CHEMISTRY FOR ENGINEERS

Abhijit Mallick, Ph.D.
Department of Chemistry
Academy of Technology
Adisaptagram, Hooghly
West Bengal, India

Viva Books
New Delhi | Mumbai | Chennai | Kolkata | Bengaluru | Hyderabad | Kochi | Guwahati

Publisher's note

Every possible effort has been made to ensure that the information contained in this book is accurate at the time of going to press, and the publisher and author cannot accept responsibility for any errors or omissions, however caused. No responsibility for loss or damage occasioned to any person acting, or refraining from action, as a result of the material in this publication can be accepted by the editor, the publisher or the author.

Every effort has been made to trace the owners of copyright material used in this book. The author and the publisher will be grateful for any omission brought to their notice for acknowledgment in the future editions of the book.

First Published 2012

Viva Books Private Limited

- 4737/23, Ansari Road, Daryaganj, New Delhi 110 002
  Tel. 42242200, 23258325, 23283121, Email: vivadelhi@vivagroupindia.net
- 76, Service Industries, Shirvane, Sector 1, Nerul, Navi Mumbai 400 706
  Tel. 27721273, 27721274, Email: vivamumbai@vivagroupindia.net
- Jamals Fazal Chambers, 26 Greams Road, Chennai 600 006
  Tel. 28294241, 28290304, Email: vivachennai@vivagroupindia.net
- B-103, Jindal Towers, 21/1A/3 Darga Road, Kolkata 700 017
  Tel. 22816713, Email: vivakolkata@vivagroupindia.net
- 7, GF, Sovereign Park Aptts., 56-58, K. R. Road, Basavanagudi, Bengaluru 560 004
  Tel. 26607409, Email: vivabangalore@vivagroupindia.net
- 101-102, Moghal Marc Apartments, 3-4-637 to 641, Narayanguda, Hyderabad 500 029
  Tel. 27564481, Email: vivahyderabad@vivagroupindia.net
- First Floor, Beevi Towers, SRM Road, Kaloor, Kochi 682 018
  Tel. 0484-2403055, 2403056, Email: vivakochi@vivagroupindia.net
- 232, GNB Road, Beside UCO Bank, Silpukhuri, Guwahati 781 003
  Tel. 0361-2666386, Email: vivaguwahati@vivagroupindia.net

www.vivagroupindia.com

ISBN: 978-81-309-2058-0

Published by Vinod Vasishtha for Viva Books Private Limited, 4737/23 Ansari Road, Daryaganj, New Delhi 110 002.

Printed and bound by Raj Press, R-3, Inderpuri, New Delhi 110 012.

E0000/E05C

Dedicated to:

My beloved wife and son

CONTENTS

ACKNOWLEDGEMENTS

I am very much obliged to my family members including my father, my wife and my beloved son for their inspiration and continuous support without which it would have been impossible to complete this uphill task.

I am indebted to Viva Books Private Limited for their continuous support and inspiration towards the completion of this book.

I must not miss the opportunity to acknowledge Mr. Digbasan Banerjee, Chairman, Mallabhum Institute of Technology and Mr. Dhurjyoti Banerjee, Director, Mallabhum Institute of Technology for their all-round cooperation and support.

I am also indebted to Dr. Sunil Kumar Roy, Principal, Mallabhum Institute of Technology, for his cooperation and support.

I must not miss the opportunity to acknowledge Mr. Indrajit Saha, Lecturer and Mr. Uttam Ganguly, Technical Assistant, Computer Science & Engineering Department of Mallabhum Institute of Technology, who have voluntarily helped me to learn computer graphics and related matters.

I would like to acknowledge Mr. Prasenjit Sikdar of Viva Books Private Limited, who encouraged me to write this book.

Abhijit Mallick

PREAMBLE

The subject Chemistry is so versed that one can understand only a part of it in one's lifetime. Perhaps Chemistry covers the widest application of science, ranging from the basic needs of people to numerous industrial applications. It also covers environmental science and biochemistry which is essential in medical science.

This book covers the chemistry courses in engineering. Physical chemistry is discussed in detail to suit the requirements for various Honours courses in Indian universities. The mathematical deductions are given to show essential relationships emerging from basic principles. This book not only meets the needs of engineering students, but promotes a fundamental understanding of engineering chemistry.

Solid state Chemistry provides the fundamental concept of electronics equipment and devices. So this chapter has been discussed in detail.

Hardness of water is a practical problem in different industries. Causes of water hardness, alkalinity of water and treatment of hard water have been discussed in detail.

Abhijit Mallick
dramallick@rediffmail.com

1

THERMODYNAMICS – I

1.1 Introduction

Definition Study of heat energy available from any thermal source is called thermodynamics. Flow of heat energy between any two states depends on temperature difference ΔT and hence the latter is called the driving force for heat flow. To study the thermodynamics the following terms are to be defined.

Universe The total material world is called universe.

System System is that part of the universe under consideration and separated from the rest by a definite boundary.

Surroundings Besides system the rest of the universe is called surroundings. Thermodynamic system is divided into three broad categories.

Thermodynamic system

Isolated system	*Closed system*	*Open system*
No heat transfer allowed No mass transfer allowed	Heat transfer allowed No mass transfer allowed	Heat transfer allowed Mass transfer allowed

Isolated system A system is said to be isolated if both mass and energy transfer with the surroundings are not allowed. In true sense no such isolated system exists. However, the universe can be called an isolated system since it has no surroundings.

Closed system A system is said to be closed if only energy transfer but not mass transfer with the surroundings is allowed.

Open system A system is said to be open if both mass transfer and energy transfer with the surroundings are allowed.

Thermodynamic parameters

Thermodynamic properties of a system are studied with the help of some parameters called thermodynamic parameters, such as pressure (P) volume (V) temperature (T) etc. All thermodynamic properties and parameters are broadly classified into two categories:

(i) Intensive property Consider of container, full of gas of volume V and pressure P.

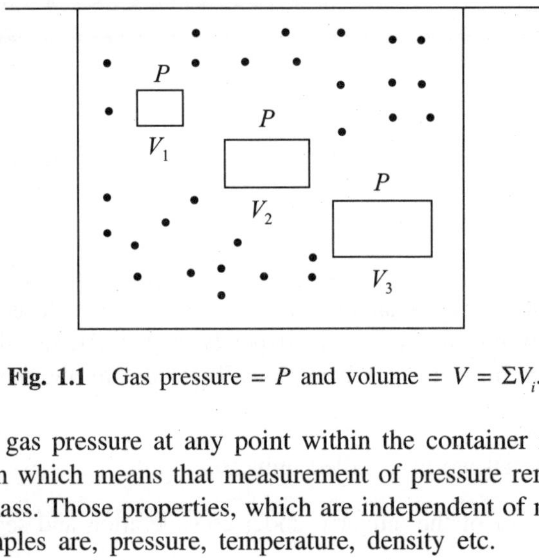

Fig. 1.1 Gas pressure = P and volume = $V = \Sigma V_i$.

Measurement of gas pressure at any point within the container indicates the pressure of the whole system which means that measurement of pressure remains same at any point, irrespective of mass. Those properties, which are independent of mass, are called intensive properties. Examples are, pressure, temperature, density etc.

(ii) Extensive property In the previous example measurement of volume at any point within the container indicates only the partial volume and does not indicate the total volume of the system, which means that measurement of volume differs from point to point and hence depends on mass. Those properties, which depend on mass, are called extensive properties. Examples are, volume, any form of energy etc.

Kinetic energy = $\frac{1}{2}(mv^2)$. So it depends on mass and so do other forms of energy.

Ratio of two extensive properties becomes an intensive property, one such example is specific volume, $\bar{v} = \left(\frac{V}{m}\right)$, where, V is the total volume and m is the mass of the gas.

1.2 Classification of thermodynamic process

Thermodynamic processes are broadly classified into two categories.

(i) Reversible process and (ii) Irreversible process.

Reversible process

Reversible process is one which proceeds through numerous steps and equilibrium is established in each and every step. Thus infinite time is required to complete a reversible process. So it is a non-flow process, also known as *Quasi-Static* process. Furthermore, at any point the physical state of a system remains same when the reaction proceeds in the forward direction as well as in the backward direction. It is well illustrated in the following figure.

A and *B* are the two states of a system and *AB* indicates the reversible path. The double arrow indicates that the forward process $A \to B$ and backward process $B \to A$ follow the same path.

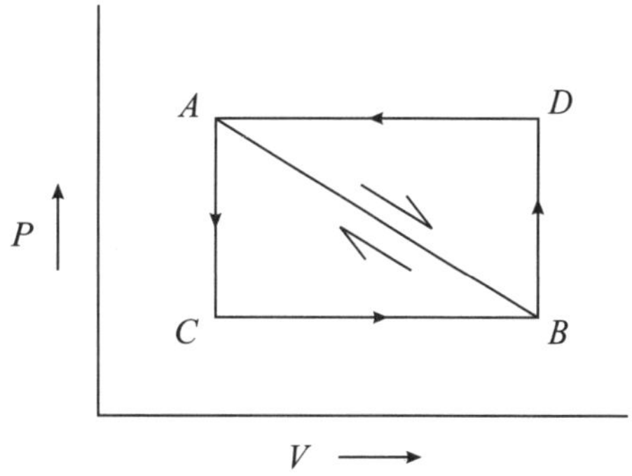

Fig. 1.2 *P-V* diagram of reversible and irreversible processes.

Irreversible process Irreversible process is one which proceeds spontaneously without establishment of any equilibrium. Furthermore at any point the physical state of a system differs for backward direction from that for forward direction as is evident from Figure 1.2, where *ACB* shows the forward process $A \to B$ and *BDA* shows the backward process $B \to A$.

Cyclic process Any process, during which a system changes its state through some changes of properties and finally returns to its original state, is called a cyclic process.

1.3 The zero-th law of thermodynamics

Statement If two systems are in thermal equilibrium with the 3rd system separately, the above two systems will also be in thermal equilibrium. Consider two systems *A* and *B*, separated by an adiabatic wall. Each of the two systems *A* and *B* are separately in equilibrium with the third system *C* through a diathermal wall. After some time thermal equilibrium (equality of temperature) is established between the systems *A* and *B*, even though they are separated by an adiabatic wall.

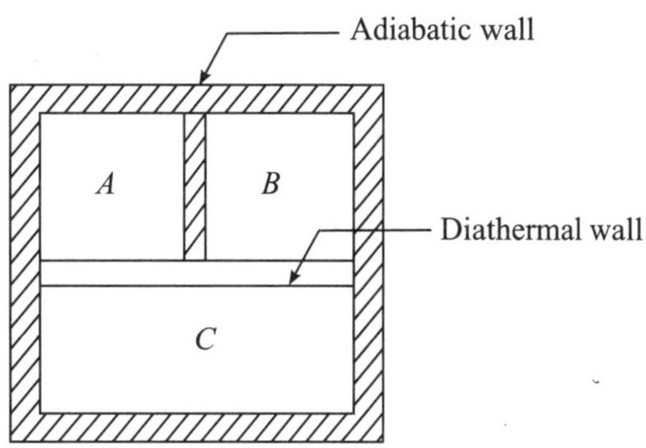

1.4 1st law of thermodynamics

Joule's experiment shows that for a cyclic process amount of work done is directly proportional to the amount of heat transfer. Mathematically it can be written as

$$\oint dW = J \oint dQ \tag{1.1}$$

Where, dW = Amount of work done in Joule and dQ = Amount of heat transfer in cal. J = proportionality constant commonly known as mechanical equivalent of heat.

When heat is absorbed by the system. dQ is +ve and hence dW is also +ve, which means that work is done *by the system.*

When heat is rejected by the system, dQ is –ve and hence dW is also –ve. which means that work is done *on the system.*

If Q and W are expressed in the same units the factor J is omitted and Equation (1.1) becomes:

$$\oint dW = \oint dQ$$

or
$$\oint dQ - \oint dW = 0 \tag{1.2}$$

Cyclic process:

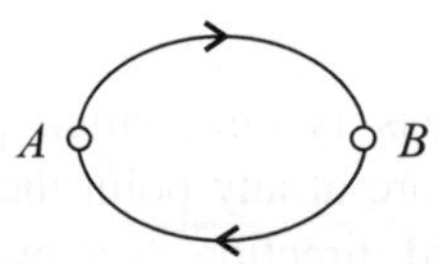

For a non-cyclic process e.g. $A \rightarrow B$ in the above figure the Equation (1.2) becomes:

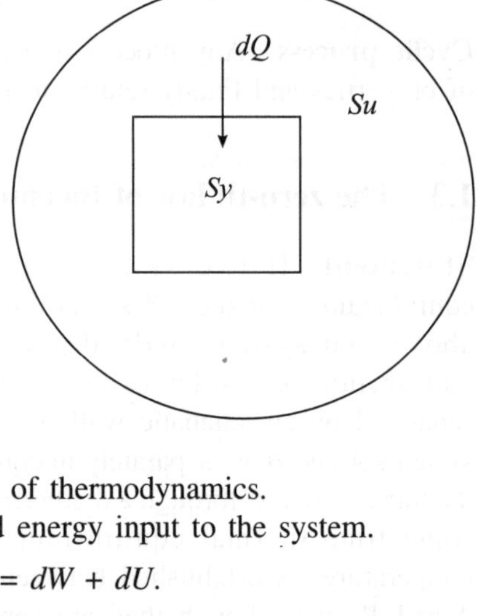

$$\int_A^B dQ - \int_A^B dW > 0$$

or,
$$\int_A^B dQ - \int_A^B dW = \int_A^B dU$$

or,
$$dQ - dW = dU$$

or,
$$dQ = dW + dU \tag{1.3}$$

Where U is called internal energy of the system.

The Equation (1.3) is the mathematical form of the 1st law of thermodynamics.

If the system gains dQ amount of heat energy, dQ is called energy input to the system.

Energy output = gross energy delivered by the system = $dW + dU$.

Net energy change of the universe = dE = Energy input – Energy output

$= dQ - (dU + dW)$

According to Equation (1.3): $dE = 0$ or E = constant.

Statement of 1st law Sum total energy of the universe is constant. If some amount of energy disappears from one part of the universe, equivalent amount of energy reappears in different form in another part of the universe.

From Equation (1.3) we have:

$$dU = dQ - dW$$

For any cyclic process we can write that:

$$\oint dU = \oint dQ - \oint dW = 0 \tag{1.4}$$

Comparing Equations (1.2) and (1.4) we get:

$$\oint dU = 0 \tag{1.5}$$

Equation (1.5) signifies that internal energy of a system has a definite value for a given state but change in internal energy is independent of path and depends only on initial and final states. This is well illustrated in the following figure:

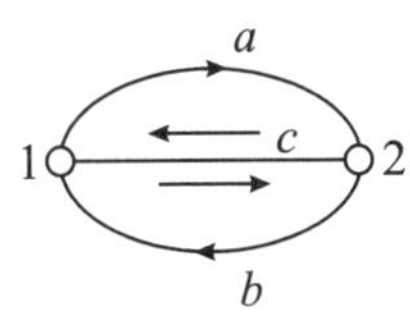

Fig. 1.3 Different routes of change of state.

Consider the two states 1 and 2 of a system. The change of state takes place through 3 different routes *a*, *b* and *c*. Along route *a* the change in internal energy is given by: $\int_1^2 dU_a$ Similarly along routes *b* and *c* the change in internal energy is given by: $\int_1^2 dU_b$ and $\int_1^2 dU_c$ respectively. Along the cyclic process 1 – *a* – 2 – *c* – 1, we can write that:

$$\int_1^2 dU_a + \int_2^1 dU_c = 0 \quad \text{[according to Equation (1.5)]}$$

Hence,

$$\int_1^2 dU_a = \int_1^2 dU_c \tag{1.6}$$

Along the cyclic process 1 – *b* – 2 – *c* – 1, we can write that:

$$\int_1^2 dU_b + \int_2^1 dU_c = 0 \quad \text{[according to Equation (1.5)]}$$

Hence,

$$\int_1^2 dU_b = \int_1^2 dU_c \tag{1.7}$$

Along the cyclic process 1 – *a* – 2 – *b* – 1, we can write that:

$$\int_1^2 dU_a + \int_2^1 dU_b = 0 \quad \text{[according to Equation (1.5)]}$$

Hence, $$\int_1^2 dU_a = \int_1^2 dU_b \tag{1.8}$$

From Equation (1.6), (1.7) and (1.8) we have:

$$\int_1^2 dU_a = \int_1^2 dU_b = \int_1^2 dU_c \tag{1.9}$$

Equation (1.9) signifies that change in internal energy is independent of path but depends only on the state of a system.

State function The function which is independent of path but depends only on state of a system is called state function. So internal energy U is a state function. Equation (1.5) holds good for each state function, i.e..

$$\oint dM = 0$$

where M is a state function.

Internal energy Each and every system possesses some inherent energy, in the form of kinetic energy, potential energy, rotational energy, vibrational energy etc., related to structure and chemical nature of the components. This inherent energy is known as internal energy. It is impossible to determine the absolute value of internal energy but only a change in internal energy can be determined.

1.5 Equation of state

Any physical state of a thermodynamic system can be expressed by thermodynamic parameters P, V and T. The general form of equation of state, relating to P, V and T is given by:

$$f(P, V, T) = 0 \tag{1.10}$$

One of the solution of the above equation is given by:

$$V = \phi(P, T)$$

The total differentiation is given by:

$$dV = \left(\frac{\partial V}{\partial T}\right)_P dT + \left(\frac{\partial V}{\partial P}\right)_T dP$$

Again differentiating with respect to T at constant volume we have:

$$0 = \left(\frac{\partial V}{\partial T}\right)_P + \left(\frac{\partial V}{\partial P}\right)_T \left(\frac{\partial P}{\partial T}\right)_V$$

Or, $$\left(\frac{\partial V}{\partial P}\right)_T \left(\frac{\partial P}{\partial T}\right)_V = -\left(\frac{\partial V}{\partial T}\right)_P$$

Or, $$\left(\frac{\partial V}{\partial P}\right)_T \left(\frac{\partial P}{\partial T}\right)_V \left(\frac{\partial T}{\partial V}\right)_P = -1$$

Or, $$\left(\frac{\partial P}{\partial V}\right)_T \left(\frac{\partial V}{\partial T}\right)_P \left(\frac{\partial T}{\partial P}\right)_V = -1 \tag{1.11}$$

Coefficient of expansion at constant pressure (α) If V_0 is the volume at temperature T_0 and pressure P_0, $\left(\frac{\partial V}{\partial T}\right)_P$ indicates the increase in volume with increase in temperature at constant pressure. Coefficient of expansion at constant pressure (α) is defined as:

$$\alpha = \frac{1}{V_0}\left(\frac{\partial V}{\partial T}\right)_P$$

Coefficient of expansion at constant volume (β) If P_0 is the pressure at temperature T_0 and volume V_0, $\left(\frac{\partial P}{\partial T}\right)_V$ indicates the increase in pressure with temperature at constant volume. Coefficient of expansion at constant volume (β) is defined as:

$$\beta = \frac{1}{P_0}\left(\frac{\partial P}{\partial T}\right)_V$$

Isothermal compressibility (k) It is the decrease in volume with increase in pressure at constant temperature. It is expressed as:

$$k = -\frac{1}{V_0}\left(\frac{\partial V}{\partial P}\right)_T$$

Putting the values of

$$\left(\frac{\partial V}{\partial T}\right)_P, \left(\frac{\partial P}{\partial V}\right)_T \text{ and } \left(\frac{\partial T}{\partial P}\right)_V$$

in Equation (1.11) we get:

$$\frac{(\alpha V_0)}{(\beta P_0)(-k V_0)} = -1$$

Or, $$\frac{\alpha}{\beta k P_0} = 1$$

Or, $$\frac{\beta k P_0}{\alpha} = 1 \tag{1.12}$$

1.6 Work

The total work delivered by the system can be divided into two parts:

(a) Expansion work.
(b) Elementary useful work.

Expansion work When work is obtained due to expansion of volume of the system the work is called expansion work. It is well illustrated in the following figure:

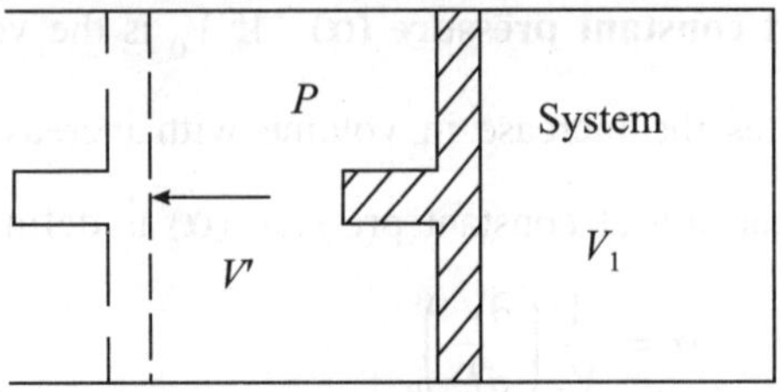

V_1 = Initial volume of the system. P = Piston.
$V_2 = V_1 + V'$ = Final volume of the system after expansion.

Fig. 1.4 Volume of the system increases due to movement of the piston (P).

Thus, mathematically the expansion work is represented as:

$$dW = PdV \tag{1.13}$$

Elementary useful work The useful work may be divided into different categories:

(i) Work due to potential energy change, represented by ($-Mgdh$),
where, Mg = weight and dh = change in height.

(ii) Work due to increase in surface area, represented by ($-\sigma\, dA$),
where, σ = surface tension and dA = change in area.

(iii) Work due to increase in charge, represented by ($-\phi . dq$),
where, ϕ = potential and dq = increment of charge.

(iv) Mechanical work, represented by ($F . dx$),
where, F = Force and dx = displacement.

1.7 Specific heat

Amount of heat required to increase the temperature of one mole of any substance by 1°K is known as molar heat capacity of that substance. In case of gases molar heat capacity can be measured either at constant volume or at constant pressure. In that case molar heat capacity is represented by C_V or C_P respectively. If dQ amount of heat is required to increase the temperature of a system by dT, C_V and C_P are represented as:

$$C_V = \left(\frac{\partial Q}{\partial T}\right)_V \quad \text{and} \quad C_P = \left(\frac{\partial Q}{\partial T}\right)_P$$

From the 1st law of thermodynamics, we have:

$$dQ = dW + dU$$

Considering only expansion work we can write that:

$$dQ = dU + P\,dV \tag{1.13a}$$

Differentiating the above equation with respect to (w.r.t.) temperature at constant volume (V), we get:

$$C_V = \left(\frac{\partial Q}{\partial T}\right)_V = \left(\frac{\partial U}{\partial T}\right)_V \tag{1.14}$$

Since U is a state function, we can write that: $U = f(V, T)$
Differentiating, we have:

$$dU = \left(\frac{\partial U}{\partial V}\right)_T dV + \left(\frac{\partial U}{\partial T}\right)_V dT \tag{1.15}$$

Again, we can write that $U = f(P, T)$

Differentiating, we have:

$$dU = \left(\frac{\partial U}{\partial P}\right)_T dP + \left(\frac{\partial U}{\partial T}\right)_P dT \tag{1.16}$$

From Equations (1.15) and (1.16) we have:

$$\left(\frac{\partial U}{\partial P}\right)_T dP + \left(\frac{\partial U}{\partial T}\right)_P dT = \left(\frac{\partial U}{\partial V}\right)_T dV + \left(\frac{\partial U}{\partial T}\right)_V dT$$

Again, $$\left(\frac{\partial U}{\partial T}\right)_P = \left(\frac{\partial Q}{\partial T}\right)_P - P\left(\frac{\partial V}{\partial T}\right)_P \qquad [\text{since, } dU = dQ - PdV]$$

And $$C_V = \left(\frac{\partial U}{\partial T}\right)_V$$

Or, $$\left(\frac{\partial U}{\partial P}\right)_T dP + \left[C_P - P\left(\frac{\partial V}{\partial T}\right)_P\right] dT = \left(\frac{\partial U}{\partial V}\right)_T dV + C_V\,dT$$

Differentiating w.r.t. T at constant P, we get:

$$C_P - P\left(\frac{\partial V}{\partial T}\right)_P = \left(\frac{\partial U}{\partial V}\right)_T \left(\frac{\partial V}{\partial T}\right)_P + C_V$$

Or, $$C_P - C_V = \left[\left(\frac{\partial U}{\partial V}\right)_T + P\right]\left(\frac{\partial V}{\partial T}\right)_P \tag{1.17}$$

Gay-Lussac-Joule's law states that in an isothermal process change in internal energy of an ideal gas is independent of its volume.

Thus for an ideal gas $\left(\frac{\partial U}{\partial V}\right)_T = 0$, and Equation (1.17) becomes:

$$C_P - C_V = P\left(\frac{\partial V}{\partial T}\right)_P \tag{1.18}$$

Again for 1 mole of an ideal gas, we know that: $PV = RT$.

Differentiating w.r.t. T at constant P, we get:

$$P\left(\frac{\partial V}{\partial T}\right)_P = R$$

Thus Equation (1.18) becomes:

$$C_P - C_V = R \tag{1.19}$$

The Joule's Experiment

In 1843 Joule performed an experiment to determine $\left(\frac{\partial U}{\partial V}\right)_T$, for a gas, assumed to behave ideally. The experiment set-up consists of two chambers A and B separated by a valve and each chamber is surrounded by an adiabatic wall. Initially chamber A is filled with a gas and chamber B is evacuated. Then valve between the chambers is opened and the gas of chamber A readily occupies the chamber B. The gas flow occurs till equilibrium is reached. The process is irreversible. Temperature change in the system is measured by a thermometer.

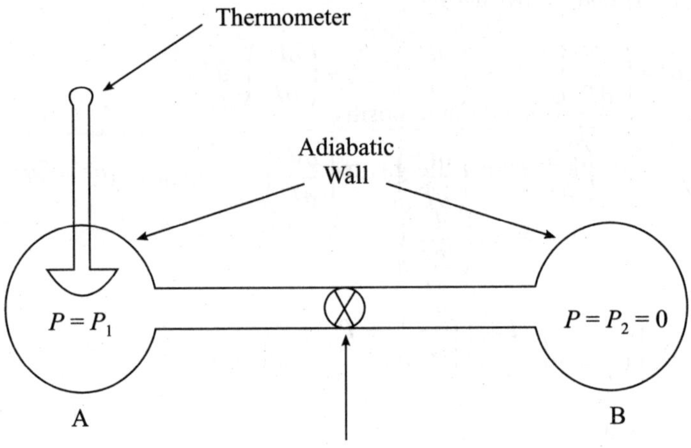

Fig. 1.5 Chamber A: Filled with gas, Chamber B: Vacuum.

There is no heat transfer between the system and the surroundings. So, $\Delta Q = 0$. The work done due to this irreversible process $= P_2(V_2 - V_1)$. As $P_2 = 0$, there is no work done. Hence from the 1st law of thermodynamics [Equation (1.3)], we have:

$$\Delta U = 0 \text{ or, } U = \text{constant}$$

At this condition Joule coefficient is defined as:

$$\mu_J = \left(\frac{\partial T}{\partial V}\right)_U$$

Since U is a state function we can write that:

$$\left(\frac{\partial T}{\partial V}\right)_U \left(\frac{\partial V}{\partial U}\right)_T \left(\frac{\partial U}{\partial T}\right)_V = -1$$

Or,
$$\left(\frac{\partial U}{\partial V}\right)_T = -\left(\frac{\partial U}{\partial T}\right)_V \left(\frac{\partial T}{\partial V}\right)_U$$

Or,
$$\left(\frac{\partial U}{\partial V}\right)_T = -C_V \mu_J \quad (1.20)$$

μ_J has been found zero for ideal gases, which means that the expansion process is an isothermal process. Hence:

$$\left(\frac{\partial U}{\partial V}\right)_T = 0$$

Conclusion In an isothermal process change in internal energy of an ideal gas does not depend on its volume. Thus for ideal gases we can write that:

$$PV = nRT \quad \text{and} \quad \left(\frac{\partial U}{\partial V}\right)_T = 0 \quad (1.21)$$

In case of real gases $\left(\frac{\partial U}{\partial V}\right)_T$ has a definite positive value. It indicates that internal pressure of gas arises due to mutual attraction among the gas molecules. Internal pressure is given by:

$$(l - P) = \left(\frac{\partial U}{\partial V}\right)_T$$

Or,
$$l = \left(\frac{\partial U}{\partial V}\right)_T + P \quad (1.22)$$

1.8 Enthalpy (H)

From the 1st law of thermodynamics, we have:

$$dQ = dU + PdV \qquad \text{[considering only expansion work]}$$

Putting the value of dU from Equation (1.15), we get:

$$dQ = \left[\left(\frac{\partial U}{\partial V}\right)_T dV + \left(\frac{\partial U}{\partial T}\right)_V dT\right] + PdV$$

$$= \left[\left(\frac{\partial U}{\partial V}\right)_T + P\right] dV + C_V\, dT$$

$$= l dV + C_V\, dT \qquad \text{[from Equation (1.22)]}$$

Or, $$dQ = l dV + C_V\, dT \tag{1.23}$$

Putting the value of dU from Equation (1.16) in the 1st law of thermodynamics, we have:

$$dQ = \left(\frac{\partial U}{\partial P}\right)_T dP + \left(\frac{\partial U}{\partial T}\right)_P dT + P dV \tag{1.24}$$

Again we know that: $V = f(P, T)$. Differentiating we have:

$$dV = \left(\frac{\partial V}{\partial P}\right)_T dP + \left(\frac{\partial V}{\partial T}\right)_P dT \tag{1.25}$$

Putting this value of dV in Equation (1.24) we get,

$$dQ = \left[\left(\frac{\partial U}{\partial P}\right)_T + P\left(\frac{\partial V}{\partial P}\right)_T\right] dP + \left[\left(\frac{\partial U}{\partial T}\right)_P + P\left(\frac{\partial V}{\partial T}\right)_P\right] dT$$

$$= h\, dP + \left[\frac{\partial (U + PV)}{\partial T}\right]_P dT$$

$$= h\, dP + \left(\frac{\partial Q}{\partial T}\right)_P dT$$

$$= h\, dP + C_P\, dT$$

where, $$h = \left[\left(\frac{\partial U}{\partial P}\right)_T + P\left(\frac{\partial V}{\partial P}\right)_T\right]$$

Or, $$dQ = h\, dP + C_P\, dT \tag{1.26}$$

Putting the value of dV from Equation (1.25) in the Equation on (1.23) we get:

$$dQ = l\left[\left(\frac{\partial V}{\partial P}\right)_T dP + \left(\frac{\partial V}{\partial T}\right)_P dT\right] + C_V\, dT$$

Or, $$dQ = l\left(\frac{\partial V}{\partial P}\right)_T dP + \left[C_V + l\left(\frac{\partial V}{\partial T}\right)_P\right] dT \tag{1.27}$$

Comparing Equations (1.26) and (1.27) we get:

$$h = l\left(\frac{\partial V}{\partial P}\right)_T \quad \text{and} \quad C_P = C_V + l\left(\frac{\partial V}{\partial T}\right)_P$$

Again $l = P$ for an ideal gas. So for an ideal gas we have:

$$h = P\left(\frac{\partial V}{\partial P}\right)_T \quad \text{and} \quad C_P = C_V + P\left(\frac{\partial V}{\partial T}\right)_P$$

For 1 mole of an ideal gas we can write that: $PV = RT$. Differentiating it w.r.t. P at constant T, we get:

$$P\left(\frac{\partial V}{\partial P}\right)_T + V = 0$$

Or, $$P\left(\frac{\partial V}{\partial P}\right)_T = -V$$

Hence $$h = -V$$

Again differentiating the ideal gas equation w.r.t. T at constant P, we get:

$$P\left(\frac{\partial V}{\partial T}\right)_P = R$$

Hence, $$C_P - C_V = P\left(\frac{\partial V}{\partial T}\right)_P = R$$

Putting the value of h in Equation (1.26), we get:

$$dQ = -VdP + C_P\,dT$$

Or, $$C_P\,dT = dQ + VdP \tag{1.28}$$

For any cyclic process the above equation can be written as:

$$\oint C_P\,dT = \oint (dQ + VdP)$$

Since for any cyclic process net change in temperature is zero, we have:

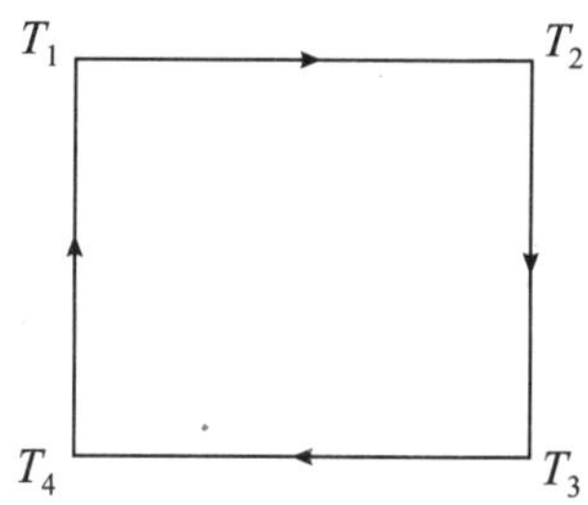

$$\oint C_P\,dT = 0$$

$$\oint dT = 0$$

Hence,

$$\oint (dQ + VdP) = 0$$

which signifies that $(dQ + VdP)$ is independent of path and a state function, represented as dH.

Or, $$dH = dQ + VdP \tag{1.29}$$

Where H as known as enthalpy or heat content of the system.

Putting the value of dQ from the 1st law of thermodynamics in Equation (1.29) we get:

$$dH = dU + \text{PdV} + VdP = d(U + PV)$$

Or, $$H = U + PV \tag{1.30}$$

Equation (1.30) represents the relation between two state functions U and H. Again from Equations (1.28) and (1.29) we have:

$$dH = C_P\,dT$$

Or, $$C_P = \left(\frac{\partial H}{\partial T}\right)_P \tag{1.31}$$

1.9 P-V-T relations in adiabatic process

From the 1st law of thermodynamics we know that:

$$dQ = dU + PdV,$$

Or, $dQ = C_V\, dT + PdV$ [since, $dU = C_V\, dT$ for 1 mole

and $dU = nC_V\, dT$ for n moles of an ideal gas]

For any adiabatic process the net heat transfer is zero or $dQ = 0$. Hence the above equation becomes:

$$C_V\, dT + PdV = 0 \quad \text{[for 1 mole of a gas]}$$

and

$$nC_V\, dT + PdV = 0 \quad \text{[for } n \text{ moles of a gas]}$$

For n moles of an ideal gas we know that:

$$PV = nRT$$

Or,

$$P = \frac{nRT}{V}$$

Putting the value of P in the above equation we get:

$$nC_V\, dT + \frac{nRT}{V}\, dV = 0$$

Or,

$$\text{dln}T = -\frac{R}{C_V}\, \text{dln}V \qquad (1.32)$$

Again applying the adiabatic condition, i.e. $dQ = 0$ in the Equation (1.28) we get:

$$C_P\, dT = VdP \quad \text{[for 1 mole of a gas]}$$

and

$$nC_P\, dT = VdP \quad \text{[for } n \text{ moles of a gas]}$$

Again for 1 mole of an ideal gas, $V = \frac{RT}{P}$.

Putting the value of V in the above equation we get:

$$C_P\, dT = \frac{RT}{P}\, dP$$

For n moles of an ideal gas:

$$nC_P\, dT = \frac{nRT}{P}\, dP$$

$$C_P\, dT = \frac{RT}{P}\, dP$$

Or,

$$\text{dln}\, T = \frac{R}{C_P}\, \text{dln}P \qquad (1.33)$$

Dividing (1.33) by (1.32) we get:

$$\frac{C_V \, d\ln P}{C_P \, d\ln V} = -1$$

Or, $$d\ln P + \frac{C_P}{C_V} \, d\ln V = 0$$

On integration we get:

$$\ln P + \gamma \ln V = \text{Constant}$$

Or, $$\ln PV^{\gamma} = \text{Constant}$$

Or, $$PV^{\gamma} = \text{Constant} \tag{1.34a}$$

Where, $$\gamma = \frac{C_P}{C_V} = \text{ratio of specific heats}$$

Again $$R = C_P - C_V \text{ for 1 mole of an ideal gas.}$$

$$\frac{R}{C_V} = \gamma - 1 \quad [\text{for 1 mole of an ideal gas}]$$

And $$\frac{nR}{nC_V} = \gamma - 1 \quad [\text{for } n \text{ moles of an ideal gas}]$$

Thus Equation (1.32) becomes: $d\ln T = -(\gamma - 1)d\ln V$

On integration we get:

$$\ln(TV^{\gamma-1}) = \text{constant. Or, } TV^{\gamma-1} = \text{constant} \tag{1.34b}$$

Again, $$\frac{R}{C_P} = 1 - \frac{C_V}{C_P} = 1 - \frac{1}{\gamma} = \left(\frac{\gamma - 1}{\gamma}\right)$$

Putting this value of $\frac{R}{C_P}$ in Equation (1.33) we get:

$$d\ln T = \left(\frac{\gamma - 1}{\gamma}\right) d\ln P$$

On integration, we get:

$$\ln\left[\text{T.P}^{\frac{1-\gamma}{\gamma}}\right] = \text{Constant}$$

Or, $$\left[\text{T.P}^{\frac{1-\gamma}{\gamma}}\right] = \text{Constant} \tag{1.34c}$$

$$\left.\begin{aligned} PV^{\gamma} &= \text{constant} \\ TV^{\gamma-1} &= \text{constant} \\ TP^{\left(\frac{1-\gamma}{\gamma}\right)} &= \text{constant} \end{aligned}\right\} \tag{1.34d}$$

So the Equation (1.34d) is valid equally for 1 mole as well as n moles of any ideal gas.

Polytropic process

Any thermodynamic process is called polytropic process, which can be divided into four categories:

(i) Adiabatic process: condition, $dQ = 0$
(ii) Isothermal process: condition, $dT = 0$
(iii) Isochoric process: condition, $dV = 0$
(iv) Isobaric process: condition, $dP = 0$

In general for an ideal gas the mathematical expression of any polytropic process is given by:

$$PV^n = \text{Constant} \tag{1.35a}$$

Or,

$$VP^{(1/n)} = \text{Constant} \tag{1.35b}$$

Different polytropic processes are summarized in the Figure 1.6.

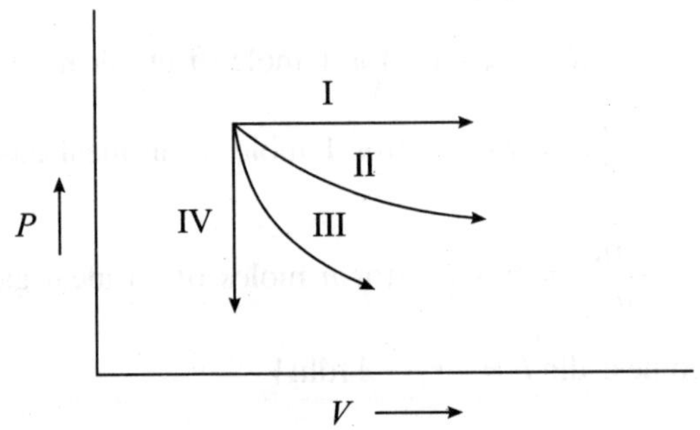

Fig. 1.6 *P-V* diagram for different polytropic processes.

Curve – I: (a) It shows an isobaric process, i.e P = constant. Hence, $n = 0$ in Equation (1.35a).

(b) Specific heat = $C = C_P = \left(\frac{\partial Q}{\partial T}\right)_P$

Curve – II: (a) It shows an isothermal process i.e., T = constant and PV = constant. Hence, $n = 1$ in Equation (1.35a).

(b) Specific heat, $C = \frac{dQ}{dT} = \infty$ since, $dT = 0$

Curve – III: (a) It shows an adiabatic process, i.e., Q = constant, and PV^γ = constant. Hence, $n = \gamma$ in Equation (1.35a).

(b) Specific heat, $C = \frac{dQ}{dT} = 0$ since, $dQ = 0$

Curve – IV: (a) It shows an isochoric process, i.e., V = constant and hence $n = \alpha$ in Equation (1.35b).

(b) Specific heat, $C = C_V = \left(\frac{\partial Q}{\partial T}\right)_V$

1.10 Calculation of work

Work Work due to expansion of gases is given by: $dW = PdV$.
For different polytropic processes work will be different.

(a) Isochoric process The condition of isochoric process is $dV = 0$.

Hence, $$dW = PdV = 0.$$

Thus no work is obtained in an isochoric process.

(b) Isobaric process Pressure is constant in an isobaric process.

Hence $$P = \text{constant and } dP = 0$$

Thus during expansion of a gas from volume V_1 to volume V_2 work done is:

$$W = \int_{V_1}^{V_2} PdV = P\int_{V_1}^{V_2} dV = P(V_2 - V_1)$$

Or, $$W = P(V_2 - V_1)$$

(c) Isothermal process In isothermal process temperature is held constant or $dT = 0$.

(i) For 1 mole of an ideal gas, we know that:

$$PV = RT$$

Or, $$P = \frac{RT}{V} \quad \text{Work done} = W = \int_{V_1}^{V_2} PdV$$

Substituting the value of P in the above equation we get:

$$W = \int_{V_1}^{V_2} \frac{RT}{V} dV = RT \int_{V_1}^{V_2} d\ln V = RT \ln \frac{V_2}{V_1}$$

Again, $$P_1 V_1 = P_2 V_2$$

Or, $$\frac{V_2}{V_1} = \frac{P_1}{P_2}$$

So, $$W = RT \ln \frac{V_2}{V_1} = RT \ln \frac{P_1}{P_2}$$

For n moles of an ideal gas:

$$W = nRT \ln \frac{V_2}{V_1} = nRT \ln \frac{P_1}{P_2}$$

(ii) For n moles of a gas, obeying Van-der-Waal's equation, we know that:

$$\left(P + \frac{an^2}{V^2}\right)(V - nb) = nRT$$

Or, $$P = \frac{nRT}{(V - nb)} - \frac{an^2}{V^2}$$

Where a and b are constants. Substituting this value of P in the expression of work we get:

$$W = \int_{V_1}^{V_2} P\,dV = \int_{V_1}^{V_2} \frac{nRT}{(V - nb)} dV - \int_{V_1}^{V_2} \frac{an^2}{V^2} dV$$

$$= nRT \ln\left[\frac{(V_2 - nb)}{(V_1 - nb)}\right] + an^2\left[\frac{1}{V_2} - \frac{1}{V_1}\right]$$

Or, $$W = nRT \ln\left[\frac{(V_2 - nb)}{(V_1 - nb)}\right] + an^2\left[\frac{1}{V_2} - \frac{1}{V_1}\right]$$

(d) Adiabatic process From the 1st law of thermodynamics we know that:

$$dQ = dU + dW$$

As the process is adiabatic $dQ = 0$ and hence,

$$dW = -dU$$

Again, $U = U(V, T)$ [since U is a state function].

Differentiating, we have:

$$dU = \left(\frac{\partial U}{\partial V}\right)_T dV + \left(\frac{\partial U}{\partial T}\right)_V dT$$

(i) For an ideal gas, as discussed in section 1.7, we have:

$$\left(\frac{\partial U}{\partial V}\right)_T = 0 \quad \text{and hence:} \quad dU = \left(\frac{\partial U}{\partial T}\right)_V = C_V\,dT$$

For n moles of an ideal gas: $dU = nC_V\,dT$.

Thus work is given by:

$$dW = -dU = -nC_V\,dT$$

Or, $$\int dW = -nC_V \int_{T_1}^{T_2} dT$$

Or, $$W = -nC_V (T_2 - T_1) \tag{1.36a}$$

Again, we know that: $dW = P\,dV$ (considering only expansion work) and $PV^\gamma = K$

Putting the value of P from adiabatic relation we get:

$$dW = \frac{K}{V^\gamma} dV \text{ where, } K \text{ is a constant.}$$

Or,
$$\int dW = \int_{V_1}^{V_2} \frac{K}{V^{\gamma}} dV = \frac{K}{(1-\gamma)}\left[\frac{1}{V^{\gamma-1}}\right]_{V_1}^{V_2}$$
$$= \frac{K}{(1-\gamma)}\left[\frac{1}{V_2^{\gamma-1}} - \frac{1}{V_1^{\gamma-1}}\right]$$
$$= \frac{1}{(1-\gamma)}\left[\frac{KV_2}{V_2^{\gamma}} - \frac{KV_1}{V_1^{\gamma}}\right]$$
$$= \frac{P_2V_2 - P_1V_1}{(1-\gamma)}\left[\text{since, } \frac{K}{V_2^{\gamma}} = P_2 \text{ and } \frac{K}{V_1^{\gamma}} = P_1\right]$$

So,
$$W = \frac{(P_2V_2 - P_1V_1)}{(1-\gamma)} \tag{1.36b}$$

Both the Equations (1.36a) and (1.36b) represent the adiabatic work for n moles of an ideal gas. Furthermore, the Equation (1.36b) represents the adiabatic work for 1 mole as well as n moles of an ideal gas.

(ii) For a real gas obeying Van-der-Waal's equation we know that:

$$\left(\frac{\partial U}{\partial V}\right)_T = (l - P) = \text{Internal pressure} \quad \text{[From Equation (1.22)]}$$

The Van-der-Waal's equation for n moles is given by:

$$\left(P + \frac{an^2}{V^2}\right)(V - nb) = \text{nRT}$$

$$\text{Internal pressure} = (l - P) = \frac{an^2}{V^2}$$

Thus dU is given by:

$$dU = \left[\left(\frac{\partial U}{\partial T}\right)_V dT + \left(\frac{\partial U}{\partial V}\right)_T dV\right]$$

$$dU = nC_V\, dT + \frac{an^2}{V^2} dV \quad \text{[for } n \text{ moles]} \tag{1.37}$$

Or,
$$\Delta U = \int dU = \int_{T_1}^{T_2} nC_V\, dT + \int_{V_1}^{V_2} \frac{an^2}{V^2} dV$$

Or,
$$\Delta U = nC_V(T_2 - T_1) - an^2\left(\frac{1}{V_2} - \frac{1}{V_1}\right)$$

$$W = -\Delta U = -nC_V\,(T_2 - T_1) + an^2\left(\frac{1}{V_2} - \frac{1}{V_1}\right) \tag{1.38a}$$

Again we know that: $dW = PdV$ [considering only expansion work]

$$\left(P + \frac{an^2}{V^2}\right)(V - nb)^{\gamma} = K\ (a \text{ constant}) \quad \text{[From Equation (1.40d)]}$$

$$\text{Thus work} = W = \int dW = \int_{V_1}^{V_2} PdV = \int_{V_1}^{V_2} \frac{K.dV}{(V - nb)^{\gamma}} - \int_{V_1}^{V_2} \frac{an^2\,dV}{V^2}$$

$$= \frac{K}{(1-\gamma)}\left[\frac{1}{(V_2 - nb)^{\gamma-1}} - \frac{1}{(V_1 - nb)^{\gamma-1}}\right] + an^2\left[\frac{1}{V_2} - \frac{1}{V_1}\right]$$

$$= \frac{1}{(1-\gamma)}\left[\frac{K(V_2 - nb)}{(V_2 - nb)^{\gamma}} - \frac{K(V_1 - nb)}{(V_1 - nb)^{\gamma-1}}\right] + an^2\left[\frac{1}{V_2} - \frac{1}{V_1}\right]$$

$$= \frac{1}{(1-\gamma)}\left[\left(P_2 + \frac{an^2}{V_2^2}\right)(V_2 - nb) - \left(P_1 + \frac{an^2}{V_1^2}\right)(V_1 - nb)\right] + an^2\left[\frac{1}{V_2} - \frac{1}{V_1}\right]$$

Or,
$$W = \frac{1}{(1-\gamma)}\left[\left(P_2 + \frac{an^2}{V_2^2}\right)(V_2 - nb) - \left(P_1 + \frac{an^2}{V_1^2}\right)(V_1 - nb)\right] + an^2\left[\frac{1}{V_2} - \frac{1}{V_1}\right] \tag{1.38b}$$

Both the equations (1.38a) and (1.38b) represent the adiabatic work for *n* moles of a real gas obeying Van-der-Waal's equation.

Adiabatic *P-V* relation for '*n*' moles of a Van-der-Waal's gas

From the 1st law of thermodynamics, we have:

$$dQ = dU + PdV \quad \text{[considering only expansion work]}$$

Adiabatic condition: $dQ = 0$ or $PdV = -dU$

Putting the value of dU from Equation (1.37) we have:

$$PdV = -nC_V\,dT - \frac{an^2}{V^2}\,dV$$

Or,
$$\left(P + \frac{an^2}{V^2}\right)dV = -nC_V\,dT$$

Again for *n* moles of a Van-der-Waal's gas we have:

$$\left(P + \frac{an^2}{V^2}\right)(V - nb) = nRT \tag{1.39}$$

Or, $$\left(P+\frac{an^2}{V^2}\right)dV = \frac{nRT}{(V-nb)}\,dV$$

Again, $$\left(P+\frac{an^2}{V^2}\right)dV = -nC_V\,dT$$

Comparing above two equations, we have:

$$\frac{dV}{(V-nb)} = -\frac{C_V}{R}\frac{dT}{T}$$

Or, $$d\ln T + (\gamma - 1)\,d\ln(V-nb) = 0$$

Or, $$T(V-nb)^{\gamma-1} = \text{constant} \qquad (1.40a)$$

From Equations (1.39) and (1.40a) we get

$$\left(P+\frac{an^2}{V^2}\right)(V-nb)^{\gamma} = \text{constant} \qquad (1.40b)$$

$$T\left(P+\frac{an^2}{V^2}\right)^{\frac{1-\gamma}{\gamma}} = \text{constant} \qquad (1.40c)$$

$$\left.\begin{aligned} T(V-nb)^{\gamma-1} &= \text{constant} \\ \left(P+\frac{an^2}{V^2}\right)(V-nb)^{\gamma} &= \text{constant} \\ T\left(P+\frac{an^2}{V^2}\right)^{\frac{1-\gamma}{\gamma}} &= \text{constant} \end{aligned}\right\} \qquad (1.40d)$$

1.11 Reversible and irreversible processes

In another angle of classification all thermodynamic processes can be divided into two broad categories: (i) Reversible process, (ii) Irreversible process.

Reversible process A process is said to be reversible if the system undergoes a change of state through numerous steps such that equilibrium holds good in each and every step. That means in a reversible process both forward and backward processes proceed through same physical states. To perform a process reversibly only infinitesimal change in pressure/temperature is sufficient and necessary to execute any chemical change slowly and reversibly. This is well illustrated in the Figure 1.7. The path *AB* indicates the reversible path since it passes through numerous steps represented by *X*. Furthermore both the forward process $A \rightarrow B$ and the backward process $(B \rightarrow A)$ follow the same path. Thus infinite time is required to complete a reversible process. So it is a non-flow process, also known as Quasi-Static process.

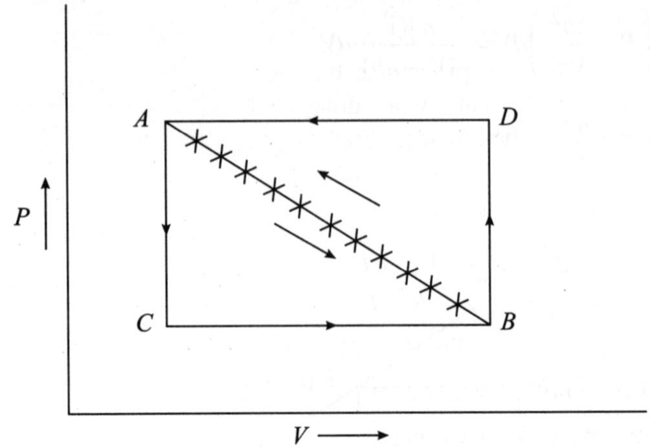

Fig. 1.7 *P-V* diagram for reversible and irreversible process.

Irreversible process A process is said to be irreversible if the system undergoes a change of state to a definite direction spontaneously and never returns to its original state following the same path. Usually irreversible process is a single-step or two-step process. In the Figure 1.7, the system changes its state $A \rightarrow B$ through one step process *ACB* but returns to its original state through another single-step process *BDA*.

Work: Maximum work is available in a reversible process

(i) Graphical approach

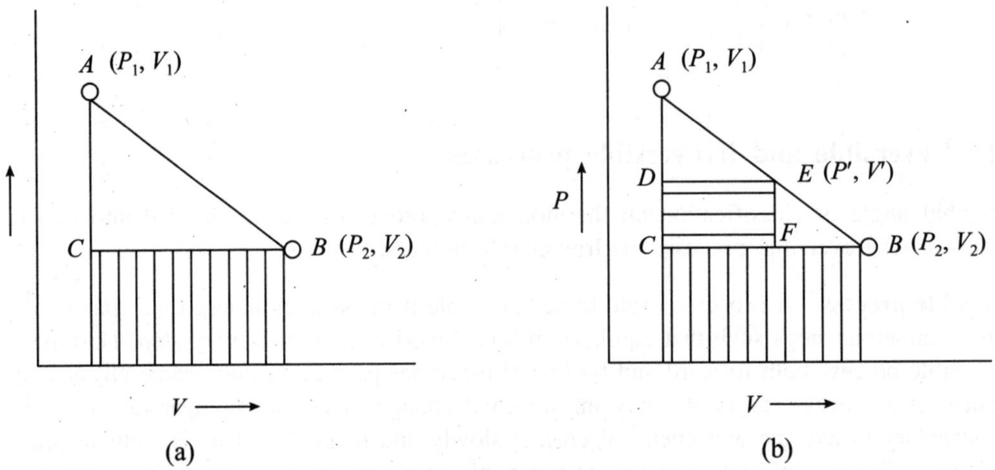

Fig. 1.8 *P-V* diagram for one step and two step processes.

Figure 1.8(a) shows a single step irreversible process *ACB*, where *AB* represents reversible path. The work done is the area under the curve *ACB*, represented by vertical hatching lines. The Figure 1.8(b) shows the two step irreversible process (*ADEFB*) where the work done is the area under the curve *ADEFB*. Compared to the Figure 1.8(a) the extra work obtained in two step process is the area of the rectangle *CDEF*, represented by horizontal hatching lines. Thus work done in two step process is more than that in one step process. Since reversible path, represented by *AB* in the Figure 1.8, consists of numerous steps, maximum work is available in a reversible process.

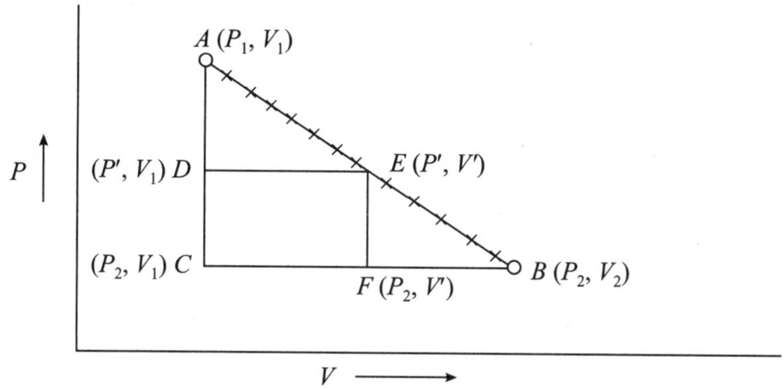

Fig. 1.9 $P - V$ diagram for one step and two step processes.

(ii) Numerical approach In the Figure 1.9 *ACB* and *ADEFB* represent single step and two step processes, respectively.

Work done along $ACB = W_1$ = Work done along *AC* + Work done along *CB*.

= Work done along $CB = P_2(V_2 - V_1)$

[since *AC* shows an isochoric process and work in an isochoric process is zero]

Work along $ADEFB = W_2$

= Work done along *AD* + Work done along *DE* + Work done along *EF* + Work done *FB*

= Work done *DE* + work done *FB* [since *AD* and *EF* represent isochoric processes]

$= P'(V' - V_1) + P_2(V_2 - V')$

Difference between two-step-work and one-step-work $= W_2 - W_1$

$$(W_2 - W_1) = [P'(V' - V_1) + P_2(V_2 - V')] - [P_2(V_2 - V_1)]$$

$$= P'V' - P'V_1 + \cancel{P_2V_2} - P_2V' - \cancel{P_2V_2} + P_2V_1$$

$$= P'V' - P_2V' - P'V_1 + P_2V_1$$

$$= V'(P' - P_2) - V_1(P' - P_2) = (P' - P_2)(V' - V_1)$$

$$(W_2 - W_1) = (P' - P_2)\,(V' - V_1)$$

From the graph it is evident that: $P' > P_2$ and $V' > V_1$.

Hence, $(W_2 - W_1) > 0$.

Or, $W_2 > W_1$

So, work obtained in a two step process is more than that obtained in a single step process. As the reversible path (*AB* in the Figure 1.9) consists of numerous steps, maximum work is available in the reversible process.

1.12 Heat engine and entropy

Heat engine An engine, which undergoes a cyclic process to convert heat energy to mechanical work, is called heat engine. Figure 1.10. shows the schematic representation of a heat engine.

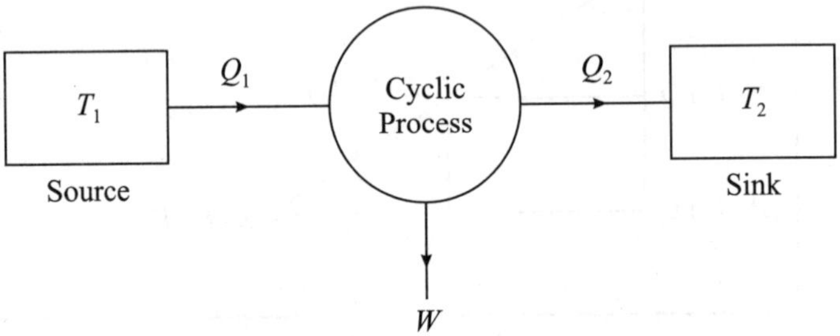

Fig. 1.10 Schematic diagram of working principle of a heat engine.

A heat engine absorbs Q_1 amount of heat from a heat source at temperature T_1 and delivers W amount of work. At the end Q_2 amount of heat is rejected to the heat sink at temperature T_2. So the efficiency of the heat engine, η is the fraction of energy input that appears as mechanical work. Thus η is given by:

$$\eta = \frac{W}{Q_1} \quad \text{[where, } Q_1 = \text{Amount of heat energy input]}$$

An engine which delivers work through a reversible cyclic process is called Carnot engine and the corresponding cycle is called Carnot cycle. A Carnot cycle is defined as a reversible cycle, consisting of two isothermal steps at two different temperatures and two adiabatic steps. It is well illustrated in the Figure 1.11.

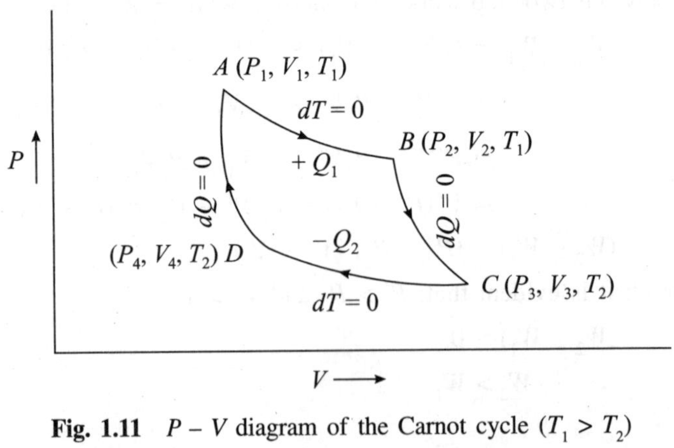

Fig. 1.11 $P - V$ diagram of the Carnot cycle ($T_1 > T_2$)

There are two isothermal steps *AB* and *CD*. Along *AB*, Q_1 amount of heat is absorbed at temperature T_1 and along *CD*, Q_2 amount of heat is rejected at temperature T_2. *BC* and *DA* are the two adiabatic steps. Applying the TV-relation [Equation (1.34)] along *BC* and along *DA* we get:

Along *BC*: $$T_1 V_2^{\gamma-1} = T_2 V_3^{\gamma-1}$$

Or, $$\frac{T_1}{T_2} = \left(\frac{V_3}{V_2}\right)^{\gamma-1}$$

Along *DA*: $$T_2 V_4^{\gamma-1} = T_1 V_1^{\gamma-1}$$

Or, $$\frac{T_1}{T_2} = \left(\frac{V_4}{V_1}\right)^{\gamma-1}$$

Hence, $$\left(\frac{V_3}{V_2}\right) = \left(\frac{V_4}{V_1}\right)$$

Or, $$\left(\frac{V_3}{V_4}\right) = \left(\frac{V_2}{V_1}\right) \tag{1.41}$$

Calculation of total work

Along the path *AB*, $dT = 0$ or T = constant.

$$\text{Work} = W_1 = \int_{V_1}^{V_2} P\,dV = nRT_1 \ln \frac{V_2}{V_1} \text{ for } n \text{ moles of an ideal gas}$$

From 1st law of thermodynamics we know that:

$$dQ = dU + dW \tag{1.42a}$$

Again we know that: $$dU = C_v\, dT$$

As $dT = 0$ along *AB*, $dU = 0$

Hence, $$Q_1 = W_1 = nRT_1 \ln \frac{V_2}{V_1}$$

As $V_2 > V_1$, $W_1 > 0$, which indicates that work is done *by* the system.

The curve *BC* shows an adiabatic path. Hence work done along *BC* is given by:

$$W_2 = -\Delta U = -nC_V \int dT \quad \text{[since } dQ = 0 \text{ in Equation (1.42)].}$$

Or, $$Q = 0 \quad \text{and} \quad W_2 = -nC_V \int dT = -nC_V (T_2 - T_1)$$

Along the curve *CD* temperature is kept constant at T_2. Hence according to Equation (1.42) we have:

$$\text{Work} = Q_2 = W_3 = nRT_2 \ln \frac{V_4}{V_3} \quad [\text{since, } dU = 0]$$

As $V_4 < V_3$, $W_3 < 0$ which means that work is done *on* the system.

The curve *DA* shows an adiabatic path. Hence work done along *DA* is given by:

$$W_4 = -\Delta U = -nC_V \int dT \quad [\text{since } dQ = 0 \text{ in Equation (1.42)}].$$

Or,
$$Q = 0 \quad \text{and} \quad W_4 = -nC_V \int dT = -nC_V (T_1 - T_2).$$

Thus the total work $= W = W_1 + W_2 + W_3 + W_4 = W_1 + W_3$ [since, $W_2 + W_4 = 0$]

$$= nRT_1 \ln \frac{V_2}{V_1} + nRT_2 \ln \frac{V_4}{V_3}$$

$$= nRT_1 \ln \frac{V_2}{V_1} - nRT_2 \ln \frac{V_2}{V_1} \quad \left[\frac{V_3}{V_4} = \frac{V_2}{V_1} \text{ equation (1.41)}\right]$$

Or,
$$W = nR(T_1 - T_2) \ln \frac{V_2}{V_1} \tag{1.42b}$$

Again,
$$Q_1 = nRT_1 \ln \frac{V_2}{V_1} \quad \text{and} \quad |Q_1| = nRT_1 \ln \frac{V_2}{V_1}$$

$$Q_2 = nRT_2 \ln \frac{V_4}{V_3} = nRT_2 \ln \frac{V_1}{V_2}$$

Or,
$$Q_2 = -nRT_2 \ln \frac{V_2}{V_1} \quad \text{and} \quad |Q_2| = nRT_2 \ln \frac{V_2}{V_1}$$

The negative sign indicates that heat is rejected to the surroundings, i.e. to the heat sink.

So,
$$W = |Q_1| - |Q_2| \tag{1.42c}$$

Thus the efficiency (η) is given by:

$$\eta = \frac{W}{Q_1} = \frac{|Q_1| - |Q_2|}{|Q_1|} = \frac{T_1 - T_2}{T_1}$$

Or,
$$\eta = 1 - \frac{T_2}{T_1} \tag{1.43}$$

Inference Efficiency of any heat engine is always less than unity.

Again,
$$\eta = \frac{W}{Q_1} = \frac{|Q_1| - |Q_2|}{|Q_1|} = \frac{T_1 - T_2}{T_1}$$

Or, $$1 - \frac{|Q_2|}{|Q_1|} = 1 - \frac{T_2}{T_1}$$

Or, $$\frac{|Q_2|}{|Q_1|} = \frac{T_2}{T_1}$$

Or, $$\frac{|Q_1|}{T_1} = \frac{|Q_2|}{T_2}$$

Or, $$\frac{|Q_1|}{T_1} - \frac{|Q_2|}{T_2} = 0$$

Thus for a complete reversible cycle we have:

$$\oint \frac{dQ}{T}\bigg|_{\text{rev}} = 0$$

So, $\left(\frac{dQ}{T}\right)_{\text{rev}}$ is a state function and is represented by dS, where S is called entropy of the system. Thus we can write that:

$$dS = \frac{dQ}{T}\bigg|_{\text{rev}} \tag{1.44a}$$

Thus for any reversible process:

$$dQ = TdS \tag{1.44b}$$

Show that: $$dS > \frac{dQ}{T}\bigg|_{\text{ir}}$$

From the definition of efficiency (η) of a heat engine we have:

$$\eta = \frac{W}{Q_1}$$

For a reversible Carnot cycle:

$$\eta_r = \frac{W_{\text{rev}}}{Q_1} = \frac{|Q_1| - |Q_2|}{|Q_1|} = \frac{T_1 - T_2}{T_1}$$

Or, $$\eta_r = \frac{T_1 - T_2}{T_1}$$

If one of the steps in a cyclic process is irreversible the whole cyclic process is considered as an irreversible cyclic process. For an irreversible cycle, efficiency is given by:

$$\eta_{\text{ir}} = \frac{W_{\text{ir}}}{Q_1} = \frac{|Q_1| - |Q_2|}{|Q_1|}$$

As discussed in the sec. 1.11, we have: $W_{ir} < W_{rev}$ Hence,

$$\frac{|Q_1| - |Q_2|}{|Q_1|} < \frac{T_1 - T_2}{T_1}$$

Or, $$1 - \frac{|Q_2|}{|Q_1|} < 1 - \frac{T_2}{T_1}$$

Or, $$-\frac{|Q_2|}{|Q_1|} < -\frac{T_2}{T_1}$$

Multiplying both sides by (–1), we get:

$$\frac{|Q_2|}{|Q_1|} > \frac{T_2}{T_1} \quad \text{Or,} \quad \left[\frac{|Q_1|}{T_1} - \frac{|Q_2|}{T_2}\right] < 0$$

Thus for a complete cycle:

$$\oint \left.\frac{dQ}{T}\right|_{ir} < 0$$

This is known as Clasius inequality but for a reversible cycle:

$$\oint dS = 0$$

Hence, $$\oint \left.\frac{dQ}{T}\right|_{ir} < \oint dS$$

Or, $$dS > \left.\frac{dQ}{T}\right|_{ir} \tag{1.44c}$$

Or, $$dQ|_{ir} < TdS \tag{1.44d}$$

Inferences

(a) For a reversible adiabatic process, $dQ = 0$ and hence $dS = 0$ according to Equation (1.44a). The process is also called isentropic process, where both S and Q are constant.

(b) For an irreversible adiabatic process, $dQ = 0$ but $dS > 0$ according to Equation (1.44c).

(c) Combining the above two conditions we can write that $dS_a \geq 0$, where dS_a is the change in entropy in an adiabatic process.

Second law of thermodynamics

(a) According to Equation (1.43), we have

$$\eta_r = \frac{|Q_1| - |Q_2|}{|Q_1|} = \frac{T_1 - T_2}{T_1}$$

So, $$\eta_r < 1$$

Statement There is no such engine which can convert absorbed heat completely into an equivalent amount of work without producing some other changes in the system.

(b) **According to Equation (1.44c), we have: $dS_a \geq 0$**
Statement All natural processes are non-equilibrium and spontaneous and hence entropy of the universe is increasing. So entropy can be created but cannot be destroyed.

(c) **Clasius Statement**
There is no such engine which can absorb heat from a cold reservoir and rejects equal amount of heat into a hot reservoir.

Zero-th law of thermodynamics

If any two systems, maintaining thermal equilibrium with each other, come in contact with the 3rd system, the latter also holds thermal equilibrium with each other.

After establishment of 1st, 2nd and 3rd law of thermodynamics the above statement is required for the development of thermodynamics but the statement logically precedes the other three. According to Zero-th law of thermodynamics temperature is a state function.

Physical significance of entropy

From the definition of entropy we have: $dS = \left.\frac{dQ}{T}\right|_{\text{rev}}$. The term $\frac{dQ}{T}$ is often called reduced heat. In adiabatic process $dQ = 0$ but in non-adiabatic process absorption of heat energy increases the temperature of the system. Only a part of absorbed heat energy is utilized to increase the temperature of the system. Thus entropy increases, since variation of temperature is small compared to amount of heat absorbed.

In an isothermal process heat is absorbed by the system at constant temperature, e.g. sublimation process, in which a solid acquires latent heat and sublimes to gas. The process is irreversible. So according to 2nd law of thermodynamics entropy increases. In a solid, molecules are fixed at their positions and remain in orderly state or compact form but after sublimation molecules are in gaseous state and execute ceaseless random motion, called Brownian motion. Thus in a sublimation process orderly state of molecules changes to the most disorderly state or random state. Thus increase in entropy signifies the disorder or randomness of a system. In other words entropy is a measure of disorder or randomness of a system.

Physical significance of internal energy (*U*)

Internal energy of a system is the amount of energy possessed inherently by that system. Any molecule of a system may possess different types of energy e.g. translational, rotational, vibrational, electronic energy and also energy due to intermolecular forces of attractions and repulsions.

Translational energy arises due to translational motion of molecules, represented by E_{tr}.

If a body changes its spatial orientation or conformation through rotation the energy associated with it is called rotational energy, represented by E_{rot}.

In the case of vibrational energy atoms of a molecule are fixed at their positions but oscillate about their equilibrium positions. The energy associated with it is called vibrational energy, represented by E_{vib}.

Electronic energy In each molecule electrons usually remain in the ground state. In a chemical reaction electronic configuration of the product molecules differ from those of the reactant molecules and hence a change in electronic energy is observed. Energy associated with this change in electronic configuration is called electronic energy represented by E_{el}.

Intermolecular forces of attraction A molecule when comes in contact with another molecule experiences an intermolecular attraction and repulsion. Intermolecular forces of attraction is maximum in a solid phase where molecules remain in most orderly state and is minimum in gaseous phase. The energy associated with it is represented by E_{interm}. Thus for a gas or liquid the total internal energy U is given by:

$$U = E_{tr} + E_{rot} + E_{vib} + E_{interm}.$$

Thermodynamic temperature scale

For a reversible Carnot engine the efficiency η_r is given by:

$$\eta_r = \frac{W_r}{Q_1} = \frac{|Q_1| - |Q_2|}{|Q_1|} = \frac{T_1 - T_2}{T_1}$$

Thus according to 2nd law of thermodynamics, $\eta_r < 1$ for all kinds of heat engines. T represents absolute temperature but there was no reference value of T with respect to which, temperature of any system can be determined. Pressure P has a reference value. The pressure of surrounding atmosphere has been universally accepted as 1 atm. Volume V is an extensive property and can be determined by using the formula: $V = \dfrac{M}{\rho}$

where M = Mass of a substance and ρ = density of that substance.

At this situation the scientist Lord Kelvin attempted to find out the reference value of absolute temperature. In the above equation $|Q_2|$ is the amount of heat rejected to the sink and T_2 is the absolute temperature of the sink. Lord Kelvin imagined a reversible cycle such that no heat is rejected to the sink and hence $|Q_2| = 0$.

Thus, $$\eta_r = \frac{|Q_1| - |Q_2|}{|Q_1|} = \frac{|Q_1|}{|Q_1|} = 1$$

Again, $$\eta_r = \frac{T_1 - T_2}{T_1} = 1$$

Or, $$T_2 = 0$$

According to 2nd law of thermodynamics the above situation is impossible to occur. According to Lord Kelvin the temperature at which the 2nd law of thermodynamics is just violated, is called absolute zero, represented by 0°K in the name of Lord Kelvin.

It was found later that, $T°K = t°C + 273.15$, where $t°C$ is the temperature expressed in °C. Thus the thermodynamic temperature scale has been set.

1.13 Third law of thermodynamics

(a) Max Planck statement (1912) At absolute zero entropy of a perfectly crystalline solid or an ideal solid is zero, where an ideal solid is a crystal possessing regularly oriented atoms or molecules without any crystal defect.

(b) Nernst-Simon statement (1907) For any isothermal process that involves only pure substances, each in internal equilibrium, the entropy change approaches to zero as absolute temperature tends to zero. Mathematically it can be expressed as:

$$\underset{\text{Lt}\, T \to 0}{\Delta S} = 0$$

From Equation (1.28) we have: $dQ = -VdP + C_P\, dT$

At constant pressure, we can write that: $dQ = C_P\, dT$

For a reversible process we know that, $dQ = TdS.$

Thus the above equation becomes: $TdS = C_P\, dT$

Or,
$$dS = \frac{C_P}{T}\, dT \tag{1.45}$$

According to 3rd law of thermodynamics, $\Delta S \to 0$ as $T \to 0$ but according to Equation (1.45), $\Delta S \to \infty$ as $T \to 0$, provided C_P is constant. Thus Equation (1.45) holds good if and only if $C_P \to 0$ as $T \to 0$.

Corollary 1 Heat capacity of an ideal solid or any pure substance approaches zero if the absolute temperature approaches zero. Thus C_P must be a function of temperature as follows:

$$C_P \propto T^a$$

(i) At $a = 1$, $\left(\dfrac{C_P}{T}\right)$ = constant. Thus according to Equation (1.45) we have:

$$\frac{dS}{dT} = \frac{C_P}{T} = \text{constant}$$

Which means that entropy does not change with temperature and has a definite value even at $T \to 0$ and hence the 3rd law of thermodynamics is violated.

(ii) At $a > 1$, $\dfrac{C_P}{T} = f(T)$ Or, $\underset{T \to 0}{\text{Lt}} \left(\dfrac{C_P}{T}\right) = \underset{T \to 0}{\text{Lt}}\; f(T) = 0$

Hence from Equation (1.45) we have:

$$\Delta S = \int \left(\underset{T \to 0}{\text{Lt}} \frac{C_P}{T}\right) dT = 0$$

Thus the 3rd law of thermodynamics holds good.

Corollary 2 Heat capacity of any crystalline substance or pure substance is directly proportional to T^a where $a > 1$

1.14 Calculation of entropy

From the 1st law of thermodynamics we have:

$$dQ = dU + PdV \quad \text{(A)}$$

[considering only expansion work]

Again from Equation (1.28) we have:

$$dQ = -VdP + C_P\, dT \quad \text{(B)}$$

From the definition of entropy, we can write that:

$$dS = \left.\frac{dQ}{T}\right|_{\text{rev}}$$

(a) In isochoric process, $dV = 0$ and the Equation (A) becomes:

$$dQ = dU = nC_V\, dT \quad \text{[for } n \text{ moles of an ideal gas]}$$

So,

$$dS_v = \left.\frac{dQ}{T}\right|_{\text{rev}} = \frac{dU}{T} = nC_v\frac{dT}{T}$$

As U is a state function, dQ also behaves as a state function in isochoric process, since in an isochoric process, $dQ_v = dU$. Thus in isochoric process:

$$dS_v = \left.\frac{dQ}{T}\right|_{\text{rev}} = \frac{dU}{T} = nC_v\frac{dT}{T}$$

On integration, within the limit T_1 and T_2 we get:

$$\Delta S_V = nC_V \ln\frac{T_2}{T_1} \quad (1.46a)$$

[assuming that C_V is independent of temperature]

(b) In an isobaric process, $dP = 0$ and Equation (B) becomes:

$$dQ_P = dH = nC_P\, dT$$

$$dS_P = \left.\frac{dQ}{T}\right|_{\text{rev}} = \left.\frac{dQ_P}{T}\right|_{\text{rev}} = \frac{dH}{T} = nC_P\frac{dT}{T}$$

Assuming that C_P is independent of temperature and integrating within the limits T_1 and T_2, we get:

$$\Delta S_P = nC_P \ln\frac{T_2}{T_1} \quad (1.46b)$$

(c) In isothermal process, $dT = 0$

So, $$dS_T = \left.\frac{dQ}{T}\right|_{rev} = \frac{PdV}{T}$$ From Equation (A)

Again, $$dS_T = \left.\frac{dQ}{T}\right|_{rev} = \frac{-VdP}{T}$$ From Equation (B)

(i) For n moles of an ideal gas we know that:

$$PV = nRT$$

Hence, $$V = \frac{nRT}{P} \quad \text{and} \quad P = \frac{nRT}{V}$$

Putting the values of P and V in the above equations we get:

$$dS_T = \left(\frac{nRT}{V}\right)\left(\frac{dV}{T}\right) \quad \text{and} \quad dS_T = -\left(\frac{nRT}{P}\right)\left(\frac{dP}{T}\right)$$

On integration, we get:

$$\Delta S_T = nR \ln \frac{V_2}{V_1} = -nR \ln \frac{P_2}{P_1}$$

Or, $$\Delta S_T = nR \ln \frac{V_2}{V_1} = nR \ln \frac{P_1}{P_2} \tag{1.46c}$$

(ii) For n moles of a Van-der-Waal's gas we have:

$$\left(P + \frac{an^2}{V^2}\right)(V - nb) = nRT$$

Or, $$P = \frac{nRT}{(V - nb)} - \frac{an^2}{V^2}$$

Hence, $$dS_T = \frac{PdV}{T} = \left[\frac{nRT}{(V - nb)} - \frac{an^2}{V^2}\right]\frac{dV}{T}$$

Or, $$dS_T = nR\,\frac{dV}{(V - nb)} - \frac{an^2}{V^2}\frac{dV}{T}$$

On integration within the limits V_1 and V_2, we get:

$$\Delta S_T = nR \ln \frac{V_2 - nb}{V_1 - nb} + \frac{an^2}{T}\left(\frac{1}{V_2} - \frac{1}{V_1}\right) \tag{1.46d}$$

Neglecting the internal pressure term, i.e. $\frac{an^2}{V^2}$ the Van-der-Waal's equation becomes:

$$P(V - nb) = nRT$$

Or, $$V = \frac{nRT}{P} + nb$$

Thus, $dS_T = -V \dfrac{dP}{T}$ Putting the value of V we get, $dS_T = -\left(\dfrac{nRT}{P} + nb\right)\dfrac{dP}{T}$

Or, $$dS_T = -nR\frac{dP}{P} - \frac{nb}{T}dP$$

On integration within the limits P_1 and P_2, we get:

$$\Delta S_T = nR \ln \frac{P_1}{P_2} + \frac{nb}{T}(P_1 - P_2) \qquad (1.46e)$$

(d) In general the total entropy of 'n' moles of an ideal gas is given by:

$$\Delta S = \Delta S_V + \Delta S_T = nC_V \ln \frac{T_2}{T_1} + nR \ln \frac{V_2}{V_1}$$

[From Equations (1.46a) and (1.46c)]

and $$\Delta S = \Delta S_P + \Delta S_T = nC_P \ln \frac{T_2}{T_1} + nR \ln \frac{P_1}{P_2}$$

[From Equations (1.46b) and (1.46c)

Or, $$S = -nR \ln P + nC_P \ln T + S_0 \qquad (1.46f)$$

$$\text{Molar entropy} = \bar{S} = \left(\frac{S}{n}\right) = -R \ln P + C_P \ln T + \bar{S}_0$$

Or, $$\bar{S} = -R \ln P + C_P \ln T + \bar{S}_0 \qquad (1.46g)$$

(e) For isochoric process we know that:

$$dS_V = \frac{dQ_V}{T} = \frac{dU}{T}$$

For n moles of a Van-der-Waal's gas, we have:

$$dU = nC_V\, dT + \frac{an^2}{V^2}\, dV \quad \text{[From Equation (1.37)]}$$

At constant volume, $dU = nC_V\, dT$

So, $$dS_V = nC_V \frac{dT}{T}$$

So, $$\Delta S_V = nC_V \ln \frac{T_2}{T_1} \quad \text{[identical with the Equation (1.46a)]}$$

So, for n moles of a Van-der-Waal's gas, the total entropy is given by:

$$\Delta S = \Delta S_V + \Delta S_T$$

Or, $$\Delta S = nC_V \ln \frac{T_2}{T_1} + nR \ln \frac{V_2 - nb}{V_1 - nb} + \frac{an^2}{T}\left(\frac{1}{V_2} - \frac{1}{V_1}\right) \quad (1.46h)$$

(f) For isobaric process, we know that:

$$dS_P = \frac{dQ_P}{T} = nC_P \frac{dT}{T} \quad \text{[From Equation (1.28)]}$$

Again, $H = f(P, T)$ [since H is a state function]

Hence, $$dH = \left(\frac{\partial H}{\partial T}\right)_P dT + \left(\frac{\partial H}{\partial P}\right)_T dP$$

At constant pressure (i.e. $dP = 0$)

$$dH_P = C_P\, dT \left[\text{Since, } C_P = \left(\frac{\partial H}{\partial T}\right)_P\right]$$

For n moles, $dH_P = nC_P\, dT$

As there is no P or V term in the above equation the equation is valid both for n moles of an ideal gas as well as n moles of a real gas.
Thus the total entropy for n moles of a Van-der-Waal's gas:

$$\Delta S = \Delta S_T + \Delta S_P = nR \ln \frac{P_1}{P_2} + nC_P \ln \frac{T_2}{T_1} + \frac{nb}{T}(P_1 - P_2)$$

Or, $$\Delta S = nR \ln \frac{P_1}{P_2} + nC_P \ln \frac{T_2}{T_1} + \frac{nb}{T}(P_1 - P_2) \quad (1.46i)$$

(g) Entropy change due to phase transition
It is an isothermal as well as isobaric process, so, $dT = 0$ and $dP = 0$.
One example is: evaporation of water:

$$H_2O(l) \rightleftharpoons H_2O(g)$$

So, $$dS_{T,P} = \frac{dQ_{T,P}}{T}$$

Or, $$\Delta S_{T,P} = \frac{L}{T} \quad (1.46j)$$

Where L is the molar latent heat.

(h) Transition of one mole of a super-cooled liquid water at (–t°C) into ice at (–t°C)

Step-I Heating of one mole of the super-cooled liquid water from (–t°C) to (0°C). Process is isobaric i.e., $dP = 0$. From Equation (1.46b), we can write that:

$$\Delta S_1 = C_{P,l} \ln \frac{T_2}{T_1}$$

where, $T_1 = (-t + 273)K$ and $T_2 = 273$ K , $C_{P,l}$ = molar specific heat of the super-cooled liquid.

Step-II Fusion of super-cooled liquid to ice at 0°C. It is an isothermal and isobaric process. So,. $dT = 0$ and $dP = 0$.

So, $$\Delta S_2 = -\frac{L}{T_2} \quad \text{[From Equation (1.46j)]}$$

The negative sign indicates that latent heat is rejected to the surroundings.

Step-III Cooling of ice from 0°C to (–t°C), i.e. from T_2 to T_1.
It is an isobaric process, i.e. $dP = 0$. So, entropy change is given by:

$$\Delta S_3 = C_{P,s} \ln \frac{T_1}{T_2} \quad \text{[From Equation (1.46b)]}$$

where, $C_{P,s}$ = molar specific heat of ice.

Thus the total entropy change for the whole process is given by:

$$\Delta S = \Delta S_1 + \Delta S_2 + \Delta S_3 = C_{P,1} \ln \frac{T_2}{T_1} - \frac{L}{T_2} + C_{P,s} \ln \frac{T_1}{T_2}$$

Or, $$\Delta S = (C_{P,1} - C_{P,s}) \ln \frac{T_2}{T_1} - \frac{L}{T_2} \quad (1.46k)$$

(i) Entropy change due to mixing of ideal gases Consider m different ideal gases, kept in m different compartments such that pressure and temperature, maintained in each compartment, are P and T respectively. At this stage, the entropy of each gas can be determined by Equation (1.46g).

Compartment	*Gases*	*No. of moles*	*Entropy*
1	g_1	n_1	$n_1 \bar{S}_1 = -n_1R \ln P + n_1 C_{P,1} \ln T + n_1 \bar{S}_1^0$
2	g_2	n_2	$n_2 \bar{S}_2 = -n_2R \ln P + n_2 C_{P,2} \ln T + n_2 \bar{S}_2^0$
3	g_3	n_3	$n_3 \bar{S}_3 = -n_3R \ln P + n_3 C_{P,3} \ln T + n_3 \bar{S}_3^0$
....	....	....	....
....	....	....	....
m	g_m	n_m	$n_m \bar{S}_m = -n_m\text{R} \ln P + n_m C_{P,m} \ln T + n_m \bar{S}_i^0$

Thus the total entropy is given by:

$$S^A = n_1 \bar{S}_1 + n_2 \bar{S}_2 + n_3 \bar{S}_3 \;\ldots\ldots\; + n_i \bar{S}_i + \ldots\ldots$$

$$= \sum_{i=1}^{m} n_i \bar{S}_i = -\sum_{i=1}^{m} n_i R \ln P + \sum_{i=1}^{m} n_i C_{P,i} \ln T + \sum_{i=1}^{m} n_i \bar{S}_i^0$$

$C_{P,i}$ = Specific heat of ith gas

n_i = Number of moles of ith gas

$\bar{S}_i$ = Molar entropy of ith gas

Now all the m different gases are allowed to mix in a single compartment where total pressure and temperature are maintained P and T respectively. After mixing entropy of ith gas is given by:

$$n_i \bar{S}_i' = -n_i R \ln p_i + n_i C_{P,i} \ln T + n_i \bar{S}_i^0$$

Where p_i = partial pressure of ith gas in the compartment = $x_i \cdot P$

x_i = Mole fraction of ith gas in the compartment = $\left(\frac{n_i}{n}\right)$

$n = \Sigma n_i$ and $P = \Sigma p_i$

Thus the total entropy of the gases after mixing is given by:

$$S^B = \sum_{i=1}^{m} n_i \bar{S}_i' = -\sum_{i=1}^{m} n_i R \ln p_i + \sum_{i=1}^{m} n_i C_{P,i} \ln T + \sum_{i=1}^{m} n_i \bar{S}_i^0$$

Thus entropy change due to mixing = ΔS^M

$$\Delta S^M = S^B - S^A = -\sum_{i=1}^{m} n_i R \ln p_i + \sum_{i=1}^{m} n_i R \ln P$$

$$= -\sum_{i=1}^{m} n_i R \ln (x_i P) + \sum_{i=1}^{m} n_i R \ln P$$

$$= -\sum_{i=1}^{m} n_i R \ln x_i$$

$$= -\frac{n}{n} \sum_{i=1}^{m} n_i R \ln x_i$$

Or, $$\Delta S^M = -nR \,\Sigma x_i \ln x_i \tag{1.47}$$

As $x_i < 1$, $\Delta S^M > 0$, which signifies that entropy is increasing on mixing of gases, which is a spontaneous non-equilibrium process since after mixing gases cannot be separated. Thus 2nd law of thermodynamics holds good in mixing of gases.

1.15 Other state functions

Besides U, H and S there are two more state functions. One is called Helmholtz energy or work function, represented by A and the other is Gibbs free energy, represented by G.

Helmholtz free energy or work function (A)

From the definition of entropy we can write that

$$dS \geq \left(\frac{dQ}{T}\right) \quad \text{or} \quad dQ \leq TdS$$

Again, from the 1st law of thermodynamics, we have

$$dW = dQ - dU$$

Putting the value of dQ in the above equation, we get

$$dW \leq TdS - dU$$

For a reversible process or equilibrium process,

$$dQ = TdS \text{ and } dW = dW_{max}$$

Or, $dW_{max} = TdS - dU$. On integration we get:

$$W_{max} = T(S_2 - S_1) - (U_2 - U_1) = (U_1 - TS_1) - (U_2 - TS_2) = A_1 - A_2 = -(A_2 - A_1)$$

Or, $$W_{max} = -\Delta A \quad (1.48)$$

Where, W_{max} = Expansion work + useful work and ΔA represents the maximum work available from a system. Hence it is called work function and is given by:

$$A = U - TS \quad (1.49)$$

Or, $$U = A + TS$$

A is the amount of energy available which can be partly or totally converted into work. Hence A is also known as Helmholtz energy. TS is called bound energy or unavailable energy since it cannot be converted into work but transformed only into heat. Entropy S of the term TS is known as capacity factor of bound energy.

Differentiating the Equation (1.49) we get:

$$dA = dU - TdS - SdT$$

$$= (dQ - dW) - TdS - SdT$$

[since from 1st law of thermodynamics $dQ = dU + dW$]

$$dA = (dQ - TdS) - dW - SdT$$

$$= (dQ - TdS) - PdV - SdT \text{ [considering only expansion work]}$$

Thus at constant V and T, we have: $(dA)_{V,T} = dQ - TdS$.

Or, $(dA)_{V,T} = 0$ for reversible process, since, $dQ = TdS$.

and $(dA)_{V,T} < 0$ for irreversible process, since, $dQ < TdS$.

In general $(dA)_{V,T} \leq 0$, or, $(\Delta A)_{V,T} \leq 0$ (1.50a)

Thus for a non-equilibrium or spontaneous process Helmholtz energy decreases but does not change for equilibrium process.

For a reversible process: $dQ = TdS$ and $dA = -PdV - SdT$ (1.50b)

At constant temperature, i.e, for isothermal process:

$$dA_T = -PdV$$

For n moles of an ideal gas, we have:

$$PV = nRT$$

Or,

$$P = \frac{nRT}{V}$$

Putting the value of P in the above equation we get:

$$dA_T = -nRT\frac{dV}{V}$$

On integration within the limits V_1 and V_2 we get:

$$A_T = nRT \ln\frac{V_1}{V_2} \quad (1.50c)$$

Gibbs free energy (*G*)

From the 1st law of thermodynamics, we have:

$$dQ = dU + dW$$

Or, $dW = dQ - dU.$

Again, $dW = \text{Expansion work} + \text{Useful work} = PdV + dW'$

We know, $dQ \leq TdS$. Thus the above equation becomes:

$$(PdV + dW') \leq (TdS - dU)$$

Or,

$$dW' \leq (TdS - dU - PdV) \quad (1.51)$$

On integration we get:

$$W' \leq T(S_2 - S_1) - (U_2 - U_1) - P(V_2 - V_1)$$

$$W' \leq (TS_2 - TS_1) - [(U_2 + PV_2) - (U_1 + PV_1)]$$

As we know $H = U + PV$, the above equation becomes:

$$W' \leq [(TS_2 - TS_1) - (H_2 - H_1)]$$

Or,

$$W'_{max} = [(TS_2 - TS_1) - (H_2 - H_1)] = [(H_1 - TS_1) - (H_2 - TS_2)]$$

$$W'_{max} = G_1 - G_2$$

Or,

$$W'_{max} = -(G_2 - G_1) = -\Delta G \quad (1.52)$$

Where, $G = H - TS$ and is known as Gibb's free energy. ΔG represents the maximum useful work available from the system:

$$G = H - TS \quad (1.53)$$

Differentiating both sides we get: $dG = dH - TdS - SdT$

As we know $H = U + PV$, we have: $dH = dU + PdV + VdP$

Putting the value of dH in the above equation we get:

$$dG = dU + PdV + VdP - TdS - SdT = (VdP - SdT) - (TdS - dU - PdV)$$

Or,
$$dG \leq (VdP - SdT - dW') \tag{1.54a}$$

[since, $dW' = dQ - dU - PdV$ and $dQ \leq TdS$. So, $(dW') \leq (TdS - dU - PdV)$ and $(-dW') \geq -(TdS - dU - PdV)$]

At constant P and T, $(dG)_{P, T} \leq (-dW')$. If there is no useful work, $dW' = 0$.

So,
$$(dG)_{P, T} \leq 0$$

Or,
$$(\Delta G)_{P, T} \leq 0 \tag{1.54b}$$

For non-equilibrium or spontaneous process $(\Delta G)_{P, T} < 0$, free energy change decreases. The physical significance of the above equation is that energy always flows from higher energy state to lower energy state but the reverse is not true. So any process, where free energy change decreases, is a spontaneous non-equilibrium process and hence the thermodynamic condition for a spontaneous process is $\Delta G < 0$.

Thus the following are true in case of any irreversible or non-equilibrium process:

(i) $dS > \left(\frac{dQ}{T}\right)_{ir}$ or $(\Delta S)_a > 0$ [from 2nd law of thermodynamics]

(ii) $(\Delta A)_{V, T} \leq 0$ [from Equation (1.50a)]

(iii) $(\Delta G)_{P, T} \leq 0$ [from Equation (1.54b)]

For any reversible or equilibrium process the following are true:

(i) $dS = \left(\frac{dQ}{T}\right)_{rev}$ or $(\Delta S)_a = 0$ [from Equation (1.44a)].

(ii) $(\Delta A)_{V, T} = 0$ [from Equation (1.50a)]

(iii) $(\Delta G)_{P, T} = 0$ [from Equation (1.54b)]

$\Delta G = 0$ signifies the thermodynamic condition of equilibrium since at this condition free energy of either side is the same.

For a reversible process Equation (1.54a) becomes:

$$dG = VdP - SdT - dW'$$

Neglecting the useful work or considering only expansion work, i.e., $dW' = 0$, the above equation becomes:

$$dG = VdP - SdT \tag{1.55a}$$

At constant temperature $dT = 0$ and hence $dG = VdP$

For n moles of an ideal gas we know that:

$$PV = nRT$$

Or,
$$V = \frac{nRT}{P}$$

Putting the value of V in the above equation, we get: $dG = nRT \dfrac{dP}{P}$

On integration we get: $G = G^\circ + nRT \ln P$ (1.55b)

Where G° is termed standard free energy, which is defined as the free energy of a system at a constant pressure of 1 atm.

1.16 Gibb's Helmholtz equation

We know from Equation (1.50b) that: $dA = -PdV - SdT$

Again, from Equation (1.49), we have: $A = U - TS$

If we consider two states having work function, internal energies and entropies are A_1 and A_2, U_1 and U_2 and S_1 and S_2 respectively, Equation (1.49) can be written as:

$$A_1 = U_1 - TS_1 \quad \text{and} \quad A_2 = U_2 - TS_2$$

Hence, $(A_2 - A_1) = (U_2 - U_1) - T(S_2 - S_1)$

Or, $\Delta A = \Delta U - T\Delta S$ (1.56)

Differentiating the Equation (1.50b) w.r.t. T at constant V we get:

$$\left(\frac{\partial A}{\partial T}\right)_V = -S$$

Or, $$S = -\left(\frac{\partial A}{\partial T}\right)_V$$

Again, $$S_1 = -\left(\frac{\partial A_1}{\partial T}\right)_V \quad \text{and} \quad S_2 = -\left(\frac{\partial A_2}{\partial T}\right)_V$$

Hence, $$(S_2 - S_1) = \Delta S = -\left[\frac{\partial (A_2 - A_1)}{\partial T}\right]_V = -\left[\frac{\partial(\Delta A)}{\partial T}\right]_V$$

Or, $$\Delta S = -\left[\frac{\partial(\Delta A)}{\partial T}\right]_V$$

Putting the value of S and ΔS in the Equations (1.49) and (1.56) respectively we get:

$$A = U + T\left(\frac{\partial A}{\partial T}\right)_V \tag{1.57a}$$

And $$\Delta A = \Delta U + T\left[\frac{\partial(\Delta A)}{\partial T}\right]_V \tag{1.57b}$$

Similarly, we know from Equation (1.53) that:

$$G = H - TS$$

If we consider two states having free energies G_1 and G_2, enthalpies H_1 and H_2 and entropies S_1 and S_2 respectively, the Equation (1.53) becomes:

$$G_1 = H_1 - TS_1 \quad \text{and} \quad G_2 = H_2 - TS_2.$$

Hence, $$(G_2 - G_1) = (H_2 - H_1) - T(S_2 - S_1)$$

Or, $$\Delta G = \Delta H - T\Delta S \tag{1.58a}$$

From Equation (1.55a) we have:

$$dG = VdP - SdT$$

Differentiating the above equation w.r.t. T at constant P, we have:

$$\left(\frac{\partial G}{\partial T}\right)_P = -S$$

Or, $$S = -\left(\frac{\partial G}{\partial T}\right)_P$$

Again, $$S_1 = -\left(\frac{\partial G_1}{\partial T}\right)_P \quad \text{and} \quad S_2 = -\left(\frac{\partial G_2}{\partial T}\right)_P$$

Hence, $$(S_2 - S_1) = \Delta S = -\left[\frac{\partial(G_2 - G_1)}{\partial T}\right]_P = -\left[\frac{\partial(\Delta G)}{\partial T}\right]_P$$

Or, $$\Delta S = -\left[\frac{\partial(\Delta G)}{\partial T}\right]_P$$

Putting the values of S and ΔS in the Equation (1.53) and (1.58a) respectively, we get:

$$G = H + T\left(\frac{\partial G}{\partial T}\right)_P \tag{1.58b}$$

And $$\Delta G = \Delta H + T\left[\frac{\partial(\Delta G)}{\partial T}\right]_P \tag{1.58c}$$

Equations 1.57a, 1.57b, 1.58b and 1.58c are known as Gibbs-Helmholtz equations:

Another form of Gibbs-Helmholtz equations

Differentiating the term $\left(\frac{G}{T}\right)$ w.r.t. T at constant P, we get:

$$\left[\frac{\partial\left(\frac{G}{T}\right)}{\partial T}\right]_P = -\frac{G}{T^2} + \frac{1}{T}\left(\frac{\partial G}{\partial T}\right)_P = -\frac{1}{T^2}\left[G - T\left(\frac{\partial G}{\partial T}\right)_P\right]$$

Comparing the above equation with (1.58b) we get:

$$\left[\frac{\partial\left(\frac{G}{T}\right)}{\partial T}\right]_P = -\frac{H}{T^2} \qquad (1.59a)$$

Similarly, we have:

$$\left[\frac{\partial\left(\frac{\Delta G}{T}\right)}{\partial T}\right]_P = -\frac{\Delta H}{T^2} \qquad (1.59b)$$

Again, differentiating the term $\left(\frac{A}{T}\right)$ w.r.t. T at constant V we get:

$$\left[\frac{\partial\left(\frac{A}{T}\right)}{\partial T}\right]_V = -\frac{A}{T^2} + \frac{1}{T}\left(\frac{\partial A}{\partial T}\right)_V = -\frac{1}{T^2}\left[A - T\left(\frac{\partial A}{\partial T}\right)_V\right]$$

Comparing the above equation with (1.57a) we get:

$$\left[\frac{\partial\left(\frac{A}{T}\right)}{\partial T}\right]_V = -\frac{U}{T^2} \qquad (1.59c)$$

Similarly, we have:

$$\left[\frac{\partial\left(\frac{\Delta A}{T}\right)}{\partial T}\right]_V = -\frac{\Delta U}{T^2} \qquad (1.59d)$$

Equations (1.59a), (1.59b), (1.59c) and (1.59d) are also known as Gibbs-Helmholtz equations.

1.17 Condition of spontaneity

The Helmholtz equation is given by:

$$\Delta G = \Delta H - T\Delta S \qquad \text{(from equation 1.58a)}$$

We have discussed in the earlier section that the thermodynamic condition for a spontaneous process is: $\Delta G < 0$

(i) For a chemical reaction if $|\Delta H| > 0$ and $|\Delta S| > 0$, the thermodynamic condition to occur the chemical reaction spontaneously is: $|T\Delta S| > |\Delta H|$

(ii) For a chemical reaction if $|\Delta H| > 0$ and $|\Delta S| < 0$, the chemical reaction cannot occur since in this case ΔG is always positive at any temperature.

(iii) For a chemical reaction if $\Delta H < 0$ and $\Delta S > 0$, the chemical reaction always occurs spontaneously, since ΔG is always negative at any temperature.

(iv) For a chemical reaction if $\Delta H < 0$ and $\Delta S < 0$, $|\Delta H| > |T\Delta S|$ is the thermodynamic condition to occur the chemical reaction spontaneously.

1.18 Phase transition Clasius-Clayperon equation

If a pure substance exists in two or more phases in equilibrium there is a possibility of transition from phase to another. Such transition takes place at constant temperature and pressure but entropy changes due to change in molecular disorderness. If phase transition takes place in a closed system, an equilibrium will be established between the phases. In that case we can write:

$$dQ = TdS.$$

From the 1st law of thermodynamics, have:

$$dU = dQ - dW = dQ - PdV \quad \text{[considering only expansion work]}$$

Assuming the process is reversible we can write that: $dU = TdS - PdV$

On integration of the above equation, we get:

$$U_2 - U_1 = T(S_2 - S_1) - P(V_2 - V_1)$$

Or, $$(U_2 + PV_2) - TS_2 = (U_1 + PV_1) - TS_1$$

Or, $$H_2 - TS_2 = H_1 - TS_1$$

Or, $$G_2 = G_1 \quad \text{Or,} \quad G_1 = G_2 \tag{1.60a}$$

If the pressure and temperature are changed slightly from P to $P + dP$ and T to $T + dT$ respectively, G_1 changes to $G_1 + dG_1$ and G_2 changes to $G_2 + dG_2$. According to La Chatelier principle the system rearranges within itself to restore the equilibrium. Thus we can write that:

$$G_1 + dG_1 = G_2 + dG_2$$

Comparing the above equation with the Equation (1.60a) we get:

$$dG_1 = dG_2 \tag{1.60b}$$

From Equations (1.55a) and (1.60b) we have:

$$V_1dP - S_1dT = V_2dP - S_2dT$$

Or, $$(V_2 - V_1)dP = (S_2 - S_1)dT$$

Or, $$\frac{dP}{dT} = \frac{(S_2 - S_1)}{(V_2 - V_1)}$$

$(S_2 - S_1)$ = Entropy change due to phase transition.

According to Equation (1.46j), we have:

$$\Delta S = (S_2 - S_1) = \frac{L}{T}$$

where L is the molar latent heat of phase transition at temperature T.

Thus the above equation becomes:

$$\frac{dP}{dT} = \frac{L}{T(V_2 - V_1)} \tag{1.60c}$$

Consider the following phase equilibrium:

$$\text{Liquid} \rightleftharpoons \text{Vapour}$$

The Equation (1.60c) can be written as:

$$\frac{dP}{dT} = \frac{L}{T(V_g - V_l)}$$

where, V_l = Volume of liquid and V_g = Volume of vapour

Since, $V_g >> V_l$, it may be approximated that: $(V_g - V_l) \approx V_g$

So, we can write that:

$$\frac{dP}{dT} = \frac{L}{T V_g}$$

Assuming that the vapour behaves ideally, we can write that: $PV_g = RT$ for one mole of vapour.

Or, $$V_g = \frac{RT}{P}$$

Putting the value of V_g in the above equation we get:

$$\frac{dP}{dT} = \frac{PL}{RT^2}$$

Or, $$\frac{dP}{P} = \frac{L}{RT^2}\, dT$$

Or, $$\int_{P_1}^{P_2} d\ln P = \int_{T_1}^{T_2} \frac{L}{RT^2}\, dT$$

Integrating the above equation within limits we get:

$$\ln \frac{P_2}{P_1} = -\frac{L}{R}\left[\frac{1}{T_2} - \frac{1}{T_1}\right] \qquad (1.60d)$$

The Equation (1.60d) is known as Classius-Clayperon equation.

Application One of the most important applications of the above equation is to determine the latent heat of vaporization of a solvent or liquid.

1.19 Maxwell's relations

From Equation (1.54b) we have:

$dG = VdP - SdT$ [Equation (1.55a)].

Consider a system, changing its state from point A to point B (Figure 1.12). The change may proceed along ACB or along ADB. The path ACB consists of two steps AC and CB. Along AC pressure is constant. So, differentiating the Equation (1.55a) w.r.t. T at constant P we get:

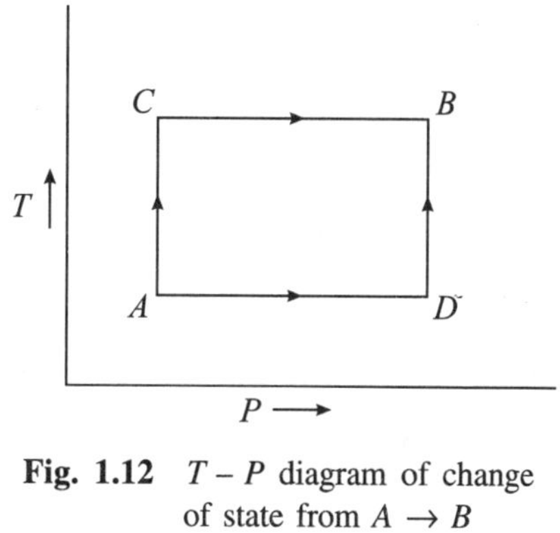

Fig. 1.12 $T - P$ diagram of change of state from $A \rightarrow B$

$$\left(\frac{\partial G}{\partial T}\right)_P = -S \tag{1.61a}$$

Along *CB* temperature remains constant. Thus differentiating the Equation (1.61a) w.r.t. *P* at constant *T*, we get:

$$\frac{\partial^2 G}{\partial T \cdot \partial P} = -\left(\frac{\partial S}{\partial P}\right)_T \tag{1.61b}$$

On the other hand, the path *ADB* consists of two steps *AD* and *DB*. Along *AD* temperature is constant and hence change in free energy can be determined by differentiating the Equation (1.55a) w.r.t., *P* at constant *T* we get:

$$\left(\frac{\partial G}{\partial P}\right)_T = V \tag{1.61c}$$

Along *DB* pressure remains constant. Thus further change in free energy can be determined by differentiating the Equation (1.61c) with respect to *T* at constant *P* we get:

$$\frac{\partial^2 G}{\partial P \cdot \partial T} = \left(\frac{\partial V}{\partial T}\right)_P \tag{1.61d}$$

As *G* is a state function, the change in free energy does not depend on path. Thus:

$$\frac{\partial^2 G}{\partial T \cdot \partial P} = \frac{\partial^2 G}{\partial P \cdot \partial T} \tag{1.61e}$$

Hence, from Equations (1.61b) and (1.61d), we can write that:

$$\left(\frac{\partial S}{\partial P}\right)_T = -\left(\frac{\partial V}{\partial T}\right)_P \tag{1.62a}$$

Again, from Equation (1.50b), we have: $dA = -PdV - SdT$ [Equation (1.50b)].

As *A* is a state function, we can write that:

$$\frac{\partial^2 A}{\partial T \cdot \partial V} = \frac{\partial^2 A}{\partial V \cdot \partial T} \quad \text{[Ch. Equation (1.61e)]}$$

$$\left(\frac{\partial A}{\partial T}\right)_V = -S \quad \text{and} \quad \frac{\partial^2 A}{\partial T \cdot \partial V} = -\left(\frac{\partial S}{\partial V}\right)_T$$

$$\left(\frac{\partial A}{\partial V}\right)_T = -P \quad \text{and} \quad \frac{\partial^2 A}{\partial V \cdot \partial T} = -\left(\frac{\partial P}{\partial T}\right)_V$$

Hence,

$$\left(\frac{\partial S}{\partial V}\right)_T = \left(\frac{\partial P}{\partial T}\right)_V \tag{1.62b}$$

From Equation (1.29) we have: $dH = dQ + VdP$ [Equation (1.29)].

As we know, $dQ = TdS$, the above equation becomes:

$$dH = TdS + VdP$$

As H is a state function we can write that:

$$\frac{\partial^2 H}{\partial P \cdot \partial S} = \frac{\partial^2 H}{\partial S \cdot \partial P} \quad \text{[Ch. Equation (1.61e)]}$$

$$\left(\frac{\partial H}{\partial P}\right)_S = V \quad \text{and} \quad \frac{\partial^2 H}{\partial P \cdot \partial S} = \left(\frac{\partial V}{\partial S}\right)_P$$

$$\left(\frac{\partial H}{\partial S}\right)_P = T \quad \text{and} \quad \frac{\partial^2 H}{\partial S \cdot \partial P} = \left(\frac{\partial T}{\partial P}\right)_S$$

Hence,

$$\left(\frac{\partial V}{\partial S}\right)_P = \left(\frac{\partial T}{\partial P}\right)_S \tag{1.62c}$$

Again from 1st law of thermodynamics, we have:

$$dQ = dU + dW$$

Or, $$dU = dQ - dW$$

Or, $$dU = dQ - PdV \quad \text{[considering only expansion work]}$$

Or, $$dU = TdS - PdV \quad \text{[since, } dQ = TdS \text{ for reversible process]}$$

As U is a state function we can write that:

$$\frac{\partial^2 U}{\partial V \cdot \partial S} = \frac{\partial^2 U}{\partial S \cdot \partial V} \quad \text{[Ch. Equation (1.61e)]}$$

$$\left(\frac{\partial U}{\partial V}\right)_S = -P \quad \text{and} \quad \frac{\partial^2 U}{\partial V \cdot \partial S} = -\left(\frac{\partial P}{\partial S}\right)_V$$

$$\left(\frac{\partial U}{\partial S}\right)_V = T \quad \text{and} \quad \frac{\partial^2 U}{\partial S \cdot \partial V} = \left(\frac{\partial T}{\partial V}\right)_S$$

Hence,

$$\left(\frac{\partial P}{\partial S}\right)_V = -\left(\frac{\partial T}{\partial V}\right)_S \tag{1.62d}$$

$$\left.\begin{array}{l} \left(\frac{\partial S}{\partial V}\right)_T = \left(\frac{\partial P}{\partial T}\right)_V \quad \text{and} \quad \left(\frac{\partial V}{\partial S}\right)_P = \left(\frac{\partial T}{\partial P}\right)_S \\ \left(\frac{\partial S}{\partial P}\right)_T = -\left(\frac{\partial V}{\partial T}\right)_P \quad \text{and} \quad \left(\frac{\partial P}{\partial S}\right)_V = -\left(\frac{\partial T}{\partial V}\right)_S \end{array}\right\} \tag{1.62e}$$

The Equation (1.62e) is known as Maxwell's equation or Maxwell's relation. It is interesting to note that the Maxwell's relation is the relation among four parameters P, V, T and S. Further, the slopes of $P - S$ curve in both isothermal and isochoric processes are negative, i.e. $\left(\frac{\partial S}{\partial P}\right)_T$ and $\left(\frac{\partial P}{\partial S}\right)_V$ terms are negative in Maxwell's relations and the slopes of $V - S$ curve in both isothermal and isobaric processes are positive, i.e., $\left(\frac{\partial S}{\partial V}\right)_T$ and $\left(\frac{\partial V}{\partial S}\right)_P$, terms are positive in Maxwell's relations.

Applications of Maxwell's relation

(i) Evaluation of ΔU As U is a state function we can write that: $U = f(V, T)$

Hence,
$$dU = \left(\frac{\partial U}{\partial V}\right)_T dV + \left(\frac{\partial U}{\partial T}\right)_V dT \tag{1.63a}$$

From, 1st law of thermodynamics, we have:

$$dU = dQ - PdV \quad \text{[considering only expansion work]}$$

Or,
$$dU = TdS - PdV \quad \text{[since, } dQ = TdS \text{ for reversible process]}$$

Differentiating w.r.t. V at constant T, we have:

$$\left(\frac{\partial U}{\partial V}\right)_T = \left[T\left(\frac{\partial S}{\partial V}\right)_T - P\right]$$

From Maxwell's relation we have:

$$\left(\frac{\partial S}{\partial V}\right)_T = \left(\frac{\partial P}{\partial T}\right)_V \quad \text{[From Equation (1.62e)]}$$

So,
$$\left(\frac{\partial U}{\partial V}\right)_T = \left[T\left(\frac{\partial P}{\partial T}\right)_V - P\right] \tag{1.63b}$$

From Equation (1.23) we have: $dQ = l\,dV + C_V\,dT$

where,
$$(l - P) = \left(\frac{\partial U}{\partial V}\right)_T \quad \text{[From Equation (1.22)]}$$

$(l - P)$ is termed internal pressure.

Thus,
$$(l - P) = \left[T\left(\frac{\partial P}{\partial T}\right)_V - P\right]$$

(a) For n moles of an ideal gas.

we know that: $PV = nRT$

Differentiating w.r.t. T at constant V, we get:

$$V\left(\frac{\partial P}{\partial T}\right)_V = nR$$

Or,
$$T\left(\frac{\partial P}{\partial T}\right)_V = \frac{nRT}{V} = P$$

Hence, $(l - P) = (P - P) = 0$

Or, $l = P$ and

$$\left(\frac{\partial U}{\partial V}\right)_T = (l - P) = 0$$

Scientist Joule found the same result in his experiment [discussed in the sec. 1.7] Thus in the case of an ideal gas, Equation (1.63a) becomes:

$$dU = \left(\frac{\partial U}{\partial T}\right)_V dT = nC_V\, dT \quad \text{[for } n \text{ moles]}$$

On integration we get:

$$\Delta U = nC_V\,(T_2 - T_1) \quad \text{and} \quad l = P \tag{1.64a}$$

(b) For n moles of a Van-der-Waal's gas, we have:

$$\left(P + \frac{an^2}{V^2}\right)(V - nb) = nRT$$

Or, $$P = \frac{nRT}{(V - nb)} - \frac{an^2}{V^2}$$

Differentiating w.r.t. T at constant V, we get:

$$\left(\frac{\partial P}{\partial T}\right)_V = \frac{nR}{(V - nb)}$$

Or, $$T\left(\frac{\partial P}{\partial T}\right)_V = \frac{nRT}{(V - nb)} = P + \frac{an^2}{V^2}$$

Thus, $$(l - P) = \left[T\left(\frac{\partial P}{\partial T}\right)_V - P\right] = \left[P + \frac{an^2}{V^2} - P\right] = \frac{an^2}{V^2}$$

Or, $$(l - P) = \left(\frac{\partial U}{\partial V}\right)_T = \frac{an^2}{V^2}$$

So, $(l - P)$ represents the internal pressure of a real gas, as represented by Equation (1.22).

Putting this value of $\left(\frac{\partial U}{\partial V}\right)_T$ in the equation on (1.63a) we get:

$$dU = \left(\frac{\partial U}{\partial T}\right)_V dT + \frac{an^2}{V^2}\, dV$$

$$= nC_V\, dT + \frac{an^2}{V^2}\, dV$$

On integration, we get:

$$\left.\begin{aligned} \Delta U &= nC_V (T_2 - T_1) - an^2\left[\frac{1}{V_2} - \frac{1}{V_1}\right] \\ \text{and} \quad l &= P + \frac{an^2}{V^2} \end{aligned}\right\} \tag{1.64b}$$

(ii) Evaluation of ΔH As H is a state function we can write that:

$$H = f(P, T)$$

$$\left.\begin{aligned} \text{Hence} \quad dH &= \left(\frac{\partial H}{\partial T}\right)_P dT + \left(\frac{\partial H}{\partial P}\right)_T dP \\ \text{Or,} \quad dH &= nC_P\, dT + \left(\frac{\partial H}{\partial P}\right)_T dP \end{aligned}\right\} \tag{1.65a}$$

From Equation (1.26), we have: $dQ = h\,dP + C_P\,dT$ [Equation (1.26)]

Where, $$h = \left(\frac{\partial Q}{\partial P}\right)_T$$

Again, we know that: $dH = dQ + VdP$ [Equation (1.29)]

Or, $$\left(\frac{\partial H}{\partial P}\right)_T = \left(\frac{\partial Q}{\partial P}\right)_T + V = h + V$$

Or, $$h = \left(\frac{\partial H}{\partial P}\right)_T - V$$

Again, $$dQ = TdS$$

Thus, $$dH = TdS + VdP$$

Differentiating w.r.t. P at constant T we have:

$$\left(\frac{\partial H}{\partial P}\right)_T = \left[T\left(\frac{\partial S}{\partial P}\right)_T + V\right]$$

From Maxwell's equation, we have:

$$\left(\frac{\partial S}{\partial P}\right)_T = -\left(\frac{\partial V}{\partial T}\right)_P \quad \text{[From Equation (1.62e)]}$$

Thus, $$\left(\frac{\partial H}{\partial P}\right)_T = \left[-T\left(\frac{\partial V}{\partial T}\right)_P + V\right] \tag{1.65b}$$

(a) For n moles of an ideal gas, we have: $PV = nRT$
Differentiating w.r.t. T at constant P, we get:

$$P\left(\frac{\partial V}{\partial T}\right)_P = nR$$

Or, $$T\left(\frac{\partial V}{\partial T}\right)_P = \frac{nRT}{P} = V$$

Or, $$\left(\frac{\partial H}{\partial P}\right)_T = -V + V = 0$$

and $$h = \left(\frac{\partial H}{\partial P}\right)_T - V = -V$$

Or, $$h + V = 0$$

Thus in the case of an ideal gas, Equation (1.65a) becomes:

$$dH = nC_P\, dT$$

On integration we get: $\Delta H = nC_P\,(T_2 - T_1)$

$$\Delta H = nC_P\,(T_2 - T_1) \quad \text{and} \quad (h + V) = 0 \tag{1.66a}$$

(b) For n moles of a Van-der-Waal's gas, we have:

$$\left(P + \frac{an^2}{V^2}\right)(V - nb) = nRT$$

Neglecting the internal pressure terms, the above equation becomes:

$$P(V - nb) = nRT$$

Or, $$V = \frac{nRT}{P} + nb$$

Differentiating w.r.t. T at constant P, we have:

$$\left(\frac{\partial V}{\partial T}\right)_P = \frac{nR}{P}$$

Or, $$T\left(\frac{\partial V}{\partial T}\right)_P = \frac{nRT}{P} = (V - nb)$$

Or, $$\left(\frac{\partial H}{\partial P}\right)_T = \left[-T\left(\frac{\partial V}{\partial T}\right)_P + V\right] = -V + nb + V = nb$$

Or, $$\left(\frac{\partial H}{\partial P}\right)_T = nb$$

and $$h = \left(\frac{\partial H}{\partial P}\right)_T - V = (nb - V) = -(V - nb)$$

Or, $$h + (V - nb) = 0$$

Putting this value of $\left(\frac{\partial H}{\partial P}\right)_T$ in the Equation (1.65a), we get:

$$dH = nC_P\, dT + nb\, dP$$

On integration we get: $\Delta H = nC_P (T_2 - T_1) + nb(P_2 - P_1)$

$$\left.\begin{aligned} &\Delta H = nC_P (T_2 - T_1) + nb(P_2 - P_1) \\ \text{and} \quad &[h + (V - nb)] = 0 \end{aligned}\right\} \quad (1.66b)$$

(iii) Evaluation of $\left(\frac{\partial S}{\partial V}\right)_T$ for an ideal gas as well as Van-der Waal's gas:

From Maxwell's relation [Equation (1.62e)], we know that:

$$\left(\frac{\partial S}{\partial V}\right)_T = \left(\frac{\partial P}{\partial T}\right)_V$$

(a) For n moles of an ideal gas, we know that: $PV = nRT$

Differentiating w.r.t. T at constant V we get:

$$V\left(\frac{\partial P}{\partial T}\right)_V = nR$$

Or, $$\left(\frac{\partial P}{\partial T}\right)_V = \frac{nR}{V} \quad (1.67a)$$

(b) For n moles of a Van-der-Waal's gas, we have:

$$\left(P + \frac{an^2}{V^2}\right)(V - nb) = nRT$$

Or, $$P = \frac{nRT}{(V - nb)} - \frac{an^2}{V^2}$$

Differentiating w.r.t. T at constant V, we get:

$$\left(\frac{\partial P}{\partial T}\right)_V = \frac{nR}{(V - nb)} \quad (1.67b)$$

(iv) Evaluation of $(C_P - C_V)$:

From the definition of specific heat we know that:

$$C_V = \left(\frac{\partial Q}{\partial T}\right)_V \quad \text{and} \quad C_P = \left(\frac{\partial Q}{\partial T}\right)_P$$

Again, we have: $dQ = TdS$

Thus, $$C_V = T\left(\frac{\partial S}{\partial T}\right)_V + dS \quad \text{and} \quad C_P = T\left(\frac{\partial S}{\partial T}\right)_P + dS$$

So, $$C_P - C_V = T\left(\frac{\partial S}{\partial T}\right)_P - T\left(\frac{\partial S}{\partial T}\right)_V$$

$$= T\left[\left(\frac{\partial S}{\partial T}\right)_P - \left(\frac{\partial S}{\partial T}\right)_V\right]$$

As S is a state function, we can write that: $S = S(V, T)$

Or, $$dS = \left(\frac{\partial S}{\partial T}\right)_V dT + \left(\frac{\partial S}{\partial V}\right)_T dV$$

Differentiating w.r.t. T at constant P, we get:

$$\left(\frac{\partial S}{\partial T}\right)_P = \left(\frac{\partial S}{\partial T}\right)_V + \left(\frac{\partial S}{\partial V}\right)_T \left(\frac{\partial V}{\partial T}\right)_P$$

Or, $$\left(\frac{\partial S}{\partial T}\right)_P - \left(\frac{\partial S}{\partial T}\right)_V = \left(\frac{\partial S}{\partial V}\right)_T \left(\frac{\partial V}{\partial T}\right)_P$$

Thus, $$C_P - C_V = T\left[\left(\frac{\partial S}{\partial T}\right)_P - \left(\frac{\partial S}{\partial T}\right)_V\right] = T\left(\frac{\partial S}{\partial V}\right)_T \left(\frac{\partial V}{\partial T}\right)_P$$

From Maxwell's relation, we have:

$$\left(\frac{\partial S}{\partial V}\right)_T = \left(\frac{\partial P}{\partial T}\right)_V \quad \text{[From Equation (1.62e)]}$$

Thus, $$C_P - C_V = T\left(\frac{\partial P}{\partial T}\right)_V \left(\frac{\partial V}{\partial T}\right)_P \tag{1.68}$$

(a) For n moles of an ideal gas, we know that: $PV = nRT$
Differentiating w.r.t. T at constant V, we get:

$$V\left(\frac{\partial P}{\partial T}\right)_V = nR$$

Or, $$\left(\frac{\partial P}{\partial T}\right)_V = \frac{nR}{V}$$

Or, $$T\left(\frac{\partial P}{\partial T}\right)_V = \frac{nRT}{V} = P$$

Again differentiating both sides of $PV = nRT$ w.r.t. T at constant P we get:

$$P\left(\frac{\partial V}{\partial T}\right)_P = nR$$

Or, $$\left(\frac{\partial V}{\partial T}\right)_P = \frac{nR}{P}$$

Putting these values of $T\left(\frac{\partial P}{\partial T}\right)_V$ and $\left(\frac{\partial V}{\partial T}\right)_P$ in the Equation (1.68), we get:

$$C_P - C_V = P\left(\frac{nR}{P}\right) = nR$$

Or, $C_P - C_V = nR$

For 1 mole of an ideal gas: $C_P - C_V = R$

(b) For n moles of a Van-der-Waal's gas, we know that:

$$\left(P + \frac{an^2}{V^2}\right)(V - nb) = nRT$$

Or, $$P = \frac{nRT}{(V - nb)} - \frac{an^2}{V^2}$$

Differentiating both sides w.r.t. T at constant V, we get:

$$\left(\frac{\partial P}{\partial T}\right)_V = \frac{nR}{(V - nb)}$$

Or, $$T\left(\frac{\partial P}{\partial T}\right)_V = \frac{nRT}{(V - nb)} = \left(P + \frac{an^2}{V^2}\right)$$

Again, $$\left(P + \frac{an^2}{V^2}\right)(V - nb) = nRT$$

So, $$PV - Pnb + \frac{an^2}{V} - \frac{an^3b}{V^2} = nRT$$

Differentiating both sides, w.r.t. T at constant P, we get:

$$P\left(\frac{\partial V}{\partial T}\right)_P - \frac{an^2}{V^2}\left(\frac{\partial V}{\partial T}\right)_P + \frac{2an^3b}{V^3}\left(\frac{\partial V}{\partial T}\right)_P = nR$$

Neglecting $\frac{1}{V^2}$ and $\frac{1}{V^3}$ terms, we get:

$$P\left(\frac{\partial V}{\partial T}\right)_P = nR$$

Or, $$\left(\frac{\partial V}{\partial T}\right)_P = \frac{nR}{P}$$

Putting these values of $T\left(\frac{\partial P}{\partial T}\right)_V$ and $\left(\frac{\partial V}{\partial T}\right)_P$, in the Equation (1.68), we get:

$$C_P - C_V = \left(P + \frac{an^2}{V^2}\right)\frac{nR}{P} = nR + \frac{an^3R}{PV^2}$$

Or, $$C_P - C_V = nR + \frac{an^3R}{PV^2}$$

For 1 mole of a Van-der-waal's gas:

$$C_P - C_V = R + \frac{aR}{PV^2}$$

(v) Joule Thomson (*J-T*) effect In 1853 Joule and William Thomson performed an experiment which involved a slow throttling of a gas through a rigid porous plug, illustrated in the Figure 1.13.

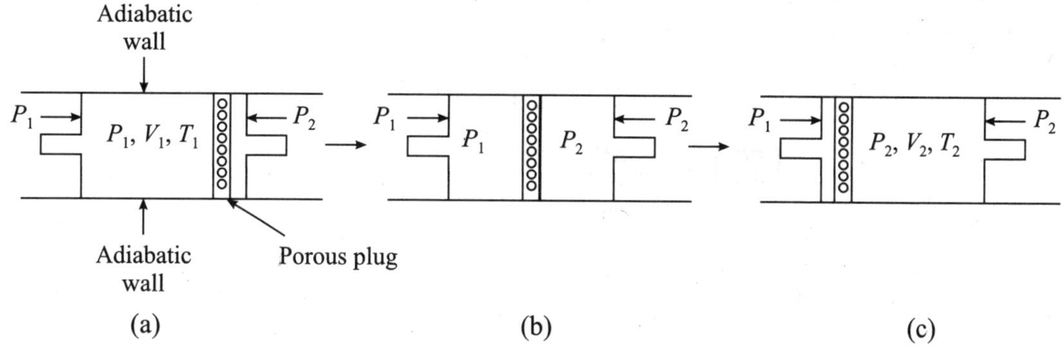

Fig. 1.13 The Joule-Thomson experiment.

There are two chambers having fixed pressures of P_1 and P_2 respectively such that $P_2 < P_1$. The chambers are separated by a porous plug, as shown in the Figure 1.13(a). Initially, the volume of left chamber is V_1 and that of right chamber is zero. The gas at the left chamber is allowed to expand through the porous plug and hence volume of the left chamber decreases while that of the right chamber increases. At the end, the volume of left chamber becomes zero and that of the right chamber is V_2.

As the system is covered by adiabatic walls there is no heat transfer between system and surroundings. So, $dQ = 0$. From the first law of thermodynamics, we have:

$$dQ = dU + dW$$

So, $$dW = -dU$$

On integration we get: $\Delta W = -(U_2 - U_1)$ = Net work done by the gas = $W_1 + W_2$

W_1 = Work done by the gas on left chamber, where pressure is fixed at P_1 and volume changes from V_1 to 0.

$$W_1 = \int_{V_1}^{0} P_1\, dV = -P_1 V_1$$

W_2 = Work done by the gas on right chamber, where pressure is fixed at P_2 and volume changes from 0 to V_2.

$$W_2 = \int_{0}^{V_2} P_2\, dV = P_2 V_2$$

$\therefore$ $$\Delta W = P_2 V_2 - P_1 V_1$$

Putting this value of ΔW in the above equation, we get:

$$P_2V_2 - P_1V_1 = -(U_2 - U_1)$$

Or, $$(U_2 + P_2V_2) = (U_1 + P_1V_1)$$

Or, $$H_2 = H_1, \quad \text{Or,} \quad dH = 0.$$

Thus Joule Thomson expansion process or throttling process is an isenthalpic process. The $T - P$ diagram of throttling process is shown in the Figure 1.14.

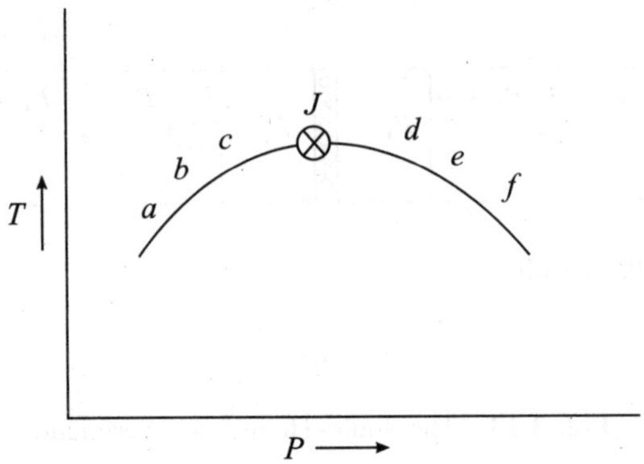

Fig. 1.14 Isenthalpic $T - P$ diagram of a gas executing throttling process for different values of P_2 at fixed P_1.

At any print on the curve, the slope is given by $\mu_J = \left(\frac{\partial T}{\partial P}\right)_H$ and μ_J is called Joule Thomson coefficient or $J - T$ coefficient. In the Figure 1.14, at the point J the curve shows a maximum and hence $\mu_J = 0$, indicating that, $dT = 0$. So no response will be observed at the point J on throttling of the gas.

To the left of the point J (i.e., *a*, *b*, *c* etc.) slope is positive, i.e., $\mu_J > 0$, indicating that $dT < 0$, since dP is always negative due to expansion of gas. Thus at these points cooling will be observed on throttling of the gas.

To the right of the point J (i.e., *d*, *e*, *f* etc.) slope is negative, i.e., $\mu_J < 0$, indicating that $dT > 0$, since dP is always negative due to expansion of gas. Thus at these points heating will be observed on throttling of the gas.

As enthalpy H is a state function, the equation of state is given by: $f(H, P, T) = 0$

Hence, we can write that:

$$\left(\frac{\partial T}{\partial P}\right)_H \left(\frac{\partial P}{\partial H}\right)_T \left(\frac{\partial H}{\partial T}\right)_P = -1 \quad \text{[Ch. Equation (1.11)]}$$

Or, $$\left(\frac{\partial T}{\partial P}\right)_H = -\frac{\left(\frac{\partial H}{\partial P}\right)_T}{\left(\frac{\partial H}{\partial T}\right)_P}$$

As we know, $\mu_J = \left(\frac{\partial T}{\partial P}\right)_H$ and $C_P = \left(\frac{\partial H}{\partial T}\right)_P$

We have, $$\mu_J = -\frac{1}{C_P}\left(\frac{\partial H}{\partial P}\right)_T \quad (1.69)$$

From Equation (1.29) we have:

$$dH = dQ + VdP \quad \text{[Equation (1.29)]}$$

Or, $$dH = TdS + VdP \quad \text{[since } dQ = TdS\text{]}$$

Differentiating w.r.t. P at constant T, we get:

$$\left(\frac{\partial H}{\partial P}\right)_T = T\left(\frac{\partial S}{\partial P}\right)_T + V$$

From Maxwell's relations, we have:

$$\left(\frac{\partial S}{\partial P}\right)_T = -\left(\frac{\partial V}{\partial T}\right)_P \quad \text{[From Equation (1.62e)].}$$

Putting the value of $\left(\frac{\partial S}{\partial P}\right)_T$ in the above equation, we get:

$$\left(\frac{\partial H}{\partial P}\right)_T = -T\left(\frac{\partial V}{\partial T}\right)_P + V$$

Putting the value of $\left(\frac{\partial H}{\partial P}\right)_T$ in the Equation (1.69), we get:

$$\mu_J = -\frac{1}{C_P}\left[-T\left(\frac{\partial V}{\partial T}\right)_P + V\right] = \frac{1}{C_P}\left[T\left(\frac{\partial V}{\partial T}\right)_P - V\right] \quad (1.70)$$

(a) In the case of an ideal gas, we have:

$$PV = nRT$$

Differentiating w.r.t. T at constant P, we get:

$$P\left(\frac{\partial V}{\partial T}\right)_P = nR$$

Or, $$T\left(\frac{\partial V}{\partial T}\right)_P = \frac{nRT}{P} = V$$

Thus, $$\mu_J = -\frac{1}{C_P}\left[-T\left(\frac{\partial V}{\partial T}\right)_P + V\right] = -\frac{1}{C_P}(-V + V) = 0$$

Thus when an ideal gas is allowed to expand through a throttle valve, the temperature of the gas remains constant. In other words, neither cooling nor heating effect is observed on execution of a throttling process of an ideal gas.

(b) For 1 mole of a Van-der-Waal's gas, we have:

$$\left(P+\frac{a}{V^2}\right)(V-b) = RT$$

$$\left(P+\frac{a}{V^2}\right) = \frac{RT}{(V-b)} \tag{1.71}$$

Differentiating w.r.t. T at constant P, we get:

$$-\frac{2a}{V^3}\left(\frac{\partial V}{\partial T}\right)_P = \frac{R}{(V-b)} - \frac{RT}{(V-b)^2}\left(\frac{\partial V}{\partial T}\right)_P$$

$$\left[\frac{RT}{(V-b)^2} - \frac{2a}{V^3}\right]\left(\frac{\partial V}{\partial T}\right)_P = \frac{R}{(V-b)}$$

Or,
$$T\left(\frac{\partial V}{\partial T}\right)_P = \frac{\dfrac{RT}{(V-b)}}{\left[\dfrac{RT}{(V-b)^2} - \dfrac{2a}{V^3}\right]} \tag{1.72}$$

Or,
$$T\left(\frac{\partial V}{\partial T}\right)_P - V = \frac{\dfrac{RT}{(V-b)}}{\left[\dfrac{RT}{(V-b)^2} - \dfrac{2a}{V^3}\right]} - V$$

$$= \frac{\dfrac{RT}{(V-b)} - \dfrac{RTV}{(V-b)^2} + \dfrac{2a}{V^2}}{\left[\dfrac{RT}{(V-b)^2} - \dfrac{2a}{V^3}\right]}$$

$$= \frac{\left[\dfrac{2a}{V^2} - \dfrac{RTb}{(V-b)^2}\right]}{\left[\dfrac{RT}{(V-b)^2} - \dfrac{2a}{V^3}\right]}$$

Or,
$$\mu_J = \frac{\left[\dfrac{2a}{V^2} - \dfrac{RTb}{(V-b)^2}\right]}{C_P\left[\dfrac{RT}{(V-b)^2} - \dfrac{2a}{V^3}\right]} \tag{1.73}$$

Inversion temperature The temperature at which $J - T$ coefficient (μ_J) of a real gas is zero, is called inversion temperature, represented by T_i. Above T_i, $\mu_J < 0$ and hence heating effect will be observed on execution of throttling process and below T_i, $\mu_J > 0$ and hence cooling effect will be observed on execution of throttling process.

Thus from Equation (1.73), $\mu_J = 0$ when $T = T_i$.

Hence,
$$\frac{RT_i b}{(V-b)^2} - \frac{2a}{V^2} = 0$$

Or,
$$T_i = \frac{2a}{Rb}\left(\frac{V-b}{V}\right)^2$$

Or,
$$\left(\frac{V-b}{V}\right)^2 = \frac{Rb}{2a}T_i \qquad (1.74)$$

Or,
$$Y^2 = 4mX$$

where,
$$Y = \left(\frac{V-b}{V}\right),\ 4m = \frac{Rb}{2a} \quad \text{and} \quad X = T_i$$

Thus the plot of $\left[\frac{(V-b)}{V}\right]$ against T is a parabola. Similar parabolic curve is also obtained by plotting T vs. P, as shown in the Figure 1.15.

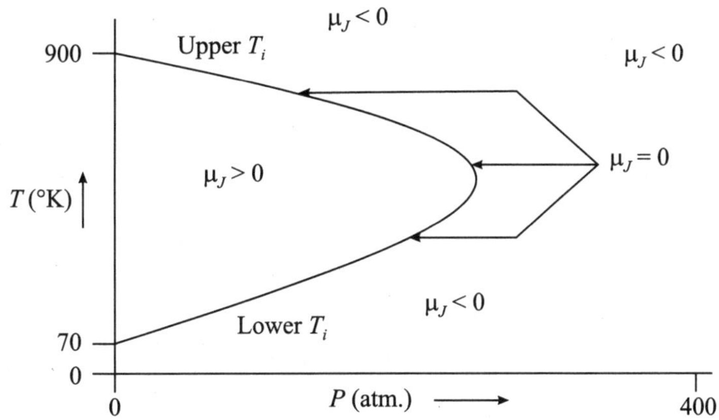

Fig. 1.15 $T - P$ diagram of a real gas undergoing a throttling process.

From the Figure 1.15 following observations can be made:

(a) If the throttling process is executed at any point within the area bounded by the curve and T-axis, the cooling effect will be observed, whereas if the same is executed at any point outside the curve, heating effect will be observed. On the line of the curve, $\mu_J = 0$ and no effect will be observed.

(b) The upper curve is called the upper T_i-curve and the lower curve is called lower T_i-curve. Thus except at nose there are two T_i-values at a particular pressure.
Usually throttling process is executed at around room temperature and all the real gases show cooling effect except hydrogen and helium. In these cases, the upper T_i value is very low e.g. the upper T_i value of hydrogen is –73°C(200°K) at 1 atm and – 92°C(181°K)

at 100 atm. So, throttling temperature (T_h) is higher than upper T_i value. Thus according to the Figure 1.15 heating effect is observed since throttling temperature (T_h) must lie within upper and lower T_i-values to get cooling effect.

If we consider $(V - b) \approx V$, the Equation (1.74) becomes:

$$1 = \frac{Rb}{2a} T_i$$

Or,
$$T_i = \frac{2a}{Rb}$$

(c) $(V - b) \approx V$, so, $\left[\frac{RT}{(V-b)^2}\right] > \left[\frac{2a}{V^3}\right]$, since volume of a gas is usually very large. Thus, denominator of Equation (1.73) is always positive. Hence, $\mu_J > 0$, when $2a > RTb$. Or, $T < \left(\frac{2a}{Rb}\right)$. Or, $T < T_i$. Thus when a real gas is subjected to throttling process at a temperature below its inversion temperature (T_i), cooling effect will be observed.

(d) $(V - b) \approx V$, so, $\left[\frac{RT}{(V-b)^2}\right] > \left[\frac{2a}{V^3}\right]$, since volume of a gas is usually very large. Thus, denominator of Equation (1.73) is always positive. Hence, $\mu_J < 0$, when $2a < RTb$. Or, $T > \left(\frac{2a}{Rb}\right)$. Or, $T > T_i$. Thus when a real gas is subjected to throttling process at a temperature above its inversion temperature (T_i), heating effect will be observed.

1.20 Fugacity

According to Equation (1.55a), we have:

$$dG = VdP - SdT \quad \text{[Equation (1.55a)]}$$

For 1 mole of an ideal gas, we know that,

$$PV = RT$$

Or,
$$V = \frac{RT}{P}$$

For an isothermal process, (i.e. $dT = 0$), the above equation becomes:

$$dG = VdP = RT\,\frac{dP}{P}$$

Or,
$$dG = RT\,d\ln P$$

On integration, we get:
$$G = G^\circ + RT\ln P$$

At a particular pressure, the value of G of a real gas differs from that calculated from the above equation. Thus for 1 mole of a real gas, the above equation can be written as:

$$G = G^\circ + RT \ln f \tag{1.75}$$

where f is known as fugacity.

As we know at very low pressure any real gas behaves ideally, we can write that:

$$\underset{P \to 0}{\text{Lt}} \left(\frac{f}{P} \right) = 1$$

Differentiating Equation (1.75), we get:

$$dG = RT \, d\ln f$$

Again we know $dG = VdP$ at constant temperature. Thus,

$$VdP = RT \, d\ln f$$

Or,
$$d\ln f = \frac{V}{RT} dP$$

Where, V = real volume of gas

= (Total volume) – (Volume of Avogadro No. of molecules)

= $(V_{id} - \alpha)$

V_{id} = Total volume = volume of one mole of an ideal gas = $\left(\frac{RT}{P} \right)$.

Or,
$$V = \left(\frac{RT}{P} - \alpha \right)$$

Putting this value of V in the above equation, we get:

$$d \ln f = \left(\frac{1}{P} - \frac{\alpha}{RT} \right) dP$$

Or,
$$d\ln f = d\ln P - \left(\frac{\alpha}{RT} \right) dP$$

On integration, we get:

$$\ln \frac{f_2}{f_1} = \ln \frac{P_2}{P_1} - \frac{1}{RT} \int_{P_1}^{P_2} \alpha \, dP$$

At,
$$P_1 \to 0, \ \frac{f_1}{P_1} = 1$$

So,
$$\ln f_2 = \ln P_2 + \ln \frac{f_1}{P_1} - \frac{1}{RT} \int_{P_1}^{P_2} \alpha \, dP$$

Or,
$$\ln f_2 = \ln P_2 + \underset{P_1 \to 0}{\text{Lt}} \left(\ln \frac{f_1}{P_1} \right) - \frac{1}{RT} \int_{P_1 \to 0}^{P_2} \alpha \, dP$$

Or, $$\ln f_2 = \ln P_2 - \frac{\alpha}{RT} \int_{P_1 \to 0}^{P_2} dP$$

[assuming the α remains constant between pressure limits P_1 and P_2]

Or, $$\ln f_2 = \ln P_2 - \frac{\alpha P_2}{RT}$$

Eliminating the subscript the above equation becomes:

Or, $$\ln f = \ln P - \frac{\alpha P}{RT}$$

Or, $$\ln \frac{f}{P} = -\frac{\alpha P}{RT}$$

Or, $$\frac{f}{P} = e^{-\frac{\alpha P}{RT}} = \gamma \tag{1.76}$$

γ is known as activity coefficient. Again:

$$\gamma = e^{-\frac{\alpha P}{RT}} = 1 - \frac{\alpha P}{RT} + \frac{1}{2!}\left(\frac{\alpha P}{RT}\right)^2 - \frac{1}{3!}\left(\frac{\alpha P}{RT}\right)^3 + \ldots$$

Neglecting the square and higher terms, we get:

$$\gamma = 1 - \frac{\alpha P}{RT} = \frac{P}{RT}\left(\frac{RT}{P} - \alpha\right) = \frac{\left(\frac{RT}{P} - \alpha\right)}{\left(\frac{RT}{P}\right)} = \frac{V}{V_{id}}$$

Or, $$\gamma = \frac{V}{V_{id}} = 1 - \frac{\alpha P}{RT} \tag{1.77}$$

V = Real volume of a gas and can be experimentally determined.

$V_{id} = \left(\frac{RT}{P}\right)$. So V_{id} can be easily calculated at a particular temperature and pressure.

Thus by measuring V, γ and hence α can easily be evaluated from Equation (1.77). Once γ is evaluated the fugacity f can easily be calculated from Equation (1.76).

1.21 Dependence of specific heat on volume and pressure

From 1st law of thermodynamics, we have:

$$dU = dQ - dW$$

Or, $$dU = dQ - PdV \quad \text{[considering only expansion work]}$$

From Equation (1.63a), we have:

$$dU = \left(\frac{\partial U}{\partial V}\right)_T dV + \left(\frac{\partial U}{\partial T}\right)_V dT \quad \text{[Equation (1.63a)]}$$

From Equation (1.63b), we have:

$$\left(\frac{\partial U}{\partial V}\right)_T = \left[T\left(\frac{\partial P}{\partial T}\right)_V - P\right] \quad \text{[Equation (1.63b)]}$$

Putting the value of $\left(\frac{\partial U}{\partial V}\right)_T$ in the above equation we get:

$$dU = \left[T\left(\frac{\partial P}{\partial T}\right)_V - P\right] dV + C_V\, dT \tag{1.78a}$$

where, $C_V = \left(\frac{\partial U}{\partial T}\right)_V$

As U is a state function, we have:

$$\frac{\partial^2 U}{\partial V \cdot \partial T} = \frac{\partial^2 U}{\partial T \cdot \partial V}$$

Thus, we have:

$$\left[\frac{\partial}{\partial T}\left\{T\left(\frac{\partial P}{\partial T}\right)_V - P\right\}\right]_V = \left[\frac{\partial}{\partial V} C_V\right]_T$$

Or, $$T\left(\frac{\partial^2 P}{\partial T^2}\right)_V = \left[\frac{\partial}{\partial V} C_V\right]_T \tag{1.78b}$$

Similarly from Equation (1.65a) and (1.65b), we have:

$$dH = \left(\frac{\partial H}{\partial P}\right)_T dP + \left(\frac{\partial H}{\partial T}\right)_P dT \quad \text{[Equation (1.65a)}$$

$$\left(\frac{\partial H}{\partial P}\right)_T = \left[-T\left(\frac{\partial V}{\partial T}\right)_P + V\right] \quad \text{[Equation (1.65b)}$$

Combining the above two equation, we get:

$$dH = \left[-T\left(\frac{\partial V}{\partial T}\right)_P + V\right] dP + C_P\, dT \tag{1.78c}$$

where, $C_P = \left(\frac{\partial H}{\partial T}\right)_P$

As H is a state function, we have:

$$\frac{\partial^2 H}{\partial P \cdot \partial T} = \frac{\partial^2 H}{\partial T \cdot \partial P}$$

$$\left[\frac{\partial}{\partial T}\left\{-T\left(\frac{\partial V}{\partial T}\right)_P + V\right\}\right]_P = \left[\frac{\partial}{\partial P}C_P\right]_T$$

Or, $$-T\left(\frac{\partial^2 V}{\partial T^2}\right)_P = \left[\frac{\partial}{\partial P}C_P\right]_T \quad (1.78d)$$

(a) For 1 mole of an ideal gas, we know that:

$$PV = RT$$

Differentiating both sides w.r.t. T at constant V, we get:

$$V\left(\frac{\partial P}{\partial T}\right)_V = R$$

Or, $$\left(\frac{\partial P}{\partial T}\right)_V = \frac{R}{V}$$

Again differentiating w.r.t. T, at constant V, we get:

$$\left(\frac{\partial^2 P}{\partial T^2}\right)_V = 0$$

Thus Equation (1.78b) becomes:

$$\left(\frac{\partial C_V}{\partial V}\right)_T = 0$$

Which indicates that in an isothermal process C_V of an ideal gas is independent of volume. Similarly, differentiating the ideal gas equation w.r.t. T at constant P, we get:

$$P\left(\frac{\partial V}{\partial T}\right)_P = R$$

Or, $$\left(\frac{\partial V}{\partial T}\right)_P = \frac{R}{P}$$

Further differentiating w.r.t. T at constant P, we get:

$$\left(\frac{\partial^2 V}{\partial T^2}\right)_P = 0$$

Thus Equation (1.78d) becomes:

$$\left(\frac{\partial C_P}{\partial P}\right)_T = 0$$

which signifies that in an isothermal process C_P of an ideal gas is independent of pressure.

(b) For 1 mole of a Van-der-Waal's gas, we have:

$$\left(P + \frac{a}{V^2}\right)(V - b) = RT$$

Or, $$P = \frac{RT}{(V-b)} - \frac{a}{V^2}$$

Differentiating w.r.t. T at constant V, we get:

$$\left(\frac{\partial P}{\partial T}\right)_V = \frac{R}{(V-b)}$$

Further differentiating w.r.t. T at constant V, we get:

$$\left(\frac{\partial^2 P}{\partial T^2}\right)_V = 0$$

Thus the Equation (1.78b) becomes:

$$\left(\frac{\partial C_V}{\partial V}\right)_T = 0$$

Which signifies that in an isothermal process C_V of a real gas is independent of volume. Neglecting the internal pressure term the Van-der-Waal's equation becomes:

$$P(V - b) = RT$$

Or, $$V = \frac{RT}{P} + b$$

Differentiating both sides w.r.t. T at constant P, we get:

$$\left(\frac{\partial V}{\partial T}\right)_P = \frac{R}{P}$$

Further differentiating w.r.t. T at constant P, we get:

$$\left(\frac{\partial^2 V}{\partial T^2}\right)_P = 0$$

Thus the Equation (1.78d) becomes:

$$\left(\frac{\partial C_P}{\partial P}\right)_T = 0$$

which signifies that in an isothermal process C_P of a real gas is independent of pressure.

1.22 Partial molar quantities and chemical potential

In the case of an intensive property sum of all partial quantities gives the total quantity. For example, total pressure (P) of a gas mixture of m different gases, is given by:

$$P = p_1 + p_2 + p_3 + \ldots\ldots\ldots + p_m$$

where, p_1, p_2, p_3, etc. are the partial pressures of the corresponding gases.

In the case of any extensive property mass or mole must appear in the additive sequence, e.g., the total volume of the same gas mixture, as described above, is given by:

$$V = n_1\bar{V}_1 + n_2\bar{V}_2 + n_3\bar{V}_3 + \ldots\ldots + ni\bar{V}_i + \ldots$$

where, $\bar{V}_i$ molar volume of ith species in the mixture and n_i is number of moles of ith species. Differentiating both sides w.r.t. n_1 at constant n_2, n_3, etc, we have:

$$\left(\frac{\partial V}{\partial n_1}\right)_{n_2} = \bar{V}_1$$

As pressure (P) and temperature (T) remain constant during mixing, we can write that:

$$\left(\frac{\partial V}{\partial n_1}\right)_{P,T,n_2} = \bar{V}_1$$

In general, $$\bar{V}_i = \left(\frac{\partial V}{\partial n_i}\right)_{P,T,n_j}$$

The subscript n_j indicates that number of moles of all species excepting the ith species are kept constant. $\bar{V}_i$ is known as partial molar volume of ith species. Similarly, we have:

Partial molar internal energy, $$\bar{U}_i = \left(\frac{\partial U}{\partial n_i}\right)_{P,T,n_j}$$

Partial molar enthalpy, $$\bar{H}_i = \left(\frac{\partial H}{\partial n_i}\right)_{P,T,n_j}$$

Partial molar entropy, $$\bar{S}_i = \left(\frac{\partial S}{\partial n_i}\right)_{P,T,n_j}$$

Partial molar work function,

$$\bar{A}_i = \left.\frac{\partial A}{\partial n_i}\right|_{P,T,n_j}$$

Partial molar Gibb's free energy,

$$\bar{G}_i = \left.\frac{\partial G}{\partial n_i}\right|_{P,T,n_j}$$

$\bar{G}_i$ is specially termed chemical potential of ith species, represented by μ_i
Thus, we may write that:

$$\mu_i = \left.\frac{\partial G}{\partial n_i}\right|_{P,T,n_j}$$

Euler's theorem For any homogeneous function of degree m e.g., $\varphi = f(x^m\ y^m\ z^m)$ we can write that:

$$x\frac{\partial\phi}{\partial x} + y\frac{\partial\phi}{\partial y} + z\frac{\partial\phi}{\partial z} = m\phi$$

In the case of a mixture of several components Gibb's free energy G is a homogeneous function of degree 1 as given below:

$$G = G(P,\ T,\ n_1,\ n_2, \ldots\ldots)$$

At constant pressure and temperature, we have:

$$G = G(n_1,\ n_2, \ldots\ldots)$$

Hence, by Euler's theorem we can write that:

$$G = n_1\frac{\partial G}{\partial n_1}\bigg|_{P,T,n_2} + n_2\frac{\partial G}{\partial n_2}\bigg|_{P,T,n_3} + \ldots\ldots + n_i\frac{\partial G}{\partial n_i}\bigg|_{P,T,n_j} + \ldots\ldots$$

From the definition of chemical potential, the above equation becomes:

$$G = n_1\mu_1 + n_2\mu_2 + \ldots + n_i\mu_i + \ldots\ldots$$

Or, $$G = \Sigma n_i\mu_i \tag{1.79a}$$

Similarly, we can write that:

$$\left.\begin{aligned} V &= \Sigma n_i \bar{V}_i \qquad U = \Sigma n_i \bar{U}_i \\ H &= \Sigma n_i \bar{H}_i \qquad S = \Sigma n_i \bar{S}_i \\ A &= \Sigma n_i \bar{A}_i \end{aligned}\right\} \tag{1.79b}$$

Again we know that:

$$G = G(n_1,\ n_1,\ \ldots) \quad \text{[at constant temperature and pressure]}$$

Differentiating we get:

$$dG = \frac{\partial G}{\partial n_1}\bigg|_{P,T,n_2} dn_1 + \frac{\partial G}{\partial n_2}\bigg|_{P,T,n_3} dn_2 + \ldots\ldots + \frac{\partial G}{\partial n_i}\bigg|_{P,T,n_j} dn_i + \ldots\ldots$$

$$= \mu_1 dn_1 + \mu_2 dn_2 + \ldots + \mu_i dn_i + \ldots$$

$$= \Sigma\mu_i dn_i$$

$$dG = \Sigma\mu_i dn_i \tag{1.79c}$$

Differentiating Equation (1.79a) we get:

$$dG = \Sigma\mu_i dn_i + \Sigma n_i d\mu_i$$

Putting the value of dG from Equation (1.79c) into the above equation we get:

$$\Sigma\mu_i dn_i = \Sigma\mu_i dn_i + \Sigma n_i d\mu_i$$

Or, $$\Sigma n_i d\mu_i = 0 \tag{1.79d}$$

Thus we have the following set of equations:

$$\left.\begin{aligned} G &= \sum_i n_i \mu_i \\ dG &= \sum_i \mu_i \, dn_i \\ \sum_i n_i \, d\mu_i &= 0 \end{aligned}\right\} \tag{1.79e}$$

The set of Equations (1.79e) is known as Gibb's – Duhem equation.

1.23 Thermodynamic properties of ideal gas mixture

Ideal gas mixture When two or more ideal gases are mixed there will be no intermolecular interaction among the gas molecules. That type of mixture is called ideal gas mixture.

Ideal solution A solution contains molecules of different components. In the case of an ideal solution, molecules of one component behave identically with those of other components and molecules of all components change their positions without changing the energy of intermolecular interactions in the solution.

Thus from the definition of an ideal solution it is clear that at constant temperature and pressure volume (V) and internal energy (U) do not change during the formation of an ideal solution from the corresponding pure components. So, mathematically we can write that: $\Delta V_{\text{Mix}} = 0$ and $\Delta U_{\text{Mix}} = 0$

As we know, $\Delta H = \Delta U + P\Delta V,$

we have: $\Delta H^M = \Delta H_{\text{Mix}} = \Delta U_{\text{Mix}} + P\Delta V_{\text{Mix}} = 0$

From Helmholtz equation, we have:

$$\Delta G = \Delta H - T\Delta S$$

In the case of an ideal solution, the above equation becomes:

$$\Delta G^M = \Delta H^M - T\Delta S^M = -T\Delta S^M \quad [\text{since, } \Delta H^M = 0]$$

Or, $\Delta G^M = -T\Delta S^M$

From the Equation (1.47), we have:

$$\Delta S^M = -nR\,\Sigma x_i \ln x_i \quad [\text{Equation (1.47)}]$$

Putting the value of ΔS^M in the above equation we get:

$$\Delta G^M = nRT\,\Sigma x_i \ln x_i \tag{1.80}$$

ΔG^M Free energy change during mixing of pure components to form an ideal solution.

As $x_i < 1$, $\Delta G^M < 0$. Thus free energy decreases during mixing. Before mixing all components are pure components. The total free energy is given by the Equation (1.79e):

$$G^* = \sum_i n_i \mu_i^*$$

μ_i^* = Chemical potential of pure liquid of ith species.

After mixing the free energy is given by:

$$G = \sum_i n_i \mu_i$$

Thus the change in free energy is given by:

$$\Delta G^M = G - G^* = \sum_i n_i \mu_i - \sum_i n_i \mu_i^*$$

$$= \sum_i n_i (\mu_i - \mu_i^*)$$

Or,

$$\Delta G^M = \sum_i n_i (\mu_i - \mu_i^*) \tag{1.81}$$

In the case of an ideal solution, Equations (1.80) and (1.81) are the same and hence we have:

$$\sum_i n_i (\mu_i - \mu_i^*) = nRT \sum_i x_i \ln x_i = RT \sum_i n_i \ln x_i$$

Where,

$$n.x_i = n\left(\frac{n_i}{n}\right) = n_i$$

So,

$$(\mu_i - \mu_i^*) = RT \ln x_i$$

Or,

$$\mu_i = \mu_i^*(T, P) + RT \ln x_i$$

Standard state In case of an ideal solution standard state of any component is nothing but the pure form of that component. So, when $x_i = 1$, the ith species is in the pure form.

So,

$$\mu_i\big|_{x_i = 1} = \mu_i^0 = \mu_i^*(T, P)$$

Thus the above equation becomes:

$$\mu_i = \mu_i^0 + \text{RT} \ln x_i \tag{1.82a}$$

This is the general expression of chemical potential of ith species for an ideal solution. Conversely, a solution is said to be ideal if the chemical potential of each component of the solution obeys the Equation (1.82a) for all solution compositions and for a range of T and P.

(i) In the case of an ideal gas mixture, we have:

$$x_i = \left(\frac{p_i}{P}\right)$$

Putting this value of x_i in the above Equation (1.82a), we get:

$$\mu_i = \mu_i^0 + RT \ln \frac{p_i}{P}$$

$$= (\mu_i^0 - RT \ln P) + RT \ln P_i$$

$$\mu_i = \mu_i^0(p) + RT \ln p_i \tag{1.82b}$$

where,

$$\mu_i^0(p) = \mu_i^0 - RT \ln P$$

P = Total pressure of the gas mixture.

p_i = partial pressure of the ith species in the mixture.

(ii) In the case of an ideal solution,

$$x_i = \left(\frac{a_i}{a}\right)$$

a_i = Activity of ith species in the solution = $m_i.\gamma_i$

a = Activity of the whole solution = $m.\gamma$

m and γ indicate the concentration (in molality) and activity coefficient respectively. Thus Equation (1.82a) becomes:

$$\mu_i = \mu_i^0 + RT \ln \frac{a_i}{a}$$

Or, $$\mu_i = (\mu_i^0 - RT \ln a) + RT \ln a_i$$

Or, $$\mu_i = \mu_i^0(a) + RT \ln a_i \tag{1.82c}$$

Again, $$\mu_i = \mu_i^0(a) + \text{RT} \ln(m_i\ \gamma_i)$$

Or, $$\mu_i = [\mu_i^0(a) + RT \ln \gamma_i] + RT \ln m_i$$

Or, $$\mu_i = \mu_i^0(m) + RT \ln m_i \tag{1.82d}$$

$$\left.\begin{aligned} \mu_i &= \mu_i^0 + RT \ln x_i \\ \mu_i &= \mu_i^0(p) + RT \ln p_i \\ \mu_i &= \mu_i^0(a) + RT \ln a_i \\ \mu_i &= \mu_i^0(m) + RT \ln m_i \end{aligned}\right\} \tag{1.82e}$$

The change in thermodynamic properties for a two-component ideal solution is given by:

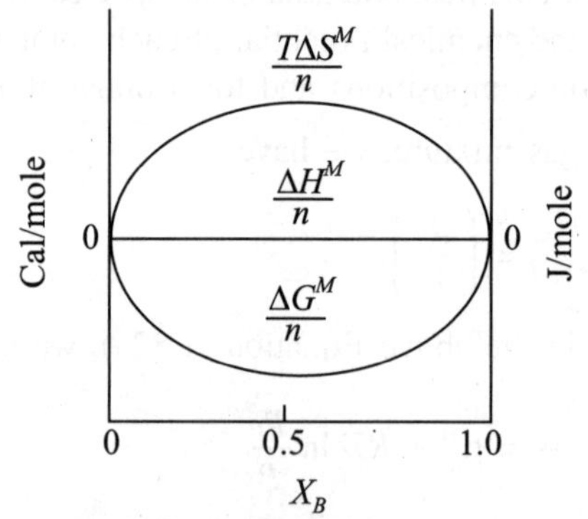

Fig. 1.16 Mixing properties of a two component system as a function of composition at 25°C. X_B = Mole fraction of the component B in the mixture.

1.24 Raoult's law and Henry's law

Other thermodynamic properties of ideal solutions: Consider a system containing a multi component ideal solution remaining in equilibrium with its vapours.

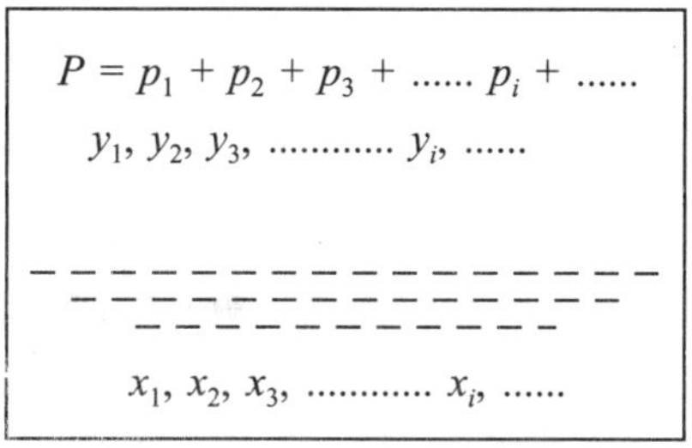

Fig. 1.17 An ideal solution in equilibrium with its vapours.
x_i = Mole fraction of *i*th species in solution
y_i = Mole fraction of *i*th species in vapour

As the system is in equilibrium we can write for *i*th component:

$$\mu_i(l) = \mu_i(v)$$

By using the Equation (1.82), we can write that:

$$\mu^0_{i,\,1}(T,\,P) + RT\ln x_i = \mu^0_{i,\,v}(T,\,P) + RT\ln y_i$$

Or, $$\mu^0_{i,\,1}(T,\,P) + RT\ln x_i = \mu^0_{i,\,v}(T,\,P) + RT\ln\frac{p_i}{P} \tag{1.83a}$$

If the above equilibrium exists between pure liquid and vapours of *i*th component, we have:

$$x_i = 1$$

$$\mu^0_{i,\,1}(T,\,P) = \mu^0_{i,\,1}(T,\,P^0)$$

Where, P^0 = vapour pressure of the pure liquid of *i*th component at temperature T.

Thus the Equation (1.83a) becomes:

$$\mu^0_{i,\,1}(T,\,P^0) = \mu^0_{i,\,v}(T,\,P) + RT\ln\frac{P^0}{P} \tag{1.83b}$$

Subtracting the Equation (1.83b) from the Equation (1.83a) we get:

$$\mu^0_{i,\,1}(T,\,P) - \mu^0_{i,\,1}(T,\,P^0) + RT\ln x_i = RT\ln\frac{p_i}{P^0}$$

As the pressure difference $(P - P^0)$ is very low:

$$\mu^0_{i,\,1}(T,\,P) = \mu^0_{i,\,1}(T,\,P^0)$$

Thus the above equation becomes:

$$RT\ln x_i = \mathrm{RT}\ln\frac{p_i}{P^0}$$

Or, $$p_i = x_i\, P^0 \qquad (1.84a)$$

Equation (1.84a) is the mathematical form of Raoult's law, which states that partial pressure of a component in an ideal solution is directly proportional to the vapour pressure of the pure liquid. From the Equation (1.83a) we have:

$$\mu_1^0(T, P) - \mu_v^0(T, P) = \text{RT} \ln p_i - \text{RT} \ln P - RT \ln x_i$$

Or, $$RT \ln \frac{p_i}{x_i} = \mu_1^0(T, P) - \mu_v^0(T, P) + RT \ln P$$

Or, $$\ln \frac{p_i}{x_i} = \frac{1}{RT}[\mu_1^0(T, P) - \mu_v^0(T, P) + RT \ln P]$$

$$= \ln K \text{ (where, } K \text{ is a constant)}$$

Or, $$p_i = K\, x_i \qquad (1.84b)$$

Equation (1.84b) is the mathematical form of Henry's law which states that partial vapour pressure of solute i is directly proportional to the mole fraction of that solute present in the ideal solution.

Inference In an ideal dilute solution the solvent obeys Raoult's law and each solute obeys Henry's law.

1.25 Application of Gibb's-Duhem equation

One of the major application of Gibb's-Duhem equation is to evaluate the thermodynamic properties of components in a solution. Consider a binary system of two components such that one component (say component 1) behaves ideally.

Thus for component 1 Equation (1.82) can be written as:

$$\mu_1 = \mu_1^0 + RT \ln p_1$$

As the component 1 behaves ideally, Raoult's law holds good. So from Equation (1.84a) we can write that, $p_1 = x_1 P_1^0$, where P_1^0 is the vapour pressure of pure liquid of component 1. Putting this value of p_1 into the above equation, we get:

$$\mu_1 = \mu_1^0 + RT \ln (x_1 P_1^0)$$

$$= \mu_1^0 + RT \ln P_1^0 + RT \ln x_1$$

Or, $$\mu_1 = \mu_{11}^0 + RT \ln x_1$$

Differentiating both sides, we get:

$$d\mu_1 = RT \,d\ln x_1$$

Or, $$d\mu_1 = RT \frac{dx_1}{x_1} \qquad (1.85a)$$

Again sum of the mole fractions of the two components is unity, i.e., $x_1 + x_2 = 1$

Differentiating both sides, we get:

$$dx_1 + dx_2 = 0$$

Or, $$dx_1 = -dx_2$$

Putting this value of dx_1 in the Equation (1.85a), we get:

$$d\mu_1 = -RT\,\frac{dx_2}{x_1} \quad (1.85b)$$

Applying Gibb's-Duhem equation in this binary system, we get:

$$n_1\, d\mu_1 + n_2\, d\mu_2 = 0 \quad \text{[Ch. Equation (1.79e)]}$$

$$d\mu_1 = -\frac{n_2}{n_1}\, d\mu_2$$

Or, $$d\mu_1 = -\frac{n_2/n}{n_1/n}\, d\mu_2$$

Or, $$d\mu_1 = -\frac{x_2}{x_1}\, d\mu_2 \quad (1.85c)$$

$$n = \text{Total number of moles} = n_1 + n_2$$

x_1 and x_2 are the mole fractions of the two components.

Combining the Equations (1.85b) and (1.85c), we get:

$$\frac{x_2}{x_1}\, d\mu_2 = RT\,\frac{dx_2}{x_1}$$

Or, $$d\mu_2 = RT\,\frac{dx_2}{x_2}$$

On integration we get:

$$\mu_2 = \mu_{22} + RT \ln x_2 \quad (1.85d)$$

The Equation (1.85d) signifies that the component 2 behaves ideally and hence according to Raoult's law, we can write that:

$P_2 = x_2\, P_2^0$, where, P_2^0 is the vapour pressure of the pure liquid of component 2.

It can be shown that in a multi-component system if one component behaves ideally the other components of the system also behave ideally.

Inference In any multi-component system all the components behave in a similar manner. Thus study of one component is enough to evaluate the thermodynamic properties of the whole system.

1.26 Van't Hoff equation

Equilibrium constant (K_a) and standard free energy change ($\Delta G°$) Consider a general chemical reaction:

$$aA + bB + cC \;........\; \rightleftharpoons lL + mM + nN \;........$$

According to Gibb's-Duhem equation [Equation (1.79e], total free energy of the reactants (G_r) is given by:

$$G_r = \sum_r n_r \mu_r = a\mu_A + b\mu_B + c\mu_C + ...$$

$$= a(\mu_A^0 + RT \ln a_A) + b(\mu_B^0 + RT \ln a_B) + c(\mu_C^0 + RT \ln a_C) + ...$$

[since we know that, $\mu_i = \mu_i^0 + RT \ln a_i$ according to Equation (1.82)]

Or, $$G_r = (a\mu_A^0 + b\mu_B^0 + c\mu_C^0 + ...) + RT \ln a_A^a + RT \ln a_B^b + RT \ln a_C^c + ...$$

Or, $$G_r = \sum_r n_r \mu_r^0 + RT \ln(a_A^a \,.\, a_B^b \,.\, a_C^c \,........)$$

Similarly the total free energy of the products (G_p) is given by:

$$G_P = \sum_P n_P \mu_P = \sum_P n_P \mu_P^0 + RT \ln(a_L^l \,.\, a_M^m \,.\, a_N^n \,........)$$

The net free energy change for the forward reaction (ΔG) is given by:

$$\Delta G = (G_P - G_r) = \sum_P n_P \mu_P - \sum_r n_r \mu_r$$

$$= \left[\sum_P n_P \mu_P^0 - \sum_r n_r \mu_r^0\right] + RT \ln \frac{a_L^l \,.\, a_M^m \,.\, a_N^n \,........}{a_A^a \,.\, a_B^b \,.\, a_C^c \,........}$$

$$\Delta G = \Delta G^0 + RT \ln Q \tag{1.86a}$$

Where, $$\Delta G^0 = \left[\sum_P n_P \mu_P^0 - \sum_r n_r \mu_r^0\right]$$ and

$$Q = \frac{a_L^l \,.\, a_M^m \,.\, a_N^n \,........}{a_A^a \,.\, a_B^b \,.\, a_C^c \,........}$$

When equilibrium is established, $\Delta G = 0$ and

$$Q = K_a = \frac{a'^l_L \,.\, a'^m_M \,.\, a'^n_N \,........}{a'^a_A \,.\, a'^b_B \,.\, a'^c_C \,........}$$

Where a'^l_L is equilibrium concentration or activity of the product, L etc.

Hence, at equilibrium, Equation (1.86a) becomes:

$$0 = \Delta G^0 + RT \ln K_a$$

Or, $$\Delta G^0 = -RT \text{ lm } K_a \tag{1.86b}$$

The Equation (1.86b) shows the relation between change in standard free energy and equilibrium constant.

Temperature dependence of the equilibrium constant (K_a)

From the Equation (1.86b), we have:

$$\ln K_a = -\frac{\Delta G^0}{RT}$$

Differentiating w.r.t. T at constant P, we get:

$$\left[\frac{\partial \ln K_a}{\partial T}\right]_P = -\frac{1}{R}\left[\frac{\partial\left(\Delta G^0/T\right)}{\partial T}\right]_P$$

We know that:

$$\left[\frac{\partial(\Delta G/T)}{\partial T}\right]_P = -\frac{\Delta H}{T^2} \quad \text{[Equation (1.59b)]}$$

Hence, we have:

$$\left[\frac{\partial\left(\Delta G^0/T\right)}{\partial T}\right]_P = -\frac{\Delta H^0}{T^2}$$

Putting this value of $\left[\frac{\partial\left(\Delta G^0/T\right)}{\partial T}\right]_P$, in the above Equation, we get:

$$\left[\frac{\partial \ln K_a}{\partial T}\right]_P = \frac{\Delta H^0}{RT^2}$$

At constant pressure we can write that:

$$\frac{d\ln K_a}{dT} = \frac{\Delta H^0}{RT^2} \tag{1.86c}$$

The Equation (1.86c) is known as Van't Hoff equation.

(a) For gas-phase reactions ΔH^0 almost remains constant within a small temperature difference. Thus on integration of the Equation (1.86c) we get:

$$\int_{K_{a,1}}^{K_{a,2}} \mathrm{d}\ln K_a = \frac{\Delta H^0}{R}\int_{T_1}^{T_2}\frac{dT}{T^2}$$

Or,

$$\ln\frac{K_{a,2}}{K_{a,1}} = -\frac{\Delta H^0}{R}\left[\frac{1}{T_2}-\frac{1}{T_1}\right] \tag{1.86d}$$

(b) In general ΔH^0 is a function of temperature as given below:

$$\Delta H^0 = A + BT + CT^2 + DT^3 + \frac{E}{T}$$

Putting this value of ΔH^0 in the Equation (1.86c) and on integration we get:

$$\ln \frac{K_{a,2}}{K_{a,1}} = -\frac{A}{R}\left[\frac{1}{T_2}-\frac{1}{T_1}\right]+\frac{B}{R}\ln\frac{T_2}{T_1}+\frac{C}{R}(T_2-T_1) + \frac{D}{2R}(T_2^2-T_1^2)-\frac{E}{2R}\left[\frac{1}{T_2^2}-\frac{1}{T_1^2}\right] \quad (1.86e)$$

1.27 Thermochemistry

1.27.1 Introduction

Study on heats of chemical reactions is called thermochemistry. Any chemical reaction may proceed either at constant temperature or with change in temperature. At constant temperature, a chemical reaction may proceed either at constant volume or at constant pressure.

(i) Reaction proceeds at constant temperature and at constant volume [$dT = 0$ and $dV = 0$]. As we know from 1st law of thermodynamics, we have:

$$dQ = dU + dW = dU + PdV \quad \text{[considering only expansion work]}$$

$$\text{As } dV = 0,\ dQ_V = dU$$

Or,
$$\Delta Q_V = \Delta U \quad (1.87a)$$

(ii) Reaction proceeds at constant temperature and at constant pressure [$dT = 0$ and $dP = 0$]. We know that:

$$H = U + PV$$

Differentiating, we have: $dH = dU + PdV + VdP$

At constant pressure the above equation becomes:

$$dH = dU + PdV$$

Or,
$$\Delta H = \Delta U + P\Delta V.$$

Considering only gas-phase reaction and assuming all the components behave ideally, we can write that:

$$PV = nRT \quad \text{and} \quad P\Delta V = \Delta n\, RT,$$

Where, Δn = (No. of moles of products) – (No. of moles of reactants).
Putting this value of $P\Delta V$ in the above equation we get:

$$\Delta H = \Delta U + \Delta n RT \quad (1.87b)$$

(a) For a chemical reaction, if $\Delta n = 0$, $\Delta H = \Delta U$
(b) For a chemical reaction, if $\Delta n > 0$, $\Delta H > \Delta U$
(c) For a chemical reaction, if $\Delta n < 0$, $\Delta H < \Delta U$

Again we know that:

$$dH = dQ + VdP \quad \text{[Equation (1.29)]}$$

Or, $dQ = dH - VdP$

At constant pressure, $dP = 0$.

Hence: $dQ_P = dH$.

Or, $\Delta Q_P = \Delta H$ (1.87c)

ΔH = Enthalpy of the products – Enthalpy of the reactants = $H_P - H_R$.

Or, $\Delta H = H_P - H_R$.

(i) If $\Delta H > 0$, the reaction is called endothermic reaction, i.e., heat is absorbed during the chemical reaction.

(ii) If $\Delta H < 0$, the reaction is called exothermic reaction, i.e., heat is released during the chemical reaction.

1.27.2 Case – I: Reactions occurring at constant temperature and pressure

Law of Lavoisier and Laplace The net enthalpy change of a given forward reaction is equal in magnitude but opposite in sign to that of the corresponding backward reaction.

Hess's law: The law of constant heat summation In any given reaction products can be obtained from the corresponding reactants by following different paths but the total heat change (ΔQ_P) of the process following one path is equal to that following any other path, i.e. heat change of a process is independent of path.

We know from Equation (1.87c) that:

$$\Delta Q_P = \Delta H$$

As H is a state function ΔH and hence ΔQ_P is independent of path.

Thus Hess's law, in other words, may be stated that, net enthalpy change of any chemical reaction is constant.

One example is formation of solid ammonium chloride from ammonia gas and hydrogen chloride gas.

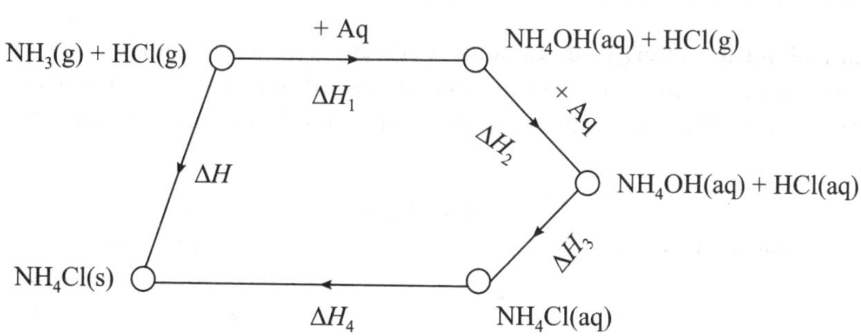

For the above cyclic process, we have:

$$\Delta H = \Delta H_1 + \Delta H_2 + \Delta H_3 + \Delta H_4$$

The algebraic value of ΔH of any process is either positive (+ve) or negative (–ve) depending on the nature of the process, i.e., endothermic or exothermic respectively.

1.27.3 Applications of Hess's law

(a) To determine heat of formation of a chemical compound

Definition Molar heat of formation of a chemical compound is the heat of reaction of one mole of that compound obtained from the corresponding elements. One example is the formation of liquid benzene from its elements.

$$6C(gr) + 3H_2(g) \rightleftharpoons C_6H_6(l)$$

The cyclic process of the reaction is shown below:

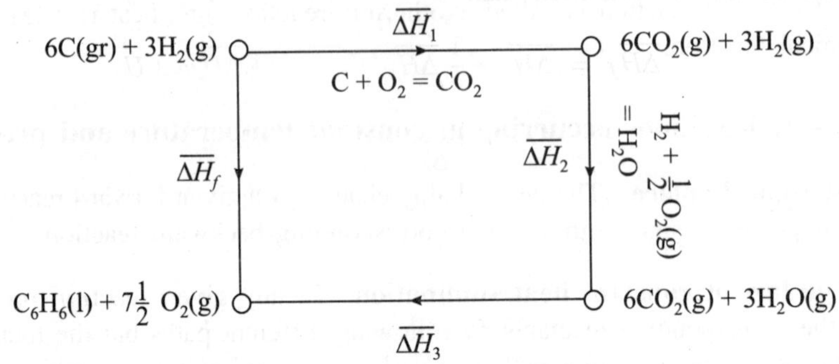

According to Hess's law we have:

$$\overline{\Delta H}_f = 6\,\overline{\Delta H}_1 + 3\,\overline{\Delta H}_2 + \overline{\Delta H}_3$$

$\overline{\Delta H}$ indicates the change in enthalpy per mole of product. Algebraic values of $\overline{\Delta H}$ should be used. Usually $\overline{\Delta H}_f$ is computed at a standard condition, i.e., at 25° (298°K) and 1 atm pressure.

$$\overline{\Delta H}_f\Big|_{298°\text{ K}, 1\text{ atm}}$$

values are frequently called standard heats of formation and are designated as $\overline{\Delta H}^0_{298}$.

(b) Calculation of lattice energy of an ionic crystal The amount of energy evolved during formation of one mole of an ionic crystal from its constituent ions is known as molar lattice energy of that crystal. The cyclic process of formation of sodium chloride crystal lattice is shown below:

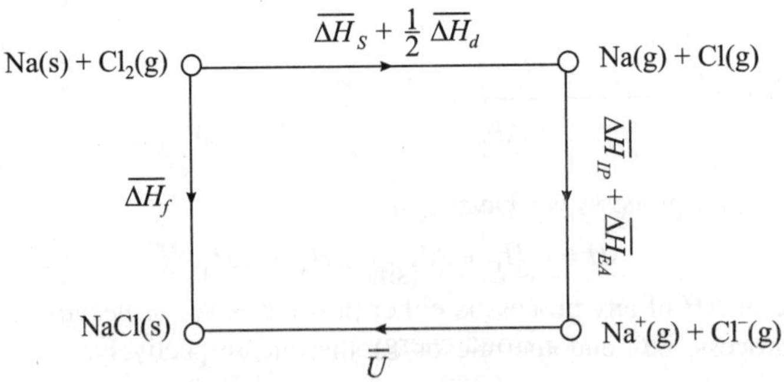

$$\overline{\Delta H}_f = \text{Molar heat of formation of NaCl(s)}$$

$$\overline{\Delta H}_s = \text{Molar heat of sublimation of Na(s)}$$

$$\overline{\Delta H}_d = \text{Molar heat of dissociation of } Cl_2(g)$$

$$\overline{\Delta H}_{IP} = \text{Molar ionization potential of Na(g)}$$

$$\overline{\Delta H}_{EA} = \text{Molar electron affinity of } Cl_2(g)$$

$$\overline{U} = \text{Molar lattice energy of NaCl(s)}$$

According to Hess's law we can write that:

$$\overline{\Delta H}_f = \overline{\Delta H}_s + \frac{1}{2}\overline{\Delta H}_d + \overline{\Delta H}_{EA} + \overline{\Delta H}_{IP} + \overline{U}$$

Or,

$$\overline{U} = \overline{\Delta H}_f - \left[\overline{\Delta H}_s + \frac{1}{2}\overline{\Delta H}_d + \overline{\Delta H}_{EA} + \overline{\Delta H}_{IP}\right]$$

(c) Calculation of bond energy Definition of Bond energy: For a compound amount of energy required to dissociate a chemical bond into its constituent atoms, is called the bond energy. It may also be defined as the amount of energy released during formation of a chemical bond .To determine bond energies of $C - H$ and $C - C$ the following cycles have been considered.

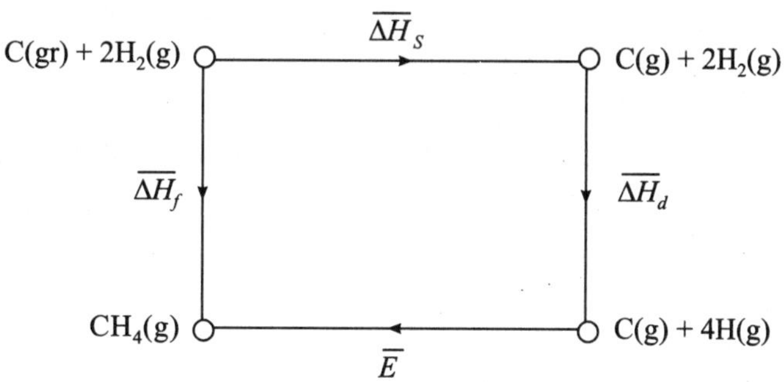

Applying Hess's law, we have:

$$\overline{\Delta H}_f = \overline{\Delta H}_s + 2\overline{\Delta H}_d + \overline{E}$$

Or,

$$\overline{E} = \overline{\Delta H}_f - (\overline{\Delta H}_s + 2\overline{\Delta H}_d)$$

$$\overline{\Delta H}_f = \text{Molar heat of formation of methane } (CH_4).$$

$$\overline{\Delta H}_s = \text{Molar heat of sublimation of solid carbon (graphite).}$$

$$\overline{\Delta H}_d = \text{Molar heat of dissociation of } H_2(g).$$

$$\overline{E} = 4\overline{E}_{C-H} \text{ (since one methane molecule has four C–H bonds).}$$

$$\overline{E}_{C-H} = \text{Bond energy of one mole C–H bond.}$$

Thus $\overline{E}_{C-H}$ can easily be calculated from the above equation.

In the next cycle let us consider the formation of ethane.

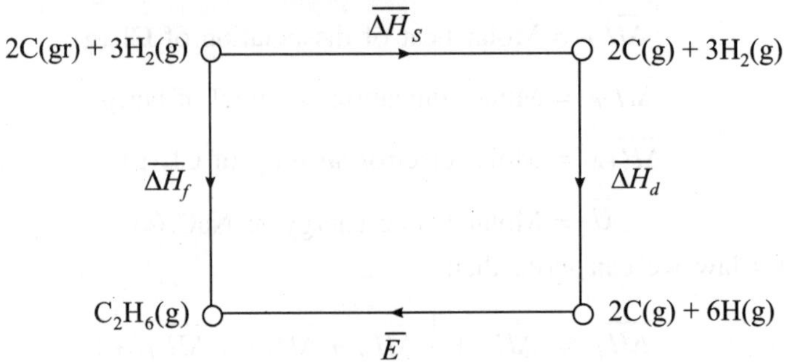

Applying Hess's law, we have:

$$\overline{\Delta H}_f = 2\overline{\Delta H}_s + 3\overline{\Delta H}_d + \overline{E}$$

Or,
$$\overline{E} = \overline{\Delta H}_f - (2\overline{\Delta H}_s + 3\overline{\Delta H}_d)$$

Where, $\overline{E} = \overline{E}_{C-C} + 6\overline{E}_{C-H}$, since one ethane molecule has one C–C bond and six C–H bonds.

$\overline{E}_{C-C}$ = Bond energy of 1 mole of C–C bond.

$\overline{E}_{C-H}$ = Bond energy of 1 mole of C–H bond and is known from the earlier cyclic process.

Thus $\overline{E}_{C-C}$ can easily be evaluated from the above equation.

1.27.4 Case – II: Reactions occurring with the change in temperature at constant pressure

If a reaction proceeds with a change in temperature the heat of reaction changes accordingly. A chemical reaction, proceeding with the change in temperature may follow in any one of the paths as shown below:

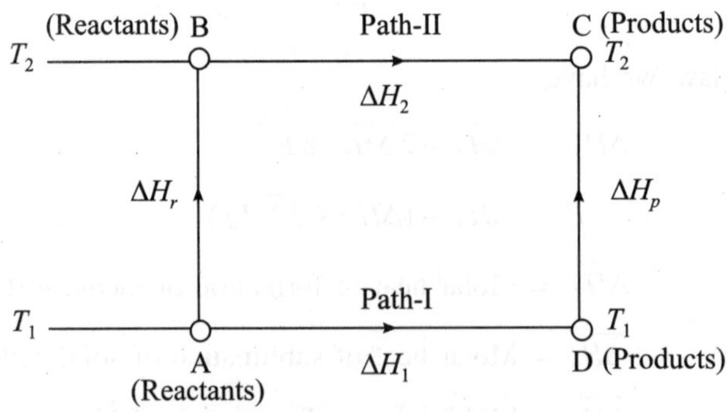

In Path-II, reactants are heated along *AB* from temperature T_1 to temperature T_2 and then they are allowed to react to form products along *BC*, at constant temperature T_2. Along path-II the net enthalpy change = $\Delta H_r + \Delta H_2$, where,

$$\Delta H_r = C_{P(r)}\ (T_2 - T_1)$$

$$C_{P(r)} = \text{Total heat capacities of the reactants.}$$

$$\Delta H_2 = \text{Heat of reaction at temperature } T_2$$

In path-I, reactants are allowed to react to form products at temperature T_1 and then the products are heated from temperature T_1 to temperature T_2.

Along Path-I, the net enthalpy change = $\Delta H_1 + \Delta H_P$, where,

$$\Delta H_P = C_{P(P)}\ (T_2 - T_1)$$

$$C_{P(P)} = \text{Total heat capacities of the products.}$$

$$\Delta H_1 = \text{Heat of reaction at temperature } T_1$$

As H is a state function, we can write that:

$$\Delta H_2 + \Delta H_r = \Delta H_1 + \Delta H_P$$

Or,
$$\Delta H_2 + C_{P(r)}\ (T_2 - T_1) = \Delta H_1 + C_{P(P)}\ (T_2 - T_1)$$

Or,
$$(\Delta H_2 - \Delta H_1) = [C_{P(P)} - (C_{P(r)}]\ (T_2 - T_1)$$

Or,
$$\left[\frac{\partial(\Delta H)}{\partial T}\right]_P = \Delta C_P$$

where,
$$\Delta C_P = [C_{P(P)} - C_{P(r)}]$$

At constant pressure we can write that:

$$\frac{d\Delta H}{dT} = \Delta C_P$$

Or,
$$d\Delta H = \Delta C_P\ dT \qquad (1.88a)$$

The Equation (1.88a) is known as differential form of Kirchaff's equation.

Let us consider the following equation:

$$xX + yY + zZ \ldots \rightleftharpoons lL + mM + nN \ldots$$

The temperature dependence of C_P is commonly expressed by a power series of the following form:

$$C_P = A + BT + CT^2 + \frac{D}{T^2}$$

Thus,
$$C_{P(P)} = (lA_L + mA_M + \ldots) + (lB_L + mB_M + \ldots)T + (lC_L + mC_M + \ldots)T^2 + \frac{(lD_L + mD_M + \ldots)}{T^2}$$

Or,
$$C_{P(P)} = \left(\sum lA_L\right) + \left(\sum lB_L\right)T + \left(\sum lC_L\right)T^2 + \frac{\left(\sum lD_L\right)}{T^2}$$

And
$$C_{P(r)} = (xA_X + yA_Y + \ldots) + (xB_X + yB_Y + \ldots)T + (xC_X + yC_Y + \ldots)T^2 + \frac{(xD_X + yD_Y + \ldots)}{T^2}$$

Or, $$C_{P(r)} = \left(\sum xA_X\right) + \left(\sum xB_X\right)T + \left(\sum xC_X\right)T^2 + \frac{\left(\sum xD_X\right)}{T^2}$$

$$\Delta C_P = [C_{P(P)} - C_{P(r)}]$$
$$= \left[\left(\sum lA_L\right) - \left(\sum xA_X\right)\right] + \left[\left(\sum lB_L\right) - \left(\sum xB_X\right)\right]T$$
$$+ \left[\left(\sum lC_L\right) - \left(\sum xC_X\right)\right]T^2 + \frac{\left[\left(\sum lD_L\right) - \left(\sum xD_X\right)\right]}{T^2}$$

Or, $$\Delta C_P = a + bT + cT^2 + \frac{d}{T^2}$$

Where, $$a = \left[\left(\sum lA_L\right) - \left(\sum xA_X\right)\right]$$
$$b = \left[\left(\sum lB_L\right) - \left(\sum xB_X\right)\right]$$
$$c = \left[\left(\sum lC_L\right) - \left(\sum xC_X\right)\right]$$
$$d = \left[\left(\sum lD_L\right) - \left(\sum xD_X\right)\right]$$

Putting this value of ΔC_P in the Equation (1.88a), we get:

$$d(\Delta H) = \left(a + bT + cT^2 + \frac{d}{T^2}\right)dT$$

On integration within the temperature limits 298°K and T°K, we get:

$$\int_{\Delta H^0_{298}}^{\Delta H_T} d(\Delta H) = a(T - 298) + \frac{b}{2}(T^2 - 298^2)$$
$$+ \frac{c}{3}(T^3 - 298^3) - d\left[\frac{1}{T} - \frac{1}{298}\right]$$

Or, $$\Delta H_T - \Delta H^0_{298} = a(T - 298) + \frac{b}{2}(T_2 - 298^2)$$
$$+ \frac{c}{3}(T^3 - 298^3) - d\left[\frac{1}{T} - \frac{1}{298}\right] \qquad (1.88b)$$

The above equation is the general form of Kirchoff's equation.

2

THERMODYNAMICS – II

2.1 Properties of fluids and equation of state

2.1.1 Triple point and critical point

Any pure material can exist in any of the three phases solid or liquid or vapour. From the knowledge of phase rule we have:

$$F = C - P + 2 \tag{2.1}$$

Where, C = Number of components.

P = Number of phases.

F = Degrees of freedom.

For a pure material $C = 1$. If it exists in a single phase, $P = 1$ and hence according to the Equation (2.1), $F = 2$. Thus the system is *bivarient*, i.e., the system can be investigated by controlling any two parameters e.g. pressure (P) and volume (V), pressure and temperature (T) or volume and temperature.

If the pure material exists in two phases in equilibrium, $P = 2$ and hence $F = 1$. The system is said to be *univariant*, i.e., the system may be investigated by controlling only one of the three parameters P, V, T and a two dimensional curve is obtained, as shown the Figure 2.1. When vapour and liquid phase are in equilibrium, the curve is called *vapourization curve*. When solid and liquid phases are in equilibrium, the curve is called *fusion curve* and when solid-vapour phases coexist the curve is called *sublimation curve*.

If three phases coexist at a point, $P = 3$ and hence $F = 0$. The system is said to be *invariant*, i.e., all the parameters are fixed. The point O is called *triple point* (Figure 2.1).

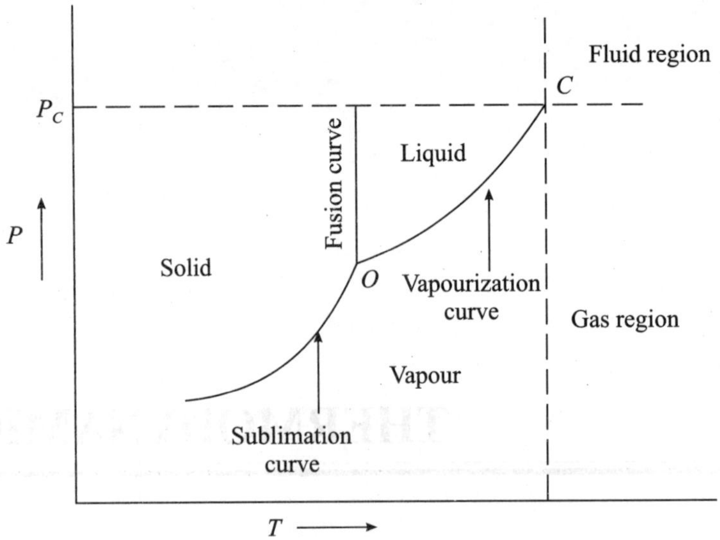

Fig. 2.1 *PT*-diagram of a pure substance.

The vapourisation curve starts from the triple point '*O*' and ends at the point '*C*', known as *critical point.*

The correspoing pressure and temperature at the critical point *C* are known as critical pressure (P_c) and critical temperature (T_c) respectively. Above the point *C* there is no demarcation between liquid and gas. So P_c and T_c are the highest pressure and temperature at which the corresponding gas can be liquefied.

Beyond the critical point *C* a wide region is left, demarcated by dashed lines. The material, existing in this region, cannot be liquefied. This region is called *fluid region.* In the Figure 2.1 a vertical dashed line, which extends beyond the critical point *C*, is drawn from 'T_c'. Above the critical point and left to the vertical dashed line, there exists *liquid region*, which can be vapourized by reduction of pressure at constant temperature. Below the critical point and left to the vertical dashed line, there exists vapour region, which can be liquefied by reduction of temperature at constant pressure.

The critical point is also known as point of inflection and the following condition holds good at the critical point.

$$\left(\frac{\partial P}{\partial V}\right)_T\bigg|_{\text{at 'C'}} = 0 \quad \text{and} \quad \left(\frac{\partial^2 P}{\partial V^2}\right)_T\bigg|_{\text{at 'C'}} = 0 \tag{2.2}$$

P_c, V_c and T_c are the critical pressure, critical volume and critical temperature respectively. Then the term reduced parameter is defined as:

$$\text{Reduced pressure} = \frac{P}{P_c} = \pi$$

$$\text{Reduced volume} = \frac{V}{V_c} = \phi$$

$$\text{Reduced temperature} = \frac{T}{T_c} = \theta$$

2.1.2 Equation of state

P, *V* and *T* are the controlling parameters to define any gas. Thus the equation of state of any gas is given by:

$$f(P, V, T) = 0 \tag{2.3}$$

One of the solution of the above equation is:

$$V = f(P, T)$$

Hence,
$$dV = \left(\frac{\partial V}{\partial T}\right)_P dT + \left(\frac{\partial V}{\partial P}\right)_T dP$$

Or,
$$\frac{dV}{V} = \frac{1}{V}\left(\frac{\partial V}{\partial T}\right)_P dT + \frac{1}{V}\left(\frac{\partial V}{\partial P}\right)_T dP$$

Or,
$$\frac{dV}{V} = \beta dT - k dP$$

$$\text{Since expansivity} = \beta = \frac{1}{V}\left(\frac{\partial V}{\partial T}\right)_P \quad \text{and}$$

$$\text{Isothermal compressibility} = k = -\frac{1}{V}\left(\frac{\partial V}{\partial P}\right)_T$$

So,
$$\frac{dV}{V} = \beta dT - k dP \tag{2.4}$$

In case of liquid, isotherms are very steep and hence the values of β and k are approximated to zero. On the basis of this assumption liquid is often called incompressible fluid.

PVT behaviour of different pure substances shows that the following power series expression holds good for all gases.

$$PV = a + bP + cP^2 + \ldots\ldots\ldots$$

Or,
$$PV = a(1 + B'P + C'P^2 + \ldots\ldots\ldots) \tag{2.5}$$

Where,
$$\left(\frac{b}{a}\right) = B', \left(\frac{c}{a}\right) = C' \ \ldots\ldots\ldots \text{ etc.}$$

As the pressure tends to zero the above equation becomes:

$$\underset{P \to 0}{\text{LT}} (PV) = a = RT \tag{2.6}$$

and the gas is known as ideal gas. Thus at very low pressure any real gas behaves ideally. Again for one mole of any ideal gas we have:

$$PV = RT \quad \text{and} \quad P = \frac{RT}{V}$$

Putting this value of P in the R. H. S. of Equation (2.5), we get:

$$PV = a\left[1 + B'\frac{RT}{V} + C'\left(\frac{RT}{V}\right)^2 + \ldots\ldots\right]$$

$$PV = a\left[1 + \frac{B}{V} + \frac{C}{V^2} + \ldots\ldots\right] \quad (2.6a)$$

Putting this value of a from the Equation (2.6) in the Equation (2.5) and (2.6a), we get:

$$PV = RT\left[1 + B'P + C'P^2 + \ldots\ldots\right]$$

and

$$PV = RT\left[1 + \frac{B}{V} + \frac{C}{V^2} + \ldots\ldots\right] \quad (2.7)$$

Or,

$$Z = \frac{PV}{RT} = \left[1 + B'P + C'P^2 + \ldots\ldots\right]$$

and

$$Z = \frac{PV}{RT} = \left[1 + \frac{B}{V} + \frac{C}{V^2} + \ldots\ldots\right] \quad (2.8)$$

where, Z is known as *compressibility factor*.

The above expansions are known as virial expansions and the coefficients are known as *virial coefficients*. For an ideal gas values of all virial coefficients are zero and

$$Z = 1 = \frac{PV}{RT}$$

In most of the cases, real gas is expressed by Van der Waal's equation of state, shown below:

$$\left(P + \frac{a}{V^2}\right)(V - b) = RT \quad \text{(for 1 mole)} \quad (2.9)$$

At the critical point the above equation becomes:

$$\left(P_c + \frac{a}{V_c^2}\right)(V_c - b) = RT_c \quad (2.10)$$

From the Equation (2.9) we have:

$$PV - Pb + \frac{a}{V} - \frac{ab}{V^2} = RT$$

Differentiating with respect to V at constant T, we get:

$$P + V\left(\frac{\partial P}{\partial V}\right)_T - b\left(\frac{\partial P}{\partial V}\right)_T - \frac{a}{V^2} + \frac{2ab}{V^3} = 0 \quad (2.11)$$

At

$$T = T_c,\ P = P_c \text{ and } \left(\frac{\partial P}{\partial V}\right)_T = 0$$

Or, $$P_c - \frac{a}{V_c^2} + \frac{2ab}{V_c^3} = 0 \qquad (2.12)$$

Further differentiating Equation (2.11) with respect to V at constant T we get:

$$\left(\frac{\partial P}{\partial V}\right)_T + V\left(\frac{\partial^2 P}{\partial V^2}\right)_T + \left(\frac{\partial P}{\partial V}\right)_T - b\left(\frac{\partial^2 P}{\partial V^2}\right)_T + \frac{2a}{V^3} - \frac{6ab}{V^4} = 0$$

At $T = T_c$, $P = P_c$ and $V = V_c$,

we have:

$$\left(\frac{\partial P}{\partial V}\right)_T = 0 \quad \text{and} \quad \frac{\partial^2 P}{\partial V^2} = 0$$

So, $$V_c = 3b,$$

Or, $$b = \left(\frac{V_c}{3}\right)$$

Putting this value of 'b' in the Equation (2.12), we get:

$$P_c - \frac{a}{V_c^2} + \frac{2aV_c}{3V_c^3} = 0$$

Or, $$\left(\frac{2a}{3} - a\right) = -P_c V_c^2$$

Or, $$a = 3P_c V_c^2$$

Or, $$V_c^2 = \frac{a}{3P_c}$$

Putting the values of V_c and V_c^2 and in the Equation (2.10), we get:

$$(P_c + 3P_c)(3b - b) = RT_c$$

Or, $$b = \frac{RT_c}{8P_c}$$

and $$a = 3P_c V_c^2 = 3P_c(3b)^2 = 27P_c b^2 = 27P_c \times \frac{R^2 T_c^2}{64P_c^2} = \frac{27R^2 T_c^2}{64P_c}$$

So, $$a = \frac{27R^2 T_c^2}{64P_c} \quad \text{and} \quad b = \frac{RT_c}{8P_c} \qquad (2.13)$$

From the concept of reduced parameter, we have:

$$P = \pi P_c,\ V = \phi V_c \quad \text{and} \quad T = \theta T_c$$

Putting these values of P, V and T in the Equation (2.9), we get:

$$\left[\pi P_c + \frac{a}{(\phi V_c)^2}\right](\phi V_c - b) = R\theta T_c$$

Again, $$V_c^2 = \frac{a}{3P_c} \quad \text{and} \quad V_c = 3b$$

Thus the above equation becomes:

$$P_c\left(\pi + \frac{3}{\phi^2}\right)(3\phi - 1)b = \theta(RT_c)$$

From Equation (2.13):

$$RT_c = 8bP_c$$

Thus the above equation becomes:

$$\left(\pi + \frac{3}{\phi^2}\right)(3\phi - 1) = 8\theta \tag{2.14}$$

Equation (2.14) is known as reduced equation of state.

Inference If π and θ of any two gases are same ϕ must be same for those two gases.

Law of corresponding states All gases, when compared at same reduced temperature and reduced pressure, have same reduced volume, hence have same compressibility factor and all of them deviate from the ideal behaviour to about the same degree.

2.2 Thermodynamic properties

2.2.1 Residual property

Free energy (G) is related with enthalpy (H) and entropy (S) as follows:

$$G = H - TS \tag{2.15}$$

and the differential form is:

$$dG = VdP - SdT \tag{2.16}$$

$$d\left(\frac{G}{RT}\right) = \frac{dG}{RT} - \frac{GdT}{RT^2} = \frac{(VdP - SdT)}{RT} - \frac{GdT}{RT^2}$$

$$= \frac{VdP}{RT} - \frac{(G+TS)dT}{RT^2} = \frac{VdP}{RT} - \frac{HdT}{RT^2}$$

Or, $$d\left(\frac{G}{RT}\right) = \frac{VdP}{RT} - \frac{HdT}{RT^2} \tag{2.17}$$

$$\left.\begin{aligned} \frac{V}{RT} &= \left[\frac{\partial(G/RT)}{\partial P}\right]_T \\ \frac{H}{RT^2} &= -\left[\frac{\partial(G/RT)}{\partial T}\right]_P \\ \frac{S}{R} &= \frac{H}{RT} - \frac{G}{RT} \end{aligned}\right\} \tag{2.18}$$

Suppose M is a parameter of any real system and M^{id} is the value of same parameter of that system assuming the latter be an ideal system. Then the difference $(M - M^{id})$ is known as the *residual property* of that system and is represented as M^r.

So we can write that:

$$M^r = M - M^{id}$$

Thus, $V^r = V - V^{id}$, $G^r = G - G^{id}$ ... etc.

Hence, Equation (2.18) can be rewritten as:

$$\left.\begin{aligned} \frac{V^r}{RT} &= \left[\frac{\partial(G^r/RT)}{\partial P}\right]_T \\ \frac{H^r}{RT^2} &= -\left[\frac{\partial(G^r/RT)}{\partial T}\right]_P \\ \frac{S^r}{R} &= \frac{H^r}{RT} - \frac{G^r}{RT} \end{aligned}\right\} \tag{2.19}$$

G^r, H^r, and S^r are the residual volume, residual free energy, residual enthalpy and residual entropy respectively of that system.

Again for 1 mole of an ideal gas:

$$V^{id} = V^{ig} = \frac{RT}{P}$$

$$V^r = V - V^{ig} = V - \frac{RT}{P}$$

Again, $Z = \dfrac{PV}{RT}$ and hence, $V = \dfrac{RTZ}{P}$

Putting this value of V in the above equation, we get:

$$V^r = V - \frac{RT}{P} = \frac{RTZ}{P} - \frac{RT}{P} = \frac{RT}{P}(Z - 1)$$

Or, $$V^r = \frac{RT}{P}(Z - 1) \tag{2.20a}$$

Comparing Equations (2.19) and (2.20a), we get:

$$\left[\frac{\partial\left(\frac{G^r}{RT}\right)}{\partial P}\right]_T = \frac{1}{P}(Z - 1)$$

Thus at constant temperature (T), we can write that:

$$\frac{G^r}{RT} = \int_0^P \frac{(Z-1)}{P}\, dP \tag{2.20b}$$

Differentiating Equation (2.20b) with respect to T at constant P we get:

$$\left[\frac{\partial\left(\frac{G^r}{RT}\right)}{\partial T}\right]_P = \int_0^P \left(\frac{\partial Z}{\partial T}\right)_P \frac{dP}{P}$$

Comparing this with Equation (2.19), we get:

$$\frac{H^r}{RT^2} = -\left[\frac{\partial\left(\frac{G^r}{RT}\right)}{\partial T}\right]_P = -\int_0^P \left(\frac{\partial Z}{\partial T}\right)_P \frac{dP}{P}$$

Or,

$$\frac{H^r}{RT} = -T\int_0^P \left(\frac{\partial Z}{\partial T}\right)_P \frac{dP}{P} \tag{2.20c}$$

Again we have:

$$\frac{S^r}{R} = \frac{H^r}{RT} - \frac{G^r}{RT} = -T\int_0^P \left(\frac{\partial Z}{\partial T}\right)_P \frac{dP}{P} - \int_0^P (Z-1)\frac{dP}{P}$$

$$\frac{S^r}{R} = -\int_0^P \left[T\left(\frac{\partial Z}{\partial T}\right)_P + (Z-1)\right] d\ln P \tag{2.20d}$$

Equations (2.20a to 2.20d) are known as fundamental residual property relations. V^r, G^r, H^r, and S^r can be evaluated by using appropriate conditions. Thus the actual property M can be determined by using the equation:

$$M = M^{ig} + M^r$$

For example: $S = S^{ig} + S^r$

S = Actual molar entropy of the pure substance.
S^{ig} = Molar entropy of the same pure substance which is assumed to be an ideal one.

$$S^{ig} = S_0^{ig} + \int_0^T C_P^{ig}\, d\ln T - R\int_0^P d\ln P$$

If C_P^{ig} and S_0^{ig} are known, S^{ig} can easily be evaluated by using the above equation.

Thus,
$$S = \left[S_0^{ig} + \int_0^T C_P^{ig}\, d\ln T - R \int_0^P d\ln P \right] + S^r \tag{2.21}$$

Similarly, the other actual thermodynamic properties of any pure substance can easily be evaluated.

2.2.2 Two-phase system

Consider a system where saturated liquid remains in equilibrium with its saturated vapour. If x and y are the mole fractions of the liquid and vapour phases respectively, we have: $x + y = 1$.

Any extensive property say volume (V) can be represented as:

$$V = xV^l + yV^g = (1 - y)V^l + yV^g$$

Or,
$$V = V^l + (V^g - V^l)y \tag{2.22}$$

For two phase system of a pure substance we have:

$$\frac{dP^{\text{sat}}}{dT} = \frac{L}{T\Delta V} = \frac{L}{T(V^g - V^l)}$$

Again $V^g >> V^l$ and assuming that the saturated vapour behaves ideally, the above equation becomes:

$$\frac{dP^{\text{sat}}}{dT} = \frac{LP^{\text{sat}}}{RT^2}$$

Or,
$$d\ln P^{\text{sat}} = \frac{L}{R} d\left(\frac{1}{T^2}\right)$$

Where, L = Molar latent heat of vapourisation.

On integration we get:

$$\text{In} P^{\text{sat}} = -\frac{L}{RT} + C \tag{2.23a}$$

When $T = T_b$ = The boiling point of the pure substance the above equation takes the following form in accordance with *Riedel*:

$$\frac{L/T_b}{R} = \frac{1.092(\ln P_c - 1.013)}{(0.930 - \theta_b)} \tag{2.23b}$$

P_c = Critical pressure in *bar*,

θ_b = Reduced temperature at T_b,

R = Molar gas constant.

Two important thermodyn̈amic properties S and H of pure substances are plotted against pressure (P) and temperature (T). There such plots are widely used, namely, (i) $T-S$ plot, (ii) ln $P-H$ plot and (iii) $H-S$ plot. The $H-S$ plot is known as *Mollier diagram.*

The tie line AB, which is parallel to S-axis, indicates the phase change from liquid to saturated vapour at constant pressure (P) and temperature (T).

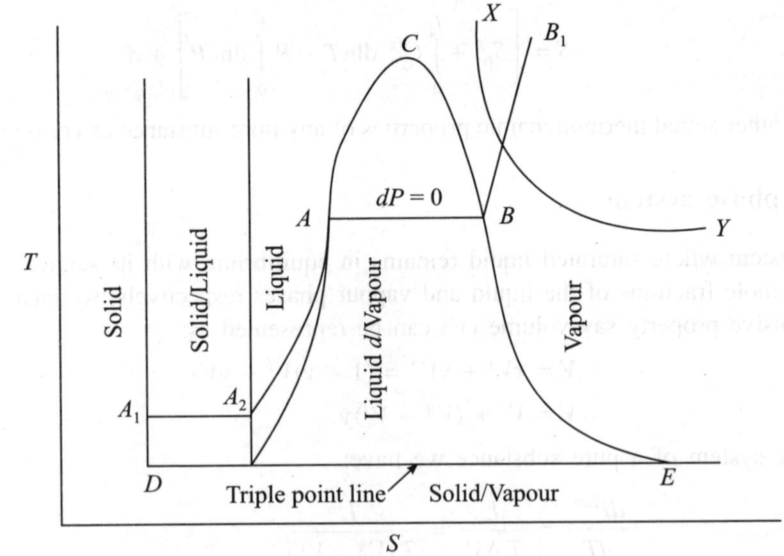

Fig. 2.2 $T - S$ diagram of a pure substance '*C*' represents the critical point.

The line *DE* represents triple point line. All three phases coexist along this line. The curve *XY* represents $T - S$ plot of vapour at constant *H*. The total phase transition of a pure substance is represented by the curve $A_1A_2BB_1$. The pure substance remains in solid state at A_1 and slowly transforms to liquid state along A_1A_2. The phase change takes place at constant temperature (*T*) and pressure (*P*). So the tie line A_1A_2 is parallel to *S*-axis. With the increase in temperature '*S*' of the pure liquid increases and reaches the point *A*, from which another phase transition (liquid to vapour) begins. Phase transition is completed at the point *B*. The line BB_1 indicates super heated vapour.

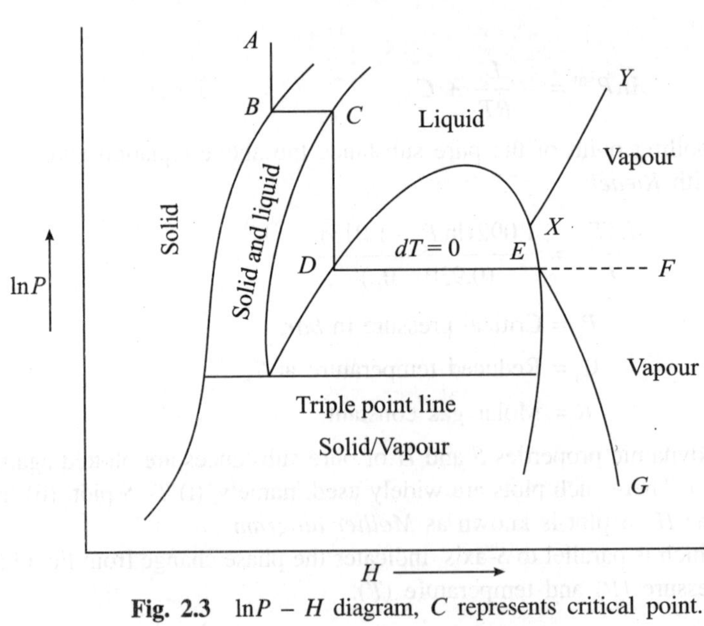

Fig. 2.3 $\ln P - H$ diagram, *C* represents critical point.

The line *XY* represents $\ln P - H$ plot of vapour at constant entropy (*S*). The total phase change of a pure substance is represented by *ABCDEG*. The dotted line *EF* represents super-heated vapour. Temperature (*T*) and pressure (*P*) remain constant during phase change, i.e., along *BC* and *DE*.

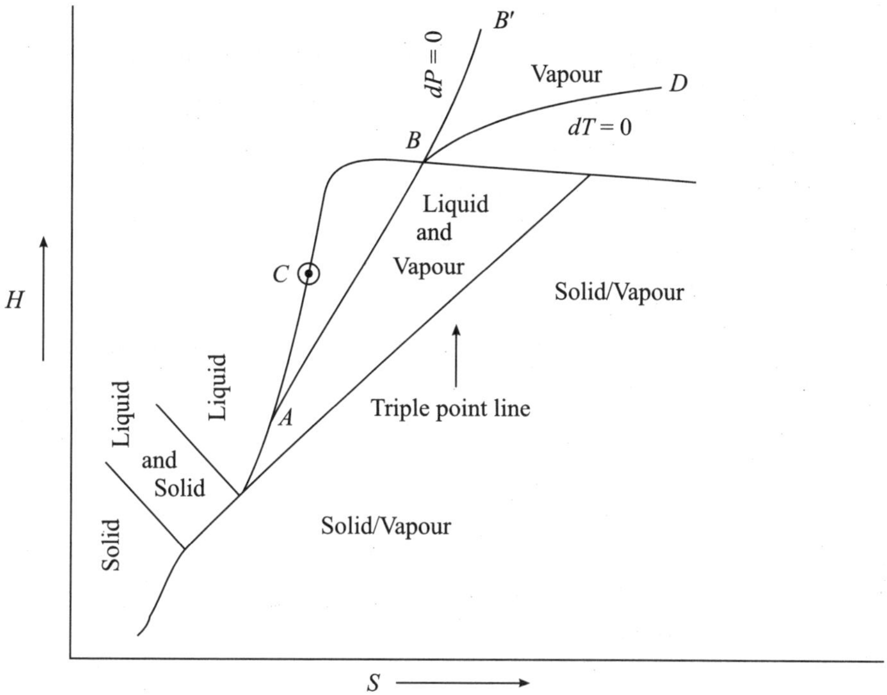

Fig. 2.4 $H-S$ diagram (Mollier diagram), *C* represents critical point.

The line *AB* represents a smooth phase change from liquid phase to vapour phase at constant temperature (*T*) and pressure (*P*). The line *BB′* represents super heating of vapour at constant pressure (*P*) whereas *BD* represents increase in entropy at constant temperature (*T*). At constant temperature *H* of an ideal gas/vapour remains constant. So the line *BD* runs almost parallel to *S*-axis. The slight deviation of *BD* to upward direction indicates that enthalpy increases slightly with pressure at constant temperature, indicating that the vapour behaves as a non-ideal gas.

2.3 Energy in flow process

2.3.1 Conservation of energy

For a system like flowing fluid through a conduit, the net energy balance of the system can be written as:

$$\frac{d(mU)}{dt} = (\dot{m}E_1 - \dot{m}E_2) + \dot{m}(Q - W) \tag{2.24}$$

$\frac{d(mU)}{dt}$ = Total change in internal energy within the system

m = Mass of the fluid within the system.

U = Specific internal energy.

E_1 and E_2 are the specific energies of the fluid at the entrance and exit respectively.

Q = Net heat flow per unit mass of the fluid within the system.

W = Net work output per unit mass of the fluid within the system.

$\dot{m}$ = Mass flow rate of the fluid.

The specific energy E consists of internal energy per unit mass, i.e., U, kinetic energy per unit mass, i.e., $\frac{1}{2}u^2$ and potential energy per unit mass, Zg, where,

Z = Height of the fluid from a reference point.

u = Velocity of the fluid.

g = Gravitational constant.

So, we can write that:

$$E = U + \frac{1}{2}u^2 + Z.g \tag{2.25}$$

At the steady state process, we can write that:

$$\frac{d(mU)}{dt} = 0$$

Hence Equation (2.24) becomes:

$$\dot{m}(E_2 - E_1) = \dot{m}(Q - W)$$

Or, $$(E_2 - E_1) = (Q - W)$$

Or, $$dE = dQ - dW$$

W = Net work output = shaft work + work due to $PdV = W_s + \int d(PV)$

Or, $$dE = dQ - dW_s - d(PV)$$

Or, $$dE + d(PV) = dQ - dW_s$$

Or, $$d[U + \left(\frac{1}{2}\right)u^2 + Zg] + d(PV) = dQ - dW_s$$

Or, $$d(U + PV) + d[\left(\frac{1}{2}\right)u^2 + Zg] = dQ - dW_s$$

Or, $$dH + d[\left(\frac{1}{2}\right)u^2 + Zg] = dQ - dW_s$$

Or, $$d[H + \left(\frac{1}{2}\right)u^2 + Zg] = dQ - dW_s$$

Or, $$\Delta[H + \left(\frac{1}{2}\right)u^2 + Zg] = \Delta Q - \Delta W_s \tag{2.26a}$$

Equation (2.26a) represents the net energy balance of a control system for a steady state flow process. Again we have

$$dH = dQ + VdP$$

Or, $$\int dH = \int dQ + \int_{P_1}^{P_2} VdP$$

Or, $$\Delta H = \Delta Q + \int_{P_1}^{P_2} VdP$$

Putting the value of ΔH in the Equation (2.26a), we get:

$$-\Delta W_s = \int_{P_1}^{P_2} VdP + \frac{1}{2}\Delta u^2 + g\Delta Z$$

Considering the frictional energy F, the above equation takes the form:

$$-\Delta W_s = \int_{P_1}^{P_2} VdP + \frac{1}{2}\Delta u^2 + g\Delta Z + F \tag{2.26b}$$

The Equation (2.26b) is the general equation of energy balance for any steady state flow process.

For a non-viscous incompressible fluid, $F = 0$ and $V =$ constant. Also assuming that the fluid does not exchange any shaft work with the surroundings, the Equation (2.26b) becomes:

$$\int_{P_1}^{P_2} VdP + \frac{1}{2}\Delta u^2 + g\Delta Z = 0$$

Or, $$V\Delta P + \frac{1}{2}\Delta u^2 + g\Delta Z = 0$$

$$V = \text{specific volume of the fluid} = \left(\frac{1}{\rho}\right),$$

where, ρ = Density of the fluid.

Hence, $$\frac{\Delta P}{\rho} + \frac{1}{2}\Delta u^2 + g\Delta Z = 0$$

Or, $$\Delta\left(\frac{P}{\rho} + \frac{u^2}{2} + gZ\right) = 0$$

Or, $$\left(\frac{P}{\rho} + \frac{u^2}{2} + gZ\right) = \text{constant} \tag{2.26c}$$

The Equation (2.26c) is known as the famous *Bernoulli's equation* for an *ideal inviscid fluid*. Consider a fluid flowing through a conduit such that $\Delta Z = 0$ and $\Delta W_S = 0$. If the flow process is assumed to be adiabatic, (i.e., $\Delta Q = 0$), Equation (2.26a) becomes:

$$\Delta H + \frac{\Delta u^2}{2} = 0$$

Or,
$$dH = -udu \tag{2.27a}$$

Again, from the law of conservation of mass, we have:

$$\dot{m} = \text{constant}$$

Hence,
$$\rho_1 u_1 A_1 = \rho_2 u_2 A_2$$

Again,
$$\rho = \frac{1}{V}$$

Where,
$$V = \text{Specific volume.}$$

So,
$$\rho_1 u_1 A_1 = \frac{u_1 A_1}{V_1} \quad \text{and} \quad \rho_2 u_2 A_2 = \frac{u_2 A_2}{V_2}$$

Or,
$$\frac{u_1 A_1}{V_1} = \frac{u_2 A_2}{V_2}$$

Or,
$$d\left(\frac{uA}{V}\right) = 0$$

Area (A) is assumed to be constant.

Or,
$$A\ d\left(\frac{u}{V}\right) = 0$$

Or,
$$-\frac{udV}{V^2} + \frac{du}{V} = 0$$

Or,
$$-\frac{udV}{V} + du = 0$$

Or,
$$udu = \frac{u^2 dV}{V} \tag{2.27b}$$

Again, we know that:

$$dH = TdS + VdP \tag{2.27c}$$

For reversible process, $dQ = TdS$ and

For irreversible adiabatic process,

$$dQ = 0 \quad \text{but} \quad dS > 0$$

Combining the Equations 2.27a, 2.27b and 2.27c, we get:

$$TdS = dH - VdP = -udu - VdP = -\frac{u^2\, dV}{V} - VdP$$

Or,
$$TdS = -\frac{u^2\, dV}{V} - VdP \tag{2.27d}$$

ΔP is the driving force for flowing a fluid through a conduit and hence it is a negative quantity. Thus 2nd term of right hand side of Equation (2.27d) is always positive. Again with the decrease in pressure volume V increases since for an ideal gas PV^{γ} = Constant for adiabatic process. Thus as the pressure of fluid at exit decreases, both V and the velocity (u) increase, although increase in volume (V) is much larger than that of velocity (u). So there is a possibility to achieve a condition such that $dS = 0$ and at that condition velocity of the fluid (u) becomes maximum, represented as u_{max}. Thus the Equation (2.27d) becomes:

$$u_{max}^2 \frac{dV}{V} = -VdP$$

Or,
$$u_{max}^2 = -V^2\left(\frac{\partial P}{\partial V}\right)_S \tag{2.27e}$$

The above expression of u_{max} is same as the expression of speed of sound and hence u_{max} is conventionally written as u_{sonic}. The ratio of fluid velocity (u) to sonic velocity (u_{sonic}) is known as *Mach number*, represented as M.

So,
$$M = \frac{u}{u_{sonic}}$$

When $M > 1$, the fluid flows with supersonic velocity and when $M < 1$, the fluid flows with subsonic velocity.

2.3.2 Flow through a nozzle

Nozzle has a converging section and a diverging section and a throat region. Thus cross-sectional area varies both in converging section and diverging section but is assumed to be constant at throat region.

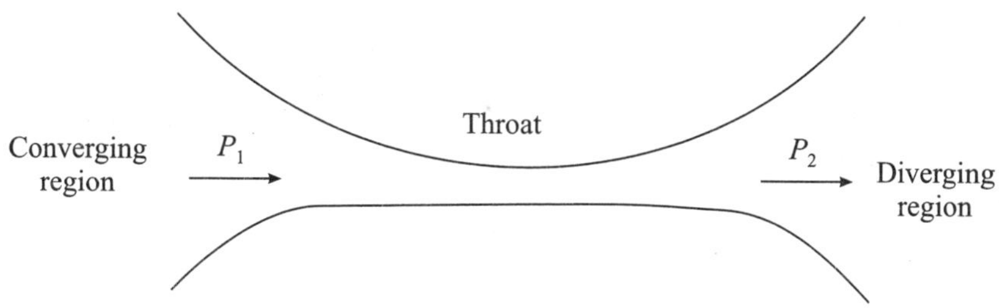

Fig. 2.5 Converging/diverging nozzle.

Fluid in the converging region is in the *subsonic state*, achieves *sonic velocity* in the throat region and emerges with *supersonic velocity* in the diverging region. It is assumed that area of cross section remains constant in the throat region. Thus for an isentropic flow (i.e., $dS = 0$) the maximum velocity achievable in the throat region is

$$u_{\text{sonic}} = -V^2\left(\frac{\partial P}{\partial V}\right)_S = u_{\text{throat}} \tag{2.28a}$$

Suppose P_1 and P_2 are the entrance and exit fluid pressures in the nozzle. So the fluid flow depends on the pressure ratio $\frac{P_2}{P_1}$. If $\frac{P_2}{P_1} = 1$, there will be no fluid flow. As the ratio decreases the fluid flow increases with fluid velocity and ultimately the ratio $\frac{P_2}{P_1}$ reaches a critical value, when the sonic flow persists at the exit of the nozzle. Further decrease in pressure ratio has no effect on fluid flow or fluid velocity. For steam this critical value $\left(\frac{P_2}{P_1}\right)_{cr}$ has been found 0.55 at moderate temperature and pressure. As the adiabatic condition is maintained in the nozzle the following relation can be conveniently used for an ideal gas flow.

$$PV^\gamma = \text{Constant} = K$$

Differentiating with respect to V at constant S, we have:

$$\left(\frac{\partial P}{\partial V}\right)_S V^\gamma + \gamma\, V^{\gamma-1} P = 0$$

Or, $$\left(\frac{\partial P}{\partial V}\right)_S = -\frac{\gamma P}{V}$$

Or, $$V^2\left(\frac{\partial P}{\partial V}\right)_S = -\gamma PV$$

Or, $$u_{\text{throat}} = -V^2\left(\frac{\partial P}{\partial V}\right)_S = \gamma P_2 V_2$$

since u_{throat} is achieved only at the exit of the nozzle where pressure and volume of fluid are P_2 and V_2 respectively.

So, $$u_{\text{throat}} = \gamma P_2 V_2 \tag{2.28b}$$

If the fluid is a gas, the frictional energy can be neglected. Further it is assumed that fluid does not transfer any shaft work with the surroundings. Again entrance and exit of a nozzle remain at the same height.

So, $$\Delta W_s = 0,\; F = 0 \quad \text{and} \quad \Delta Z = 0$$

Hence, the Equation (2.26b) reduces to:

$$\frac{1}{2}\Delta u^2 = -\int_{P_1}^{P_2} VdP$$

Or, $$u_2^2 - u_1^2 = -2\int_{P_1}^{P_2} \frac{KdP}{P^{\frac{1}{\gamma}}} \quad \text{[since } PV^{\gamma} = \text{Constant} = K' \text{ and } K = K'^{\frac{1}{\gamma}}\text{]}$$

Or, $$u_2^2 - u_1^2 = -2K\left[\frac{P^{-\frac{1}{\gamma}+1}}{-\frac{1}{\gamma}+1}\right]_{P_1}^{P_2}$$

$$= -\frac{2\gamma}{(\gamma-1)}\left[K P_2 P_2^{-\frac{1}{\gamma}} - K P_1 P_1^{-\frac{1}{\gamma}}\right]$$

$$= -\frac{2\gamma}{(\gamma-1)} K P_2 P_2^{-\frac{1}{\gamma}}\left[1-\left(\frac{P_1}{P_2}\right)^{-\frac{1}{\gamma}+1}\right]$$

$$= \frac{2\gamma P_2 V_2}{(\gamma-1)}\left[\left(\frac{P_1}{P_2}\right)^{\frac{(\gamma-1)}{\gamma}} - 1\right]$$

To achieve the sonic velocity at the throat of a nozzle assume $u_1 = 0$.

So, $u_{\text{throat}} = u_{\text{sonic}} = \gamma P_2 V_2$ (according to Equation 2.28b). Putting these conditions in the above equation we get:

$$u_{\text{throat}}^2 = \frac{2\gamma P_2 V_2}{(\gamma-1)}\left[\left(\frac{P_1}{P_2}\right)^{\frac{(\gamma-1)}{\gamma}} - 1\right]$$

Or, $$\gamma P_2 V_2 = \frac{2\gamma P_2 V_2}{(\gamma-1)}\left[\left(\frac{P_1}{P_2}\right)_{cr}^{\frac{(\gamma-1)}{\gamma}} - 1\right]$$

Or, $$\left[\left(\frac{P_1}{P_2}\right)_{cr}^{\frac{(\gamma-1)}{\gamma}} - 1\right] = \frac{(\gamma-1)}{2}$$

[Where, $\left(\frac{P_1}{P_2}\right)_{cr}$ = Critical Pressure Ratio]

Or, $$\left(\frac{P_2}{P_1}\right)_{cr} = \left[\frac{2}{(\gamma+1)}\right]^{\frac{\gamma}{(\gamma-1)}} \tag{2.28c}$$

Thus for any ideal fluid critical pressure ratio can easily be evaluated just by knowing the γ-value.

For varying cross-sectional area the equation of conservation of mass is given by:

$$\frac{d(m)}{dt} = 0$$

Or, $$d(m) = 0$$

Or, $$d(\rho uA) = 0$$

Or, $$d\left(\frac{uA}{V}\right) = 0$$

Or, $$\frac{udA + Adu}{V} - \frac{uA}{V^2}dV = 0$$

Or, $$\frac{udA + Adu}{uA} = \frac{dV}{V}$$

Or, $$\frac{dA}{A} + \frac{du}{u} = \frac{dV}{V}$$

The above equation holds good at any position of converging or diverging section of the pipe. At the throat of the nozzle sonic velocity is achieved.

Thus at the throat $\dfrac{udu}{u_{\text{sonic}}^2} = \dfrac{dV}{V}$ [From Equation (2.27b), since at nozzle, $dA = 0$]

Putting this value of $\left(\dfrac{dV}{V}\right)$ in the above equation, we get:

$$\frac{dA}{A} + \frac{du}{u} = \frac{dV}{V} = \frac{udu}{u_{\text{sonic}}^2}$$

Or, $$\frac{dA}{A} = \frac{udu}{u_{\text{sonic}}^2} - \frac{du}{u} = \left[\frac{u^2}{u_{\text{sonic}}^2} - 1\right]\frac{du}{u}$$

Or, $$\frac{dA}{A} = (M^2 - 1)\frac{du}{u} \tag{2.28d}$$

For the converging section of the nozzle dA is negative and hence $M < 1$. So fluid flow is subsonic in the converging section. For the diverging section of the nozzle, dA is positive and hence $M > 1$, so fluid flow is supersonic in the diverging section.

2.3.3 Expanders and Compressors

Expanders If a gas is allowed to expand through a nozzle, internal energy of the gas gets converted into shaft work. The device, through which the shaft work is realised through expansion, is known as expanders. In different chemical or petrochemical plants *ammonia or ethylene* is used as working fluid in expanders. In power plants usually *steam* is used as a working fluid and then the expander is specially called *turbine*.

The expansion process is fully adiabatic and hence $dQ = 0$. Neglecting changes in kinetic and potential energies the Equation (2.26a) becomes:

$$\Delta H = -\Delta W_S$$

Maximum shaft work, $(W_S)_{max}$, is available if and only if the expansion process is reversible adiabatic, i.e., isentropic ($dS = 0$). So for isentropic expansion process:

$$(\Delta W_S)_{max} = (\Delta H)_S$$

and hence efficiency of any expander is given by:

$$\eta = \frac{\Delta W_S}{(\Delta W_S)_{max}} = \frac{\Delta H}{(\Delta H)_S} \tag{2.29}$$

The $H - S$ diagram of an expander is shown in the Figure 2.6.

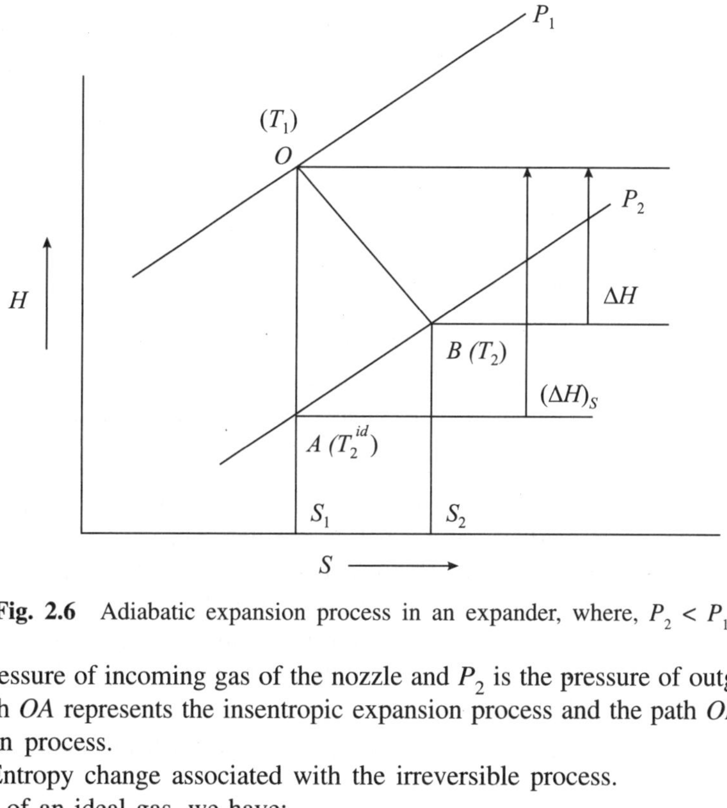

Fig. 2.6 Adiabatic expansion process in an expander, where, $P_2 < P_1$.

P_1 is the pressure of incoming gas of the nozzle and P_2 is the pressure of outgoing gas of the nozzle. The path *OA* represents the insentropic expansion process and the path *OB* represents the actual expansion process.

$S_2 - S_1$ = Entropy change associated with the irreversible process.

For 1 mole of an ideal gas, we have:

$$\Delta S = C_P \ln \frac{T_2}{T_1} - R \ln \frac{P_2}{P_1}$$

For isentropic process, i.e. along the path *OA*, $T_2 = T_2^{id}$ and $\Delta S = 0$.

Hence,
$$T_2^{id} = T_1 \left(\frac{P_2}{P_1}\right)^{\frac{R}{C_P}} = T_1 \left(\frac{P_2}{P_1}\right)^{\frac{C_P - C_V}{C_P}} = T_1 \left(\frac{P_2}{P_1}\right)^{1 - \frac{1}{\gamma}}$$

Or, $$T_2^{id} = T_1\left(\frac{P_2}{P_1}\right)^{\frac{(\gamma-1)}{\gamma}} \tag{2.30}$$

Thus, $$W_s\big|_{max} = -(\Delta H)_S = -C_P(T_2^{id} - T_1)$$

$$= -C_P\, T_1\left[\left(\frac{P_2}{P_1}\right)^{\frac{(\gamma-1)}{\gamma}} - 1\right]$$

So, $$W_s\big|_{max} = -C_P(T_2^{id} - T_1) \tag{2.31a}$$

and $$W_S\big|_{max} = C_P\, T_1\left[1-\left(\frac{P_2}{P_1}\right)^{\frac{(\gamma-1)}{\gamma}}\right] \tag{2.31b}$$

Again, actual shaft work = $W_S = -\Delta H = -C_P(T_2 - T_1)$

Hence, $$\eta = \frac{W_S}{W_S\big|_{max}} = \frac{-C_P(T_2 - T_1)}{-C_P(T_2^{id} - T_1)} = \frac{(T_2 - T_1)}{(T_2^{id} - T_1)}$$

$$T_2 = T_1 + \eta(T_2^{id} - T_1) = \eta\, T_2^{id} + (1 - \eta)T_1$$

Or, $$T_2 = \eta\, T_2^{id} + (1 - \eta)T_1 \tag{2.32}$$

T_2 can easily be determined. T_2^{id} can easily be calculated by using Equation (2.30) and hence the efficiency (η) of an expander can easily be evaluated by using Equation (2.32).

Compressors The operating system of a compressor is just reverse to that of an expander. In an expander shaft work is realised at the cost of internal energy of working fluid. Whereas, in the case of a compressor, shaft work is introduced at one end and very high kinetic energy of working fluid is received at the other end. This kinetic energy is mostly used in rotating blades in different equipment like pumps, blowers, compressors etc. Thus it is obvious that volume of working fluid decreases or pressure increases during compression process but in expansion process pressure of working fluid decreases or volume increases. Like expanders, in a compressor also the process may be visualized in two ways:

(i) Isentropic process, i.e. $dS = 0$. This is the ideal process where minimum shaft work is required to achieve the final pressure.

(ii) Irreversible adiabatic process, i.e. $dQ = 0$ but $dS > 0$. This is the real process where a definite shaft work, which is greater than that required in the isentropic process, to achieve the final pressure. Thus the efficiency of compressor can be written as:

$$\eta = \frac{W_S\big|_{isentropic}}{W_S} = \frac{(\Delta H)_S}{(\Delta H)} \quad \ldots \tag{2.33}$$

$$W_S\big|_{isentropic} = -(\Delta H)_S = -C_P(T_2^{id} - T_1)$$

$$W_S = -(\Delta H) = -C_P(T_2 - T_1)$$

T_1 = Initial temperature

T_2 = Final temperature achieved in a real process

T_2^{id} = Final temperature achieved in the isentropic process.

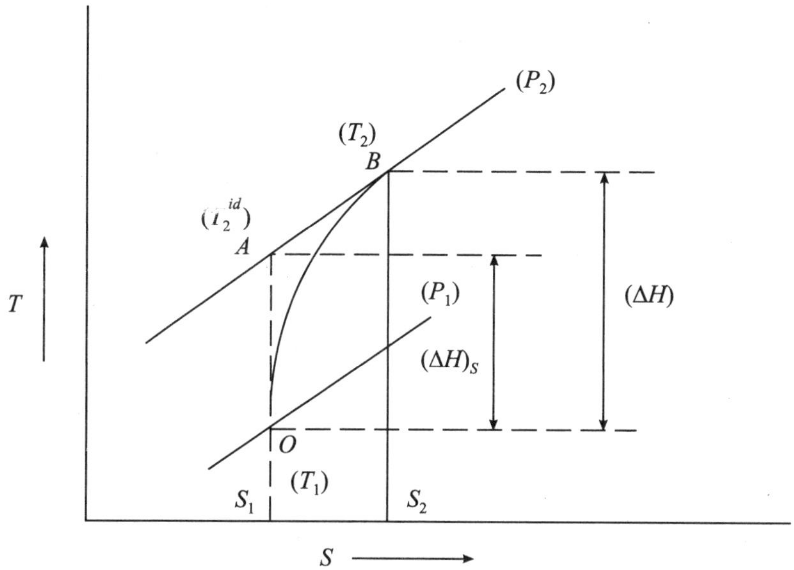

Fig. 2.7 Adiabatiac compression process in a compressor, where $P_2 > P_1$.

Efficiency of a compressor is given by:

$$\eta = \frac{W_S\big|_{\text{isentropic}}}{W_S} = \frac{C_P(T_2^{id} - T_1)}{C_P(T_2 - T_1)} = \frac{(T_2^{id} - T_1)}{(T_2 - T_1)}$$

$$(T_2 - T_1) = \frac{(T_2^{id} - T_1)}{\eta}$$

Or,
$$T_2 = \frac{T_2^{id}}{\eta} - \left(\frac{1}{\eta} - 1\right)T_1 \tag{2.34}$$

In the Figure 2.7, *OA* represents the ideal adiabatiac compression process, where enthalpy change is $(\Delta H)_S$. The curve *OB* represents the actual or real adiabatic compression process, where enthalpy change is ΔH.

During compression process the fluid is sufficiently compressed with increase in temperature. The working fluid is air in the cases of blowers, fans etc. but it is a liquid in the case of pumps.

$(\Delta H)_S$ may be evaluated by using thermodynamic property relations.

$$dH = dQ + VdP \quad \text{and} \quad dQ = TdS \text{ for reversible process.}$$

Or,
$$dH = TdS + VdP$$

$$\left(\frac{\partial H}{\partial P}\right)_T = V + T\left(\frac{\partial S}{\partial P}\right)_T$$

Again, from Maxwell's relations, we have:

$$\left(\frac{\partial S}{\partial P}\right)_T = -\left(\frac{\partial V}{\partial T}\right)_P = -\beta V$$

Where, $\beta = \text{volume expansivity} = \frac{1}{V}\left(\frac{\partial V}{\partial T}\right)_P$

Hence, $\left(\frac{\partial H}{\partial P}\right)_T = V - \beta TV = V(1 - \beta T)$

Again we can write that: $H = f(P, T)$

Or, $dH = \left(\frac{\partial H}{\partial T}\right)_P dT + \left(\frac{\partial H}{\partial P}\right)_T dP = C_P\, dT + (V - \beta TV)\, dP$

$\Delta H|_{rev}$ = Enthalpy change for a reversible process. As $(\Delta H)_S$ is a reversible isentropic process, we can write that:

$$(\Delta H)_S = C_P\, \Delta T + V(1 - \beta T)\Delta P \tag{2.35a}$$

Again, we know that for a reversible process:

$$dS = \frac{dQ}{T} = \frac{dH - VdP}{T} = \frac{[C_P\, dT + (V - \beta TV)dP] - VdP}{T}$$

$$= \frac{C_P\, dT}{T} - \beta VdP$$

On integration, we get:

$$\Delta S = C_P \ln \frac{T_2}{T_1} - \beta V \Delta P \tag{2.35b}$$

$$\text{Shaft work} = W_S = -\frac{(\Delta H)_S}{\eta},$$

where, η = efficiency of the compressor.

Or, $$W_S = -\frac{(\Delta H)_S}{\eta} = -\frac{1}{\eta}[C_P\, \Delta T + (V - \beta TV)\Delta P] \tag{2.35c}$$

The Equations 2.35b and 2.35c represent the entropy change and actual shaft work respectively of any real adiabatic compression process. However, Equations 2.35a, 2.35b and 2.35c are mostly used in those adiabatic compression processes, where liquid is used as working fluid since in many cases enthalpy values of compressed liquids are not available. Thus the above equations are very much useful in the case of pumps.

2.4 Power cycles

2.4.1 Introduction

In any conventional power plant heat energy is realized in either through combustion of fossil fuel or through nuclear fission or fusion processes. The function of heat engine is to convert part of this heat energy into work. In all work producing devices conversion of heat energy into work takes place through cyclic process, which are commonly known as power cycles. The heat engines can be divided into two broad categories.

(a) Steam engine
(b) Internal-combustion (IC) engine

In the case of steam engine, working fluid, i.e., steam is enclosed in a cylinder and heat is supplied from outside the cylinder. So to reach the working fluid heat has to cross the metallic boundary. The working fluid after receiving the heat energy, executes a cyclic process, delivering a definite amount of work.

In the case of *IC-engine* the source of heat energy is available within heat engine itself. Chemical energy of fuel is converted into heat energy through combustion and combustion products serve as the working medium.

Disadvantages of steam engine

(a) Huge amount of heat loss is inevitable during conduction process.
(b) Since metallic wall has to withstand high temperature and pressure, temperature, at which heat absorption occurs, must not be very high.

2.4.2 Steam engine

The most ideal and efficient heat engine is the Carnot heat engine, where water gets converted into steam in the boiler. The steam then undergoes an expansion process in a turbine and finally condensed into water, which is then pumped to the boiler. The schematic diagram of the power plant and the corresponding $T - S$ diagram are shown in the Figures 2.8 and 2.9 respectively. Steps $2 \rightarrow 3$ and $4 \rightarrow 1$ are isentropic processes.

(a) At stage 2 the saturated vapour enters into the turbine and the exhaust contains high liquid content, bringing about severe damage to the turbine.
(b) In step $4 \rightarrow 1$ pump accepts liquid/vapour mixture and delivers saturated liquid to the boiler. It is next to impossible to design such an engine to execute the step $4 \rightarrow 1$ isentropically.

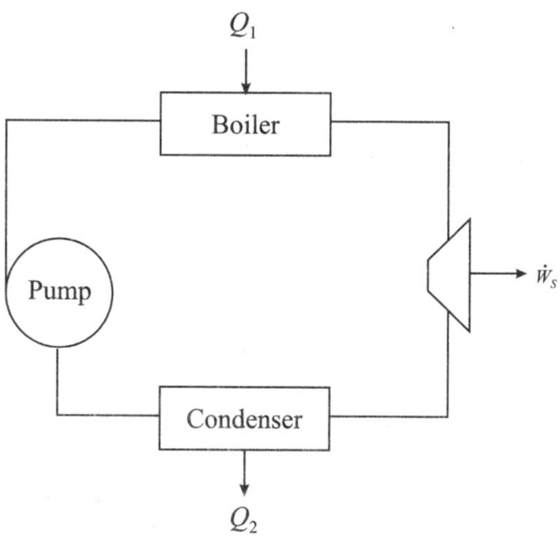

Fig. 2.8 Schematic diagram of a steam power plant.

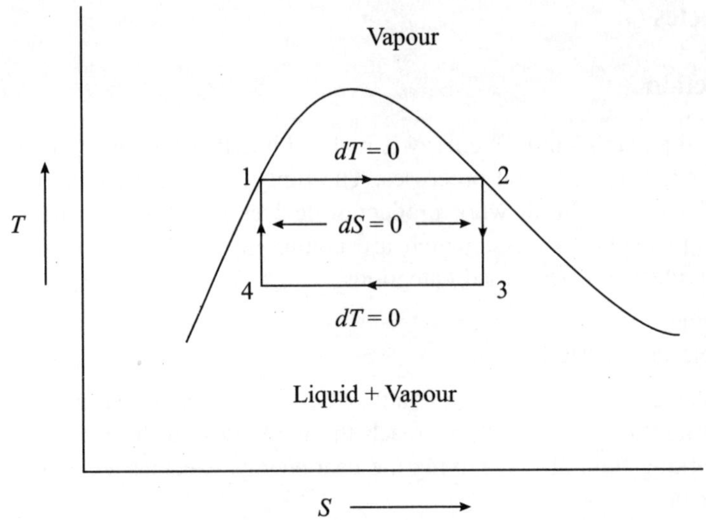

Fig. 2.9 $T - S$ diagram of Carnot heat engine.
1 → Saturated liquid, 2 → Saturated vapour, 3 → Liquid/vapour mixture
4 → Liquid/vapour mixture.

The above two problems can be avoided by modifying the cycle as follows:

(a) Super heated vapour is used in the turbine, which delivers works isentropically.
(b) Saturated liquid is delivered to the pump, which absorbs work isentropically.

Thus the corresponding $T - S$ diagram is given by:

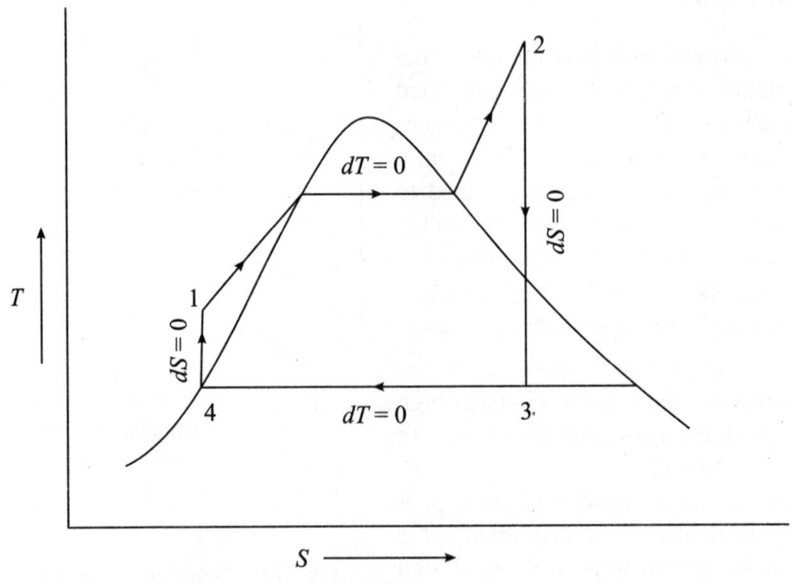

Fig. 2.10 *TS*-diagram of Rankine cycle.

The cycle, for which the above $T - S$ diagram (Figure 2.10) is shown, is known as Rankine cycle and is accepted as a standard cycle for heat engines.

In actual practice the expansion process in turbine is an irreversible process and hence the process is adiabatic but not isentropic. Furthermore, work requiring step, i.e. step 4 → 1 is also irreversible and hence the process is also not isentropic. The $T - S$ diagram of actual power cycle is shown in the Figure 2.11.

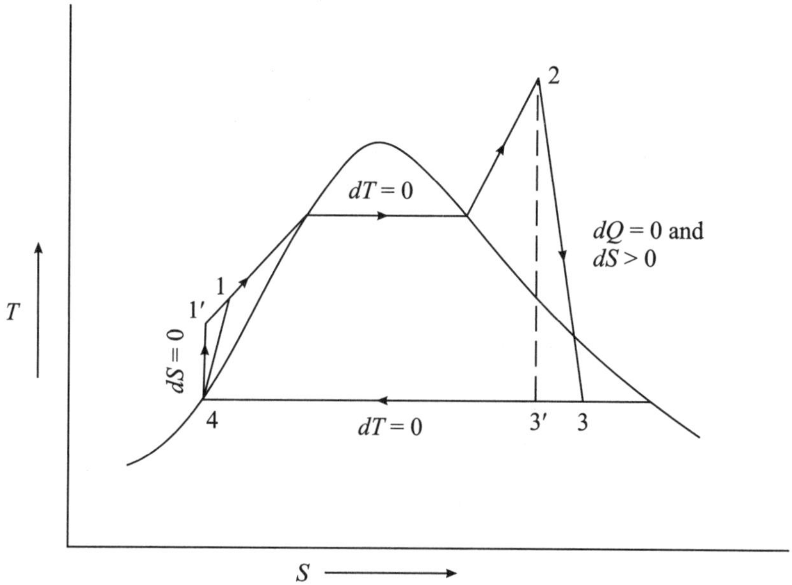

Fig 2.11 *TS*-diagram of actual power cycle (1-2-3-4-1) (1′-2-3′-4-1′) indicates Rankine cycle.

Work absorbed at step 4 → 1′ = $W_1' = -(\Delta H_1)_S$

Work delivered at step 2 → 3′ = $W_2' = -(\Delta H_2)_S$

Work delivered at step 2 → 3 = $W_2 = -(\Delta H_2)$

$$|\Delta H_2| = \phi|(\Delta H)_S| \quad \text{and} \quad \phi < 1$$

Hence, $$W_2 = \phi W_2'$$

Net work available at Rankine cycle = $(W_1' + W_2') = W_S'$

$$(\Delta H_1)_S = (H_1' - H_4) > 0 \quad \text{and hence,} \quad W_1' < 0$$

$$(\Delta H_1) = (H_1 - H_4) > 0 \quad \text{and hence,} \quad W_1 < 0,$$

since, $$W_1 = -(\Delta H_1)$$

$$(\Delta H_2)_S = (H_3' - H_2) < 0 \quad \text{and hence,} \quad W_2' > 0$$

$$(\Delta H_2) = (H_3 - H_2) < 0 \quad \text{and hence,} \quad W_2 > 0$$

Net work available at actual cycle = $(W_2 + W_1) = W_S$,

Where, $W_1 = \frac{W_1'}{\zeta}$ and $\zeta < 1$

Hence, it is obvious that:

$$W_S = (W_2 + W_1) = \left(\phi W_2' + \frac{W_1'}{\zeta}\right) < (W_2' + W_1')$$

Or, $W_S < W_S'$

As $|W_1| > |W_1'|$, $\Delta H_1 > (\Delta H_1)_S$, where ΔH_1 is the enthalpy change at step 4 → 1 in actual power cycle and $(\Delta H_1)_S$ is the enthalpy change in 4 → 1′ in Rankine Cycle.

So, heat absorbed in the boiler in Rankine cycle = $(H_2 - H_1') = Q_1'$

and heat absorbed in the boiler in actual cycle = $(H_2 - H_1) = Q_1$

$$\Delta H_1 = (H_1 - H_4) \quad \text{and} \quad (\Delta H_1)_S = (H_1' - H_4)$$

As, $\Delta H_1 > (\Delta H_1)_S$, $H_1 > H_1'$

and hence, $(H_2 - H_1) < (H_2 - H_1')$

Or, $Q_1 < Q_1'$

So, thermal efficiency of a Rankine cycle = η_R

$$\eta_R = \frac{W_S'}{Q_1'}$$

and thermal efficiency of an actual power cycle = η

$$\eta = \frac{W_S}{Q_1}$$

As $H_1 \approx H_1'$, $Q_1 \approx Q_1'$, but $W_S << W_S'$ and hence, $\eta < \eta_R$

2.4.3 Internal combustion engines

Two commercially important IC-engines are:

(a) The Otto engine and (b) the Diesel engine.

(a) The Otto engine:

Working principle The combustion chamber consists of a cylinder with a movable piston. In the 1st step the piston moves outward with the intake of air/fuel mixture into the cylinder.

In the 2nd step mixture is compressed almost adiabatically up to the pressure P_2. The mixture is then ignited and combustion occurs very rapidly so that pressure P_2 rises to P_3 at constant volume.

In the 3rd step the combustion products expand almost adiabatically till the pressure falls to P_4 at which point valve opens and pressure falls rapidly to P_1 at constant volume.

In the 4th and final step the piston moves inward and disposes all combustion products to the atmosphere.

The total cycle is shown in the Figure 2.12.

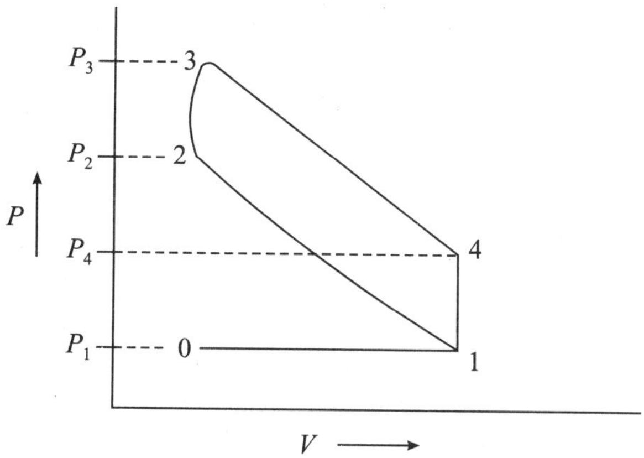

Fig. 2.12 Otto cycle for internal combustion engine

$0 \rightarrow 1$ indicates step-I

$1 \rightarrow 2 \rightarrow 3$ indicates step-II

$3 \rightarrow 4 \rightarrow 1$ indicates step-III

$1 \rightarrow 0$ indicates step-IV

The working fluid is petrol. As the behaviour of combustion products is not known, an idealized standard cycle is considered where air is used as working medium. The efficiency and other characteristics of that idealized standard cycle is equivalent to that of Otto cycle. That idealized standard cycle is known as air-standard Otto cycle. The *PV*-diagram of air-standard Otto cycle is shown in the Figure 2.13.

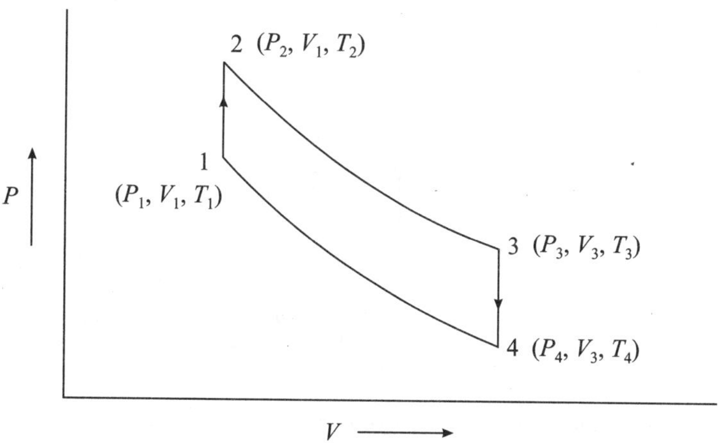

Fig. 2.13 Air-standard Otto cycle.

Step 1 → 2 Absorption of heat at constant volume = $|Q_1|$

Step 2 → 3 Reversible adiabatic expansion of air.

Step 3 → 4 Rejection of heat or cooling at constant volume = $|Q_2|$

Step 4 → 1 Reversible adiabatic compression of air.

The thermal efficiency of the above cycle is given by:

$$\eta = \frac{W_{net}}{|Q_1|} = \frac{|Q_1| - |Q_2|}{|Q_1|} = 1 - \frac{|Q_2|}{|Q_1|}$$

Again $|Q_1| = C_V(T_2 - T_1)$ for 1 mole of an ideal gas. Assume that air, used in the above cycle, behaves as an ideal gas.

$$Q_2 = C_V(T_4 - T_3) = -C_V(T_3 - T_4)$$

Or, $$|Q_2| = C_V(T_3 - T_4) \quad [\text{since } T_3 > T_4]$$

Or, $$\frac{|Q_2|}{|Q_1|} = \frac{(T_3 - T_4)}{(T_2 - T_1)}$$

Again, along the adiabatic curve 2 → 3, we have:

$$T_2 V_1^{\gamma-1} = T_3 V_3^{\gamma-1}$$

$$\frac{V_3}{V_1} = r = \text{compression ratio}$$

$$r = \frac{\text{volume of gas at the beginning of compression stroke}}{\text{volume of gas at the end of compression stroke}}$$

$$T_2 = T_3\left(\frac{V_3}{V_1}\right)^{\gamma-1}$$

Or, $$T_2 = T_3(r)^{\gamma-1}$$

Along the adiabatic curve 4 → 1, we have:

$$T_4 V_3^{\gamma-1} = T_1 V_1^{\gamma-1}$$

Or, $$T_1 = T_4\left(\frac{V_3}{V_1}\right)^{\gamma-1} = T_4(r)^{\gamma-1}$$

Hence, $$(T_2 - T_1) = T_3(r)^{\gamma-1} - T_4(r)^{\gamma-1} = (T_3 - T_4)(r)^{\gamma-1}$$

$$\eta = 1 - \frac{|Q_2|}{|Q_1|} = 1 - \frac{(T_3 - T_4)}{(T_2 - T_1)} = 1 - \left(\frac{1}{r}\right)^{\gamma-1}$$

Thus the efficiency of an air standard Otto cycle = η

$$\eta = 1 - \left(\frac{1}{r}\right)^{\gamma-1} \tag{2.36}$$

It is evident that with the increase in compression ratio (r) the efficiency of the Otto cycle increases.

(b) The diesel engine In contrast to the Otto engine high compression ratio has been set in the case of Diesel engine in order to achieve high temperature so that combustion occurs spontaneously. Furthermore, in the case of the Diesel engine fuel is injected at the end of compression step but slow enough to proceed heat absorption at constant pressure.
At the same compression ratio efficiency of the Otto engine is higher than that of the Diesel engine but the Otto engine cannot be operated at high compression ratio due to pre-ignition problem. As the Diesel engine can be operated at very high compression ratio safely higher efficiency can be obtained in the case of the Diesel engine compared to that of the Otto engine.

In order to evaluate efficiency of the Diesel engine an idealized standard cycle is considered where air is used as working medium. The PV-diagram of air-standard Diesel cycle is shown in the Figure 2.14.

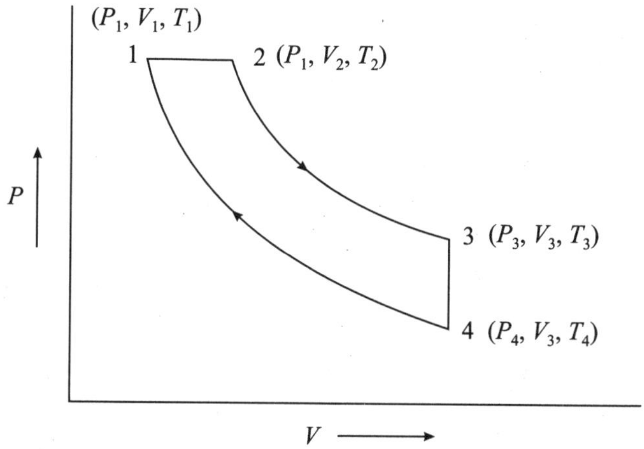

Fig. 2.14 Air standard Diesel cycle.

Step 1 → 2 Absorption of heat at constant pressure = $|Q_P|$
Step 2 → 3 Reversible adiabatic expansion of air
Step 3 → 4 Rejection of heat at constant volume = $|Q_V|$
Step 4 → 1 Reversible adiabatic compression of air.

The internal efficiency of the above cycle is given by:

$$\eta = \frac{W_{\text{net}}}{|Q_P|} = \frac{|Q_P| - |Q_V|}{|Q_P|} = 1 - \frac{|Q_V|}{|Q_P|}$$

Again, $|Q_P| = C_P(T_2 - T_1)$ for 1 mole of an ideal gas and assume that air, used in the above cycle, behaves as an ideal gas.

Similarly, $Q_V = C_V(T_4 - T_3)$ for 1 mole of an ideal gas

$$\gamma = \left(\frac{C_P}{C_V}\right) \quad \text{and} \quad |Q_V| = C_V(T_3 - T_4) \quad [\text{since } T_3 > T_4]$$

So, $$\eta = 1 - \frac{|Q_V|}{|Q_P|} = 1 - \frac{C_V(T_3 - T_4)}{C_P(T_2 - T_1)} = 1 - \frac{(T_3 - T_4)}{\gamma(T_2 - T_1)} \quad (2.37a)$$

Along the adiabatic path 2 → 3, we have:

$$P_1 V_2^{\gamma} = P_3 V_3^{\gamma}$$

Or, $$\frac{P_1}{P_3} = (r_e)^{\gamma}$$

Along the adiabatic path 4 → 1, we have:

$$P_4 V_3^{\gamma} = P_1 V_1^{\gamma}$$

Or, $$\frac{P_1}{P_4} = (r_c)^{\gamma}$$

$\frac{V_3}{V_1}$ = Compression ratio = r_c and $\frac{V_3}{V_2}$ = Expansion ratio = r_e

Hence, $$\frac{r_c}{r_e} = \frac{V_2}{V_1}$$

$$\frac{P_1}{P_3} = (r_e)^{\gamma}, \quad \frac{P_1}{P_4} = (r_c)^{\gamma}, \quad \frac{P_3}{P_4} = \left(\frac{r_c}{r_e}\right)^{\gamma} = \left(\frac{V_2}{V_1}\right)^{\gamma} \quad (2.37b)$$

At the point 3, $P_3 V_3 = RT_3$ for 1 mole of air and at the point 4, $P_4 V_3 = RT_4$ for 1 mole of air.

Hence, $$\frac{T_3}{T_4} = \frac{P_3}{P_4} = \left(\frac{r_c}{r_e}\right)^{\gamma}$$

Or, $$T_3 = T_4\left(\frac{r_c}{r_e}\right)^{\gamma} \quad (2.37c)$$

Again, along the adiabatic path 2 → 3, we have:

$$T_2 V_2^{\gamma-1} = T_3 V_3^{\gamma-1}$$

Or, $$T_2 = T_3\left(\frac{V_3}{V_2}\right)^{\gamma-1} = T_3(r_e)^{\gamma-1}$$

Along the adiabatic path 4 → 1, we have:

$$T_4 V_3^{\gamma-1} = T_1 V_1^{\gamma-1}$$

Or, $$T_1 = T_4\left(\frac{V_3}{V_1}\right)^{\gamma-1} = T_4(r_c)^{\gamma-1}$$

So, $$T_2 = T_3(r_e)^{\gamma-1} \text{ and } T_1 = T_4(r_c)^{\gamma-1} \quad (2.37d)$$

Thus combining Equations (2.37a), (2.37c) and (2.37d), we have:

$$\eta = 1 - \frac{T_4\left(\frac{r_c}{r_e}\right)^{\gamma} - T_4}{\gamma\left[T_3\,(r_e)^{\gamma-1} - T_4\,(r_c)^{\gamma-1}\right]} = 1 - \frac{T_4\left[\left(\frac{r_c}{r_e}\right)^{\gamma} - 1\right]}{\gamma\left[T_4\left(\frac{r_c}{r_e}\right)^{\gamma}(r_e)^{\gamma-1} - T_4\,(r_c)^{\gamma-1}\right]}$$

$$= 1 - \frac{\left[\left(\frac{1}{r_e}\right)^{\gamma} - \left(\frac{1}{r_c}\right)^{\gamma}\right]}{\gamma\left(\frac{1}{r_e} - \frac{1}{r_c}\right)}$$

Thus the efficiency of an air standard diesel cycle is given by:

$$\eta = 1 - \frac{\left[\left(\frac{1}{r_e}\right)^{\gamma} - \left(\frac{1}{r_c}\right)^{\gamma}\right]}{\gamma\left(\frac{1}{r_e} - \frac{1}{r_c}\right)} \tag{2.38}$$

2.4.4 Gas turbine power plant

The advantage of internal combustion engine is the direct use of energy at high temperature and pressure without any heat transfer. Simultaneously, the *IC*-engine has the disadvantage of loss of energy due to friction between reciprocating piston and cylinder but the turbine does not have this disadvantage.

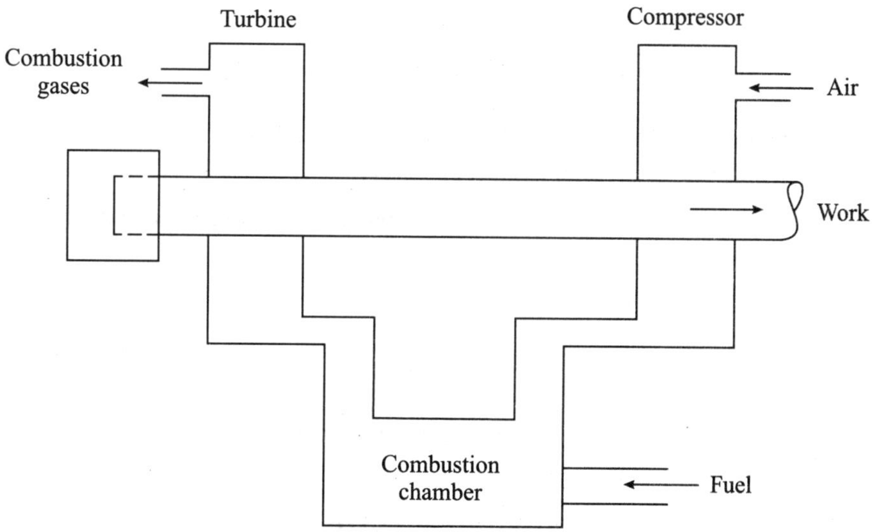

Fig. 2.15 Schematic diagram of a gas-turbine power plant.

The gas turbine engine enjoys both the advantages of *IC*-engine and turbine. The schematic diagram of a gas turbine power plant is shown in the Figure 2.15.

Working principle

(a) Air is compressed to several atmospheres in a compression chamber and then allowed to pass into combustion chamber where fuel is injected.
(b) The fuel is burnt in the combustion chamber and the combustion gases enter into the turbine, where expansion occurs due to which mechanical work is obtained. Part of this mechanical work is used to compress air in the compressor.
(c) Efficiency of the power plant increases with the increase in inlet temperature of the combustion gases in the turbine.

The *PV*-diagram of (air-standard idealized) gas-turbine cycle, popularly known as Brayton cycle, is shown in the Figure 2.16.

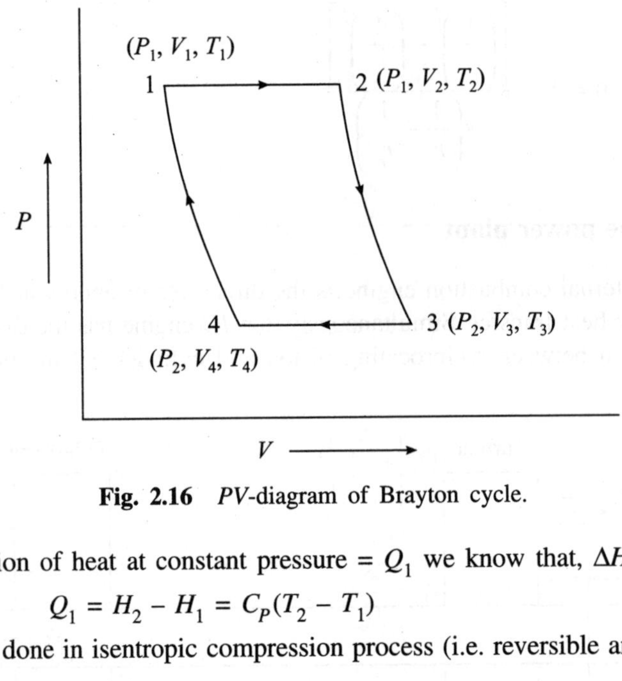

Fig. 2.16 *PV*-diagram of Brayton cycle.

Step 1 → 2 Addition of heat at constant pressure = Q_1 we know that, $\Delta H_P = \Delta Q_P$

Thus: $Q_1 = H_2 - H_1 = C_P(T_2 - T_1)$

Step 4 → 1 Work done in isentropic compression process (i.e. reversible and adiabatic process) $= W_{4\to 1}$

We know from Equation 2.26a that:

$$\Delta\left(H + \frac{u^2}{2} + zg\right) = \Delta Q - \Delta W_S$$

As the process is adiabatic $\Delta Q = 0$. Neglecting potential energy and kinetic energy, we get:

$$-W_S = \Delta H$$

Thus, $$W_{4\to 1} = -(\Delta H)_{4\to 1} = -C_P(T_1 - T_4)$$

Step 2 → 3 Work done in isentropic expansion process = $W_{2\to 3} = -(\Delta H)_{2\to 3} = -C_P(T_3 - T_2)$

$$W = \text{Total work} = W_{2\to 3} + W_{4\to 1} = -C_P(T_3 - T_2) - C_P(T_1 - T_4)$$
$$= C_P(T_2 - T_3) - C_P(T_1 - T_4)$$

Along the adiabatic curve 2 → 3, we have:

$$T_2 P_1^{\frac{1-\gamma}{\gamma}} = T_3 P_2^{\frac{1-\gamma}{\gamma}}$$

Or,
$$T_3 = T_2\left(\frac{P_1}{P_2}\right)^{\frac{1-\gamma}{\gamma}}$$

Along the adiabatic curve 4 → 1, we have:

$$T_4 P_2^{\frac{1-\gamma}{\gamma}} = T_1 P_1^{\frac{1-\gamma}{\gamma}}$$

Or,
$$T_4 = T_1\left(\frac{P_1}{P_2}\right)^{\frac{1-\gamma}{\gamma}}$$

$$W = C_P\left[T_2 - T_2\left(\frac{P_1}{P_2}\right)^{\frac{1-\gamma}{\gamma}}\right] - C_P\left[T_1 - T_1\left(\frac{P_1}{P_2}\right)^{\frac{1-\gamma}{\gamma}}\right]$$

$$= C_P(T_2 - T_1)\left[1 - \left(\frac{P_1}{P_2}\right)^{\frac{1-\gamma}{\gamma}}\right]$$

$$= Q_1\left[1 - \left(\frac{P_1}{P_2}\right)^{\frac{1-\gamma}{\gamma}}\right]$$

$$\text{Efficiency} = \eta = \frac{W}{Q_1} = \left[1 - \left(\frac{P_1}{P_2}\right)^{\frac{1-\gamma}{\gamma}}\right]$$

Or,
$$\eta = \left[1 - \left(\frac{P_2}{P_1}\right)^{\frac{\gamma-1}{\gamma}}\right] \tag{2.39}$$

It is evident that higher the value of P_1, i.e., higher the inlet pressure, higher is the efficiency of the Brayton cycle, since $\left(\frac{\gamma-1}{\gamma}\right) > 0$.

2.5 Refrigeration cycle

2.5.1 Carnot cycle

Refrigeration means maintenance of temperature below the temperature of surroundings. This is reverse of the heat engine cycle. In heat engine cycle heat is absorbed at higher temperature and rejected at lower temperature but in refrigeration cycle heat is absorbed at lower temperature and rejected at higher temperature. Efficiency of a refrigeration cycle is expressed as coefficient of performance (COP).

$$\text{COP} = \frac{\text{Amount of heat absorbed}}{\text{Net work}}$$

Different types of refrigeration cycles are available, e.g., (i) vapour-compression cycle, (ii) Absorption refrigeration cycle.

Ideal refrigeration cycle is the Carnot refrigeration cycle, where Q_2 amount of heat is absorbed at lower temperature T_2 and Q_1 amount of heat is rejected at higher temperature T_1, followed by two adiabatic steps. According to 2nd law of thermodynamics this cycle cannot be operated without use of any external work. So some work is required to operate the cycle. Let the work be W.

According to the 1st law of thermodynamics we know that:

$$\Delta Q = \Delta U + W$$

For a cyclic process: $\Delta U = 0$ and $W = \Delta Q = |Q_1| - |Q_2|$., since $|Q_1| > |Q_2|$

Hence,

$$\text{COP} = \frac{|Q_2|}{W} = \frac{|Q_2|}{|Q_1| - |Q_2|} = \frac{1}{\dfrac{|Q_1|}{|Q_2|} - 1}$$

Again from Carnot cycle we know that:

$$\frac{|Q_1|}{|Q_2|} = \frac{T_1}{T_2}$$

where, $T_1 > T_2$

Or,

$$\text{COP} = \frac{1}{\dfrac{T_1}{T_2} - 1} = \frac{T_2}{T_1 - T_2} \tag{2.40}$$

Carnot refrigerator is independent of working medium.

2.5.2 Vapour-compression cycle

The most useful commercial refrigeration cycle is the vapour-compression cycle. In this cycle liquid refrigerant absorbs heat from the surroundings and gets converted to saturated vapour, which is then compressed sufficiently to superheat the refrigerant vapour. That superheated vapour is allowed to pass through a condenser where heat is rejected at higher temperature and vapour is converted to saturated liquid, which is then passed through a throttle valve where expansion occurs in isenthalpic process, as a result of which, a sharp drop in temperature is observed. During evaporation process liquid refrigerant absorbs heat from the surroundings, as a result of which, surrounding atmosphere gets cooled. The whole process is shown in the Figure 2.17 and the corresponding $T-S$ diagram is shown in the Figure 2.18.

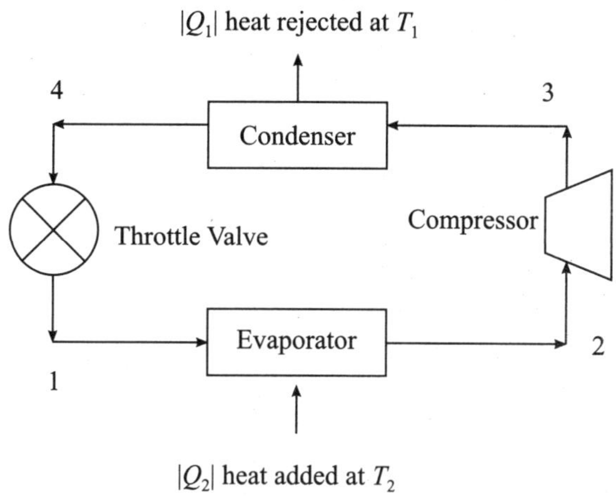

Fig. 2.17 Schematic diagram of a vapour-compression cycle

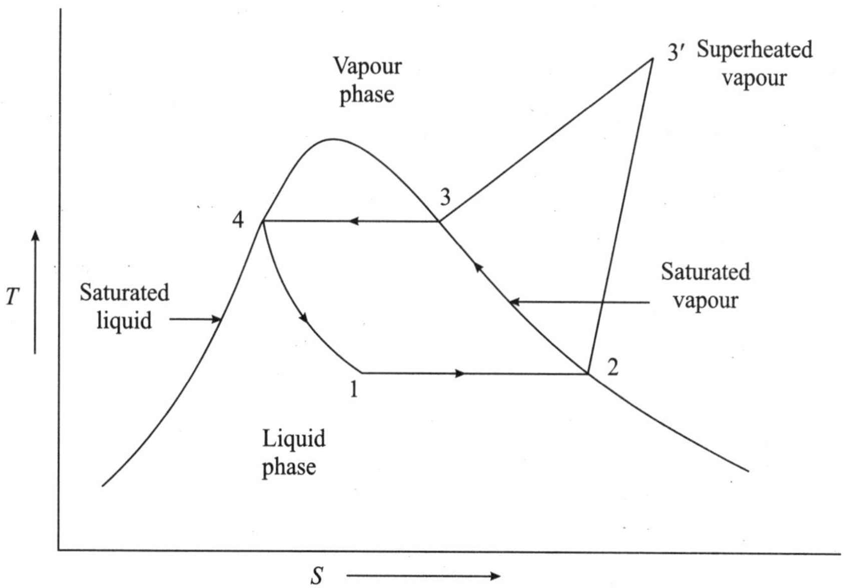

Fig. 2.18 $T-S$ diagram of a typical vapour-compression cycle.

Step 1 → 2 Heat is absorbed from the surroundings by the refrigerant to form saturated vapour. Amount of heat added = $|Q_2|$

Step 2 → 3′ → 3 Saturated vapour is heated to form superheated vapour (3′) which is then cooled to saturated vapour (state 3).

Step 3 → 4 The saturated vapour is allowed to condense to form saturated liquid (state 4) Amount of heat rejected = $|Q_1|$

Step 4 → 1 The saturated liquid is then passed through a throttle valve due to which a sharp drop of temperatures is observed. The process is isenthalpic but entropy increases, as shown in the Figure 2.18.

Condenser Heat is rejected at constant pressure and temperature. H_1, H_2, H_3 and H_4 are the enthalpy values per unit mass flow rate at the states 1, 2, 3 and 4.

We know that: $dH = dQ + VdP$

At constant pressure, $dH = dQ$

Or, $\Delta Q = \Delta H$

Thus at condenser: $Q_1 = (H_4 - H_3)$ and $|Q_1| = (H_3 - H_4)$,

since, $H_3 > H_4$

Evaporator Heat is absorbed from the surroundings at constant pressure and temperature. So,

$$Q_2 = (H_2 - H_1) \quad \text{and} \quad |Q_2| = (H_2 - H_1)$$

since, $H_2 > H_1$

Throttle valve Isenthalpic process.

So, $H_1 = H_4$

Net work required to operate the cycle = $|Q_1| - |Q_2|$

$$|Q_1| - |Q_2| = (H_3 - H_4) - (H_2 - H_1) = (H_3 - H_2) \quad [\text{since } H_1 = H_4]$$

Hence,

$$\text{COP} \doteq \frac{|Q_2|}{W} = \frac{H_2 - H_1}{H_3 - H_2} \tag{2.41}$$

It is evident that, lower the value of H_3 higher the value of COP.

Again, $|Q_2| = (H_2 - H_1)$, where H_1 and H_2 are the enthalpy values per unit mass flow rate. If $\dot{m}$ is the mass flow rate of refrigerant, we can write that:

$$|Q_2| = \dot{m}(H_2 - H_1)$$

Or,

$$\dot{m} = \frac{|Q_2|}{(H_2 - H_1)} \tag{2.42}$$

Ton of refrigeration 1 ton of refrigeration implies that amount of heat to be removed to freeze 1 ton of water to 1 ton of ice at 0°C in 24 hrs.

Latent heat of fusion of ice = 80 cal/g = 80 kcal/kg. = 336 kJ/kg.

Total heat to be removed to freeze 1 ton of water = 336 × 1000 = 336000 kJ.

This amount of heat is to be removed in 24 hrs.

So, amount of heat removed = $\frac{336000}{24}$ kJ/hr = 14000 kJ/hr

Thus tons of refrigeration = $\frac{\dot{m}(H_2 - H_1)}{14000}$

2.5.3 Absorption refrigeration

In absorption refrigeration process refrigerant is absorbed in a relatively non-volatile solvent and refrigerant is regenerated as vapour, which is then executed as vapour-compression cycle. The schematic diagram of an absorption refrigeration cycle is shown in the Figure 2.19.

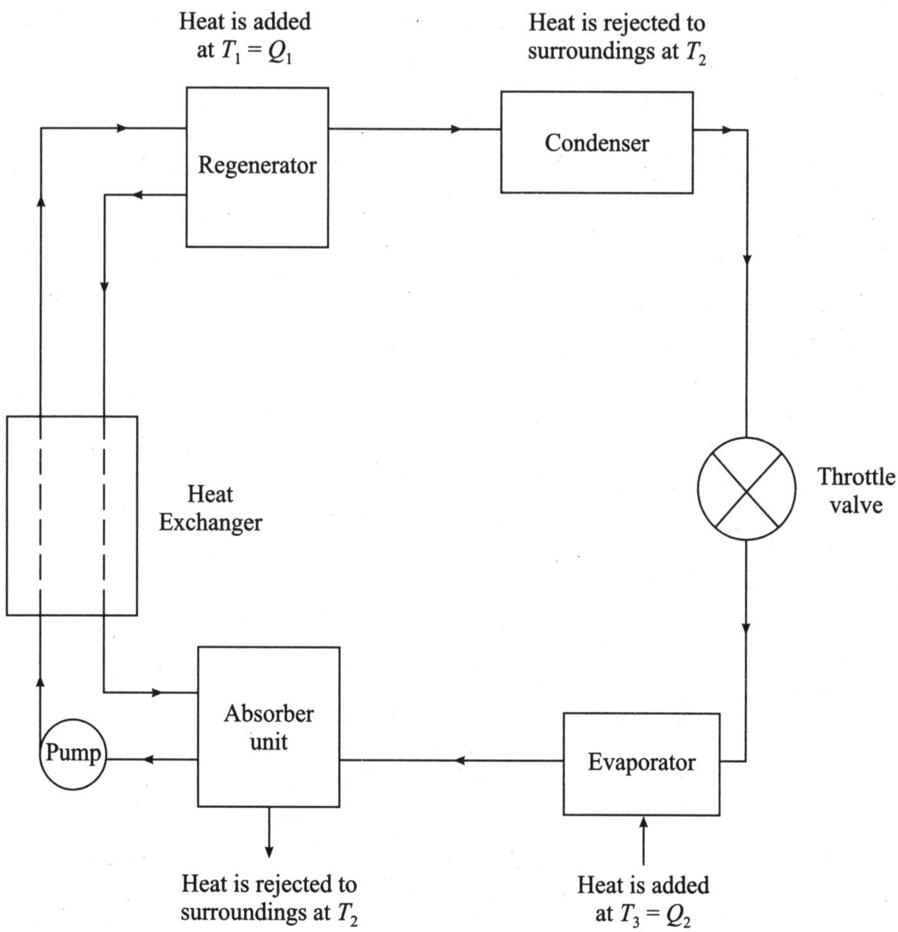

Fig. 2.19 Schematic diagram of an absorption refrigerator.

Evaporator At this unit liquid refrigerant is evaporated with the absorption of heat $|Q_2|$ at temperature T_3.

Absorber Vapour of the refrigerant is absorbed by a relatively non-volatile solvent and heat of absorption is rejected to the surroundings at temperature T_2.

Pump and heat exchanger Solvent with refrigerant is compressed and pumped to the regenerator through heat exchanger. In the heat exchanger two streams are flowing. In the upstream, solvent, enriched with refrigerant, absorbs heat. In the downstream, solvent, having low concentration of refrigerant and flowing from the regenerator, rejects heat in the heat exchanger and enters into the absorber unit.

Regenerator Pressure in the regenerator is the same as that in the condenser and $|Q_1|$ amount of heat is added isothermally at temperature T_1. In this unit, solvent, enriched with refrigerant, is heated to reach the temperature T_1 such that refrigerant gets evaporated and leaves the regenerator, moving towards the condenser unit. So, the concentration of refrigerant in the solvent falls and reaches a minimum. This solvent, with low concentration of refrigerant, is sent to the absorber unit through heat exchanger. Net work available can be computed by using Carnot engine efficiency equation:

$$\eta = \frac{W}{|Q_1|} = \frac{T_1 - T_2}{T_1}$$

Or,
$$W = |Q_1| \left(\frac{T_1 - T_2}{T_1} \right) \tag{2.43}$$

This work is used in the next section of the refrigeration cycle, consisting of condenser and evaporator units.

Condenser In this unit refrigerant vapour gets condensed with the release of latent heat to the surroundings, at temperature T_2. After condensation the saturated liquid refrigerant is allowed to pass through the throttle valve, as a result of which, temperature of the liquid refrigerant drops to T_3 and the liquid is sent to the evaporator unit. Amount of work required for refrigeration is given by Equation (2.43).

$$\text{COP} = \frac{|Q_2|}{W} = \frac{T_3}{T_2 - T_3} \tag{2.44a}$$

Or,
$$|Q_2| = W \left(\frac{T_3}{T_2 - T_3} \right) \tag{2.44b}$$

Putting the value of W from Equation (2.43) in the Equation (2.44b), we get:

$$|Q_2| = |Q_1| \left(\frac{T_1 - T_2}{T_1} \right) \left(\frac{T_3}{T_2 - T_3} \right)$$

Or,
$$\frac{|Q_1|}{|Q_2|} = \left(\frac{T_1}{T_1 - T_2} \right) \left(\frac{T_2 - T_3}{T_3} \right) \tag{2.45}$$

All the processes considered in the above cycle are reversible and hence COP becomes maximum for this cycle. As COP is maximum $|Q_2|$ is also maximum according to Equation (2.44a). Hence, $\frac{|Q_1|}{|Q_2|}$ is minimum for Carnot absorption refrigerator.

Example 2.1 A Carnot engine is coupled to a Carnot refrigerator so that all of the work produced by the engine is used by the refrigerator in extraction of heat from a heat reservoir at 270 K at the rate of 4 kJ/s. The source of energy for the Carnot engine is a heat reservoir at 500 K. If both devices discard heat to the surroundings at 300 K, how much heat does the engine absorb from the 500 K-reservoir?

If the actual coefficient of performance of the refrigerator is $\omega = [\omega_{\text{Carnot}}/1.5]$ and if the thermal efficiency of the engine is $\eta = [\eta_{\text{Carnot}}/1.5]$, how much heat does the engine absorb from the 500 K-reservoir?

Sol. We know that

$$\frac{|Q_1|}{|Q_2|} = \left(\frac{T_1}{T_1 - T_2}\right)\left(\frac{T_2 - T_3}{T_3}\right)$$

$$|Q_2| = \text{Heat absorbed at } T_3 = 4 \text{ kJ/s}$$

where, $T_3 = 270$ K

$$|Q_1| = \text{Heat absorbed at } T_1$$

where, $T_1 = 500$ K

T_2 = Temperature at which heat is rejected to the surroundings = 300 K.

$$|Q_1| = 4 \times \left(\frac{500}{500 - 300}\right)\left(\frac{300 - 270}{270}\right) = 4 \times \frac{500}{200} \times \frac{30}{270} = 1.11 \text{ kJ/s}$$

Amount of heat absorbed at 500 K = 1.11 kJ/s
The work obtained from Carnot engine = W

$$W = |Q_1| \left(\frac{T_1 - T_2}{T_1}\right)$$

$$\text{Actual work} = W_S = \frac{W}{1.5} = \frac{|Q_1|}{1.5}\left(\frac{T_1 - T_2}{T_1}\right)$$

$$\text{COP of carnot refrigerator} = \frac{|Q_2|}{W} = \left(\frac{T_3}{T_2 - T_3}\right)$$

$$\text{COP of actual refrigerator} = \frac{1}{1.5}\left(\frac{T_3}{T_2 - T_3}\right)$$

So, $$\frac{|Q_2|}{W_S} = \frac{T_3}{1.5(T_2 - T_3)}$$

Or, $$|Q_2| = \frac{W_S T_3}{1.5(T_2 - T_3)}$$

Or, $$|Q_2| = \frac{|Q_1|}{1.5}\left(\frac{T_1 - T_2}{T_1}\right)\frac{T_3}{1.5(T_2 - T_3)}$$

Or, $$|Q_1| = 2.25\,|Q_2|\,\frac{T_1(T_2 - T_3)}{T_3(T_1 - T_2)} = \frac{4\times 2.25\times 500(300-270)}{270(500-300)}$$

Or, $$|Q_1| = 2.5 \text{ kJ/s}$$

So the actual amount of heat absorbed by the refrigerator = 2.5 kJ/s.

Refrigerants In most cases methane and ethane based halogenated hydrocarbons are used as refrigerants. Nomenclature of different refrigerants is based on number of hydrogen atoms and halogen atoms present in the compound. Halogenated hydrocarbon is usually chlorofluoro carbon, which is written as CFC.
For CFC – 12, add 90 with 12, i.e. 90 + 12 = 102

1	0	2
↑	↑	↑
No. of C-atoms	No. of H-atoms	No. of F-atoms

So, two fluorine atoms are attached with one carbon atom. The remaining two positions must be occupied by chlorine atoms. Thus the compound is CCl_2F_2. Technically it is represented as R-12.
Similarly for R-143: add 90 with 143, i.e. 90 + 143 = 233

2	3	3
↑	↑	↑
No. of C-atoms	No. of H-atoms	No. of F-atoms

Thus the compound will be $C_2H_3Cl_2F_3$.
CFCs are very much stable, readily move up to stratosphere and destroy the ozone layer. That is why the use of CFC is very much limited. Ammonia is also a good refrigerant, particularly in absorption refrigeration system but it is toxic, inflammable and very much unfriendly to environment. Sulphur-di-oxide (SO_2), methyl chloride (CH_3Cl), propane (C_3H_8), butane (C_4H_{10}) etc. are also used as refrigerants.

Example 2.2 An engine working on Otto cycle is supplied with air at 0.1 MPa and 35°C. The compression ratio is 8. Heat supplied is 2100 KJ/kg. Calculate the maximum pressure and temperature of the cycle, the cycle efficiency and the mean effective pressure (mep).
(For air, C_P = 1.005, C_V = 0.718 and R = 0.287 kJ/kg. °K)

Sol. The *PV*-diagram of Otto-cycle is shown below:

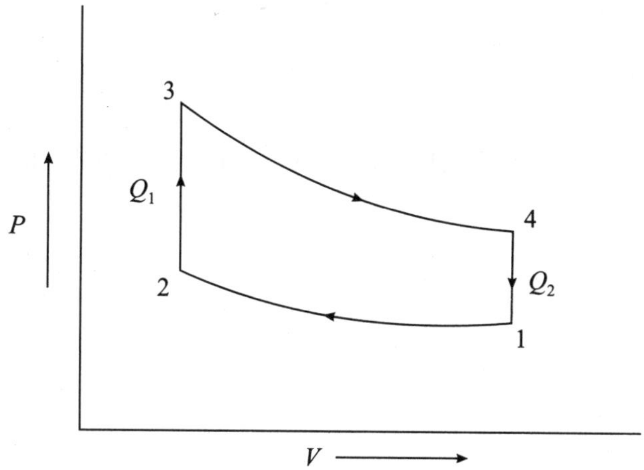

PV-diagram of the Otto cycle.

At the state 1, P_1 = 0.1 MPa, T_1 = (273 + 35) K = 308 K, Q_1 = 2100 kJ/kg.
Efficiency of the cycle = η

$$\eta = 1 - \left(\frac{1}{r}\right)^{\gamma - 1}, \; r = \frac{V_1}{V_3} = \frac{V_1}{V_2} = 8$$

$$\gamma = \frac{C_P}{C_V} = \frac{1.005}{0.718} = 1.4$$

Hence, $$\eta = 1 - \left(\frac{1}{8}\right)^{0.4} = 0.565$$

Heat is absorbed at constant volume.

So, $$Q_1 = C_V(T_3 - T_2) \tag{2.46}$$

$$2100 = 0.718\,(T_3 - T_2)$$

Process 1 → 2 is an adiabatic process.

So, $$T_1 V_1^{\gamma-1} = T_2 V_2^{\gamma-1}$$

Or, $$\frac{T_2}{T_1} = \left(\frac{V_1}{V_2}\right)^{\gamma - 1} = (8)^{0.4}$$

Or, $$T_2 = 308 \times (8)^{0.4} = 708.4 \text{ K}$$

Putting the value of T_2 in Equation (2.46) we get:

$$(T_3 - 708.4) = \frac{2100}{0.718} = 2925$$

Or, $T_3 = 3633\ \text{K}$

T_3 is the maximum temperature of this cycle.
Again, applying adiabatic expression in the process 1 → 2 we have:

$$P_1 V_1^{\gamma} = P_2 V_2^{\gamma}$$

Or, $$P_2 = P_1\left(\frac{V_1}{V_2}\right)^{\gamma} = (0.1) \times (8)^{1.4} = 1.837\ \text{MPa}$$

Again, along 2 → 3, we have:

$$\frac{P_2 V_2}{T_2} = \frac{P_3 V_3}{T_3}$$

$$V_2 = V_3$$

$$T_3 = 3633\ \text{K}$$

$$T_2 = 708.4\ \text{K}$$

$$P_2 = 1.837\ \text{MPa}$$

Or, $$P_3 = 1.837\left(\frac{3633}{708.4}\right) = 9.421\ \text{MPa}$$

P_3 is the maximum pressure of this cycle.

$$\text{Work} = W = (\eta \times Q_1) = 0.565 \times 2100 = 1186.5\ \text{kJ/kg.}$$

If P_m is the mean effective pressure,

$$W = P_m(V_1 - V_2) \quad \text{and} \quad V_1 = \left(\frac{RT_1}{P_1}\right)$$

So, $$V_1 = (0.287 \times 308/100) = 0.884\ \text{m}^3$$

$$P_1 = 0.1\ \text{MPa} = 100\ \text{KPa} = 100\ \text{kJ/m}^3$$

$$R = 0.287\ \text{kJ/kg.°K}$$

$$T_1 = 308\ \text{K.}$$

$$\left(\frac{V_1}{V_2}\right) = 8.$$

Or, $$V_2 = \left(\frac{V_1}{8}\right) = \left(\frac{0.884}{8}\right) = 0.11\ \text{m}^3/\text{kg}$$

Thus, $$W = 1186.5\ \text{kJ/kg}$$

$$V_1 = 0.884\ \text{m}^3/\text{kg}$$

$$V_2 = 0.11\ \text{m}^3/\text{kg}$$

Hence, $$1186.5 = P_m(0.884 - 0.11) = 0.774\ P_m$$

Or, $$P_m = \left(\frac{1186.5}{0.774}\right) \text{ kPa} = 1533 \text{ kPa} = 1.533 \text{ MPa}$$

Problem 2.1 In an air standard Diesel cycle the compression ratio is 16. At the beginning of compression the temperature and pressure are 15°C and 0.1 MPa respectively. Heat is added at constant pressure, until the temperature reaches to 1480°C. Calculate (a) the cut-off ratio, (b) the heat supplied per kg of air, (c) the cycle efficiency and (d) the mean effective pressure.

Problem 2.2 An air standard Diesel cycle absorbs 1500 J/mole of heat at constant pressure. The pressure and temperature at the beginning of the compression step are 1 bar and 20°C. The pressure at the end of the compression step is 4 bar. Assuming air to be an ideal gas for which $C_P = \frac{7}{2}R$ and $C_V = \frac{5}{2}R$, what are the compression ratio and expansion ratio of the cycle?

Example 2.3 An air-standard dual cycle has a compression ratio of 16 and the compression begins at 1 bar, 50°C. The maximum pressure is 70 bar. The heat transferred to air at constant pressure is equal to that at constant volume. Estimate (a) the pressures and temperatures at the cardinal points of the cycle, (b) the cycle efficiency and (c) the mean effective pressure of the cycle.

$$C_V = 0.718 \text{ kJ/kg°K}$$
$$C_P = 1.005 \text{ kJ/kg°K}$$

Sol. The *PV*-diagram of dual cycle is shown below:

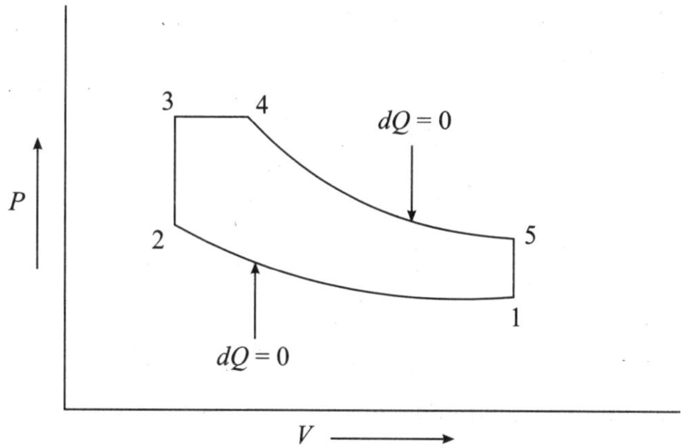

Heat is absorbed partly at constant volume (process 2 → 3) and partly at constant pressure (process 3 → 4). So, net heat absorbed = $Q_1 = m C_V (T_3 - T_2) + m C_P (T_4 - T_3)$

Heat is rejected at the end of the cycle (process 5 → 1) at constant volume.

So, $$Q_2 = m C_V (T_1 - T_5) \; [T_5 > T_1]$$

Or, $$|Q_2| = m C_V (T_5 - T_1)$$

$$|Q_1| = Q_1 = m\,C_V\,(T_3 - T_2) + m\,C_P\,(T_4 - T_3)\ [T_3 > T_2 \text{ and } T_4 > T_3].$$
$$= 2\,m\,C_V\,(T_3 - T_2) \quad \text{[according to the condition given in the problem]}$$

Thus efficiency of the cycle $= \eta = 1 - \dfrac{|Q_2|}{|Q_1|}$

Or,
$$\eta = 1 - \frac{m\,C_V\,(T_5 - T_1)}{m\,C_V\,(T_3 - T_2) + m\,C_P\,(T_4 - T_3)}$$

$$= 1 - \frac{(T_5 - T_1)}{(T_3 - T_2) + \gamma(T_4 - T_3)}$$

$$T_1 = (273 + 50) = 323 \text{ K},\ P_1 = 1 \text{ bar},\ P_3 = P_4 = P_{max} = 70 \text{ bar}$$

$$\text{Compression ratio} = r = \frac{V_1}{V_2} = \frac{V_1}{V_3} = 16,\ \gamma = \frac{C_P}{C_V} = 1.4$$

Along 1 → 2

$$P_1 V_1^{\gamma} = P_2 V_2^{\gamma}$$

Or,
$$P_2 = P_1 \left(\frac{V_1}{V_2}\right)^{\gamma} = 1 \times (16)^{1.4} = 48.5 \text{ bar}$$

$$T_1 V_1^{\gamma-1} = T_2 V_2^{\gamma-1}$$

Or,
$$T_2 = T_1 \left(\frac{V_1}{V_2}\right)^{\gamma-1} = 323 \times (16)^{0.4} = 979 \text{ K}$$

Again, along 2 → 3

$$\frac{P_2 V_2}{T_2} = \frac{P_3 V_3}{T_3}$$

$$V_2 = V_3,\ P_3 = 70 \text{ bar}$$
$$T_2 = 979 \text{ K},\ P_2 = 48.5 \text{ bar}$$

So,
$$T_3 = 979 \left(\frac{70}{48.5}\right) = 1413 \text{ K}$$

Again, according to given condition:

$$Q_{2 \to 3} = Q_{3 \to 4}$$

Or,
$$m\,C_V\,(T_3 - T_2) = m\,C_P\,(T_4 - T_3)$$

Or,
$$T_4 = \frac{(1+\gamma)T_3 - T_2}{\gamma} = \frac{2.4 \times 1413 - 979}{1.4} = 1723 \text{ K}$$

$$V_1 = \frac{RT_1}{P_1} = \frac{0.287 \times 323}{100} = 0.927 \text{ m}^3/\text{kg}$$

[since, 1 bar = 10^5Pa = 100 KPa = 100 kJ/m^3]

$$V_2 = \left(\frac{0.927}{16}\right) = 0.0579 \text{ m}^3/\text{kg}$$

$$V_3 = V_2 = 0.0579 \text{ m}^3/\text{kg}$$

$$P_3 = P_4 = 70 \text{ bar}$$

Along 3 → 4

$$\frac{P_3 V_3}{T_3} = \frac{P_4 V_4}{T_4}, \; V_4 = \frac{V_3 T_4}{T_3} = \frac{0.0579 \times 1723}{1413} = 0.0706 \text{ m}^3/\text{kg}$$

$$P_1 = 1 \text{ bar} = 100 \text{ kPa}$$

$$V_5 = V_1 = \frac{RT_1}{P_1} = 0.927 \text{ m}^3/\text{kg}$$

Along 4 → 5

$$T_5 V_5^{\gamma-1} = T_4 V_4^{\gamma-1}$$

Or,
$$T_5 = T_4\left(\frac{V_4}{V_5}\right)^{\gamma-1} = 1723 \times \left(\frac{0.0706}{0.927}\right)^{0.4} = 615 \text{ K}$$

$$\eta = 1 - \frac{(T_5 - T_1)}{(T_3 - T_2) + \gamma(T_4 - T_3)}$$

$$= 1 - \frac{(615 - 323)}{(1413 - 979) + 1.4(1723 - 1413)} = 0.664$$

$$Q_1 = 2C_V (T_3 - T_2) = 0.718 \ (1413 - 979) \times 2 = 623.2 \text{ kJ/kg}$$

$$W = \eta \times Q_1 = 0.664 \times 623.2 \text{ kJ/kg} = 413.82 \text{ kJ/kg}$$

Mean effective pressure = P_m

So, $$P_m(V_1 - V_2) = W = 413.82$$

Or, $$P_m(0.927 - 0.0579) = 413.82$$

Or, $$P_m = 476.15 \text{ kPa} = 4.7615 \text{ bar.}$$

Problem 3 A refrigerated space is maintained at –12°C and cooling water is available at 21°C. The evaporator and condenser are of sufficient size that –12°C temperature difference can be maintained for heat transfer in each case. The refrigeration capacity is 120,000 kJ/hr.

(a) What is the value of COP(ω) for a Carnot refrigerator?

(b) Calculate ω and mass flow rate ($\dot{m}$) for the vapour compression cycle where expansion occurs isentropically.

(c) Calculate ω and $\dot{m}$ for the vapour compression cycle where expansion occurs at constant enthalpy with compressor efficiency 80%.

3

CHEMICAL KINETICS

3.1 Introduction

To occur a spontaneous chemical reaction the required thermodynamic condition is: $\Delta G < 0$ but this condition is not enough to bring about a chemical reaction unless we have sufficient knowledge on the following:

(a) Rate at which the molecules interact
(b) Factors controlling the reaction rate
(c) Mechanism through which the reactants are converted into products.

On the basis of the above information, reaction conditions, popularly known as kinetic conditions, (e.g. temperature, pressure, catalyst, solvent etc.) are set. Thus to occur a chemical reaction the reactants must satisfy the thermodynamic condition as well as kinetic condition. Chemical kinetics or reaction kinetics deals with the study on the rates and mechanisms of chemical reactions.

Chemical reactions are broadly classified into two categories:

(a) Homogeneous reaction, where all the reacting species are in same phase.
(b) Heterogeneous reaction, where the reacting species exist in two or more phases.

In another way of classification chemical reactions can be divided into two categories:

(a) One way reaction, which proceeds along one direction only, represented by a single arrow ($\rightarrow$), e.g.

$$aA + bB \rightarrow lL + mM$$

It is also called irreversible (non-equilibrium) reaction.

(b) Two way reaction, which proceeds both in the forward and backward directions, represented by a double headed arrow ($\rightleftharpoons$) e.g.

$$aA + bB \rightleftharpoons lL + mM$$

It is also called reversible (equilibrium) reaction.

Assumption The most important assumption in studying the chemical kinetics of chemical reactions is that the volume of the whole system (V) remains constant throughout the reaction.

Fundamental postulates Let us consider the following irreversible homogeneous reaction:

$$aA + bB + \ldots\ldots \rightarrow lL + mM + \ldots\ldots$$

(a) The rate of consumption per mole is same for each reactant and also equal to the rate of formation per mole of any of the products. If, n_A, n_B ...etc. are the number of moles of reactants A, B ...etc. present at time t, we have:

$$-\frac{1}{a}\frac{dn_A}{dt} = -\frac{1}{b}\cdot\frac{dn_B}{dt} = \ldots\ldots$$

The negative sign indicates that the reactants are consumed. Again, if n_L, n_M ...etc. are the number of moles of L, M, ...etc. present at time t, we have:

$$\frac{1}{l}\cdot\frac{dn_L}{dt} = \frac{1}{m}\cdot\frac{dn_M}{dt} = \ldots\ldots$$

Combing the above two equations, we get,

$$-\frac{1}{a}\frac{dn_A}{dt} = -\frac{1}{b}\cdot\frac{dn_B}{dt} = \ldots\ldots = \frac{1}{l}\cdot\frac{dn_L}{dt} = \frac{1}{m}\cdot\frac{dn_M}{dt} = \ldots\ldots$$

Hence, we have:

$$\frac{|dn_A|}{a} = \frac{|dn_B|}{b} = \ldots\ldots = \frac{|dn_L|}{l} = \frac{|dn_M|}{m} = \ldots\ldots = d\varepsilon$$

$d\varepsilon$ is known as reaction coordinate. It indicates percent (%) conversion of reactants to products.

However, the above relation need not hold good for a reaction occurring in more than one step. Nevertheless, the above relation holds good for multi-step reactions if the concentrations of all reaction intermediates are small so that their effect on stoichiometry can be neglected.

(b) Rate of the reaction is given by:

$$J = \left(\frac{1}{V}\right)\left[-\frac{1}{a}\left(\frac{dn_A}{dt}\right)\right] = k.C_A^{n_1}\, C_B^{n_2} \tag{3.1a}$$

Again, $\left(\frac{n_A}{V}\right) = C_A$ and

$$J = \left[-\frac{1}{a}\left(\frac{dC_A}{dt}\right)\right] = \left[-\frac{1}{b}\left(\frac{dC_B}{dt}\right)\right] = \left[\frac{1}{l}\left(\frac{dC_L}{dt}\right)\right] = \left[\frac{1}{m}\left(\frac{dC_M}{dt}\right)\right]$$

J = Velocity or rate of the chemical reaction

k = Velocity constant or rate constant or specific reaction rate. It is defined as the rate of a reaction at unit concentrations of reactants.

C_A, C_B etc. are the concentrations of reactants left at time t. n_1, the exponent to C_A, is termed order of the reaction with respect to the reactant A. Similarly n_2 is the order of the reaction with respect to the reactant B etc. n_1, n_2 ...etc. are also called partial orders. The overall order or simply order of the whole reaction is the sum of the partial orders and given by:

$$n = n_1 + n_2 + \ldots \tag{3.1b}$$

The Equation (3.1a) is called the rate law.

3.2 Molecularity and order

Consider the following reaction:

$$aA + bB \rightarrow lL + mM$$

The rate is given by:

$$J = k.C_A^{n_1}\, C_B^{n_2}$$

Definition of molecularity Total number of molecules involved in a stoichiometrically balanced equation is called molecularity of that reaction. In other words, molecularity of a reaction is the total number of molecules that are taking part in that given reaction. Thus molecularity of the above reaction = $(a + b)$. As it is a sum of stoichiometric coefficient, molecularity is a theoretical quantity.

Definition of order Order of a reaction is the sum of exponents of concentrations of reactants to which the rate of that chemical reaction depends. In the above example order of a reaction is given by: $n = n_1 + n_2$. As n_1 and n_2 are determined experimentally, the order of a chemical reaction is an experimental quantity.

Differences between molecularity and order

Molecularity	*Order*
(1) It is a theoretical concept.	(1) It is an experimentally determined quantity.
(2) It is always a positive whole number and cannot be zero or fraction.	(2) It may be positive whole number/zero/ fractional number/negative.
(3) Molecularity has no limit since it is obtained from stoichiometrically balanced equation e.g. $KClO_3 + 6FeCl_2 + 6HCl = KCl + 6FeCl_3 + 3H_2O$ Molecularity is 13 but the order is only 3.	(3) The order of a chemical reaction is usually limited within +3. Only a very few reactions, known to us, have order greater than 3.

(Contd.)

Molecularity	*Order*
(4) Rate of a reaction does not depend on molecularity.	(4) Rate of a reaction depends on order.
(5) It provides idea about reaction mechanism, through which product is formed.	(5) It does not throw any light on reaction mechanism.

3.3 Rate equations for different types of reactions

3.3.1 Irreversible 1st order reaction

Case – I Consider the reaction:

$$A \rightarrow P + Q + R + \ldots \text{ etc.}$$

time (t)	*No. of moles of reactant 'A'*	*No. of moles of any product (say P)*
0	a	0
t	$(a - x)$	x
$t_{\frac{1}{2}}$	$\left(\frac{a}{2}\right)$	$\left(\frac{a}{2}\right)$

x = Number of moles of reactant A consumed during time t.

V = Volume of the system = constant.

C_A = Concentration of reactant A left at time $t = \dfrac{a - x}{V}$

$C_P = \dfrac{x}{V}$ Concentration of product P at time t

Boundary conditions: (i) at $t = 0$, $(a - x) = a$ and $x = 0$

(ii) at $t = t$, $(a - x) = (a - x)$ and $x = x$

The rate equation is given by:

$$J = -\left(\frac{dC_A}{dt}\right) = kC_{\text{A}}$$

Again,

$$J = \frac{dC_P}{dt} = \frac{d\left(\frac{x}{V}\right)}{dt} = \left(\frac{1}{V}\right)\frac{dx}{dt}$$

Putting the value of C_A in the above equation, we get:

$$-\frac{d}{dt}\left(\frac{a-x}{V}\right) = k\left(\frac{a-x}{V}\right)$$

Or, $$-\frac{1}{V}\frac{d(a-x)}{dt} = k\left(\frac{a-x}{V}\right)$$

Or, $$-\frac{d(a-x)}{dt} = k(a-x)$$

Method 1 Using the boundary conditions of $(a - x)$, we get:

$$-\int_a^{(a-x)} \frac{d(a-x)}{(a-x)} = k\int_0^t dt$$

On integration, we get:

$$\ln\frac{a}{(a-x)} = k.t$$

Or, $$t = \frac{1}{k}\ln\frac{a}{(a-x)} \qquad (3.2)$$

Equation (3.2) is the rate equation of first order reaction of type shown above.

Method 2 Using the boundary conditions of x, we get:

$$-\frac{d(a-x)}{dt} = \frac{dx}{dt} = k(a-x)$$

Or, $$\int_0^x \frac{dx}{(a-x)} = k\int_0^t dt$$

On integration, we get:

$$[-\ln(a-x)]_0^x = [k.t]_0^t$$

Or, $$\ln\frac{a}{(a-x)} = k.t$$

Or, $$t = \frac{1}{k}\ln\frac{a}{(a-x)}$$

Case – II Consider the reaction:

$$lL \rightarrow pP + qQ + rR + \ldots \text{ etc.,}$$

Where, l, p, q, r etc. are the stoichiometric coefficients.

time (t)	*No. of moles of reactant*	*No. of moles of any product (say P)*
0	a	0
t	$(a-x)$	$\left(\frac{p}{l}\right)\cdot x$
$t_{\frac{1}{2}}$	$\left(\frac{a}{2}\right)$	$\left(\frac{p}{l}\right)\left(\frac{a}{2}\right)$

x = No. of moles of reactant A consumed during time t.

V = Volume of the system = constant

C_A = Concentration of the reactant A left at time $t = \frac{(a-x)}{V}$ moles/litre.

$C_P = \left(\frac{p}{l}\right)\left(\frac{x}{V}\right)$ = Concentration of product P at time t.

Boundary conditions: (i) at $t = 0$, $(a-x) = a$ and $x = 0$

(ii) at $t = t$, $(a-x) = (a-x)$ and $x = x$

The rate equation is given by:

$$J = -\left(\frac{1}{l}\right)\frac{dC_A}{dt} = kC_A$$

Or,
$$-\left(\frac{1}{l}\right)\frac{d\left[\frac{(a-x)}{V}\right]}{dt} = k\frac{(a-x)}{V}$$

Or,
$$-\frac{d(a-x)}{(a-x)} = k.l.dt = k_L\,dt$$

where, $k_L = k.l$

On integration of the above equation w.r.t. $(a-x)$ within the boundary conditions we get:

Or,
$$t = \frac{1}{k_L}\ln\frac{a}{(a-x)} \tag{3.3}$$

Again,

$$J = \left(\frac{1}{p}\right)\frac{dC_P}{dt} = kC_A = k\frac{(a-x)}{V}$$

Or,
$$J = \left(\frac{1}{p}\right)\frac{d\left(\frac{px}{lV}\right)}{dt} = \left(\frac{1}{lV}\right)\frac{dx}{dt}$$

Or, $$\left(\frac{1}{lV}\right)\frac{dx}{dt} = k\frac{(a-x)}{V}$$

Or, $$\frac{dx}{(a-x)} = k.l.dt = k_L\, dt$$

On integration of the above equation w.r.t. x within the boundary conditions we get:

$$\ln\frac{a}{(a-x)} = k_L t$$

Or, $$t = \frac{1}{k_L}\ln\frac{a}{(a-x)}$$

The Equation (3.3) is the rate equation of 1st order reaction of type shown above. The only difference between Equation (3.2) and Equation (3.3) is in the value of rate constant, k and k_L, which are related as follows:

$$k_L = l.k$$

Half life period The time taken to consume a reactant to half of its amount is called half-life period of that reactant and it is designated by $t_{\frac{1}{2}}$.

Thus at $$t = t_{\frac{1}{2}},\quad x = \left(\frac{a}{2}\right)$$

Hence, $$(a-x) = \left(a - \frac{a}{2}\right) = \left(\frac{a}{2}\right)$$

Putting the value of $(a - x)$ at $t_{\frac{1}{2}}$ in the Equation (3.3), we get:

$$t_{\frac{1}{2}} = \frac{1}{k_L}\ln\frac{a}{\frac{a}{2}} = \frac{2.303\log 2}{k_L} = \frac{0.693}{k_L}$$

Or, $$t_{\frac{1}{2}} = \frac{0.693}{k_L} \qquad (3.4)$$

Unit of specific reaction rate (k_L) From the Equation (3.4), we get:

$$k_L = \frac{0.693}{t_{\frac{1}{2}}}$$

The unit of $t_{\frac{1}{2}}$ is sec and hence the unit of k_L is sec^{-1}.

Average life period Consider the total number of molecules present in a moles of reactant A is N_0 and N is the number of molecules present in $(a - x)$ moles of reactant A.

Thus the Equation (3.3) becomes:

$$\ln \frac{N_0}{N} = k_L t$$

Or,
$$\ln \frac{N}{N_0} = -k_L t$$

Or,
$$N = N_0\, e^{-k_L t}$$

Differentiating both sides w.r.t. t, we get:

$$\frac{dN}{dt} = -N_0\, k_L\, e^{-k_L t}$$

Or,
$$dN = -N_0\, k_L\, e^{-k_L t}\, dt \qquad (3.5)$$

The N_0 molecules are divided into numerous windows, such that first window contains dN_1 number of molecules, each of which has a life-time t_1, which means that each of dN_1 molecules is converted into products after time t_1.

Thus the total life-time of dN_1 molecules = $t_1(-dN_1)$.

The negative sign indicates that the reactant molecules are consumed.

Similarly, number of molecules in the 2nd window = dN_2 and life-time of each molecule = t_2

Total life-time of all dN_2 molecules = $t_2(-dN_2)$

1	2	3			$(i + 1)$	
dN_1 t_1	dN_2 t_2	dN_3 t_3	....	....	dN_{i+1} t_{i+1}	....
$t_1(-dN_1)$	$t_2(-dN_2)$	$t_3(-dN_3)$	....	....	$t_{i+1}(-dN_{i+1})$	....

So,
$$N_0 = dN_1 + dN_2 + dN_3 + \ldots\ldots + dN_{i+1} + \ldots\ldots$$

$$N_0 = \sum_{i=0}^{\infty} dN_{i+1}$$

Total life-time of all N_0 molecules = T

$$T = t_1(-dN_1) + t_2(-dN_2) + t_3(-dN_3) + \ldots\ldots + t_{i+1}(-dN_{i+1}) + \ldots\ldots$$

$$T = \sum_{i=0}^{\infty} t_{i+1}\,(-dN_{i+1})$$

As the series is continuous, we can write that:

$$T = \int_0^{\infty} t.(-dN)$$

Putting the value of dN [from Equation (3.5)] into the above equation, we get:

$$T = \int_0^\infty t\, N_0\, k_L\, e^{-k_L t}\, dt$$

Thus the average life-period per molecule = τ

$$\tau = \frac{T}{N_0} = \int_0^\infty t\, k_L\, e^{-k_L t}\, dt = k_L \int_0^\infty t\, e^{-k_L t}\, dt$$

$$= k_L \left[\frac{t\, e^{-k_L t}}{(-k_L)}\right]_0^\infty - k_L \int_0^\infty \frac{e^{-k_L t}}{(-k_L)} = \int_0^\infty e^{-k_L t}\, dt$$

$$= \left[\frac{e^{-k_L t}}{(-k_L)}\right]_0^\infty = \frac{1}{k_L}$$

Or, $$\tau = \frac{1}{k_L} \quad (3.6)$$

Examples of 1st order reaction:

(i) Dissociation of di-methyl-ether: $CH_3\,O\,CH_3 \rightarrow CH_4 + H_2 + CO$

(ii) Dissociation of nitrogen pentoxide: $2N_2O_5 \rightarrow 4NO_2 + O_2$

(iii) Decomposition of hydrogen peroxide in presence of Pt catalyst.

$$H_2O_2 \xrightarrow{[Pt]} H_2O + O$$

(iv) Conversion of N-chloroacetanilide to p-chloroacetanilide:

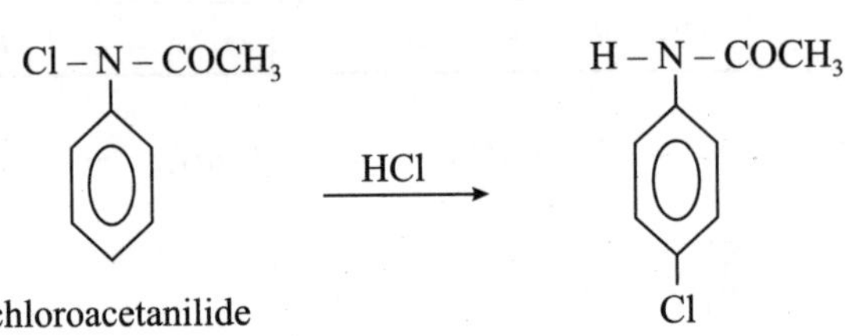

N-chloroacetanilide

p-chloroacetanilide

3.3.2 Irreversible 2nd order reactions

Case – I Consider the reaction

$$lL \rightarrow pP + qQ + rR + \ldots\ldots \text{ etc.,}$$

Where l, p, q, r etc. are stoichiometric coefficients. x moles of the reactant L gets reacted producing $\left(\frac{px}{l}\right)$ moles of product P, $\left(\frac{qx}{l}\right)$ moles of Q etc.

Time (t)	*No. of moles of reactant (A)*	*No. of moles of any product (say P)*
0	a	0
t	$(a - x)$	$\left(\frac{p \cdot x}{l}\right)$
$t_{\frac{1}{2}}$	$\left(\frac{a}{2}\right)$	$\left(\frac{p}{l}\right)\left(\frac{a}{2}\right)$

x = Number of moles of reactant L consumed during time t.

V = Volume of the system = constant.

$C_L = \left[\frac{(a-x)}{V}\right]$ moles/litre = concentration of reactant A left at time t.

$C_P = \left(\frac{p}{l}\right)\left(\frac{x}{V}\right)$ moles/litre = concentration of product P at time t.

Boundary conditions: (i) at $t = 0$: $(a - x) = a$ and $x = 0$

(ii) at $t = t$: $(a - x) = (a - x)$ and $x = x$

The rate is given by:

$$J = -\left(\frac{1}{l}\right)\frac{dC_L}{dt} = -\left(\frac{1}{l}\right)\frac{d\left[\frac{(a-x)}{V}\right]}{dt}$$

$$= -\left(\frac{1}{lV}\right)\frac{d(a-x)}{dt} = \left(\frac{1}{lV}\right)\frac{dx}{dt}$$

Again,

$$J = \left(\frac{1}{p}\right)\frac{d\,C_P}{dt} = \left(\frac{1}{p}\right)\frac{d\left(\frac{p.x}{l.V}\right)}{dt} = \left(\frac{1}{lV}\right)\frac{dx}{dt}$$

The rate equation is given by:

$$J = k.C_L^2$$

Or,
$$-\left(\frac{1}{lV}\right)\frac{d(a-x)}{dt} = k\left[\frac{(a-x)}{V}\right]^2$$

$$-\frac{d(a-x)}{(a-x)^2} = \left(\frac{k.l}{V}\right)dt = \left(\frac{k_L}{V}\right)dt$$

where, $k_L = k.l.$

Or, $$-\int_0^{(a-x)} \frac{d(a-x)}{(a-x)^2} = \frac{k_L}{V}\int_0^t dt$$

On integration, we get:

$$\left[\frac{1}{(a-x)} - \frac{1}{a}\right] = \frac{k_L t}{V}$$

Or, $$t = \frac{V}{k_L}\frac{x}{a(a-x)} \tag{3.7a}$$

Equation (3.7a) is the rate equation of 2nd order reaction of type shown above.

Case – I(a) If $l = 1$, i.e. the stoichiometric coefficient of reactant L is unity, $k_L = k$ and Equation (3.7a) becomes:

$$t = \frac{V}{k}\frac{x}{a(a-x)} \tag{3.7b}$$

Half-life period At $t = t_{\frac{1}{2}}$, $x = \left(\frac{a}{2}\right)$

Thus at half-life period Equation (3.7a) becomes:

$$t_{\frac{1}{2}} = \frac{V}{k_L}\frac{\left(\frac{a}{2}\right)}{a\left(\frac{a}{2}\right)} = \frac{V}{k_L a}$$

Or, $$t_{\frac{1}{2}} = \frac{V}{k_L a} \tag{3.8}$$

Unit of specific reaction rate k_L From Equation (3.8), we have:

$$k_L = \frac{V}{a\,t_{\frac{1}{2}}}$$

The unit of a is moles, V is litre and $t_{\frac{1}{2}}$ is sec. Thus the unit of k_L is litre/mole-sec.

Case – II Consider the reaction:

$$lL + mM \rightarrow pP + qQ + rR + \text{........ etc.,}$$

where l, m, p, q, n etc.are stoichiometric coefficients.

Time (*t*)	*No. of moles of reactant L*	*No. of moles of reactant M*	*No. of moles of any product (say P)*
0	a	b	0
t	$(a-x)$	$\left[b-\frac{mx}{l}\right]$	$\left(\frac{p}{l}\right).x$
$\left(t_{\frac{1}{2}}\right)_L$	$\frac{a}{2}$	$\left[b-\left(\frac{m}{l}\right)\left(\frac{a}{2}\right)\right]$	$\left(\frac{p}{l}\right)\left(\frac{a}{2}\right)=\left(\frac{pa}{2l}\right)$
$\left(t_{\frac{1}{2}}\right)_M$	$\left[a-\left(\frac{b}{2}\right)\left(\frac{l}{m}\right)\right]$	$\frac{b}{2}$	$\left[\left(\frac{b}{2}\right)\left(\frac{l}{m}\right)\right]\left(\frac{p}{l}\right)=\left(\frac{pb}{2m}\right)$

x = Number of moles of reactant L consumed during time t

V = Volume of the system = constant

$C_L=\left[\frac{a-x}{V}\right]$ = Concentration of the reactant L at time t.

$C_M=\left[\frac{b-(mx/l)}{V}\right]$ = Concentration of the reactant M at time t.

$C_P=\left(\frac{px}{lV}\right)$ = Concentration of the product P at time t.

Boundary conditions: (i) at t = 0: $(a-x)=a$ and $x=0$

(ii) at $t=t$: $(a-x)=(a-x)$ and $x=x$

The rate is given by:

$$J=-\frac{1}{l}\frac{dC_L}{dt}=-\frac{1}{lV}\frac{d[(a-x)]}{dt}=\frac{1}{lV}\frac{dx}{dt}$$

$$J=-\frac{1}{m}\frac{dC_M}{dt}=-\frac{1}{mV}\frac{d\left[b-\left(\frac{mx}{l}\right)\right]}{dt}=\frac{1}{lV}\frac{dx}{dt}$$

$$J=\frac{1}{p}\frac{dC_P}{dt}=\frac{1}{pV}\frac{d\left(\frac{px}{l}\right)}{dt}=\frac{1}{lV}\frac{dx}{dt}$$

The rate equation is given by:

$$J=k\,.\,C_L\,.\,C_M$$

This equation holds good when order of the reaction w.r.t. each reactant is unity.

Or, $$\frac{1}{lV}\frac{dx}{dt}=k\left[\frac{(a-x)}{V}\right]\left[\frac{\left\{b-\left(\frac{mx}{l}\right)\right\}}{V}\right]$$

Or, $$\frac{dx}{dt} = \left(\frac{k.l}{V}\right)(a-x)\left(\frac{m}{l}\right)\left\{\left(\frac{lb}{m}\right)-x\right\}$$

Or, $$\frac{dx}{dt} = \frac{k.m}{V}(a-x)(b'-x)$$

where, $$b' = \left(\frac{bl}{m}\right)$$

Or, $$\frac{dx}{dt} = \frac{k_M}{V}(a-x)(b'-x)$$

where, $$k_M = k.m$$

Or, $$\int_0^x \frac{dx}{(a-x)(b'-x)} = \frac{k_M}{V}\int_0^t dt$$

Or, $$\frac{1}{(a-b')}\int_0^x \left[\frac{1}{(b'-x)} - \frac{1}{(a-x)}\right]dx = \frac{k_M}{V}\int_0^t dt$$

On integration, we get:

$$\frac{k_M t}{V}(a-b') = -\ln\frac{(b'-x)}{b'} + \ln\frac{(a-x)}{a} = \ln\frac{b'(a-x)}{a(b'-x)}$$

Or, $$t = \frac{V}{k_M(a-b')}\ln\frac{b'(a-x)}{a(b'-x)} \qquad (3.9a)$$

Equation (3.9a) is the rate equation of 2nd order reaction of type shown above.

Case – II(a) $l = m = 1$ but $a \neq b$. So, $k_M = k$ and $b' = b$.
Hence, Equation (3.9a) becomes:

$$t = \frac{V}{k(a-b)}\ln\frac{b(a-x)}{a(b-x)} \qquad (3.9b)$$

Case – II(b) $a = b$. So, $b' = \left(\frac{lb}{m}\right) = \left(\frac{la}{m}\right) = a'$ (say).

Thus Equation (3.9a) becomes:

$$t = \frac{V}{k_M(a-a')}\ln\frac{a'(a-x)}{a(a'-x)} \qquad (3.9c)$$

Half-life period on the basis of Equation (3.9a):

At $t = \left(t_{\frac{1}{2}}\right)_L$, $x = \left(\frac{a}{2}\right)$ and hence,

$$C_L = \frac{a}{2V} \quad \text{and} \quad C_M = \left(\frac{b - \frac{ma}{2l}}{V} \right)$$

$$\left(t_{\frac{1}{2}}\right)_L = \frac{V}{k_M (a - b')} \ln \frac{b'\left(a - \frac{a}{2}\right)}{a\left(b' - \frac{a}{2}\right)} = \frac{V}{k_M (a - b')} \ln \frac{b'}{(2b' - a)} \qquad (3.10a)$$

At $t = \left(t_{\frac{1}{2}}\right)_M$, $x = \left(\frac{l}{m}\right)\left(\frac{b}{2}\right) = \frac{b'}{2}$ and hence,

$$C_L = \frac{\left(a - \frac{bl}{2m}\right)}{V} \quad \text{and} \quad C_M = \frac{b}{2V}$$

$$\left(t_{\frac{1}{2}}\right)_M = \frac{V}{k_M (a - b')} \ln \frac{b'\left(a - \frac{b'}{2}\right)}{a\left(b' - \frac{b'}{2}\right)} = \frac{V}{k_M (a - b')} \ln \frac{(2a - b')}{a} \quad \ldots (3.10b)$$

Examples of 2nd order reaction

(i) Formation of hydrogen iodide from its elements.

$$H_2 + I_2 \rightarrow 2HI$$

(ii) Saponification of an ester by an alkali.

$$CH_3COOC_2H_5 + NaOH \rightarrow CH_3COONa + C_2H_5OH$$

3.3.3 Irreversible 3rd order reactions

In the case of 3rd order reaction there are three possible cases:

Case – I $lL \rightarrow pP + qQ + rR + \ldots$ etc.

Rate equation is given by:

$$J = -\frac{1}{l}\frac{dC_L}{dt} = k\, C_L^3$$

Case – II $lL + mM \rightarrow pP + qQ + rR + \ldots$ etc.

Rate equation is given by:

$$J = -\frac{1}{l}\frac{dC_L}{dt} = k\, C_L^2\, C_M$$

Case – III $lL + mM + nN \rightarrow pP + qQ + rR + \ldots$ etc.

The rate equation is given by:

$$J = -\frac{1}{l}\frac{dC_L}{dt} = k\,C_L\,C_M\,C_N$$

It is left as an exercise for the readers to find out ultimate rate equation in all the three cases.

Examples of 3rd order reaction

(i) Oxidation of stannous chloride by ferric chloride.

$$2FeCl_3 + SnCl_2 \rightarrow SnCl_4 + 2FeCl_2$$

(ii) Reactions between nitric oxide with one of the elements, chlorine, bromine, oxygen or hydrogen. e.g.

$$NO + NO \rightarrow (NO)_2$$

$$(NO)_2 + O \rightarrow 2NO_2$$

(iii) Reaction between sulphurdioxide and oxygen in presence of nitric oxide.

$$2SO_2 + O_2 \xrightarrow{NO} 2SO_3$$

3.3.4 Zero order irreversible reactions

Consider the following reaction:

$$lL \rightarrow pP + qQ + rR + \ldots \text{ etc.}$$

Where l, p, q, r etc. are stoichiometric coefficients.

Time (t)	*No. of moles of reactant L*	*No. of moles of any product (say P)*
0	a	0
t	$(a - x)$	$\left(\frac{px}{l}\right)$
$t_{\frac{1}{2}}$	$\left(\frac{a}{2}\right)$	$\left(\frac{pa}{2l}\right)$

x = No. of moles of reactant L consumed during time t.

V = Volume of the system = constant

$$C_L = \frac{(a-x)}{V} \text{ moles/litre = Concentration of reactant } L \text{ at time } t.$$

Boundary conditions: (i) at $t = 0$: $(a - x) = a$ and $x = 0$

(ii) at $t = t$: $(a - x) = (a - x)$ and $x = x$

The rate equation is given by:

$$J = -\frac{1}{l}\frac{dC_L}{dt} = -\frac{1}{l}\frac{d}{dt}\left[\frac{(a-x)}{V}\right] = \frac{1}{lV}\frac{dx}{dt}$$

$$\frac{1}{lV}\frac{dx}{dt} = k\,C_L^0 = k$$

Or, $$\frac{dx}{dt} = k\,lV\,dt$$

Or, $$\int_0^x dx = k_L V \int_0^t dt$$

where, $$k_L = l.k$$

On integration, we get:

$$x = k_L V.t$$

Or, $$t = \frac{x}{k_L V} \quad (3.11)$$

The Equation (3.11) is the rate equation for zero-order reaction.

Half-life period At $t = t_{\frac{1}{2}}$, $x = \left(\frac{a}{2}\right)$.

Thus, $$t_{\frac{1}{2}} = \frac{a}{2k_L V} \quad (3.12)$$

Unit of specific reaction rate From Equation (3.12), we have:

$$k_L = \frac{a}{2V\,t_{\frac{1}{2}}}$$

Unit of a is moles, unit of V is litre and that of $t_{\frac{1}{2}}$ is sec.

Thus the unit of k_L is moles/(litre-sec).

Examples of zero-order reactions:

(i) Thermal decomposition of hydrogen iodide on gold surface.
(ii) Decomposition of ammonia over molybdenum surface.
(iii) Iodination of acetone in acid medium

$$CH_3COCH_3 + I_2 \xrightarrow{H^+} CH_3CH_2I + HI$$

The reaction is zero order w.r.t. iodine.

3.3.5 Examples of negative and fractional order reactions and catalyst

(a) Examples of negative order reaction:

(i) Oxidation of iodide with hypochlorite in presence of an alkali:

$$OCl^- + I^- \xrightarrow{OH^-} OI^- + Cl^-$$

The rate is given by:

$$J = k \cdot \frac{[I^-][OCl^-]}{[OH^-]}$$

Thus the order of the reaction w.r.t. $[OH^-]$ is -1. Increase in concentration of $[OH^-]$ results in decrease in rate and hence $[OH^-]$ is said inhibitor to the above reaction.

(ii) Oxidation of mercurous ion by the thalium ion,

$$Hg_2^{2+} + Tl^{3+} \rightarrow 2Hg^{2+} + Tl^+$$

The rate is given by:

$$J = k \cdot \frac{[Hg_2^{2+}][Tl^{3+}]}{[Hg^{2+}]}$$

The presence of mercuric ion decreases the rate and hence it acts as an inhibitor. Order of the reaction w.r.t. Hg^{2+} is (–1)
Thus any species for which the partial order is negative is called inhibitor.

(b) Example of catalytic reaction:

Oxidation of sulphur-di-oxide in presence of nitric oxide:

$$2SO_2(g) + O_2(g) \xrightarrow{NO} 2SO_3(g)$$

The rate is given by:

$$J = k[O_2][SO_2]^2$$

The presence of NO speeds up the reaction and hence is called catalyst.

(c) Example of reaction, having non-integer order:

$$CH_3CHO \rightarrow CH_4 + CO$$

The rate is given by:

$$J = k[CH_3CHO]^{3/2}$$

3.3.6 Irreversible n^{th}-order reaction

Case – I Consider the following reaction:

$$lL \rightarrow pP + qQ + rR + \dots \text{ etc.,}$$

Where l, p, q, r etc. are the stoichiometric coefficients.

x = Number of moles of reactant L consumed at time t.

V = Volume of the system = constant

$C_L = \dfrac{(a-x)}{V}$ moles/litre = Concentration of the reactant L at time t.

$C_P = \left(\dfrac{px}{l}\right)$ moles/litre = Concentration of the reactant P at time t.

Boundary conditions: (i) at $t = 0$: $(a - x) = a$ and $x = 0$

(ii) at $t = t$: $(a - x) = (a - x)$ and $x = x$

Time (t)	*No. of moles of reactant L*	*No. of moles of any product (say P)*
0	a	0
t	$(a - x)$	$\left(\dfrac{px}{l}\right)$
$t_{\frac{1}{2}}$	$\left(\dfrac{a}{2}\right)$	$\left(\dfrac{pa}{2l}\right)$

The rate equation is given by:

$$J = k\ C_L^n$$

$$J = -\frac{1}{l}\frac{dC_L}{dt} = -\frac{1}{l}\frac{d\left[\dfrac{(a-x)}{V}\right]}{dt} = \frac{1}{lV}\frac{dx}{dt}$$

Or,

$$J = \frac{1}{p}\frac{dC_P}{dt} = \frac{1}{p}\frac{d\left(\dfrac{px}{lV}\right)}{dt} = \frac{1}{lV}\frac{dx}{dt}$$

Thus,

$$-\frac{1}{lV}\frac{d(a-x)}{dt} = k.\left(\frac{a-x}{V}\right)^n$$

Or,

$$-\frac{d(a-x)}{(a-x)^n} = \frac{l.k}{V^{n-1}}dt = \frac{k_L}{V^{n-1}}dt$$

where,

$$k_L = lk$$

Or,

$$-\int_0^{(a-x)}\frac{d(a-x)}{(a-x)^n} = \frac{k_L}{V^{n-1}}\int_0^t dt$$

On integration, we get:

$$\frac{k_L t}{V^{n-1}} = \frac{1}{(n-1)}\left[\frac{1}{(a-x)^{n-1}} - \frac{1}{a^{n-1}}\right]$$

Or, $$t = \frac{V^{n-1}}{k_L(n-1)}\left[\frac{1}{(a-x)^{n-1}} - \frac{1}{a^{n-1}}\right] \tag{3.13}$$

The Equation (3.13) is the rate equation for *n*-th order reaction.

Half-life period At $t = t_{\frac{1}{2}}$, $x = \left(\frac{a}{2}\right)$

Thus Equation (3.13) becomes:

$$t_{\frac{1}{2}} = \frac{V^{n-1}}{k_L(n-1)}\left[\frac{1}{\left(a-\frac{a}{2}\right)^{n-1}} - \frac{1}{a^{n-1}}\right]$$

$$t_{\frac{1}{2}} = \frac{V^{n-1}}{k_L(n-1)}\left[\frac{2^{n-1}-1}{a^{n-1}}\right] \tag{3.14}$$

Unit of specific reaction rate (k_L) From Equation (3.14) we have:

$$t_{\frac{1}{2}} = \frac{V^{n-1}}{k_L(n-1)}\left[\frac{2^{n-1}-1}{a^{n-1}}\right]$$

The unit of a is moles, V is litre and $t_{\frac{1}{2}}$ is sec.

Thus, the unit of k_L is (Litre)$^{n-1}$/[(moles)$^{n-1}$.sec].

Case – II Consider the following reaction:

$$lL + mM \rightarrow pP + qQ + rR + \ldots \text{ etc.,}$$

Where l, m, p, q, r etc. are stoichiometric coefficients.

Time (t)	*No. of moles of reactant L*	*No. of moles of reactant M*	*No. of moles of any product (say P)*
0	a	b	0
t	$(a - x)$	$\left[b - \left(\frac{mx}{l}\right)\right]$	$\left(\frac{px}{l}\right)$
$\left(t_{\frac{1}{2}}\right)_L$	$\left(\frac{a}{2}\right)$	$\left[b - \left(\frac{ma}{2l}\right)\right]$	$\left(\frac{pa}{2l}\right)$
$\left(t_{\frac{1}{2}}\right)_M$	$\left[a - \left(\frac{bl}{2m}\right)\right]$	$\left(\frac{b}{2}\right)$	$\left(\frac{pb}{2m}\right)$

x = No. of moles of reactant L consumed, during time t.

V = Volume of the system = constant

$$C_L = \frac{(a-x)}{V} \text{ moles/litre = Concentration of the reactant } L \text{ at time } t.$$

$$C_M = \left(\frac{1}{V}\right)\left[b-\left(\frac{mx}{l}\right)\right] \text{ moles/litre = Concentration of the reactant } M \text{ at time } t.$$

$$C_P = \left(\frac{px}{l}\right) \text{ moles/litre = Concentration of the reactant } P \text{ at time } t.$$

Boundary conditions: (i) at $t = 0$: $(a - x) = a$ and $x = 0$

(ii) at $t = t$: $(a - x) = (a - x)$ and $x = x$

The rate is given by:

$$J = -\frac{1}{l}\frac{dC_L}{dt} = -\frac{1}{l}\frac{d\left[\frac{(a-x)}{V}\right]}{dt} = \frac{1}{lV}\frac{dx}{dt}$$

Or,

$$J = -\frac{1}{m}\frac{dC_M}{dt} = -\frac{1}{m}\frac{d\left[\frac{\{b-(mx/l)\}}{V}\right]}{dt} = \frac{1}{lV}\frac{dx}{dt}$$

Or,

$$J = \frac{1}{p}\frac{dC_P}{dt} = \frac{1}{p}\frac{d\left(\frac{px}{lV}\right)}{dt} = \frac{1}{lV}\frac{dx}{dt}$$

The rate equation is given by:

$$J = k\ C_L^{n_1}\ C_M^{n_2}$$

where n_1 and n_2 are partial orders and the total order = $n = n_1 + n_2$.

Or,

$$\frac{1}{lV}\frac{dx}{dt} = k\left[\frac{(a-x)}{V}\right]^{n_1}\left[\frac{\left\{b-\left(\frac{mx}{l}\right)\right\}}{V}\right]^{n_2}$$

$$= \frac{k}{V^n}(a-x)^{n_1}\left[b-\left(\frac{mx}{l}\right)\right]^{n_2}$$

Or,

$$\frac{dx}{(a-x)^{n_1}\left[b-\left(\frac{mx}{l}\right)\right]^{n_2}} = \frac{k\,.\,l}{V^{n-1}}\cdot dt = \frac{k_L}{V^{n-1}}\cdot dt \tag{3.15}$$

where, $k_L = k\,.\,l$

According to stoichiometric equation we have, a minimum l moles of L and m moles of M are required to bring about the chemical reaction, where l and m are positive integers. Reactant L, having number of moles less than l, cannot undergo reaction. Similar explanation is true for the reactant M also. Thus the reaction takes place between

either (l) moles of L and (m) moles of M
or ($2l$) moles of L and ($2m$) moles of M
or ($3l$) moles of L and ($3m$) moles of M
.................................etc.

Hence, x = Number of moles of reactant L consumed

$$= (s \,.\, l),$$

where, $s = 1, 2, 3, \ldots$ etc.

So,
$$(a - x) = (a - sl) \quad \text{and} \quad \left(b - \frac{mx}{l}\right) = \left(b - \frac{m}{l} sl\right) = (b - sm)$$

a and b are the number of moles of reactants L and M taken to bring about the chemical reaction. To make the Equation (3.15) simpler, a and b are so chosen that,

$$\frac{\text{No. of moles of reactant } L}{\text{No. of moles of reactant } M} = \frac{a}{b} = \frac{l}{m} = w$$

Hence, $a = (w \,.\, l)$ and $b = (w \,.\, m)$

Thus, $(a - x) = (a - sl) = (wl - sl) = l(w - s)$

and
$$\left(b - \frac{mx}{l}\right) = (b - sm) = (wm - sm) = m(w - s)$$

Putting these values of $(a - x)$ and $\left(b - \frac{mx}{l}\right)$ in the Equation (3.15) we get:

$$\frac{dx}{[l(w - s)]^{n_1}\,[m(w - s)]^{n_2}} = \frac{k_L}{V^{n-1}}\, dt$$

$$\frac{dx}{(w - s)^{n_1 + n_2}} = \frac{k_L}{V^{n-1}}\, l^{n_1}\, m^{n_2}\, dt$$

Again, $x = sl$

Or, $dx = lds$

When (i) $x = 0$, $s = 0$, since $l \neq 0$ and when (ii) $x = x$, $s = s$.

$$\frac{lds}{(w - s)^n} = \frac{k_L}{V^{n-1}}\, l^{n_1}\, m^{n_2}\, dt \quad [\text{since, } n = n_1 + n_2]$$

$$\frac{ds}{(w - s)^n} = \frac{k_L}{V^{n-1}}\, (l^{n_1 - 1})\, (m^{n_2}) dt$$

$$\int_0^s \frac{ds}{(w - s)^n} = \frac{k_{LM}}{V^{n-1}} \int_0^t dt$$

where, $k_{LM} = k_L\, (l^{n_1 - 1})\, (m^{n_2})$

$$\frac{k_{LM}}{V^{n-1}} t = -\int_{w}^{(w-s)} \frac{d(w-s)}{(w-s)^n} = \left(\frac{1}{n-1}\right)\left[\frac{1}{(w-s)^{n-1}} - \frac{1}{(w)^{n-1}}\right]$$

$$= \left(\frac{1}{n-1}\right)\left[\frac{1}{\left\{\frac{(a-x)}{l}\right\}^{n-1}} - \frac{1}{\left(\frac{a}{l}\right)^{n-1}}\right] \quad \left[\text{Since, } (w-s) = \frac{a-x}{l} \text{ and } w = \frac{a}{l}\right]$$

$$k_L t \frac{l^{n_1 - 1} m^{n_2}}{V^{n-1}} = \frac{l^{n-1}}{(n-1)}\left[\frac{1}{(a-x)^{n-1}} - \frac{1}{a^{n-1}}\right]$$

$$\frac{k_L t \left(\frac{m}{l}\right)^{n_2}}{V^{n-1}} = \frac{1}{(n-1)}\left[\frac{1}{(a-x)^{n-1}} - \frac{1}{a^{n-1}}\right]$$

$$t = \frac{V^{n-1}}{k_{\frac{M}{L}}(n-1)}\left[\frac{1}{(a-x)^{n-1}} - \frac{1}{a^{n-1}}\right] \tag{3.16}$$

$$\left[\text{where, } k_{\frac{M}{L}} = k_L \left(\frac{m}{l}\right)^{n_2}\right]$$

Equation (3.16) is similar to Equation (3.13).

3.4 Methods of determination of order of a reaction

(a) Half-life period method Consider the following irreversible reaction:

$$lL + mM \rightarrow pP + qQ + rR + \ldots \text{ etc.}$$

The rate equation is given by:

$$J = k\, C_L^{n_1}\, C_M^{n_2}$$

n_1 and n_2 are partial orders and $n = n_1 + n_2$ = total order of the reaction.

To find out n_1, the reaction is carried out in large excess of the reactant M such that C_M remains almost constant throughout the reaction. Thus the above rate equation becomes:

$$J = k'\, C_L^{n_1}$$

where, $$k' = k\, C_M^{n_2}$$

The half-life period of the reactant L is given by Equation (3.14):

$$t_{\frac{1}{2}} = \frac{V^{n_1 - 1}}{k_L'(n_1 - 1)}\left[\frac{2^{n_1 - 1} - 1}{a^{n_1 - 1}}\right]$$

where, $$k_L' = lk'$$

Thus we have:

$$t_{\frac{1}{2}} \propto \left(\frac{1}{a^{n_1-1}}\right)$$

Where a is the initial number of moles of the reactant L.

Let us carry out the reaction in two phases, starting with the two initial concentrations of the reactant L, namely $\left(\frac{a_1}{V}\right)$ moles/litre and $\left(\frac{a_2}{V}\right)$ moles/litre. Thus we have two different half-life periods $\left(t_{\frac{1}{2}}\right)_1$ and $\left(t_{\frac{1}{2}}\right)_2$ for the two phases. In both the cases we can write that:

$$\left(t_{\frac{1}{2}}\right)_1 \propto \left(\frac{1}{a_1^{n_1-1}}\right) \quad \text{and} \quad \left(t_{\frac{1}{2}}\right)_2 \propto \left(\frac{1}{a_2^{n_1-1}}\right)$$

$$\frac{\left(t_{\frac{1}{2}}\right)_2}{\left(t_{\frac{1}{2}}\right)_1} = \left(\frac{a_1}{a_2}\right)^{n_1-1}$$

Taking $\log_{10}$ on both sides we get:

$$\log \frac{\left(t_{\frac{1}{2}}\right)_2}{\left(t_{\frac{1}{2}}\right)_1} = (n_1 - 1) \log \frac{a_1}{a_2} \tag{3.17}$$

a_1 and a_2 are known, $\left(t_{\frac{1}{2}}\right)_1$ and $\left(t_{\frac{1}{2}}\right)_2$ are measurable quantities. Thus the partial order n_1 can easily be determined by Equation (3.17).

The total order of the said reaction can be determined from the overall rate equation:

$$t = \left[\frac{V^{n-1}}{k_{\frac{L}{M}}(n-1)}\right]\left[\frac{1}{(a-x)^{n-1}} - \frac{1}{(a)^{n-1}}\right] \tag{3.16}$$

At, $t = \left(t_{\frac{1}{2}}\right)$, $x = \left(\frac{a}{2}\right)$. Thus Equation (3.16) becomes:

$$t_{\frac{1}{2}} = \frac{V^{n-1}}{k_{\frac{M}{L}}(n-1)}\left[\frac{2^{n-1}-1}{a^{n-1}}\right]$$

Or,

$$t_{\frac{1}{2}} \propto \frac{1}{a^{n-1}}$$

Now proceeding in the same manner as described earlier we have two values of $t_{\frac{1}{2}}$ namely $\left(t_{\frac{1}{2}}\right)_1$ and $\left(t_{\frac{1}{2}}\right)_2$ at two different values of a namely a_1 and a_2

Thus we have,

$$\left(t_{\frac{1}{2}}\right)_1 \propto \left(\frac{1}{a_1^{n-1}}\right) \quad \text{and} \quad \left(t_{\frac{1}{2}}\right)_2 \propto \left(\frac{1}{a_2^{n-1}}\right)$$

$$\frac{\left(t_{\frac{1}{2}}\right)_2}{\left(t_{\frac{1}{2}}\right)_1} = \left(\frac{a_1}{a_2}\right)^{n-1}$$

Taking $\log_{10}$ on both sides we get:

$$\log \frac{\left(t_{\frac{1}{2}}\right)_2}{\left(t_{\frac{1}{2}}\right)_1} = (n-1) \log \frac{a_1}{a_2} \tag{3.18}$$

$\left(t_{\frac{1}{2}}\right)_1$ and $\left(t_{\frac{1}{2}}\right)_2$ are measurable quantities and a_1 and a_2 are known quantities. Thus the total order of the reaction (n) can easily be evaluated from Equation (3.18).

As we know, $$n = n_1 + n_2 \tag{3.19}$$

Both n_1 and n are determined. So the partial order n_2 can be evaluated easily from the Equation (3.19)

(b) Graphical method For the same reaction mentioned in the earlier section we have:

$$t = \left[\frac{V^{n-1}}{k_{\frac{M}{L}}(n-1)}\right]\left[\frac{1}{(a-x)^{n-1}} - \frac{1}{(a)^{n-1}}\right] \quad \text{(from equation 3.16)}$$

Or, $$t = mX - C, \text{ where,}$$

$$m = \frac{V^{n-1}}{k_{\frac{M}{L}}(n-1)} = \text{Constant}$$

$$C = \frac{m}{a^{n-1}} = \text{Constant} \quad \text{and} \quad X = \frac{1}{(a-x)^{n-1}}$$

Or, we can write that:

$$X = \frac{t}{m} + \frac{C}{m} \tag{3.20}$$

Thus $X - t$ plot is a straight line with slope $\left(\frac{1}{m}\right)$ and intercept $\left(\frac{C}{m}\right)$ as shown in the Figure 3.1.

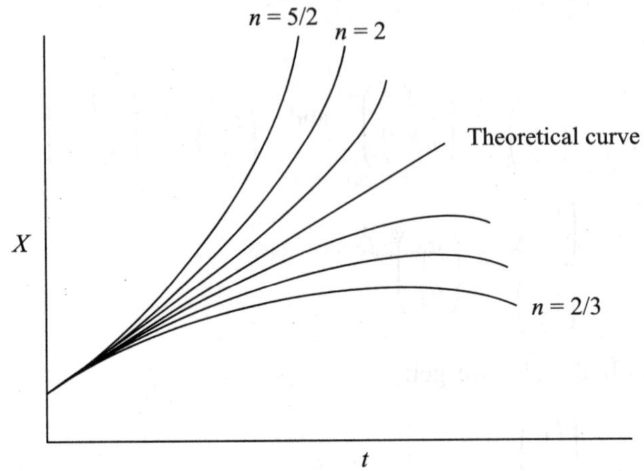

Fig. 3.1 X vs. t plot at different values of n Equation (3.16) is valid for all values of n excepting $n = 1$.

At different intervals of time amount of product formed (x) can be measured. a, the initial concentration of the reactant L, is known. Thus $(a - x)$ can be calculated at different values of t. All these values are listed in the Table 2.1.

Table 2.1 List of parameters x, $(a - x)$ etc. at different values of 't'

t	X	$(a-x)$	X_b at $n = \frac{1}{2}$	X_b at $n = \frac{2}{3}$	X_b at $n = \frac{3}{2}$	X_b at $n = 2$	X_b at $n = \frac{5}{2}$
t_1	x_1	$(a-x_1)$	X_1	X_1	X_1	X_1	X_1
t_2	x_2	$(a-x_2)$	X_2	X_2	X_2	X_2	X_2
t_3	x_3	$(a-x_3)$	X_3	X_3	X_3	X_3	X_3
t_4	x_4	$(a-x_4)$	X_4	X_4	X_4	X_4	X_4
t_5	x_5	$(a-x_5)$	X_5	X_5	X_5	X_5	X_5
...	...	...	...	...	...	...	...
...	...	...	...	...	...	...	...
...	...	...	...	...	...	...	...
t_{15}	x_{15}	$(a-x_{15})$	X_{15}	X_{15}	X_{15}	X_{15}	X_{15}

b: $X = \dfrac{1}{(a-x)^{n-1}}$.

The values of X are calculated at different values of 'x' by taking different values of n. The values of X are plotted against t-values and a separate curve for each value of n is obtained as shown in the Figure 3.1. For a particular value of n, the curve becomes straight line and matches with the theoretical curve. That value of n is called the order of the reaction. Slope of the straight line $= \left(\frac{1}{m}\right) = \frac{\left\{k_{\frac{M}{L}} \cdot (n-1)\right\}}{V^{n-1}}$.

The value of $\left(\frac{1}{m}\right)$ can be evaluated from the curve. V is known and n has been evaluated.

Thus from the above equation the rate constant $k_{\frac{M}{L}}$ of the said reaction can easily be calculated.

Proceeding in the same manner the partial order n_1 can be evaluated and the another partial order n_2 can be evaluated from the equation: $n = n_1 + n_2$

3.5 Significances of half-life period $\left(t_{\frac{1}{2}}\right)$

(a) For a reactant, participating in a particular chemical reaction, $t_{\frac{1}{2}}$ is the time taken by that reactant to decay to half-of its amount.

(b) For any reactant $\left(t_{\frac{1}{2}}\right)$ depends on rate constant (k) of the chemical reaction. A reactant may participate in different chemical reactions. As k-values for different chemical reactions are different the $\left(t_{\frac{1}{2}}\right)$ of that reactant varies from reaction to reaction.

(c) The value of $\left(t_{\frac{1}{2}}\right)$ of any reactant depends on the initial concentration of that reactant, e.g., $\left(t_{\frac{1}{2}}\right) \propto a$ for zero order reaction, $\left(t_{\frac{1}{2}}\right)$ is independent of a for 1st order reaction and $\left(t_{\frac{1}{2}}\right) \propto \left(\frac{1}{a}\right)$ for 2nd order reaction ... etc.

(d) For 1st order reaction we have:

$$t_{\frac{1}{2}} = \frac{0.693}{k_L} \quad \text{[Equation (3.4)]}$$

Thus by measuring $t_{\frac{1}{2}}$ one can easily evaluate the rate constant of that reaction.

(e) By measuring half-life period at different initial concentration of a reactant both partial orders and total order of any irreversible chemical reaction can be evaluated.

3.6 Elementary reactions

There are many examples, where partial orders are totally different from the corresponding stoichiometric coefficients. This is because those reactions occur in multi-steps and reaction in each step is known as elementary reaction. An elementary reaction is said to be unimolecular or bimolecular depending on the number of species involved in that reaction. The elementary reaction $A \rightarrow products$ is unimolecular since one molecule is involved in the reaction but the elementary reaction $2A \rightarrow products$ or $A + B \rightarrow products$ is bimolecular since two molecules are involved in the elementary reaction.

According to collision theory molecules are involved in random collisions through which the reactant molecules gain necessary activation energy to form products. Usually, collision comes about between two molecules of same reacting species (i.e., unimolecular) or between two molecules of different reacting species (i.e. bimolecular). Trimolecular elementary reaction of type $A + B + C \rightarrow products$ is uncommon since simultaneous three-body collision rarely occurs. Obviously no elementary reaction, involving more than three molecules, is known.

Let us consider a bimolecular reaction: $A + B \rightarrow$ products.

If Z_{A-B} is the rate of $A - B$ collisions per unit volume, rate of reaction (J) is is given by:

$$J \propto Z_{A-B}$$

Again from the kinetic theory of ideal gas, we know that:

$$Z_{A-B} \propto \left(\frac{n_A}{V}\right)\left(\frac{n_B}{V}\right)$$

where n_A and n_B are the number of moles of the reactants A and B respectively and V is the volume of the system.

Hence, $$J = k\,C_A \,.\, C_B$$

Thus for any ideal elementary reaction the partial orders are nothing but the corresponding stoichiometric coefficients.

For an ideal elementary reaction of type $aA + bB \rightarrow products$, rate is given by:

$$J = k\,C_A^a\,C_B^b$$

where, $$(a + b) = 1 \text{ or } 2 \text{ or } 3.$$

For a reversible elementary reaction of type

$$lL + mM \rightleftharpoons pP + qQ$$

At equilibrium, rate of forward reaction = rate of backward reaction

Or, $$k_f\,C_L^l\,C_M^m = k_b\,C_P^p\,C_Q^q$$

Or, $$\text{Equilibrium constant} = K_c = \frac{k_f}{k_b} = \frac{C_P^p\,C_Q^q}{C_L^l\,C_M^m}$$

3.7 Study of reaction mechanism

There are two concepts to study a reaction mechanism:

(i) Rate-determining-step approach
(ii) Steady-state approach.

(i) Rate-determining-step approach According to this concept a reaction takes place in multiple steps and one of them is the slowest one. The rate of the slowest step actually determines the rate of the whole reaction. One of such examples is:

Example 3.1 Decomposition of hydrogen peroxide by iodide ion in acid medium.
The overall reaction is:

$$H_2O_2 + 2H^+ + 2I^- \rightarrow I_2 + 2H_2O$$

The elementary reactions are:

$$H^+ + I^- \underset{k_{-1}}{\overset{k_1}{\rightleftharpoons}} HI \qquad \text{[Rapid equilibrium]}$$

$$HI + H_2O_2 \xrightarrow{k_2} H_2O + HOI \qquad \text{[Slow]}$$

$$HOI + I^- \xrightarrow{k_3} I_2 + OH^- \qquad \text{[Fasy]}$$

$$OH^- + H^+ \xrightarrow{k_4} H_2O \qquad \text{[Fast]}$$

The second step is the slowest step and hence rate of the whole reaction is given by:

$$J = k_2 \,.\, [HI].[H_2O_2] \tag{3.21}$$

As equilibrium holds in the first step we can write that:

$$K_c = \frac{[HI]}{[H^+][I^-]}$$

Or, $$[HI] = K_c[H^+]\,[I^-]$$

Putting the value of [HI] in the above equation we get,

$$J = k_2 \,.\, K_C \,.\, [H^+]\,[I^-]\,[H_2O_2]$$

Or, $$J = k\,[H^+]\,[I^-]\,[H_2O_2] \tag{3.22}$$

where, $$k = k_2 \,.\, K_C$$

Example 3.2 Nitration of benzene in acid medium:

$$HNO_3 + C_6H_6 \xrightarrow{H^+} C_6H_5NO_2 + H_2O$$

The elementary reactions are:

$$H^+ + HO - NO_2 \underset{k_{-1}}{\overset{k_1}{\rightleftharpoons}} H_2O^+ - NO_2 \qquad \text{[Rapid equilibrium]}$$

$$H_2O^+ - NO_2 \xrightarrow{k_2} H_2O + NO_2^+ \quad \text{[Slow]}$$

$$NO_2^+ + C_6H_6 \xrightarrow{k_3} C_6H_5NO_2 + H^+ \quad \text{[Fast]}$$

The second step is the slowest step and hence the rate of the whole reaction is given by:

$$J = k_2[H_2O^+ - NO_2] \quad (3.23)$$

As equilibrium exists in the 1st step we can write that:

$$K_c = \frac{[H_2O^+ - NO_2]}{[H^+][HNO_3]}$$

Or, $$[H_2O^+ - NO_2] = K_c [H^+] [HNO_3]$$

Putting the value of $[H_2O^+ - NO_2]$ in the above equation we get:

$$J = k_2 \,.\, K_C [H^+] [HNO_3]$$

Or, $$J = k \,.\, [H^+] [HNO_3] \quad (3.24)$$

where, $$k = k_2 \,.\, K_C$$

(ii) Steady-state approach According to this concept rate of formation of any intermediate species, formed in elementary reactions, is equal to the rate of consumption of that intermediate species since intermediate species are very much reactive and do not appear in the final phase of the reaction.

By using this concept rate of reactions, discussed in the earlier section, can be determined.

Example 3.3 Decomposition of hydrogen peroxide by iodide ion in acid medium.

The overall reaction is:

$$H_2O_2 + 2H^+ + 2I^- \rightarrow I_2 + 2H_2O$$

Elementary reactions are:

$$H^+ + I^- \underset{k_{-1}}{\overset{k_1}{\rightleftharpoons}} HI \quad \text{[Rapid equilibrium]}$$

$$HI + H_2O_2 \xrightarrow{k_2} H_2O + HOI \quad \text{[Slow]}$$

$$HOI + I^- \xrightarrow{k_3} I_2 + OH^- \quad \text{[Fast]}$$

$$OH^- + H^+ \xrightarrow{k_4} H_2O \quad \text{[Fast]}$$

Rate of formation of product

$$J = k_3 [HOI] [I^-] \quad (3.25)$$

According to steady-state concept, we have:

$$\frac{d}{dt}[HOI] = 0 = k_2[HI] [H_2O_2] - k_3[HOI] [I^-]$$

Or, $$[HOI] = \frac{k_2 [HI] [H_2O_2]}{k_3 [I^-]}$$

Again, $$\frac{d}{dt}[\text{HI}] = 0 = k_1[\text{H}^+][\text{I}^-] - k_{-1}[\text{HI}] - k_2[\text{HI}][\text{H}_2\text{O}_2]$$

Or, $$[\text{HI}] = \frac{k_1[\text{H}^+][\text{I}^-]}{k_{-1} + k_2[\text{H}_2\text{O}_2]}$$

Hence, $$[\text{HOI}] = \frac{k_1 k_2}{k_3[\text{I}^-]}\frac{[\text{H}^+][\text{I}^-][\text{H}_2\text{O}_2]}{\{k_{-1} + k_2[\text{H}_2\text{O}_2]\}}$$

Putting this value of [HOI] in the rate equation, we get:

$$J = \frac{k_1 k_2[\text{H}^+][\text{I}^-][\text{H}_2\text{O}_2]}{\{k_{-1} + k_2[\text{H}_2\text{O}_2]\}} \tag{3.26}$$

Equations (3.22) and (3.26) resembles with each other but not the same. These two equations will be the same if $k_{-1} >> k_2[H_2O_2]$

Example 3.4 Nitration of benzene in acid medium:

$$\text{HNO}_3 + \text{C}_6\text{H}_6 \xrightarrow{\text{H}^+} \text{C}_6\text{H}_5\text{NO}_2 + \text{H}_2\text{O}$$

Elementary reactions are:

$$\text{H}^+ + \text{HO–NO}_2 \underset{k_{-1}}{\overset{k_1}{\rightleftharpoons}} \text{H}_2\text{O}^+ - \text{NO}_2 \quad \text{[Rapid equilibrium]}$$

$$\text{H}_2\text{O}^+ - \text{NO}_2 \xrightarrow{k_2} \text{H}_2\text{O} + \text{NO}_2^+ \quad \text{[Slow]}$$

$$\text{NO}_2^+ + \text{C}_6\text{H}_6 \xrightarrow{k_3} \text{C}_6\text{H}_5\text{NO}_2 + \text{H}^+ \quad \text{[Fast]}$$

Rate of formation of the product:

$$J = k_3 \,.\, [\text{NO}_2^+][\text{C}_6\text{H}_6] \tag{3.27a}$$

According to steady-state concept, we have:

$$\frac{d[\text{NO}_2^+]}{dt} = 0$$

Or, $$0 = k_2[H_2\text{O}^+ - \text{NO}_2] - k_3[\text{NO}_2^+][\text{C}_6\text{H}_6]$$

Or, $$k_2[\text{H}_2\text{O}^+ - \text{NO}_2] - J = 0 \quad \text{[using Equation (3.27a)]}$$

$$J = k_2[\text{H}_2\text{O}^+ - \text{NO}_2] \tag{3.27b}$$

Again,

$$\frac{d[\text{H}_2\text{O}^+ - \text{NO}_2]}{dt} = 0$$

Or, $$k_1[\text{H}^+][\text{HNO}_3] - k_{-1}\,.\,[\text{H}_2\text{O}^+\text{–NO}_2] - k_2[\text{H}_2\text{O}^+\text{–NO}_2] = 0$$

Or, $$[\text{H}_2\text{O}^+ - \text{NO}_2] = \frac{k_1[\text{H}^+][\text{HNO}_3]}{(k_{-1} + k_2)}$$

Putting this value of $[H_2O^+–NO_2]$ in the above equation we get:

$$J = \frac{k_1 k_2 [H^+][HNO_3]}{(k_{-1} + k_2)}$$

Or, $$J = k'[H^+]\,[HNO_3] \tag{3.28}$$

Equations (3.24) and (3.28) are same but differ with the magnitude of rate constant.

$$k' = k \quad \text{if} \quad k_{-1} >> k_2$$

Free radical mechanism

Besides ionic mechanism there are many reactions which proceed through free radical mechanism. One such example is: gas-phase reaction between hydrogen and bromine.

The overall reaction is:

$$H_2 + Br_2 \rightarrow 2HBr$$

The elementary reactions are:

$$Br_2 \xrightarrow[k_1]{h\nu} 2Br^* \qquad \text{Rate} = J_1 = k_1[Br_2]$$

$$Br^* + H_2 \xrightarrow{k_2} HBr + H^* \qquad \text{Rate} = J_2 = k_2[Br^*]\,[H_2]$$

$$H^* + Br_2 \xrightarrow{k_3} HBr + Br^* \qquad \text{Rate} = J_3 = k_3[H^*]\,[Br_2]$$

$$Br^* + Br^* \xrightarrow{k_4} Br_2 \qquad \text{Rate} = J_4 = k_4[Br^*]^2$$

$$H^* + HBr \xrightarrow{k_5} H_2 + Br^* \qquad \text{Rate} = J_5 = k_5[H^*]\,[HBr]$$

The elementary reactions show that reactive intermediates, generated from the reactants, are converted into products and regenerated. Thus the whole process is repeating continuously. These types of reactions are called chain reactions. Usually chain reactions proceed through free radical mechanism.

Rate of formation of product is given by:

$$\frac{1}{2}\frac{d[HBr]}{dt} = J = \frac{1}{2}(J_2 + J_3 - J_5) \tag{3.29a}$$

According to the steady-state concept, we can write that:

$$\frac{d[H^*]}{dt} = 0 = (J_2 - J_3 - J_5) \tag{3.29b}$$

$$\frac{d[Br^*]}{dt} = 0 = (J_1 + J_3 + J_5 - J_4 - J_2) \tag{3.29c}$$

Adding the above two equations, we get:

$$J_1 = J_4$$

Or, $$k_1[Br_2] = k_4[Br^*]^2$$

Or, $$[\text{Br}^*] = \left(\frac{k_1}{k_4}\right)^{\frac{1}{2}} [\text{Br}_2]^{\frac{1}{2}} \qquad (3.29d)$$

Again from Equation (3.29b) we have:

$$J_3 + J_5 = J_2$$

Or, $$k_3[\text{H}^*][\text{Br}_2] + k_5[\text{H}^*][\text{HBr}] = k_2[\text{Br}^*][\text{H}_2]$$

Or, $$[\text{H}^*] = \frac{k_2[\text{Br}^*][\text{H}_2]}{k_3[\text{Br}_2] + k_5[\text{HBr}]}$$

Putting the value of $[\text{Br}^*]$ in the above equation we get:

$$[\text{H}^*] = \frac{k_2\left(\frac{k_1}{k_4}\right)^{\frac{1}{2}} [\text{Br}_2]^{\frac{1}{2}} [\text{H}_2]}{k_3[\text{Br}_2] + k_5[\text{HBr}]}$$

Or, $$k_3[\text{H}^*]\,[\text{Br}_2] = \frac{k[\text{H}_2][\text{Br}_2]^{\frac{1}{2}}}{1 + k'\frac{[\text{HBr}]}{[\text{Br}_2]}}$$

where, $$k = k_2\left(\frac{k_1}{k_4}\right)^{\frac{l}{2}} \quad \text{and} \quad k' = \frac{k_5}{k_3}$$

We know from Equation (3.29a):

$$J = \frac{1}{2}\frac{d[\text{HBr}]}{dt} = \frac{1}{2}(J_2 + J_3 - J_5)$$

Again from Equation (3.29b) we have:

$$J_5 = (J_2 - J_3)$$

Putting the value of J_5 in the above equation we get:

$$J = \frac{1}{2}(2J_3) = J_3 = k_3.[\text{H}^*].[\text{Br}_2]$$

Thus, $$J = \frac{k[\text{H}_2][\text{Br}_2]^{\frac{1}{2}}}{1 + k'\frac{[\text{HBr}]}{[\text{Br}_2]}}$$

Here it is interesting to note that an increase in [HBr] decreases the rate J and hence the product [HBr] acts as an inhibitor.

3.8 Mechanism to explain unimolecular reaction

Most of the reactions are bimolecular, where two molecules A and B collide with each other to form products. But in the case of unimolecular reactions only one species is involved and hence apparently collision theory fails to explain the unimolecular reactions. In 1922 Lindemann proposed a scheme to explain unimolecular reaction on the basis of collision theory.

Isomerization reactions and decomposition reactions are very good examples of unimolecular reactions, e.g. cis-CHCl=CHCl to trans-CHCl=CHCl,

$$CH_3CH_2I \rightarrow CH_2 = CH_2 + HI, \text{ etc.}$$

Consider the following unimolecular reaction: $A \rightarrow$ Products

The corresponding elementary reactions are:

Step-I: $A + A \xrightarrow{k_1} A^* + A$ (activation)

Step-II: $A^* + A \xrightarrow{k_2} A + A$ (deactivation)

Step-III: $A^* \xrightarrow{k_3}$ Products

1 mole of the reacting species A contains Avogadro number of molecules. According to Lindemann collisions continuously take place within the molecules of the same species and these collisions lead to the above elementary reactions. The species A^* is an activated or energized molecule, that means in A^* the molecule A remains in high vibrational energy level.

The rate of formation of product = $J = k_3\ C_A^*$

According to the steady-state approach, we can write that:

$$\frac{d[C_A^*]}{dt} = 0 = k_1\ C_A^2 - k_2\ C_A^*\ C_A - k_3\ C_A^*$$

Or,

$$C_A^* = \frac{k_1\ C_A^2}{k_2\ C_A + k_3}$$

Putting the value of C_A^* in the above rate equation we get,

$$J = \frac{k_1\ k_3\ C_A^2}{k_2\ C_A + k_3} \tag{3.30}$$

Case – I $k_3 >> k_2\ C_A$ and $k_2\ C_A + k_3 \approx k_3$, which means that only a few molecules get deactivated through step-II.

Thus Equation (3.30) becomes:

$$J = \frac{k_1\ k_3\ C_A^2}{k_2\ C_A + k_3} = \frac{k_1\ k_3\ C_A^2}{k_3} = k_1\ C_A^2 \tag{3.31a}$$

So, the reaction follows 2nd order kinetics.

Again for a unimolecular reaction the rate is given by:

$$J = k_U \cdot C_A \tag{3.31b}$$

Comparing Equations (3.31a) and (3.31b) we get,

$$k_U = k_1 C_A \tag{3.31c}$$

The Figure 3.2 shows the plot of k_u vs. p_A^0 (initial pressure of A):

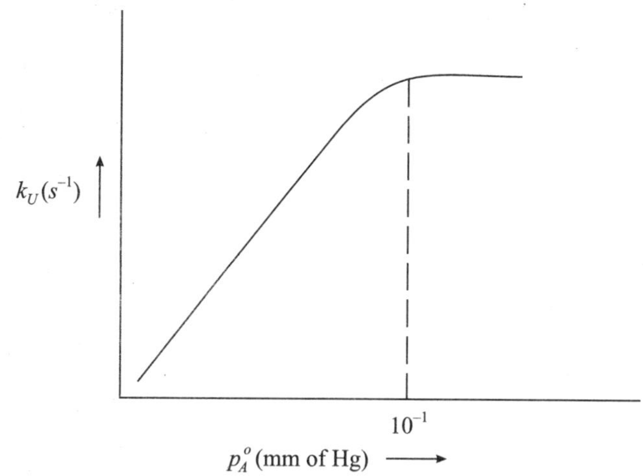

Fig. 3.2 Plot of k_u vs. initial pressure (p_A^0) of decomposition of N_2O_5

From Figure 3.2 it is evident that the Equation (3.31c) holds good only at very low pressure ($p^0_{N_2O_5}$ < 0.1 mm of Hg)

Case – II $k_2 C_A >> k_3$, which means that maximum number of molecules undergo deactivation through step-II. Hence, $k_2 C_A + k_3 \approx k_2 C_A$

Thus the Equation (3.30) becomes:

$$J = \frac{k_1 k_3 C_A^2}{k_2 C_A + k_3} = \frac{k_1 k_3 C_A^2}{k_2 C_A} = k\, C_A \tag{3.32a}$$

Where, $$k = \frac{k_1 k_3}{k_2}$$

So the reaction follows 1st order kinetics. Comparing Equations (3.31b) and (3.32a), we get:

$$k_U = k \tag{3.32b}$$

Thus k_U is constant. It is reflected in the Figure 3.2. It has been found that the value of k_U and k are independent of $p^0_{N_2O_5}$ in wide range of pressure (0.6 mm of Hg to 70 mm of Hg).

Nitrogen pentoxide decomposes into nitrogen-di-oxide through the following elementary reactions:

$$N_2O_5 \rightarrow N_2O_3 + O_2$$
$$N_2O_3 + N_2O_5 \rightarrow 4NO_2$$

The value of rate constant k has been investigated over a wide pressure range (0.0002 to 70 mm of Hg).

At very low pressure $p^0_{N_2O_5} < 0.1$ mm of Hg, the above reaction follows 2nd order kinetics and at high pressure $p^0_{N_2O_5} > 0.6$ mm of Hg, the above reaction follows 1st order kinetics.

Inferences

(i) Order of a unimolecular reaction depends on initial concentration of the reactant.

(ii) Collision theory explains unimolecular, bimolecular and trimolecular reactions satisfactorily. Thus the mechanism of formation of product is independent of molecularity of a reaction.

(iii) The time lag between two successive collisions is called collision time. If the collision time is greater than the life time of activated or energized complex the latter splits into the products before colliding with another molecule. Thus step – III is dominating over step-II. So, $k_3 >> k_2 C_A$ and hence, $k_2 C_A + k_3 \approx k_3$. The reaction follows 2nd order kinetics. This usually comes true at low concentration of reactant.

(iv) If the initial concentration of reactant is very high, the collision time is very short, as a result of which, molecules prefer colliding with each other to form products. So, $k_2 C_A >> k_3$ and hence, $k_2 C_A + k_3 \approx k_2 C_A$. The reaction follows 1st order kinetics.

3.9 Kinetic study of some special reactions

3.9.1 Pseudo 1st order reactions

There are some reactions where more than one reactant is involved but rate depends on one reactant only. One such example is: Inversion of cane sugar.

$$\underset{\substack{\text{(sucrose)} \\ \text{(+) dextro}}}{C_{12}H_{22}O_{11}} + H_2O \xrightarrow{H^+} \underset{\substack{\text{(glucose)} \\ \text{(+) dextro}}}{C_6H_{12}O_6} + \underset{\substack{\text{(fructose)} \\ \text{(–) leveo}}}{C_6H_{12}O_6}$$

The rate equation is given by: $J = k[C_{12}H_{22}O_{11}]\,[H_2O]\,[H^+]$

The concentration of the catalyst i.e. $[H^+]$ is fixed since it is regenerated at the end of the reaction. As the reaction is carried out in profuse amount of water, concentration of water, i.e $[H_2O]$ is fixed. Thus the above rate equation becomes:

$$J = k'[C_{12}H_{22}O_{11}] \tag{3.33}$$

where, $k' = k[H_2O]\,[H^+]$

Thus the apparent order of the above reaction is 1. Hence it is called pseudo first order reaction. Again the rate depends only on the concentration of one reactant though more than one reactant is involved in actual reaction. Thus pseudo first order reaction is also called pseudo unimolecular reaction.

Determination of rate constant '*k*' of the above reaction

Sucrose is an optically active substance and shows dextro-rotatory (+) characteristics. It, on hydrolysis, gives glucose (dextro-rotatory) and fructose (laveo-rotatory).

When a plane polarized light is passed through sucrose solution, the net angle of rotation of the plane polarized light decreases with the increase in fructose concentration. Thus the concentration of sucrose is directly proportional to the net angle of rotation. As the reaction is 1st order the rate equation is given by:

$$t = \frac{1}{k}\ln\left[\frac{a}{(a-x)}\right] \quad \text{[Equation (3.2)]}$$

θ_0 = Angle of rotation of plane polarized light at $t = 0$, i.e. at the beginning of hydrolysis.

θ_t = Angle of rotation measured at any time t.

θ_∞ =Angle of rotation measured at ∞, i.e., at the end of hydrolysis.

We have: $\theta_0 > \theta_t >> \theta_\infty$

Thus the initial concentration of sucrose = $\left(\frac{a}{V}\right) \propto [(\theta_0 - \theta_\infty)]$

where V = volume of the total solution = constant.

Concentration of the product formed at any time t = $\left(\frac{x}{V}\right) \propto [(\theta_0 - \theta_t)]$

where x is the number of moles of product.

Hence, $$\left(\frac{a-x}{V}\right) \propto [(\theta_0 - \theta_\infty) - (\theta_0 - \theta_t)]$$

Or, $$\left(\frac{a-x}{V}\right) \propto (\theta_t - \theta_\infty)$$

Thus, $$\left(\frac{a}{a-x}\right) = \left(\frac{\theta_0 - \theta_\infty}{\theta_t - \theta_\infty}\right)$$

Putting this value of $\left(\frac{a}{a-x}\right)$ in the above rate equation, we get,

$$t = \frac{1}{k}\ln\left[\frac{\theta_0 - \theta_\infty}{\theta_t - \theta_\infty}\right] = \frac{2.303}{k}\log\left[\frac{\theta_0 - \theta_\infty}{\theta_t - \theta_\infty}\right]$$

Or, $$t = \frac{2.303}{k}\log\left[\frac{\theta_0 - \theta_\infty}{\theta_t - \theta_\infty}\right] \tag{3.34a}$$

θ_0 and θ_∞ can easily be measured with the help of polarimeter and the values of θ_t are measured at different time t. By plotting $\log\left[\frac{\theta_0 - \theta_\infty}{\theta_t - \theta_\infty}\right]$ versus t a straight line is obtained. The rate constant k can easily be evaluated from the slope of the straight line.

Another example of pseudo order reaction is hydrolysis of ester in acid medium.

$$\text{RCOOR}' + H_2O \xrightarrow{H^+} \text{RCOOH} + \text{R}'\text{OH}$$

The products are acid and alcohol and the amount of acid produced can be estimated through titration with a base say caustic soda. As the reaction proceeds amount of acid increases and the corresponding titre value also increases. At the beginning of hydrolysis acid concentration is the least and the corresponding titre value is V_0(say). At the end of hydrolysis the acid concentration is the maximum and the corresponding titre value is V_∞(say). At any time t the titre value is V_t (say). The rate equation of the above reaction is given by:

$$J = k[\text{RCOOR}'] \, [\text{H}^+] \, [\text{H}_2\text{O}]$$

The acid acts as a catalyst and is regenerated at the end of reaction. Thus the concentration of acid, i.e., $[H^+]$ is constant. As the reaction is carried out in profuse amount of water, $[H_2O]$ is also constant. Hence the above rate equation becomes:

$$J = k'[\text{RCOOR}']$$

Thus it is a pseudo-unimolecular 1st order reaction and the rate equation is given by:

$$t = \frac{2.303}{k} \log \frac{a}{(a - x)}$$

a = Initial concentration of the ester.

From the knowledge of V_0 and V_t , we can write that: $a \propto (V_\infty - V_0)$

x = Concentration of acid produced at any time t.

So, $x \propto (V_t - V_0)$

Hence, $(a - x) \propto (V_\infty - V_t)$

Putting the values of a and $(a - x)$ in the above equation we get:

$$t = \frac{2.303}{k} \log \left[\frac{(V_\infty - V_0)}{(V_\infty - V_t)} \right] \qquad (3.34b)$$

V_0 and V_t are the titre values at $t = 0$ and $t = t$ respectively. Different values of V_t are obtained at different time intervals and hence different values of $\log \left[\frac{(V_\infty - V_0)}{(V_\infty - V_t)} \right]$ are obtained. A straight line is obtained by plotting $\log \left[\frac{(V_\infty - V_0)}{(V_\infty - V_t)} \right]$ against t. The rate constant k can easily be obtained from the slope of the straight line.

3.9.2 Competing reactions/parallel reactions/side reactions

There are many reactions where a reactant gives two products at different rates, e.g. dissociation of potassium chlorate as shown below:

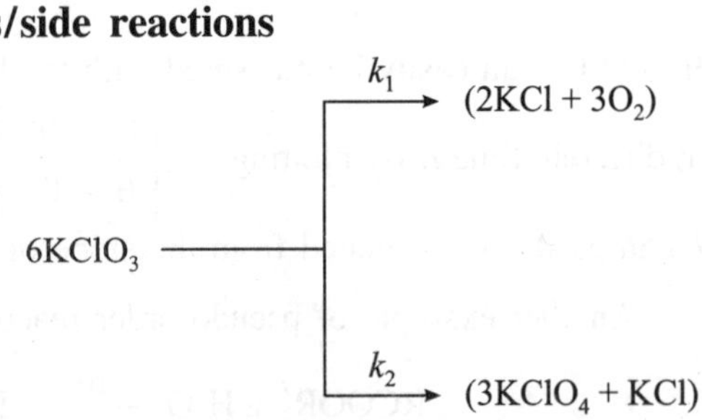

Another example is chlorination of toluene:

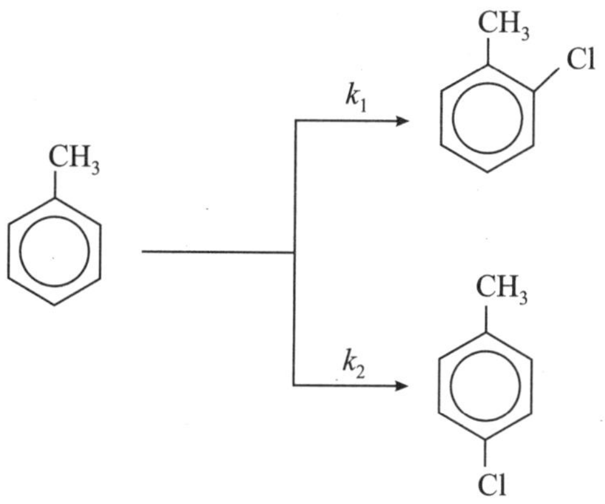

(o-chlorotoluene)

(p-chlorotoluene)

These type of reactions are known as parallel reactions.

Rate equation of a competing reaction: Let us consider a parallel reaction:
k_1 and k_2 are the rate constants for the two products P and Q respectively.

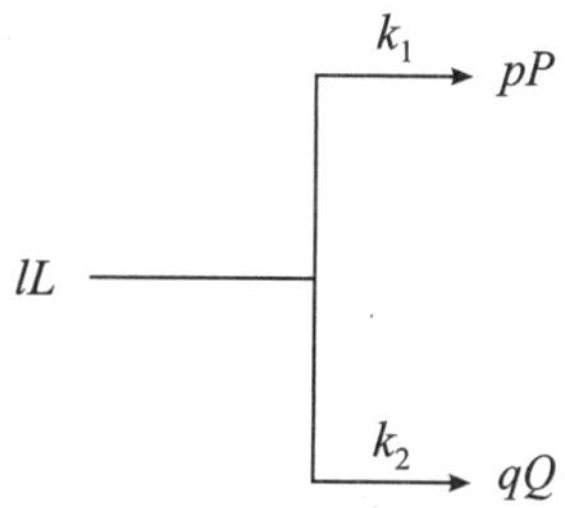

Time (t)	*No. of moles of reactant L*	*No. of moles of product (P)*	*No. of moles of product (Q)*
0	a	0	0
t	$(a-x)$	$\left(p.\frac{x}{l}\right)$	$\left(q.\frac{x}{l}\right)$
$t_{\frac{1}{2}}$	$\left(\frac{a}{2}\right)$	$\left(p.\frac{a}{2l}\right)$	$\left(q.\frac{a}{2l}\right)$

x = Number of moles of reactant L consumed at time t.

V = volume of the system = constant

$$C_L = \left[\frac{(a-x)}{V}\right] \text{ moles/litre} = \text{concentration of the reactant } L \text{ at time } t.$$

$$C_P = \left[\frac{(px)}{(lV)}\right] \text{ moles/litre} = \text{concentration of the product } P \text{ at time } t.$$

$$C_Q = \left[\frac{(qx)}{(lV)}\right] \text{ moles/litre} = \text{concentration of the product } Q \text{ at time } t.$$

Boundary conditions: (i) at $t = 0$, $(a - x) = a$ and $x = 0$

(ii) at $t = t$ $(a - x) = (a - x)$ and $x = x$

The rate equation is given by:

$$J = -\frac{1}{l}\frac{d\left[\dfrac{(a-x)}{V}\right]}{dt} = \frac{1}{l.V}\frac{dx}{dt}$$

Again, $$J = \frac{1}{p}\frac{dC_P}{dt} = \frac{1}{p}\frac{d\left(\dfrac{p.x}{lV}\right)}{dt} = \frac{1}{l.V}\frac{dx}{dt}$$

Again, $$J = \frac{1}{q}\frac{dC_Q}{dt} = \frac{1}{q}\frac{d\left(\dfrac{q.x}{lV}\right)}{dt} = \frac{1}{l.V}\frac{dx}{dt}$$

For the product P, the rate equation is given by:

$J_1 = k_1 \, . \, C_L$, assuming the parallel reaction is of 1st order

and for the product Q, the rate equation is given by:

$J_2 = k_2 \, . \, C_L$, assuming the parallel reaction is of 1st order

Thus the total rate of decomposition of the reactant L is given by:

$$J = k_1 \, . \, C_L + k_2 \, . \, C_L = (k_1 + k_2) \, . \, C_L = k \, . \, C_L$$

where, $$k = k_1 + k_2$$

$$-\frac{1}{l}\frac{d\left[\dfrac{(a-x)}{V}\right]}{dt} = \frac{k\,(a-x)}{V}$$

$$-\frac{1}{l.V}\frac{d\,(a-x)}{(a-x)} = \frac{k}{V}\,dt$$

$$-\int_a^{(a-x)}\frac{d\,(a-x)}{(a-x)} = lk\int_0^t dt$$

On integration, we get:

$$\ln\frac{a}{(a-x)} = k_L t$$

where, $$k_L = lk$$

Or, $$t = \frac{2.303}{k_L}\log\left(\frac{a}{a-x}\right) \qquad (3.35a)$$

Again, we have:

$$\frac{\text{Amount of product '}P\text{' formed at time '}t\text{'}}{\text{Amount of product '}Q\text{' formed at time '}t\text{'}} = \frac{\left(\dfrac{px}{l}\right)}{\left(\dfrac{qx}{l}\right)} = \frac{p}{q}$$

Or,

$$\frac{p}{q} = \frac{k_1 C_L}{k_2 C_L} = \frac{k_1}{k_2} \tag{3.35b}$$

As p and q are known, $\dfrac{k_1}{k_2}$ is known.

Again k_L can easily be evaluated by plotting $\log\left[\dfrac{a}{(a-x)}\right]$ versus t.

$$k_L = lk = l(k_1 + k_2)$$

Thus from the above two equations k_1 and k_2 can easily be determined.

3.9.3 Consecutive reactions

There are many reactions which occur through multiple steps and each step is called elementary reaction. e.g.:

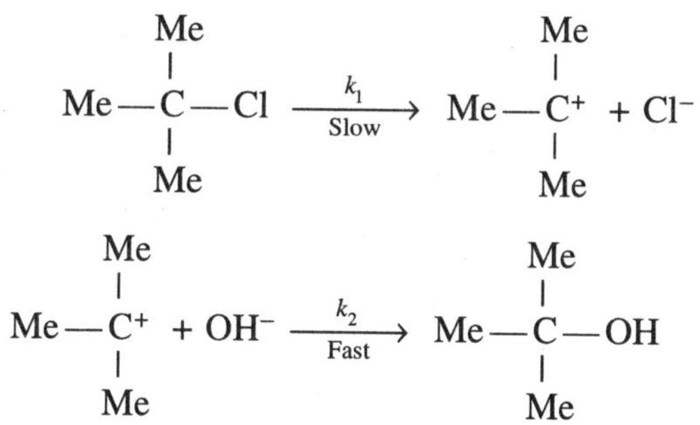

In that case, the slowest step is considered as rate-determining step and the rate of the rate-determining step is the rate of the whole reaction. The detailed discussion has been made in sec. 3.6 and 3.7. This type of multi-step reaction is known as consecutive reaction.

Radioactive nuclear reactions occur through multi-steps, e.g.,

$${}_{92}U^{239} \rightarrow {}_{93}Np^{239} + {}_{-1}e^{0}$$

$${}_{93}Np^{239} \rightarrow {}_{94}Pu^{239} + {}_{-1}e^{0}$$

This type of nuclear reaction can also be termed consecutive reaction.

3.9.4 Opposing reactions

1st order reversible reactions:

1st order reversible reaction of type

$$A \underset{k_2}{\overset{k_1}{\rightleftharpoons}} B \quad \text{Total volume} = V$$

Time (*t*)	*No. of moles of reactant A*	*No. of moles of product B*
0	*a* moles	*b* moles
t	(*a* − *x*) moles	(*b* + *x*) moles

x = Number of moles of reactant A converted into product

$$C_A = \frac{(a-x)}{V} = \text{Concentration of reactant } A \text{ at time } t$$

$$C_B = \frac{(b+x)}{V} = \text{Concentration of reactant } B \text{ at time } t$$

If several reactions are occurring simultaneously in a system, each of them is independent of others and the rate of each reaction is directly proportional to the concentrations of the corresponding reactants only.

Thus, the rate of formation of $B = -\dfrac{d\,C_A}{dt}$

$$-\frac{d\,C_A}{dt} = k_1 C_A = k_1 \frac{(a-x)}{V}$$

where,
$$C_A = \frac{(a-x)}{V}$$

$$-\frac{d\left[\dfrac{(a-x)}{V}\right]}{dt} = k_1 \frac{(a-x)}{V}$$

Or,
$$\frac{dx}{dt} = k_1(a-x) \qquad (3.36a)$$

Thus, the rate of disappearance of $B = -\dfrac{d\,C_B}{dt}$

$$-\frac{d\,C_B}{dt} = k_2 C_B = k_2 \frac{(b+x)}{V}$$

where,
$$C_B = \frac{(b+x)}{V}$$

$$-\frac{d\left[\dfrac{(b+x)}{V}\right]}{dt} = k_2 \frac{(b+x)}{V}$$

Or, $$\frac{dx}{dt} = -k_2(b + x) \tag{3.36b}$$

Combining the above two equations, we get:

$$\begin{aligned}\frac{dx}{dt} &= k_1(a - x) - k_2(b + x)\\ &= (k_1a - k_2b) - (k_1 + k_2)x\\ &= (k_1 + k_2)\left[\frac{(k_1a - k_2b)}{(k_1 + k_2)} - x\right]\\ &= (k_1 + k_2)(L - x)\end{aligned}$$

where, $$L = \frac{(k_1a - k_2b)}{(k_1 + k_2)} = \frac{\left[\left(\frac{k_1}{k_2}\right)a - b\right]}{\left[\left(\frac{k_1}{k_2}\right) + 1\right]} = \frac{(Ka - b)}{(K + 1)}$$

where, $$K = \frac{k_1}{k_2}$$

Or, $$\frac{dx}{dt} = (k_1 + k_2)(L - x) \tag{3.36c}$$

$$\frac{dx}{(L - x)} = (k_1 + k_2)dt$$

$$-\frac{d(L - x)}{(L - x)} = (k_1 + k_2)dt$$

Boundary conditions: (i) at $t = 0$: $x = 0$ and $(L - x) = L$

(ii) at $t = t$: $x = x$, and $(L - x) = (L - x)$.

$$-\int_L^{(L-x)} \frac{d(L - x)}{(L - x)} = (k_1 + k_2)\int_0^t dt$$

Or, $$(k_1 + k_2)t = \ln\left[\frac{L}{(L - x)}\right]$$

$$t = \frac{1}{(k_1 + k_2)}[\ln L - \ln(L - x)] \tag{3.36d}$$

where, $$L = \frac{(Ka - b)}{(K + 1)} \quad \text{and} \quad K = \frac{k_1}{k_2} \tag{3.36e}$$

At equilibrium, $\frac{dx}{dt} = 0$ and $x = x_e$. Thus the Equation (3.36c) becomes:

$$(k_1 + k_2)(L - x_e) = 0$$

Or, $$(L - x_e) = 0$$

Hence, $$L = x_e$$

Putting the value of L we get:

$$\frac{(Ka - b)}{(K+1)} = x_e$$

Or, $$(Ka - b) = Kx_e + x_e$$

Or, $$K = \frac{(b + x_e)}{(a - x_e)} \tag{3.36f}$$

The value of x_e can be determined experimentally. Hence K can be calculated from equation (3.36f). Using this value of K the value of L can be evaluated from Equation (3.36e). By measuring x at different time t, one can plot t versus log $(L - x)$. The plot will be a straight line according to Equation (3.36d).

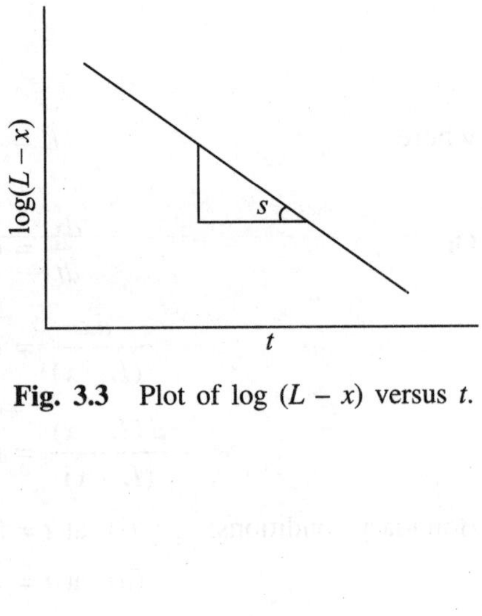

Fig. 3.3 Plot of log $(L - x)$ versus t.

Rearranging Equation (3.36d) we get:

$$2.303 \log(L - x) = -(k_1 + k_2) . t + 2.303 \log L$$

Or, $$\log(L - x) = -\frac{(k_1 + k_2)t}{2.303} + \log L$$

$$\text{The slope} = s = -\left(\frac{k_1 + k_2}{2.303}\right)$$

Thus, $$\frac{k_1}{k_2} = K \quad \text{and} \quad (k_1 + k_2) = -2.303 \text{ s}$$

Thus from the above two equations k_1 and k_2 can easily be calculated.

3.10 Theory of reaction rate

Potential energy For each element or compound there is a potential well, which contains numerous energy levels. Each energy level contains a definite number of molecules, which are continuously vibrating. A typical potential energy (PE) diagram is shown in the Figure 3.4. Consider the reaction:

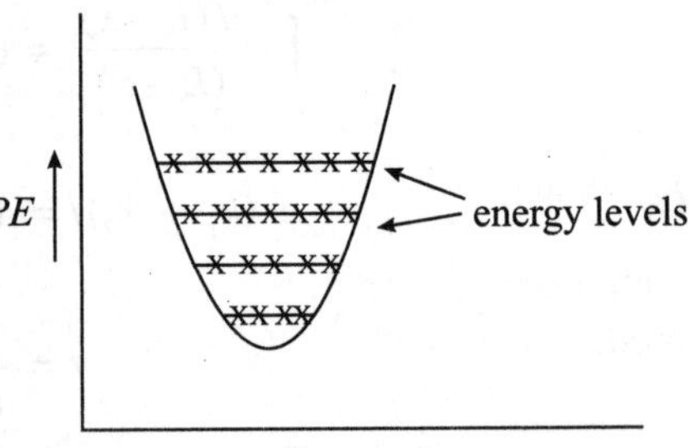

Fig. 3.4 Potential energy diagram of an element or compound, r = Internuclear distance.

$$X - Y + Z \rightleftharpoons X + Y - Z$$

As atom Z approaches $X - Y$, the bond between Y and Z becomes stronger and stronger, while the bond between X and Y becomes weaker and weaker, leading to the formation of an activated complex, or transition state (TS).

$$X - Y + Z \rightarrow [X \cdots\cdot Y \cdots\cdot Z] \text{ (TS)}$$

$$[X \cdots\cdot Y \cdots\cdot Z] \rightarrow X + Y - Z$$

Both the bonds $X - Y$ and $Y - Z$ are weak and hence TS is placed at high energy level, as shown in the Figure 3.5. Thus activated complex is unstable and readily dissociates into products.

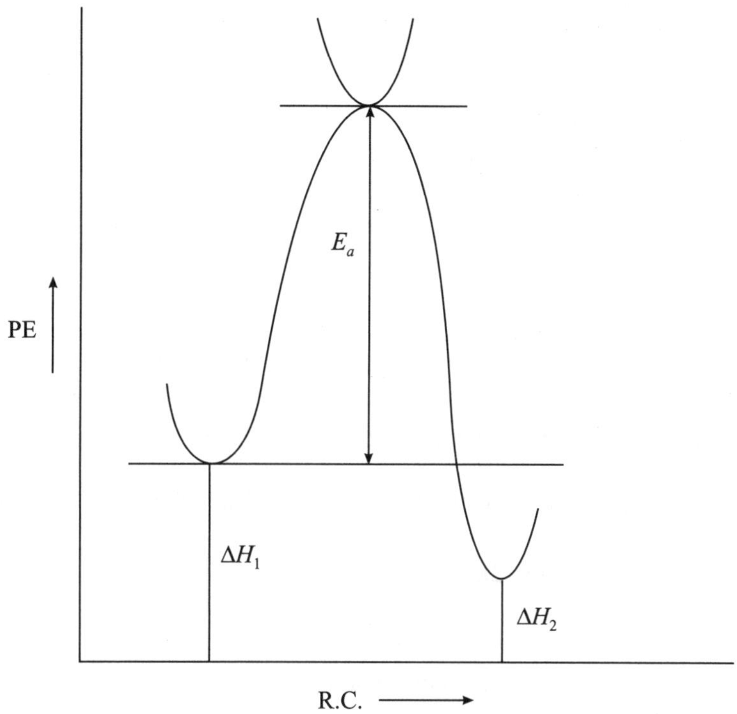

Fig. 3.5 Reaction path profile for the reaction: $X - Y + Z \rightarrow X + Y - Z$
R.C. = Reaction coordinate = $d\varepsilon$,

E_a, shown in the Figure 3.5, is called the activation energy.

Activation energy Amount of energy required to convert a reactant into the corresponding product is called activation energy of that reaction.

$$|\Delta H_1| = \text{Enthalpy of the reactants.}$$

$$|\Delta H_2| = \text{Enthalpy of the products.}$$

If ($|\Delta H_2| - |\Delta H_1|$) is negative some amount of energy is released in the form of heat. The reaction is then said to be an *exothermic reaction.*

If ($|\Delta H_2| - |\Delta H_1|$) is positive some amount of energy is absorbed during chemical reaction. The reaction is then said to be an *endothermic reaction.*

Collision theory According to collision theory all the molecules are colliding with each other. Each and every molecule possesses a secluded volume (see Figure 3.6), into which no other molecule is allowed to enter.

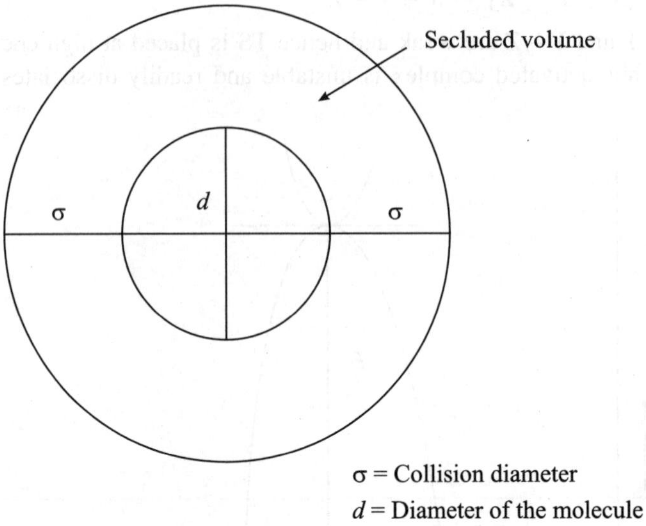

Fig. 3.6 Secluded volume and collision diameter of a molecule.

For a reaction of type $2A \rightarrow$ *Products*, the total number of collisions (Z_0) given by:

$$Z_0 = 2n^2\,\sigma_T^2\,\sqrt{\frac{\pi RT}{M}} \tag{3.37a}$$

n = Number of molecules/c.c.

M = Molecular mass.

σ_T = Collision diameter at temperature T°K.

According to Arrheneous approach, in each collision one molecule gains energy and goes to higher energy level. But only a fraction of total number of activated molecules gains the necessary activation energy E_a. Only those activated molecules, gaining the activation energy E_a, form the activated complex and ultimately break into the products.

$$A + A \rightarrow A^* \text{ (T.S)} \rightarrow \text{Products.}$$

$$A + B \rightarrow (AB)^* \text{ (T.S.)} \rightarrow \text{Products.}$$

Thus only those collisions, by which reactant molecules gain necessary activation energy, will lead to a chemical reaction. These collisions are called effective collisions represented as Z_a which is related to Z_0 by Boltzman's distribution factor $e^{-\left(\frac{E_a}{RT}\right)}$ as follows:

$$Z_a = Z_0 \; e^{-\left(\frac{E_a}{RT}\right)} \tag{3.37b}$$

Inferences

1. Only a fraction of collisions is effective in bringing about a chemical change. Those collisions are called effective collisions.
2. With the increase in number of reactant molecules the number of effective collisions increases and hence concentration of activated complex increases and so does reaction rate (J). Thus rate of a chemical reaction is directly proportional to the concentration of the reactants.
3. Since the rate (J) of a chemical reaction depends on the fraction of effective collisions, i.e., $\left(\dfrac{Z_a}{Z_0}\right)$, Arrheneous proposed that rate constant (k) is directly proportional to $\left(\dfrac{Z_a}{Z_0}\right)$.

   Thus, we can write that, $k \propto \left(\dfrac{Z_a}{Z_0}\right)$.

   Or, $$k \propto e^{-\left(\frac{E_a}{RT}\right)} \quad \text{[From Equation (3.37b)]}$$

   Or, $$k = A\, e^{-\left(\frac{E_a}{RT}\right)}$$

   where, 'A' is the Arrheneous factor.

   So, $$k = A\, e^{-\left(\frac{E_a}{RT}\right)} \tag{3.37c}$$

   Taking $\log_e$ or *ln* on both sides, we get:

   $$\ln k = \ln A - \frac{E_a}{RT}$$

   At $T = T_1$, $k = k_1$ and at $T = T_2$, $k = k_2$.

   Using these boundary conditions, we have:

   $$\ln k_1 = \ln A - \frac{E_a}{RT_1} \quad \text{and} \quad \ln k_2 = \ln A - \frac{E_a}{RT_2}$$

   $$\ln \frac{k_2}{k_1} = -\frac{E_a}{R}\left[\frac{1}{T_2} - \frac{1}{T_1}\right]$$

   $$\log \frac{k_2}{k_1} = -\frac{E_a}{2.303\,R}\left[\frac{1}{T_2} - \frac{1}{T_1}\right] \tag{3.37d}$$

   For a chemical reaction if k_1 and k_2 are known at two different temperatures, activation energy E_a can easily be evaluated from Equation (3.37d).

3.11 Factors influencing the reaction rate

(a) Activation energy As we have seen in the earlier discussion, the rate constant (k) is related to the activation energy E_a according to the following equation:

$$k = A\, e^{-\left(\frac{E_a}{RT}\right)} \qquad \text{(from equation 3.37c)}$$

Thus it is evident that with the decrease in activation energy the rate constant (k) increases and hence the reaction rate increases.

(b) Catalyst In presence of a catalyst, the path of a chemical reaction changes and also the activation energy. The Figure 3.7 shows how the reaction path changes in presence of a catalyst.

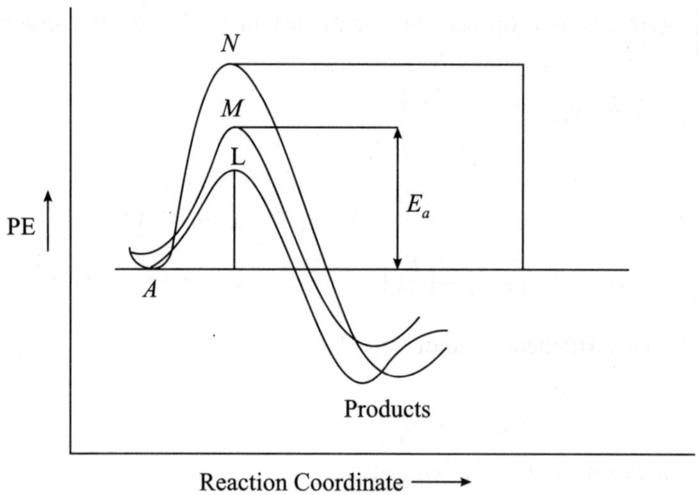

Fig. 3.7 PE diagram of reaction of type $2A \rightarrow$ Products. E_a = Activation energy of uncatalysed reaction.

The curve M indicates the uncatalysed reaction path. In presence of a catalyst if the reaction path changes from M to L the activation energy decreases and the catalyst is called positive catalyst. The catalyst is said to be negative catalyst if its presence changes the reaction path from M to N such that the E_a value increases.
Uncatalysed reaction may be written as:

$$A + A \rightarrow A^* + A$$

$$A^* \rightarrow \text{products.}$$

Whereas in the catalysed reaction collision occurs between reactant (A) and catalyst (C):

$$A + C \rightarrow (AC)^*$$

$$(AC)^* \rightarrow \text{Products} + C$$

It the catalyst is a positive catalyst the potential energy of $(AC)^*$ is much lower than that of A^* and the reaction occurs at much faster rate than that of uncatalysed reaction, e.g., esterification reaction in acid medium.

$$RCOOH + R'OH \xrightarrow{H^+} RCOOR' + H_2O$$

Here acid acts as a positive catalyst.

In the case of negative catalysed reaction the potential energy of $(AC)^*$ is much higher than that of A^* and hence the reaction occurs at much slower rate than that of uncatalysed reaction. One example is oxidation of hydrochloric acid by potassium permanganate. The reaction rate is sufficiently (or significantly) retarded in presence of managanous sulphate.

$$KMnO_4 + HCl \rightarrow MnCl_2 + Cl_2$$

$MnSO_4$ acts as negative catalyst.

(c) Temperature From Equation (3.37c) it is evident that the rate constant (k) increases with the increase in temperature. It has been found that with just 10° rise in temperature the rate of reaction is nearly doubled.

(d) Concentration of reactants Rate of a reaction is directly related to the concentration of the reactants. The reactant to which order of a reaction is positive the reaction rate increases with the increase in the reactant concentration but the same decreases with the increase in reactant concentration, to which order of the reaction is negative. That type of reactant is called inhibitor. One example is:

$$OCl^- + I^- \xrightarrow{OH^-} OI^- + Cl^-$$

The rate of reaction (J) is given by:

$$J = k.\frac{[I^-][OCl^-]}{[OH^-]}$$

The order of reaction with respect to $[OH^-]$ is (-1) and hence the rate J decreases with the increase in $[OH^-]$. Thus OH^- is called inhibitor to the above reaction.

(e) State of subdivision If one of the reactant or catalyst remains in solid state, the reaction rate depends on the particle size of the solid. Smaller the particle size larger the area and hence faster the reaction rate.

(f) Radiation If the reactants of a reaction are irradiated with a monochromatic beam of light the reaction proceeds through free radical mechanism. One such example is the formation of hydrochloric acid from its elements.

$$Cl_2 \xrightarrow{h\nu} 2Cl\cdot \qquad \text{(Initiation)}$$

$$Cl\cdot + H_2 \rightarrow HCl + H\cdot \qquad \text{(Propagation)}$$

$$H\cdot + Cl_2 \rightarrow HCl + Cl\cdot \qquad \text{(Propagation)}$$

$$Cl\cdot + Cl\cdot \rightarrow Cl_2 \qquad \text{(Termination)}$$

4

CATALYSIS

4.1 Definition

Studies on several chemical reactions under different circumstances reveal that rate of a chemical reaction may be significantly altered in presence of a specific foreign substance, although the latter is not at all consumed and remains unchanged chemically at the end of the reaction. The phenomenon is known as catalysis and the foreign substance is called catalyst.

Characteristic features of a catalyst:

(i) It remains unchanged chemically and quantitatively at the end of reaction.
(ii) It cannot start a chemical reaction but can only change the velocity of the reaction.
(iii) Only a trace amount (of the order of 10^{-10} gm/litre or less) of catalyst is enough to produce a massive impact on reaction rate.
(iv) It speeds up both forward and backward reactions equally such that equilibrium constant (K) remains unaltered and hence also the position of equilibrium since we know that:

$$\Delta G^0 = -RT \ln K,$$

where, ΔG^0 = Standard free energy change and
K = Equilibrium constant

(v) A catalyst can change the nature of product since action of catalyst on chemical reactions is highly specific. For example, ethanol gives acetaldehyde in presence of *Ni*-catalyst but gives ethylene in presence of alumina (Al_2O_3).

4.2 Classification

(a) On the basis of *activity* catalysts can be divided into two main categories:

(i) Positive catalyst and
(ii) Negative catalyst.

(i) Positive catalyst A catalyst is said to be positive catalyst when it accelerates the reaction rate appreciably, e.g., in the manufacture of sulphuric acid by contact process, platinum is used as a positive catalyst in the conversion of sulphur-di-oxide to sulphur-tri-oxide.

$$2SO_2(g) + O_2(g) \xrightarrow{[Pt]} 2SO_3(g)$$

(ii) Negative catalyst A catalyst is said to be negative catalyst when it retards the reaction rate noticeably, e.g., decomposition of hydrogen peroxide is sufficiently restricted in presence of orthophosphoric acid.

(b) On the basis of *physical state* catalysts can be broadly divided into two main categories:

(i) Homogeneous catalyst
(ii) Heterogeneous catalyst.

(i) Homogeneous catalyst When catalyst, reactants and products remain in the same phase, the catalyst is called homogeneous catalyst, e.g., in the chamber process of manufacturing sulphuric acid gaseous nitric oxide (NO) acts as a catalyst, which remains in the same phase with reactants (SO_2 and O_2) and product (SO_3).

$$2SO_2(g) + O_2(g) \xrightarrow{[NO(g)]} 2SO_3(g)$$

(ii) Heterogeneous catalyst When catalyst remains in different phase with reactants or products or both the catalyst is called heterogeneous catalyst, e.g., iron oxide (Fe_2O_3) acts as a catalyst in the Haber process of manufacturing ammonia.

$$N_2(g) + 3H_2(g) \xrightarrow{Fe_2O_3(s)} 2NH_3(g)$$

4.3 Other type of classification

4.3.1 Auto catalysis

There are chemical reactions where one of the reaction products acts as a catalyst. The phenomenon is known as autocatalysis.

Example 4.1 Hydrolysis of ester:

$$\underset{\text{(Ester)}}{RCOOR'} + H_2O \rightleftharpoons \underset{\text{(Acid)}}{RCOOH} + \underset{\text{(Alcohol)}}{R'OH}$$

Acid, produced, during the course of hydrolysis, accelerates the rate of hydrolysis appreciably.

Example 4.2 Redox reaction between oxalic acid and potassium permanganate in acid medium.

$$MnO_4^- + \begin{array}{c} COOH \\ | \\ COOH \end{array} + 6H^+ \rightleftharpoons \underset{\text{(Catalyst)}}{Mn^{2+}} + 2CO_2 + 4H_2O$$

Initially the forward reaction is very slow but as manganous ion concentration increases the reaction rate becomes faster and faster. The reaction is self-catalysed by manganous ion (Mn^{2+}).

4.3.2 Acid-base catalysis

There are some reactions which are catalysed by acids as well as by bases. One such example is hydrolysis of ester:

(i) Uncatalysed reaction

$$RCOOR' + H_2O \rightleftharpoons RCOOH + R'OH$$

$$\text{Rate} = -\frac{d[\text{Ester}]}{dt} = k_0'[\text{Easter}]\,[H_2O] = k_0[\text{Easter}]$$

Where $k_0 = k_0' \cdot [H_2O]$, since the reaction is carried out in huge excess of water.

(ii) Acid-catalysed reaction

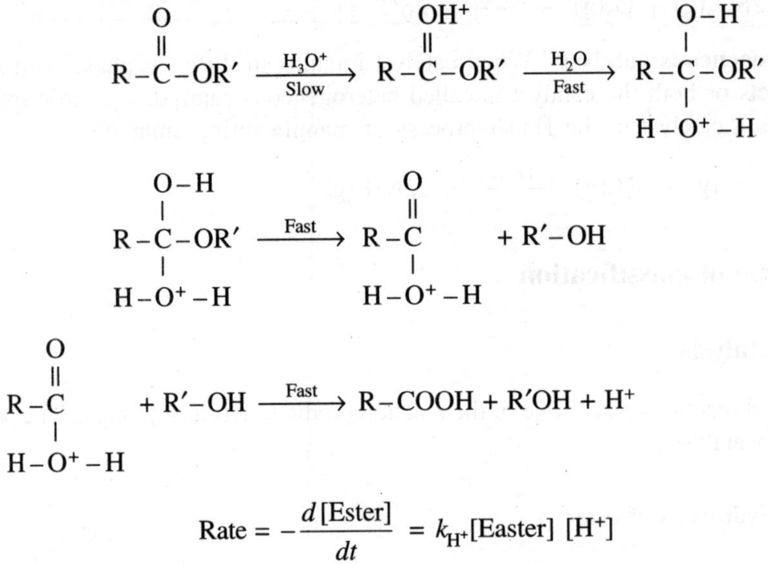

$$\text{Rate} = -\frac{d[\text{Ester}]}{dt} = k_{H^+}[\text{Easter}]\,[H^+]$$

(iii) Base-catalysed reaction

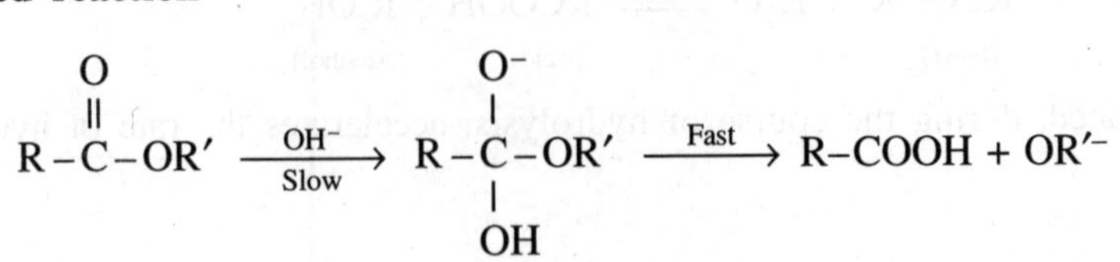

$$OR'^{-} + H_2O \xrightarrow{\text{Fast}} R'\text{–}OH + OH^{-}$$

$$\text{Rate} = -\frac{d\,[\text{Ester}]}{dt} = k_{OH^-}[\text{Easter}]\,[OH^-]$$

Considering acid-catalysed reaction, we have:

$$k = k_{H^+} \,.\, C_{H^+}, \text{ where } k \text{ is the overall rate constant}$$

Or,
$$\log_{10} k = \log_{10} k_{H^+} + \log_{10} C_{H^+}.$$

$$\log_{10} k = \log_{10} k_{H^+} - \text{pH} \quad [\text{since, pH} = -\log_{10} C_{H^+}{}^{+}] \qquad (4.1)$$

k_{H^+} is constant but the overall rate constant (k) decreases linearly with the increase in pH. From the intercept the value of k_{H^+} can easily be evaluated.
Considering base-catalysed reaction, we have:

$$k = k_{OH^-} \,.\, C_{OH^-}\,, \text{ where } k \text{ is the overall rate constant.}$$

Or,
$$\log_{10} k = \log_{10} k_{OH^-} + \log_{10} C_{OH^-}$$

$$\log_{10} k = \log_{10} k_{OH^-} - \text{pOH} \quad [\text{since, pOH} = -\log_{10} C_{OH^-}]$$

$$= \log_{10} k_{OH^-} - (pk_w - \text{pH})$$

Or,
$$\log_{10} k = \log_{10}(k_{OH^-} \cdot k_w) + \text{pH} \quad [\text{since, } pk_w = -\log_{10} k_w] \qquad (4.2)$$

k_{OH^-} is constant but the overall rate constant (k) increases with increase in pH. From the intercept the value of k_{OH^-} can easily be evaluated since k_w is known and its value is 10^{-14}.

4.3.3 Enzyme catalysis

Enzymes are complex protein substances. They usually reside in living cells in colloidal form. Enzymes are essential to execute numerous biochemical reactions, constantly occurring in living cells of human body. In most of the reactions enzymes are acting as catalyst but their action as catalyst is highly specific in nature. One enzyme can catalyse only one chemical reaction and remains inert in other reactions, e.g., enzyme urease hydrolyses only urea but cannot hydrolyse even methyl urea or protein.

Mechanism In an enzyme catalysed reaction usually one substrate undergoes chemical change. At first enzyme (E) reacts with the substrate (S) to form an intermediate unstable complex [ES], which further dissociates into products (P) with the regeneration of enzyme. Schematically it is represented as:

$$E + S \underset{k_2}{\overset{k_1}{\rightleftharpoons}} [\text{ES}] \xrightarrow{k_3} E + P$$

Rate of formation of product:

$$= \frac{d\,[\text{P}]}{dt} = k_3[ES] = k_3 C_{ES} \qquad (4.3)$$

As the complex [ES] is unstable, at steady state we have:

Rate of formation of $[ES]$ = Rate of disappearance of $[ES]$

Or, $$k_1[E]\,[S] = k_2[ES] + k_3[ES]$$

Again, if C_E^0 is the initial concentration of enzyme, at time 't', $C_E^0 = C_E + C_{ES}$.

Where C_E is the concentration of enzyme at time 't'.

$$k_1 \,.\, C_E \,.\, C_S = (k_2 + k_3) \,.\, C_{ES} \tag{4.4a}$$

$$C_E = C_E^0 - C_{ES} \tag{4.4b}$$

Combining the above two equations, we get:

$$k_1(C_E^0 - C_{ES})C_S = (k_2 + k_3)C_{ES}$$

Or, $$C_{ES} = \frac{k_1\, C_E^0\, C_S}{k_1\, C_S + k_2 + k_3} = \frac{C_E^0}{1+\dfrac{k_2+k_3}{k_1\, C_S}} = \frac{C_E^0}{1+\dfrac{k_0}{C_S}}$$

Or, $$C_{ES} = \frac{C_E^0}{1+\dfrac{k_0}{C_S}} \tag{4.5}$$

[where, $k_0 = \dfrac{k_2+k_3}{k_1}$]

Combining Equations (4.3) and (4.5) we get:

$$\frac{d[\mathrm{P}]}{dt} = \frac{k_3\, C_E^0}{\left[1+\dfrac{k_0}{C_S}\right]} \tag{4.6}$$

Case – I If the concentration of the substrate is very small, i.e. $\left[1+\left(\dfrac{k_0}{C_S}\right)\right] \approx \left(\dfrac{k_0}{C_S}\right)$.

Thus Equation (4.6) becomes:

$$\frac{d[\mathrm{P}]}{dt} = \frac{k_3\, C_E^0}{\left(\dfrac{k_0}{C_S}\right)} = \left(\frac{k_3}{k_0}\right) C_E^0\, C_S = kC_S$$

$$\frac{d[\mathrm{P}]}{dt} = kC_S \tag{4.7}$$

Where, $k = \left(\dfrac{k_3}{k_0}\right) .\, C_E^0$ = constant for a given initial concentration of enzyme.

Conclusion The enzyme-catalysed reaction follows first order kinetics.

Case – II If the concentration of the substrate is very large, i.e. $\left[1+\left(\frac{k_0}{C_S}\right)\right] = 1.$

Thus Equation (4.6) becomes:

$$\frac{d[\mathrm{P}]}{dt} = k_3\, C_E^0 = k'$$

Or,
$$\frac{d[\mathrm{P}]}{dt} = k' \qquad (4.8)$$

Where, $k' = (k_3 \,.\, C_E^0)$ for a given initial concentration of enzyme.

Conclusion The enzyme-catalysed reaction follows zero-order kinetics.

Characteristic features of enzyme catalysed reactions:

(i) Catalytic activity of enzyme is highly specific and selective. One enzyme can catalyse only one reaction. The enzyme, invertase breaks sucrose into glucose and fructose but unable to break another disaccharide maltose, dissociation of which needs another enzyme, known as maltase.

(ii) Catalytic activity of an enzyme is worst affected in presence of a trace amount of HCN, H_2S, etc.

(iii) Catalytic activity of an enzyme very much depends on pH of the medium, usually enzymes are the most active near its *isoelectric point*, e.g., the enzyme trypsin is the most active at pH around 8.0, which is also the isoelectric point of trypsin.

(iv) Catalytic activity of enzymes is strongly pronounced within 30°C to 50°C. Reaction rate of an enzyme catalysed reaction remarkably decreases below and above the temperature range.

4.4 Promoters and inhibitors

There are substances, which promote the catalytic activity of catalysts. These substances are known as promoters. For example, molybdenum (Mo) acts as promoter in the Haber's process of manufacturing ammonia.

$$N_2(g) + 3H_2(g) \xrightarrow[\text{[Promoter]}]{\text{[Catalyst]}} 2NH_3(g)$$

Catalyst is Fe_2O_3 and promoter is Mo.

In the contact process of manufacturing sulphuric acid (H_2SO_4) platinum catalyst is used and V_2O_5 is used as promoter.

$$2SO_2(g) + O_2(g) \xrightarrow[\text{[Promoter]}]{\text{[Catalyst]}} 2SO_3(g)$$

Catalyst is Pt and promoter is V_2O_5.

Sometimes presence of a trace amount of foreign substance retard or inhibit the reaction rate. The function of retardar is to slow down the reaction rate appreciably, e.g., decomposition of hydrogen peroxide (H_2O_2) is sufficiently slowed down in presence of a trace amount of orthophosphoric acid.

Inhibitors are those substances presence of which totally stops the reaction. The reaction commences only after the inhibitor is totally consumed. The period during which the reaction does not occur at all is called inhibition period. After the inhibition period the reaction starts at the same rate as that observed without inhibitor. For example, potassium cyanide (KCN) acts as inhibitor in auto-oxidation of sodium sulphite solution, catalysed by cupric ion [Cu^{2+}]. In presence of KCN, Cu^{2+} forms stable complex $[Cu(CN)_4]^{2-}$ and loses its catalytic activity. Under this circumstances auto oxidation of Na_2SO_3 solution ceases to occur and the same resumes only after the total consumption of KCN.

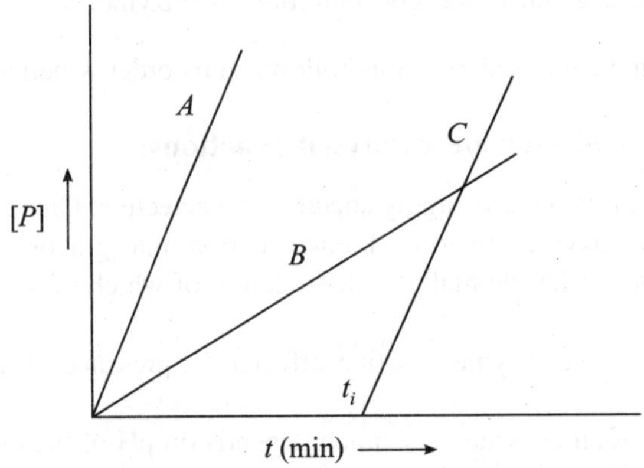

Fig. 4.1 Amount of product vs. time plot of a reaction under different conditions.

In the above figure, Curve-*A* represents rate of formation of product without any inhibitor or retardar. Curve-*B* shows that the same reaction proceeds with much reduced rate in presence of a retardar. In the case of Curve-*C* it is clear that the reaction is inhibited up to time t_i and after that the reaction proceeds with the same rate as that of curve-*A*. The time t_i, is known as inhibition period and the substance, that causes inhibition, is called inhibitor.

4.5 Mechanism of catalytic action in homogeneous catalysis system

From the knowledge of chemical kinetics we know that each and every reaction has a definite potential energy barrier. It is believed that in presence of a catalyst the energy barrier is sufficiently diminished such that reactants easily cross the energy barrier and form products. This is well illustrated in the Figure 4.2.

Consider the following 1st order reaction:

$$A \rightarrow \text{Products}$$

In presence of catalyst the above reaction is believed to follow two-step mechanism as shown below:

(i) $$A + C(\text{catalyst}) \underset{k_2}{\overset{k_1}{\rightleftharpoons}} [AC] \quad \text{(Transition state complex)}$$

(ii) $$[AC] \xrightarrow{k_3} \text{Products} + C$$

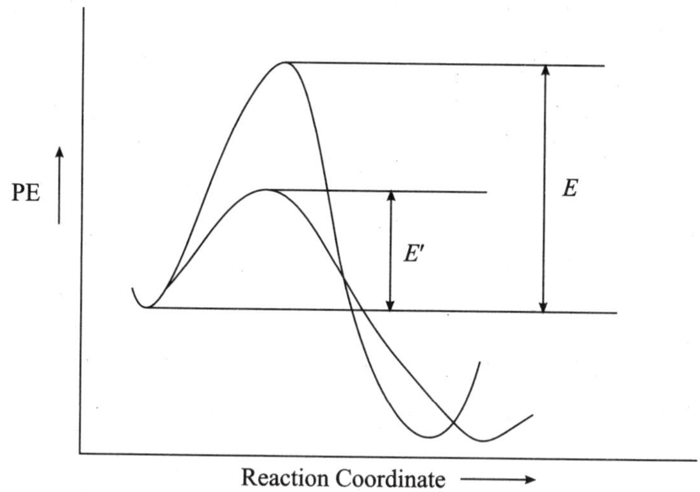

Fig. 4.2 Activation energy curve of both uncatalysed and catalysed reactions.
E = Activation energy of uncatalysed reaction.
E' = Activation energy of catalysed reaction.

The rate of formation of product (r), is given by:

$$r = k_3 C_{[AC]} \tag{4.9}$$

As the transition state (*T.S.*) complex is short lived, steady state concept can be applied on it. So, we have:

Rate of formation of *T.S. – complex* = Rate of disappearance of *T.S. – complex*.

Or, mathematically we can write that:

$$k_1 C_A C_C = k_2\, C_{[AC]} + k_3\, C_{[AC]}$$

Or,
$$C_{[AC]} = \frac{k_1\, C_A\, C_C}{k_2 + k_3}$$

Putting this value of $C_{[AC]}$ in Equation (4.9) we have:

$$r = \frac{k_1 k_3\, C_A\, C_C}{k_2 + k_3} = k\, C_A C_C$$

where,
$$k = \frac{k_1 k_3}{k_2 + k_3} \tag{4.10a}$$

The above equation clearly shows that reaction rate depends on catalyst concentration. However, for a given catalyst concentration C_C remains constant since catalyst remains unchanged at the end of the reaction. In that case equation (4.10a) becomes.

$$r = \frac{k_1 k_3\, C_A\, C_C}{k_2 + k_3} = k' C_A$$

where,
$$k' = \frac{k_1 k_3 C_C}{k_2 + k_3} \tag{4.10b}$$

Example 4.1 Preparation of sulphur-tri-oxide (SO_3) in presence of catalyst nitric oxide (NO).

$$2SO_2(g) + O_2(g) \xrightarrow{[NO]} 2SO_3(g)$$

Step-I $2NO(g) + O_2(g) \rightarrow 2NO_2(g)$

Step-II $[SO_2 + NO_2(g) \rightarrow SO_3(g) + NO(g)] \times 2$

$$2SO_2(g) + O_2(g) \rightarrow 2SO_3(g)$$

This is a gas phase homogeneous reaction.

Example 4.2 Oxidation of thiosulphate by ceric ion, catalysed by iodide ion

$$2Ce^{4+} + 2S_2O_3^{2-} \xrightarrow{[I^-]} 2Ce^{3+} + S_4O_6^{2-}$$

Step-I $2Ce^{4+} + 2I^- \rightarrow 2Ce^{3+} + I_2$

Step-II $I_2 + 2S_2O_3^{2-} \rightarrow 2I^- + S_4O_6^{2-}$

$$2Ce^{4+} + 2S_2O_3^{2-} \rightarrow 2Ce^{3+} + S_4O_6^{2-}$$

This is an example of liquid phase homogeneous reaction.

4.6 Theory of heterogeneous catalysis system

In any heterogeneous catalyst system, usually metal or metal oxide is chosen as catalyst, i.e., catalyst remains in solid phase. Reactant molecules are adsorbed on the catalyst surface and undergo chemical reactions. The product, thus formed, is desorbed from the catalyst surface. If the reactants are gases, the following processes may occur:

Step-I Diffusion of reactants to solid catalyst surface. The process is very fast.

Step-II Adsorption of reactants on the catalyst surface.

Step-III Reaction takes place between reactants.

Step-IV Desorption and diffusion of the products from the catalyst surface.

Diffusion of gases occurs very fast and (adsorption $\rightleftharpoons$ desorption) equilibrium sets up very quickly. So the overall reaction mostly depends on the fraction of catalyst surface covered by reactants. If p and θ be the reactant gas pressure and fraction of surface covered by reactants, we have:

$$-\frac{dp}{dt} \propto \theta$$

Or,

$$-\frac{dp}{dt} = k'\theta$$

According to Langmuir concept, reactant molecules form unimolecular film thickness on catalyst surface. So, θ (fraction of surface covered by reactant molecules) is related to gas pressure (p) according to the following equation:

$$\theta = \frac{ap}{1+ap}, \quad \text{where, } a \text{ is a constant.}$$

Thus the rate equation becomes:

$$-\frac{dp}{dt} = k'\theta = \frac{k'ap}{1+ap}$$

Or,
$$-\frac{dp}{dt} = \frac{k'ap}{1+ap} \tag{4.11}$$

Case – I If adsorption is small, gas pressure (p) on catalyst surface is also very low and hence, $(1 + ap) \approx 1$. Thus Equation (4.11) becomes:

$$-\frac{dp}{dt} = k'ap \tag{4.12}$$

The catalysed reaction follows 1st order kinetics.

Examples (i) Decomposition of hydrogen iodide (HI) on platinum surface.
(ii) Decomposition of phosphine on glass.

Case – II If the adsorption is strong, gas pressure (p) on catalyst surface is very high and it may be approximated that $(1 + ap) \approx ap$. Thus Equation (4.11) becomes:

$$-\frac{dp}{dt} = \frac{k'ap}{ap} = k' \tag{4.13}$$

The catalysed reaction follows Zero-order Kinetics.

Examples (i) Dissociation of hydrogen iodide (HI) on gold surface.
(ii) Dissociation of ammonia on tungsten.

4.6.1 Bimolecular reaction: both reactants are adsorbed

Bimolecular reaction: In the above section unimolecular heterogeneous reaction is discussed. Heterogenously catalysed bimolecular reaction is more complicated.

$$A + B \xrightarrow{\text{Catalyst}} \text{Products + Catalyst}$$

Mechanism Both the reactants are believed to be adsorbed on catalyst surface and readily form a transition state complex, which is ultimately desorbed and breaks into products.

The rate equation is given by:

$$-\frac{dp}{dt} \propto \theta_1\theta_2 \tag{4.14}$$

where, θ_1 = Fraction of the surface covered by the reactant A

θ_2 = Fraction of the surface covered by the reactant B

p = Total pressure = partial pressure of the reactant $A(p_1)$
+ partial pressure of the reactant $B(p_2)$.

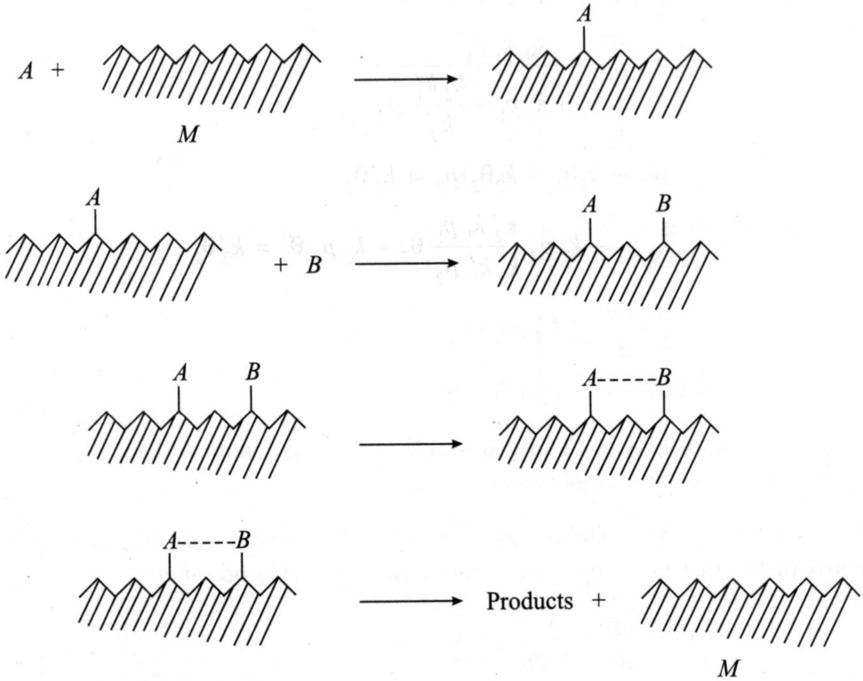

According to the principle of adsorption:

Rate of adsorption of reactant 'A', i.e., $r_a \propto (1 - \theta_1 - \theta_2)p_1$
= Rate of desorption of reactant 'A', i.e., $r_d \propto \theta_1$

Rate of adsorption of reactant 'B', i.e., $r'_a \propto (1 - \theta_1 - \theta_2)p_2$
= Rate of desorption of reactant 'B', i.e., $r'_d \propto \theta_2$

$$\left.\begin{aligned} k_1(1-\theta_1-\theta_2)p_1 &= k_1'\theta_1 \\ \text{and} \quad k_2(1-\theta_1-\theta_2)p_2 &= k_2'\theta_2 \end{aligned}\right\} \tag{4.15}$$

where, $(1 - \theta_1 - \theta_2)$ = Fraction of the surface still vacant for further adsorption.

$$\frac{k_1 p_1}{k_2 p_2} = \frac{k_1'\theta_1}{k_2'\theta_2}$$

Or,
$$\theta_1 = \frac{k_2' k_1 p_1}{k_2 k_1' p_2}\theta_2$$

and
$$\theta_2 = \frac{k_2 k_1' p_2}{k_2' k_1 p_1}\theta_1$$

Putting these values of θ_1 and θ_2 in the 1st and 2nd equations of Equation (4.15), we have:

$$k_1\left(1-\theta_1-\frac{k_1' k_2 p_2}{k_2' k_1 p_1}\theta_1\right)p_1 = k_1'\theta_1$$

$$\theta_1 = \frac{k_1 p_1}{k_1' + k_1 p_1 + \frac{k_2 k_1'}{k_2'} p_2} \tag{4.16}$$

$$(k_2 - k_2\theta_1 - k_2\theta_2)p_2 = k_2'\theta_2$$

$$k_2 p_2 - k_2 p_2 \frac{k_2' k_1 p_1}{k_2 k_1' p_2}\theta_2 - k_2 p_2 \theta_2 = k_2'\theta_2$$

Or,
$$\theta_2 = \frac{k_2 p_2}{k_2' + k_2 p_2 + \frac{k_2'}{k_1'} k_1 p_1} \tag{4.17}$$

Thus the rate equation [Equation (4.14)] becomes:

$$-\frac{dp}{dt} \propto \theta_1\theta_2 = k\,\theta_1\theta_2$$

Or,
$$-\frac{dp}{dt} = \frac{k(k_1 p_1)(k_2 p_2)}{\left[k_1' + k_1 p_1 + \frac{k_2 k_1'}{k_2'} p_2\right]\left[k_2' + k_2 p_2 + \frac{k_2' k_1}{k_1'} p_1\right]}$$

Case – I If adsorption of both the gases are very weak, i.e., p_1 and p_2 are very low:

$$k_1' + k_1 p_1 + \frac{k_2 k_1'}{k_2'} p_2 \approx k_1' \quad \text{and} \quad k_2' + k_2 p_2 + \frac{k_2' k_1}{k_1'} p_1 \approx k_2'$$

Thus the rate equation becomes:

$$-\frac{dp}{dt} = \frac{k\,k_1 k_2}{k_1' k_2'} p_1 p_2 \tag{4.18}$$

Thus the reaction follows 1st order kinetics with respect to each reactant.

Examples (a) Hydrogenation of ethylene on copper surface.
(b) Formation of nitrogen pentoxide from nitric oxide and oxygen on glass.

Case – II Reactant *A* is more strongly adsorbed than reactant *B*, i.e. p_2 can be neglected with respect to p_1: Thus the rate equation becomes:

$$-\frac{dp}{dt} = \frac{k\,k_1 k_2 p_1 p_2}{(k_1' + k_1 p_1)\left[k_2' + \frac{k_2' k_1}{k_1'} p_1\right]} = \frac{k\,k_1 k_2}{k_1' k_2'} \frac{p_1 p_2}{\left[1 + \frac{k_1}{k_1'} p_1\right]^2}$$

Or, $$-\frac{dp}{dt} = \frac{k_0 p_1 p_2}{(1+ap_1)^2} \tag{4.19}$$

where, $$k_0 = \frac{k\,k_1\,k_2}{k_1'\,k_2'} \quad \text{and} \quad a = \frac{k_1}{k_1'}$$

Example Reaction between carbon-di-oxide and hydrogen on platinum surface.

Case – III Reactant *A* is very strongly adsorbed and $\theta_1 \approx 1$ hence, $p_1 >> p_2$

Thus the rate equation becomes:

$$-\frac{dp}{dt} = k\theta_2 = \frac{k\,k_2\,p_2}{\left[k_2' + \frac{k_2'\,k_1}{k_1'}p_1\right]} \approx \frac{k\,k_2\,k_1'}{k_2'\,k_1}\frac{p_2}{p_1} \quad \left[\text{since, } \frac{k_2'\,k_1}{k_1'}p_1 >> k_2'\right]$$

Or, $$-\frac{dp}{dt} = k_0'\frac{p_2}{p_1}$$

where, $$k_0' = \frac{k\,k_2\,k_1'}{k_2'\,k_1} \tag{4.20}$$

Example Reaction between carbon monoxide and oxygen on platinum surface.

4.6.2 Bimolecular reaction: one reactant is absorbed

The reaction kinetics is same as that of unimolecular reaction since the rate of reaction is independent of the unabsorbed reactant *B*.

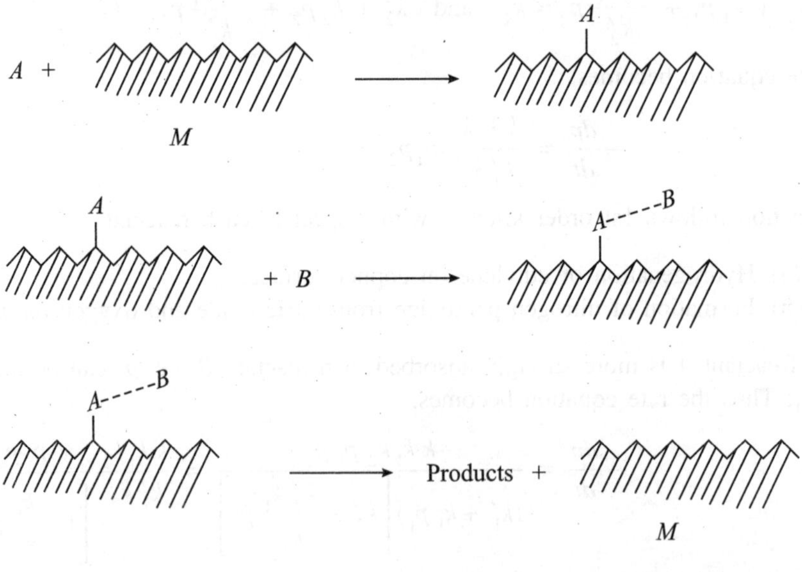

4.7 Primary Kinetic salt effect

In the case of ionic reactions the rate constant depends on the ionic strength (I) of the solution. This effect of ionic strength (I) on rate constant is known as primary kinetic salt effect, established by Brönsted and Bjerrum.

Consider the following bimolecular reaction:

$$A^{Z_1} + B^{Z_2} \underset{k_2}{\overset{k_1}{\rightleftharpoons}} [AB]^{Z_1 + Z_2} \xrightarrow{k_3} \text{Products}$$

If the rate equation follows 1st order kinetics with respect to each reactant, the rate of the reaction is given by:

$$\text{Rate} = \frac{d\,[\text{Products}]}{dt} = k\, C_A z_1\, C_B z_2 \tag{4.21}$$

Again, from the steady-state concept, we have:
Rate of formation of T.S. complex $\{[AB]\}$ = Rate of dissociation of T.S. complex.

Thus we have:

$$k_1\, a_A z_1\, a_B z_2 = (k_2 + k_3)\, a_{[AB]}$$

Or,
$$C_{[AB]} = \frac{k_1}{(k_2 + k_3)} \frac{(C_A z_1\, f_A z_1)(C_B z_2\, f_B z_2)}{f_{[AB]}}$$

$$\text{Rate of formation of product} = \frac{d\,[\text{Products}]}{dt} = k_3 C_{[AB]}$$

$a_A z_1$, $a_B z_2$ and $a_{[AB]}$ are the activity terms of reactants A, B and T.S. complex respectively.

$$\frac{d\,[\text{Products}]}{dt} = k_3\, C_{[\text{AB}]} = \frac{k_1 k_3}{(k_2 + k_3)} (C_A z_1\, C_B z_2) \frac{f_A z_1\, f_B z_2}{f_{[AB]}} \tag{4.22}$$

Comparing Equations (4.21) and (4.22), we have:

$$k = k' \frac{f_A z_1\, f_B z_2}{f_{[AB]}\, z_1 + z_2}$$

where,
$$k' = \frac{k_1 k_3}{(k_2 + k_3)}$$

Taking $\log_{10}$ on both sides, we get:

$$\log k = \log k' + [\log f_A z_1 + \log f_B z_2 - \log f_{[AB]} z_1 + z_2] \tag{4.23}$$

Again from Debye-Hückel relation, we have:

$$\log f = -0.509\ Z^2 \sqrt{I} \ \text{(at 25°C)}.$$

where, I = Ionic strength of the solution and Z = valency of the ion.

$f_A z_1 \cdot f_B z_2$ and $f_{[A+B]} z_1 + z_2$ are the activity coefficients of reactants A, B and T.S. complex respectively. So, we can write that:

$$\log f_A z_1 = -0.509\ z_1^2 \sqrt{I}$$

$$\log f_B z_2 = -0.509\ z_2^2 \sqrt{I} \quad \text{and}$$

$$\log f_{[AB]} z_1 + z_2 = -0.509\ (z_1 + z_2)^2 \sqrt{I}$$

where, Z_1 = valency of the cation and Z_2 = valency of the anion.

$$[\log f_A z_1 + \log f_B z_2 - \log f_{[AB]} z_1 + z_2] = -0.509\ [z_1^2 + z_2^2 - (z_1 + z_2)^2] \sqrt{I}$$

$$= 1.018\ z_1 z_2 \sqrt{I}$$

Valency of the T.S. complex is $(Z_1 + Z_2)$.

So, $[AB] = [AB]^{z_1 + z_2}$, $f_{[AB]} = f_{[AB]} z_1 + z_2$ and $a_{[AB]} = a_{[AB]} z_1 + z_2$

Thus the Equation (4.23) becomes:

$$\log k = \log k' + 1.018\ Z_1 Z_2 \sqrt{I} \tag{4.24}$$

The Equation (4.24) is known as Brönsted-Bjerrum equation.

The plot of log k against $\sqrt{I}$ yields a straight line of slope $(1.018\ Z_1 Z_2)$.

Case – I Both the reactants (A and B) are positively charged, or negatively charged.
Slope = $(1.018\ Z_1 Z_2)$ is positive since Z_1 and Z_2 are of same sign.

Example $[NiCl_4]^{2-} + 4CN^- \rightarrow [Ni(CN)_4]^{2-} + 4Cl^-$

Case – II The reactants (A and B) are of oppositely charged, i.e. $(Z_1 \times Z_2)$ is negative and hence the slope $(1.018\ Z_1 Z_2)$ is also negative.

Example $I_2 + 2S_2O_3^{2-} \rightarrow 2I^- + S_4O_6^{2-}$

If one of the reactants (A or B) is neutral, Z_1 or Z_2 is zero and the magnitude of the slope $(1.018\ Z_1 Z_2)$ is zero, which means that the rate constant of the reaction is independent of ionic strength [I].

Example $\underset{\text{(Sucrose)}}{C_{12}H_{22}O_{11}} + H^+ \rightarrow \underset{\text{(Glucose)}}{C_6H_{12}O_6} + \underset{\text{(Fructose)}}{C_6H_{12}O_6}$

5

ELECTROCHEMISTRY

5.1 Theory of electrolytes

For an electrolytic solution the osmotic pressure π is given by:

$$\pi = icRT \tag{5.1}$$

where c = molarity of the solution and i = Van't Hoff factor.

S. Arrhenius proposed that total number ions and neutral molecules present per mole of a solute constitute the factor i. Consider an electrolyte AB which dissociates as follows:

$$\underset{(1-\alpha)}{A - B} \rightleftharpoons \underset{(\nu_+\alpha)}{\nu_+ A^{z+}} + \underset{(\nu_-\alpha)}{\nu_- B^{z-}}$$

Where α is the degree of dissociation. Then the factor i is given by:

$$i = (1 - \alpha) + \nu_+\alpha + \nu_-\alpha = 1 + (\nu - 1)\alpha,$$

where, $\nu = \nu_+ + \nu_-$

Or, $$i = 1 + (\nu - 1)\alpha \tag{5.2}$$

Where α is the degree of dissociation of electrolyte AB.

5.1.1 Shortcomings of Arrhenius theory

(1) According to Arrhenius theory, ions in a solution behave as ideal gas molecules. Thus ions execute chaotic motion in solution. Interionic attraction is not considered. In actual practice strong electrostatic force of attraction exists among the ions, as a result of which, each ion is surrounded by a definite number of oppositely charged ions, forming an ionic

atmosphere. This ionic atmosphere, surrounding each ion, restricts the chaotic motion of that ion.

(2) According to Arrhenius theory the dissociation constant of any strong electrolyte is not constant but changes with concentration. In actual practice every strong electrolyte remains in completely dissociated form at any concentration.

(3) Actual degree of dissociation of any electrolyte is much higher than that calculated from Equation (5.2).

5.1.2 Activity (*a*) and activity coefficient (γ) of electrolytes

Electrolytes are dissociated into ions in solution. The chemical potentials of a positive and a negative ions are given by:

$$\left.\begin{aligned} \mu_+ &= \mu_+^0 + RT\ln a_+ = \mu_+^0 + RT\ln m_+ + RT\ln\gamma_+ \\ \mu_- &= \mu_-^0 + RT\ln a_- = \mu_-^0 + RT\ln m_- + RT\ln\gamma_- \end{aligned}\right\} \quad (5.3)$$

Where, $a_+ = m_+ . \gamma_+$ and $a_- = m_- . \gamma_-$. m_+ and m_- are the molalities of positive ion and negative ion respectively.

Consider an electrolyte *AB*, which is dissociated in solution as follows:

$$AB \rightarrow \nu_+ A^{z+} + \nu_- B^{z-}$$

Molality → (m) $(\nu_+ m)$ $(\nu_- m)$

So, $m_+ = (\nu_+ m)$ and $m_- = (\nu_- m)$

If n_1 and n_2 are the number of moles of solvent and electrolyte respectively, we can write that:

$$n_1 d\mu_1 + n_2 d\mu_2 = 0$$

Again we have: $d\mu_1 = RT\,\mathrm{d}\ln a_1$ and $d\mu_2 = RT\,\mathrm{d}\ln a_2$

Or, $n_1\,\mathrm{d}\ln a_1 + n_2\,\mathrm{d}\ln a_2 = 0$

For 1000 g of solvent, $n_1 = \left(\frac{1000}{M_1}\right)$ and $n_2 = m$, where, M_1 = Molecular weight of the solvent.

Putting the values of n_1 and n_2 in the above equation, we get:

$$\frac{1000}{M_1}\mathrm{d}\ln a_1 + m\,\mathrm{d}\ln a_2 = 0 \quad (5.4)$$

As the electrolyte remains in completely dissociated form in solution, we can write that:

$$\frac{1000}{M_1}\mathrm{d}\ln a_1 + m_+\mathrm{d}\ln a_+ + m_-\mathrm{d}\ln a_- = 0 \quad (5.5)$$

From Equations (5.4) and (5.5) we get:

$$m_+\mathrm{d}\ln a_+ + m_-\mathrm{d}\ln a_- - m\,\mathrm{d}\ln a_2 = 0$$

Or, $(\nu_+ m)\mathrm{d}\ln a_+ + (\nu_- m)\mathrm{d}\ln a_- - m\,\mathrm{d}\ln a_2 = 0$

Or, $\nu_+\,\mathrm{d}\ln a_+ + \nu_-\mathrm{d}\ln a_- - \mathrm{d}\ln a_2 = 0$

Or, $$\text{dln}\frac{(a_+^{\nu_+})(a_-^{\nu_-})}{a_2} = 0$$

On integration, we get:

$$\ln\frac{(a_+^{\nu_+})(a_-^{\nu_-})}{a_2} = \text{Constant}$$

Or, $$\frac{(a_+^{\nu_+})(a_-^{\nu_-})}{a_2} = \text{Constant} = K$$

The unit is so chosen that the value of the constant is unity.

So, $$a_2 = (a_+^{\nu_+})\ (a_-^{\nu_-}) \tag{5.6}$$

If $a_\pm$ is the mean activity, we have:

$$(a_\pm)^\nu = (a_+^{\nu_+})\ (a_-^{\nu_-})$$

Or, $$a_2 = (a_\pm)^\nu$$

Similarly, we can write that:

$$m_2 = m = (m_\pm)^\nu \quad \text{and} \quad \gamma_2 = (\gamma_\pm)^\nu$$

$$a_2 = (a_\pm)^\nu,\ m = (m_\pm)^\nu \quad \text{and} \quad \gamma_2 = (\gamma_\pm)^\nu \tag{5.7}$$

$$m = (m_+^{\nu_+})\ (m_-^{\nu_-}) = (\nu_+ m)^{\nu_+}\ (\nu_- m)^{\nu_-} = (\nu_+^{\nu_+})(\nu_-^{\nu_-})\, m^\nu$$

Again, $$m = (m_\pm)^\nu$$

Hence, $$(m_\pm)^\nu = (\nu_+^{\nu_+})\ (\nu_-^{\nu_-})\, m^\nu$$

Or, $$m_\pm = (\nu_+^{\nu_+}\ \nu_-^{\nu_-})^{\frac{1}{\nu}}\ \text{m}$$

Or, $$m_\pm = \text{L.m} \tag{5.8}$$

where, $$L = (\nu_+^{\nu_+}\ \nu_-^{\nu_-})^{\frac{1}{\nu}}$$

Thus we can write that:

$$a_\pm = m_\pm\ \gamma_\pm = L\ \gamma_\pm m \tag{5.9}$$

or uni-uni-valent salt like NaCl, $\nu_+ = 1$, $\nu_- = 1$ and $\nu = 2$. Hence, $L = 1$.
For di-uni-valent salt like, $CaCl_2$, $\nu_+ = 2$, $\nu_- = 1$ and $\nu = 3$.

Hence, $$L = (2^2 . 1^1)^{\frac{1}{3}} = 1.588$$

Thus for a solution containing a mixture of salts, it is difficult to express the concentration of solution by activity term. Usually the concentration of an electrolytic solution containing a mixture of salts, is expressed by *ionic strength* (I), which is expressed as:

$$I = \frac{1}{2}\sum m_i\, z_i^2 \tag{5.10}$$

where, m_i = molality of ith ionic species and

z_i = valency ith ionic species.

5.2 Electrolysis

5.2.1 Conductor

Any medium, which conducts electricity, is called conductor.

Type of conductors There are two types of conductors:
(a) Electronic conductor, (b) Electrolytic conductor.

(a) Electronic conductor The material which conducts electricity through flow of electrons, is called electronic conductor, e.g., copper, silver etc.

(b) Electrolytic conductor The material, which conducts electricity in solution, where flow of ions is the origin of current, is called electrolytic conductor, e.g., sodium chloride.

Electrochemical reaction When a chemical reaction takes place at the interface of an electrode (electronic conductor) and electrolytic solution, the reaction is called an electrochemical reaction.

For example: $Cu^{2+} + 2e \rightarrow Cu$

The process is known as electrolysis.

5.2.2 Faraday's Laws

Laws of electrolysis If current is passed through an electrolytic solution metals are deposited on the respective electrodes.

1st Law Amount of substance deposited at an electrode is directly proportional to the quantity of electricity passed through it. Mathematically it can be written as:

$$Q \propto m$$

Or, $$Q = F.m, \text{ where,}$$

Q = Quantity of electricity passed.

F = Proportionality constant, also known as Faraday constant.

m = mass of substance deposited in terms of gm-equivalent.

If m = 1 gm-equiv. $Q = Q_0 = F$. Thus F is the amount of electricity required to deposit 1 gm-eqiv. of substance. It is constant for any electrolyte.

2nd Law Same quantity of electricity is required to deposit 1 gm-equiv. of any substance at any electrode. It is called 1 Faraday (1F) of electricity. The value of 'F' is given by:

$$1\,\text{F} = 96{,}500 \text{ Coulomb/gm-equiv.}$$

Electrochemical Equivalent (ECE)

As we know, $Q = \text{Fm}$

If $Q = 1\,\text{C},\ m = \left(\frac{1}{F}\right)$ gm.equiv. $= \left(\frac{1000\,E}{F}\right)$mg,

where, E = Equivalent weight of electrolyte.

Amount of substance deposited at one electrode by passing 1 *coulomb of electricity into its solution, is called the electrochemical equivalent* (ECE) *of that substance.*

ECE values of silver, copper and hydrogen have been found 1.118 mg, 0.3293 mg and 0.010446 mg respectively. In the case of silver, 1.118 mg silver is deposited for 1 Coulomb of charge. Thus,

$$1.118 = \frac{1000 \times E_{\text{Ag}}}{F} = \frac{1000 \times 107.87}{F}$$

Or, $$F = \frac{1000 \times 107.87}{1.118} \text{ C/gm.eqiv.} = 96{,}484 \text{ C/gm.eqiv.}$$

In the case of copper, 0.3293 mg copper is deposited for 1 Coulomb of charge.

Thus, $$0.3293 = \frac{1000 \times E_{\text{Cu}}}{F} = \frac{1000 \times 31.77}{F}$$

Or, $$F = \frac{1000 \times 31.77}{0.3293} \text{ C/gm.eqiv.} = 94{,}481 \text{ C/gm.eqiv.}$$

In the case of hydrogen, 0.010446 mg hydrogen is deposited for 1 Coulomb of charge.

Thus, $$0.010446 = \frac{1000 \times E_{\text{H}_2}}{F} = \frac{1000 \times 1.00797}{F}$$

Or, $$F = \frac{1000 \times 1.00797}{0.010446} \text{ C/gm.eqiv.} = 94{,}494 \text{ C/gm.eqiv.}$$

The average value of F is taken as 96,500 C/gm-equiv.

For uni-valent ion like H^+ equivalent weight is equal to molecular weight. Thus 1 mole of a uni-valent ion carries 96,500 Coulomb. As 1 mole uni-valent ion contains 6.023×10^{23} number of ions, which is equivalent 6.023×10^{23} number of electrons, since charge carried by a uni-valent ion is equal to that carried by an electron.

Thus, 6.023×10^{23} number of electrons carry 96,500 C of charge.

Or, 1 electron carries $(96{,}500/6.023 \times 10^{23})$ Coulomb of charge

$= 1.602 \times 10^{-19}$ C of charge $= 4.8063 \times 10^{-12}$ e.s.u. of charge.

5.3 Electrical conductance of electrolytes

For any conductor relation between resistance and resistivity is given by:

$$R = \rho \frac{l}{A}$$

R = Resistance of the conductor (Ω)

ρ = Resistivity of the conductor (Ω-cm)

A = Area of the electrode or conductor (cm^2)

l = Length of the conductor or distance of separation of two electrodes (cm)

$$\frac{1}{R} = G = \text{Conductance } \Omega^{-1}$$

Similarly, $\left(\frac{1}{\rho}\right) = \kappa$ = Conductivity or specific conductance (Ω^{-1} cm^{-1})

Conductivity data of some selected substances:

$$\kappa_{Ag} = 615{,}000\ \Omega^{-1}.\text{cm}^{-1}$$

$$\kappa_{Al} = 360{,}000\ \Omega^{-1}.\text{cm}^{-1}$$

$$\kappa_{NaCl(l)} = 3.40\ \Omega^{-1}.\text{cm}^{-1} \quad \text{[at 750°C]}$$

$$\kappa_{KCl(1N)} = 9.789 \times 10^{-2}\ \Omega^{-1}.\text{cm}^{-1}$$

$$\kappa_{water} = 4.3 \times 10^{-8}\ \Omega^{-1}.\text{cm}^{-1}$$

It is evident from the above table that, conductivity of an electronic conductor is much higher than that of an electrolytic conductor.

5.3.1 Conductivity

As we know

$$G = \kappa \frac{A}{l}$$

Or,

$$\kappa = G \frac{l}{A}$$

When l = 1 cm and A = 1 cm^2, κ = G. Thus specific conductance or conductivity of an electrolytic solution is nothing but the conductance of that electrolytic solution containing electrodes of area 1 cm^2 separated by a distance of 1cm.

Conductivity of an electrolyte depends on concentration of that electrolyte. Figure 5.3.1 shows the dependence of conductivity on concentration of different electrolytes.

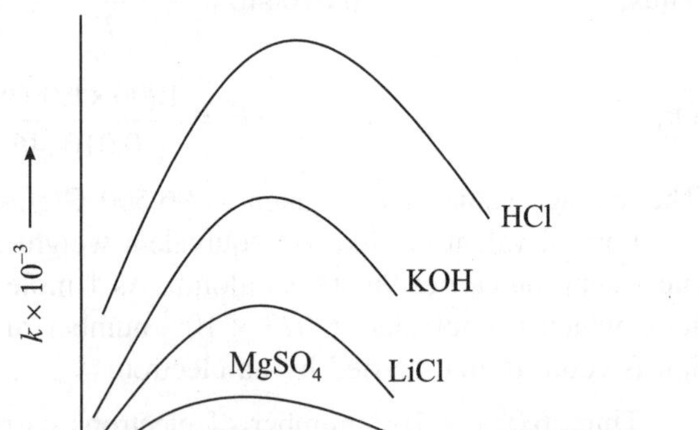

Fig. 5.3.1 Conductivity versus concentration plot of different electrolytes

From the Figure 5.3.1, it is evident that conductivity of a strong electrolyte increases with the increase in concentration and reaches a maximum. As we know strong electrolyte remains in completely dissociated form in solution, thus with the increase in electrolyte concentration number of ions *increases* and hence conductivity *increases*. Again each positive ion is surrounded by a negative ion atmosphere and vice versa. This ionic atmosphere appreciably *reduces* the velocity of the central ion. At higher concentration the effect of ionic atmosphere dominates over the effect of increase in ionic concentration and hence conductivity shows a decrease at higher concentration.

In the case of a weak electrolyte, with the increase in electrolytic concentration increase in ionic concentration is only marginal, since weak electrolyte is partially dissociated in solution. Furthermore the effect of ionic atmosphere is negligible due to low concentration of ions. Thus the concentration of electrolyte has negligible effect on its conductivity as is evident from the Figure 5.3.1.

5.3.2 Equivalent conductivity

Equivalent conductivity of an electrolyte is the conductance of such a volume of solution that contains 1 gm-equivalent of solute and the electrodes are 1 cm apart. It is represented as Λ_N or simply Λ. If v ml of an electrolytic solution of strength $c(N)$ contains 1 gm-equiv. of that electrolyte, Λ is given by:

$$\Lambda = \kappa . v \tag{5.11}$$

where κ is the conductivity of the solution.

Again from the definition of Normality, we can write that,

1000 ml of the solution contains c gm-equiv. of electrolyte.

or c gm-equiv. of electrolyte is present in 1000 ml of solution.

1 gm-equiv. of electrolyte is present in $\frac{1000}{c}$ ml of solution.

Hence, $$v = \frac{1000}{c} \text{ ml}$$

Thus Equation (5.11) becomes:

$$\Lambda = \frac{1000\,\kappa}{c} \; \frac{\Omega^{-1}\,\text{cm}^2}{\text{gm.eqiv.}} \tag{5.12}$$

Molar conductivity Molar conductivity of an electrolyte is the conductivity of a solution containing 1 mole of solute and the electrodes are 1 cm apart. It is represented as Λ_M, which is expressed as:

$$\Lambda_M = \frac{1000\,\kappa}{M} \; \frac{\Omega^{-1}\,\text{cm}^2}{\text{mole}} \tag{5.13}$$

Where M is the molarity of electrolyte solution.

As we have discussed earlier conductivity of strong electrolyte depends on concentration. So, equivalent conductivity (Λ) also depends on concentration. Figure (5.3.2) shows how Λ changes with concentration (c).

From the Figure 5.3.2, it is evident that the equivalent conductivity of a strong electrolyte decreases linearly but only slightly on increasing the concentration. The mathematical expression of this linear relationship was first proposed by *Kohlraush* as follows:

$$\Lambda_c = \Lambda_0 - A\sqrt{c} \qquad (5.14)$$

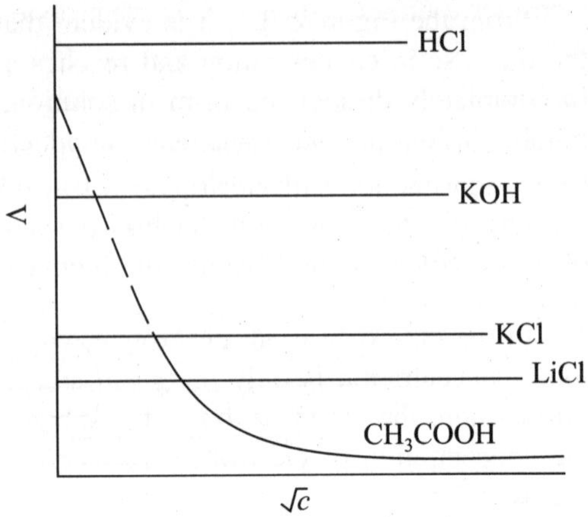

Fig. 5.3.2 Λ versus $\sqrt{c}$ plots of some selected electrolytes.

Where Λ_0 is the equivalent conductivity at infinite dilution, i.e. at $c \to 0$.

As we know from Equation (5.12) that:

$$v = \frac{1000}{c}$$

So, $$v \to \infty \quad \text{as } c \to 0$$

So, Λ_0 may also be represented as Λ_∞ i.e., equivalent conductivity at infinite dilution. A is a constant and Λ_c is the equivalent conductivity at any concentration c. The Equation (5.14) is valid only for strong electrolytes.

The above equation does not hold good for weak electrolyte due to partial dissociation of the latter. However, it is evident from the Figure 5.3.2 that, equivalent conductivity of weak electrolyte sharply increases with the increase in dilution and reaches a very high value, equal to that of a strong electrolyte, at infinite dilution.

Thus at infinite dilution both strong and weak electrolytes behave in a similar manner.

5.3.3 Kohlrausch's law

Bearing this observation in mind Kohlrausch proposed the following law:

At infinite dilution all interionic effects vanish and ions are moving independently. At this condition equivalent conductivity of any electrolyte is the sum of equivalent conductivities of cation and anion.

Mathematically it can be expressed as:

$$\Lambda_\infty = \Lambda_0 = \lambda_+ + \lambda_- \qquad (5.15)$$

Where, λ_+ and λ_- are the equivalent conductivities of cation and anion respectively. As all interionic effects vanish at infinite dilution, the total current (I) carried by an electrolyte is the sum of currents carried by cation and anion, i.e., I_+ and I_- respectively.

Thus we can write that:

$$I = I_+ + I_- \text{ (at infinite dilution)} \qquad (5.16a)$$

Consider a uni-uni-valent strong electrolyte AB, which dissociates in solution as follows:

$$AB \rightarrow A^+ + B^-$$

Concentration c(N) c(N) c(N)

Thus concentration of cation (c_+) in gm-equiv./c.c = concentration of anion (c_-) in gm-equiv/c.c. = concentration of electrolyte (c) in gm-equiv/c.c.

Let, q_+ and q_- are the velocities of cation and anion respectively in cm/s. Let A be the area of cross-section through which the ions are passing [see Figure 5.3.3].

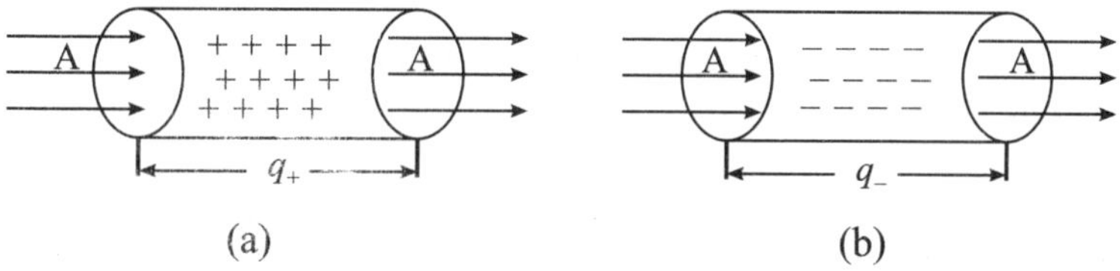

Figure 5.3.3 (a) volumetric flow rate of cation, (b) volumetric flow rate of anion.

Let us consider two cylinders of length q_+ and q_- respectively. Thus $(A \,.\, q_+)$ is the volume of cations (in c.c.or ml) flowing per sec through the area of cross-section A [see Figure 5.3.3]. Similarly volumetric flow rate of anion through the area of cross-section A is $(A \,.\, q_-)$ c.c/s.

So, mass flow rate of cation = (Aq_+) c.c/s $\times$ c_+ gm-equiv/c.c. = $(Aq_+ \, c_+)$ gm-equiv./s.

and mass flow rate of anion = (Aq_-) c.c/s $\times$ c_- gm-equiv/c.c. = $(Aq_- \, c_-)$ gm-equiv./s.

We know that 1 gm-equiv. of ions carry F Coulomb of charge (C).

Thus current carried by cation = $I_+ = (Aq_+ \, c_+)$ (gm.-equiv./s) $\times$ F (C/gm.-equiv.)

$= (Aq_+ \, c_+ . F)\ C/s = (Aq_+ \, c_+ . F)$ Amp.

$= (Aq_+ \, c . F)$ Amp [since, $c_+ = c$]

Similarly, current carried by anion = $I_- = (Aq_- \, c_-)$ (gm.-equiv./s) $\times$ F (C/gm.-equiv.)

$= (Aq_- \, c_- .\mathrm{F})\ C/s = (Aq_- \, c_- . F)$ Amp.

$= (Aq_- \, c . F)$ Amp. [since $c_- = c$]

$$I_+ = (Aq_+ \, c . F) \text{ Amp. and } I_- = (Aq_- \, c . F) \text{ Amp.} \tag{5.16b}$$

If E is the e.m.f. of electrolytic cell and l is the distance of separation of two electrodes, $\left(\frac{E}{l}\right)$ is the field gradient of the cell. The velocity of any ion largely depends on field gradient. Higher the value of $\left(\frac{E}{l}\right)$ velocity of an ion increases proportionately. Thus velocity of any ion is directly proportional to the field gradient. Hence, we can write that:

$$q_+ \propto \frac{E}{l} \quad \text{and} \quad q_- \propto \frac{E}{l}$$

Or, $$q_+ = u_+\frac{E}{l} \quad \text{and} \quad q_- = u_-\frac{E}{l}$$

Or, $$u_+ = \frac{q_+}{\left(\frac{E}{l}\right)} \quad \text{and} \quad u_- = \frac{q_-}{\left(\frac{E}{l}\right)} \qquad (5.16c)$$

The proportionality constants u_+ and $u\_$ are called the absolute mobilities of cation and anion respectively.

Definition of mobility The velocity of an ion under unit field gradient is called mobility of that ion. Combining Equations (5.16b) and (5.16c), we get:

$$I_+ = AFc\left(\frac{E}{l}\right)u_+$$

and $$I_- = AFc\left(\frac{E}{l}\right)u_- \qquad (5.16d)$$

Combining Equations (5.16a) and (5.16d), we get:

$$I = AFc\left(\frac{E}{l}\right)(u_+ + u_-) = \left(\frac{AcE}{l}\right)(Fu_+ + Fu_-)$$

$$U_+^\infty = u_+F$$

Unit: $$\frac{(\text{cm/s})}{(\text{Volt/cm})}\frac{C}{\text{gm-eqiv.}} = \text{cm}^2\frac{\text{Amp}}{\text{Volt}}\frac{1}{\text{gm-eqiv.}} = \frac{\Omega^{-1}\,\text{cm}^2}{\text{gm-eqiv.}}$$

Thus U_+^∞ is nothing but the equivalent conductivity of the cation, i.e. λ_+, or, $U_+^\infty = \lambda_+$ Similarly, $U_-^\infty = u_-F = \lambda_-$. U_+^∞ and U_-^∞ are the limiting mobilities of cation and anion respectively. The limiting mobility is the mobility at infinite dilution. Thus the above equation becomes:

$$I = \left(\frac{AcE}{l}\right)(U_+^\infty + U_-^\infty) \qquad (5.16e)$$

Again, from Ohm's law, we have:

$$I = \frac{E}{R} = E.G = E.\kappa.\frac{A}{l} = \left(\frac{EA}{l}\right)\left(\frac{\Lambda}{v}\right) \quad [\text{since, } \Lambda = k.v]$$

Putting this value of I in the Equation (5.16e), we have:

$$\left(\frac{EA}{l}\right)\left(\frac{\Lambda}{v}\right) = \left(\frac{AcE}{l}\right)(U_+^\infty + U_-^\infty)$$

Or, $$\Lambda_\infty = (v.c)\,(U_+^\infty + U_-^\infty)$$

Or, $$\Lambda_\infty = (U_+^\infty + U_-^\infty) \qquad (5.16f)$$

since, v = Volume occupied by 1 gm-equiv. of ions, unit = cc/gm-equiv.
and c = Amount in gm-equiv. per unit volume, unit = gm-equiv/c.c.

Thus, $$v = \frac{1}{c}$$

Or, $$vc = 1$$

The Equation (5.16f) is same as Equation (5.15). Thus we have:

$$U_+^\infty = \lambda_+ \quad \text{and} \quad U_-^\infty = \lambda_-$$

The Equation (5.16f) is the mathematical expression of *Kohlrausch's law* for uni-uni-valent electrolyte like NaCl, CH_3COOH etc.

(a) As strong electrolyte always remains in dissociated form in solution, the Equation (5.16f) is valid at any concentration *c*.
Thus for strong electrolyte, we have:

$$\Lambda_c = (U_+ + U_-) \tag{5.16g}$$

(b) For weak electrolyte the dissociation is incomplete as shown below:

$$AB \rightleftharpoons A^+ + B^-$$

Concentration $\quad c(N) \quad \alpha c \quad \alpha c$

where α is the degree of dissociation.

Hence, $$c_+ = \alpha c \quad \text{and} \quad c_- = \alpha c$$

Putting these values of c_+ and c_- in the following equations, we get:

$$I_+ = AFc_+ q_+ \quad \text{and} \quad I_- = AFc_- q_- \tag{5.16h}$$

$$I_+ = AF(\alpha c)q_+ = (AF\alpha c)q_+ \quad \text{and} \quad I_- = AF(\alpha c)q_- = (AF\alpha c)q_-$$

Using these values of I_+ and I_- the Equation (5.16g) becomes:

$$\Lambda_c = \alpha(U_+ + U_-) \tag{5.16i}$$

(c) If *AB* is a multi-multi-valent strong electrolyte the dissociation of *AB* is given by:

$$AB \rightleftharpoons \nu_+ A^{z_+} + \nu_- B^{z_-}$$

Concentration $\quad c(N) \quad (\nu_+ c) \quad (\nu_- c)$

where ν_+ and ν_- are the number of moles of cations and anions respectively. Thus, we have:

$$c_+ = \nu_+ c \quad \text{and} \quad c_- = \nu_- c$$

Putting these values of c_+ and c_- in the Equation (5.16h), we get:

$$I_+ = AF(\nu_+ c)q_+ = (AFc)\nu_+ q_+ \quad \text{and} \quad I_- = AF(\nu_- c)q_- = (AFc)\nu_- q_-$$

Using these values of I_+ and I_- the Equation (5.16g) becomes:

$$\Lambda_c = (\nu_+ U_+ + \nu_- U_-) \tag{5.16j}$$

5.3.4 Mobility

Mobility of an ion is the velocity of that ion at unit field gradient. As an ion is moving, its motion is restricted by the surrounding oppositely charged ionic atmosphere. Thus the motion of an ion in an electrolytic solution is very much similar to the motion of a macroscopic sphere in a viscous medium. So Stoke's formula may be applied to correlate the ionic mobility and its radius as given below:

$$u_+ = \frac{Ze}{6\pi\eta r_+} \tag{5.17a}$$

Z = valency of the cation.

e = Electronic charge = 1.602×10^{-19} C.

η = Viscosity factor or resistance offered by the ionic atmosphere.

r_+ = radius of the cation.

u_+ = Mobility of the cation = $\left[\dfrac{q_+}{\left(\dfrac{E}{l}\right)}\right]$

Putting this value of u_+ in Equation (5.17a), we get:

$$q_+ = \frac{Ze}{6\pi\eta r_+}\left(\frac{E}{l}\right) \tag{5.17b}$$

Similarly for an anion, we have:

$$q_- = \frac{Ze}{6\pi\eta r_-}\left(\frac{E}{l}\right) \tag{5.17c}$$

To calculate absolute mobilities of several ions the value of driving force, i.e. $\left(\dfrac{E}{l}\right)$ is taken as unity. So, $\left(\dfrac{E}{l}\right)$ = 1 Volt/cm.

Hence, $$u_+ = q_+ = \frac{Ze}{6\pi\eta r_+}$$

and $$u_- = q_- = \frac{Ze}{6\pi\eta r_-} \tag{5.17d}$$

From Equation (5.17d) it is evident that:

$$u_+ \propto \frac{1}{r_+} \quad \text{and} \quad u_- \propto \frac{1}{r_-}$$

Or, $$U_+^\infty \propto \frac{1}{r_+} \quad \text{and} \quad U_-^\infty \propto \frac{1}{r_-}$$

Thus with the increase in ionic radius, mobility decreases. As ionic radius increases down the group, it is expected that:

$$U_{Rb^+}^\infty < U_{K^+}^\infty < U_{Na^+}^\infty < U_{Li^+}^\infty$$

but in actual practice the reverse is true, i.e.,

$$U^{\infty}_{Li^+} < U^{\infty}_{Na^+} < U^{\infty}_{K^+} < U^{\infty}_{Rb^+}$$

The explanation of the fact lies within the radius. In aqueous solution all the ions get hydrated. Thus the radius, given in the Equation (5.17d) is the hydrated radius, represented as r_H and in general mobility is given by:

$$U = u.F = \frac{ZeF}{6\pi\eta\, r_H} \tag{5.17e}$$

Smaller the ionic radius higher the hydration and hence higher the value of r_H. Thus it is expected that:

$$(r_H)_{Li^+} > (r_H)_{Na^+} > (r_H)_{K^+} > (r_H)_{Rb^+} \quad \text{and hence,}$$
$$U^{\infty}_{Li^+} < U^{\infty}_{Na^+} < U^{\infty}_{K^+} < U^{\infty}_{Rb^+}$$

Mobility and hydration number of ions

The actual radius of an ion is its crystal radius, represented as r_c, which can easily be computed from crystal unit cell dimension, e.g. unit cell of NaCl crystal is Face-Centred-Cubic or FCC structure. Thus we have, $r_c = \frac{a}{2\sqrt{2}}$, where a is the edge length of the unit cell cube. Thus r_c can be evaluated from the crystallographic analysis. r_H can be calculated from the Equation (5.17e).

$$\text{Volume of hydrated ionic cell} = \left(\frac{4}{3}\right)\pi r^3$$

$$\text{Volume of actual ionic cell} = \left(\frac{4}{3}\right)\pi r_c^3$$

$$\text{So, volume of hydration} = \left[\left(\frac{4}{3}\right)\pi r^3 - \left(\frac{4}{3}\right)\pi r_c^3\right] = V_H \times 10^{-30}\ \text{m}^3$$

Volume of one molecule of water is 30×10^{-30} m^3

$$\text{So, the hydration number} = n = \frac{V_H}{30}$$

r is the true radius of the ion and it is related to r_H as follows:

$$r = \frac{ZeF}{6\pi\eta U}\left(\frac{r_H}{r_c}\right) \tag{5.17f}$$

$F = 96.500$ C/gm.equiv.,

$e = 1.602 \times 10^{-19}$ C and $\left(\frac{E}{l}\right) = 1$ V/cm.

So, $e.\left(\frac{E}{l}\right) = 1.602 \times 10^{-19}$ J/cm $= 1.602 \times 10^{-12}$ ergs/cm.

$$r = \frac{1.602\times10^{-12}\times ZF}{6\times3.14\times U\eta}\left(\frac{r_H}{r_c}\right)\text{cm} = \frac{1.602\times10^{-4}}{6\times3.14}\left(\frac{ZF}{U\eta}\right)\left(\frac{r_H}{r_c}\right)\text{Å}$$

$$= \frac{0.820Z}{U\eta}\left(\frac{r_H}{r_c}\right)\text{Å}$$

Using these values of r and r_c, V_H can easily be calculated and hence also the hydration number n.

$$\eta \text{ of water} = 1.005\ cp = 1.005\times10^{-2}\ p.$$

Thus,

$$r = \frac{0.820Z}{U\eta\times10^{-2}}\left(\frac{r_H}{r_c}\right) = \left(\frac{82Z}{U\eta}\right)\left(\frac{r_H}{r_c}\right)\text{Å}$$

For Na^+:

$$U^{\infty}_{Na^+} = 50.10\ \Omega^{-1}\ \text{cm}^2$$

$$r = 3.3\ \text{Å},\quad r_c = 0.97\ \text{Å},\quad r_H = 1.83\ \text{Å}$$

$$V_H = \frac{4\pi}{3}(r^3 - r_c^3) \approx 150\ \text{Å}^3$$

So,

$$n = \frac{V_H}{30} = \frac{150}{30} = 5$$

Hydration number of other ions may be evaluated in the same way [see Table 5.3.7].

5.3.5 Effect of solvent on conductivity

Mobility or equivalent conductivity of an ion largely depends on the polarity of solvent. It is expected that higher the polarity of the solvent higher will be the degree of dissociation and hence higher will be the concentration of ions. Hence higher will be the mobility. Furthermore, higher the polarity of the solvent higher will be the effective ionic radius and hence lower will be the mobility. In any polar solvent these two opposing factors come into effect. Among the usual solvents water is the most polar solvent and it has been found that mobility of an ion is less in non-aqueous solution than in aqueous solution. However, it has been established that mobility of an ion (expect H^+ and OH^-) is much higher in acetone and hydrogen cyanide than any other solvent [see Table 5.3.6]. In a dilute solution the Equation (5.17e) becomes:

$$U_0\eta_0 = \frac{ZeF}{6\pi r} = \text{Constant} \tag{5.18}$$

Provided the effective radius r is constant for different solvents.

Thus the product of mobility (or equivalent conductivity) and viscosity of an ion or salt is constant. This is known as *Walden's rule*.

5.3.6 Mobility of hydroxonium (H_3O^+) and hydroxyl (OH^-) ions

Mobilities of cations and anions usually fall within 30 – 70 Ω^{-1} cm^2 with two exceptions, hydrogen ion (H^+) and hydroxyl ion (OH^-).

$$U^{\infty}_{H^+} = 350\ \Omega^{-1}\ \text{Cm}^2 \quad\text{and}\quad U^{\infty}_{OH^-} = 200\ \Omega^{-1}\ \text{cm}^2$$

In aqueous solution H^+ readily gets hydrated to form H_3O^+. Water acts as an excellent carrier to transfer H^+ ion from one part to another part as illustrated in the following:

$$H^+ + H_2O \rightarrow H_3O^+$$

$$H_3O^+ + H_2O \rightarrow H_2O + H_3O^+$$

$$H_3O^+ + H_2O \rightarrow H_2O + H_3O^+$$

$$\begin{array}{ccccccccccccc} H_2O^+ & & HO & & HO & & HO & & HO & & HO & & HO \\ | & \longrightarrow & | & \longrightarrow & | & \longrightarrow & | & \longrightarrow & | & \longrightarrow & | & \longrightarrow & | \\ H & & H & & H & & H & & H & & H & & H \end{array}$$

Thus in aqueous medium H^+ ion has very high mobility. This type of conductance is called relay or chain conductance. In nitrobenzene the relay conductance is hardly possible and hence mobility to H^+ ion is very less (only 23 Ω^{-1} cm^2) in nitrobenzene.

The high mobility of hydroxyl ion can be explained in the same way as shown below:

$$OH^- + H_2O \rightarrow [OH\cdots H{-}O{-}H]^-$$

Ion complex (A)

$$[OH\cdots H{-}O{-}H]^- + H_2O \rightarrow H_2O + [OH\cdots H{-}O{-}H]^-$$

(A)

In this case hydroxyl ion first forms an *ion complex* (*A*), which is then transferred to the next water molecule and hence the relay process continues. So here also the chain conductance is the primary reason for high mobility of hydroxyl ion. However, the size of H_3O^+ is much smaller than the ion complex '*A*', and hence the transfer of H_3O^+ from one water molecule to another is much easier and faster than the ion complex '*A*'. That is why, the mobility of hydroxonium ion is much higher then that of hydroxyl ion.

5.3.7 Relation between ionic mobility and concentration

The equivalent conductivity Λ_c is related to concentration (c) as follows:

$$\Lambda_c = \Lambda_0 - A\sqrt{c} = \Lambda_\infty - A\sqrt{c} \quad \text{[Equation (5.14)]}$$

For uni-uni-valent strong electrolyte:

$$\Lambda_c = U_+ + U_- \quad \text{and} \quad \Lambda_\infty = U_+ + U_- \quad \text{[Kohlrausch's law]}$$

Thus Equation (5.14) becomes:

$$\left.\begin{array}{ll} & U_+ = U_+^\infty - B_1\sqrt{c}, \quad U_- = U_-^\infty - B_2\sqrt{c}, \\ \text{and} & B_1 + B_2 = A \end{array}\right\} \qquad (5.19)$$

Thus it is evident that with the increase in ionic concentration mobility decreases. P. Debye and E.Hückel explained it with the help of ionic atmosphere theory, according to which each ion is surrounded by oppositely charged ionic atmosphere. This oppositely charged ionic atmosphere substantially reduces the velocity of the central ion and hence the absolute mobility of the latter decreases. The phenomenon is known as *electrophoretic effect*. Higher the concentration higher the density of ionic atmosphere and hence lower the mobility of the ion.

While moving an ion leaves the ionic atmosphere but a new one immediately develops surrounding the central ion. This destruction and creation of ionic atmosphere occurs due to simultaneous movement of ionic atmosphere and central ion in opposite direction. The time-lag between destruction and creation of ionic atmosphere is very short. The phenomenon is called *relaxation or asymmetry effect*. Thus the retardation of an ion occurs mainly due to *electrophoretic effect and asymmetry effect.*

Wien effect According to Deybe-Hückel theory the mobility decreases with the increase in concentration due to increase in density of ionic atmosphere, but at the same time the stability of ionic atmosphere depends on field intensity, i.e. on $\left(\frac{E}{l}\right)$. With the increase in field intensity the ionic atmosphere becomes more and more unstable and at a certain value of $\left(\frac{E}{l}\right)$, called critical field intensity, the ionic atmosphere no longer exists. The ions are moving freely as if there is no interionic attraction and under this circumstances, they have maximum mobility. The phenomenon was first established by M.Wien in 1927 and hence the name. It has been found that at a field gradient of 2,00,000 V/cm the equivalent conductivity of a salt at any concentration c is equal to Λ_∞ of that salt. Thus Wien effect merely substantiates the Dubye-Hückel theory.

Debye-Falkenhagen effect/Frequency effect/Dispersion of electrical conductivity

As we have discussed earlier in asymmetry effect there is a definite time-lag (τ) between destruction and creation of ionic atmosphere surrounding the central ion. If a periodic electric field is applied in an electrolytic solution, the existence of ionic atmosphere depends on frequency of *A/C* supply (ν). If $\tau < \left(\frac{1}{\nu}\right)$ the asymmetry effect exists but there is a frequency, called *critical frequency* ν_{cr}, at which point $\tau > \left(\frac{1}{\nu_{cr}}\right)$ and hence after destruction of one ionic atmosphere no new ionic atmosphere develops. Thus asymmetry effect vanishes at ν_{cr}. The phenomenon is known as *Debye-Falkenhagen effect*. Thus at $\nu > \nu_{cr}$, the equivalent conductivity of a salt increases sharply. The frequency ν_{cr} and τ are related according to the following equation:

$$\nu_{cr}\tau = 1 \quad \text{and} \quad \nu_{cr} = \frac{1}{\tau} = \frac{cz\Lambda_c}{71.3} \times 10^{10} \tag{5.20}$$

where, c = concentration of ion in (N), z = valency of ion and Λ_c = equivalent conductivity of electrolyte at concentration c.

For a uni-uni-valent salt at c = 0.001 (N) asymmetry effect vanishes at 10^9 Hz. Considering both electrophoretic effect and asymmetry effect L.Onsagar deduced an equation correlating equivalent conductivity (Λ_c) with concentration (c) as given below:

$$\Lambda_c = \Lambda_\infty - \left[\frac{8.2\times10^5}{(\varepsilon_r T)^{\frac{3}{2}}}\Lambda_\infty + \frac{82.4}{(\varepsilon_r T)^{\frac{1}{2}}\eta}\right]\sqrt{c} \tag{5.21}$$

where, ε_r = Relative permittivity, η = Coefficient of viscosity, and

T = Absolute temperature.

The Equation (5.21) is very much similar with that proposed by Kohlrausch [Equation (5.14)].

$$\Lambda_c = \Lambda_\infty - A\sqrt{c}$$

The first term in the bracket of Equation (5.21) appears due to asymmetry effect and the second term indicates the electrophoretic effect. Experimental studies reveal that the factor due to electrophoretic effect is almost double to that due to asymmetry effect.

5.3.8 Ionic mobility and temperature

In the case of electronic conductor conductivity decreases with the increase in temperature. In the case of electrolytic conductor with the increase in temperature kinetic energy of ions increases and hence velocity (or mobility) of ions increases. The temperature coefficient of mobility is given by:

$$\mu_M = \frac{1}{U_{298\,\text{K}}}\left(\frac{\Delta U}{\Delta T}\right)$$

For most ions the value of μ_M in aqueous solution is about 0.02, i.e. mobility and hence conductivity of an ion increases by 2% with increase in temperature by just 1 °K. Again we know that, with the increase in temperature, viscosity of a solution decreases. If the relative increase in mobility is comparable to relative decrease in viscosity with increase in temperature, the quantity $U_0\eta_0$ should be independent of temperature, which means that Equation (5.18) is true for any electrolytic solution. Or, in other words *Walden's rule* is applicable for all electrolytic solutions. Experimental studies reveal that *Walden's rule* is true for most of the ions with a little deviation. However, a large deviation has been observed in the cases of H^+ and OH^- ions.

5.3.9 Applications of Kohlrausch's law

(a) **Determination of equivalent conductivity of weak electrolyte at infinite dilution** As strong electrolytes always remain in dissociated form Λ_c of any strong electrolyte is the sum of equivalent conductivities or (mobilities) of cation and anion but dissociation of a weak electrolyte is only partial. Only at infinite dilution complete dissociation is assumed to take place, but infinite dilution cannot be achieved in actual practice. Thus Kohlrausch's law [Equation (5.16f)] is applicable at any concentration for strong electrolytes but applicable only at infinite dilution for weak electrolytes. Thus Λ_0 and Λ_∞ of a weak electrolyte can be computed by using Λ_∞ values of two or more suitable strong electrolytes. The following example illustrates how Λ_∞ of acetic acid (CH_3COOH) is calculated.

$$\Lambda^\infty_{CH_3COOH} = U^\infty_{H^+} + U^\infty_{CH_3COO^-}$$

$$= (U^\infty_{H^+} + U^\infty_{Cl^-}) - (U^\infty_{Na^+} + U^\infty_{Cl^-}) + (U^\infty_{Na^+} + U^\infty_{CH_3COO^-})$$

Or,

$$\Lambda^\infty_{CH_3COOH} = \Lambda^\infty_{HCl} - \Lambda^\infty_{NaCl} + \Lambda^\infty_{CH_3COONa}$$

$$= \Lambda^c_{HCl} - \Lambda^c_{NaCl} + \Lambda^c_{CH_3COONa}$$

Thus by measuring equivalent conductivities of strong electrolytes HCl, NaCl and CH_3COONa at any concentration c one can easily calculate the $\Lambda^{\infty}_{CH_3COOH}$. Its value has been calculated to be around 391 Ω^{-1} cm^2. As we know:

$$U^{\infty}_{H^+} = 350\ \Omega^{-1}\ cm^2$$

$$U^{\infty}_{CH_3COO^-} = \Lambda^{\infty}_{CH_3COOH} - U^{\infty}_{H+} = (391 - 350)\ \Omega^{-1}\ cm^2$$

Or,
$$U^{\infty}_{CH_3COO^-} = 41\ \Omega^{-1}\cdot cm^2 \text{ (at 25°C)}$$

(b) Determination of degree of dissociation (α) of a weak electrolyte: Equivalent conductivity of a weak electrolyte at any concentration c is given by:

$$\Lambda_c = \alpha(U_+ + U_-) \quad \text{[Equation (5.16i)]}$$

Where α is the degree of dissociation.

Again, at infinite dilution equivalent conductivity is given by:

$$\Lambda_0 = \Lambda_\infty = (U_+^\infty + U_-^\infty) \quad \text{[Equation (5.16f)]}$$

Or,
$$\Lambda_\infty = (U_+^\infty + U_-^\infty) = (U_+ + U_-)$$

Dividing Λ_c by Λ_∞, we get:

$$\frac{\Lambda_c}{\Lambda_\infty} = \alpha \tag{5.22}$$

Λ_∞ can be computed easily by using Kohlrausch's law as discussed above. Λ_c can be determined experimentally. Thus the degree of dissociation can easily be calculated from Equation (5.22).

(c) Determination of solubility of a sparingly soluble salt Any sparingly soluble salt has a limited solubility in water and the solubility does not change even at infinite dilution. Thus equivalent conductivity of a sparingly soluble salt remains constant at any concentration even at infinite dilution. So, we can write that:

$$\Lambda_c = \Lambda_\infty = (U_+^\infty + U_-^\infty) \quad \text{[for uni-uni-valent salt]}$$

Again, we have, $\Lambda_c = \Lambda_\infty = \kappa v$, where κ is the conductivity and v is the volume in cm^3, containing 1 gm-equiv. of the salt.

So, v c.c of salt solution contains 1 gm-equiv. of the salt.

1000 c.c. of salt solution contains $\frac{1000}{v}$ gm – equiv. of the salt

So, the concentration of the salt solution $= \frac{1000}{v} \frac{\text{gm-eqivalent}}{\text{litre}}$

$$= \frac{1000 \times (EW)}{v} \frac{\text{gms}}{\text{litre}}$$

Where, *EW* is the equivalent weight of the salt.

Again, concentration of salt in aqueous solution is nothing but the solubility (S) of the salt. So, we can write that:

$$S = \frac{1000 \times (EW)}{v}$$

Or,
$$v = \frac{1000 \times (EW)}{S}$$

Putting this value of v in $\Lambda_\infty = \kappa v$, we get:

$$\Lambda_\infty = \frac{\kappa 1000 \times (EW)}{S}$$

Or,
$$S = \frac{\kappa 1000 \times (EW)}{\Lambda_\infty} = \frac{\kappa . 1000 \times (EW)}{(U_+^\infty + U_-^\infty)} \frac{\text{gms}}{\text{litre}}$$

For multi-multi-valent electrolyte:

$$S = \frac{\kappa . 1000 \times (EW)}{(\nu_+ U_+^\infty + \nu_- U_-^\infty)} \quad \text{[From Equation (5.16j)]}$$

Where ν_+ and ν_- are the number of moles of cation and anion respectively.

So,
$$S = \frac{\kappa . 1000 \times (EW)}{(\nu_+ U_+^\infty + \nu_- U_-^\infty)} \frac{\text{gms}}{\text{litre}} \tag{5.23}$$

All the quantities of R.H.S. of Equation (5.23) are known except κ, which can easily be determined experimentally. Thus by measuring the conductivity κ of a sparingly soluble salt one can easily evaluate the solubility (S) of that salt.

(d) Determination of transport number It will be discussed in detail in the following section.

5.3.10 Transport numbers of ions

In 1854, J. Hittorf noticed that during electrolysis concentration of electrolyte changed at the electrodes. Quantity of electricity carried by any ion depends on the concentration and mobility of that ion. As an electrolytic solution is electrically neutral, the concentrations of cation and anion are same but mobilities of cation and anion differ with each other. If the cationic mobility is higher than anionic mobility, concentration of electrolyte should be higher at cathode than that at anode. Similarly if the anionic mobility is higher than cationic mobility, concentration of electrolyte is higher at anode than that at cathode. Thus transport number of a given ion in an electrolyte is the fraction of electricity carried by that ion through a given cross-section. If I_+ and I_- are the current carried by the cation and anion respectively, transport number (also called transference number or migration number) of cation (t_+) and anion (t_-) are given by:

$$t_+ = \frac{I_+}{I} \quad \text{and} \quad t_- = \frac{I_-}{I}$$

Where I is the total current passed through the solution.

Putting the values of I_+ , I_- and I from Equations (5.16d) and (5.16e) in the above equations, we get:

$$t_+ = \frac{U_+}{U_+ + U_-} \quad \text{and} \quad t_- = \frac{U_-}{U_+ + U_-} \tag{5.24a}$$

Thus the transport number of a given ion is the ratio of velocity of that ion to the sum of velocities of the cation and anion. For a uni-uni-valent weak electrolyte:

$$\left.\begin{aligned} t_+ &= \frac{\alpha U_+}{\alpha(U_+ + U_-)} = \frac{U_+}{U_+ + U_-} \quad \text{and} \\ t_- &= \frac{\alpha U_-}{\alpha(U_+ + U_-)} = \frac{U_-}{U_+ + U_-} \end{aligned}\right\} \tag{5.24b}$$

For a strong electrolyte of the following type:

$$AB \rightarrow \nu_+ A^{z+} + \nu_- A^{z-}$$

$$t_+ = \frac{\nu_+ U_+}{\nu_+ U_+ + \nu_- U_-} \quad \text{and} \quad t_- = \frac{\nu_- U_-}{\nu_+ U_+ + \nu_- U_-} \tag{5.24c}$$

From the above equation it is evident that transport number of an ion depends on ν_+ , ν_-, U_+ and U_-. Thus transport number of a given ion varies from salt to salt e.g., t_+ of Ca^{2+} in $Ca_3(PO_4)_2$ solution is given by:

$$t_+ = \frac{3U_{Ca^{2+}}}{3U_{Ca^{2+}} + 2U_{Po_4^{3-}}}$$

but t_+ of Ca^{2+} in $CaCl_2$ solution is given by:

$$t_+ = \frac{U_{Ca^{2+}}}{U_{Ca^{2+}} + 2U_{Cl^-}}$$

True transport number

During electrolysis cations are leaving anode and moving towards cathode while anions are leaving cathode and moving towards anode [see Figure 5.3.4]. Thus change in concentration of electrolyte takes place in the vicinity of both the electrodes. So it may be concluded that, loss of electrolyte at cathode is directly proportional to the mobility of anion and loss of electrolyte at anode is directly proportional to the mobility of cation.

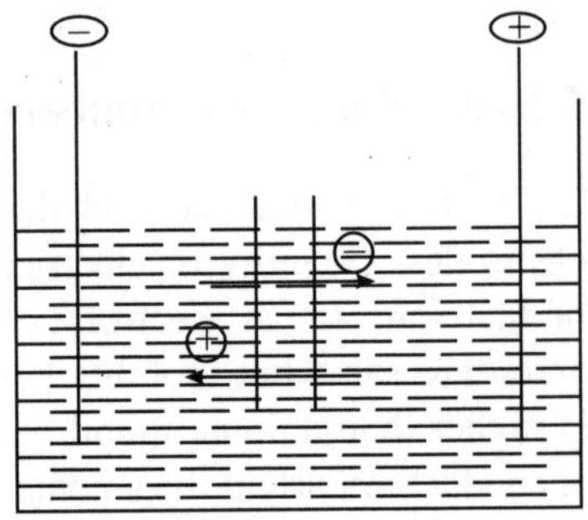

Fig. 5.3.4 Movement of ions during electrolysis.

Hence, we can write that:

$$\frac{\text{Loss around cathode}}{\text{Loss around anode}} = \frac{U_-}{U_+}$$

Or,

$$1 + \frac{\text{Loss around cathode}}{\text{Loss around anode}} = 1 + \frac{U_-}{U_+}$$

Or, $$t_+ = \frac{U_+}{U_+ + U_-} = \frac{\text{Loss around anode}}{\text{Total loss around both the electrodes}}$$

and $$t_- = \frac{U_-}{U_+ + U_-} = \frac{\text{Loss around cathode}}{\text{Total loss around both the electrodes}}$$

Thus by measuring individual loss of electrolyte at each electrode one can easily calculate the transport numbers of ions. This is the basic principle adopted by J. Hittorf to determine transport numbers but in these losses he did not consider the loss of water molecules, which are also transferred with the ions, since the ions always remain in hydrated form in aqueous solution. Thus t_+ and t_- are called apparent transport numbers. Suppose τ_+ and τ_- are taken to be true transport numbers and n_+ and n_- are the number of moles of water associated with one mole of cation and one mole of anion respectively.

At cathode Gain of water in moles = $\tau_+ n_+$, since cation moves towards the cathode.
Loss of water in moles = $\tau_- n_-$, since anion leaves the cathode.

$$\text{Net Gain} = (\tau_+ n_+ - \tau_- n_-) = y \text{ moles of water., if } y > 0$$

At anode Gain of water in moles = $\tau_- n_-$, since anion moves towards the anode.
Loss of water in moles = $\tau_+ n_+$, since cation leaves the anode.

$$\text{Net loss} = (\tau_+ n_+ - \tau_- n_-) = y \text{ moles of water, if } y > 0$$

Net gain of water at cathode = Net loss of water at anode

If the electrolytic solution contains n_1 moles of solvent and n_2 moles of solute, we can write that:
n_1 moles of solvent (water) is associated with n_2 moles of solute.

So, y moles of solvent (water) is associated with $\left(\frac{n_2}{n_1}\right)y$ moles of solute

$$= \frac{\dfrac{n_2}{n_1 + n_2}}{\dfrac{n_1}{n_1 + n_2}} y \text{ moles of solute} = \frac{x_2}{x_1} y \text{ moles of solute}$$

Where, x_1 and x_2 are the mole fractions of solvent and solute respectively.
As, net gain of water at cathode = net loss of water at anode, we can write that:
net gain of solute at cathode = net loss of solute at anode

$$= \left(\frac{x_2}{x_1}\right)y \text{ moles.}$$

Therefore the true transport number of the cation = τ_+

$$\tau_+ = \tau_+ + \left(\frac{x_2}{x_1}\right)y$$

and the true transport number of the anion = τ_-

$$\tau_- = \tau_- - \left(\frac{x_2}{x_1}\right)y$$

$$\tau_+ = \tau_+ + \left(\frac{x_2}{x_1}\right)y \quad \text{and} \quad \tau_- = \tau_- - \left(\frac{x_2}{x_1}\right)y \tag{5.24d}$$

If $y > 0$, $\left(\frac{x_2}{x_1}\right)y > 0$ and $\tau_+ > \tau_-$. If $y < 0$, $\left(\frac{x_2}{x_1}\right)y < 0$ and $\tau_+ < \tau_-$.

5.3.11 Method of determination of transport number

The most widely used method to determine transport number is the moving boundary method. In this method two solutions of different densitities are usually taken, such that the two solutions are separated by a boundary *ab* as shown in the Figure 5.3.5.

The electrolytic solutions are so chosen that they meet the following criteria.

(a) The electrolytes have a common anion A^-
(b) The mobility of one cation L^+ is higher than that of other cation M^+
(c) Electrolyte having higher mobility cation i.e. *LA* must have higher density than the other electrolyte, i.e, *MA*, such that *MA*-solution is placed above the *LA*-solution, i.e. above the boundary ab. *LA*-solution is called indicator solution.

Now the current *I* Amp. is passed through the system for *t* sec. So, the amount of electrocity passed through the system = $q = I \cdot t$ Coulomb. Current is passed in such a direction that cations move upward. As the velocity of L^+ is higher than that of M^+ the boundary *ab* moves to *cd* (see the Figure 5.3.5) in time *t*. It is obvious that as the boundary moves from *ab* to *cd* volume of M^+ ions within the region [*abcd*] must cross *cd*, since M^+ ions always remain above the boundary.

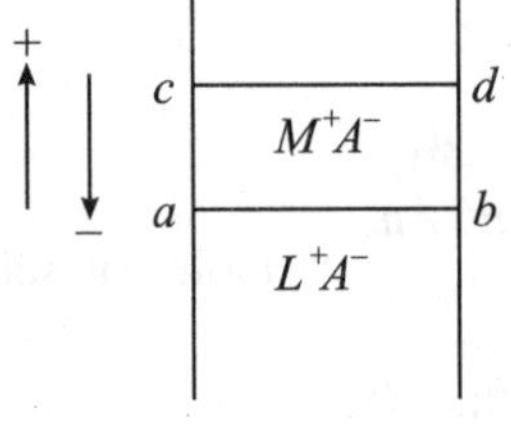

Fig. 5.3.5 Determination of t_+ by moving boundary method. Both the electrolytes are uni-uni-valent.

If *V* c.c. is the volume of M^+ ions within the boundary [*abcd*] the amount of M^+ ions within the boundary [*abcd*] is *cV* gm-equiv., where *c* is the concentration of the electrolyte *MA* in gm-equiv./c.c.

Thus the total electricity carried by M^+ ions in time *t*
= *cV* gm-equiv. × *F* Coulomb./gm-equiv. = *cVF* Coulomb.

From the definition of transport number, we have:

$$t_+ = \frac{\text{Electricity carried by } M^+ \text{ ions}}{\text{Total electricity passed}} = \frac{cVF}{q} \tag{5.25}$$

c, F and q are known. Thus by measuring the volume V, one can easily calculate the transport number of M^+. As we know, $t_+ + t_- = 1$, transport number of A^- can easily be determined.

The Table 5.3.1. shows the difference between true (τ_+) and apparent (t_+) transport numbers of some selected electrolytes.

Table 5.3.1 Values of τ_+ and t_+ of some selected electrolytes at 25°C [c = 1.3 N].

Salt	y^a	t_+	τ_+
HCl	0.24	0.820	0.844
LiCl	1.50	0.278	0.304
NaCl	0.76	0.366	0.383
KCl	0.60	0.482	0.495
NaCl	0.33	0.485	0.491

a: y has the same meaning as given in Equation (5.24d) i.e., it indicates the net gain in water (in moles) at cathode.

It is evident from the Table 5.3.1 that the difference between t_+ and τ_+ is small even at high concentration and this difference is negligible on further dilution. Thus universally accepted transport number is t_+ and hence no correction factor is introduced into the values of t_+ determined experimentally by moving boundary method.

5.3.12 Dependence of transport number on concentration of electrolyte and temperature

Usually transport number responds only marginally with the change in concentration as shown in the Table 5.3.2. It is evident that transport number, having value < 0.5 decreases slightly with the increase in concentration but transport number, having value > 0.5 increases slightly with the increase in concentration. In some exceptional cases transport number decreases drastically with increase in concentration and may be equal to or less than zero, e.g. t_+ of CdI_2 has a negative value at high concentration of CdI_2. This is because of the formation of complex ion $[CdI_4]^{2-}$ at high concentration.

Table 5.3.2 Differences between transport number and equivalent conductivity

Sl. No.	*Transport Number*	*Equivalent conductivity*
1	It has no unit.	It has a definite unit, $\Omega^{-1}.cm^2/gm.eq.$
2	It is a fraction, having value lies within 0 and 1.	It has a large value, much greater than 1.
3	Transport number of a cation or an anion varies from one salt to the other.	Equivalent conductivity of a cation or anion is independent of type of salt.
4	Effect of concentration and temperature on transport number is marginal.	Effect of concentration and temperature on equivalent conductivity is remarkable.

Table 5.3.3 Dependence of transport numbers of cations (t_+) on concentrations of selected electrolytes in aqueous solutions at 25°C.

Concentration gm-equiv./lit.	t_+			
	HCl	KNO_3	LiCl	$BaCl_2$
0.01	0.8251	0.5084	0.3289	0.440
0.02	0.8263	0.5087	0.3261	0.4375
0.05	0.8292	0.5093	0.3211	0.4317
0.10	0.8314	0.5103	0.3168	0.4253
0.20	0.8337	0.5120	0.3112	0.4162
0.50	—	—	0.3000	0.3986
1.00	—	—	0.2870	0.3792

Temperature has marginal effect on transport number. Usually transport number approaches to 0.5 with the increase in temperature [see Table 5.3.3].

Table 5.3.4 Dependence of transport number on temperature at infinite dilution i.e., $c \to 0$

Temperature (°C)	t_+	
	KCl	NaCl
15	0.4928	0.3929
25	0.4905	0.3962
35	0.4889	0.4002
45	0.4872	0.4039

Table 5.3.5 Limiting mobilities of ions in water at 25°C. Unit of mobility is Ω^{-1} cm²/gm.equiv.

Ion	U_+^∞	*Ion*	U_+^∞	*Ion*	U_-^∞	*Ion*	U_-^∞
H_3O^+	349.8	Sr^{2+}	59.5	OH^-	197.6	$[Fe(CN)_6]^{3-}$	99.1
Li^+	38.6	Ba^{2+}	63.6	F^-	55.4	$[Fe(CN)_6]^{4-}$	111
Na^+	50.1	Cu^{2+}	56.6	Cl^-	76.4	HCO_3^-	44.5
K^+	73.5	Zn^{2+}	52.8	Br^-	78.1	CNS^-	66.5
Rb^+	77.8	Cd^{2+}	54.0	I^-	76.8		
Cs^+	77.2	Co^{2+}	49.0	N_3^-	69.0		
Ag^+	61.9	Fe^{2+}	53.5	NO_3^-	71.4		
Tl^+	74.7	Fe^{3+}	68.0	ClO_3^-	64.6		
NH_4^+	73.5	Al^{3+}	63.0	BrO_3^-	55.8		
$[N(Me)_4]^+$	44.9			ClO_4^-	67.3		
$[N(Et)_4]^+$	32.6			IO_4^-	54.5		
$[N(Pr)_4]^+$	23.5			SO_4^{2-}	80.0		
$[N(Bu)_4]^+$	19.4			$C_2O_4^{2-}$	74.1		
Be^{2+}	45.0			CO_3^{2-}	69.3		
Mg^{2+}	53.0			$HCOO^-$	54.6		
Ca^{2+}	59.5			CH_3COO^-	40.9		

Table 5.3.6 Limiting mobilities of ions in selected solvents at 25°C.

Ion	U_+^∞ (Ω^{-1} *cm²/gm.equiv.*)				
	Water	*Methyl alcohol*	*Ethyl alcohol*	*Acetone*	*Nitrobenzene*
H^+	349.8	143.0	59.5	88.0	23.0
Li^+	38.6	39.8	*a*	*a*	*a*
Na^+	50.1	45.2	18.7	80.0	17.2
K^+	73.5	52.4	22.0	82.0	19.2
NH_4^+	73.5	57.9	19.3	98.0	*a*
Ag^+	61.9	50.3	17.5	88.0	18.5
Cl^-	76.4	52.9	24.3	111.0	17.3
Br^-	78.1	55.5	25.8	113.0	19.6
I^-	76.8	62.7	*a*	*a*	*a*
ClO_4^-	67.3	70.9	33.8	117.0	19.9
NO_3^-	71.4	60.8	27.9	120.0	*a*
OH^-	197.6	53.0	22.5	*a*	*a*
CNS^-	66.5	61.0	29.2	123.0	*a*

a: Date not available.

Table 5.3.7 Hydration number of ions (n). Formula $n = \frac{V_H}{30}$,

$$V_H = \frac{4}{3}\pi(r^3 - r_c^3)\text{Å}^3$$

Ion	U_+^∞	r_H(Å)	r_c(Å)	r(Å)	V_H(Å³)	n
Na^+	50.10	1.83	0.97	3.3	147	5
Li^+	36.68	2.37	0.60	3.7	211	7
Be^{2+}	45.00	4.08	0.55	4.6	407	13 – 14
Mg^{2+}	53.05	3.46	0.65	4.4	355	12
Ca^{2+}	59.50	3.09	0.99	4.2	306	10
Sr^{2+}	59.45	3.09	1.13	4.2	304	10
Ba^{2+}	63.63	2.88	1.35	4.1	278	9 – 10
Zn^{2+}	53.00	3.46	0.74	4.4	355	12
La^{3+}	69.75	3.95	1.15	4.6	401	13 – 14

5.3.13 Conductometric titrations

Conductivity of an electrolytic solution changes noticeably in presence of another electrolytic solution. By measuring the change in conductivity one can easily evaluate the strength or concentration of an electrolytic solution. The method is called conductometric titration. Conductometric titration can be divided into two broad categories:

(a) Acid-base titration, (b) Precipitation titration.

(a) Acid-base titration

(i) Titration of a strong acid by a strong base The conductivity of an acid solution is very high due to the presence of H^+ ion, which has very high mobility [see Table 5.3.5]. Let us take a hydrochloric acid solution. If we add sodium hydroxide solution dropwise, the following reaction takes place:

$$(H^+ + Cl^-) + (Na^+ + OH^-) \rightarrow (Na^+ + Cl^-) + H_2O$$

As NaOH is consumed, Na^+ ion replaces H^+ ion. As the mobility of Na^+ ion is much lower than that of H^+ ion [see Table 5.3.5], conductivity of HCl solution decreases sharply with the addition of NaOH solution, till the end point is reached.

At the end point there is no more H^+ ion present in the solution. Thus further addition of NaOH solution increases the conductivity sharply due to the presence of high mobile OH^- ion. The nature of the curve is shown in the Figure 5.3.6.

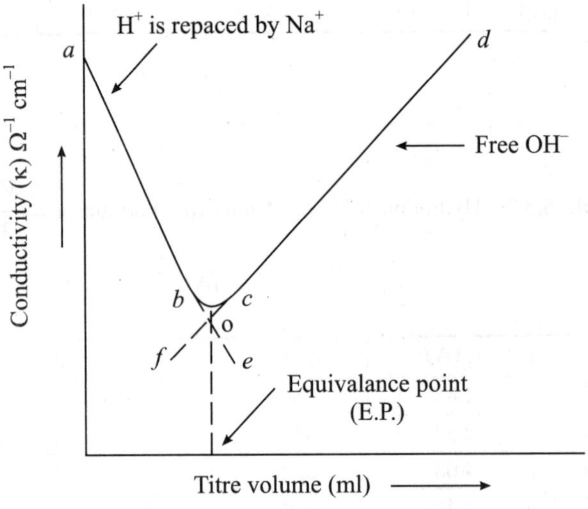

Fig. 5.3.6 Conductometric titration of HCl solution by a strong base NaOH solution.

Two straight lines are obtained before and after the end point. The straight line before the end point is steeper than that obtained after the end point.

Along *ab* conductivity decreases. $(U^{\infty}_{H^+} - U^{\infty}_{Na^+}) \approx 300\ \Omega^{-1}\ cm^2/gm.equiv.$

Along *cd* conductivity increases. $(U^{\infty}_{Na^+} + U^{\infty}_{OH^-}) \approx 250\ \Omega^{-1}\ cm^2/gm.equiv.$

Thus rate of decrease in conductivity along *ab* is greater than the rate of increase in conductivity along *cd*. That is why, curve *ab* is steeper than the curve *cd*. Intersecting region between the two curves *ab* and *cd*, i.e., *bc* is usually flat due to dilution effect. To minimize the dilution effect the titre (base) is taken as approximately five times stronger than the titrant (acid). The equivalence point *o* can easily be found out by extrapolating the curves *ab* and *cd* [see Figure 5.3.6].

(ii) Titration of a weak acid by a strong base If a weak acid (say CH_3COOH) is titrated by a strong base (say NaOH), the nature of the curve is shown in the Figure 5.3.7.

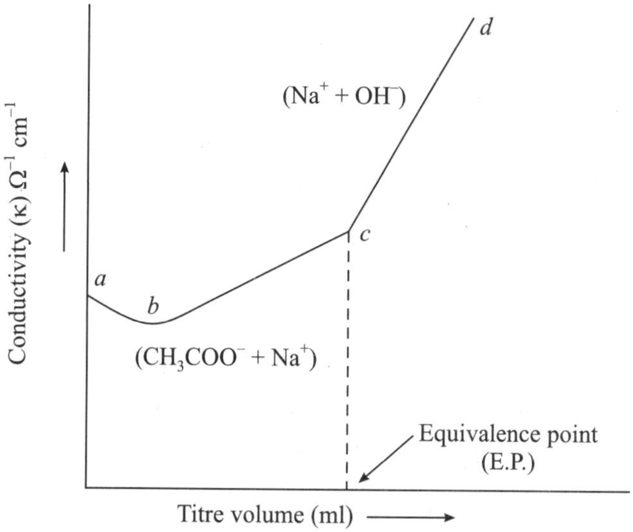

Fig. 5.3.7 Conductometric titration of CH_3COOH solution by a strong base NaOH solution.

The overall reaction, occurring during titration is given below:

$$CH_3COOH \rightleftharpoons H^+ + CH_3COO^- \quad \text{[partially dissociated]}$$

$$(CH_3COOH) + (Na^+ + OH^-) \rightarrow (CH_3COO^- + Na^+) + H_2O$$

Acetic acid is weakly dissociated. So initially H^+ is replaced by Na^+ and hence an initial drop in conductivity is observed. (*ab*) of the curve shown in Figure 5.3.7 represents the initial drop in conductivity. The neutralization reaction shows that acetate ion and sodium ion are produced from undissociated acetic acid. Hence conductivity increases on titration [(*bc*) of the curve shown in Figure 5.3.7]. After complete neutralization further addition of NaOH solution increases the conductivity sharply due to the presence of OH^- ion [(*cd*) of the curve shown in the Figure 5.3.7]. The point *c*, intersection of *bc* and *cd*, indicates the end point of titration.

(iii) Titration of a strong acid by a weak base If a strong acid (say HCl) solution is titrated with a weak base (say NH_4OH) solution, sharp drop in conductivity is observed due to the replacement of H^+ ion by NH_4^+ ion according to the following reaction:

$$(H^+ + Cl^-) + NH_4OH \rightarrow (NH_4^+ + Cl^-) + H_2O$$

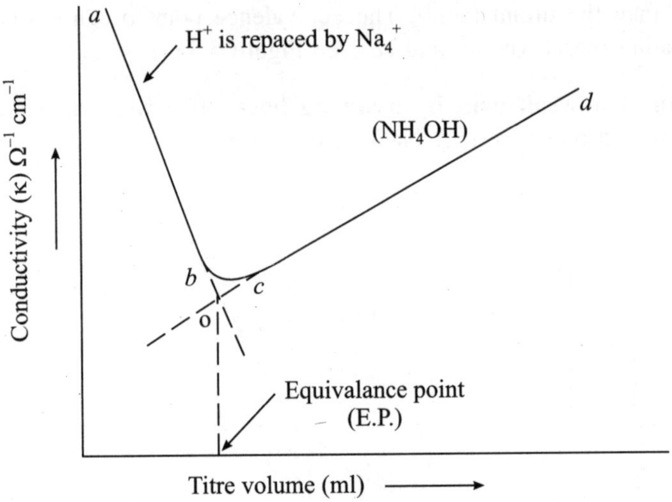

Fig. 5.3.8 Conductometric titration of a strong acid (HCl) by a weak base (NH_4OH)

After the end point the conductivity increases slowly since only small amount of OH^- ions is available from weakly dissociated NH_4OH [curve (*cd*) in the Figure 5.3.8]. The intersecting point '*o*' indicates the end point of titration.

(iv) Titration of a weak acid by a weak base If a weak acid (say CH_3COOH) solution is titrated by a weak base (say NH_4OH) solution, an initial drop in conductivity is observed due to the replacement of H^+ ion by NH_4^+ ion [Curve (*ab*) in the Figure 5.3.9]. After that conductivity increases due to salt formation according to the following reaction:

$$(CH_3COOH) + (NH_4OH) \rightarrow (CH_3COO^- + NH_4^+) + H_2O$$

[curve (*cd*) in the Figure 5.3.9]

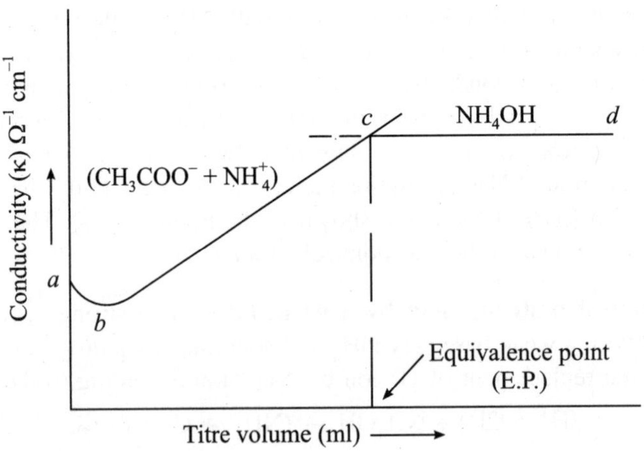

Fig. 5.3.9 Conductometric titration of a weak acid (CH_3COOH) by a weak base (NH_4OH).

After neutralization any further addition of NH_4OH solution increases the conductivity only marginally since NH_4OH is a weak base and almost remains in undissociated form. This is shown in the Figure 5.3.9.

Intersecting point '*c*' of the two curves *bc* and *cd* indicates the end point of titration. Weak acid-weak base titration cannot be performed by volumetric method.

(b) Precipitation titration

(i) Titration of potassium chloride solution by silver nitrate solution The reaction, taking place during titration, is given below:

$$(K^+ + Cl^-) + (Ag^+ + NO_3^-) \rightarrow AgCl \downarrow + (K^+ + NO_3^-)$$

The reaction shows that nitrate ion replaces chloride ion but they have almost same mobility [see Table 5.3.5]. Thus conductivity almost remains constant till the end point is reached [curve (*ab*) in the Figure 5.3.10]. After the end point there is no KCl in the system. So conductivity increases sharply with the addition of $AgNO_3$ solution due to increase in Ag^+ and NO_3^- concentration [curve (*cd*) in the Figure 5.3.10]. Intersecting point of the two curves indicates the end point of titration.

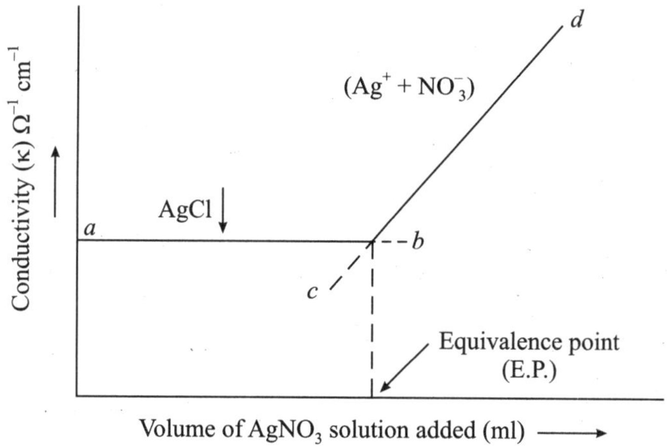

Fig. 5.3.10 Precipitation titration of KCl solution by $AgNO_3$ solution.

(ii) Titration of $BaCl_2$ solution by Na_2SO_4 solution The following reaction is taking place during titration:

$$(Ba^{2+} + 2Cl^-) + (2Na^+ + SO_4^{2-}) \rightarrow BaSO_4 \downarrow + (2Na^+ + 2Cl^-)$$

The reaction shows that 1 mole of Ba^{2+} is replaced by 2 moles of Na^+.

Equivalent conductivity of $Ba^{2+} = 63.6\ \Omega^{-1}\ cm^2$ [see Table 5.3.5]

Molar conductivity of $Ba^{2+} = 2 \times 63.6 = 127.2\ \Omega^{-1}\ cm^2$, since Ba^{2+} is bi-valent.

Equivalent conductivity of Na^+ = Molar conductivity of $Na^+ = 50.1\ \Omega^{-1}\ cm^2$

[see Table 5.3.5]

Thus the decrease in conductivity during titration $= (127.2 - 100.2)\ \Omega^{-1}\ cm^2$.

Thus the titration shows a decrease in conductivity, [see Figure 5.3.11] till the end point is reached. After the end point conductivity shows an increase due to increase in Na^+ and SO_4^{2-} concentrations. The point of intersection *o* of two curves (*ab*) and (*cd*) indicates the end point.

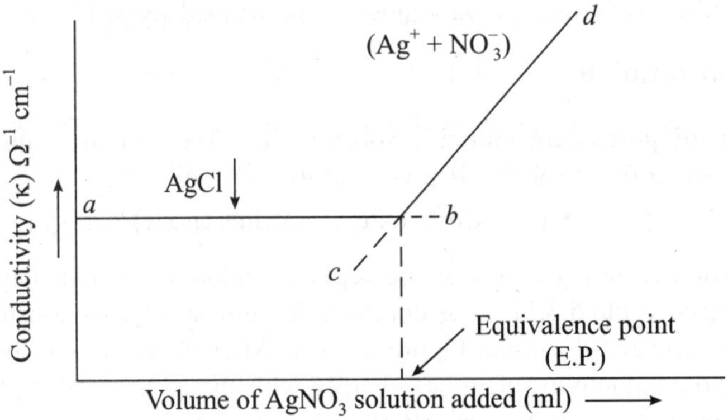

Fig. 5.3.11 Precipitation titration of $BaCl_2$ solution by Na_2SO_4 solution.

Advantages of conductometric titration over volumetric titration

1. In volumetric titration an indicator is required to detect the end point but incase of conductometric titration no indicator is required to detect the end point.
2. Detection of end point in conductometric titration is far more accurate than that in volumetric titration since end point in conductometric titration is detected graphically without any personal error.
3. Weak acid-weak base type of titration can be performed by conductometric method only.
4. Precipitation type of titration can be performed by conductometric method only.

5.4 Electrochemical cells and electromotive force

5.4.1 Electrochemical cell

Definition The cell, in which electrical energy is produced at the expense of chemical energy or vice versa, i.e. chemical reaction takes place at the expense of electrical energy, is called electrochemical cell.

There are two types of electrochemical cells:

(a) Galvanic cell The cell, which generates current at the expense of spontaneous chemical reaction occurring in the cell, is called a galvanic cell.

(b) Electrolytic cell Any cell, where current is passed into the cell to bring about chemical reaction within the cell, is called an electrolytic cell.

Electrochemical cells can also be classified in the following way:

(a) Reversible cell If the cell reaction exactly reverses when the direction of current is reversed, the cell is called reversible cell, e.g., *Daniell-Jacobi cell*, represented as:

$$\text{Cu} \,|\, \overset{\ominus}{\text{Zn}} \,|\, \text{ZnSO}_4\,(\text{aq}) \,||\, \text{CuSO}_4\,(\text{aq}) \,|\, \overset{\oplus}{\text{Cu}}$$

The overall cell reaction is given by:

$$Zn + Cu^{2+} \rightleftharpoons Zn^{2+} + Cu$$

When current is drawn from the cell the forward reaction occurs, i.e.,

$$Zn \rightarrow Zn^{2+} + 2e \quad \text{(at the cathode)}$$

and $$Cu^{2+} + 2e \rightarrow Cu \quad \text{(at the anode)}$$

When current is passed into the cell the reverse reaction occurs, i.e.

$$Zn^{2+} + 2e \rightarrow Zn \quad \text{(at the cathode)}$$

and $$Cu \rightarrow Cu^{2+} + 2e \quad \text{(at the anode)}$$

(b) Irreversible cell If the cell reaction does not reverse when the direction of current is reversed, the cell is called irreversible cell, e.g., the Voltaic cell, represented as:

$$\overset{\ominus}{Zn} \,|\, H_2SO_4\,(aq) \,|\, \overset{\oplus}{Cu}$$

When current is drawn from the cell, the following reaction takes place:

$$Zn \rightarrow Zn^{2+} + 2e \quad \text{(at the cathode)}$$

and $$2H^+ + 2e \rightarrow H_2 \quad \text{(at the anode)}$$

When current is passed into the cell the following reaction takes place:

$$2H^+ + 2e \rightarrow H_2 \quad \text{(at the cathode)}$$

$$Cu \rightarrow Cu^{2+} + 2e \quad \text{(at the anode)}$$

Two important characteristics of Galvanic cell

(i) Current is drawn from the cell as shown in the Figure 5.4.1.

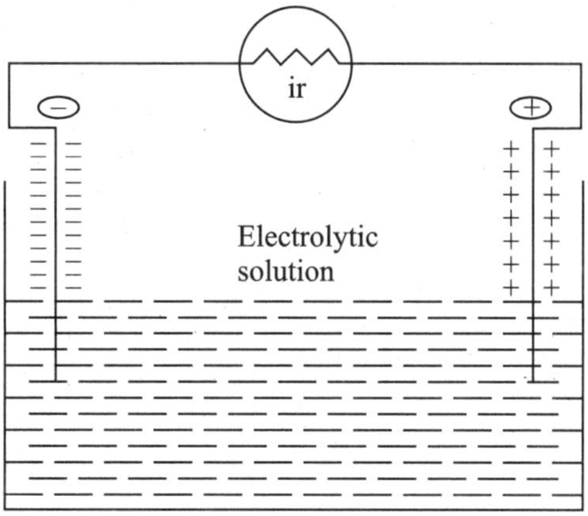

Fig. 5.4.1 An electrochemical cell where electric current is drawn from the cell.

(ii) Oxidation takes place at the cathode plate of voltaic cell e.g., $Zn \rightarrow Zn^{2+} + 2e$ and hence cathode plate is negatively charged. Reduction takes place at the anode, e.g., $2H^+ + 2e \rightarrow H_2$ and hence anode plate is positively charged.

Two important characteristics of electrolytic cell

(i) Current is sent to the cell as shown in the Figure 5.4.2.

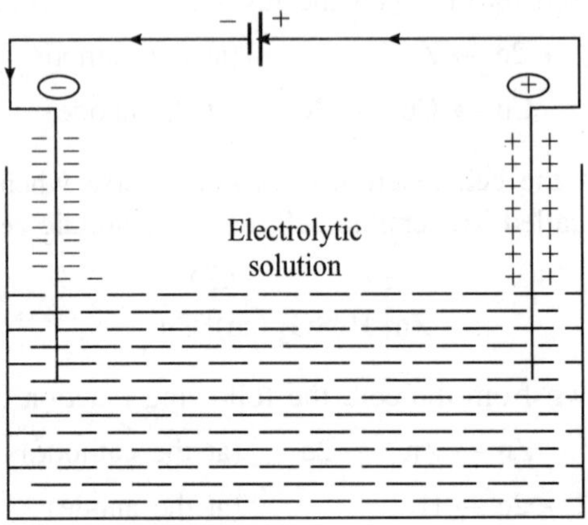

Fig. 5.4.2 An electrochemical cell where current is sent to the cell.

Cathode plate is made negative by external battery. Similarly anode plate is made positive by the same external battery.

(ii) Reduction takes place at the cathode, e.g., $2H^+ + 2e \rightarrow H_2$ in the Voltaic cell. Oxidation takes place at the anode, e.g., $Cu \rightarrow Cu^{2+} + 2e$ in the Voltaic cell.

e.m.f. and p.d. In an electrochemical cell, cathode plate becomes negative and anode plate becomes positive. Cathode plate is called negative electrode and anode plate is called positive electrode. When the cell is externally disconnected, the potential difference between the two electrodes is called e.m.f., represented as '*E*'. When the electrodes are externally connected, as shown in the Figure 5.4.1, the net potential difference between the two electrodes is $V = E - ir$, where i is the current drawn from the cell and r is the external resistance. This '*V*' is called potential difference or p.d. of the cell.

Sign of e.m.f. It is universally accepted that in any electrochemical cell negative end, i.e., cathode is always situated at the left and the positive end, i.e., anode is always situated at the right. When current is drawn from the cell, i.e., chemical energy is converted into electrical energy the cell e.m.f. is taken as positive. On the other hand when current is sent to the cell to bring about chemical reaction in the cell electrolyte, the cell e.m.f. is taken as negative.

5.4.2 Thermodynamics of electrochemical cells

For any electrochemical cell two parts are clearly distinguished:

(a) External source or battery or output voltage, as shown in Figures 5.4.1 and 5.4.2. The free energy change = ΔG_{out}

(b) Internal source of current within the cell, i.e. chemical reaction, continuously occurring within the cell. The free energy change = ΔG_{cell}

The net free energy change = $\Delta G = \Delta G_{cell} + \Delta G_{out}$

At equilibrium the net current produced within the cell is equal to the current drawn out externally. Thus applying thermodynamic condition of equilibrium, we have:

$$\Delta G = 0 = \Delta G_{cell} + \Delta G_{out}$$

Or,
$$\Delta G_{cell} = -\Delta G_{out}$$

$\Delta G_{out} = nFE$ Joules, where, E = e.m.f. of the cell in volt, F = 96,500 C/gm.equiv. and n = Number of gm.equiv. of electrons exchanged within the electrodes.

Thus,
$$\Delta G_{cell} = -nFE$$

Usually, ΔG_{cell} is written as ΔG. So the above equation becomes:

$$\Delta G = -nFE \tag{5.4.1a}$$

Differentiating the above equation w.r.t. T at constant P, we get:

$$\left[\frac{\partial(\Delta G)}{\partial T}\right]_P = -nF\left(\frac{\partial E}{\partial T}\right)_P \tag{5.4.2}$$

$\left(\frac{\partial E}{\partial T}\right)_P$ is known as temperature coefficient of the cell.

From Gibb's Helmholtz relation, we have:

$$\Delta S = -\left[\frac{\partial(\Delta G)}{\partial T}\right]_P = +nF\left(\frac{\partial E}{\partial T}\right)_P$$

Or,
$$\Delta S = nF\left(\frac{\partial E}{\partial T}\right)_P \tag{5.4.1b}$$

Again, from Helmholtz equation, we have:

$$\Delta G = \Delta H - T\Delta S. \quad \text{Hence,}$$

$$\Delta H = \Delta G + T\Delta S = -nFE + nFT\left(\frac{\partial E}{\partial T}\right)_P$$

Or,
$$\Delta H = -nF\left[E - T\left(\frac{\partial E}{\partial T}\right)_P\right] \tag{5.4.1c}$$

$$\left.\begin{aligned}\Delta G &= -nFE, \quad \Delta S = nF\left(\frac{\partial E}{\partial T}\right)_P \\ \text{and} \qquad \Delta H &= -nF\left[E - T\left(\frac{\partial E}{\partial T}\right)_P\right]\end{aligned}\right\} \tag{5.4.1}$$

The Equation (5.4.1) shows the relation between thermodynamic properties and e.m.f.

Significance of $\left(\frac{\partial E}{\partial T}\right)_P$

For any cell $\left(\frac{\partial E}{\partial T}\right)_P$ signifies the change in e.m.f. with temperature. The cell for which $\left(\frac{\partial E}{\partial T}\right)_P$ is close to Zero, the cell e.m.f. remains constant within a wide range of temperature. That type of cell is called standard cell. One of such standard cells is Weston-Cadmium cell [Figure 5.4.3].

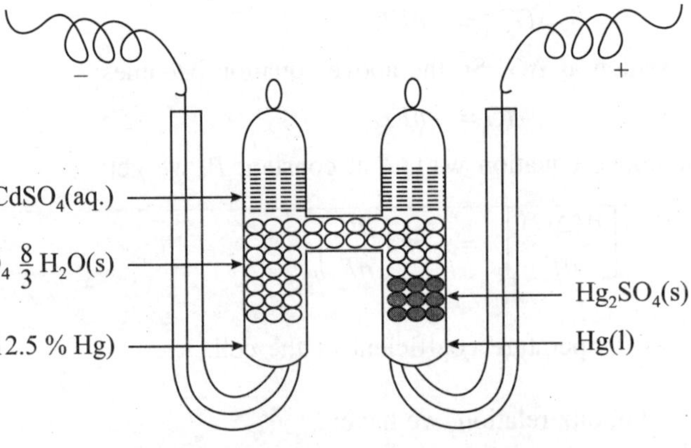

Fig. 5.4.3 Standard Weston-Cadmium cell.

Chemical reaction occurring at cathode (LHE)

The LHE consists of 12.5% cadmium amalgam followed by crystals of $CdSO_4 . \left(\frac{8}{3}\right)H_2O$. Electrolyte is saturated solution of $CdSO_4$.

$$Cd \rightarrow Cd^{2+} + 2e$$

$$Cd^{2+} + SO_4^{2-} + \left(\frac{8}{3}\right)H_2O \rightarrow CdSO_4 . \frac{8}{3}H_2O(s)$$

$$Cd + SO_4^{2-} + \frac{8}{3}H_2O \rightarrow CdSO_4 . \frac{8}{3}H_2O(s) + 2e$$

Chemical reaction occurring at anode (RHE):

The RHE consists of mercury followed by crystals of mercurous sulphate and $CdSO_4 \cdot \left(\frac{8}{3}\right)H_2O$. The electrolyte is saturated solution of $CdSO_4$.

$$2Hg(l) \rightarrow Hg_2^{2+} + 2e$$
$$Hg_2^{2+} + SO_4^{2-} \rightarrow Hg_2SO_4(s)$$
$$2Hg(l) + SO_4^{2-} \rightarrow Hg_2SO_4(s) + 2e$$

The overall cell reaction = LHE – RHE:

$$Cd + Hg_2SO_4(s) + \frac{8}{3}H_2O \rightarrow CdSO_4 \cdot \frac{8}{3}H_2O(s) + 2Hg(l)$$

The cell e.m.f. at 20°C = 1.0180 V and $\left(\frac{\partial E}{\partial T}\right)_P = -5.0 \times 10^{-5}$ V/°K.

5.4.3 Nernst equation for single electrode potential and electrochemical cell

Consider a metal/metal ion electrode of type M/M^{n+}. The electrode reaction is:

$$M \rightarrow M^{n+} + ne$$

For 1 mole of metal the free energy change is given by:

$$\Delta G = (\mu_{M^{n+}} - \mu_M)$$

Since, $$G = \sum \mu_i n_i$$

Where n_i is the number of moles of ith species and μ_i is the chemical potential of ith species. Again, we know that, $\mu_i = \mu_i^0 + RT \ln a_i$, where a_i is the activity of ith species.

Thus the above equation becomes:

$$\Delta G = \mu_{M^{n+}}^0 + RT \ln a_{M^{n+}} - \mu_M$$
$$\Delta G = (\mu_{M^{n+}}^0 - \mu_M) + RT \ln a_{M^{n+}} \qquad (5.4.3a)$$

Again, we know, $\Delta G = -nF\phi$ [Equation (5.4.1a)].

Combining the above two equations, we get:

$$-nF\phi = (\mu_{M^{n+}}^0 - \mu_M) + RT \ln a_{M^{n+}}$$

Or, $$\phi = -\left(\frac{\mu_{M^{n+}}^0 - \mu_M}{nF}\right) - \frac{RT}{nF} \ln a_{M^{n+}}$$

$$\mu_M = \mu_M^0 + RT \ln a_M$$

For solid substance activity is taken as unity.

So, $a_M = 1$ and $\mu_M = \mu_M^0$.

Or, $$\phi_{ox} = \phi^0_{ox} - \frac{RT}{nF} \ln a_{M^{n+}} \tag{5.4.3b}$$

Where ϕ_{ox} and ϕ^0_{ox} are the oxidation potential and standard oxidation potential of the single electrode M/M^{n+}. Equation (5.4.3b) is the *Nernst equation* for the single electrode potential.

$$\phi^0_{ox} = -\left(\frac{\mu^0_{M^{n+}} - \mu_M}{nF}\right)$$

It is defined as the *oxidation potential* at unit activity of metal ion. The electrode reaction is oxidation reaction since electrons are released. If the electrode reaction is reduction type, i.e., $M^{n+} + ne \rightarrow M$ (electrons are accepted), the corresponding potential is written as ϕ_{red}, which is expressed as: $\phi_{red} = -\phi_{ox}$ and $\phi^0_{red} = -\phi^0_{ox}$.

Putting these values of ϕ_{ox} and ϕ^0_{ox} in the Equation (5.4.3b) we have:

$$\phi_{red} = \phi^0_{red} + \frac{RT}{nF} \ln a_{M^{n+}} \tag{5.4.3c}$$

Nernst equation for a cell

An electrochemical cell consists of two electrodes, namely, cathode and anode.

Let us consider Doniell-Jacobi cell.

$$\overset{\ominus}{\text{Cu}}\,|\,\text{Zn}\,|\,\text{ZnSO}_4\,(\text{aq})\,||\,\text{CuSO}_4\,(\text{aq})\,|\,\overset{\oplus}{\text{Cu}}$$

L.H.E. $Zn \rightarrow Zn^{+2} + 2e$

Applying Equation (5.4.3b) we get:

$$\phi_{\text{Zn}} = \phi^0_{\text{Zn}} - \frac{RT}{2F} \ln a_{\text{Zn}^{2+}}$$

R.H.E. $Cu \rightarrow Cu^{+2} + 2e$

Applying Equation (5.4.3b) we get:

$$\phi_{\text{Cu}} = \phi^0_{\text{Cu}} - \frac{RT}{2F} \ln a_{\text{Cu}^{2+}}$$

E_{ox} = The cell e.m.f. = L.H.E. e.m.f. – R.H.E. e.m.f. = $\phi_{\text{Zn}} - \phi_{\text{Cu}}$.

$$E_{ox} = \phi_{\text{Zn}} - \phi_{\text{Cu}} = (\phi^0_{\text{Zn}} - \phi^0_{\text{Cu}}) - \frac{RT}{2F} \ln \frac{a_{\text{Zn}^{2+}}}{a_{\text{Cu}^{2+}}}$$

$$= E^0_{ox} - \frac{RT}{2F} \ln \frac{a_{\text{Zn}^{2+}}}{a_{\text{Cu}^{2+}}}$$

where, E^0 is the standard cell e.m.f., i.e. e.m.f. of the cell at unit concentration of ions.

So, $$E_{ox} = E^0_{ox} - \frac{RT}{2F} \ln \frac{a_{Zn^{2+}}}{a_{Cu^{2+}}} \tag{5.4.4a}$$

The overall cell reaction = L.H.E. – R.H.E.

Or, $$Zn + Cu^{2+} \rightarrow Zn^{2+} + Cu$$

Thus in the Equation (5.4.4a) the numerator of *ln* contains the activity terms of right hand side of the cell reaction and the denominator of *ln* consists of activity terms of left hand side.

Let us consider a general cell reaction of the following type:

$$lL + mM + \ldots \rightarrow pP + qQ + \ldots$$

The oxidation potential of the cell is given by:

$$E_{ox} = E^0_{ox} - \frac{RT}{nF} \ln \frac{a_P^p \,.\, a_Q^q \ldots}{a_L^l \,.\, a_M^m \ldots} \tag{5.4.4b}$$

Where n is number of electron transfer in the above chemical reaction.

The reduction potential of the cell is given by:

$$E_{red} = -E_{ox} \quad \text{and} \quad E^0_{red} = -E^0_{ox}$$

Or, $$E_{red} = E^0_{red} + \frac{RT}{nF} \ln \frac{a_P^p \,.\, a_Q^q \ldots}{a_L^l \,.\, a_M^m \ldots} \tag{5.4.4c}$$

Equations (5.4.4b) and (5.4.4c) are known as Nernst equation for an electrochemical cell. E^0_{ox} is called the standard cell e.m.f., which is the cell e.m.f. at unit concentration of the participants.

$$E_{ox} = E^0_{ox} - \frac{RT}{nF} \ln \frac{a_P^p \,.\, a_Q^q \ldots}{a_L^l \,.\, a_M^m \ldots}$$

$$= E^0_{ox} - \frac{2.303\,RT}{nF} \log \frac{a_P^p \,.\, a_Q^q \ldots}{a_L^l \,.\, a_M^m \ldots}$$

$$\frac{2.303\,RT}{F} = \frac{2.303 \times 8.314 \times 298}{96,500} \left(\frac{J}{C}\right) \quad \text{[at 25°C]}$$

$$= 0.0592 \text{ V} \quad [\text{since } 1\,J = 1\,V - C]$$

Thus, $$E_{ox} = E^0_{ox} - \frac{0.0592\,RT}{n} \log \frac{a_P^p \,.\, a_Q^q \ldots}{a_L^l \,.\, a_M^m \ldots} \quad \text{[at 25°C]} \tag{5.4.4d}$$

Equation (5.4.4d) is the Nernst equation of an electrochemical cell at 25°C.

5.4.4 Classification of electrodes

Electrodes can be broadly classified into three categories:

(a) The 1st kind, which is metal/metal ion electrode. If a metal plate is immersed into a solution of same metal salt, a potential is developed surrounding the metal plate. The electrode of this type is called 1st kind electrode, e.g. $Zn|Zn^{+2}$ (aq.)
The electrode reaction is:

$$Zn \rightarrow Zn^{2+} + 2e$$

The electrode potential is given by:

$$\phi_{Zn} = \phi^0_{Zn} - \frac{RT}{2F} \ln a_{Zn^{2+}} \quad [a_{Zn(s)} = 1] \tag{5.4.5a}$$

(b) The 2nd kind, which is a metal/sparingly soluble salt/saturated salt solution. A metal plate, coated with its sparingly soluble salt, is immersed in a solution containing the same sparingly soluble salt (soluble) and an easily soluble salt containing the same anion. One example is:

$$\text{Ag}|\text{AgCl(s), KCl}$$

The electrode reaction is:

$$\text{Ag} \rightarrow \text{Ag}^+ + \text{e}$$

$$\text{Ag}^+ + \text{Cl}^- \rightarrow \text{AgCl(s)}$$

$$\text{Ag} + \text{Cl}^- \rightarrow \text{AgCl(s)} + \text{e}$$

The electrode potential is given by:

$$\phi_{\text{Ag}|\text{AgCl(s)}} = \phi^0_{\text{Ag}|\text{AgCl(s)}} - \frac{RT}{F} \ln \frac{a_{\text{AgCl(s)}}}{a_{\text{Ag(s)}} \cdot a_{\text{Cl}^-}}$$

Or,

$$\phi_{\text{Ag}|\text{AgCl(s)}} = \phi^0_{\text{Ag}|\text{AgCl(s)}} + \frac{RT}{F} \ln a_{\text{Cl}^-} \tag{5.4.5b}$$

(c) The 3rd kind, which is also known as redox electrode. If a platinum plate is immersed into a solution containing a multivalent metal ion, redox electrode is formed. *Pt*-plate is chosen since it does not participate in chemical reaction but serves only as a carrier of electrons. One example of redox electrode is *Pt*-plate dipped into a solution containing both Fe^{2+} and Fe^{3+} ions. The electrode reaction is:

$$\text{Fe}^{2+} \rightarrow \text{Fe}^{3+} + \text{e}$$

The electrode potential is given by:

$$\phi_{\text{Fe}^{2+}|\text{Fe}^{3+}} = \phi^0_{\text{Fe}^{2+}|\text{Fe}^{3+}} - \frac{RT}{F} \ln \frac{a_{\text{Fe}^{3+}}}{a_{\text{Fe}^{2+}}} \tag{5.4.5c}$$

5.4.5 Reference electrodes

An electrode of known potential is called reference electrode since with the help of a reference electrode the potential of any unknown electrode can easily be determined. The basic reference electrode is the hydrogen electrode, also known as Standard Hydrogen Electrode (SHE) since the value of standard potential (ϕ^0) of hydrogen electrode is universally accepted as Zero.

So, $$\phi^0_{H_2|H^+} = 0$$

Preparation of hydrogen electrode

A platinised *Pt*-plate is immersed in a solution containing sulphuric acid or hydrohloric acid. Pure hydrogen gas is passed into the solution under a constant pressure. Hydrogen gas is first adsorbed on the *Pt*-surface and then diffuses into the solution. *Pt*-surface is further platinised to increase the efficiency of hydrogen-adsorption.

The overall electrode reaction is given by:

$$\frac{1}{2} H_2(g) \rightarrow H^+ + e$$

The electrode potential is given by:

$$\phi_{H_2|H^+} = \phi^0_{H_2|H^+} - \frac{RT}{F} \ln \frac{a_{H^+}}{p_{H_2}^{\frac{1}{2}}} \qquad (5.4.6a)$$

[since $a_{H_2} = p_{H_2}$ for gaseous state]

Again, we know that, $$\phi^0_{H_2|H^+} = 0$$

So, $$\phi_{H_2|H^+} = -\frac{RT}{F} \ln \frac{a_{H^+}}{p_{H_2}^{\frac{1}{2}}} \qquad (5.4.6a)$$

a_{H^+} = Activity of H^+ ion in aqueous solution. It is maintained at 1 unit.

p_{H_2} = Gas pressure of hydrogen on the electrode.

Thus the above equation becomes:

$$\phi_{H_2|H^+} = \frac{RT}{F} \ln p_{H_2}^{\frac{1}{2}} = \frac{RT}{2F} \ln p_{H_2} = \frac{0.0592}{2} \log p_{H_2} \quad \text{[at 25°C]}$$

Disadvantages of hydrogen electrode

1. Calculation shows that a deviation of gas pressure by 0.1 atm results in a change in potential by 0.0012 V. As the gas pressure cannot be maintained accurately, the electrode potential of standard hydrogen electrode ($\phi_{H_2|H^+}$) is often fluctuating. Obviously the potential of an unknown electrode, evaluated with the help of hydrogen electrode, is erroneous.
2. Presence of impurities changes the potential significantly.
3. Hydrogen electrode is susceptible to corrosion under strongly acidic or alkaline media.

Due to above disadvantages hydrogen electrode is seldom used as a reference electrode.

Other reference electrodes

(i) Calomel electrode Mercurous chloride (Hg_2Cl_2) is known as *calomel*. When a mercury electrode is dipped in a solution of potassium chloride (KCl) of a definite concentration saturated with calomel, i.e., Hg_2Cl_2, the resultant electrode is called *calomel electrode*. It is represented as:

$$Hg(l)\big| Hg_2Cl_2(s), KCl(a)$$

The electrode reaction is given by:

$$2Hg(l) \rightarrow Hg_2^{2+} + 2e$$
$$Hg_2^{2+} + 2Cl^- \rightarrow Hg_2Cl_2(s)$$
$$\overline{2Hg(l) + 2Cl^- \rightarrow Hg_2Cl_2(s) + 2e}$$

The electrode potential is given by:

$$\phi_{cal} = \phi^0_{cal} - \frac{RT}{2F} \ln \frac{a_{Hg_2Cl_2(s)}}{a^2_{Hg(l)} \cdot a^2_{Cl^-}} = \phi^0_{cal} + \frac{RT}{F} \ln a_{Cl^-}$$

$$[\text{since, } a_{Hg_2Cl_2}(s) = a_{Hg(l)} = 1]$$

So, $$\phi_{cal} = \phi^0_{cal} + \frac{RT}{F} \ln a_{Cl^-} \tag{5.4.6b}$$

The value of ϕ_{cal} depends on KCl concentration.
Thus it has a definite potential at a definite concentration of KCl, e.g.,

(a) $a_{Cl^-} = m_{Cl^-} = \left(\frac{N}{10}\right)$, $\phi_{cal} = -0.3338$ V at 25°C.

(b) $a_{Cl^-} = m_{Cl^-} = $ (N), $\phi_{cal} = -0.2800$ V at 25°C.

(c) $a_{Cl^-} = m_{Cl^-} = $ Saturated KCl solution, $\phi_{cal} = -0.2415$ V at 25°C.

(ii) Silver-silver chloride electrode This electrode is similar to the calomel electrode. The electrode is represented as:

$$Ag\big| AgCl(s), KCl(a)$$

A silver electrode is immersed into a solution of potassium chloride (KCl), saturated with silver chloride (AgCl). The electrode reaction is given by:

$$Ag(s) \rightarrow Ag^+ + e$$
$$Ag^+ + Cl^- \rightarrow AgCl(s)$$
$$\overline{Ag(s) + Cl^- \rightarrow AgCl(s) + e}$$

The electrode potential is given by:

$$\phi_{Ag|AgCl(s),\ KCl(a)} = \phi^0_{Ag|AgCl(s),\ KCl(a)} - \frac{RT}{F} \ln \frac{AgCl(s)}{a_{Ag(s)} \cdot a_{Cl^-}}$$

Or, $$\phi_{Ag|AgCl(s),\ KCl(a)} = \phi^0_{Ag|AgCl(s),\ KCl(a)} + \frac{RT}{F} \ln a_{Cl^-} \tag{5.4.6c}$$

Table 5.4.1 Standard electrode potential of different electrodes in aqueous solution.

Electrode	ϕ^0 (V)	*Electrode*	ϕ^0 (V)	*Electrode*	ϕ^0 (V)
Li/Li^+	+3.045	Ni/Ni^{2+}	+0.250	Au/Au^{3+}	−1.290
K/K^+	+2.925	Pb/Pb^{2+}	+0.126	H_2/H_3O^+	0.000
Na/Na^+	+2.714	Cu/Cu^{2+}	−0.337	OH^-/O_2	−0.401
Mg/Mg^{2+}	+2.37	Cu/Cu^+	−0.521	I^-/I_2	−0.536
Al/Al^{3+}	+1.66	Pb/Pb^{4+}	−0.700	Br^-/Br_2	−1.065
Zn/Zn^{2+}	+0.763	Hg/Hg_2^{2+}	−0.789	Cl^-/Cl_2	−1.360
Fe/Fe^{2+}	+0.440	Ag/Ag^+	−0.799	F^-/F_2	−2.650
Cd/Cd^{2+}	+0.403	Hg/Hg^{2+}	−0.854		

Table 5.4.2 Some standard redox potentials in aqueous solution

Electrode	*Electrode Process*	ϕ^0 (V)
$Cr^{2+}, Cr^{3+}\|Pt$	$Cr^{2+} \rightarrow Cr^{3+} + e$	+0.410
$Sn^{2+}, Sn^{4+}\|Pt$	$Sn^{2+} \rightarrow Sn^{4+} + 2e$	−0.150
$(ClO_3 + 2OH^-), ClO_4^-\|Pt$	$(ClO_3 + 2OH^-) \rightarrow ClO_4^- + H_2O + e$	−0.360
$MnO_4^{2-}, MnO_4^-\|Pt$	$MnO_4^{2-} \rightarrow MnO_4^- + e$	−0.564
$Fe^{2+}, Fe^{3+}\|Pt$	$Fe^{2+} \rightarrow Fe^{3+} + e$	−0.771
$Hg_2^{2+}, Hg^{2+}\|Pt$	$Hg_2^{2+} \rightarrow 2Hg^{2+} + 2e$	−0.920
$Mn^{2+}, Mn^{4+}\ (4OH^-)\|Pt$	$Mn^{2+} + 2H_2O \rightarrow MnO_2 + 4H^+ + 2e$ $4H^+ + 4OH^- \rightarrow 4H_2O$	−1.230
$Pb^{2+}, Pb^{4+}\ (4OH^-)\|Pt$	$Pb^{2+} + 2H_2O \rightarrow PbO_2 + 4H^+ + 2e$ $4H^+ + 4OH^- \rightarrow 4H_2O$	−1.455
$Mn^{2+}, Mn^{3+}\|Pt$	$Mn^{2+} \rightarrow Mn^{3+} + e$	−1.510
$PbSO_4, PbO_2\|Pt$	$PbSO_4 + 2H_2O \rightarrow PbO_2 + SO_4^{2-} + 4H^+ + 2e$	−1.685
$Pb^{2+}, Pb^{4+}\|Pt$	$Pb^{2+} \rightarrow Pb^{4+} + 2e$	−1.700
$Co^{2+}, Co^{3+}\|Pt$	$Co^{2+} \rightarrow Co^{3+} + e$	−1.820
$Cr^{3+}, Cr^{6+}\|Pt$	$2Cr^{3+} + 7H_2O \rightarrow Cr_2O_7^{2-} + 14H^+ + 6e$	−1.330
$Mn^{2+}, Mn^{7+}\|Pt$	$Mn^{2+} + 7H_2O \rightarrow MnO^{4-} + 8H^+ + 5e$	−1.520
$Ce^{3+}, Ce^{4+}\|Pt$	$Ce^{3+} \rightarrow Ce^{4+} + e$	−1.610

5.4.6 Applications of single electrode potential

(i) Determination of pH of a buffer solution

Let us consider the following cell:

$$\overset{\ominus}{\text{Pt}|\text{H}_2\text{(g)}} \left| \begin{array}{l}\text{Buffer solution} \\ \text{containing H}^+ \text{ and Cl}^- \\ \text{of molality } m\end{array} \right| \text{AgCl(s)}, \left| \overset{\oplus}{\text{Ag(s)}} \right.$$

L.H.E. $\frac{1}{2}H_2 \rightarrow H^+ + e.$ Electrode potential $= \phi_L$.

$$\phi_L = \phi^0_{H_2|H^+} - \frac{RT}{F} \ln \frac{a_{H^+}}{p_H^{\frac{1}{2}}} \quad [\phi^0_{H_2|H^+} = 0.00 \text{ V}, p_{H_2} = 1 \text{ atm}]$$

Or, $\phi_L = -0.0592 \log a_{H^+}$ [at 25°C]

R.H.E.

$$\text{Ag(s)} \rightarrow \text{Ag}^+ + \text{e}$$

$$\text{Ag}^+ + \text{Cl}^- \rightarrow \text{AgCl(s)}$$

$$\overline{\text{Ag(s)} + \text{Cl}^- \rightarrow \text{AgCl(s)} + \text{e}} \quad \text{Electrode potential} = \phi_R$$

$$\phi_R = \phi^0_R + \frac{RT}{F} \ln a_{Cl^-}$$

Or, $\phi_R = \phi^0_R + 0.0592 \log m_{Cl^-}$ [at 25°C and assuming that $\gamma_{Cl^-} = 1$]

$$\text{Cell e.m.f.} = E = \phi_L - \phi_R$$

$$E = \phi_L - \phi_R = -0.0592 \log a_{H^+} - \phi^0_R - 0.0592 \log m_{Cl^-}$$

$$= E^0 - 0.0592 \log m_{Cl^-} + 0.0592\, pH_a$$

[where, $E^0 = -\phi^0_R$ and $pH_a = -\log a_{H^+}$]

Or, $$\left(\frac{E - E^0}{0.0592}\right) = pH_a - \log m_{Cl^-}$$

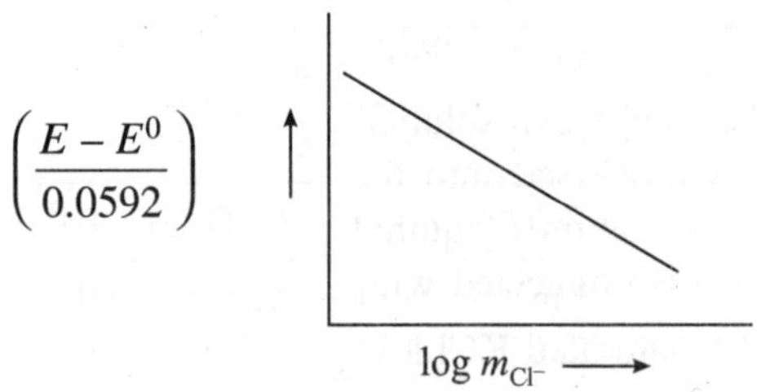

Fig. 5.4.4 Plot of $\left(\frac{E - E^0}{0.0592}\right)$ Vs. log m_{Cl^-} of a buffer solution. Intercept = pH_a

E^0 is known and constant and E is a measurable quantity. Thus different E-values are recorded at different chloride ion concentrations (m_{Cl^-}). Thus by plotting $\left(\frac{E - E^0}{0.0592}\right)$ against $\log m_{Cl^-}$ a straight line of negative slope is obtained. The intercept, obtained at m_{Cl^-} = 1 m, gives the value of pH of the buffer solution.

Two electrodes are widely used for the determination of pH of an unknown solution.

1. Quinhydrone electrode
2. Glass electrode.

1. Quinhydrone electrode

Quinol or hydroquinone (often represented as H_2Q) is oxidized reversibly in presence of $FeCl_3$ to produce p-benzoquinone or simply quinone (often represented as Q) as shown below

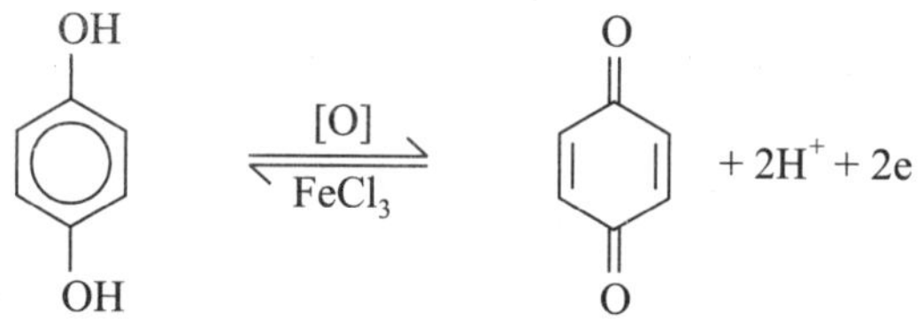

Quinol/Hydroquinone *p-Benzoquinone*

Symbolically, the above equation may be represented as

$$H_2Q \rightleftharpoons Q + 2H^+ + 2e$$

Thus, using Nernst's equation, we can write that

$$E_Q = E_Q^0 - \frac{RT}{2F} \ln \frac{a_Q}{a_{H_2Q}} . a_{H^+}^2 = E_Q^0 - \frac{RT}{2F} \ln \frac{a_Q}{a_{H_2Q}} - \frac{RT}{F} \ln a_{H^+}$$

The concentrations of quinone and hydroquinone are so maintained that $a_Q/a_{H_2Q} = 1$. Hence, the above equation becomes

$$E_Q = E_Q^0 - \frac{RT}{F} \ln a_{H^+}^2 = E_Q^0 - \frac{2.303RT}{F} \log a_{H^+} = E_Q^0 + \frac{2.303RT}{F} \text{pH}$$

At 25°C, the value of $2.303RT/F$ is 0.0592. Putting this value in the above equation, we get

$$E_Q = E_Q^0 + 0.0592\text{pH}$$

To measure the pH of an unknown solution, quinhydrone system is added to the unknown solution. A gold foil is immersed into the solution. Platinum foil cannot be used since platinum catalyses the oxidation of quinol and hence the equilibrium is disturbed. The quinhydronone electrode is connected with a standard calomel electrode as shown below

$$\text{Hg|Hg}_2\text{Cl}_2\text{(S), saturated KCl || Unknown solution + quinohydrone|Au}$$

Calomel electrode is used as cathode and quinhydrone electrode is used as anode. The cell e.m.f. can be written as

$$E_Q^0 = E_{cal} - E_Q = E_{cal} - E_Q^0 - 0.0592\text{pH} \quad \text{(at 25°C)}$$

Hence, $\text{pH} = \dfrac{E_{cal} - E_Q^0 - E}{0.0592}$ (at 25°C)

At 25ºC, the value of $E_Q^0 = -0.699$ V and $E_{cal}|_{sat} = -0.241$ V. E is measurable quantity and hence pH can be calculated from the above equation.

Advantages of quinhydrone electrode

(a) The electrode is simple and easy to handle.
(b) The equilibrium between quinone and hydroquinone is quickly attained.
(c) A little amount of solution is enough to determine pH of an unknown solution.
(d) The electrode is stable under corrosive environment.
(e) The electrode can be safely used in presence of other ions or organic substances.

Disadvantages of quinhydrone electrode

(a) Platinum foil or wire cannot be used in this electrode since it catalyses oxidation of hydroquinone, thereby disturbing the equilibrium.
(b) The electrode can be safely used in alkaline solution of pH up to 8 but cannot be used for solutions having pH greater than 8.5.

2. Glass electrode

Principle

Glass electrode is widely used in determination of pH of an unknown solution. If two electrodes of different pH are separated by a thin glass membrane, a potential is developed across the membrane due to concentration difference. If pH of one solution is kept constant while pH of other solution varies, the electrode potential can be represented as

$$E_G = E_G^0 - \frac{RT}{F}\ln a_{H^+}^2 = E_G^0 - \frac{2.303RT}{F}\log a_{H^+} = E_G^0 + 0.0592\,\text{pH} \qquad \text{(at 25ºC)}$$

In a glass electrode, a glass membrane is used to separate Ag|AgCl(s), 0.1 (N)HCl solution and the unknown solution, whose pH is to be determined. To measure the pH of an unknown solution, glass electrode is connected with a standard calomel electrode as shown below

Ag|AgCl(s), 0.1 (N)HCl|Glas|Unknown solution (a_{H^+})|Standard Calomel electrode

The e.m.f. of the cell is given by

$$E = E_G - E_{cal} = E_G^0 + 0.0592\text{ pH} - E_{cal} \qquad \text{(at 25ºC)}$$

E_G^0 is not known. Thus pH of unknown solution is determined against known pH of a standard solution. Suppose, pH(S) is the pH of the standard solution and E_S is the e.m.f. of the corresponding cell.

$$E_S = E_G^0 + 0.0592\text{ pH(S)} - E_{cal} \qquad \text{(at 25ºC)}$$

From the above two equations, we get

$$\Delta E = E - E_S = 0.0592\,[\text{pH} - \text{pH(S)}] \qquad \text{(at 25ºC)}$$

pH(S) is known, E and E_S are measurable quantities. So pH of unknown solution can be calculated from the above equation.

Advantages of glass electrode

(a) It is simple and easy to handle.
(b) Equilibrium across the glass membrane is quickly achieved.
(c) Accurate pH value can be obtained.
(d) It can be used to measure very high pH quite accurately.
(e) The glass electrode is not affected in presence of impurities, ions, acid, alkali, oxidizing agents and reducing agents.

Disadvantages of glass electrode

(a) Special glass membranes are required to measure high pH especially for pH > 10.
(b) Ordinary potentiometer cannot be used due to high resistance of the glass membrane.
(c) The glass electrode is safe to measure pH up to 12 but for solutions, having pH > 12, cations affect the glass membrane and the results are erroneous.

(ii) Determination of solubility product of a sparingly soluble salt

Let us find out the solubility product of silver chloride [AgCl(s)], which dissociates in aqueous solution as follows:

$$AgCl(s) \rightleftharpoons Ag^+ + Cl^-$$

Solubility product = $K_{aP} = a_{Ag^+} . a_{Cl^-}$. If S is the solubility, $K_{aP} = S^2$
Let us consider the following cell:

$$\overset{\ominus}{Ag(s)} \Big| AgCl(s),\ 0.1\ N\ KCl,\ Hg_2Cl_2(s) \Big| \overset{\oplus}{Hg(l)}$$

The R.H.E. is the calomel electrode. The e.m.f. of calomel electrode at 0.1 N KCl, is −0.334 V. The cell e.m.f. is given by:

$$E = \phi_{Ag} - \phi_{cal} = \phi_{Ag} + 0.334$$

The cell e.m.f. E is measured and its value is 0.0455 V. Hence,

$$\phi_{Ag} = E - 0.334 = 0.0455 - 0.334 = -0.289\ V$$

Again, L.H.E.: $\quad Ag(s) \rightarrow Ag^+ + e$

$$\phi_{Ag} = \phi^0_{Ag} - \frac{RT}{F} \ln a_{Ag^+}$$

$$= \phi^0_{Ag} - 0.0592 \log \frac{a_{Ag^+}\, a_{Cl^-}}{a_{Cl^-}} \quad \text{[at 25°C]}$$

$$= -0.799 - 0.0592 \log \frac{K_{aP}}{a_{Cl^-}}$$

[since, $\phi^0_{Ag} = -0.799$ V (Table 5.4.1)]

$$a_{Cl^-} = m_{Cl^-}\, \gamma_{Cl^-} = 0.1 \times 0.769 = 0.0769$$

[since, $\gamma_{Cl^-} = 0.769$ at $m_{Cl^-} = 0.1$ N]

Putting all these values in the above equation and solving K_{ap} we get:

$$K_{ap} = 1.75 \times 10^{-10}$$

Again, $\quad K_{ap} = S^2$

Hence, $S^2 = 1.75 \times 10^{-10}$

Or, $S = 1.323 \times 10^{-5}$ unit.

(iii) Determination of ionic product of water

Let us consider the following cell:

$$\underset{(1\text{ atm.})}{\overset{\ominus}{H_2(g)}} \Big| NaOH(m_1) \| NaCl(m_2), AgCl(s) \Big| \overset{\oplus}{Ag(s)}$$

L.H.E. $\frac{1}{2} H_2(g) \rightarrow H^+ + e$

$$\phi_L = \phi^0_{H_2|H^+} - \frac{RT}{F} \ln \frac{a_{H^+}}{a^{\frac{1}{2}}_{H_2}}$$

$$\phi_L = \phi^0_{H_2|H^+} - \frac{RT}{F} \ln \frac{a_{H^+}}{p^{\frac{1}{2}}_{H_2}} \quad [\text{since, } a_{H_2} = p_{H_2}]$$

$$\phi^0_{H_2|H^+} = 0.00 \text{ V} \quad \text{and} \quad p_{H_2} = 1 \text{ atm}$$

So, $$\phi_L = -\frac{RT}{F} \ln a_{H^+}$$

R.H.E.

$$Ag(s) \rightarrow Ag^+ + e$$
$$Ag^+ + Cl^- \rightarrow AgCl(s)$$

$$Ag(s) + Cl^- \rightarrow AgCl(s) + e \quad \text{Electrode potential} = \phi_R$$

$$\phi_R = \phi^0_R - \frac{RT}{F} \ln \frac{a_{AgCl(s)}}{a_{Ag(s)} \cdot a_{Cl^-}}$$

Or, $$\phi_R = \phi^0_R + \frac{RT}{F} \ln a_{Cl^-} \quad [\text{since, } a_{Ag(s)} = a_{AgCl(s)} = 1]$$

The cell e.m.f. = E

$$E = \phi_L - \phi_R = -\phi^0_R - \frac{RT}{F} \ln a_{H^+} a_{Cl^-}$$

Or, $$E = E^0 - \frac{RT}{F} \ln a_{H^+} a_{Cl^-} \quad [\text{where, } E^0 = -\phi^0_R]$$

Or, $$E = E^0 - 0.0592 \log \frac{a_{Cl^-} (a_{H^+} a_{OH^-})}{a_{OH^-}} \quad [\text{at } 25°\text{C}]$$

Or, $$E = E^0 - 0.0592 \log \frac{a_{Cl^-} K_w}{a_{OH^-}} \quad [\text{at } 25°\text{C}]$$

[since, $K_w = (a_{H^+} a_{OH^-})$]

Or, $$E = E^0 - 0.0592 \log K_w - 0.0592 \log \frac{m_2}{m_1} \quad \text{[at 25°C]}$$

[Putting the values of a_{Cl^-} and a_{OH^-}]

$$\log K_w = -\left(\frac{E - E^0}{0.0592}\right) - \log \frac{m_2}{m_1} \quad \text{[at 25°C]} \qquad (5.4.7)$$

m_1 and m_2 are the known concentrations of caustic soda (NaOH) and sodium chloride (NaCl). E^0 is also a known quantity. Thus simply by measuring the cell e.m.f. (E) the ionic product of water K_w can easily be determined by using Equation (5.4.7).

(iv) Determination of equilibrium constant

For any reversible chemical reaction, we know that,

$$\Delta G^0 = -RT \ln K_a \qquad (5.4.8a)$$

Where, ΔG^0 = Standard free energy change and K_a = equilibrium constant of that reversible reaction.

Again, we know that, $\Delta G = -nFE$ [Equation (5.4.1a)]. Hence, $\Delta G^0 = -nFE^0$, where, n is the number of electrons exchange in the chemical reaction.

Or, $$\Delta G^0 = -nFE^0 \qquad (5.4.8b)$$

Comparing Equations (5.4.8a) and (5.4.8b), we get:

$$E^0 = \frac{RT}{nF} \ln K_a = \frac{0.0592}{n} \log K_a \quad \text{[at 25°C]}$$

Or, $$\log K_a = \frac{nE^0}{0.0592} \quad \text{[at 25°C]} \qquad (5.4.8c)$$

For any reversible cell, n and E^0 are known. Hence the equilibrium constant K_a can easily be determined from the Equation (5.4.8c).

Example 5.1 One example is Daniell-Jacobi-cell. The cell reaction is given by:

$$Zn(s) + Cu^{2+} \rightleftharpoons Zn^{2+} + Cu(s) \quad (n = 2)$$

$$E^0 = \phi^0_{Zn} - \phi^0_{Cu} = 0.763 - (-0.337) = 1.1 \text{ V} \quad \text{[From Table 5.4.1]}$$

The equilibrium constant K_a is given by:

$$K_a = \frac{a^{eq.}_{Zn^{2+}}\, a^{eq.}_{Cu(s)}}{a^{eq.}_{Cu^{2+}}\, a^{eq.}_{Zn(s)}} = \frac{a^{eq.}_{Zn^{2+}}}{a^{eq.}_{Cu^{2+}}}$$

where, a_i^{eq} = activity of the ith species at equilibrium condition.

From Equation (5.4.8c), we have:

$$\log K_a = \frac{2 \times 1.1}{0.0592} = 37.162$$

Or, $$K_a = 1.45 \times 10^{37} \quad \text{[at 25°C]}$$

Example 5.2 Consider the following redox reaction:

$$Sn^{2+} + 2Fe^{3+} \rightleftharpoons Sn^{4+} + 2Fe^{2+}$$

The equilibrium constant K_a is given by:

$$K_a = \frac{a^{\text{eq.}}_{Sn^{4+}} \left(a^{\text{eq.}}_{Fe^{2+}}\right)^2}{a^{\text{eq.}}_{Sn^{2+}} \left(a^{\text{eq.}}_{Fe^{3+}}\right)^2}$$

$$E^0 = \phi^0_{Sn} - \phi^0_{Fe} = -0.150 - (-0.771) = 0.621 \text{ V}$$

Putting these values of n and E^0 in the Equation (5.4.8c), we get:

$$\log K_a = \frac{2 \times 0.621}{0.0592} = 20.98$$

Or,

$$K_a = 9.54 \times 10^{20} \quad [\text{at } 25°\text{C}]$$

(v) Potentiometric titration

At first a suitable indicator electrode is placed in the titrant (the solution to be titrated). So a potential is developed around the electrode, denoted by ϕ_a. Then this electrode is connected with a reference electrode (ϕ_R) (say, calomel electrode) to form a complete cell. The cell e.m.f. = $E = (\phi_a - \phi_R)$.

As ϕ_R is fixed, any change in ϕ_a results in a change in E. The titre is added drop-wise from outside and the cell e.m.f (E) is measured continuously. A slight change in E is observed up to the end point. After the end point titrant does not exist but only titre exists, and hence the nature of the electrode also changes. The potential developed around the electrode is ϕ_b not ϕ_a and the cell e.m.f. is given by: $E = (\phi_b - \phi_R)$.

Thus at the end point a sharp change in potential occurs as shown in the Figure 5.4.5.

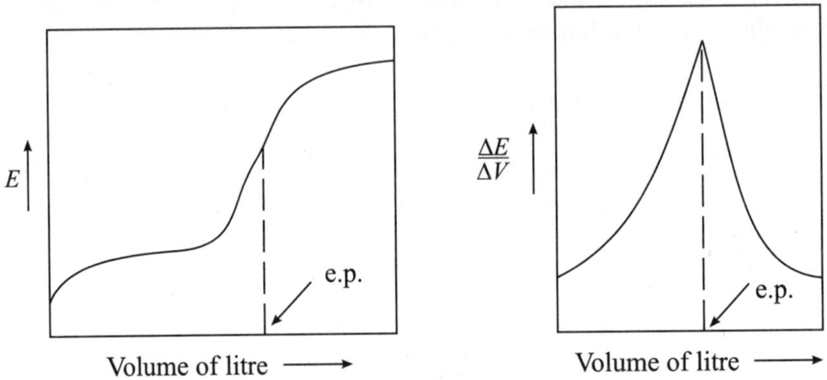

Fig. 5.4.5 **(a)** Cell e.m.f. (E) Vs. volume of titre plot of a typical redox reaction.

(b) $\left(\frac{\Delta E}{\Delta V}\right)$ Vs. volume plot of the same redox reaction.

The titre value at which the last trace of titrant is consumed by the titre, is known as the *equivalence point* and the corresponding potential is called *equivalence potential*, denoted by E_p. The equivalence point is difficult to identify in Figure 5.4.5a but it is very sharp in Figure 5.4.5b.

Selection of indicator electrode depends on types of titration, e.g.,

(1) in the case of *redox titration* the indicator electrode is usually *Pt-electrode*.
(2) in the case of *precipitation titration* like titration of sodium chloride (NaCl) by silver nitrate ($AgNO_3$), *silver electrode* is chosen as an indicator electrode.
(3) In the case of *acid-base titration* a suitable *hydrogen electrode* is used as an indicator electrode.

Calculation of equivalence potential (E_p) of a redox reaction:
A redox reaction can be written as:

$$a\,\mathrm{Ox}_1 + b\,\mathrm{Red}_2 \rightleftharpoons a\,\mathrm{Red}_1 + b\,\mathrm{Ox}_2$$

Ox_1 = Oxidizing agent, converted to corresponding reduced form (Red_1)

Red_2 = Reducing agent, converted to corresponding oxidizing form (Ox_2)

At cathode: $\mathrm{Red}_2 \rightarrow \mathrm{Ox}_2 + \mathrm{be}$

$$\phi_c = \phi_c^0 - \frac{RT}{bF}\ln\frac{[\mathrm{Ox}_2]}{[\mathrm{Red}_2]}$$

At anode: $\mathrm{Red}_1 \rightarrow \mathrm{Ox}_1 + ae$

$$\phi_a = \phi_a^0 - \frac{RT}{aF}\ln\frac{[\mathrm{Ox}_1]}{[\mathrm{Red}_1]}$$

At the equivalence point: $\phi_c = \phi_a = E_P$

$$E_p = \phi_c^0 - \frac{RT}{bF}\ln\frac{[\mathrm{Ox}_2]}{[\mathrm{Red}_2]} \quad \text{and} \quad E_p = \phi_a^0 - \frac{RT}{aF}\ln\frac{[\mathrm{Ox}_1]}{[\mathrm{Red}_1]}$$

Or,

$$bE_p = b\phi_c^0 - \frac{RT}{F}\ln\frac{[\mathrm{Ox}_2]}{[\mathrm{Red}_2]} \quad \text{and} \quad aE_p = a\phi_a^0 - \frac{RT}{F}\ln\frac{[\mathrm{Ox}_1]}{[\mathrm{Red}_1]}$$

Again, at the equivalence point: $a[\mathrm{Ox}_1] = b[\mathrm{Red}_2]$ and $a[\mathrm{Red}_1] = b[\mathrm{Ox}_2]$

Hence,

$$\frac{[\mathrm{Ox}_1]}{[\mathrm{Red}_1]} = \frac{[\mathrm{Red}_2]}{[\mathrm{Ox}_2]}$$

Or,

$$\frac{[\mathrm{Ox}_1]}{[\mathrm{Red}_1]} \times \frac{[\mathrm{Ox}_2]}{[\mathrm{Red}_2]} = 1$$

Thus,

$$(a+b)E_p = (b\phi_c^0 + a\phi_a^0) - \frac{RT}{F}\ln\left\{\frac{[\mathrm{Ox}_1]}{[\mathrm{Red}_1]} \times \frac{[\mathrm{Ox}_2]}{[\mathrm{Red}_2]}\right\}$$

Or, $(a + b)E_p = (b\phi_c^0 + a\phi_a^0)$ since $\frac{[Ox_1]}{[Red_1]} \times \frac{[Ox_2]}{[Red_2]} = 1$

Or, $$E_p = \left[\frac{(b\phi_c^0 + a\phi_a^0)}{(a+b)}\right] \quad (5.4.9)$$

5.5 Concentration cell and accumulator

5.5.1 Concentration cell

An important type of electrochemical cell is concentration cell. In this special type of cell, electrolyte is the same at both the electrodes. There is no net chemical reaction in the cell. So the current produced in this cell is not due to chemical reaction but due to concentration difference of electrolyte at two electrodes. Let us consider a concentration cell, where two similar electrodes are separated by a reference electrode.

$$\ominus \qquad\qquad\qquad\qquad \oplus$$
$$\underset{(1\ \text{atm.})}{\text{Pt}, H_2(g)} \Big| HCl(a_1), AgCl(s)|Ag(s)|AgCl(s), HCl(a_2), \Big| \underset{(1\ \text{atm.})}{H_2(g), \text{Pt}}$$

Each electrode is an elementary cell:

L.H.E.

$$\frac{1}{2} H_2(g) \rightarrow H^+ + e$$

$$AgCl(s) + e \rightarrow Ag(s) + Cl^-$$

$$\frac{1}{2} H_2(g) + AgCl(s) \rightarrow Ag(s) + H^+ + Cl^-$$

Or, $$\frac{1}{2} H_2(g) + AgCl(s) \rightarrow Ag(s) + HCl(a_1) \quad (5.5.1a)$$

Thus 1 mole of HCl of activity a_1 is produced at the L.H.E.

R.H.E.

$$H^+ + e \rightarrow \frac{1}{2} H_2(g)$$

$$Ag(s) + Cl^- \rightarrow AgCl(s) + e$$

$$Ag(s) + H^+ + Cl^- \rightarrow \frac{1}{2} H_2(g) + AgCl(s)$$

Or, $$Ag(s) + HCl(a_2) \rightarrow \frac{1}{2} H_2(g) + AgCl(s) \quad (5.5.1b)$$

Thus, 1 mole of HCl of activity a_2 is consumed at the R.H.E.

The net cell reaction is obtained by adding the Equations (5.5.1a) and (5.5.1b):

$$HCl(a_2) \rightarrow HCl(a_1)$$

Free energy change (ΔG) is given by:

$$\Delta G = \mu_{HCl(a_1)} - \mu_{HCl(a_2)} = (\mu^0_{HCl} + RT \ln a_1) - (\mu^0_{HCl} + RT \ln a_2)$$

Or,
$$\Delta G = RT \ln \frac{a_1}{a_2}$$

Again, we have:
$$\Delta G = -FE,$$

where, E is the cell e.m.f.. Thus,

$$-FE = RT \ln \frac{a_1}{a_2}$$

Or,
$$E = \frac{RT}{F} \ln \frac{a_2}{a_1}$$

Again, HCl remains in dissociated form, i.e., $HCl \rightarrow H^+ + Cl^-$

So,
$$a_{HCl} = a_{H^+} \cdot a_{Cl^-} \quad \text{and} \quad a_1 = (a_+)_1\,(a_-)_1$$

If $a_\pm$ is the mean activity, we can write that:

$$(a_\pm)_1^2 = (a_+)_1\,(a_-)_1 \quad \text{and} \quad (a_+)_1 = (a_-)_1 = (a_\pm)_1$$

$$a_1 = (a_\pm)_1^2 \quad \text{and similarly} \quad a_2 = (a_\pm)_2^2 \qquad (5.5.2)$$

Putting these values of a_1 and a_2 in the above equation we get:

$$E = \frac{RT}{F} \ln \frac{(a_\pm)_2^2}{(a_\pm)_1^2}$$

Or,
$$E = \frac{2RT}{F} \ln \frac{(a_\pm)_2}{(a_\pm)_1} \qquad (5.5.3a)$$

Now we consider another cell where the reference electrode is absent.

$$\ominus \qquad\qquad M \qquad\qquad \oplus$$

$$\underset{(1\ \text{atm.})}{Pt, H_2(g)} \;\Bigg|\; \begin{array}{c} HCl(a_1) \mid HCl(a_2) \\ \xrightarrow{t_+} H^+ \\ Cl^- \xleftarrow{t_-} \end{array} \;\Bigg|\; \underset{(1\ \text{atm.})}{H_2(g), Pt}$$

The two hydrochloric acid solutions are separated by a membrane, which allows ions to diffuse from one compartment to another. So, t_+ gm-equiv. of H^+ moves towards the anode and t_- gm-equiv. of Cl^- moves towards the cathode. The transfer of ions at two electrodes are given below:

L.H.E	R.H.E.
$\frac{1}{2}H_2 \rightarrow H^+ + e$	$H^+ + e \rightarrow \frac{1}{2}H_2$
$+1$ mole of H^+ (appears) $-t_+$ mole of H^+ (disappears) $+t_-$ mole of Cl^- (appears)	-1 mole of H^+ (disappears) $+t_+$ mole of H^+ (appears) $-t_-$ mole of Cl^- (disappears)
$+(1 - t_+)$ mole of H^+ (appears) $+t_-$ mole of Cl^- (appears)	$-(1 - t_+)$ mole of H^+ (disappears) $-t_-$ mole of Cl^- (disappears)
$+t_-$ mole of H^+ (appears) $+t_-$ mole of Cl^- (appears)	$-t_-$ mole of H^+ (disappears) $-t_-$ mole of Cl^- (disappears)
$+t_-$ mole of HCl of activity a_1 (appears)	$-t_-$ mole of HCl of activity a_2 (disappears)

Thus free energy change (ΔG) is given by:

$$\Delta G = t_- \mu_{HCl(a_1)} - t_- \mu_{HCl(a_2)} = t_- [\mu_{HCl(a_1)} - \mu_{HCl(a_2)}]$$
$$= t_- [(\mu^0_{HCl} + RT \ln a_1) - (\mu^0_{HCl} + RT \ln a_2)]$$

$$\Delta G = t_- RT \ln \frac{a_1}{a_2}$$

If E_T is the cell e.m.f., we have: $\Delta G = -FE_t$

Or, $$-FE_T = t_- RT \ln \frac{a_1}{a_2}$$

Or, $$E_T = \frac{t_- RT}{F} \ln \frac{a_2}{a_1}$$

Putting the values of a_1 and a_2 from Equation (5.5.2) in the above equation we get:

$$E_T = \frac{t_- RT}{F} \ln \frac{(a_\pm)_2^2}{(a_\pm)_1^2} = \frac{2t_- RT}{F} \ln \frac{(a_\pm)_2}{(a_\pm)_1} \quad (5.5.3b)$$

From Equations (5.5.3a) and (5.5.3b), we have:

$$\frac{E_T}{E} = t_- \quad (5.5.3c)$$

Thus transport number can easily be determined by measuring E_T and E.

5.5.2 Liquid junction potential or diffusion potential

If the electrolytic solutions are separated by a membrane such that ions are not allowed to pass through, the above cell becomes:

$$\ominus \qquad\qquad \text{M} \qquad\qquad \oplus$$

$$\underset{(1\text{ atm.})}{\text{Pt}, H_2(g)}\Big| \text{HCl}(a_1) \Big\| \text{HCl}(a_2) \Big| \underset{(1\text{ atm.})}{H_2(g), \text{Pt}}$$

The above cell is called concentration cell without transference.

L.H.E. $\frac{1}{2} H_2(g) \rightarrow H^+ + e$

R.H.E. $H^+ + e \rightarrow \frac{1}{2} H_2(g)$

So, 1 mole of H^+ is consumed at R.H.E. and 1 mole of H^+ is generated at L.H.E.

So, free energy change (ΔG) is given by:

$$\Delta G = (\mu_{H^+})_1 - (\mu_{H^+})_2 = [\mu^0_{H^+} + RT \ln(a_{H^+})_1] - [\mu^0_{H^+} + RT \ln(a_{H^+})_2]$$

$$= RT \ln \frac{(a_{H^+})_1}{(a_{H^+})_2} = \text{RT} \ln \frac{(a_\pm)_1}{(a_\pm)_2}$$

$$[\text{since, } (a_{H^+})_1 = (a_\pm)_1 \quad \text{and} \quad (a_{H^+})_2 = (a_\pm)_2]$$

Again, we know that, $\Delta G = -FE_w$, where E_w is the cell e.m.f. Thus we have:

$$-FE_w = \text{RT} \ln \frac{(a_\pm)_1}{(a_\pm)_2}$$

Or,

$$E_w = \frac{RT}{F} \ln \frac{(a_\pm)_2}{(a_\pm)_1} \tag{5.5.4}$$

The same cell with transference has the cell e.m.f. E_T given by Equation (5.5.3b).

$$E_T = \frac{2t_- RT}{F} \ln \frac{(a_\pm)_2}{(a_\pm)_1} \quad \text{[Equation (5.5.3b)]}$$

The difference $(E_T - E_w)$ is known as diffusion potential or liquid junction potential. This potential difference arises due to transference of ions or diffusion of ions through the membrane.

$$E_l = (E_T - E_w) = (2t_- - 1) \frac{RT}{F} \ln \frac{(a_\pm)_2}{(a_\pm)_1} \tag{5.5.5}$$

Liquid junction potential If the two electrolytes, dissolved in two immiscible solvents come in contact, a liquid junction potential sets in at the boundary of the solutions. Let us consider the following cell.

$$\ominus \qquad\qquad\qquad\qquad \oplus$$
$$\underset{(1\text{ atm.})}{Pt, H_2(g)}\,\Big|\,HCl(a_1)\,(aq.)\,\Big|\,HCl(a_2)\,(phenol)\,\Big|\,\underset{(1\text{ atm.})}{H_2(g), Pt}$$

Hydrochloric acid is dissolved in both water and phenol. If these two solutions are separated by a membrane a potential difference is automatically developed at the junction of two solutions. This potential is called liquid junction potential (E_l). The net e.m.f. of the above concentration cell is given by:

$$E = (\phi_L + E_l) - \phi_R$$

When the diffusion is near to complete there is no transfer of solute from one phase to another phase. So equilibrium is reached and hence $\Delta G = 0$. As we know, $\Delta G = -FE$ for 1 mole of solute transfer, we have: $E = 0$.

Thus we can write that: $(\phi_L + E_l) - \phi_R = 0$ (at equilibrium)

Or, $$E_l = \phi_R - \phi_L$$

$$\phi_R = \phi_R^0 - \frac{RT}{F} \ln a_{H^+}(\text{phenol}) \text{ and } \phi_L = \phi_L^0 - \frac{RT}{F} \ln a_{H^-}(\text{aq.})$$

$$E_l = \phi_R - \phi_L = (\phi_R^0 - \phi_L^0) - \frac{RT}{F} \ln \frac{a_{H^+}(\text{phenol})}{a_{H^+}(\text{aq.})}$$

$$E_l = E_l^0 - \frac{RT}{F} \ln \frac{a_{H^+}(\text{phenol})}{a_{H^+}(\text{aq.})} \tag{5.5.6}$$

As the solvents ϕ_L^0 and ϕ_R^0 are different and unknown. As ϕ_R^0 is unknown, one cannot evaluate E_l by using Equation (5.5.6). To solve the problem different kind of cell is chosen as given below:

$$\ominus \quad \text{I} \quad \text{II} \quad \text{III} \quad \text{IV} \quad \text{V} \quad \text{VI} \quad \oplus$$
$$Hg(l)\,\Big|\,Hg_2Cl_2(s), \underset{(aq.)}{HCl(a_1)}\,\Big|\,\underset{(phenol)}{HCl(a_2)}\,\Big\|\,\underset{(phenol)}{HCl(a_3)}\,\Big|\,\underset{(aq.)}{HCl(a_4)}\,\Big|\,\underset{(aq.)}{HCl(a_1)}\; Hg_2Cl_2(s)\,\Big|$$
$$E_l^1 \qquad E_d^1 \qquad E_l^2 \qquad E_d^2$$

The net e.m.f. of the above cell is given by:

$$E = (\phi_L + E_l^1 + E_d^1) + E_l^1 - E_d^2 - \phi_R$$

The phases I and VI are equal. So, $\phi_L = \phi_R$

E_d^1 and E_d^2 are the diffusion potentials, i.e., potential at the junction of two similar solutions, phenol-phenol solutions and aqueous-aqueous solutions respectively. E_l^1 and E_l^2 are the liquid junction potentials, i.e. potential at the boundary of two dissimilar solutions, i.e., phenol water solutions. The values of E_d^1 and E_d^2 are very small in compared to E_l^1 and E_l^2. Thus, cell e.m.f. is given by: $E = E_l^1 + E_l^2$

$$E_l^1 = E_{l(\text{W/Ph})}^0 - \frac{RT}{F} \ln \frac{a_{H^+(W)}}{a_{H^+(Ph)}} = E_{l(\text{W/Ph})}^0 - \frac{RT}{F} \ln \frac{a_1}{a_2}$$

Or, $$E_l^1 = E^0_{l(\text{W/Ph})} + \frac{RT}{F} \ln \frac{a_2}{a_1}$$

and $$E_l^2 = E^0_{l(\text{Ph/W})} - \frac{RT}{F} \ln \frac{a_{\text{H}^+(\text{Ph})}}{a_{\text{H}^+(\text{W})}} = -E^0_{l(\text{W/Ph})} - \frac{RT}{F} \ln \frac{a_3}{a_4}$$

$$E = E_l^1 + E_l^2 = \frac{RT}{F}\left[\ln \frac{a_2}{a_1} - \ln \frac{a_3}{a_4}\right] = \frac{RT}{F} \ln \frac{a_2\, a_4}{a_1\, a_3}$$

Or, $$E = 0.0592 \log \frac{a_2\, a_4}{a_1\, a_3} \quad \text{(at 25°C)} \qquad (5.5.7)$$

Putting the values of a_1, a_2 , a_3 and a_4 in the above equation E, the liquid junction potential, can easily be calculated.

Definition of liquid junction potential The potential, developed at the boundary of any two electrolytic solutions is called liquid junction potential. If the solvent of both the solutions is same the liquid junction potential is specially called diffusion potential.

5.5.3 Determination of activity and activity coefficient

Let us consider the cell

$$\overset{\ominus}{\underset{(1\text{ atm.})}{\text{Pt, H}_2\text{(g)}}} \Big| \text{HCl}(a_1)\ \text{AgCl(s)} \,|\, \text{Ag(s)} \,|\, \text{AgCl(s), HCl}(a_2) \Big| \overset{\oplus}{\underset{(1\text{ atm.})}{\text{H}_2\text{(g), Pt}}}$$

The e.m.f. of the cell is given by:

$$E = \frac{2RT}{F} \ln \frac{(a_\pm)_2}{(a_\pm)_1} \qquad \text{[Equation (5.5.3a)]}$$

$$E = 2 \times 0.0592\, [\log(a_\pm)_2 - \log(m_\pm)_1 - \log(\gamma_\pm)_1] \quad \text{[at 25°C]}$$

[since, $(a_\pm)_1 = (m_\pm)_1\ (\gamma_\pm)_1$]

Where $(a_\pm)_1$ and $(a_\pm)_2$ are the mean activities of HCl solutions of cathode and anode respectively.

$$[E + 0.1184\ \log(m_\pm)_1] = 0.1184\ \log(a_\pm)_2 - 0.1184\ \log(\gamma_\pm)_1$$

Or, $$Y = I - 0.1184\ \log(\gamma_\pm)_1 \qquad (5.5.8a)$$

Plot of Y vs. $\sqrt{(m_\pm)_1}$ is shown in the Figure 5.5.1.

$$I = \underset{(m_\pm)\to 0}{\text{Lt}}\ [E + 0.1184\ \log(m_\pm)_1] \qquad (5.5.8b)$$

Extrapolation of the graph on y-axis gives the value of I. Again, $I = 0.1184\ \log(a_\pm)_2$

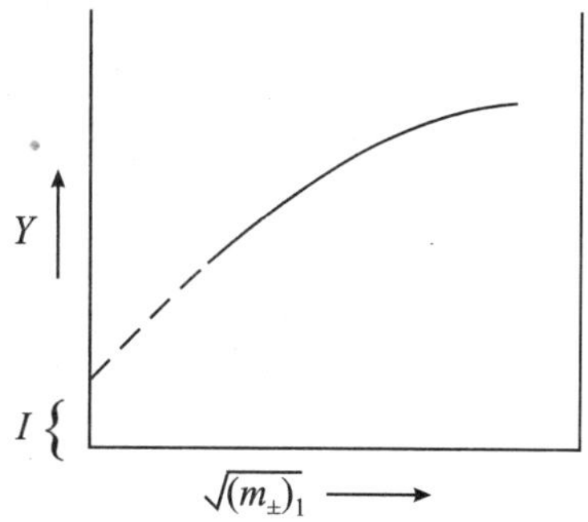

Fig. 5.5.1 Plot Y vs. $\sqrt{(m_\pm)_1}$

Thus $(a_{\pm})_2$ can easily be determined, and hence $(\gamma_{\pm})_2$ can be obtained from the following relation:

$$(a_{\pm})_2 = (m_{\pm})_2\ (\gamma_{\pm})_2$$

$(m_{\pm})_2$ is known. Note that $(\gamma_{\pm})_2$ varies with $(m_{\pm})_2$. Again, from Equation (5.5.8a), we have:

$$\log(\gamma_{\pm})_1 = \frac{(I - Y)}{0.1184}$$

I is known. At a particular value of $(m_{\pm})_1$, Y can be determined from the Figure 5.5.1. Thus $(\gamma_{\pm})_1$ can be easily computed from the above equation. Hence $(a_{\pm})_1$ can be determined from the following relation:

$$(a_{\pm})_1 = (m_{\pm})_1\ (\gamma_{\pm})_1$$

So here also, $(\gamma_{\pm})_1$ varies with $(m_{\pm})_1$.
Thus activity coefficient is not constant but varies with concentration.

5.5.4 Decomposition voltage and over voltage

Exchange currents If an electrode is immersed in an electrolytic solution, exchange of charged particles takes place between electrodes and solution. Due to this exchange some quantity of electricity is passed from electrode to the solution and vice versa. This quantity of exchange electricity per unit time is called exchange current.

Concentration polarization

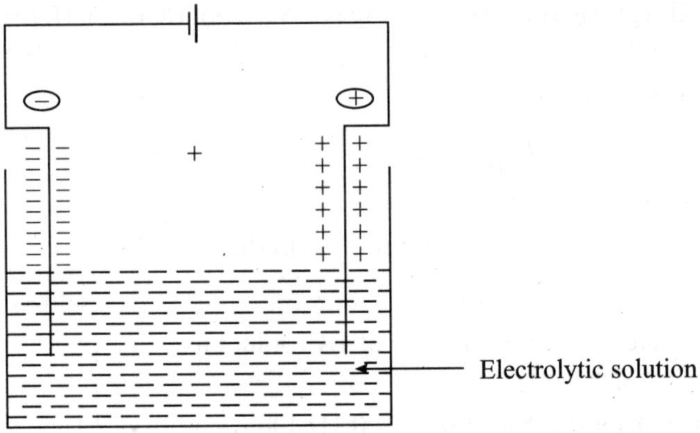

Fig. 5.5.2 Electrolysis of an electrolytic solution C_0 = concentration of electrolyte at the beginning of electrolysis. C_S = concentration of electrolyte near cathode and on the surface of cathode.

During electrolysis cations get deposited on cathode and hence concentration of electrolyte decreases in the vicinity of cathode, though the concentration of cation remains unchanged in the bulk of the solution. Let C_0 and C_S be the concentrations of cation in the bulk and near the cathode respectively. Thus a concentration gradient sets up, resulting in a flow of current in the opposite direction. Due to the existence of concentration gradient cations diffuse from bulk of the solution to cathode. If (n_i) is the number of gm-equiv. of cations diffused per unit time, we can write in accordance with the Fick's law that:

$$n_i = DA\left(\frac{C_0 - C_s}{l}\right)$$

Where, D = Diffusion coefficient of discharging ion

A = surface area of electrode

l = Thickness of the diffused layer.

As we know F is the electricity carried by 1 gm-equiv. of ions, the total current carried due to mass flow rate of cation n_i is:

$$I = n_i F = FDA\left(\frac{C_0 - C_s}{l}\right) \tag{5.5.9a}$$

The value of diffusion current (I) becomes maximum when $C_s = 0$, i.e. all cations get deposited at cathode.

$$I = I_0 = FDA\frac{C_0}{l}$$

Or, $$C_0 = \frac{I_0 l}{FDA}$$

Again, from Equation (5.5.9a), we have:

$$(C_0 - C_s) = \frac{I.l}{FDA}$$

Or, $$C_0\left(1 - \frac{C_s}{C_0}\right) = \frac{I.l}{FDA}$$

Putting the value of C_0 in the above equation we get:

$$\frac{I_0 l}{FDA}\left(1 - \frac{C_s}{C_0}\right) = \frac{I.l}{FDA}$$

Or, $$\left(1 - \frac{C_s}{C_0}\right) = \frac{I}{I_0}$$

Or, $$\left(1 - \frac{I}{I_0}\right) = \frac{C_s}{C_0}$$

Or, $$C_s = C_0\left(1 - \frac{I}{I_0}\right) \tag{5.5.9b}$$

According to Nernst equation we can write that:

$$\phi_{\text{bulk}} = \phi_0 - \frac{RT}{Z_+F}\ln C_0 \quad \text{and} \quad \phi_{\text{surface}} = \phi_0 - \frac{RT}{Z_+F}\ln C_s$$

Where Z_+ is the valency of the cation. Thus,

$$\Delta\phi_c = \phi_{\text{bulk}} - \phi_{\text{surface}} = -\frac{RT}{Z_+F}\ln\frac{C_0}{C_s}$$

$$= \frac{RT}{Z_+F}\ln\frac{C_S}{C_0} = \frac{RT}{Z_+F}\ln\left(I - \frac{I}{I_0}\right) \quad \text{[From Equation (5.5.9b)}$$

Or,

$$\Delta\phi_c = \frac{RT}{Z_+F}\ln\left(I - \frac{I}{I_0}\right) \tag{5.5.9c}$$

$\Delta\phi_c$ is called potential due to concentration polarization at cathode. Similarly, we have $\Delta\phi_a$, which is the potential due to concentration polarization at anode. Thus e.m.f. of concentration polarization is given by:

$$E_{\text{conc.}} = \Delta\phi_c + \Delta\phi_a$$

As the direction of $E_{\text{conc.}}$ is opposite to the applied voltage ($E_{\text{appl.}}$), the net e.m.f. required to execute the electrolysis = E

$$E = E_{\text{appl.}} + E_{\text{conc.}}$$

In industrial application the value of $E_{\text{conc.}}$ is sufficiently high and hence considerable amount of electrical energy is required to overcome the concentration e.m.f.

To avoid this loss of electrical energy continuous stirring is employed to avoid setting up concentration gradient.

Electrochemical polarization In the case of electronic conductor current density is directly proportional to the applied voltage, i.e. $I = \frac{E}{R}$ [Ohm's law]

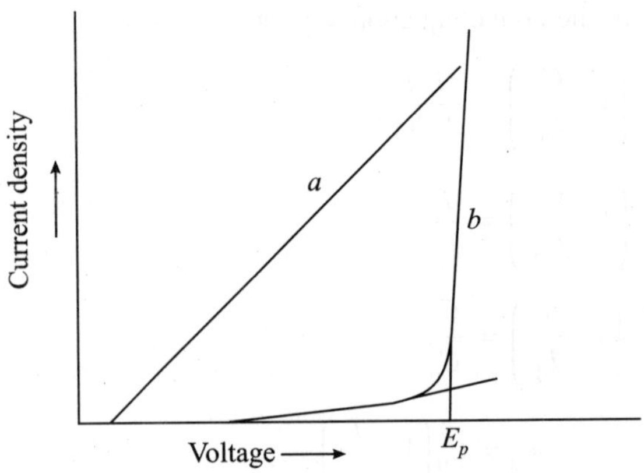

Fig. 5.5.3 Current density Vs. voltage curves for electronic conductors (Curve *a*) and for electrolytic conductors (Curve *b*)

In the case of electrolytic conductor, the curve characteristics shows that the current realized in the circuit remains insignificant until a certain voltage is reached. After that growth of current with voltage is very fast and the curve rises very steeply. The cross-over voltage is known as polarization voltage, denoted by E_P (shown in the Figure 5.5.3). Thus the current (I) is related to the applied voltage (E) in the following way:

$$I = \frac{(E - E_P)}{R} \tag{5.5.10}$$

Where, E_P is the polarization voltage.

In absence of any external e.m.f., ionic equilibrium is established at both the electrodes. On application of an external e.m.f. cathode receives extra negative charge and anode receives extra positive charge. As a result of this equilibrium is disturbed and a potential difference sets in against the direction of applied e.m.f. This potential difference is our well-known polarization voltage (E_P), which depends on electrolytic solution. The extra negative charge received by cathode is shifted to the surrounding cations, e.g. in the case of HCl solution the following reaction takes place at the cathode:

$$2H_3O^+ + 2e \rightarrow 2H_2O + H_2 \text{ (absorbed on Pt)} \rightarrow 2H_2O + H_2(g)\uparrow$$

The process is called discharging of ions, which is the origin of polarization. Hence discharging of ions is also known as electrochemical polarization, and the potential due to discharging of ions, is called discharge potential. Thus discharge potential is the potential required to discharge an ion to the corresponding element. The sum of discharge potentials of ions at two electrodes is known as decomposition voltage of the electrolyte. Thus the decomposition voltage of an electrolyte is the minimum potential difference required to begin electrolysis. Thus it is clear that the decomposition voltage of an electrolyte is equal to the e.m.f. of polarization.

5.5.5 Rule of discharging of ion

If more than one cation is present at cathode that cation having the lowest oxidation potential (or the highest discharge potential) will get deposited first. When cathode potential increases and reaches next higher oxidation potential (or lower discharge potential) the next cation, having second lowest oxidation potential will start depositing and so on. For example, in acidic aqueous solution of KCl two species of cations namely K^+ and H_3O^+ are present, but discharge potential of K^+ ion is –2.925 V and that of H_3O^+ ion is around zero volt.

$$K^+ + e \rightarrow K \quad \phi = -2.925 \text{ V (discharge of } K^+ \text{ ion)} \quad \text{[see Table 5.4.1]}$$

Hence liberation of hydrogen always takes place at cathode during electrolysis of KCl solution according to the following reaction.

$$2H_3O^+ + 2e \rightarrow 2H_2O + H_2(g)\uparrow$$

If the concentration of hydroxonium ion (H_3O^+) is sufficiently low the following reaction takes place at the cathode.

$$2H_2O + 2e \rightarrow 2OH^- + H_2(g)\uparrow$$

If more than one anion is present at anode, that anion having highest oxidation potential will get deposited first. For example if OH^-, H_2O and SO_4^{2-} are present, following anodic reactions may occur:

$$4OH^- \rightarrow 2H_2O + O_2(g) + 4e \quad \ldots(a)\ \phi^0 \text{ is around } -1.23 \text{ V}$$

$$6H_2O \rightarrow 4H_3O^+ + O_2(g) + 6e \quad \ldots(b)\ \phi^0 \text{ is around } -1.00 \text{ V}$$

$$[2SO_4^{2-} \rightarrow 2SO_4 + 4e$$

$$2SO_4 + 6H_2O \rightarrow 2SO_4^{2-} + 4H_3O^+ + O_2(g)] \quad \ldots(c)\ \phi^0 \text{ is around } -2.00 \text{ V}$$

Thus in alkaline solution hydroxyl ion will get deposited first and in acidic solution hydroxonium ion (H_3O^+) will be produced first but in no case sulfate ion (SO_4^{2-}) will be deposited at anode.

Thus during electrolysis of an electrolyte only selected ions are deposited at cathode and anode and the remaining ions maintain sufficient electrical conductance to continue the electrolysis.

Electrolysis of hydrochloric acid solution At anode both OH^- and Cl^- are present. Following two reactions may proceed at anode.

$$4OH^- \rightarrow 2H_2O + O_2(g) + 4e \qquad \phi^0 = -1.23 \text{ V}$$

$$2Cl^- \rightarrow Cl_2(g) + 2e \qquad \phi^0 = -1.36 \text{ V}$$

$$\phi_{Cl^-/Cl_2} = \phi^0_{Cl^-/Cl_2} - \frac{RT}{F} \ln \frac{p_{Cl_2}^{\frac{1}{2}}}{a_{Cl^-}}$$

The electrode is so set that $p_{Cl_2} = 1$ atm

So $$\phi_{Cl^-/Cl_2} = \phi^0_{Cl^-/Cl_2} + \frac{RT}{F} \ln a_{Cl^-} = -1.36 + 0.0592 \log a_{Cl^-} \quad \text{[at 25°C]}$$

At high concentration of Cl^- ions, a_{Cl^-} is high and hence net electrode potential ϕ_{Cl^-/Cl_2} is higher then -1.23 V and hence chlorine is liberated at the anode, whereas in dilute solutions a_{Cl^-} is low enough so that the value of ϕ_{Cl^-/Cl_2} falls below -1.23 V and hence oxygen is liberated at the anode.

Decomposition voltage depends on several factors:

(a) shape and dimension of the electrodes.
(b) Nature of the electrode surface.
(c) Removal of products formed during electrode processes.
(d) Concentration of ions.

The nature of electrode process depends on the net electrode potential of ions. At cathode that electrode process, having lowest oxidation potential, takes place but at anode that electrode process, having highest oxidation potential, takes place.

Consider a cell where the following electrode processes take place:

At cathode: $M^{p+} + pe \rightarrow M(s)$ (discharge of cation)

At anode: $N^{q-} \rightarrow N + qe$ (discharge of anion)

Where, p, q are the valencies of cation and anion respectively.

Discharge potential of cation = ϕ_c

$$\phi_c = \phi_c^0 + \frac{RT}{pF} \ln a_+$$

where, a_+ = activity of M^{p+}, and $a_M = 1$.

Discharge potential of anion = ϕ_a

$$\phi_a = \phi_a^0 + \frac{RT}{qF} \ln a_-$$

where, a_- = Activity of N^{q-} and $a_{N(g)} = p_{N(g)} = 1$ atm.

Decomposition voltage = $E_{dc} = \phi_c + \phi_a$

$$E_{dc} = (\phi_c^0 + \phi_a^0) + \frac{RT}{pF} \ln a_+ + \frac{RT}{qF} \ln a_-$$

$$E_{dc} = E_{dc}^0 + \frac{RT}{pF} \ln a_+ + \frac{RT}{qF} \ln a_- \qquad (5.5.11)$$

where, $E_{dc}^0 = (\phi_c^0 + \phi_a^0)$

It is to be noted here that ϕ_c and ϕ_c^0 are the reduction potentials and ϕ_a and ϕ_a^0 are the oxidation potentials.

5.5.6 Over-voltage

According to the knowledge of decomposition voltage, to begin an electrolysis the minimum potential difference (p.d.) applied from an external source is equal to the decomposition voltage. In actual practice the minimum p.d. required to begin an electrolysis is always greater than the decomposition voltage. This excess voltage, required to begin an electrolysis, is called over-potential or over-voltage.

If η_c and η_a are the over potentials at cathode and anode respectively, we can write that:

$$\eta_c = \phi_c - \phi_{dc} \quad \text{and} \quad \eta_a = \phi_a - \phi_{da} \qquad (5.5.12a)$$

ϕ_{dc} and ϕ_{da} are the decomposition potentials of cathode and anode respectively.

The total over voltage of the cell = $(\eta_c + \eta_a) = (\phi_c + \phi_a) - (\phi_{dc} + \phi_{da})$

It is to be noted that the excess voltage, i.e., $[(\phi_c + \phi_a) - (\phi_{dc} + \phi_{da})]$ is the sum of over-voltage (η) and Ohmic voltage drop in solution.

The Ohmic voltage drop is small enough to be neglected in compared to over voltage. Hence we can write that:

$$\eta = (\phi_c + \phi_a) - (\phi_{dc} + \phi_{da}) \qquad (5.5.12b)$$

The magnitude of the over voltage on an electrode depends on several factors: (a) Nature of the electrode, (b) Density of the current (j), (c) Composition of the solution.

I. Tafel (1905) was the first scientist who proposed the following relation between over voltage and current density (j).

$$\eta = a + b \log j \quad (5.5.13a)$$

Where, a and b are the constants. The value of a for any electrochemical process, largely depends on the materials of electrode, but the value of b slightly depends on materials and is approximately equal to $2 \times 0.0592 = 0.1184$ at 25°C. Again, at very low current density, $j \rightarrow 0$ or, $j \rightarrow e^{-\infty}$. So, $\log j \rightarrow -\infty$. Hence, $\eta \rightarrow -\infty$. In actual practice $\eta \rightarrow 0$ while $j \rightarrow 0$. Thus the Equation (5.5.13a) is not applicable at very low current density.

5.5.7 Theories of hydrogen over-voltage

Discharge of an ion usually proceeds through multiple steps. If all the steps proceed at comparable rate there will be no over voltage but if one of the steps proceeds at relatively low rate, over voltage is resulted. The discharge of hydroxonium ion consists of five steps:

Step-I Diffusion In this step hydroxonium ion (H_3O^+) diffuses from aqueous solution to electrode (cathode).

Step-II Dehydration In this step dehydration of hydroxonium ion takes place according to the following reaction:

$$H_3O^+ \rightarrow H^+ + H_2O$$

Step-III Discharge and adsorption In this step H^+ ion gets converted to atomic hydrogen which is then adsorbed on the electrode surface.

$$H^+ + e \rightarrow H \text{ (nascent)}$$

$$H\text{(nascent)} + Me \rightarrow H(Me), \text{ where, } Me = \text{a metal or glass electrode.}$$

Step-IV Recombination In this step adsorbed atomic hydrogen recombines to molecular hydrogen which still remains adsorbed on the surface.

$$2H(Me) \rightarrow H_2(Me)$$

Step-V Gas liberation step In this step adsorbed hydrogen is liberated from the electrode as gas.

$$H_2(Me) \rightarrow H_2(g)\uparrow + Me$$

Among the five steps it is very difficult to identify the slowest one. According to Taffel step-IV is the slowest one but according to Volmer et al. discharge of ion, i.e. step-III is the slowest step.

Recombination Theory According to this theory step-IV is the slowest step.

$$2H(Me) \rightarrow H_2(Me)$$

$$J = \text{Rate} = k\,.\,[H]^2 = \frac{dn}{dt}$$

Or,
$$\frac{dn}{dt} = k\,.\,[H]^2$$

where, $\frac{dn}{dt}$ = number of moles of hydrogen produced per second.

If current intensity (j) is applied for time dt, we can write that:
$j\,dt = zF\,dn$. Where z is the number of electrons transferred. In this case, $z = 2$.

Hence, $$\frac{j}{2F} = \frac{dn}{dt} \quad \left[\text{since, } 2\text{H}^+ + 2\text{e} \rightarrow 2\text{H} \text{ and } j = \frac{I}{A}\right]$$

Combining the above two equations we get:

$$k\,.\,[\text{H}]^2 = \frac{j}{2F}$$

Or, $$[\text{H}]^2 = \frac{j}{2kF}$$

Or, $$[\text{H}] = k_1\sqrt{j} \quad (5.5.13b)$$

where, $$k_1 = \left(\frac{1}{2kF}\right)^{\frac{1}{2}}$$

Thus the amount of hydrogen adsorbed is always greater than the amount of hydrogen recombined, since the latter is the *slowest* process. The over-voltage appears due to increase in adsorbed hydrogen on the electrode.

If $[H_c]$ is the amount of adsorbed hydrogen converted to molecular hydrogen, the over-voltage is given by:

$$\eta = \phi_\text{H} - \phi_{\text{H}_c} = \left\{\phi^0 + \frac{RT}{F}\ln[\text{H}]\right\} - \left\{\phi^0 + \frac{RT}{F}\ln[\text{H}_c]\right\}$$

$$= \frac{RT}{F}\ln\frac{[\text{H}]}{[\text{H}_c]}$$

Putting the value of [H] from Equation (5.5.13b) in the above equation we get:

$$\eta = \frac{RT}{F}\ln\frac{k_1\sqrt{j}}{[\text{H}_c]} = a + \frac{RT}{2F}\ln j$$

Or, $$\eta = a + 0.0296 \log j \quad (\text{at } 25°\text{C}) \quad (5.5.13c)$$

The above equation is similar to the Equation (5.5.13a) but the value of 'b' is 0.0296 in Equation (5.5.13c). In actual practice the value of 'b' is around 0.1184. This is the serious drawback of Tafel's recombination theory.

Theory of slow discharge of ions According to this theory discharge of ions is the slowest step. The discharge of hydroxonium ion consists of two steps.

$$\text{H}_3\text{O}^+ \xrightarrow{\text{slow}} \text{H}^+ + \text{H}_2\text{O} \quad (\text{dehydration})$$

$$\text{H}^+ + \text{e} \xrightarrow{\text{fast}} \text{H} \quad (\text{discharge})$$

Dehydration step is the slowest step and hence rate = J

$$J = k\,.\,[H_3O^+] = A.e^{-\frac{E}{RT}}\,[H_3O^+] \quad \text{[since we know that, } k = A.e^{-\frac{E}{RT}}\text{]}$$

$E = E^0 - \alpha F\eta$, where, E^0 = Activation energy of an unpolarized electrode and $(\alpha F\eta)$ is the activation energy due to over-voltage and α is a constant.

Again, we have:

$$J = \frac{dn}{dt} = \frac{j}{2F}$$

where j is the current density $\left(\frac{I}{A}\right)$ in $\frac{\text{Amp}}{\text{cm}^2}$.

Hence we can write that:

$$J = \frac{j}{2F} = A\,.\,[H_3O^+]\,e^{-\frac{(E^0 - \alpha F\eta)}{RT}} = A\,.\,[H_3O^+]\,e^{-\frac{E^0}{RT}}\,e^{\frac{\alpha F\eta}{RT}}$$

$$= C\,.\,e^{\frac{\alpha F\eta}{RT}}\text{, where, } C = A[H_3O^+]\,e^{-\frac{E^0}{RT}}$$

Or,

$$j = 2FC\,.\,e^{\frac{\alpha F\eta}{RT}}$$

Taking *ln* on both sides we get:

$$\ln j = \ln 2FC + \frac{\alpha F\eta}{RT}$$

Or,

$$\eta = a + \frac{RT}{\alpha F}\ln j = a + \frac{0.0592}{\alpha}\log j \quad \text{(at 25°C)} \qquad (5.5.13d)$$

The coefficient of log j in the above equation is the constant b in Equation (5.5.13a).

Or,

$$b = \frac{0.0592}{\alpha}\quad \text{At } \alpha = \frac{1}{2},\quad b = 0.1184$$

The value of b in Equation (5.5.13d) is equal to that found in actual practice.

5.5.8 Accumulators

In an electrochemical cell chemical energy is the source of current or electricity. In many cells electricity is obtained only once due to irreversible nature of chemical reactions. These type of cells are called galvanic cells. In other type of cells energy is stored and current is drawn from the cell with repeated action. This type of cell is called storage cell or accumulator.

Principle of accumulator At first electric current is passed from an external source to execute electrochemical reactions. The free energy of reaction is stored within the cell. The process is known as charging.

During charging electrons are accepted from cathode plate and electrons are rejected to anode plate. In the process of discharge current is withdrawn from the cell. So in the discharging process electrons are rejected to cathode plate and electrons are accepted from anode plate. Some widely used accumulators are:

(i) Lead-acid accumulator
(ii) Cadmium-Nickel alkaline accumulator
(iii) Silver-Zinc alkaline accumulator

(i) Lead-acid accumulator The cell is given by:

$$\ominus \ \mathrm{Pb, PbSO_4(s)} \,|\, \mathrm{H_2SO_4\ (32-34\ \%)} \,|\, \mathrm{PbO_2, Pb} \ \oplus$$

Charging at cathode:

$$\mathrm{PbSO_4(s) \rightarrow Pb^{2+} + SO_4^{2-}}$$
$$\mathrm{Pb^{2+} + 2e \rightarrow Pb(s)}$$
$$\mathrm{PbSO_4(s) + 2e \rightarrow Pb(s) + SO_4^{2-}}$$

Discharging at cathode:

$$\mathrm{Pb(s) \rightarrow Pb^{2+} + 2e}$$
$$\mathrm{Pb^{2+} + SO_4^{2-} \rightarrow PbSO_4(s)}$$
$$\mathrm{Pb(s) + SO_4^{2-} \rightarrow PbSO_4(s) + 2e}$$

Charging at anode:

$$\mathrm{PbSO_4(s) \rightarrow Pb^{2+} + SO_4^{2-}}$$
$$\mathrm{Pb^{2+} + 2H_2O \rightarrow PbO_2(s) + 4H^+ + 2e}$$
$$\mathrm{PbSO_4(s) + 2H_2O \rightarrow PbO_2(s) + SO_4^{2-} + 4H^+ + 2e}$$

Discharging at anode: $\mathrm{PbO_2(s) + 4H^+ + 2e \rightarrow Pb^{2+} + 2H_2O}$

$$\mathrm{Pb^{2+} + SO_4^{2-} \rightarrow PbSO_4(s)}$$
$$\mathrm{PbO_2(s) + 4H^+ + + SO_4^{2-} + 2e \rightarrow PbSO_4(s) + 2H_2O}$$

The cell reaction during charging process is given by:

$$\mathrm{2PbSO_4(s) + 2H_2O \rightarrow Pb(s) + PbO_2(s) + 4H^+ + 2SO_4^{2-}}$$

During discharging process the overall cell reaction is given by:

$$\mathrm{Pb(s) + PbO_2(s) + 4H^+ + 2SO_4^{2-} \rightarrow 2PbSO_4(s) + 2H_2O}$$

The combined cell reaction is:

$$\mathrm{Pb(s) + PbO_2(s) + 4H^+ + 2SO_4^{2-} \underset{\text{Charge}}{\overset{\text{Discharge}}{\rightleftharpoons}} 2PbSO_4(s) + 2H_2O}$$

Calculation of cell e.m.f. during discharging process:

$$\phi_c^0 = \phi^0_{\mathrm{Pb|Pb^{2+}}} = +0.126 \text{ V} \quad \text{[see Table 5.4.1]}$$

e.m.f. at cathode: $\phi_c = \phi_c^0 - \dfrac{RT}{2F} \ln a_{\mathrm{Pb^{2+}}}$

Again, $\mathrm{PbSO_4(s) \rightleftharpoons Pb^{2+} + SO_4^{2-}}$

Or, $$K_{aP} = a_{Pb^{2+}}\, a_{SO_4^{2-}} = 2.2 \times 10^{-8}$$

$$\phi_c = +0.126 - \frac{0.0592}{2}\log\frac{a_{Pb^{2+}}\, a_{SO_4^{2-}}}{a_{SO_4^{2-}}} \quad \text{(at 25°C)}$$

$$= +0.126 - \frac{0.0592}{2}\log\frac{K_{aP}}{a_{SO_4^{2-}}} \quad \text{(at 25°C)}$$

$$= +0.126 + 0.226 + \frac{0.0592}{2}\log a_{SO_4^{2-}} \quad \text{(at 25°C)}$$

$$\phi_c = 0.352 + \frac{0.0592}{2}\log a_{SO_4^{2-}} \quad \text{(at 25°C)}$$

e.m.f. at anode: $$\phi_a = \phi_a^0 - \frac{RT}{2F}\ln\frac{a_{H^+}^4\, a_{SO_4^{2-}}}{a_{H_2O}^2}$$

$$\phi_a^0 = -1.685 \text{ V}$$

Or, $$\phi_a = -1.685 - \frac{0.0592}{2}\log\frac{a_{H^+}^4\, a_{SO_4^{2-}}}{a_{H_2O}^2} \quad \text{(at 25°C)}$$

$$\text{Cell e.m.f.} = E = \phi_c - \phi_a$$

$$= (0.352 + 1.685) + \frac{0.0592}{2}\log\frac{a_{H^+}^4\, a_{SO_4^{2-}}^2}{a_{H_2O}^2} \quad \text{(at 25°C)}$$

Or, $$E = 2.037 + 0.0592\log\frac{a_{H^+}^2\, a_{SO_4^{2-}}}{a_{H_2O}} \quad \text{(at 25°C)}$$

Again $$\underset{(m)}{H_2SO_4} \rightarrow \underset{(2\,m)}{2H^+} + \underset{(m)}{SO_4^{2-}}$$

Hence, $$a_{H_2SO_4} = a_{H^+}^2\, a_{SO_4^{2-}}$$

If $a_\pm$ and m are the mean activity and molality, we can write that:

$$a_\pm^3 = \gamma_\pm^3\, 4m^3$$

where, $\gamma_\pm$ is the mean activity coefficient. Thus the above equation becomes:

$$E = 2.037 + 0.0592\log\frac{(\gamma_\pm^3)(4m^3)}{a_{H_2O}} \qquad (5.5.14)$$

(ii) Nickel-Cadmium alkaline accumulator The cell is given by:

$$\ominus\ Cd\,|\,Cd(OH)_2,\ KOH(20\,\%)\,||\,KOH(20\,\%),\ Ni(OH)_2,\ Ni(OH)_3\,|\,Ni\ \oplus$$

At cathode: $Cd \rightarrow Cd^{2+} + 2e$ (discharge)

$$Cd^{2+} + 2OH^- \rightarrow Cd(OH)_2$$

$$Cd + 2OH^- \rightarrow Cd(OH)_2 + 2e$$

At anode: $Ni^{3+} + e \rightarrow Ni^{2+}$ (discharge)

Or, $Ni(OH)_3 + e \rightarrow Ni(OH)_2 + OH^-$

$$2Ni(OH)_3 + 2e \rightarrow 2Ni(OH)_2 + 2OH^-$$

The overall cell reaction is given by:

$$Cd + 2Ni(OH)_3 \xrightarrow{\text{Discharge}} Cd(OH)_2 + 2Ni(OH)_2$$

and $Cd(OH)_2 + 2Ni(OH)_2 \xrightarrow{\text{Charge}} Cd + 2Ni(OH)_3$

Combining above two equations we can write that:

$$Cd + 2Ni(OH)_3 \underset{\text{Charge}}{\overset{\text{Discharge}}{\rightleftharpoons}} Cd(OH)_2 + 2Ni(OH)_2$$

The e.m.f. of Nickel-Cadmium accumulator is about 1.36 V.

(iii) Zinc-Silver alkaline accumulator

Porous Zn-plate forms the cathode and (silver and silver oxide) form the anode. The cell is given by:

$$\ominus\ Zn \mid K_2ZnO_2 + KOH(40\ \%) \mid AgO\ (\text{or } Ag_2O) \mid Ag\ \oplus$$

At cathode: $Zn \rightarrow Zn^{2+} + 2e$ (discharge)

$$Zn^{2+} + 2OH^- \rightarrow ZnO + H_2O$$

$$Zn + 2OH^- \rightarrow ZnO + H_2O + 2e$$

At anode: $2AgO + H_2O + 2e \rightarrow Ag_2O + 2OH^-$ (discharge)

$$Ag_2O + H_2O + 2e \rightarrow 2Ag + 2OH^-$$

$$2AgO + 2H_2O + 4e \rightarrow 2Ag + 4OH^-$$

Or, $AgO + H_2O + 2e \rightarrow Ag + 2OH^-$

The cell reaction, at discharge state, is:

$$Zn + AgO \xrightarrow{\text{Discharge}} ZnO + Ag$$

and at charge state the cell reaction is:

$$ZnO + Ag \xrightarrow{\text{Charge}} Zn + AgO$$

Combining the above two equations, we can write that:

$$Zn + AgO \underset{\text{Charge}}{\overset{\text{Discharge}}{\rightleftharpoons}} ZnO + Ag$$

If only Ag_2O/Ag constitutes the anode, the overall cell reaction can be written as:

$$Zn + Ag_2O \underset{\text{Charge}}{\overset{\text{Discharge}}{\rightleftharpoons}} ZnO + 2Ag$$

The cell e.m.f. with anode AgO/Ag is 1.86 V but the same is around 1.60 V if the anode is Ag_2O/Ag. The speciality of Zn-Ag accumulator is that: unlike lead and nickel accumulators, the electrolyte of Zn-Ag accumulator does not precipitate either in the charge or in the discharge state. This criterion makes it possible to place the electrodes very close to each other and the electrodes are separated by a thin layer of cellophane.

The electrolyte is freely flowing through the porous electrodes. So, the Zn-Ag accumulator is available in a very small size also. Furthermore, it has high capacity, high energy and high power per unit mass and volume. That is why Zn-Ag accumulator has wide application area particularly where small size is desired.

5.6 Ionic equilibrium in aqueous solutions

5.6.1 Debye-Hückel limiting law

An ion remains in solvated form in aqueous solution. Thus the effective radius of an ion is always greater than the actual ionic radius. Furthermore, each ion is surrounded by oppositely charged ions. Thus the potential around an ion consists of the potential of the central ion and the potential of the surrounding ionic atmosphere. So, we can write that:

$$\phi = \phi_i + \phi_a, \text{ where,}$$

ϕ_i and ϕ_a are the potentials of central ion and ionic atmosphere respectively. Considering this concept Debye-Hückel deduced an expression of activity coefficient (γ_i) of any ionic species i as given below:

$$\log(\gamma_i) = -A\, Z_i^2 \sqrt{I} \tag{5.6.1a}$$

Where A a is constant and has a value of 0.5117 at 25°C for pure water. Z_i is the valency of the ion i and I is the Ionic strength of the solution.

Let us consider an electrolyte MN which dissociates in aqueous solution as follows:

$$\underset{(m)}{MN} = \underset{(\nu_+ m)}{\nu_+ M^{z+}} + \underset{(\nu_- m)}{\nu_- M^{z-}}$$

The mean activity ($a_\pm$) is given by:

$$a_\pm^\nu = (\gamma_+^{\nu_+} \gamma_-^{\nu_-}) m_+^{\nu_+} m_-^{\nu_-} \quad [m_+ = (\nu_+ m) \quad \text{and} \quad m_- = (\nu_- m)]$$

$$= \gamma_\pm^\nu m_\pm^\nu \quad \text{where,} \quad \nu = \nu_+ + \nu_-$$

Hence,

$$\gamma_\pm^\nu = \gamma_+^{\nu_+} \gamma_-^{\nu_-}$$

Or,

$$\nu \log \gamma_\pm = \nu_+ \log \gamma_+ + \nu_- \log \gamma_-$$

$$\log \gamma_\pm = \frac{\nu_+ \log \gamma_+ + \nu_- \log \gamma_-}{(\nu_+ + \nu_-)} \tag{5.6.1b}$$

where, $a_{\pm}$ = Mean activity of the electrolytic solution.

$m_{\pm}$ = Mean molality of the electrolytic solution.

$\gamma_{\pm}$ = Mean activity coefficient of the electrolytic solution.

Combining Equations (5.6.1a) and (5.6.1b), we have:

$$\log \gamma_{\pm} = -A\left(\frac{\nu_+ z_+^2 + \nu_- z_-^2}{\nu_+ + \nu_-}\right)\sqrt{I}$$

As the solution is electrically neutral, we have: $\nu_+ z_+ = \nu_- z_-$

Or,
$$\frac{\nu_+ z_+^2 + \nu_- z_-^2}{\nu_+ + \nu_-} = \frac{(\nu_- z_-) z_+ + (\nu_+ z_+) z_-}{\nu_+ + \nu_-} = \frac{(\nu_+ + \nu_-) z_+ z_-}{(\nu_+ + \nu_-)}$$

$$= z_+ z_-$$

Thus the above equation becomes:

$$\log(\gamma_{\pm}) = -A\, z_+ z_- \sqrt{I} \tag{5.6.1c}$$

Considering the radius of ionic atmosphere the above equation becomes:

$$\log \gamma_{\pm} = \frac{-A\, z_+ z_- \sqrt{I}}{1+\sqrt{I}} \quad \text{(in aqueous solution)} \tag{5.6.1d}$$

As we know, each ion remains in hydrated form, the polarization of solvent molecules must be taken into account. The ultimate equation of mean activity coefficient becomes:

$$\log \gamma_{\pm} = \frac{-A\, z_+ z_- \sqrt{I}}{1+\sqrt{I}} + CI \tag{5.6.1e}$$

where C is a constant.

The Equations (5.6.1c), (5.6.1d) and (5.6.1e) are the Debye-Hückel equations for mean activity coefficient.

5.6.2 Acids and bases

Old definition A substance, capable of releasing proton(s) in aqueous solution, is called acid and a base is that substance, capable of releasing hydroxyl ion(s) in aqueous solution.

Later on J. Brönsted and T. Lowry proposed a new definition. According to them acid is a substance which is capable of giving up a proton and a base is a substance, capable of accepting a proton. Consider the following reaction:

$$HA \rightleftharpoons H^+ + A^-$$

According to the above definition, HA is an acid since it releases H^+ ion. The corresponding anion formed is A^-, which accepts H^+ ion to form HA. Thus A^- is a base. As A^- is originating from HA, A^- is often called a conjugate base of HA and the latter (HA) is called conjugate acid of A^-.

Thus conjugate base of HCl is Cl^- and conjugate acid of NH_3 is NH_4^+, since NH_3 accepts H^+ ion to form NH_4^+ according to the following equation:

$$NH_3 + H^+ \rightleftharpoons NH_4^+ \quad \text{(conjugate acid)}$$

$$HCl \rightleftharpoons H^+ + Cl^- \quad \text{(conjugate base)}$$

Solvent plays an important role in acid-base characteristics. On this basis solvents can be divided into four categories:

(a) Protophilic solvent Any solvent, having the property of accepting a proton, is called protophilic solvent, e.g., water, alcohol, ammonia, amine, ether, acetone etc.

(b) Protogenic (protophobic) solvent Any solvent, having the property of giving up a proton, is called protogenic (or protophobic) solvent, e.g., sulphuric acid, formic acid, acetic acid etc.

(c) Amphoprotic (amphoteric) solvent Any solvent, capable of either accepting or giving up a proton, is called amphoprotic or amphoteric solvent, e.g., water, alcohol etc.

(d) Aprotic solvent Any solvent, incapable of accepting or giving up a proton, is called aprotic solvent. One example of an aprotic solvent is benzene.

Lewis modified the definition of acid and base. According to him an acid is a substance that can accept an electron pair from other molecule and a base is a substance that can donate a pair of electron to another molecule. According to this definition BF_3 is an acid and NH_3 is a base.

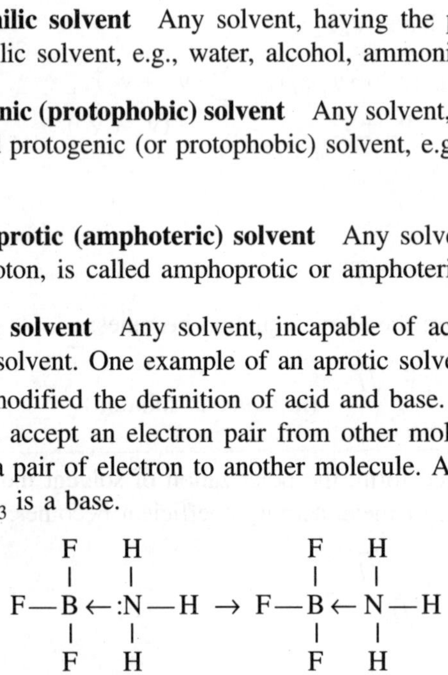

(acid) (base) (salt)

5.6.3 Ionic product of water

Water is a unique amphoteric solvent, which dissociates into hydroxonium ion and hydroxyl ion according to the following equation:

$$H_2O + H_2O \rightleftharpoons H_3O^+ + OH^-$$

Or,

$$H_2O \rightleftharpoons H^+ + OH^-$$

Thus water contains equimolar quantities of H^+ and OH^- ions and hence water is a perfectly neutral solvent. The dissociation constant K_{d, H_2O} is given by:

$$K_{d, H_2O} = \frac{a_{H^+} \, a_{OH^-}}{a_{H_2O}}$$

Or, $$K_w = K_{d,\,H_2O} \times a_{H_2O} = a_{H^+}\, a_{OH^-}$$

As water molecules mostly remain in undissociated form a_{H_2O} may be taken as constant. K_w is known as ionic product of water.

$$K_w = a_{H^+}\, a_{OH^-} \tag{5.6.2a}$$

Neglecting the activity coefficient terms, we can write that:

$$a_{H^+} = C_{H^+} = [H^+] \quad \text{and} \quad a_{OH^-} = C_{OH^-} = [OH^-]$$

Thus, $$K_w = [H^+]\,[OH^-] \tag{5.6.2b}$$

K_w can easily be measured by e.m.f. method.

The value of K_w is 1.0×10^{-14} at 25°C.

Again, $$a_{H^+} = a_{OH^-}$$

So, $$K_w = a_{H^-}\, a_{OH^-} = a^2_{H^+} = 1.0 \times 10^{-14}$$

Or, $$a_{H^+} = a_{OH^-} = 10^{-7}$$

Or, $$[H^+] = [OH^-] = 10^{-7}$$

5.6.4 pH

Strong acids and bases always remain in dissociated form in aqueous solution. Thus strength of acids or bases can be represented by the corresponding concentrations or acids or bases. For example, HCl in aqueous solution dissociates as follows:

$$\underset{(m)}{HCl} \rightarrow \underset{(m)}{H^+} + \underset{(m)}{Cl^-}$$

So, HCl solution of molality m has the concentration of $H^+ = [H^+] = m$ molal.

In case of weak acids the concentration of acid solution does not represent the concentration of H^+ since weak acids are partly dissociated in aqueous solution, e.g.

$$\underset{(m)}{CH_3COOH} \rightleftharpoons \underset{(\alpha\, m)}{H^+} + \underset{(\alpha\, m)}{CH_3COO^-}$$

If α be the degree of dissociation of acetic acid in aquaous solution, $[H^+] = \alpha m$, where m is the molality of acid solution. The value of α is different for different acids. Furthermore it varies with dilution. To avoid this difficulty S. Sorensen in 1909 introduced a term pH, to express hydrogen ion concentration in case of weak acids.

Definition of pH It is a measure of acid strength in aqueous solution of any weak acid.

Mathematically it is expressed as: $$pH = -\log_{10}[H^+]$$

In terms of activity: $$pH_a = -\log_{10} a_{H^+}$$

In case of pure water we know that: $$[H^+] = 10^{-7}$$

Hence, $pH = -\log_{10}[H^+] = 7$. So pH of pure demineralized water is 7. As pure water is neutral in nature, pH of any neutral solution is 7. In acid solution $[H^+]$ is greater than 10^{-7} and hence pH value falls below 7 whereas in case of basic solution $[OH^-]$ is greater than 10^{-7}. As we know:

$$K_w = [H^+][OH^-] = 10^{-14},\ [H^+] < 10^{-7}$$

since, $[OH^-] > 10^{-7}$

Or, $\log_{10}[H^+] < -7$

Or, $-\log_{10}[H^+] > 7$

Or, $pH > 7$

Thus pH value is greater than 7 in basic solution. When, $[OH^-] = 10^\circ$, $[H^+] = 10^{-14}$. Thus maximum value of pH is 14 and minimum value of pH is 1. Negative pH has no meaning.

A solution is said to be acidic when pH of that solution lies within 1 to 6 and a solution behaves basic in nature when pH lies within 8 to 14. Like pH, pOH may be expressed as follows:

$$pOH = -\log_{10}[OH^-]$$

We know that, $K_w = [H^+][OH^-] = 10^{-14}$

Taking $\log_{10}$ on both sides, we get: $\log_{10} K_w = \log_{10}[H^+] + \log_{10}[OH^-]$

Or, $-\log_{10}[H^+] - \log_{10}[OH^-] = -\log_{10} K_w = pK_w = 14$

Or, $pK_w = pH + pOH = 14$

Hence, $pOH = 14 - pH$ (5.6.3)

where, $pK_w = -\log_{10} K_w$

5.6.5 (a) Dissociation of weak acid

As discussed earlier any weak acid reversibly dissociates in aqueous solution at a low degree. Let us consider the dissociation of acetic acid (HOAC):

$$HOAC + H_2O \rightleftharpoons H_3O^+ + OAC^-$$

The equilibrium constant K is given by:

$$K = \frac{[H_3O^+][OAC^-]}{[HOAC][H_2O]}$$

As the concentration of water does not change, we can write that:

$$K_a = K[H_2O] = \frac{[H_3O^+][OAC^-]}{[HOAC]}$$

Or, in other way:

$$\underset{(1-\alpha)c}{HOAC} \rightleftharpoons \underset{(\alpha c)}{H^+} + \underset{(\alpha c)}{OAC^-}$$

c is the molarity of HOAC solution and α is the degree of dissociation.

Thus, $$K_c = \frac{[H^+][OAC^-]}{[HOAC]} = \frac{(\alpha c)(\alpha c)}{(1-\alpha)c} = \frac{\alpha^2 c}{(1-\alpha)}$$

Or, $$K_c = \frac{\alpha^2 c}{(1-\alpha)} \tag{5.6.4a}$$

The Equation (5.6.4a) is known as Ostwald's dilution law, first formulated by W. Ostwald in 1888. It is evident from Equation (5.6.4a) that, greater the value of K_c, higher the value of α and hence higher the acid strength. Thus dissociation constant K_c is a measure of acid strength.

Considering that α is very low, we can assume that $(1 - \alpha) \approx 1$. Thus Equation (5.6.4a) becomes:

$$K_c = \alpha^2 c$$

Or, $$(\alpha c)^2 = K_c c$$

Or, $$[H^+] = \alpha c = \sqrt{K_c c}$$

Taking $\log_{10}$ on both sides we get:

$$\log_{10}[H^+] = \frac{1}{2}\log_{10} K_c + \frac{1}{2}\log_{10} c$$

Or, $$\text{pH} = \frac{1}{2} pK_c - \frac{1}{2}\log_{10} c \quad [pK_c = -\log_{10} K_c] \tag{5.6.4b}$$

The above equation is the general expression of pH of any monobasic weak acid solution.

Again, $$K_c = \alpha^2 c$$

Or, $$\alpha = \sqrt{\frac{K_c}{c}}$$

α can easily be determined by measuring conductivity and using Kohlransch's law:

$$\alpha = \frac{\Lambda_c}{\Lambda_\infty}$$

where, Λ_c and Λ_∞ are the equivalent conductivities of weak acid solution at concentration c and at infinite dilution respectively. Using this value of α one can easily calculate K_c of that weak acid solution.

(b) Dissociation of weak base

Let us consider a weak base ammonium hydroxide (NH_4OH), which dissociates reversibly as follows:

$$\underset{(1-\alpha)c}{NH_4OH} \rightleftharpoons \underset{(\alpha c)}{NH_4^+} + \underset{(\alpha c)}{OH^-}$$

$$\text{Equilibrium constant} = K_c = \frac{[NH_4^+][OH^-]}{[NH_4OH]}$$

Or, $$K_c = \frac{(\alpha c)(\alpha c)}{(1-\alpha)c} = \frac{\alpha^2 c}{(1-\alpha)}$$

Considering α to be very low, we can write that: $K_c = \alpha^2 c$.

Or, $$K_c.c = (\alpha.c)^2$$

Thus, $$[OH^-] = \alpha.c = \sqrt{K_c \cdot c}$$

Taking $\log_{10}$ on both sides, we get:

$$\log_{10}[OH^-] = \frac{1}{2}\log_{10}K_c + \frac{1}{2}\log_{10}c.$$

Or, $$pOH = \frac{1}{2}pK_c - \frac{1}{2}\log_{10}c \ [pK_c = -\log_{10}K_c] \quad (5.6.4c)$$

Again, $$pOH = pK_w - pH$$

Putting this value of pOH in the above equation, we get:

$$pK_w - pH = \frac{1}{2}pK_c - \frac{1}{2}\log_{10}c$$

Or, $$pH = pK_w - \frac{1}{2}pK_c + \frac{1}{2}\log_{10}c \quad (5.6.4d)$$

The Equation (5.6.4d) is the general expression of pH of any monoacidic weak base solution.

(c) Dissociation of polybasic acids

A polybasic acid dissociates in steps, e.g., phosphoric acid. It dissociates in three steps.

Step-I $$H_3PO_4 + H_2O \rightleftharpoons H_3O^+ + H_2PO_4^-$$

$$K_{d1} = \frac{[H_3O^+][H_2PO_4^-]}{[H_3PO_4]} = 7.6\times10^{-3}\Big|_{18°C}$$

Step-II $$H_2PO_4^- + H_2O \rightleftharpoons H_3O^+ + HPO_4^{2-}$$

$$K_{d2} = \frac{[H_3O^+][HPO_4^{2-}]}{[H_2PO_4^-]} = 5.9\times10^{-8}\Big|_{18°C}$$

Step-III $$HPO_4^{2-} + H_2O \rightleftharpoons H_3O^+ + PO_4^{3-}$$

$$K_{d3} = \frac{[H_3O^+][PO_4^{3-}]}{[HPO_4^{2-}]} = 3.5\times10^{-13}\Big|_{18°C}$$

The value of first dissociation constant is maximum, since presence of H_3O^+ shifts the equilibrium towards the left in step-II. In step-III the concentration of H_3O^+ further increases and hence the value of K_{d3} decreases sharply.

5.6.6 Salt hydrolysis

5.6.6.1 Salt of strong acid and strong base

This type of salt does not hydrolyze at all. It remains in completely dissociated form and the corresponding ions get hydrated in aqueous solution, e.g. sodium chloride (NaCl).

NaCl is a salt produced from HCl (strong acid) and NaOH (strong base).

$$(H^+ + Cl^-) + (Na^+ + OH^-) \rightarrow (Na^+ + Cl^-) + H_2O$$

$$Na^+ + nH_2O \rightarrow [Na(H_2O)_n]^+ \quad \text{and} \quad Cl^- + nH_2O \rightarrow [Cl(H_2O)_n]^-$$

5.6.6.2 Salt of weak acid and strong base

This type of salt readily hydrolyzes in aqueous solution and the corresponding salt solution is alkaline in nature, e.g. sodium acetate (NaOAC). It is produced from acetic acid (weak acid) and sodium hydroxide (NaOH) according to the following equation:

$$(Na^+ + OH^-) + HOAC \rightarrow (Na^+ + OAC^-) + H_2O$$

$$(Na^+ + OAC^-) + H_2O \rightleftharpoons (Na^+ + OH^-) + HOAC$$

As HOAC is a weak acid, it remains in undissociated form and the solution is alkaline due to free OH^- ions. It is to be noted that salt hydrolyses reversibly. The salt NaOAC is continuously produced during titration of HOAC by NaOH. At the beginning of titration only HOAC is present and the following equilibrium exists.

$$HOAC \rightleftharpoons H^+ + OAC^- \tag{5.6.5a}$$

The concentration of acid = a

a = conc. of undissociated acid (HOAC) + conc. of dissociated acid which is equivalent to C_{OAC^-}.

$$a = C_{HOAC} + C_{OAC^-} \tag{5.6.5b}$$

Now, NaOH solution is added in little quantity. NaOH readily reacts with undissociated acid to form salt according to the following reaction:

$$(Na^+ + OH^-) + HOAC \xrightarrow{\text{step-I}} (Na^+ + OAC^-) + H_2O$$

$$OAC^- + H_2O \underset{}{\overset{\text{step-II}}{\rightleftharpoons}} OH^- + HOAC$$

(OAC^-), produced in step-I, undergoes hydrolysis in step-II to form undissociated HOAC. Otherwise equilibrium in Equation (5.6.5a) will be disturbed. Thus the solution contains two types of cations Na^+, H^+ and two types of anions OH^-, OAC^-. As the solution is electrically neutral, we can write that:

Concentration of cation = concentration of anion.

$$C_{Na^+} + C_{H^+} = C_{OH^-} + C_{OAC^-}$$

NaOH remains fully dissociated form in aqueous solution. So, $C_{NaOH} = C_{Na^+}$. If b be the concentration of sodium hydroxide solution, $C_{Na^+} = b$.

Thus, $$C_{OAC^-} = b + C_{H^+} - C_{OH^-} \tag{5.6.5c}$$

Combining Equations (5.6.5b) and (5.6.5c), we get:

$$C_{HOAC} = a - C_{OAC^-} = (a - b - C_{H^+} + C_{OH^-})$$

Or, $$C_{HOAC} = (a - b - C_{H^+} + C_{OH^-}) \tag{5.6.5d}$$

The equilibrium of Equation (5.6.5.a) is still maintained in aqueous solution. The equilibrium constant (K_c) is given by:

$$K_c = \frac{C_{H^+}\, C_{OAC^-}}{C_{HOAC}}$$

Or, $$C_{H^+} = \frac{K_c\, C_{HOAC}}{C_{OAC^-}}$$

Putting the values of C_{OAC^-} and C_{HOAC} from Equations (5.6.5c) and (5.6.5d) in the above equation, we get:

$$C_{H^+} = \frac{K_c\,(a - b - C_{H^+} + C_{OH^-})}{(b + C_{H^+} - C_{OH^-})} \tag{5.6.5e}$$

(i) At the beginning of titration or before adding any NaOH solution, b = concentration of NaOH solution added = 0 and C_{OH^-} = Concentration of OH^- in the mixture = 0.

Thus Equation (5.6.5e) becomes:

$$C_{H^+} = K_c\left(\frac{a - C_{H^+}}{C_{H^+}}\right)$$

Or, $$C_{H^+}^2 + K_c\, C_{H^+} - K_c a = 0$$

Or, $$C_{H^+} = \frac{-K_c \pm \sqrt{K_c^2 + 4aK_c}}{2} = -\frac{1}{2}K_c \pm \sqrt{\frac{1}{4}K_c^2 + aK_c}$$

Or, $$C_{H^+} = -\frac{1}{2}K_c + \sqrt{\frac{1}{4}K_c^2 + aK_c}$$

The other root of C_{H^+} is negative and hence is not considered.

Again, a = concentration of acid and is far greater than K_c or $a >> K_c$.

$$\frac{K_c^2}{4} + aK_c \approx aK_c$$

Or, $$C_{H^+} = \frac{1}{2}K_c + \sqrt{aK_c}$$

Again, $$a >> K_c$$

Or, $$aK_c >> K_c^2$$

Or, $\sqrt{(aK_c)} >> K_c$

Hence, $C_{H^+} = -\frac{1}{2}K_c + \sqrt{aK_c} \approx \sqrt{aK_c}$

Or, $C_{H^+} = \sqrt{aK_c}$

Taking $\log_{10}$ on both sides, we get:

$$\log_{10} C_{H^+} = \frac{1}{2}\log K_c + \frac{1}{2}\log a$$

Or $$-\log_{10} C_{H^+} = -\frac{1}{2}\log K_c - \frac{1}{2}\log a$$

Or, $$\text{pH} = \frac{1}{2}pK_c - \frac{1}{2}\log a \qquad (5.6.5f)$$

The Equation (5.6.5.f) is identical with the Equation (5.6.4.b).

(ii) Near the end point of titration or near the equivalence point: $C_{H^+} \approx C_{OH^-}$ and $b < a$. So, all base is consumed. Thus the Equation (5.6.5e) becomes:

$$C_{H^+} = K_c\left(\frac{a-b}{b}\right)$$

Taking $\log_{10}$ on both sides we get:

$$\log C_{H^+} = \log K_c + \log\left(\frac{a-b}{b}\right)$$

$(a - b)$ = Concentration of acid remaining after consumption of the base

b = Concentration of base added = Concentration of salt since all base is converted into salt.

So, when the salt (Na^+ OAC^-) hydrolyses to produce free OH^-, the acid HOAC readily consumes that free OH^-.

$$(Na^+ + OAC^-) + H_2O \rightleftharpoons (Na^+ + OH^-) + HOAC$$

$$HOAC + OH^- \rightarrow H_2O + OAC^-$$

So $$-\log C_{H^+} = -\log K_c - \log\left(\frac{a-b}{b}\right)$$

Or, $$\text{pH} = pK_c - \log\left(\frac{a-b}{b}\right)$$

Or, $$\text{pH} = pK_c + \log\left(\frac{b}{a-b}\right) \qquad (5.6.5g)$$

The above equation represents the pH of a buffer solution.

5.6.6.3 Buffer solution

Definition of a buffer solution Buffer solution is a solution which possesses a definite pH and offers strong resistance to change the pH whenever a strong acid or a strong base is added to it. Buffer solution is usually prepared by mixing a weak acid and its salt or by mixing a weak base and its salt. The Equation (5.6.5g) represents the pH of a buffer solution prepared by mixing weak acid and its salt.

Mechanism of buffer action pH of a buffer solution, prepared from weak acid and strong base, is given by:

$$\text{pH} = pK_c + \log \frac{[\text{Salt}]}{[\text{Acid}]} \tag{5.6.5h}$$

Where, K_c is the dissociation constant of weak acid.

K_c of HOAC = 1.76×10^{-5} and $pK_c = -\log_{10} K_c = 4.76$. Salt is sodium acetate (NaOAC).

Consider a buffer solution containing both acid and salt having equal concentration of 1 mole/litre, i.e., [Salt] = [Acid] = 1 mole/lit.

So, $$\text{pH} = pK_c = 4.76$$

(i) Now HCl solution containing 0.9 mole of HCl is added into the buffer solution. 1 mole HCl reacts with 1 mole salt to form 1 mole undissociated acid according to the following equation:

$$(H^+ + Cl^-) + (Na^+ + OAC^-) \rightarrow (Na^+ + Cl^-) + HOAC$$

So, H^+ ions are consumed and hence pH changes very little. Thus 0.9 mole HCl produces 0.9 mole HOAC.

Thus, $$[\text{Acid}] = (1 + 0.9) \text{ moles/l} = 1.9 \text{ moles/l and}$$
$$[\text{Salt}] = (1 - 0.9) \text{ moles/l} = 0.1 \text{ moles/l}$$

Thus pH is given by:

$$\text{pH} = pK_c + \log \frac{0.1}{0.9} = 3.48$$

Thus 0.9 mole HCl is required to change the pH from 4.76 to 3.48.

Let us consider an acid solution of pH = 4.76.

Or, $$\text{pH} = -\log_{10} C_{H^+} = 4.76$$

Or, $$C_{H^+} = 10^{-4.76} \text{ moles/l}$$

pH of the acid solution is changed to 3.48 after addition of HCl solution.

So, $$\text{pH} = -\log_{10} C_{H^+} = 3.48$$

Or, $$C_{H^+} = 10^{-3.48} \text{ moles/l}$$

Thus moles of HCl added = $(10^{-3.48} - 10^{-4.76})$ mole = 0.00031 mole = 3.1×10^{-4} mole.

Thus only 3.1×10^{-4} mole of HCl is sufficient to change pH from 4.76 to 3.48 but in case of buffer solution 0.9 mole HCl is required to change pH from 4.76 to 3.48.

(ii) Instead of HCl, NaOH solution containing 0.9 mole of NaOH, is added to the buffer solution. Then free OH^- reacts with undissociated acid to form equivalent amount of salt according to the following equation:

$$(Na^+ + OH^-) + HOAC \rightarrow (Na^+ + OAC^-) + H_2O$$

Thus OH^- ions are consumed and hence pH of the buffer solution changes very little. So, 0.9 mole of NaOH reacts with 0.9 mole of HOAC to form 0.9 mole of salt.

So, [Salt] = (1 + 0.9) mole/l = 1.9 moles/l and [Acid] = (1 – 0.9) mole/l = 0.1 mole/l.

So, $$\text{pH} = pK_c + \log \frac{[\text{Salt}]}{[\text{Acid}]} = 4.76 + \log \frac{0.1}{0.9} = 5.039$$

[since, $pK_c = 4.76$]

Thus 0.9 mole NaOH is required to change pH of buffer solution from 4.76 to 5.039. If we take a weak acid solution of pH = 4.76, $[H^+] = C_{H^+} = 10^{-4.76}$ mole/l.
Now pH of that solution is changed to 5.039 after addition of NaOH solution.

So, $$C_{H^+} = 10^{-5.039} \text{ mole/l.}$$

Thus the amount of NaOH required to change the pH from 4.76 to 5.039 is = $(10^{-4.76} - 10^{-5.039})$ mole = (0.0000174 – 0.00000914) mole = 8.26×10^{-6} mole.
So only 8.26×10^{-6} mole of NaOH is enough to change the pH from 4.76 to 5.039.

Buffer capacity or buffer index Amount of strong acid or base, in moles, required to change the pH of a buffer solution by unity, is called buffer capacity or buffer index of that buffer solution. Let us take a buffer solution, containing acetic acid (0.8 mole) and its salt sodium acetate (0.2 mole). Thus pH of the buffer solution is given by:

$$\text{pH} = pK_c + \log \frac{[\text{Salt}]}{[\text{Acid}]} = 4.76 + \log \frac{0.2}{0.8} = 4.158$$

[since, $pK_c = 4.76$]

Now pH of buffer solution decreases by unity upon addition of x moles of HCl or increases by unity upon addition of y moles of NaOH.

So, $$\text{pH} = 3.158 = 4.76 + \log \frac{0.2 - x}{0.8 + x}$$

and $$\text{pH} = 5.158 = 4.76 + \log \frac{0.2 + y}{0.8 - y}$$

Or, $$\log \frac{0.2 - x}{0.8 + x} = 3.158 - 4.76$$

Or, $$\log \frac{0.8 + x}{0.2 - x} = 4.76 - 3.158 = 1.602$$

Or, $$\frac{0.8 + x}{0.2 - x} = 10^{1.602} = 39.994 \approx 40$$

Or, $$\frac{0.8 + x}{0.2 - x} = 40$$

Or, $$(0.8 + x) = 40(0.2 - x)$$

Or, $$x = 0.176 \text{ mole}$$

Thus 0.176 mole of HCl is required to decrease the pH of that buffer solution by unity.

So, the acidic buffer capacity = 0.176.

Again, $$\log \frac{0.2 + y}{0.8 - y} = 5.158 - 4.76 = 0.398$$

Or, $$\frac{0.2 + y}{0.8 - y} = 10^{0.398} = 2.5$$

Or, $$(0.2 + y) = 2.5(0.8 - y)$$

Or, $$y = 0.514 \text{ mole}$$

So the alkaline buffer capacity = 0.514.

Expression of buffer index (β)

The pH of a weak acid/salt buffer solution is given by:

$$\text{pH} = pK_c + \log\left(\frac{b}{a - b}\right) \quad \text{[Equation (5.6.5g)]}$$

a = Number of moles of weak acid present in the solution at the beginning.

b = Number of moles of base added in the acid solution = No. of moles of salt present in the buffer solution.

Assume that *(db)* moles of base is added to a buffer solution to increase the pH by *(dpH)*.

So, *(dpH)* is positive. Addition of *(db)* moles of base means buffer solution contains $(b + db)$ moles of salt and $[a - (b + db)]$ moles of acid. So, base content or salt content of buffer solution increases by *(db)* moles.

Definition of buffer index For any buffer solution the change of pH w.r.t. base content, *b*, is represented as α. So, $\alpha = \frac{dpH}{db}$. The inverse of α is known as buffer index, represented by β. In other words, amount of base required to change the pH of a buffer solution by unity is called buffer index of that buffer solution.

$$\alpha = \frac{dpH}{db}$$

Or, $$\beta = \frac{1}{\alpha} = \frac{db}{dpH}$$

For weak acid/salt buffer solution, we know that:

$$\text{pH} = pK_c + \log\left(\frac{b}{a-b}\right) \quad \text{[Equation (5.6.5g)]}$$

Differentiating both sides of the above equation w.r.t. b we get:

$$\frac{dpH}{db} = \left(\frac{a-b}{b}\right)\frac{(a-b+b)}{(a-b)^2} = \frac{a}{b(a-b)}$$

$$\beta = \frac{db}{dpH} = \frac{b(a-b)}{a} \tag{5.6.6}$$

a, the initial acid content, is constant, So, β of a buffer solution depends on the base content, b. The $\beta - b$ curve shows a maximum, where the value of β is β_{max}. To determine the value of β_{max}, 1st derivative of β i.e., $\frac{d\beta}{db}$ is set to zero, i.e., $\frac{d\beta}{db} = 0$. Differentiating the Equation (5.6.6) with respect to b, we get:

$$\frac{d\beta}{db} = \frac{[(a-b)-b]}{a} = \frac{a-2b}{a} \quad \text{and} \quad \frac{d^2\beta}{db^2} = -\frac{2}{a} < 0 \quad [\text{since, a} > 0]$$

At $\beta = \beta_{max}$, $\frac{d\beta}{db} = 0$

Or, $\frac{a-2b}{a} = 0$

Or, $a - 2b = 0$

Or, $b = \frac{a}{2}$

Table 5.6.1 Some buffer solutions with their compositions and pH range

Sl. No.	*Composition*	*pH range*
1.	Hydrochloric acid and calcium chloride	1.0 – 2.2
2.	Potassium hydrogen phthalate and hydrochloric acid	2.0 – 3.8
3.	Acetic acid and sodium acetate	3.7 – 5.6
4.	Potassium hydrogen phthalate and sodium hydroxide	4.0 – 6.2
5.	Dibasic sodium citrate and sodium hydroxide	5.0 – 6.3
6.	Monobasic potassium phosphate and sodium hydroxide	5.8 – 8.0
7.	Dibasic sodium phosphate and sodium hydroxide	11.0 – 12.0
8.	Borax and hydrochloric acid	7.6 – 9.2
9.	Borax and sodium hydroxide	9.2 – 11.0
10.	Boric acid and sodium hydroxide	7.8 – 10.0

Putting the value of b in Equation (5.6.6), we get:

$$\beta_{max} = \frac{\left(\frac{a}{2}\right)\left(a - \frac{a}{2}\right)}{a}$$

Or, $$\beta_{max} = \frac{a}{4}$$

Thus buffer index reaches to maximum value of $\frac{a}{4}$ at $b = \frac{a}{2}$.

5.6.6.4 Expression of pH: Salt of weak acid and strong base

We have seen in the earlier section that buffer solution is formed near the end point of titration of weak acid (HOAC) by a strong base (NaOH) and $b < a$. At the end point of titration or just beyond the end point, all acid is consumed and there is no free H^+ ion. So, $a = b$ and $C_{H^+} = 0$. Hydrolysis of salt results in the formation of free OH^- and the solution becomes alkaline:

$$(Na^+ + OAC^-) + H_2O \rightleftharpoons (Na^+ + OH^-) + HOAC$$

Putting the values of b and C_{H^+} in the Equation (5.6.5.e), we get:

$$C_{H^+} = K_c \frac{[(a - a) - 0 + C_{OH^-}]}{(a + 0 - C_{OH^-})} = K_c \frac{C_{OH^-}}{(a - C_{OH^-})}$$

Again, we know that:

$$C_{H^+}\, C_{OH^-} = K_w$$

Or, $$C_{OH^-} = \frac{K_w}{C_{H^+}}$$

Putting this value of C_{OH^-} in the above equation we get:

$$C_{H^+} = K_c \frac{\left(\frac{K_w}{C_{H^+}}\right)}{a - \left(\frac{K_w}{C_{H^+}}\right)} = \frac{K_c\, K_w}{(a\, C_{H^+} - K_w)}$$

Or, $$a\, C_{H^+}^2 - K_w C_{H^+} - K_c K_w = 0$$

Or, $$C_{H^+} = \frac{K_w \pm \sqrt{K_w^2 + 4a\, K_c\, K_w}}{2a}$$

As $$K_c >> K_w,\ K_c K_w >> K_w^2$$

Or, $$K_w << \sqrt{K_c\, K_w}$$

So, $$K_w^2 + 4a K_c K_w \approx 4a K_c K_w$$

Thus, $$C_{H^+} = \frac{K_w \pm \sqrt{4a\, K_c\, K_w}}{2a} = \frac{K_w}{2a} \pm \sqrt{\frac{K_c\, K_w}{a}}$$

Again, $K_w << \sqrt{K_c K_w}$ and $\frac{K_w}{2a} << \sqrt{\frac{K_c K_w}{a}}$

Hence, $C_{H^+} = \pm\sqrt{\frac{K_c K_w}{a}}$

Or, $C_{H^+} = \sqrt{\frac{K_c K_w}{a}}$ [since C_{H^+} cannot be negative]

Taking $\log_{10}$ on both sides we get:

$$\log C_{H^+} = \frac{1}{2}\log K_c + \frac{1}{2}\log K_w - \frac{1}{2}\log a$$

Or, $-\log C_{H^+} = -\frac{1}{2}\log K_c - \frac{1}{2}\log K_w + \frac{1}{2}\log a$

Or, $$\text{pH} = \frac{1}{2}pK_c + \frac{1}{2}pK_w + \frac{1}{2}\log a \qquad (5.6.7)$$

This is the expression of pH of solution of a salt of weak acid and strong base, a is the concentration of weak acid. As $\frac{1}{2}pK_w = 7$, pH > 7 and hence the solution is alkaline.

5.6.6.5 Salt of strong acid and weak base

One example of this type of salt is ammonium chloride (NH_4Cl), which hydrolyses in aqueous solution according to the following equation:

$$NH_4OH + (H^+ + Cl^-) \rightarrow (NH_4^+ + Cl^-) + H_2O$$

$$(NH_4^+ + Cl^-) + H_2O \rightleftharpoons NH_4OH + (H^+ + Cl^-)$$

Thus the aqueous solution of ammonium chloride is acidic in nature. The salt NH_4Cl is continuously produced during titration of NH_4OH by HCl. At the beginning of titration only NH_4OH is present and the following equilibrium exists:

$$NH_4OH \rightleftharpoons NH_4^+ + OH^- \qquad (5.6.8a)$$

$b = C_{NH_4OH}$ + Concentration of dissociated base which is equivalent to $C_{NH_4^+}$

$$b = C_{NH_4OH} + C_{NH_4^+} \qquad (5.6.8b)$$

Now, a little quantity of HCl is added to this solution. HCl reacts with NH_4OH to form NH_4Cl, which on hydrolysis gives:

$$(NH_4^+ + Cl^-) + H_2O \rightleftharpoons NH_4OH + (H^+ + Cl^-)$$

As the solution is electrically neutral, we can write that:

Concentration of cations = concentration of anions. Thus, we have:

$$C_{H^+} + C_{NH_4^+} = C_{OH^-} + C_{Cl^-}$$

$$C_{Cl^-} = C_{HCl} = \text{Amount of acid added} = a$$

[since HCl is completely dissociated in aqueous solution]

$$C_{NH_4^+} = (a + C_{OH^-}) - C_{H^+} \tag{5.6.8c}$$

Again from Equation (5.6.8.b), we have:

$$C_{NH_4OH} = b - C_{NH_4^+} = b - (a + C_{OH^-}) + C_{H^+}$$

Or,
$$C_{NH_4OH} = (b - a) - C_{OH^-} + C_{H^+} \tag{5.6.8d}$$

The equilibrium constant K_b of Equation (5.6.8a) is given by:

$$K_b = \frac{C_{NH_4^+} C_{OH^-}}{C_{NH_4OH}}$$

Putting the values of $C_{NH_4^+}$ and C_{NH_4OH} in the above equation, we get:

$$K_b = \frac{[(a + C_{OH^-}) - C_{H^+}] C_{OH^-}}{[(b - a) - C_{OH^-} + C_{H^+}]}$$

Or,
$$C_{OH^-} = K_b \left[\frac{(b - a - C_{OH^-} + C_{H^+})}{(a + C_{OH^-} - C_{H^+})} \right] \tag{5.6.8e}$$

(i) At the begining of titration there is no HCl in the mixture. So, $a = 0$ and $C_{H^+} = 0$. Thus Equation (5.6.8e) becomes:

$$C_{OH^-} = K_b \frac{(b - C_{OH^-})}{C_{OH^-}}$$

Or,
$$C_{OH^-}^2 + K_b\, C_{OH^-} - bK_b = 0$$

Solving C_{OH^-} we get:

$$C_{OH^-} = \frac{-K_b \pm \sqrt{K_b^2 + 4bK_b}}{2}$$

As $b >> K_b$, $bK_b >> K_b^2$ and hence, $K_b^2 + 4bK_b \approx 4bK_b$

So,
$$C_{OH^-} = \frac{-K_b \pm \sqrt{4bK_b}}{2}$$

Again,
$$K_b^2 << bK_b \quad \text{and} \quad K_b << \sqrt{bK_b}$$

So,
$$-K_b \pm \sqrt{4bK_b} \approx \pm \sqrt{4bK_b} = \pm 2\sqrt{bK_b}$$

Or,
$$C_{OH^-} = \frac{\pm 2\sqrt{bK_b}}{2} = \pm \sqrt{bK_b}$$

Or,
$$C_{OH^-} = \sqrt{bK_b} \quad \text{[since } C_{OH^-} \text{ cannot be negative]}$$

Taking $\log_{10}$ on both sides, we get:

$$\log_{10} C_{OH^-} = \frac{1}{2}\log_{10} K_b + \frac{1}{2}\log_{10} b$$

Or, $$pOH = -\log_{10} C_{OH^-} = -\frac{1}{2}\log_{10} K_b - \frac{1}{2}\log_{10} b = \frac{1}{2}pK_b - \frac{1}{2}\log_{10} b$$

Or, $$pOH = \frac{1}{2}pK_b - \frac{1}{2}\log_{10} b$$

Again, we know that, pH + pOH = pK_w.

Or, $$pH = pK_w - pOH = pK_w - (\frac{1}{2}pK_b - \frac{1}{2}\log b) = pK_w - \frac{1}{2}pK_b + \frac{1}{2}\log b$$

Or, $$pH = pK_w - \frac{1}{2}pK_b + \frac{1}{2}\log b \quad (5.6.8f)$$

This is the expression of pH of solution of any monoacidic weak base like NH_4OH.
It is to be noted that the Equation (5.6.8f) is the same as the Equation (5.6.4d).

(ii) Near the end point of titration, $C_{OH-} \approx C_{H^+}$ and $a < b$. So all acid is consumed to form salt.
So moles of acid added = moles of salt formed = a moles.
Moles of remaining weak base in the mixture = $(b - a)$ moles.
Thus the Equation (5.6.8e) becomes:

$$C_{OH^-} = K_b\left(\frac{b-a}{a}\right) = K_b\frac{[\text{base}]}{[\text{salt}]}$$

Taking $\log_{10}$ on both sides we get:

$$\log_{10} C_{OH^-} = \log_{10} K_b + \log_{10}\frac{[\text{base}]}{[\text{salt}]}$$

Or, $$pOH = pK_b + \log_{10}\frac{[\text{salt}]}{[\text{base}]} \quad (5.6.8g)$$

Again, we know that: pH + pOH = pK_w.

Or, $$pH = pK_w - pOH$$

Putting the value of pOH in the above equation, we get:

$$pH = pK_w - \left(pK_b + \log_{10}\frac{[\text{salt}]}{[\text{base}]}\right)$$

Or, $$pH = pK_w - pK_b - \log\frac{[\text{salt}]}{[\text{base}]} \quad (5.6.8h)$$

Equation (5.6.8h) is the expression of pH of a buffer solution prepared from weak base and strong acid.

(iii) At the end point of titration or just beyond the end point, all base is consumed and there is no free OH^- ion. So, $b = a$ and $C_{OH^-} = 0$. Hydrolysis of salt results in the formation of free H^+ ion and the solution becomes acidic, e.g. NH_4Cl on hydrolysis forms free H^+ ion according to the following equation:

$$(NH_4^+ + Cl^-) + H_2O \rightleftharpoons NH_4OH + (H^+ + Cl^-)$$

Putting these values of b and C_{OH^-} in the Equation (5.6.8e), we get:

$$C_{OH^-} = K_b \frac{(b - b - 0 + C_{H^+})}{(b - C_{H^+})} = \frac{K_b C_{H^+}}{(b - C_{H^+})}$$

Again, $$C_{H^+} = \frac{K_w}{C_{OH^-}}$$

Putting this value of C_{H^+} in the above equation we get:

$$C_{OH^-} = \frac{K_b \dfrac{K_w}{C_{OH^-}}}{b - \dfrac{K_w}{C_{OH^-}}} = \frac{K_w K_b}{b C_{OH^-} - K_w}$$

Or, $$b C_{OH^-}^2 - K_w C_{OH^-} - K_w K_b = 0$$

Solving C_{OH^-} we get:

$$C_{OH^-} = \frac{K_w \pm \sqrt{K_w^2 + 4b K_b K_w}}{2b}$$

As $K_b >> K_w$, $K_b K_w >> K_w^2$ and hence, $K_w^2 + 4bK_bK_w \approx 4bK_bK_w$

Or, $$C_{OH^-} = \frac{K_w \pm \sqrt{4b K_b K_w}}{2b}$$

Again, $$K_b K_w >> K_w^2$$

Or, $$\sqrt{K_b K_w} >> K_w$$

So, $$K_w \pm \sqrt{4b K_b K_w} \approx \pm \sqrt{4b K_b K_w}$$

Or, $$C_{OH^-} = \frac{\pm \sqrt{4b K_b K_w}}{2b} = \frac{\sqrt{4b K_b K_w}}{2b}$$

[since C_{OH^-} cannot be negative]

Or, $$C_{OH^-} = \sqrt{\frac{K_b K_w}{b}}$$

Taking $\log_{10}$ on both sides, we get:

$$\log_{10} C_{OH^-} = \frac{1}{2} \log_{10} K_w + \frac{1}{2} \log_{10} K_b - \frac{1}{2} \log_{10} b$$

Or, $$pOH = \frac{1}{2} pK_w + \frac{1}{2} pK_b + \frac{1}{2} \log_{10} b \qquad (5.6.8i)$$

Again, we know that: $\mathrm{pH} + p\mathrm{OH} = pK_w$.

Or, $$\mathrm{pH} = pK_w - p\mathrm{OH} = pK_w - \left[\frac{1}{2}pK_w + \frac{1}{2}pK_b + \frac{1}{2}\log_{10} b\right]$$

Or, $$\mathrm{pH} = \frac{1}{2}pK_w - \frac{1}{2}pK_b - \frac{1}{2}\log_{10} b \quad [\text{usually } |b| > 1] \qquad (5.6.8j)$$

This is the expression of pH of solution of a salt of weak base and strong acid.

$\frac{1}{2}pK_w = 7$ Hence pH < 7 and the solution is acidic.

5.6.6.6 Salt of weak acid and weak base

NH_4OH is a weak base and HOAC is a weak acid. So ammonium acetate (NH_4OAC) is formed during the reaction of NH_4OH and HOAC according to the following equation:

$$NH_4OH + HOAC \rightarrow (NH_4^+ + OAC^-) + H_2O$$

$$(NH_4^+ + OAC^-) + H_2O \rightleftharpoons NH_4OH + HOAC$$

Thus the hydrolysis of salt gives rise the corresponding weak acid and weak base. In aqueous solution NH_4OH and HOAC dissociate as follows:

$$NH_4OH \rightleftharpoons NH_4^+ + OH^- \quad [\text{Equilibrium constant} = K_b]$$

$$HOAC \rightleftharpoons H^+ + OAC^- \quad [\text{Equilibrium constant} = K_a]$$

The expressions of K_b and K_c are given as:

$$K_b = \frac{C_{NH_4^+}\, C_{OH^-}}{C_{NH_4OH}} \quad \text{and} \quad K_a = \frac{C_{H^+}\, C_{OAC^-}}{C_{HOAC}} \qquad (5.6.9a)$$

The salt NH_4OAC is continuously formed during titration of weak acid (HOAC) by weak base (NH_4OH). At the beginning of titration only acetic acid is present.

a = concentration of acid = concentration of undissociated acid (HOAC) + concentration of dissociated acid, which is equivalent to concentration of OAC^-

Or, $$a = C_{HOAC} + C_{OAC^-} \qquad (5.6.9b)$$

Now a little quantity of NH_4OH solution is added to this. The mixture contains H^+, NH_4^+ and OAC^- ions. As the solution is electrically neutral, concentration of cations is equal to concentration of anions. So, we have:

$$C_{H^+} + C_{NH_4^+} = C_{OAC^-} \qquad (5.6.9c)$$

$$C_{HN_4OH} = \text{Amount of base added} = b \text{ moles}$$

Again, from Equation (5.6.9b), we have:

$$C_{HOAC} = a - C_{OAC^-}$$

Or, $$C_{HOAC} = a - C_{H^+} - C_{NH_4^+} \qquad (5.6.9d)$$

Again, the equilibrium constant K_a is given by:

$$K_a = \frac{C_{H^+}\, C_{OAC^-}}{C_{HOAC}}$$

Putting the values of C_{OAC^-} and C_{HOAC} in the above equation we get:

$$K_a = \frac{C_{H^+}\,(C_{H^+} + C_{NH_4^+})}{(a - C_{H^+} - C_{NH_4^+})}$$

Again, from Equation (5.6.9a) we have:

$$C_{HN_4^+} = \frac{K_b\, C_{NH_4OH}}{C_{OH^-}} = \frac{b K_b}{C_{OH^-}}$$

Putting the value $C_{HN_4^+}$ in the above equation, we get:

$$K_a = \frac{C_{H^+}\left[C_{H^+} + \dfrac{bK_b}{C_{OH^-}}\right]}{\left[a - C_{H^+} - \dfrac{bK_b}{C_{OH^-}}\right]} = \frac{C_{H^+}\left[K_w + bK_b\right]}{\left[a\, C_{OH^-} - K_w - bK_b\right]}$$

$$K_a = \frac{\left(\dfrac{K_w}{C_{OH^-}}\right)\left[K_w + bK_b\right]}{\left[a\, C_{OH^-} - (K_w + bK_b)\right]} = \frac{K_w\left[K_w + bK_b\right]}{\left[a\, C_{OH^-}^2 - (K_w + bK_b)C_{OH^-}\right]}$$

Or, $$K_a\, a\, C_{OH^-}^2 - K_a(K_w + bK_b)C_{OH^-} - (K_w^2 + bK_bK_w) = 0$$

Solving C_{OH^-} we get:

$$C_{OH^-} = \frac{K_a(K_w + bK_b) \pm \sqrt{K_a^2(K_w + bK_b)^2 + 4aK_a(K_w^2 + bK_b K_w)}}{2aK_a}$$

As $K_b >> K_w$, $K_b K_w >> K_w^2$, $K_a K_b K_w >> K_a K_w{}^2$, $K_b{}^2 << K_b$

Again, $K_a^2(K_w + bK_b)^2 << 4aK_a(K_w^2 + bK_b K_w)$ and

$4aK_a(K_w^2 + bK_bK_w) \approx 4ab\, K_a K_b K_w$

Thus the above equation reduces to:

$$C_{OH^-} = \frac{(K_a K_w + bK_a K_b) \pm \sqrt{4ab\, K_a K_b K_w}}{2aK_a} \approx \pm\frac{\sqrt{4ab\, K_a K_b K_w}}{2aK_a}$$

[since, $\sqrt{4ab\, K_a K_b K_w} >> (K_aK_w + bK_aK_b)$]

$$= \sqrt{\frac{4ab\, K_a K_b K_w}{4a^2 K_a{}^2}}$$ [since C_{OH^-} cannot be negative]

$$= \sqrt{\frac{b}{a}\frac{K_b K_w}{K_a}}$$

Or, $$C_{OH^-} = \sqrt{\frac{b K_b K_w}{a K_a}}$$

Taking $\log_{10}$ on both sides, we get:

$$\log_{10} C_{OH^-} = \frac{1}{2}\log_{10}\frac{b}{a} + \frac{1}{2}\log_{10}\frac{K_b}{K_a} + \frac{1}{2}\log_{10} K_w$$

At the end of titration or just beyond the end point all acid is consumed by all base. Hence, $b = a$. Thus the above equation becomes:

$$\log_{10} C_{OH^-} = \frac{1}{2}\log_{10} K_w + \frac{1}{2}\log_{10} K_b - \frac{1}{2}\log_{10} K_a$$

Or, $$pOH = \frac{1}{2} pK_w + \frac{1}{2} pK_b - \frac{1}{2} pK_a$$

Again, we know that: $pH + pOH = pK_w$

Or, $$pH = pK_w - pOH = pK_w - \left(\frac{1}{2} pK_w + \frac{1}{2} pK_b - \frac{1}{2} pK_a\right)$$

Or, $$pH = \frac{1}{2} pK_w + \frac{1}{2} pK_a - \frac{1}{2} pK_b$$

This is the expression of pH of a salt solution produced from weak acid and weak base. $\frac{1}{2} pK_w = 7$. It is interesting to note that the above expression does not contain any concentration term. So, pH of weak acid/weak base salt solution is independent of concentration of either acid or base.

(i) If acid is stronger than base, $K_a > K_b$ and hence $pK_a < pK_b$. Thus pH < 7 hence the salt solution is acidic.

(ii) If acid is weaker than base, $K_a < K_b$ and hence $pK_a > pK_b$. Thus pH > 7 and hence the salt solution is basic.

5.6.7 Indicators

In any acid-base titration it is difficult to identify the end point. So some specific substances are used to identify the end point of titration. These substances have special characteristics to change their colour within a certain pH range. These substances are called indicators. Thus an indicator changes its colour depending on the concentration of hydrogen ion in the solution. Some indicators change their colour in the acidic pH-range and the others change their colour in the alkaline pH-range [see Table 5.6.2]. Several researchers tried hard to establish the theory according to which an indicator changes its colour. In 1891 W. Ostwald first established a theory according to which in alkaline medium each indicator forms an anion, which co-exists with its tautomeric form with different colours. Concentration of one tautomeric form dominates over the other in acid medium and the solution appears in the colour of that tautomeric form. Conversely, in alkaline medium the other tautomeric form dominates and hence the colour of the solution changes near the end point of titration. For example phenolphthalein dissociates in aqueous solution according to the following equation:

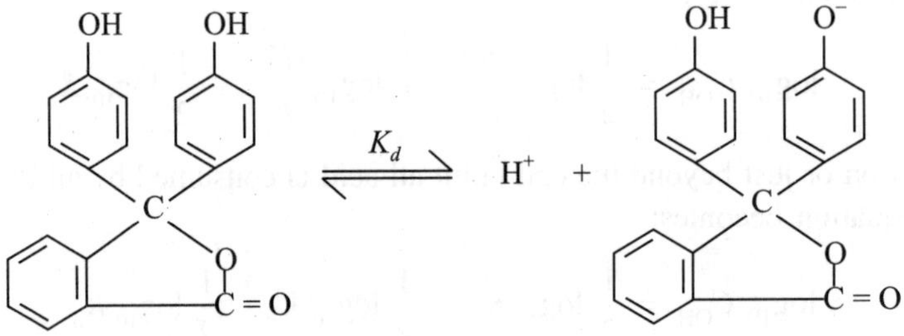

Phenolpthalein (Hln) $\quad\quad$ (ln_I^-)

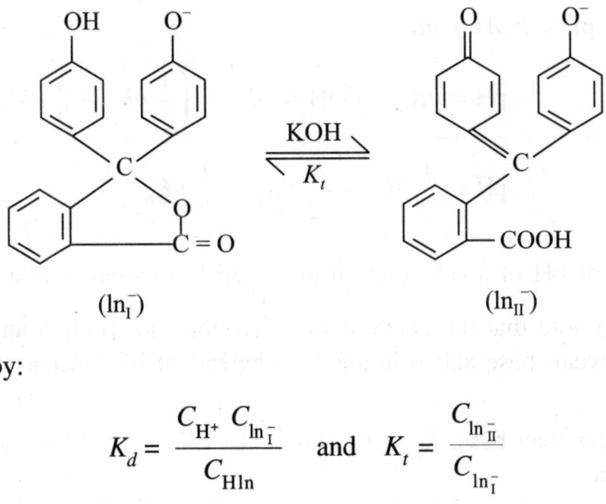

(ln_I^-) $\quad\quad$ (ln_II^-)

K_d and K_t are given by:

$$K_d = \frac{C_{\text{H}^+}\, C_{\text{ln}_\text{I}^-}}{C_{\text{Hln}}} \quad \text{and} \quad K_t = \frac{C_{\text{ln}_\text{II}^-}}{C_{\text{ln}_\text{I}^-}}$$

Multiplying K_d with K_t gives the indicator constant K_{In}.

Or,
$$K_{\text{ln}} = K_d\, K_t = \frac{C_{\text{H}^+}\, C_{\text{ln}_\text{II}^-}}{C_{\text{Hln}}}$$

Or,
$$C_{\text{H}^+} = \frac{K_{\text{ln}}\, C_{\text{Hln}}}{C_{\text{ln}_\text{II}^-}}$$

Taking $\log_{10}$ on both sides we get:

$$\log C_{\text{H}^+} = \log K_{\text{ln}} + \log \frac{C_{\text{Hln}}}{C_{\text{ln}_\text{II}^-}}$$

Or,
$$pH = pK_{In} + \log \frac{[In_{II}^-]}{[HIn]} \tag{5.6.10}$$

where,
$$pK_{In} = -\log_{10} K_{In}$$

It has been established experimentally that when $[In_{II}^-] = 10[HIn]$, the colour of In_{II}^- predominates over that of [HIn] and vice versa. Thus at pH = $pK_{In} + \log_{10} 10$, the colour of In_{II}^- predominates and at pH = $pK_{In} - \log_{10} 10$, the colour of HIn predominates. Thus the colour change takes place within the pH-range $[(pK_{In} - 1)$ to $(pK_{In} + 1)]$.

5.6.8 pH-metric titration

Selection of indicator in acid-base titration is an important factor but the selection will be easier if one considers the pH-curve. So we first consider pH-curve of different acid-base titrations.

(a) Titration of strong acid by a strong base pH of strong acid lies around 1 and it slowly increases upon addition of strong base. Near equivalence point pH sharply increases to around 11 [Figure 5.6.1a]. Any indicator can be used to determine the end point of titration.

Table 5.6.2 Selected indicators and their characteristic properties

Indicator	*pH-range* pK_{In}	*for colour change*	*Colour*	
			In acid	*In alkali*
Thymol blue	1.51	1.2 – 2.8	Red	Yellow
Methyl orange	3.70	3.1 – 4.4	Light Pink	Yellow
Methyl red	5.10	4.4 – 6.0	Light Pink	Yellow
Bromocresol green	4.67	3.8 – 5.4	Yellow	Blue
Bromocresol red	5.10	4.2 – 6.3	Red	Yellow
Bromocresol purple	6.30	5.2 – 6.8	Yellow	Purple
Bromocresol blue	3.98	3.0 – 4.6	Yellow	Blue
Bromophenol	6.16	5.2 – 6.8	Yellow	Red
Chlorophenol red	5.98	4.8 – 6.4	Yellow	Red
Phenol red	7.9	6.8 – 8.4	Yellow	Red
Cresol red	8.3	7.2 – 8.8	Yellow	Red
Cresolphthalein	9.4	8.2 – 9.8	Colourless	Red
Phenolphthalein	9.4	8.3 – 10.0	Colourless	Dark Pink
Thymolphthalein	9.4	9.2 – 11.6	Colourless	Blue

The change in pH is instantaneous and the curve runs almost vertically [see Figure 5.6.1a]

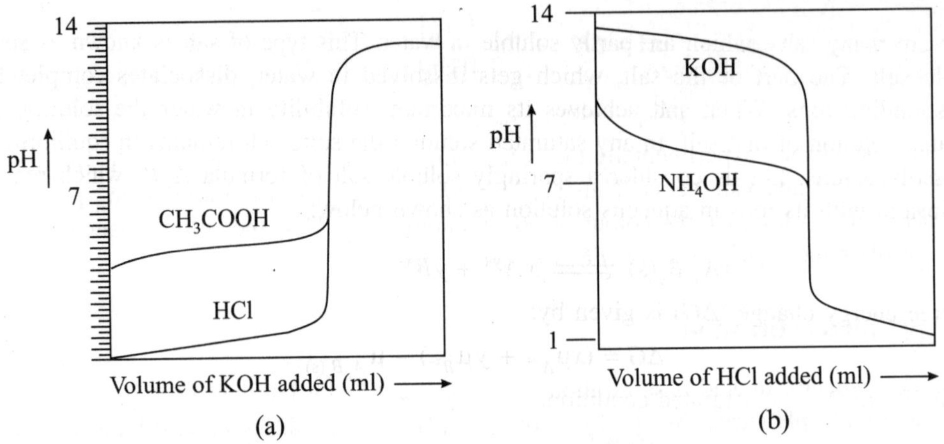

Fig. 5.6.1 Titration curves: (a) Titration of strong acid and weak acid by a strong base (KOH), (b) Titration of strong base and weak base by a strong acid (HCl).

(b) Titration of strong base by a strong acid (HCl) A sharp and wide change of pH (3 – 11) is observed near the equivalence point [Figure 5.6.1b]. Thus any indicator, having pH-range within 3 – 11, can be used to determine the end point of titration, that means any indicator from methyl orange to phenolphthalein can be used in strong acid-strong base titration.

(c) Titration of weak acid by strong base During titration of a weak acid (CH_3COOH) by a strong base (KOH) a narrow change in pH (7 – 11) is observed near the equivalence point [Figure 5.6.1a] since solution becomes alkaline on addition of strong base. Thus any indicator, which changes its colour within the pH-range 7 – 11, is suitable to detect the end point of titration. So, phenolphthalein is the right choice in this case. If methyl orange is used instead of phenolphthalein, the colour change will be observed even at pH 5, i.e. well before the end point. So use of methyl orange as an indicator in the weak acid-strong base titration only misguides us in detecting the end point.

(d) Titration of weak base by strong acid During titration of a weak base by a strong acid the solution becomes slowly acidic due to salt hydrolysis and hence a decrease of pH is observed [Figure 5.6.1b]. Near the end point a narrow change of pH (5 – 2) is observed. Thus any indicator, which changes its colour within the pH-range (5 – 2), is suitable to detect the end point of titration. So methyl orange is the right choice in this case.

(e) Titration of weak acid by weak base During titration of a weak acid by a weak base the pH changes very slowly and usually lies near about 7. The pH-curve does not have any vertical section, that means, there is no sharp change in pH near the end point of titration. Thus practically no indicator is suitable to detect the end point. So, weak acid-weak base titration cannot be performed accurately by volumetric method. In this case usually conductometric method is used to detect the end point of titration.

5.6.9 Solubility product

There are many salts, which are partly soluble in water. This type of salt is known as sparingly soluble salt. The part of the salt, which gets dissolved in water, dissociates completely into corresponding ions. When salt achieves its maximum solubility in water the solution is said saturated solution of that salt. In any saturated solution the solid salt remains in equilibrium with its dissolved ions. Let us consider a sparingly soluble salt of formula A_xB_y which remains in equilibrium with its ions in aqueous solution as shown below:

$$A_x B_y(\mathrm{s}) \rightleftharpoons xA^{y+} + yB^{x-}$$

The free energy change (ΔG) is given by:

$$\Delta G = (x\mu_{A^{y+}} + y\mu_{B^{x-}}) - \mu_{A_xB_y(\mathrm{s})}$$

At equilibrium, i.e. at saturated condition:

$$\Delta G = 0$$

Or, $$\mu_{A_xB_y(\mathrm{s})} = (x\mu_{A^{y+}} + y\mu_{B^{x-}})$$

Again we know that:

$$\mu_i = \mu_i^0 + RT \ln a_i$$

Hence, $$\mu_{A_xB_y(\mathrm{s})} = \mu^0_{A_xB_y(\mathrm{s})} + \mathrm{RT} \ln a_{A_xB_y(\mathrm{s})}$$

$$a_{A_xB_y(\mathrm{s})} = 1$$

Or, $$\mu_{A_xB_y(\mathrm{s})} = \mu^0_{A_xB_y(\mathrm{s})} = \text{Constant} \quad \text{[at a particular temperature]}$$

$$\mu_{A^{y+}} = \mu^0_{A^{y+}} + RT \ln a_{A^{y+}} \quad \text{and} \quad \mu_{B^{x-}} = \mu^0_{B^{x-}} + RT \ln a_{B^{x-}}$$

Thus, we have:

$$\text{Constant} = (x\,\mu^0_{A^{y+}} + y\,\mu^0_{B^{x-}}) + RT \ln a^x_{A^{y+}}\, a^y_{B^{x-}}$$

Or, $$a^x_{A^{y+}}\, a^y_{B^{x-}} = \text{Constant at a particular temperature} = K_{\mathrm{aP}}$$

[since, $(x\mu^0_{A^{y+}} + y\mu^0_{B^{x-}}) = \text{Constant}$]

Or, $$K_{\mathrm{aP}} = a^x_{A^{y+}}\, a^y_{B^{x-}} \quad (5.6.11a)$$

K_{aP} is known as activity product.

$$K_{\mathrm{aP}} = a^x_{A^{y+}}\, a^y_{B^{x-}} = (C^x_{A^{y+}}\, C^y_{B^{x-}})\,(f^x_{A^{y+}}\, f^y_{B^{x-}})$$

Or, $$K_{\mathrm{aP}} = K_{\mathrm{SP}}(f^x_{A^{y+}}\, f^y_{B^{x-}})$$

The solution of a partly soluble salt may be considered as a dilute salt solution. So, we can write that:

$$K_{\mathrm{aP}} = K_{\mathrm{SP}} = [A^{y+}]^x\,[B^{x-}]^y \quad (5.6.11b)$$

K_{SP} is known as the solubility product of a sparingly soluble salt.

If S is the solubility of the salt A_xB_y, $[A^{y+}] = xS$ and $[B^{x-}] = yS$, since 1 mole of A_xB_y produces x moles of A^{y+} and y moles of B^{x-}. Thus the Equation (5.6.11b) becomes:

$$K_{\mathrm{SP}} = (x\,S)^x\,(y\,S)^y = S^{x+y}\,x^x\,y^y \quad (5.6.11c)$$

(a) Consider a uni-uni-valent sparingly soluble salt AgCl which dissociates in aqueous solution according to the following equation:

$$AgCl(s) \rightleftharpoons Ag^+ + Cl^-$$

$$x = 1 \quad \text{and} \quad y = 1 \quad \text{and} \quad K_{sP} = S^2$$

Consider another sparingly soluble salt (Ag_2S), which dissociates in aqueous solution according to the following equation:

$$Ag_2S(s) \rightleftharpoons 2Ag^+ + S^{2-}$$

$$x = 2 \text{ and } y = 1 \quad \text{and} \quad K_{SP} = S^{(2+1)} \,.\, 2^2 \,.\, 1^1 = 4S^3$$

(b) **Common ion effect** sparingly soluble salt A_xB_y dissociates in aqueous solution as follows:

$$A_xB_y(s) \rightleftharpoons xA^{y+} + yB^{x-}$$

If S is the solubility of the salt, the solubility product K_{sP} is given by:

$$K_{SP} = S^{x+y}\, x^x y^y$$

Now a metal salt M_xB having a common ion B^{x-} is added to the solution. Thus according to the La Chatelier principle the backward reaction is favoured and hence the solubility S decreases to S'. If the solution contains C gms of the salt M_xB, concentration of B^{x-} will be = $[B^{x-}] = (S' + C)$. Concentration of $[A^{y+}] = S'$. Thus the above equation becomes:

$$K_{SP} = (xS')^x\,(yS' + C)^y$$

The value of C is far greater than S' and hence, $(yS' + C) \approx C$. Thus the above equation becomes:

$$K_{SP} = (S')^x\, C^y\, x^x$$

Combining this with above equation we get:

$$(S')^x\, C^y\, \cancel{x^x} = S^{x+y}\, \cancel{x^x}\, y^y \qquad (5.6.11d)$$

Equation (5.6.11d) shows how the solubility of a sparingly soluble salt is effected in presence of a common ion.

(i) In the case of Ag_2S, if Na_2S is added in the saturated solution of Ag_2S, the Equation (5.6.11d) becomes:

$$(S')^2\, C^1\, 2^2 = S^{(2+1)} \,.\, 2^2 \,.\, 1^1$$

Or, $$CS'^2 = S^3$$

Or, $$S' = \left(\frac{S^3}{C}\right)^{\frac{1}{2}} = \frac{S^{\frac{3}{2}}}{\sqrt{C}}$$

S = solubility of the salt Ag_2S before adding Na_2S and

S' = solubility of the salt Ag_2S after adding Na_2S.

(ii) If $AgNO_3$ is added to the solution of Ag_2S, the concentration of $[Ag^+] = (2S' + C)$ and the concentration of $S^{2-} = [S^{2-}] = S'$, C = concentration of $AgNO_3$ in the solution.

Thus, $K_{sP} = 4S^3 = (2S' + C)^2 . S'$

Or, $C^2 . S' = 4S^3$

Or, $S' = \frac{4S^3}{C^2}$ since, $C >> S'$ and hence, $2S' + C \approx C$

(iii) If $Na_2S_2O_3$ solution is added to Ag_2S solution, Ag^+ readily forms a complex ion $[Ag(S_2O_3)_2]^{3-}$ according to the following equation:

$$Ag^+ + 2S_2O_3^{2-} \rightleftharpoons [Ag(S_2O_3)_2]^{3-}$$

$$[Ag(S_2O_3)_2]^{3-} \rightleftharpoons Ag^+ + 2S_2O_3^{2-}$$

Instability constant K_{IC} is given by:

$$K_{IC} = \frac{[Ag^+][S_2O_3^{2-}]^2}{[Ag(S_2O_3)_2]^{3-}} = 1.02 \times 10^{-13}$$

$$[Ag^+] = \frac{[Ag(S_2O_3)_2]^{3-}}{[S_2O_3^{2-}]^2} \times 1.02 \times 10^{-13}$$

Thus all Ag^+ ions remain as $[Ag(S_2O_3)_2]^{3-}$ and hence the new solubility is S'

$$2S' = [Ag^+] = [Ag(S_2O_3)_2]^{3-} \quad \text{and} \quad [S^{2-}] = S'$$

$$K_{SP} = [Ag^+]^2 \, [S^{2-}]$$

Or, $$K_{SP} = \frac{\left([Ag(S_2O_3)_2]^{3-}\right)^2}{[S_2O_3^{2-}]^4} (1.02 \times 10^{-13})^2 \, [S^{2-}]$$

$$K_{SP} = (2S')^2 \, (S') \, \frac{(1.02 \times 10^{-13})^2}{[S_2O_3^{2-}]^4}$$

Or, $$(4S')^3 = \frac{K_{SP}}{(1.02 \times 10^{-13})^2} \times [S_2O_3^{2-}]^4$$

Hence, S' can easily be computed since K_{sP} is known and $[S_2O_3^{2-}]$ is the free thiosulphate ion concentration = $(C - S')$, where C is the concentration of $Na_2S_2O_3$ in the solution. As $C >> S'$, $(C - S') \approx C$. Thus the above equation becomes:

$$(4S')^3 = \frac{K_{SP} \, C^4}{(1.02 \times 10^{-13})^2}$$

Hence, S' can easily be calculated.

(iv) The solubility product of Ag_2S is given by:

$$K_{SP} = (2S_0)^2 \, S_0 = 4S_0^3$$

The activity product is given by:

$$K_{aP} = (2S_0 \, \gamma_0)^2 \, (S_0 \, \gamma_0) = 4S_0^3 \, \gamma_0^3$$

where, γ_0 is the mean ionic activity coefficient.
Now an electrolyte, having no common ion, is added. Solubility S_0 changes to S and the mean ionic activity γ_0 changes to $\gamma_\pm$, but K_{aP} remains constant. Thus we can write that:

$$K_{aP} = (2S\ \gamma_\pm)^2\ (S\ \gamma_\pm) = 4S^3\ \gamma_\pm^3$$

Combining the above two equations, we get:

$$4S_0^3\ \gamma_0^3 = 4S_3\ \gamma_\pm^3$$

Or, $$\frac{S}{S_0} = \frac{\gamma_0}{\gamma_\pm}$$

Taking $\log_{10}$ on both sides, we get:

$$\log_{10}\frac{S}{S_0} = \log_{10}\frac{\gamma_0}{\gamma_\pm} = \log_{10}(\gamma_0) - \log_{10}(\gamma_\pm) \qquad (5.6.12a)$$

Again, we have:

$$\log \gamma_\pm = -A\ Z_+\ Z_-\ \sqrt{I} \quad \text{[Equation (5.6.1c)]}$$

Thus, $$\log \gamma_0 = -A\ Z_+\ Z_-\ \sqrt{I_0}$$

and $$\log \gamma_0 - \log \gamma_\pm = -A\ Z_+\ Z_-\left(\sqrt{I_0} - \sqrt{I}\right) \qquad (5.6.12b)$$

With the addition of an electrolyte *ionic strength* of the solution increases and hence $I > I_0$. Thus $(\log \gamma_0 - \log \gamma_\pm) > 0$ and hence $S > S_0$ according to Equation (5.6.12a).

Inference Solubility of a sparingly soluble salt *increases* with the addition of a foreign electrolyte having no *common ion*.
The variation of solubility of a sparingly soluble salt with the addition of an electrolyte is well illustrated in the Figure 5.6.2.

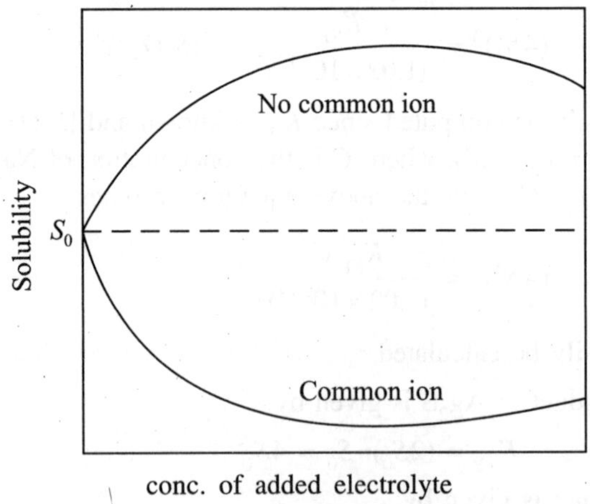

Figure 5.6.2 Effect of addition of an electrolyte to the solution of a sparingly soluble salt.

6

SOLID STATE CHEMISTRY

6.1 Crystallography

Study of crystal structures and properties of crystalline solids is known as crystallography.

Crystal A solid is said to be a crystal when constituent atoms are orderly arranged in a definite sequence. To define a three-dimensional sequence three axes are required. These three axes are known as crystallographic axes.

Space lattice It is an arrangement of indefinite number of points in three dimensions such that environment of any atom or point is the reflection of that of any other point in the array. Studies on several crystal structures show that *fourteen* different kinds of arrangements or fourteen space lattices are available to describe all crystal structures. These fourteen space lattices are known as *Bravis lattices*. These fourteen Bravis lattices are divided into *seven* crystal systems. To describe any crystal lattice six parameters are essential: three crystallographic axes, a, b, c and angles between them, α, β, γ as shown in the Figure 6.1.

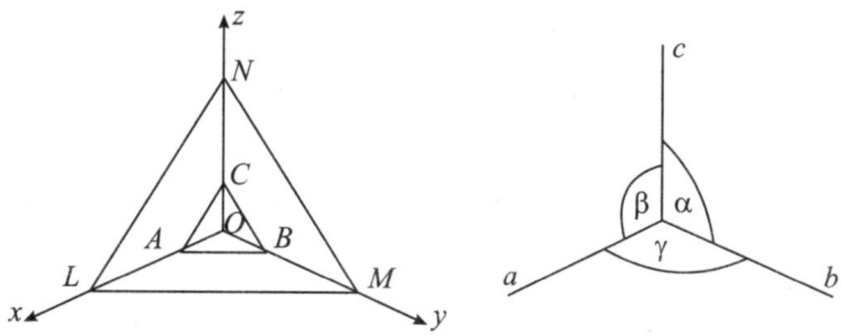

Fig. 6.1 Lattice parameters of a crystal. Crystallographic axes and angles. Intercepts cut by unit cell and a crystal plane.

Seven crystal systems are:

(i) Cubic system It has three space lattices, namely, Simple Cubic (SC), Body Centred Cubic (BCC) and Face Centred Cubic (FCC). In all the cases $a = b = c$ and $\alpha = \beta = \gamma = 90°$. Example: NaCl etc.

(ii) Trigonal or Rhombohedral system It has one space lattice, i.e., Simple Trigonal with lattice parameter: $a = b = c$ and $\alpha = \beta = \gamma \neq 90°$. Examples are As, Sb and Bi crystals.

(iii) Tetragonal system It has two space lattices, namely, Simple Tetragonal and Body Centred Tetragonal with lattice parameters: $a = b \neq c$ and $\alpha = \beta = \gamma = 90°$. One example is SnO_2 crystal.

(iv) Hexagonal system It has one space lattice, i.e., Simple Hexagonal with lattice parameter: $a = b \neq c$ and $\alpha = \beta = 90°$ and $\gamma = 120°$. One example is SiO_2 crystal.

(v) Orthorhombic system It has four space lattices: Simple Orthorhombic, End Centred Orthorhombic, Body Centred Orthorhombic and Face Centred Orthorhombic with lattice parameters: $a \neq b \neq c$ and $\alpha = \beta = \gamma = 90°$. One example is $BaSO_4$ crystal.

(vi) Monoclinic system It has two space lattices, namely, Simple Monoclinic and End Centred Monoclinic with lattice parameters: $a \neq b \neq c$ and $\alpha = \gamma = 90° \neq \beta$. One example is $CaSO_4$, $2H_2O$ (Gypsum).

(vii) Triclinic system It has one space lattice, Simple Triclinic with lattice parameters: $a \neq b \neq c$ and $\alpha \neq \beta \neq \gamma \neq 90°$. One example is $K_2Cr_2O_7$.

6.2 Unit cell

It is the smallest unit of a crystal, expansion of which gives rise a 3-D real crystal. Geometry of unit cell represents the geometry of the real crystal. Structure of some space lattices:

(a) Simple Cubic (SC) eight corners of the cube constitute eight space sites.

(b) Body Centred Cube (BCC) (i) eight corners of the cube constitute eight space sites, (ii) center of the cube constitutes one space site.

(c) Face Centred Cube (FCC) (i) eight corners of the cube constitute eight space sites, (ii) each center of six faces of the cube constitutes one space site.

(d) End Centred Orthorhombic/Monoclinic (i) eight corners of the unit cell constitute eight space sites, (ii) centers of two opposite faces constitute two space sites.

(e) Hexagonal (i) twelve corners of the two hexagonal faces constitute twelve space sites, (ii) centers of two hexagonal faces constitute two space sites. Three space sites lie within the crystal. Hexagonal and cubic crystals are shown in the Figure 6.2.

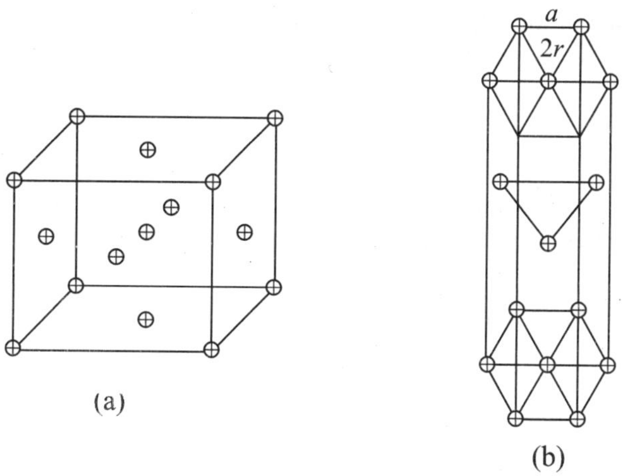

Fig. 6.2 (a) Cubic crystal, (b) HCP crystal.

6.3 Atomic radius

In a unit cell half of the distance between centers of two nearest neighbouring atoms is called atomic radius, represented by r.

Lattice constant For cubic system edge length of a cube is called lattice constant, represented by a.

Relation between 'a' and 'r' This is well illustrated in the Figure 6.3.

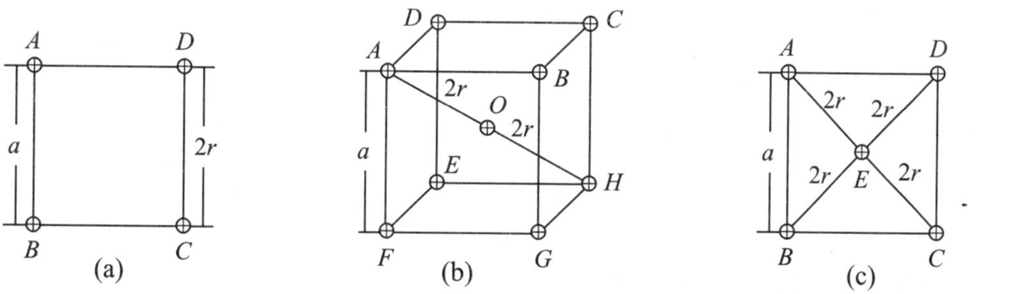

ABCD = Face of a cube
a = Edge length of the cube

H = Corner of bottom face *EFGH*
A = Opposite corner of top face *ABCD*
O = Centre of the cube
a = Edge length of the cube
AH = Cube diagonal

BD = Face diagonal of *ABCD*
E = Centre of face diagonal *BD*
a = Edge length of the cube

Fig. 6.3 (a) Atomic radius and edge length of a SC crystal, (b) Atomic radius and edge length of a BCC crystal, (c) Atomic radius and edge length of a FCC crystal.

(i) SC crystal One of the six faces of a simple cube is shown in the Figure 6.3(a). According to Figure 6.3(a) we have: $AB = BC = a = 2r$.

Or, $$r = \frac{a}{2} \tag{6.1a}$$

(ii) FCC crystal One of the six faces of a FCC crystal is shown in the Figure 6.3(c), according to which we can write that: (Face diagonal)^2 = $AC^2 = AB^2 + BC^2 = CD^2 + DA^2$.

Or, $$(2r + 2r)^2 = a^2 + a^2$$

Or, $$r = \frac{a}{2\sqrt{2}} \tag{6.1.b}$$

(iii) BCC crystal In the Figure 6.3(b) AOH represents a cube diagonal, connecting a corner of one face to another corner of opposite face through the centre of the cube. Thus we can write that:

$$AH^2 = AF^2 + FG^2 + GH^2$$

Or, $$(2r + 2r)^2 = a^2 + a^2 + a^2$$

Or, $$r = \frac{a\sqrt{3}}{4} \tag{6.1c}$$

(iv) HCP crystal According to Figure 6.2(b), we have:

$$2r \cos 60° = \frac{a}{2}$$

Or, $$r = \frac{a}{2} \tag{6.1d}$$

Coordination number The total number of nearest neighbouring atoms, surrounding a lattice atom, is known as coordination number of that lattice atom. It is 6 for SC crystal, 8 for BCC crystal and 12 for both FCC and HCP crystals.

6.4 Effective number of atoms

Consider the lattice site C in the Figure 6.4.

The point C is common to four faces, $DABC$, $BJLC$, $NMLC$ and $DINC$. Again each face is common to two cubes. Thus the lattice site C is common to eight cubes and hence its contribution to each cube is $\frac{1}{8}$.

Each face has four corners and one centre. Each corner is common to eight cubes (as discussed above) and the centre is obviously common to two cubes. Thus the contribution of the centre of each face to a cube is $\frac{1}{2}$.

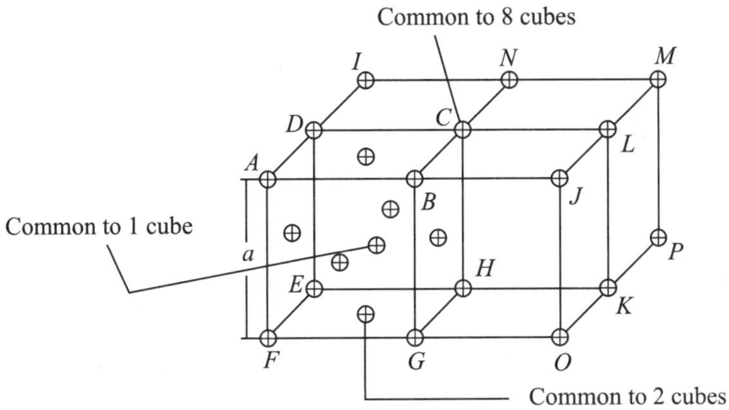

Fig. 6.4 Structure of a cubic crystal.

There is a centre inside each cube and the centre is not shared by any other cube. Thus the contribution of the cube centre to a cube is 1.

Simple Cubic (SC) Crystal The eight corners of the unit cell possess eight lattice points or atoms. Thus the total number of effective atoms is $\left(\frac{1}{8}\right) \times 8 = 1$.

Body Centred Cubic (BCC) Crystal Eight corners of the unit cell possess eight atoms and the centre of the unit cell possesses one atom. Thus the total number of effective atoms = $\left(\frac{1}{8}\right) \times 8 + 1 = 1 + 1 = 2$.

Face Centred Cubic (FCC) Crystal Eight corners of the unit cell possess eight atoms and the centre of each face possesses one atom. Thus the total number of effective atoms = $\left(\frac{1}{8}\right) \times 8 + \left(\frac{1}{2}\right) \times 6$ (since there are six faces) $= 1 + 3 = 4$.

Hexagonal Close Packed (HCP) Crystal The unit cell of a HCP crystal has two hexagonal faces. Each hexagonal face contains six lattice points at six corners, each of which is common to six unit cells. The centre of each hexagonal face possesses one lattice point and the centre of the face is common to two unit cells. Furthermore, the unit cell has three interior centre atoms and each centre atom is not shared by any other unit cell [see the Figure 6.2(b)].

Thus the total number of effective atoms = $(6 + 6) \times \left(\frac{1}{6}\right) + (1 + 1) \times \left(\frac{1}{2}\right) + (1 + 1 + 1) \times 1$
$= 2 + 1 + 3 = 6$.

6.5 Atomic packing factor (APF)

It is obvious that the total volume of the unit cell cannot be occupied by lattice points or atoms. Effective number of atoms occupies only a fraction of the total volume of the unit cell. That fraction of occupied volume is known as Atomic Packing Factor (APF). In the case of cubic system the volume of the unit cell is a^3, where a is the lattice constant.

(a) SC crystal The effective number of atoms in the unit cell of a SC crystal is 1. Atomic radius is r and hence volume of the atom is $\left(\frac{4}{3}\right)\pi r^3$. Thus %APF is given by:

$$\%\text{APF} = \frac{\%\text{ Volume occupied by effective number of atoms}}{\text{Volume of the unit cell}}$$

$$= \frac{\frac{4}{3}\pi r^3}{a^3} \times 100 = \frac{\frac{4}{3}\pi\left(\frac{a}{2}\right)^3}{a^3} \times 100 \quad \left[\text{since, } r = \frac{a}{2}\right]$$

$$= \frac{\pi}{6} \times 100 = 52\ \%$$

(b) BCC crystal The effective number of atoms in the unit cell of BCC crystal is 2. Volume occupied by 2 atoms is $2 \times \left(\frac{4}{3}\right)\pi r^3$, where r is the atomic radius. Thus % APF is given by:

$$\%\text{APF} = \frac{\%\text{ Volume occupied by effective number of atoms}}{\text{Volume of the unit cell}}$$

$$= \frac{2\times\frac{4}{3}\pi r^3}{a^3} \times 100 = \frac{1}{a^3}\times\frac{8}{3}\pi\left(\frac{a\sqrt{3}}{4}\right)^3 \times 100$$

$$\left[\text{since, } r = \frac{a\sqrt{3}}{4}\right]$$

$$= \frac{\pi\sqrt{3}\times 100}{8} = 68\ \%$$

(c) FCC crystal The effective number of atoms in the unit cell of a FCC crystal is 4. Volume occupied by 4 effective atoms = $4 \times \left(\frac{4}{3}\right)\pi r^3$, where r is the atomic radius. Thus %APF is given by:

$$\%\text{APF} = \frac{\%\text{ Volume occupied by effective number of atoms}}{\text{Volume of the unit cell}}$$

$$= \frac{4 \times \frac{4}{3}\pi r^3}{a^3} \times 100 = \frac{1}{a^3} \times \frac{16}{3}\pi\left(\frac{a}{2\sqrt{2}}\right)^3 \times 100$$

$$\left[\text{since, } r = \frac{a}{2\sqrt{2}}\right]$$

$$= \frac{\pi \times 100}{3\sqrt{2}} = 74\%$$

(d) HCP crystal The effective number of atoms in the unit cell of a HCP crystal is 6. The hexagonal face consists of six equilateral triangles of angle 60°. Height of the unit cell = c

Total area of the hexagonal face $= 6 \times \frac{1}{2} \times a \times a \sin 60°$

$$= 6 \times \frac{1}{2} \times a \times a\frac{\sqrt{3}}{2} = \frac{3\sqrt{3}}{2}a^2$$

$$\text{Volume of the unit cell} = \frac{3\sqrt{3}}{2}a^2 \times c = \frac{3\sqrt{3}}{2}a^2 \times \sqrt{\frac{8}{3}}a$$

$$= 3\sqrt{2}\,a^3 \quad \left[\text{since, } c = \sqrt{\frac{8}{3}}a\right]$$

Volume occupied by six effective atoms $= 6 \times \frac{4}{3}\pi r^3$

$$= 6 \times \frac{4}{3}\pi\left(\frac{a}{2}\right)^2 \quad \left[\text{since, } r = \frac{a}{2}\right]$$

Thus %APF is given by:

$$\%\,\text{APF} = \frac{\%\text{ Volume occupied by effective number of atoms}}{\text{Volume of the unit cell}}$$

$$= \frac{6 \times \frac{4}{3}\pi\left(\frac{a}{2}\right)^3}{3\sqrt{2}\,a^3} \times 100 = 74\%$$

$\frac{c}{a}$ **ratio in HCP crystal** Consider the Figure 6.5. The top face of the HCP crystal is divided into six equilateral triangles. One such equilateral triangle AOB is drawn separately. GN is the half of the vertical distance, i.e., $\frac{c}{2}$. Thus the angle $\angle(AGN)$ is 90°. So we can write that:

$$AN^2 = NG^2 + GA^2$$

$$AN = a,\ NG = \frac{c}{2},\ BC = CO = \frac{a}{2}.$$

$$GA = \frac{2}{3}\ CA = \frac{2}{3}\ OA\ \cos 30° = \frac{a}{\sqrt{3}}$$

Putting these values of *AN*, *NG* and *GA* we get:

$$a^2 = \left(\frac{c}{2}\right)^2 + \frac{a^2}{3}$$

Or,
$$\frac{c}{a} = \sqrt{\frac{8}{3}} = 1.633$$

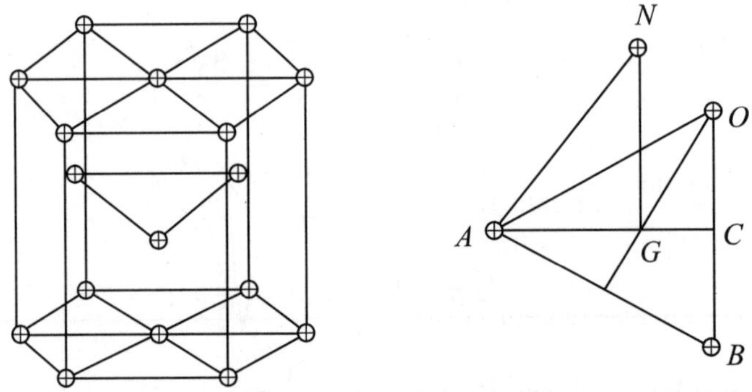

Fig. 6.5 $\frac{c}{a}$ ratio of HCP crystal.

Table 6.1 Characteristics of different crystal structures

Crystal Structure	*Effective no. of atoms*	*Atomic radius (r)*	*Coordination Number*	*APF (%)*
Simple Cubic	1	$\left(\frac{a}{2}\right)$	6	52
Body centred cubic	2	$\left(\frac{a\sqrt{3}}{4}\right)$	8	68
Face centred cubic	4	$\left(\frac{a}{2\sqrt{2}}\right)$	12	74
Hexagonal close packed crystal	6	$\left(\frac{a}{2}\right)$	12	74

6.6 Density of a cubic crystal

From the definition of density, we have: density (ρ) = (Mass)/(Volume). Volume of the unit cell is (a^3) and Mass = mass of effective number of atoms present in the unit cell. If Z be the effective number of atoms present in the unit cell, Mass = $\left(\frac{Z \times M}{N}\right)$, where, M is the molar mass, i.e., mass of Avogadro number of atoms and N is the Avogadro number. Thus the density of the crystal = ρ = $\left(\frac{Z \times M}{N \times a^3}\right)$ gms/cc. $Z = 1$ for SC crystal, $Z = 2$ for BCC crystal and $Z = 4$ for FCC crystal and a = lattice parameter, i.e, edge length of the cube.

6.7 Fundamental laws of crystallography

Each and every crystal possesses a definite geometrical shape and follows the following laws of crystallography: (a) law of constancy of interfacial angle, (b) law of rationality of indices, (c) law of symmetry.

(a) **Law of constancy of interfacial angle** Each and every crystal is surrounded by a definite number of plane surfaces, which are often termed faces. These faces intersect among themselves at a definite angle, known as interfacial angle. The interfacial angle is constant for a particular type of crystal, e.g., it is 90° for NaCl type of crystal structure.

Statement A crystal may appear in different shape or size depending on the condition of crystallization but the interfacial angle between the corresponding faces remains always constant.

(b) **Law of rationality of indices** Different lattice planes in a crystal intersect the crystallographic axes to form intercepts but in all cases ratio of intercepts, made by a lattice plane, to fundamental vectors bears a simple ratio of rational numbers. In the Figure 6.1, *OA*, *OB* and *OC* are the three fundamental vectors, represented by *a*, *b* and *c* respectively. A lattice plane makes the three intercepts *OL*, *OM* and *ON* on the three axes. Then $\frac{OL}{a}$, $\frac{OM}{b}$ and $\frac{ON}{c}$ are all integers and known as *Weiss indices*. If the lattice plane is parallel to *z*-axis, *ON* becomes ∝ and in that case Weiss indices are difficult to handle. The scientist Miller proposed that the inverse of the ratio should be taken as indices, $\frac{a}{OL}$, $\frac{b}{OM}$ and $\frac{c}{ON}$. These are known as *Miller indices*, represented by (h k l). Miller indices are easy to handle. If $OL = 2a$, $OM = 3b$ and $ON = 4c$, (2 3 4) are the Weiss indices of the plane and the corresponding Miller indices are $\left(\frac{1}{2}, \frac{1}{3}, \frac{1}{4}\right)$. LCM of 2, 3, 4 is 12. Multiplying with LCM the above figure becomes (6 4 3). So the plane is designated as (6 4 3).

Characteristics of Miller Indices (i) If a plane meets a crystallographic axis at infinity, the corresponding Miller index is 0. (ii) Two equidistant parallel planes from the origin have same Miller indices. (iii) If Miller indices of two planes bear the same ratio, the planes are parallel, e.g., Miller indices of (8 4 4) plane and (4 2 2) plane bear the same ratio (2 : 1 : 1).

(c) Law of symmetry Statement: Each and every crystal exhibits some sort of symmetry about an axis or a plane or the centre. For a particular substance all crystals show the same element of symmetry irrespective of their size and shape.

(i) Centre of symmetry If any two opposite faces or planes are equidistant from a point, the latter is called centre of symmetry. A crystal may possess one centre of symmetry.

(ii) Plane of symmetry If one part of a crystal appears as a mirror image of the other part with respect to a plane, the latter is called plane of symmetry. Two types of plane of symmetry are possible in a cubic crystal as shown in the Figure 6.6(c).

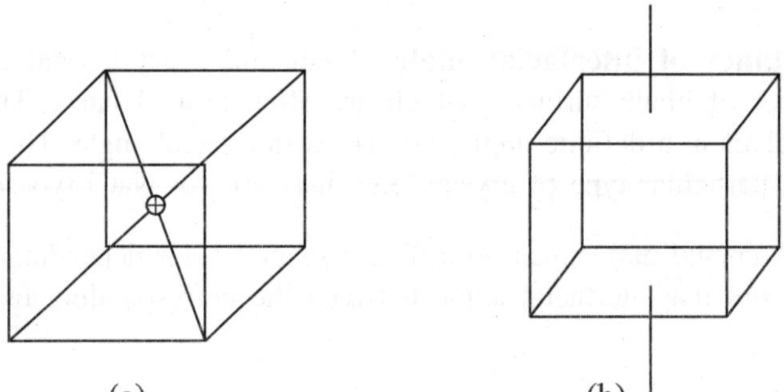

(a) (b)

Fig. 6.6 (a) Centre of symmetry and (b) Axis of symmetry of a cubic crystal.

One is called *straight plane symmetry* and the other is called *diagonal plane symmetry*. A cubic crystal has *three* straight plane symmetry and *six* diagonal plane symmetry, a total of *nine* plane of symmetery.

(iii) Axis of symmetry If a crystal is rotated by 360° about an axis such that the original configuration of the crystal returns more than once, the axis is called axis of symmetry. If the original configuration returns twice, the axis is called 2-fold axis of symmetry or *diad axis*. If the original configuration returns thrice, the axis is called 3-fold axis of symmetry or *triad axis*. If the original configuration returns four times, the axis is called 4-fold axis of symmetry or *tetrad axis*. If the original configuration returns six times, the axis is called 6-fold axis of symmetry or *hexad axis*. A crystal cannot have 5-fold axis of symmetry or more than 6-fold axis of symmetry.

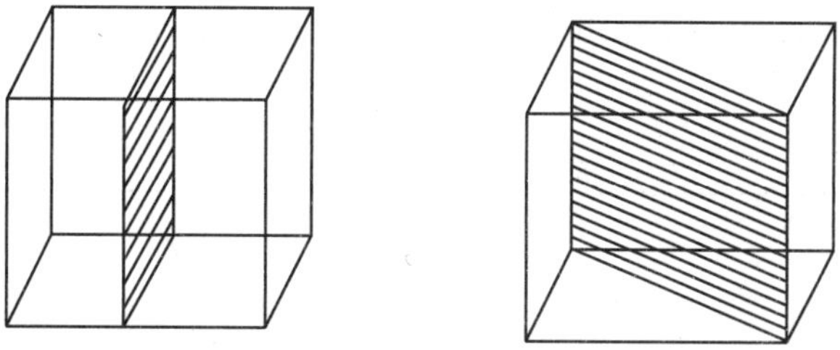

Straight plane symmetry Diagonal plane symmetry

Fig. 6.6 (c) Plane of symmetry of a cubic crystal.

Proof In the Figure 6.7, *MA*, *AB* and *BN* are the lattice points in the same plane. The crystal is rotated to an angle φ such that the point *M* takes the position *P*, the point *N* takes the position *Q*. If the original configuration returns the line *PQ* must be parallel to the line *MN* and the length *PQ* is an integral multiple of *AB*. Thus, $PQ = m\ AB = ma$, where *a* is the lattice parameter and *m* is an integer. Again from the Figure 6.7, we have:

$$PQ = PR + RS + SQ \quad \text{and} \quad PR = AP \cos \varphi = a \cos \varphi$$

$$SQ = BQ \cos \varphi = a \cos \varphi,\ RS = a$$

Thus, $$PQ = 2a \cos \varphi + a$$

Again, $$PQ = ma$$

Hence, $$2a \cos \varphi + a = ma.$$

Or, $$2 \cos \varphi = (m - 1) = N$$

n-fold axis of symmetry $= n = \left(\dfrac{360°}{\phi}\right).$

Table 6.2 Angle of rotation (φ) and corresponding value of '*n*'.

(*m*)	(*N*)	cos φ	φ	(*n*)
2	1	$\frac{1}{2}$	60°	6
1	0	0	90°	4
0	−1	$-\frac{1}{2}$	120°	3
−1	−2	−1	180°	2
3	2	1	360°	1

Thus it is evident that a crystal cannot have 5-fold axis of symmetry or more than 6-fold axis of symmetry. A cubic crystal possesses 6-diad axis of symmetry, 4-triad axis of symmetry and 3-tetrad axis of symmetry, a total of 13 axis of symmetry. Thus a cubic crystal possesses 1 centre of symmetry, 9 plane of symmetry and 13 axis of symmetry, a total of 23 elements of symmetry.

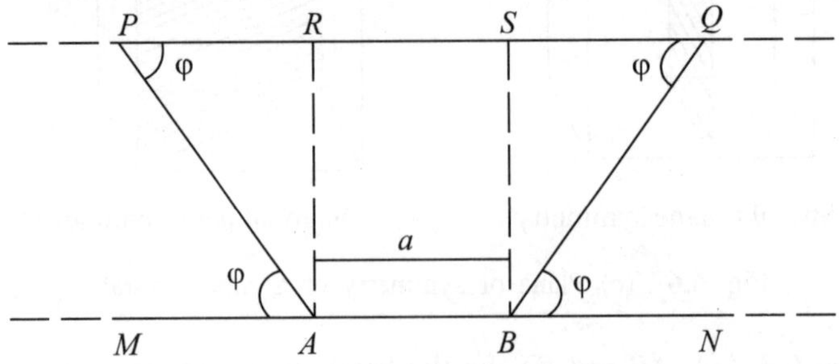

Fig. 6.7 Rotation of axis.

6.8 Miller indices and interplanar spacing

Miller indices of lattice plane is usually designated by (hkl). $h = 0$ signifies that the plane meets the corresponding crystallographic axis at infinity. $\{hkl\}$, represents the family of all possible planes. $\bar{h}$ signifies that the plane makes the intercept at negative corresponding crystallographic axis. Thus {210} contains the following planes: (210), (120), (201), (102), (012), (021), $(\bar{2}\ 1\ 0)$, $(\bar{2}\ 0\ 1)$, $(1\ \bar{2}\ 0)$, $(1\ 0\ \bar{2})$, $(0\ 1\ \bar{2})$, $(0\ \bar{2}\ 1)$, $(2\ \bar{1}\ 0)$, $(2\ 0\ \bar{1})$, $(\bar{1}\ 2\ 0)$, $(\bar{1}\ 0\ 2)$, $(0\ \bar{1}\ 2)$, $(0\ 2\ \bar{1})$, $(\bar{2}\ \bar{1}\ 0)$, $(\bar{2}\ 0\ \bar{1})$, $(\bar{1}\ \bar{2}\ 0)$, $(\bar{1}\ 0\ \bar{2})$, $(0\ \bar{1}\ \bar{2})$, $(0\ \bar{2}\ \bar{1})$. Total number of planes is 24.

Interplanar spacing Consider two parallel planes one of which is passing through the origin and the other plane ABC has the Miller indices (hkl). A normal ON is drawn from the origin to the plane ABC, which makes the intercepts OA, OB and OC on the crystallographic axes respectively. Let the normal ON makes angles α, β and γ with the corresponding crystallographic axes as shown in the Figure 6.8. Thus we can write that:

$$\cos\alpha = \frac{d}{OA}, \cos\beta = \frac{d}{OB} \quad \text{and} \quad \cos\gamma = \frac{d}{OC}$$

[since any straight line in the plane ABC is perpendicular to the normal ON and d is the interplanar distance between the two parallel planes = ON].

Again from the Figure 6.8 it is evident that:

$$ON^2 = d^2 = x^2 + y^2 + z^2 \quad \text{and} \quad x = d\cos\alpha,\ y = d\cos\beta \quad \text{and} \quad z = d\cos\gamma$$

Putting these values of x, y and z in the above equation, we have:

$$\cos^2\alpha + \cos^2\beta + \cos^2\gamma = 1 \quad [x = OL,\ y = OM,\ z = OK]$$

Again, we know from the definition of Miller indices that:

$$OA = \frac{a}{h}, OB = \frac{b}{k} \quad \text{and} \quad OC = \frac{c}{l} \quad [a = b = c \text{ for cubic system}]$$

ON is perpendicular to the plane *ABC*. So, $\angle NOA = \alpha$, $\angle NOB = b$, $NOC = \gamma$, $\angle ONA = \angle ONB = \angle ONC = 90°$. Thus we can write that:

$$\cos\alpha = \frac{d}{OA} = \frac{dh}{a},\ \cos\beta = \frac{d}{OB} = \frac{dk}{a} \text{ and } \cos\gamma = \frac{d}{OC} = \frac{dl}{a}$$

Hence, $$\cos^2\alpha + \cos^2\beta + \cos^2\gamma = \frac{d^2}{a^2}[h^2 + k^2 + l^2]$$

Or, $$\frac{d^2}{a^2}[h^2 + k^2 + l^2) = 1$$

Or, $$d = \frac{a}{\sqrt{h^2 + k^2 + l^2}}$$

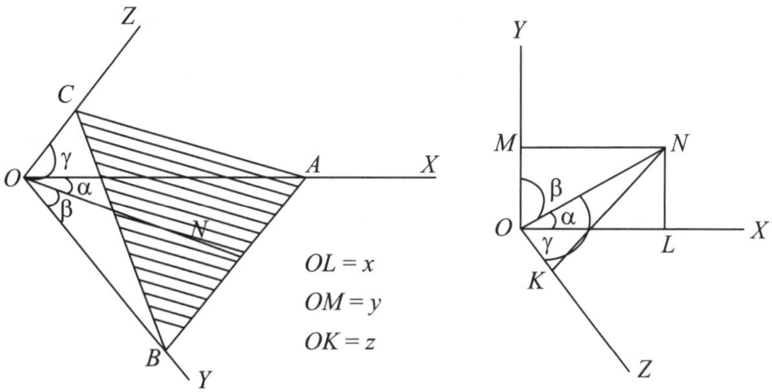

Fig. 6.8 Schematic diagram of the distance between two immediate parallel planes.

Miller indices for HCP crystals In the case of hexagonal crystal, a lattice plane is usually designated by four coordinate axes. The unit cell of a HCP crystal is shown in the Figure 6.9. The hexagonal base is defined by three coordinate axes a_1, a_2 and a_3, making an angle 120° with each other. The height of the crystal is defined by fourth coordinate axis, represented by *c*-axis, making an angle 90° with the hexagonal base. The three axes a_1, a_2 and a_3 form a vector triangle such that $\hat{a}_1 + \hat{a}_2 + \hat{a}_3 = 0$. Hence, $\hat{a}_1 = -(\hat{a}_2 + \hat{a}_3)$, $\hat{a}_2 = -(\hat{a}_1 + \hat{a}_3)$ and $\hat{a}_3 = -(\hat{a}_1 + \hat{a}_2)$. Thus, if two Miller indices are known the third Miller index is the negative sum of the first two. Thus the Miller indices of any plane in a hexagonal crystal can be represented as $(a_1\ a_2\ a_3\ c)$ and $a_3 = -(a_1 + a_2)$. If a plane has directions lying only in the basal plane, $c = 0$ and Miller indices are $(a_1\ a_2\ a_3\ 0)$.

Type-I basal plane The basal plane has direction parallel to a_1-axis as shown in the Figures 6.9. The Miller indices of the plane is $(h\ \bar{k}\ l\ 0)$, where $l = (-h + k)$.

Type-II basal plane The basal plane is perpendicular to the a_2-axis. Thus the basal plane is represented by $(h\ 0\ \bar{h}\ 0)$.

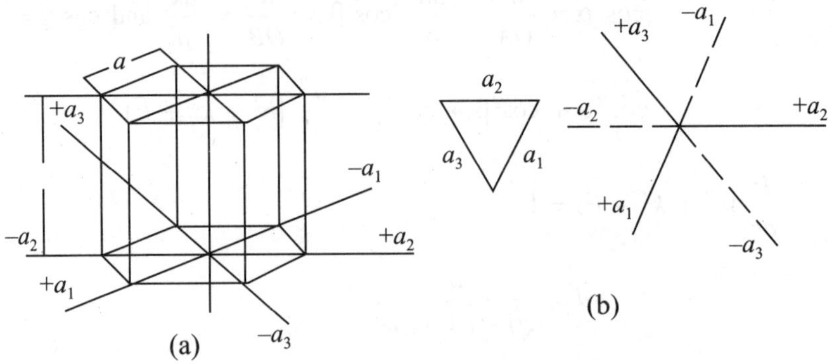

Fig. 6.9 (a) Unit cell of a HCP crystal, (b) Miller indices of a HCP crystal

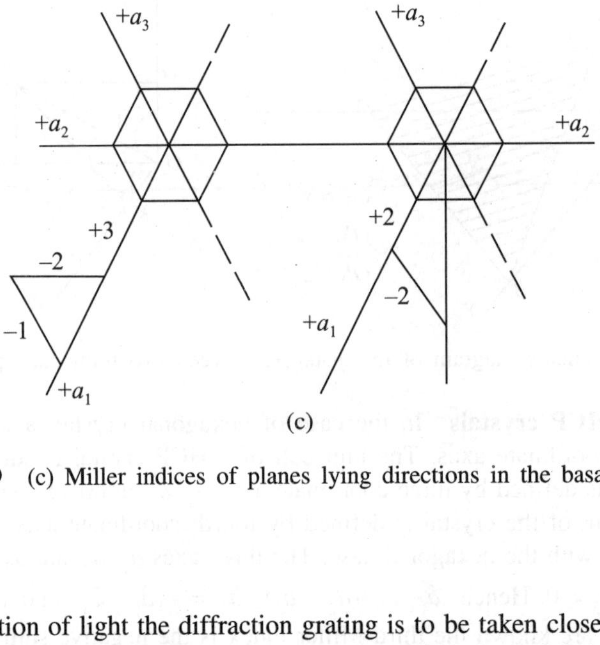

Fig. 6.9 (c) Miller indices of planes lying directions in the basal plane.

6.9 Bragg's law

In the case of diffraction of light the diffraction grating is to be taken close to the wavelength of light source. This is the primary condition to observe diffraction of light. The scientist Laue suggested that the parallel lattice planes can be used as diffraction grating. The interplanar distance between the lattice planes is of the order of 2 – 3 Å. Thus *X*-rays can be used as light source to observe diffraction in the crystal since the wavelength of *X*-ray is of the order of 1 Å. W. L. Bragg and W. H. Bragg showed that if a crystal is irradiated with *X*-rays, electrons within the crystal are excited and start vibrating with the frequency same as that of the incident *X*-rays. As a result of these vibrations, *X*-rays are reradiated with the same frequency. This interaction between *X*-rays and electrons can be treated as reflection of *X*-rays by atomic planes. According to Bragg's proposition, any two reflected *X*-rays will reinforce with each other if and only if the path difference between two *X*-rays is an integral multiple of wavelength or $n\lambda$, where λ is the wavelength of *X*-ray and n is an integer, 1, 2, 3, etc. As it is a case of reflection angle of incidence equals the angle of reflection. Let us consider two consecutive parallel lattice planes *AB* and *CD* in a crystal and a parallel beam of *X*-rays are striking on them (see the Figure 6.10a). *NOO′* is the normal to the planes *AB* and *CD*. *RO* and *SO′* are the two incident *X*-rays striking on *AB* and *CD* respectively. *OR′* and *O′S′* are the corresponding reflected *X*-rays. Angle of incidence is θ. From the point *O* two perpendiculars *OP* and *OQ* are drawn on *SO′* and *O′S′* respectively.

Thus, $\angle ROA = \angle SO'C = \angle PO'C = \theta$ and $\angle R'OB = \angle S'O'D = \angle QO'D = \theta$.

Again, $\angle PO'C + \angle PO'O = 90°$ and $\angle PO'O + \angle POO' = 90°$.

Hence, $\angle PO'C = \angle POO' = \theta$.

Similarly, $\angle QOO' = \theta$. $OO' = d$ = Distance between two parallel planes, *AB* and *CD*.

Hence, $PO' = QO' = OO' \sin\theta = d \sin\theta$.

Path difference between two sets of *X*-rays *ROR′* and *SO′S′* is $= PO' + QO' = d\sin\theta + d\sin\theta = 2d \sin\theta$. W. L. Bragg and W. H. Bragg proposed that path difference ia an integral multiple of wavelength, i.e.,

$$2d \sin\theta = n\lambda \tag{6.2}$$

where n is the order of reflection.

The value of n is positive integer only. For a given values of d and λ, the value of n increases with the increase in the value of θ. When $n = 1$, θ has the smallest value and the reflection is called 1st order reflection. Similarly, when $n = 2$, it is called 2nd order reflection etc. This is well illustrated in the Figure. 6.10(b).

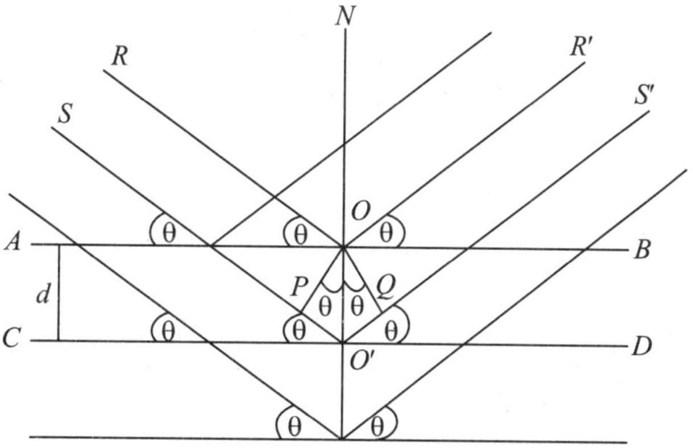

Fig. 6.10(a) Reflection of *X*-rays from parallel crystal planes.

Inference: $n\lambda = 2d \sin\theta$

Or, $$d = \frac{n\lambda}{2\sin\theta}$$

For, $n = 1,\ d = \dfrac{\lambda}{2\sin\theta}$

Theoretically, $\theta_{max} = 90°$ and $\sin\theta = 1$, which is practically impossible. So, $\lambda_{max} = 2d$ and hence wavelength of X-rays is always less than $2d$ or $\lambda < 2d$.

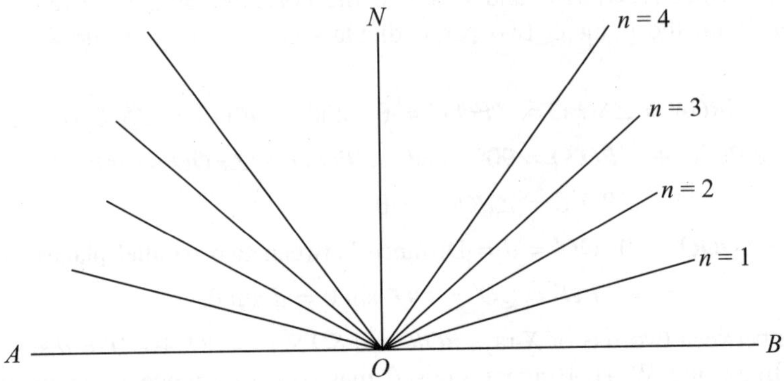

Fig. 6.10(b) Reflection of X-rays from a crystal plane AB at different angles.

6.10 Crystal imperfections/defects

In an ideal crystal regularity or periodicity of arrangement of atoms or ions is maintained throughout the crystal but in real crystal disruption in periodicity is very common within and on the surface of the crystal. These disruptions arise due to some defects, known as *crystal imperfections.* On the basis of stoichiometry crystal defects or imperfections can be classified into two broad categories: (a) Stoichiometric defects and (b) Non-stoichiometric defects.

6.10.1 Stoichiometric defects

When defects are formed in a crystal such that overall electrical neutrality is maintained, the defects are called stoichiometric defects. Stoichiometric defects can further be classified into two categories: (i) Schottky defect, (ii) Frenkel defect.

(i) **Schottky defect** Schottky defect is observed both in ionic crystals and metallic crystals. In the case of ionic crystal it may be possible that one or more pairs of ions are missing from their regular positions, leaving behind vacancies within the crystal but the overall electrical neutrality is maintained. Such defects are called Schottky defects in an ionic crystal. It is well illustrated in the Figure 6.11.

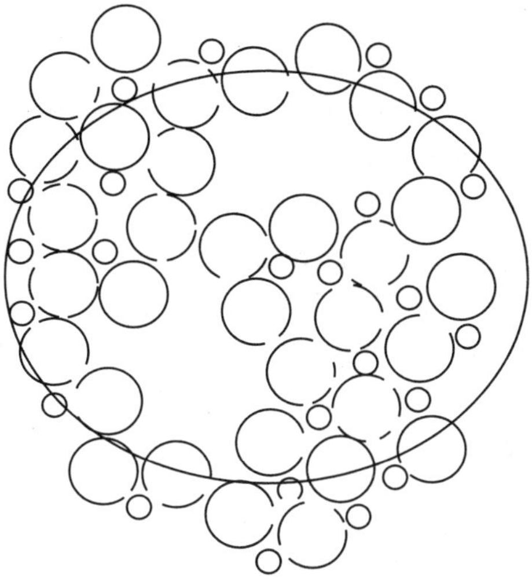

Fig. 6.11 Schottky defect: Vacancy created due to missing of pairs of ions.

In the case of metallic crystals the irregularity on the surface is the source of Schottky defect. Atoms, arranged periodically within a metallic crystal, are always vibrating. Some of them are able to cross the energy barrier and jump on to the surface, leaving behind vacancies within the crystal. With the increase in temperature, probability of jumping increases and hence number of vacancies increases. Vacancies, thus created, are moving inside the crystal, opposite to the direction of jumping. Creation of vacancies is illustrated in the Figure 6.12. It is expected that with the increase in temperature vacancy motion increases.

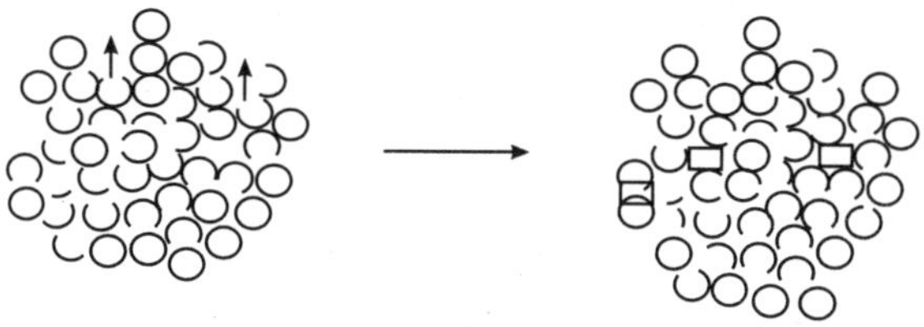

(a)

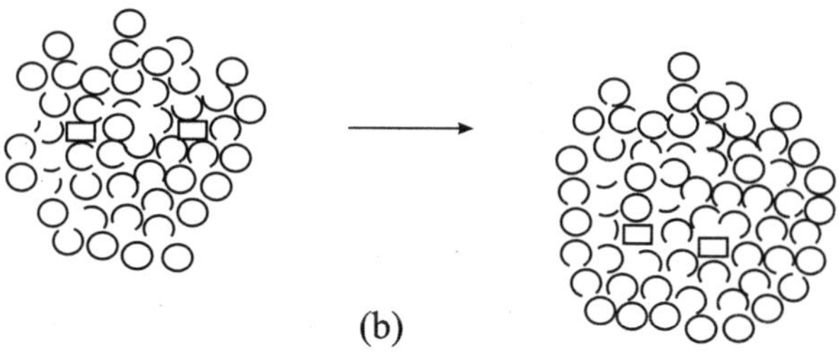

(b)

□ ≡ Vacancy

Fig. 6.12 (a) Creation of vacancy, (b) Motion of vacancy.

Thus a real crystal contains two constituents, *lattice point* and *vacancy* and they are intermixed. So free energy of mixing is given by ΔG^M.

n_0 = Number of moles of lattice points in a crystal

n_v = Number of moles of vacancies in that crystal.

w = Amount of work required to create one vacancy.

Nw = Amount of work required to create one mole of vacancies or, enthalpy of formation of one mole of vacancies.

$Nw \times n_v$ = Total enthalpy of formation of n_v moles of vacancies.

N = Avogadro Number = 6.023×10^{23}/gm-mole.

Total number of moles = $n_0 + n_v$

Mole fraction of lattice points = $x_0 = \dfrac{n_0}{n_0 + n_v}$

Mole fraction of vacancies = $x_v = \dfrac{n_v}{n_0 + n_v}$

Free energy of mixing ΔG^M is given by:

$$\Delta G^M = \Delta H^M - T\Delta S^M \tag{6.3}$$

$$\Delta H^M = Nw\, n_v$$

$$\Delta S^M = -R\, [n_0 \ln x_0 + n_v \ln x_v]$$

Putting these values of ΔH^M and ΔS^M in the Equation (6.3), we get:

$$\Delta G^M = Nw\, n_v + RT\left[n_0 \ln \frac{n_0}{n_0 + n_v} + n_v \ln \frac{n_v}{n_0 + n_v}\right] \tag{6.4}$$

The plot of ΔG^M versus n_v curve is shown in the Figure 6.13.

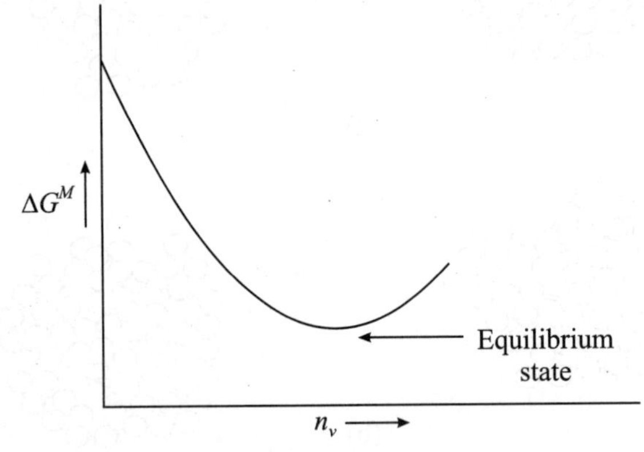

Fig. 6.13 Schematic diagram of variation of ΔG^M with n_v

As a crystal always remains in equilibrium state, unless it is disturbed, the ΔG^M assumes its minimum value, i.e., $(\Delta G^M)_{min}$. To find out $(\Delta G^M)_{min}$, differentiate the Equation (6.4) with respect to n_v and set it to zero.

$$\Delta G^M = Nw\, n_v + RT[n_0 \ln n_0 + n_v \ln n_v - n_0 \ln(n_0 + n_v) - n_v \ln(n_0 + n_v)]$$

$$\left(\frac{\partial \Delta G^M}{\partial n_v}\right)_{P,T,n_0} = Nw + RT[\ln n_v - \ln(n_0 + n_v)]$$

$$= Nw + RT \ln \frac{n_v}{n_0 + n_v}$$

At the minimum, $$\left(\frac{\partial \Delta G^M}{\partial n_v}\right)_{P,T,n_0} = 0$$

Or, $$Nw + RT \ln \frac{n_v}{n_0 + n_v} = 0$$

Again, $$n_0 + n_v \approx n_0$$

since, $$n_0 >> n_v$$

So, $$RT \ln \frac{n_v}{n_0} = -Nw$$

Or, $$n_v = n_0\, e^{-\frac{Nw}{RT}}$$

Or, $$n_v = n_0\, e^{-\frac{\Delta H_f}{RT}} \tag{6.5}$$

where, ΔH_f = Enthalpy of formation of 1 mole of vacancies = Nw

AT $T = 0$, $n_v = 0$, which means that a real crystal tends to be ideal one as temperature approaches to absolute zero.

The Equation (6.5) gives us an idea about number of vacancies in any real crystal, e.g. ΔH_f for Cu-crystal at 27°C is 20 kcal/mole. So, molar vacancies = n_v

$$n_v = n_0\, e^{-\frac{\Delta H_f}{RT}} = n_0 e^{-\frac{20{,}000}{2 \times 300}}$$

Or, $$\frac{n_v}{n_0} = e^{-33.34} = 4.45 \times 10^{-15}$$

where, $R = 2$ cal/(mole-°K) and $T = (273 + 27)$°K = 300°K

Thus for every 10^{15} atoms or lattice points there are 5 vacancies in the crystal.

(ii) Frenkel defect It is usually observed in an ionic crystal. An ionic crystal is constituted by cations and anions. As we know any real crystal possesses interstitial voids and due to small in size cations have a tendency to jump into interstitial voids from their regular positions. The defect, thus produced, is known as Frenkel defect. This is well illustrated in the Figure 6.14.

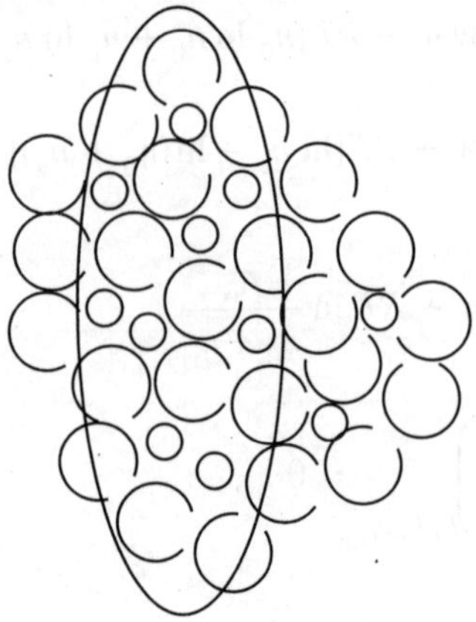

Fig. 6.14 Schematic representation of Frenkel defect in an ionic crystal.

Characteristic features of Schottky defect

(a) It arises due to jumping of atoms or missing of pair of ions, leading to the formation of vacancies within the crystal lattice.

(b) As fraction of lattice sites is unoccupied density of the crystal decreases. Thus the observed density (ρ_{obs}) is always less than the ideal density (ρ_{id}), and this density difference helps us evaluate percent unoccupied sites in a crystal lattice.

$$\% \text{ Unoccupied sites} = \frac{\rho_{id} - \rho_{obs}}{\rho_{id}} \times 100$$

ρ_{obs} is an experimentally determined quantity. For cubic crystal, we have:

$$\rho_{id} = \frac{MZ}{N_A a^3}$$

M = Molecular mass of the crystal,

Z = 1 for SC, 2 for BCC, 4 for FCC

N_A = Avogadro number = 6.023×10^{23}/g.mole

a = lattice parameter of the crystal (edge length of the cube)

(c) Ionic crystals, having cations and anions of comparable sizes, show this type of defect. Examples are: NaCl, KBr etc.

(d) This defect is observed both in ionic and metallic crystals.

Characteristic features of Frenkel defect

(a) This type of defect arises due to jumping of positive ions from their regular sites to the interstitial voids.

(b) As there is no pair of missing ions within the crystal, density of the crystal does not decrease due to the presence of Frenkel defect.
(c) This defect is observed mainly in the ionic crystal.
(d) Ionic crystals, having cationic size much smaller than anionic size, show this type of defect. Examples are: AgCl, ZnS etc.

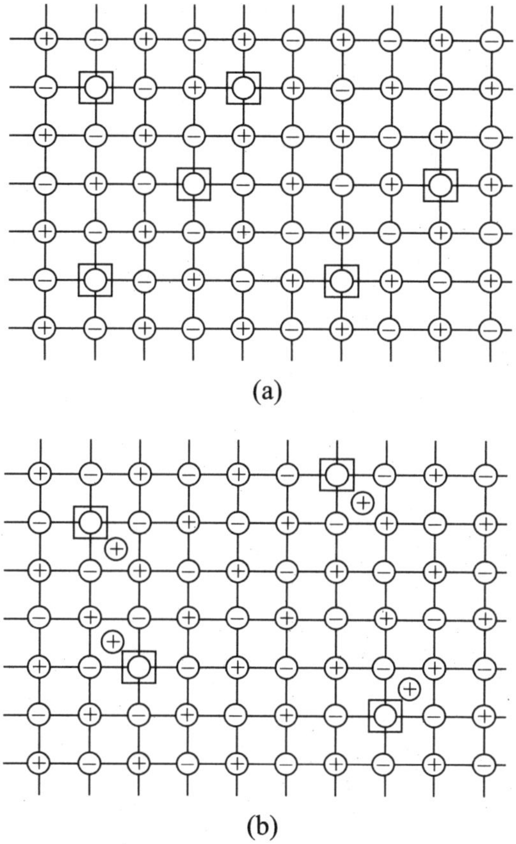

Fig. 6.15 Schematic representation of (a) Schottky defect and (b) Frenkel defect.

6.10.2 Non-stoichiometric defects

Non-stoichiometric defect

If crystal imperfections are so created in an ionic crystal that the composition of the ionic crystal does not follow the exact stoichiometric formula. Such defects are called non-stoichiometric defects, as a result of which, cation or anion excess always occurs. These type of defects are called non-stiochiometric defects, which can be divided into three types:

(i) Metal excess defects.
(ii) Metal deficiency defects.
(iii) Impurity defects.

(i) Metal excess defect

This may occur in either of the following two ways:

(a) Anion vacancies

In case of an ionic solid crystal, some anions are missing from their regular sites, generating vacancies or positive holes. To maintain electrical neutrality, electrons from outside diffuse into the crystal lattice and are trapped by the positive holes. In presence of sunlight or heat energy, these trapped electrons absorb energy and jump to the nearby vacancies. Due to this absorption of energy the crystal appears coloured. Hence the trapped electrons are often called F-centres or colour centres as they are responsible for imparting colour to the crystal.

When NaCl crystal is heated in presence of Na vapour, Cl^- diffuses out of the crystal and electrons absorb energy due to which the crystal appear yellow. Due to same reason, LiCl appears pink colour when it is heated in Li vapour and KCl appears violet in presence of K vapour. As the ions are missing from their regular sites, the defect is similar to *Schottky defects*.

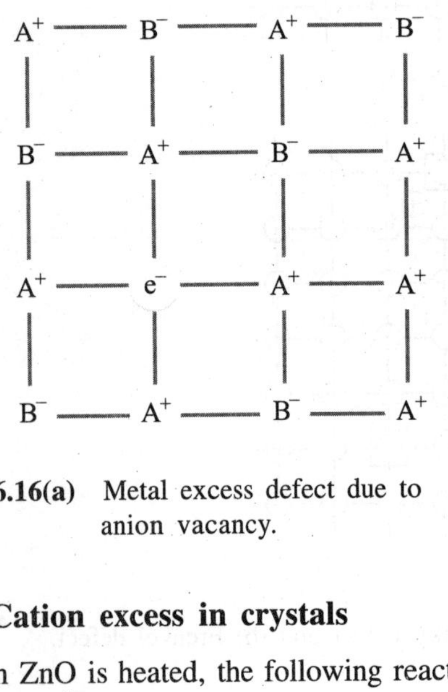

Fig. 6.16(a) Metal excess defect due to anion vacancy.

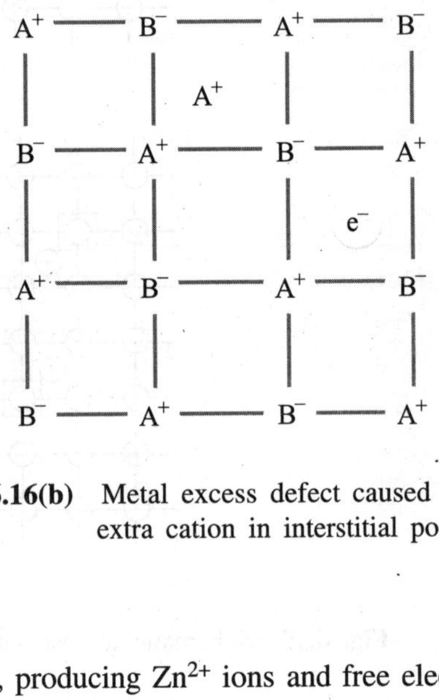

Fig. 6.16(b) Metal excess defect caused by extra cation in interstitial position.

(b) Cation excess in crystals

When ZnO is heated, the following reactions take place, producing Zn^{2+} ions and free electrons as shown below:

$$ZnO \rightleftharpoons Zn^{2+} + \frac{1}{2}O_2 + 2e$$

These Zn^{2+} ions and electrons diffuse into the crystal lattice and occupy the interstitial sites. The trapped electrons absorb colour when the ZnO crystal is heated and hence the crystal appears yellow. Again the crystal becomes white when it is cooled to room temperature. As Zn^{2+} ions occupy the interstitial voids, the defect is similar to ***Frenkel defects***. Due to the presence of excess Zn^{2+} ions the formula of the crystal should be Zn_xO with $x > 1$.

Crystals with either type of metal excess act as semiconductors.

(ii) Metal deficiency defect

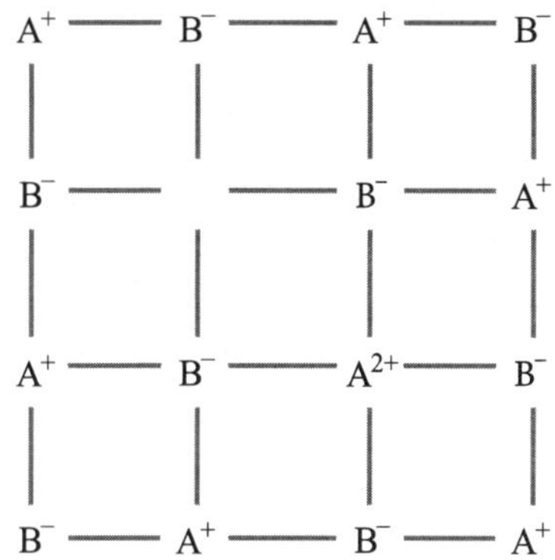

Fig. 6.16(c) Metal deficiency defect due to missing of a cation of lower valency and presence of a cation of higher valency.

This type of defect is observed when metal shows variable valency. Examples are: FeO, FeS, NiO. When FeO crystal is heated, some Fe^{2+} ions are oxidized to Fe^{3+} ions and hence one vacancy is created for every two Fe^{3+} ions in order to maintain the electrical neutrality. Free electrons, generated during oxidation, are trapped within these vacancies and give colour to the crystal. So, FeO is better represented as FegO, where, $y < 1$.

Applications of metal excess/metal deficiency defects

1. The free electrons are often trapped in the interstitial sites. These electrons are responsible for characteristic colours of different ionic crystals.
2. The characteristic colours, generated due to these defects, are practically used to identify the elements, e.g., golden yellow colour indicates the presence of Na in the sample or purple colour indicates the presence of K etc.
3. The free electrons, trapped in the interstitial sites, impart conductivity to the crystal, behaving the latter as semiconductor.
4. As the trapped electrons reflect light, the crystals appear metallic lustrous.

(iii) Impurity defect in covalent solids

The element Si has valency 4. If Si crystal is treated with Group 13 elements like Al, Ga etc., some Si atoms are replaced by Group 13 elements. The phenomenon is called doping. As valency of Group 13 element is 3, doped Si crystal becomes electron deficient, generating some positive holes within the crystal. These positive holes are responsible for conductivity of doped Si crystal. These are called p-type semiconductors.

In another case if the Si crystal is treated with Group 15 elements like P, As etc., some Si atoms are replaced by Group 15 elements. The phenomenon is also called doping. As valency of Group 15 element is 5, doped Si crystal possesses excess electrons, generating some excess valence electrons within the crystal. These excess valence electrons are responsible for conductivity of doped Si crystal. These are called n-type semiconductors.

The above type of defect is called impurity defect.

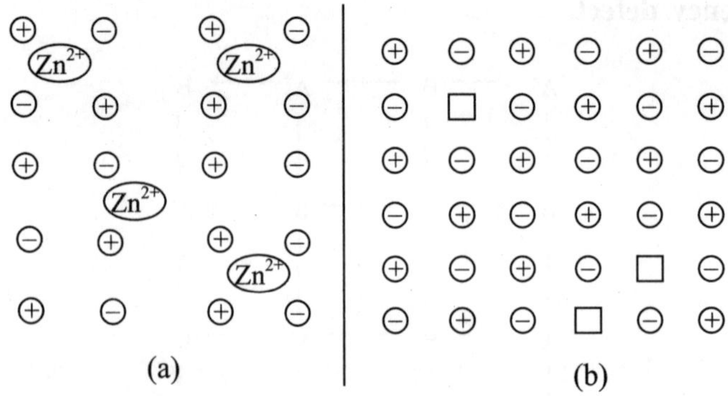

Fig. 6.17 (a) Defects in ZnO crystal due to excess Zn^{2+} ions, (b) Defects in FeO crystal due to vacant cation sites.

6.11 Schrödinger wave equation, energy states and Fermi energy

6.11.1 Schrödinger wave equation

According to Heisenberg's uncertainty principle, position and velocity of any small particle, like electron, cannot be determined at the same time. Thus a small particle is believed to have both particulate and wavy characteristics. Schrödinger introduced probability density function (ψ^2) to describe electron density in a given energy state. ψ depends on both time and distance. It is defined as:

$$\psi = \psi_0 \sin(\omega t - kx) \tag{6.6}$$

ψ is known as probability of finding electron in a given energy state and ψ^2 is the probability density function in a given volume $(dx.dy.dz)$ such that:

$$\int_{-\infty}^{+\infty}\int_{-\infty}^{+\infty}\int_{-\infty}^{+\infty} \psi^2 \, dx\, dy\, dz = 1$$

In one dimension the above equation becomes:

$$\int_{-\infty}^{+\infty} \psi^2 \, dx = 1$$

Differentiating the Equation (6.6) with respect to x we get:

$$\frac{\partial \psi}{\partial x} = -k\, \psi_0 \cos(\omega t - kx) \quad \text{and} \quad \frac{\partial^2 \psi}{\partial x^2} = -k^2\, \psi_0 \sin(\omega t - kx) = -k^2 \psi$$

Or,
$$\frac{\partial^2 \psi}{\partial x^2} + k^2 \psi = 0 \tag{6.7}$$

where, $$k = \frac{2\pi}{\lambda}, \quad \omega = 2\pi f,$$

λ = wave length and f = frequency of electronic wave.

In three dimensions the Equation (6.7) takes the form:

$$\frac{\partial^2\psi}{\partial x^2} + \frac{\partial^2\psi}{\partial y^2} + \frac{\partial^2\psi}{\partial z^2} + k^2\psi = 0$$

Or, $$\nabla^2\psi + k^2\psi = 0 \tag{6.8}$$

According to De Broglie's concept, matter waves possess both particulate and wavy characteristics and hence we have: Total energy = $T = mv^2$ and $T = h\nu$

Or, $$mv^2 = h\nu = \frac{h\nu}{\lambda}$$

Or, $$\lambda = \frac{h}{mv} \quad \text{and} \quad k = \frac{2\pi}{\lambda} = \frac{2\pi\, mv}{h}$$

So, $$k^2 = \left(\frac{2\pi}{h}\right)^2 (mv)^2 = \frac{4\pi^2}{h^2}(2m)\left(\frac{1}{2}\, mv^2\right) = \frac{8\pi^2\, m\, E}{h^2}$$

Or, $$k^2 = \frac{8\pi^2\, m\, E}{h^2} \quad \text{and} \quad E = \frac{h^2\, k^2}{8\pi^2\, m} \tag{6.9}$$

E = Kinetic energy of electron = $(T - V) = \frac{1}{2}mv^2$

= Total energy (T) – potential energy (V)

m = Mass of an electron, which is assumed to be constant = 9.11×10^{-31} kg

h = Planck's constant = 6.625×10^{-34} J – s.

Combining Equations (6.8) and (6.9), we get:

$$\nabla^2\psi + \frac{8\pi^2\, m\, E}{h^2}\psi = 0$$

Or, $$\nabla^2\psi + \frac{8\pi^2\, m}{h^2}(T - V)\psi = 0 \tag{6.10}$$

The Equation (6.10) is known as time-independent Schrödinger wave equation.

According to Equation (6.9) $E - k$ curve is a parabola as shown in the Figure 6.18.

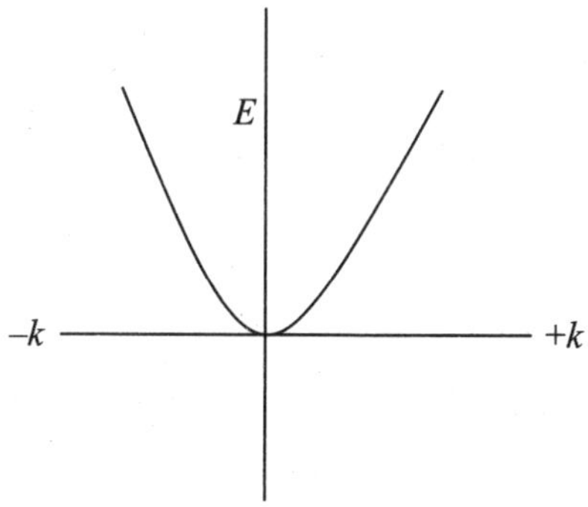

Fig. 6.18 $E - k$ curve of an electron wave

6.11.2 Sommerfeld theory

Quantum concept of free electron theory

In 1900, Drude and Lorentz first proposed free electron concept of metallic bonding. Later on Sommerfeld modified the free electron theory by introducing quantum concept.

Postulates

(a) Electrons in a metal are free to move throughout the crystal lattice but not allowed to leave the crystal.

(b) Electrons possess both particulate and wavy characteristics. So they are believed to follow De Broglie concept of wave-particle duality. Electron waves are often considered as *matter waves*.

(c) The electron wave follows Schrödinger wave equation given below:

$$\nabla^2\psi + \frac{8\pi^2 m}{h^2}(T - V)\psi = 0$$

$T = E + V$, E = Kinetic energy T = Total energy. V = Potential energy, ψ = Electron wave function, m = Mass of electron and h = Planck's constant

(d) At absolute zero electrons do not have any kinetic energy and they are confined in different energy states.

Energy, associated with the highest energy state, where electron exists at absolute zero, is known as *Fermi energy*. However, at higher temperature electrons gain kinetic energy and have a tendency to jump to the vacant energy states. In that case probability of finding electron at any given energy state is given by *Fermi-Dirac formula*, as given below:

$$F(E) = \frac{1}{1 + e^{\frac{(E - E_F)}{kT}}}$$

where, $F(E)$ is the probability of finding electron at any given energy state of energy E. E_F is known as Fermi energy, k is the Boltzman constant and T is the absolute temperature in °K.

The following conclusions can be drawn from the Fermi-Dirac formula:

(i) At $T = 0$°K, $E = 0$ and $F(E) = 1$.

(ii) At $T > 0$°K, and $E < E_F$, the value of $e^{\frac{(E - E_F)}{kT}}$ increases with increase in temperature and hence $F(E)$ decreases.

(iii) At $T > 0$°K and $E > E_F$, the value of $e^{\frac{(E - E_F)}{kT}}$ decreases with increase in temperature and hence $F(E)$ increases.

(iv) At $T > 0$°K and $E = E_F$, $F(E) = \frac{1}{2}$ at all temperatures.

These are well illustrated in the following figure:

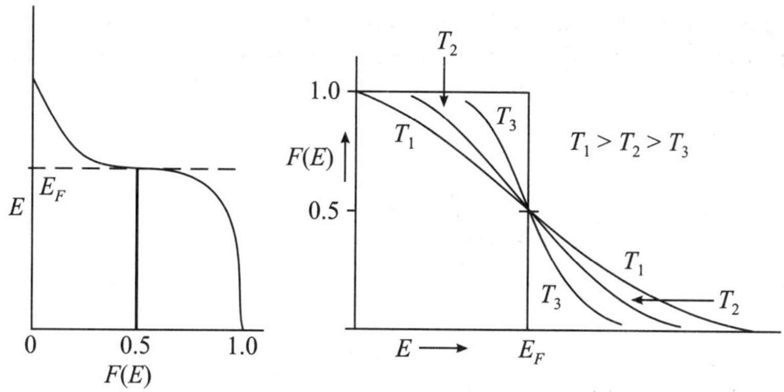

Fig. 6.19 $F(E)$ versus E curve

Fermi level

(a) It is the highest energy state where electron exists even at absolute zero.
(b) It is that energy level where probability of finding electron is 1 at absolute zero.
(c) It is that energy level where probability of finding electron is $\frac{1}{2}$ at any temperature other than 0°K.
(d) Although electrons are moving in a metal in a complicated potential field, the potential energy of each election is constant within the metal.
Consider a one-dimensional potential well as shown in the Figure 6.19, with the following boundary conditions:

(i) Inside the potential well $V = 0$ and outside the potential well $V = \infty$.
(ii) At $x = 0$, $\psi = 0$ and at $x = L$, $\psi = 0$.

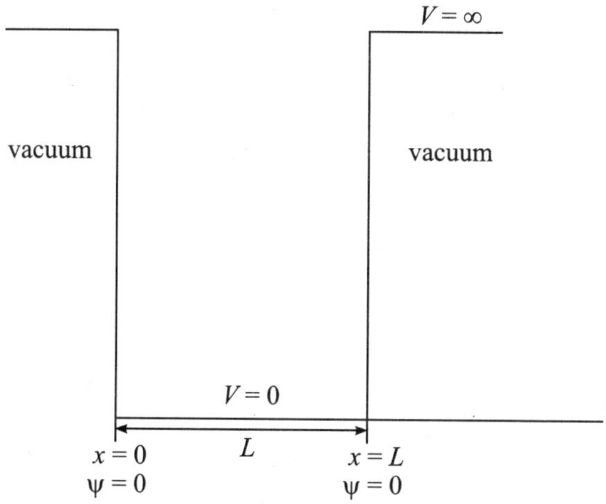

Fig. 6.20 One-dimensional potential well. L = Length of the potential well.

The length of the potential well is so chosen that the electron density vanishes at the boundary walls, that means, electrons are behaving as stationary waves within the potential well, as shown in the Figure 6.21.

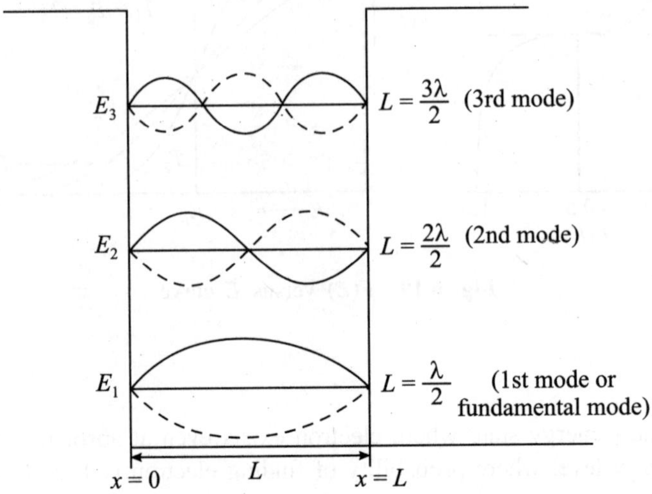

Fig. 6.21 Electron waves in a potential well.

$$L = \frac{n\lambda}{2} \quad \text{or} \quad \lambda = \frac{2L}{n} \quad \text{or} \quad n = \frac{2L}{\lambda}$$

At the 1st mode of vibration: $L = \dfrac{\lambda}{2}$

Or, $\lambda = \dfrac{2L}{1}$

At the 2nd mode of vibration: $L = \lambda$

Or, $\lambda = \dfrac{2L}{2}$

At the 3rd mode of vibration: $L = \dfrac{3\lambda}{2}$

Or, $\lambda = \dfrac{2L}{3}$

At the *n*th mode of vibration: $L = \dfrac{n\lambda}{2}$

Or, $\lambda = \dfrac{2L}{n}$

Considering the one-dimensional potential well, $n = n_x$

Or, $$\lambda = \frac{2L}{n_x}, \quad k = \frac{2\pi}{\lambda} = \frac{2\pi\, n_x}{2L} = \frac{\pi\, n_x}{L}$$

Or, $$k = \frac{\pi\, n_x}{L}$$

Substituting this value of k in the Equation (6.9), we get:

$$E = \left(\frac{\pi\, n_x}{L}\right)^2 \frac{h^2}{8\pi^2 m} = \frac{h^2 n_x^2}{8m\, L^2}$$

Or, $$E_x = \frac{h^2 n_x^2}{8m\, L^2} \quad \text{(along } x\text{-axis)}$$

along y-axis: $E_y = \dfrac{h^2 n_y^2}{8m\, L^2}$ and along z-axis: $E_z = \dfrac{h^2 n_z^2}{8m\, L^2}$

Thus the total kinetic energy $= E = E_x + E_y + E_z$

$$E = E_x + E_y + E_z = \frac{h^2}{8m\, L^2}\,(n_x^2 + n_y^2 + n_z^2) \tag{6.11}$$

where, n_x, n_y and n_z are the energy states along x–, y– and z– axes and $n = \dfrac{2L}{\lambda}$

So, λ cannot be zero or infinity. It has only positive values.

Thus, $n_x > 0$, $n_y > 0$ and $n_z > 0$. Thus, the values of n_x, n_y and n_z are positive integers only.

So, $n_x = 1, 2, 3, \ldots$, $n_y = 1, 2, 3, \ldots$ and $n_z = 1, 2, 3, \ldots$

(i) Configuration of the lowest energy state is (111) , i.e., $n_x = 1$, $n_y = 1$ and $n_z = 1$:

$$E = \frac{h^2}{8mL^2}(1^2 + 1^2 + 1^2) = \frac{3h^2}{8mL^2}$$

(ii) Configuration of next higher energy state is,

(121), i.e., $n_x = 1$, $n_y = 2$ and $n_z = 1$ or,

(211), i.e., $n_x = 2$, $n_y = 1$ and $n_z = 1$ or,

(112), i.e., $n_x = 1$, $n_y = 1$ and $n_z = 2$.

So, the energy state {112} has three degenerate states, each of which has kinetic energy:

$$E = \frac{h^2}{8mL^2}(1^2 + 1^2 + 2^2) = \frac{6h^2}{8mL^2}$$

(iii) For the energy state {123}, we have:

{132} and {123} $\Rightarrow n_x = 1$, $n_y = 3$, $n_z = 2$ and $n_x = 1$, $n_y = 2$, $n_z = 3$

$\{213\}$ and $\{231\} \Rightarrow n_x = 2, n_y = 1, n_z = 3$ and $n_x = 2, n_y = 3, n_z = 1$

$\{321\}$ and $\{312\} \Rightarrow n_x = 3, n_y = 2, n_z = 1$ and $n_x = 3, n_y = 1, n_z = 2$

So in each case: $(n_x)^2 + (n_y)^2 + (n_z)^2 = 14$. So there are six degenerate states, each of which has kinetic energy:

$$E = \frac{h^2}{8mL^2}(1^2 + 2^2 + 3^2) = \frac{14h^2}{8mL^2}$$

Number of energy states In two dimensions, the expression of energy (E) is given by:

$$E = \frac{h^2}{8mL^2}(n_x^2 + n_y^2)$$

Or,
$$E \propto (n_x^2 + n_y^2)$$

Or,
$$E \propto r^2$$

$$r = (n_x^2 + n_y^2)^{\frac{1}{2}} \tag{6.12}$$

For any value of r there are number of energy states available in the metal. If the value of r is small, there is only a few number of energy states and the total number can be counted easily, but for large value of r it is difficult to find out total number of energy states by counting method. Graphical approach obviates the tedious counting method and helps us evaluate the total number of energy states very easily. Equation (6.12) is an equation of a circle the 1st quadrant of which is shown in Figure 6.22. As $n_x > 0$ and $n_y > 0$, all energy states are confined in the 1st quadrant only.

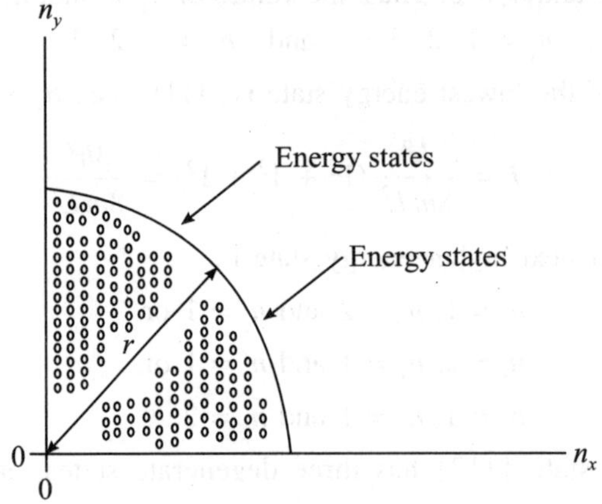

Fig. 6.22 A two-dimensional quantum number space. The total number of energy states within the arc = Area of the quadrant

$$\text{Area of the quadrant} = \frac{\pi}{4}r^2 = \frac{\pi}{4}(n_x^2 + n_y^2)$$

In three dimensional space radius of the sphere = r, which is expressed as:

$$r^2 = (n_x^2 + n_y^2 + n_z^2)$$

Or,
$$r = (n_x^2 + n_y^2 + n_z^2)^{\frac{1}{2}} \quad (6.13)$$

As $n_x > 0$, $n_y > 0$ and $n_z > 0$, in the case of sphere, all energy states are concentrated in an octant i.e., $\frac{1}{8}$th of the volume of a sphere. The volume of octant indicates the total number of energy states. Thus the total number of energy states within the radius r is equal to the volume of octant.

$$\text{Volume of octant} = \eta = \frac{1}{8} \times \frac{4}{3}\pi r^3 = \frac{\pi}{6} r^3$$

$$= \frac{\pi}{6}(n_x^2 + n_y^2 + n_z^2)^{\frac{3}{2}}$$

Or,
$$\eta = \frac{\pi}{6}(n_x^2 + n_y^2 + n_z^2)^{\frac{3}{2}}$$

Putting the value of $(n_x^2 + n_y^2 + n_z^2)$ from Equation (6.11) into the above equation, we get:

$$\eta = \frac{\pi}{6}\left(\frac{8m L^2 E}{h^2}\right)^{\frac{3}{2}} \quad (6.14)$$

Filling up of energy states by electrons follows Aufbau principle, Pauli's exclusion principle and Hund's rule of maximum multiplicity. 'η' indicates total number of energy states available within the radius r. According to Pauli's exclusion principle each energy state can possess a maximum of two electrons. At absolute zero, total number of electrons (N) in a piece of metal requires $\frac{N}{2}$ number of energy states since each energy state possesses two electrons. Furthermore, at absolute zero $E = E_F$, where E_F is known as Fermi energy .

So, at $E = E_F$, $\eta = \left(\frac{N}{2}\right)$. Thus Equation (6.14) becomes:

$$\frac{N}{2} = \frac{\pi}{6}\left(\frac{8m L^2 E_F}{h^2}\right)^{\frac{3}{2}}$$

$$E_F = \left(\frac{3N}{\pi}\right)^{\frac{2}{3}}\left(\frac{h^2}{8m L^2}\right)$$

$$= \left(\frac{h^2}{8m}\right)\left(\frac{3N}{\pi L^3}\right)^{\frac{2}{3}} = \left(\frac{h^2}{8m}\right)\left(\frac{3N}{\pi \bar{V}}\right)^{\frac{2}{3}}$$

Or,
$$E_F = \left(\frac{h^2}{8m}\right)\left(\frac{3N}{\pi\bar{V}}\right)^{\frac{2}{3}} \tag{6.15}$$

where, $\bar{V} = L^3$ = volume of the assumed metal box = Molar volume of the metal.

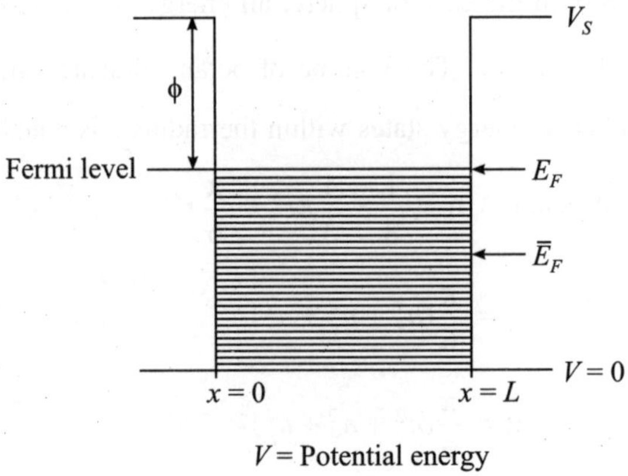

Fig. 6.23 Fermi energy according to Sommerfeld theory.

Fermi level At absolute zero electron-occupied highest energy state is known as Fermi level and the corresponding energy is known as Fermi energy. So E_F is the highest energy level where electron can be found at absolute zero. Thus at absolute zero energy associated with electrons lies between 0 and E_F. Thus average energy possessed by electrons in the metal box = $\bar{E}_F$.

$$\bar{E}_F = \frac{1}{N}\int_0^N \left(\frac{h^2}{8m}\right)\left(\frac{3N}{\pi\bar{V}}\right)^{\frac{2}{3}} dN$$

$$= \frac{1}{N}\left[\left(\frac{h^2}{8m}\right)\left(\frac{3}{\pi\bar{V}}\right)^{\frac{2}{3}} \frac{N^{\frac{2}{3}+1}}{\left(\frac{2}{3}+1\right)}\right]$$

$$= \frac{3}{5}\left(\frac{h^2}{8m}\right)\left(\frac{3N}{\pi\bar{V}}\right)^{\frac{2}{3}} = \frac{3}{5}E_F$$

Potential energy inside the metal box is zero and that outside the metal box, i.e., in vacuum is V_S. In the Figure 6.23, ϕ represents amount of energy required to remove an electron from the Fermi level and take it out of the metal box.

6.12 Bragg's law for electron waves

In a metal positive nuclei or kernels form parallel planes and the electrons are free to move throughout the crystal. Electrons are moving in the crystal lattice like waves, having similar optical properties like light waves. These free electrons, while travelling, are reflected by several planes, formed by kernels. [See. Figure 6.24]

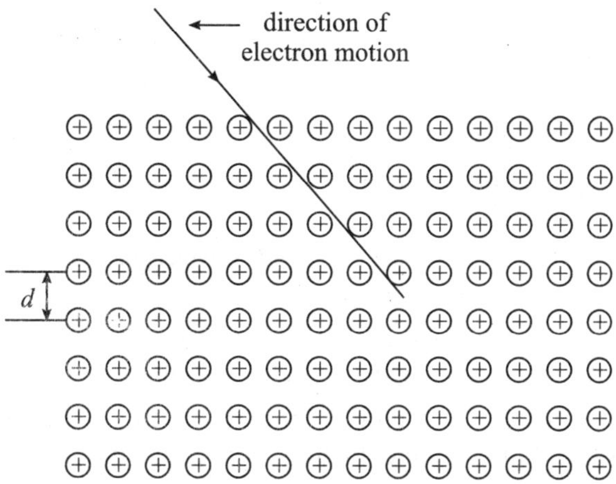

Fig. 6.24 Motion of electrons in a metallic crystal. '*d*' is called interplanar spacing.

For a wavelength λ, electron waves are reflected by a kernel-plane at a particular angle of incidence (θ). This reflection of electron waves is believed to follow Bragg's equation of reflection of light waves, i.e. Equation (6.2):

$$n\lambda = 2d \sin \theta \quad (6.2)$$

where n is the order of reflection.

$n = 1$ for two immediate parallel planes and is called the 1st order reflection.

Thus Bragg's equation is the condition of reflection of electron waves, by internal planes. Again we know that:

$$k = \frac{2\pi}{\lambda} = \frac{2\pi}{2d \sin\theta} = \frac{\pi}{d \sin\theta}, \text{ since}$$

$$\frac{1}{\lambda} = \frac{1}{2d \sin\theta} \text{ for 1st order reflection}$$

Or, $$k \sin\theta = \frac{\pi}{d} \quad (6.16)$$

As d is constant for a particular metallic crystal, k will change with the incident angle θ, shown in the Figure 6.25. From the figure it is evident that electrons with different wavelengths make different incident angles on the same lattice plane but the perpendicular distance remains the same. So we can write that:

$$k \sin\theta = k' \sin\theta' = \frac{\pi}{d} \quad (6.17)$$

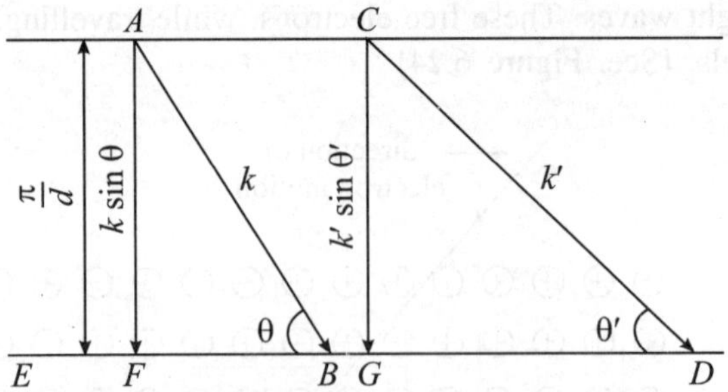

Fig. 6.25 Variation of incident angle (θ) with wavelength (λ) or wave number (*k*).

Electrons, having wave vector $\vec{k}$, are travelling along the path *AB* and get reflected by the lattice plane *EFBGD* since the incident angle (θ) satisfies the Bragg's Equation (6.17). Similarly electrons, having wave vector $\vec{k}'$, are travelling along the path *CD* and also get reflected by the same lattice plane *EFBGD* since the incident angle (θ′) satisfies the Bragg's Equation (6.17). If we consider two-dimensional space, $\vec{k}$-vector can be resolved along the axes shown in the Figure 6.26. $\vec{k}$-vector is reflected, while travelling in any of the four quadrants, provided it satisfies the Bragg's equation (Equation 6.17). If we consider four lattice planes in four quadrants at a distance of $\frac{\pi}{d}$ each, we get a square of arm of length $\frac{2\pi}{d}$.

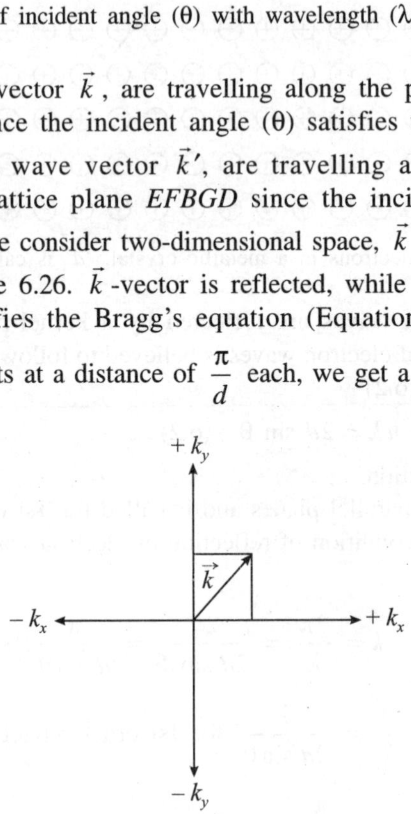

Fig. 6.26 Two-dimensional wave number space.

The movement of an electron wave remains uninterrupted until it touches any of the four lattice planes, (e.g., $\vec{k}'''$ vector in the Figure 6.27) but any electronic wave, emanating from the source *o*, gets reflected when it touches any of the four arm of the square *ABCD*, (e.g., $\vec{k}'$, $\vec{k}''$ vectors etc.). In other words, we can say that motion of any electron is confined within the square *ABCD*, which is also known as 1st Brillouin zone, (Figure 6.27). In a similar manner 2nd Brillouin zone (*EFGH*) is formed by straight lines, touching the corners of the 1st Brillouin zone (*ABCD*).

Only the shaded area is considered as the 2nd Brillouin zone (Figure 6.28). Electrons, confined in the 2nd Brillouin zone cannot pass through the 1st or the 3rd Brillouin zones and vice versa.

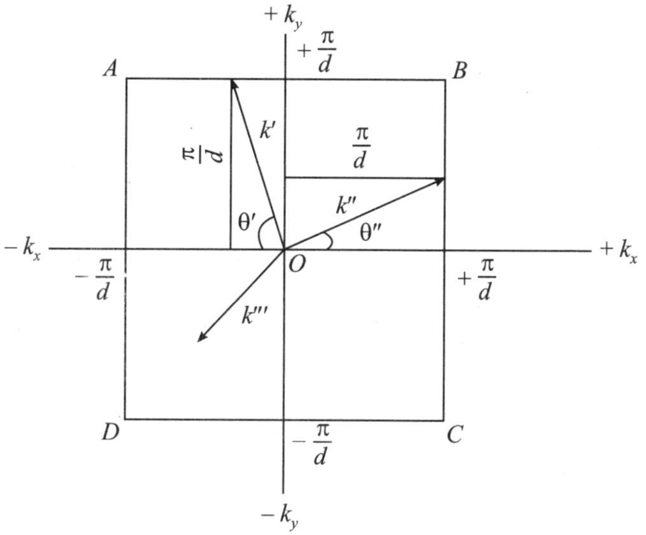

Fig. 6.27 The first Brillouin Zone in a two-dimensional simple cubic lattice.

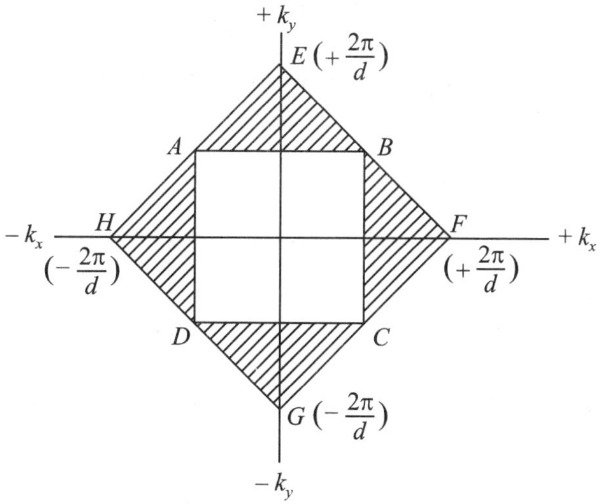

Fig. 6.28 The 1st and 2nd Brillouin Zones in a simple cubic lattice. The 2nd zone is crosshatched.

Thus a definite energy gap exists between any two successive Brillouin Zones. Thus the $E - k$ curve, which is apparently a continuous parabola according to Equation (6.9), is actually a discontinuous parabola as shown in the Figure 6.29(b).

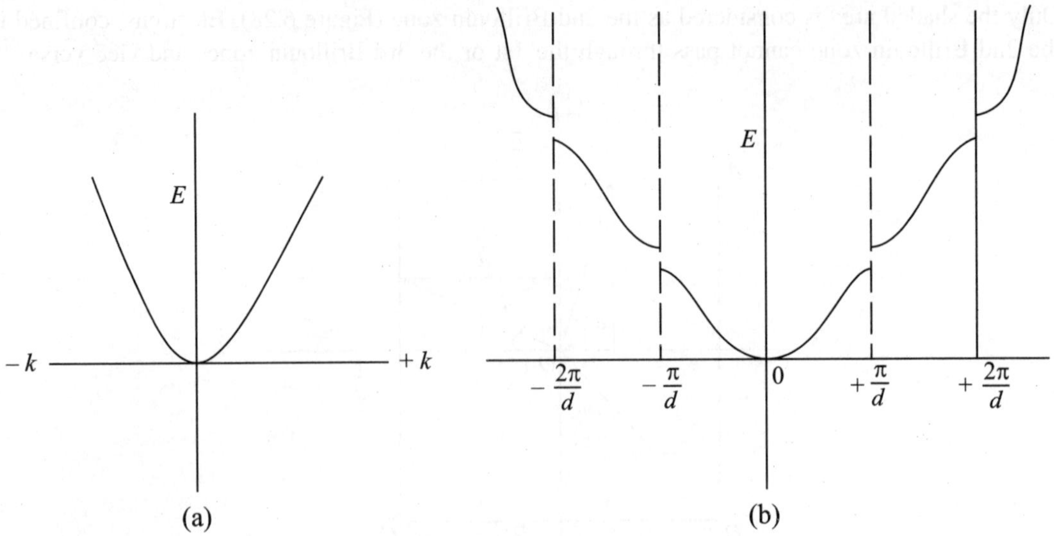

Fig. 6.29 (a) Variation of energy with k according to Equation (6.9).
(b) Actual variation of energy with k according to zone theory.

Electrons, having higher values of $\vec{k}$-vector, face the boundary wall at longer distance [Figure 6.30a]. As the magnitude of $\vec{k}$-vector increases corresponding energy E increases sharply since $E \propto k^2$ (Equation 6.9). Obviously appearance of energy gap or break point energy (E_b) depends on $\vec{k}$-value of electron wave. Higher the magnitude of $\vec{k}$-vector higher the value of E_b as is evident from the Figure 6.30b.

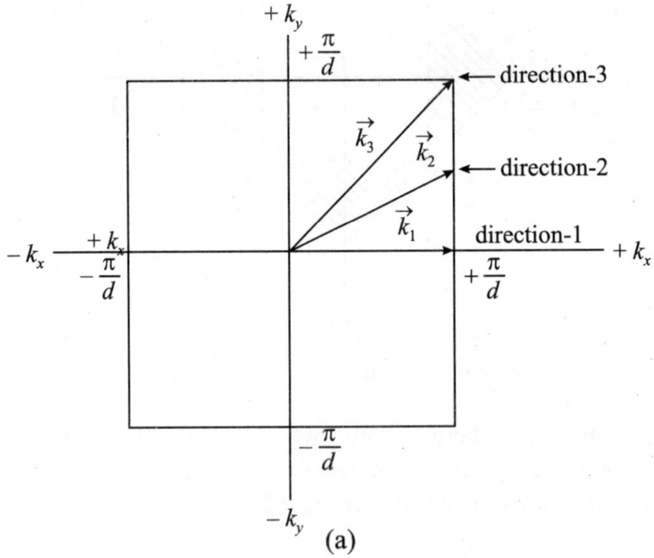

Fig. 6.30(a) Direction of motion of different electron waves.

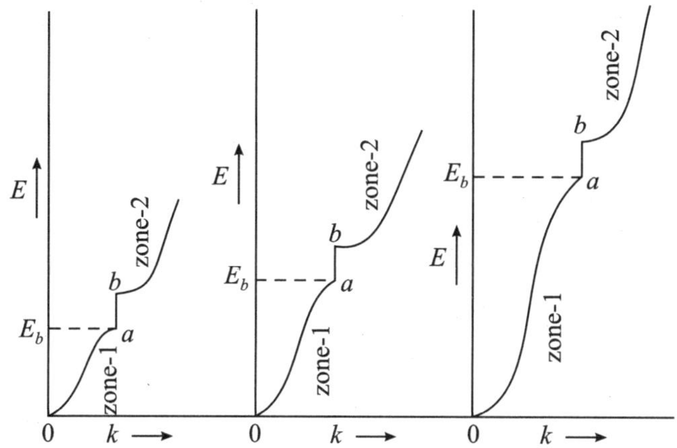

(b)

Fig. 6.30(b) Break-point energy (E_b) varies with the direction of motion of electron waves. *a* and *b* are the two points infinitesimally close to the boundary, indicating the energy gap between Zone 1 and Zone 2.

So, electrons, having wave number $\vec{k}_1$, face the energy gap (E_b) at the lowest energy level whereas electrons, travelling along the direction-3, face the energy gap (E_b) at the highest energy level. From the Figure 6.30b it is evident that there is a definite energy gap between Zone-1 and Zone-2 and the energy gap is popularly known as forbidden energy gap, represented by E_g.

6.13 Conductors, semi-conductors and insulators

Considering all *k*-vectors the $E - k$ curve takes the following form:

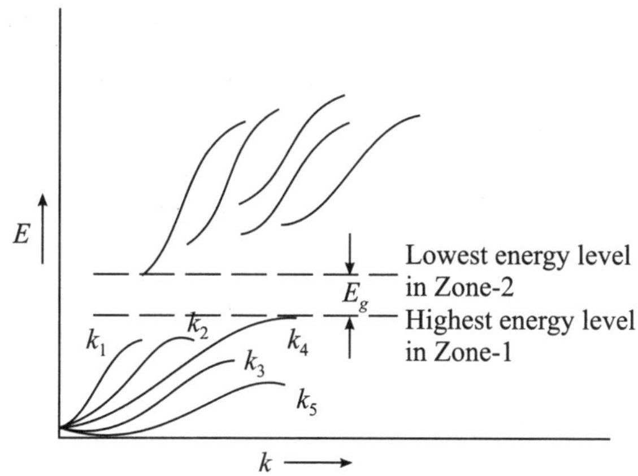

Fig. 6.31 Energy levels of Zones 1 and 2.

In any metal, forbidden energy gap is actually the energy difference between the highest energy level of Zone-1 and the lowest energy level of Zone-2 (See Figure 6.31).

Conductors Mechanism of electrical conduction in any metallic conductor.

There are two types of metallic conductors:

Type-I Each metal atom has single electron in its valence shell. As each energy state may possess two electrons, only half of the total energy states of Zone-1 is filled up with electrons and the energy states of other half of Zone-1 remain empty (See Figure 6.32).

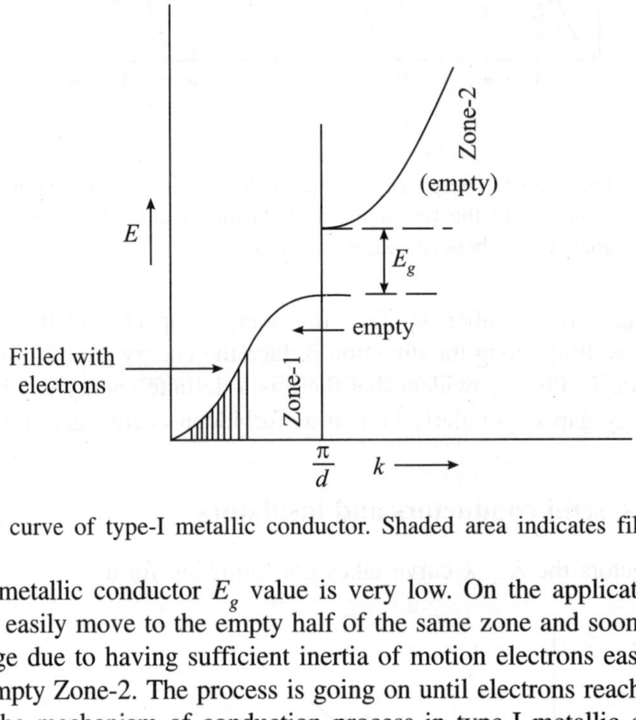

Fig. 6.32 $E - k$ curve of type-I metallic conductor. Shaded area indicates filled energy states.

In the case of metallic conductor E_g value is very low. On the application of electric field electrons in Zone-1 easily move to the empty half of the same zone and soon reach to the end of Zone-1. At this stage due to having sufficient inertia of motion electrons easily cross the energy barrier and reach empty Zone-2. The process is going on until electrons reach to the other end of the metal. This is the mechanism of conduction process in type-I metallic conductors.

Type-II Each metal atom has two electrons in its valence shell. Hence Zone-1 is totally filled up with electrons. The corresponding $E - k$ curve shows that there is a definite overlap between lower and higher energy states, eliminating energy barrier between zones [see Figure 6.33].

Thus on the application of electric field, electrons of a filled zone readily move up to the empty Zone-2 of another plane but of equal energy. The process continues and electrons soon reach to another end of metallic conductor. Thus in type-II metallic conductor overlapping of energy states plays the vital role in conduction process.

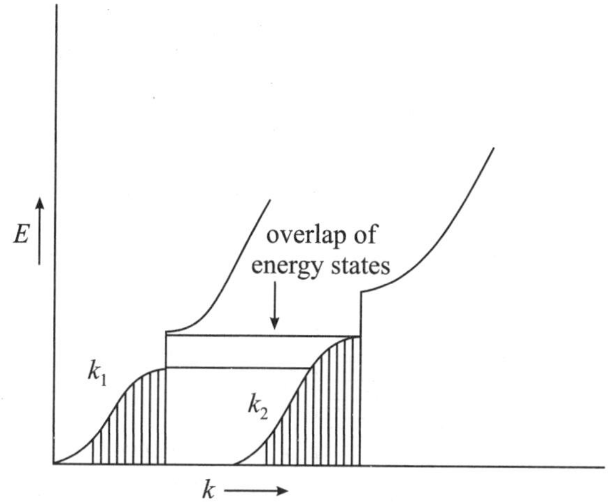

Fig. 6.33 Overlap of energy states.

Insulators In the case of an insulator Zone-1 is totally filled up with electrons and forbidden energy gap is sufficiently high so that on the application of electric field electrons cannot move in Zone-1, nor will they be able to jump to Zone-2 due to high value of E_g. As a result of this, electrons are confined in Zone-1 only and the material behaves as an insulator. One example of insulator is diamond, E_g-value of which is around 6 eV.

Semi-Conductors In the case of semi-conductors Zone-1 is filled up with electrons but E_g-value lies within 2 eV. Thus on the application of an electric field some electrons of Zone-1 gain sufficient kinetic energy through collision and cross the energy barrier to move into the empty Zone-2. Furthermore, fraction of those successful electrons, moving into Zone-2, is able to cross the energy barrier to move into Zone-3 etc. So ultimately only a few electrons is able to reach the other end of semi-conducting material. Hence semi-conductors are weak conductor of electricity. Silicon and germanium are two widely used semiconductors.

$$E_g\text{-value of silicon} = 1.1 \text{ eV}, \quad E_g\text{-value of germanium} = 0.74 \text{ eV}.$$

Filled Zone-1 is called valence band and empty Zone-2 is called conduction band.

Semiconductors are of two types: (a) Intrinsic type and (b) Extrinsic type.

6.13.1 Intrinsic semiconductor

As discussed earlier intrinsic semiconductor processes a filled Zone-1 with a narrow gap between valence band and conduction band [see Figure 6.34].

On application of an electric field electrons in the valence band undergo *random collision*. Some of them gain sufficient energy through collision and *jump* to the empty conduction band. At this stage the material becomes a weak conductor of electricity. An electron while jumping to conduction band leaves behind a '*positive hole*' in the valence band. So, as electrons move forward by jumping, holes are moving backward direction. This is why, a hole is considered as a carrier of positive charge.

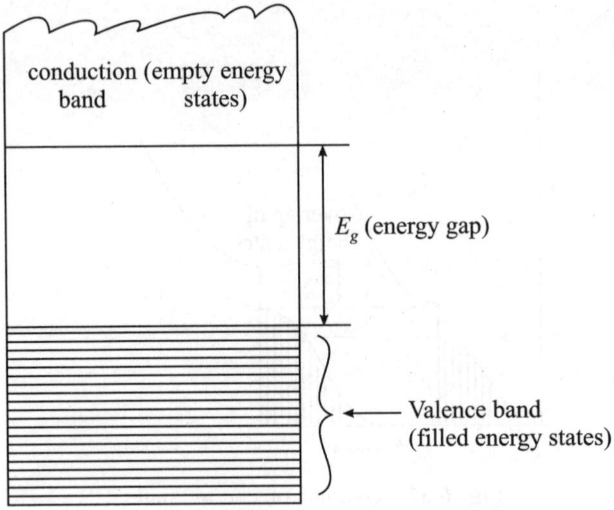

Fig. 6.34 Valence band and conduction band in an intrinsic semiconductor.

Characteristic features of intrinsic semiconductors

(1) Fraction of electrons excited and jumped to the second Brillouin zone is nothing but the probability of finding an electron in the second Brillouin zone. Thus we can write that:

$$F(E) = \frac{n}{N}$$

where, n = Number of electrons jumped to the conduction band.

N = Number of electrons available in the valance band.

Second Brillouin zone is often called conduction band, and 1st Brillouin zone is known as valence band. $F(E)$ is calculated from Fermi-Dirac probability distribution, given below:

$$F(E) = \frac{1}{1 + e^{\frac{(E - E_F)}{kT}}} \tag{6.18}$$

Where E_F = Fermi energy, k = Boltzmann's constant = 8.614×10^{-5} eV/°K. E = Any energy state. For conduction band $E = E_C$ and for valence band $E = E_V$

The Figure 6.35 shows that probability of finding an electron in Fermi level is 50%. So actual forbidden energy gap = $(E_C - E_F) = \frac{E_g}{2}$, as shown in the Figure 6.35. Usually, E_g-value of semiconductors lies within 1 to 2 eV, e.g., for silicon E_g = 1.1 eV. So, $\frac{E_g}{2} = \left(\frac{1.1}{2}\right)$ = 0.55 eV. At room temperature, T = 298 K and the value of $kT = 8.614 \times 10^{-5} \times 298$ eV = 0.0257 eV.

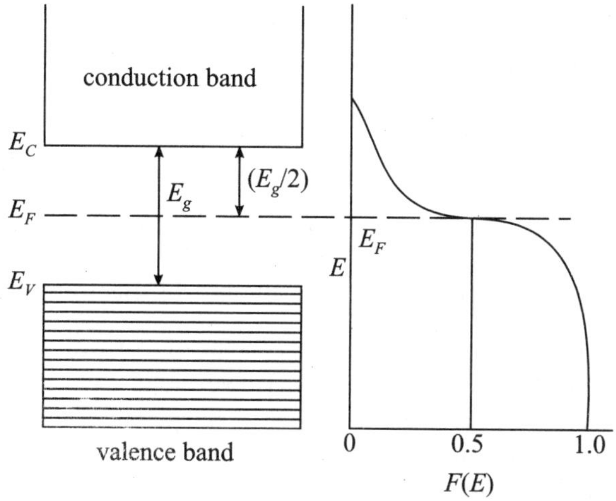

Fig. 6.35 Position of Fermi level and energy gap in a semiconductor, $(E_C - E_F) = \dfrac{E_g}{2}$

Thus, $$\frac{\left(\frac{E_g}{2}\right)}{kT} = \left(\frac{0.55}{0.0257}\right) = 21.4 \quad \text{and} \quad e^{\frac{E_g}{2kT}} = e^{21.4} >> 1$$

So, Equation (6.18) becomes:

$$F(E) = \frac{1}{1 + e^{\frac{(E_C - E_F)}{kT}}} = \frac{1}{1 + e^{\frac{E_g}{2kT}}} \approx \frac{1}{e^{\frac{E_g}{2kT}}} = e^{-\frac{E_g}{2kT}}$$

Again, $F(E) = \dfrac{n}{N}$, where,

n = Number of electrons excited to the empty second zone, also known as carrier concentration.

N = Total number of electrons available for excitation from the top of the valence band.

Thus above equation becomes:

$$\frac{n}{N} = e^{-\frac{E_g}{2kT}} \tag{6.19}$$

(i) If the magnitude of $\left(\dfrac{E_g}{2kT}\right)$ (where T = 298 K) lies below 35 the material is said to be semiconductor.

(ii) If the magnitude of $\left(\frac{E_g}{2kT}\right)$ (where $T = 298$ K) is very high the material is said to be an insulator, e.g., E_g of diamond = 7.0 eV and $\left(\frac{E_g}{2kT}\right) = 136.2$ eV.

(2) The Equation (6.19) shows that with the increase in temperature kT increases and hence $\left(\frac{n}{N}\right)$ also increases, which means that, number of electrons in the conduction band increases with temperature. Thus it may be concluded that with the increase in temperature conductivity of a semiconductor increases. However, in case of conductors, conductivity decreases with increase in temperature. At low temperature, electrons in a conductor are moving orderly on application of an electric field. There is a little or no intermixing of electrons during their motion. However, at higher temperature thermal vibration of lattice atoms increases, which ultimately destroys the orderly motion of electrons, facilitating intermixing of electrons, thereby, increasing in resistance to the flow of current. That is why, conductivity of a conductor decreases with increase in temperature.

(3) As discussed earlier, conductivity of a semiconductor increases with temperature and ultimately a temperature is reached when $\left(\frac{n}{N}\right) = \left(\frac{1}{e}\right)$ or, $E_g = 2kT$. At this temperature the number of electrons in the conduction band is so high that the semiconductor is assumed to behave as a conductor.

For $\quad E_g = 1.0$ eV

$$T = \frac{1.0\times1.602\times10^{-19}\text{ J}}{2\times1.38\times10^{-23}\text{ J/K}} = 5805\text{ K}$$

Thus a semiconductor, having $E_g = 1.0$ eV, can behave as a conductor at around 5800 K, the temperature at which all known materials melt and vaporize. Thus it may be concluded that a semiconductor cannot behave as a conductor.

(4) According to Equation (6.19) the value of $\left(\frac{n}{N}\right) = 0$ at $T = 0$°K, which means that there is no electron in the conduction band at 0°K. Thus it may be inferred that a semiconductor behave as an insulator at 0°K.

6.13.2 Extrinsic semiconductor

The forbidden energy gap (E_g) of a semiconductor is drastically reduced in presence of a foreign element which assumes one or more lattice point in the crystal. Two cases may arise:

Case – I If some lattice points of tetravalent silicon (Si) are replaced by a trivalent element (e.g., B, Al, Ga, In or Tl) part of Zone-1 remains vacant, resulting in the formation of an acceptor level just above the valence band as shown in the Figure 6.36.

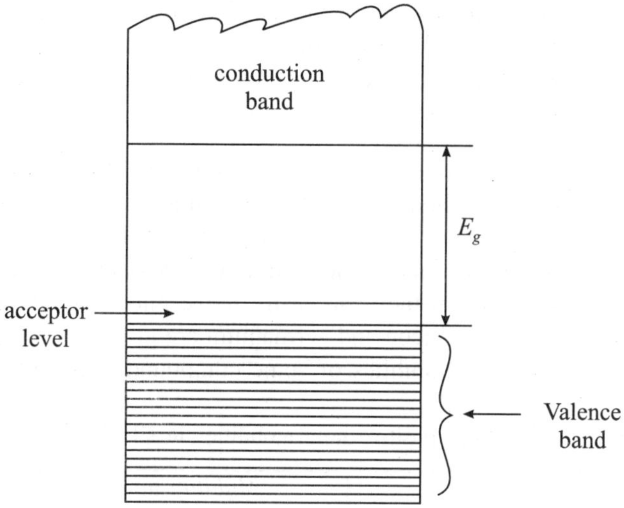

Fig. 6.36 Silicon, doped with Group-III element, gives a *p*-type semiconductor.

The energy gap between acceptor level and valence band is equivalent to kT at room temperature, i.e. of about 0.0257 eV and thus thermal vibration in crystal lattice is enough to excite electrons from valence band to acceptor level and consequently holes are created in the valence band. These holes are propagating throughout the crystal and hence appear as the majority charge carriers. This type of semiconductor is called *p*-type semiconductor.

Case – II If some lattice points of tetravalent silicon (Si) are replaced by a pentavalent element (e.g., P, As, Sb or Bi), Zone-1 will be in a state of electron excess. These excess electrons form donor level just below the conduction band, as shown in the Figure 6.37.

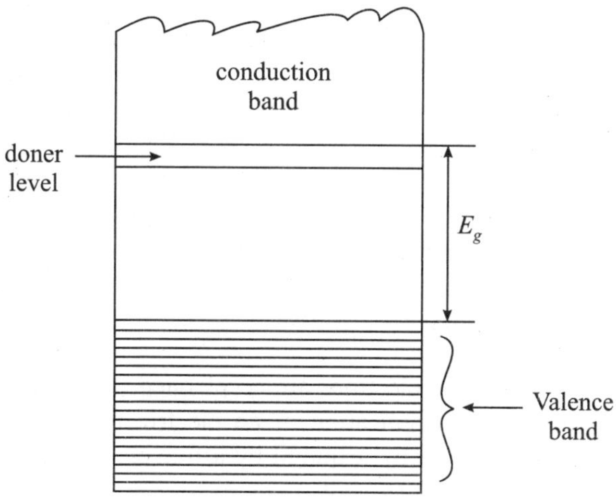

Fig. 6.37 Silicon doped with Group-*V* element gives an *n*-type semiconductor.

The energy gap between donor level and conduction band is small and only thermal vibration in crystal lattice is enough to excite electrons from donor level to conduction band. These electrons in the conduction band move forward freely and hence are the majority charge carriers. This type of semiconductor is called *n*-type semiconductor.

Formation of donor level When one silicon atom is replaced by one phosphorus atom the 5th electron is loosely bound since it cannot take part in the direct covalent bonding. This electron is moving in the silicon crystal. The orbit radius is very large about 80Å and hence ionization energy is very low compared to the energy gap. These loosely bound electrons, form donor level close to the conduction band, so that they can easily move into conduction band.

Electron and hole mobilities in silicon and germanium are given in Table 6.3. Table 6.4 shows energy gap and electron and hole mobilities of some compound semiconductors.

Table 6.3 Electron and positive hole mobilities in silicon and germanium.

Species	*Mobility at room temperature* (m^2/v.s)	
	Silicon	*Germanium*
Electrons	0.14	0.39
Positive holes	0.05	0.19

Table 6.4 Properties of some compound semiconductors.

Compound	*Energy gap* (eV)	*Mobility* (m^2/V–s)	
		Electrons	*Holes*
GaP	2.26	—	0.002
AlSb	1.52	0.02	0.02
GaAs	1.43	0.34	0.02
InP	1.29	0.46	0.015
GaSb	0.78	0.40	0.07
InAs	0.35	2.30	0.024
InSb	0.18	6.50	0.100

6.14 Mobility and conductivity

On the application of an electric field electrons are accelerated in a given direction. The velocity acquired by a free electron on the application of an electric field is known as *drift velocity*, represented by v_d. Again the drift velocity acquired by a free electron under unit field gradient is known as *mobility*, represented by the symbol μ.

Thus,
$$\text{mobility } (\mu) = \frac{\text{Drift velocity (m/s)}}{\text{Field gradient (V/m)}} = \frac{v_d}{E_x} \ (\text{m}^2/\text{V–s})$$

where, E_x = Field gradient

Conductivity Inverse of resistivity of a material is known as conductivity. If ρ and σ represent resistivity and conductivity respectively, we can write that:

$$\sigma = \frac{1}{\rho}$$

Again ρ is related with the resistance (R) as follows:

$$\rho = R \, . \, \frac{A}{l}$$

where, A = Area of cross-section and l = Length of the material. The unit of ρ is Ω–m.

So, $$\sigma = \frac{1}{\rho} = \frac{1}{R}\frac{l}{A}$$

Or, $$\sigma = \frac{l}{AR} \ \Omega^{-1} \ \text{m}^{-1} \tag{6.20a}$$

As we know in case of a semiconductor, electrons in valence band are excited by the application of an external field and jump to the conduction band, leaving behind positive holes in the valence band. Thus on application of an electric field electrons and positive holes start moving in the opposite direction. So net conductivity of a semiconductor (σ) is the sum of conductivity due to electrons and conductivity due to positive holes. Mathematically we can write that:

$$\sigma = \sigma_e + \sigma_h \tag{6.20b}$$

where σ_e and σ_h are the conductivities due to electron and positive hole respectively.

Consider a material of constant cross-section A and length dx. If dq amount of charge is crossing the length dx in time dt, the current I is given by:

$$I = \frac{dq}{dt}$$

n = Number of electrons per unit volume. Volume of the section = $A \, . dx$.

Charge per electron = e. So, $dq = n \, eAdx$.

$$I = \frac{dq}{dt} = n.e.A\left(\frac{dx}{dt}\right), \ \frac{dx}{dt} = \text{drift velocity} = v_d$$

The *current flux* (J) flowing through the cross-section A is given by:

$$\text{Current flux } (J) = \frac{I}{A}$$

Or, $$J = n \, . \, e \, . \, v_d$$

Again from the definition of mobility, we have:

$$v_d = \mu E_x$$

Replacing v_d in the above equation, we get:

$$\text{Conductivity } (\sigma) = \frac{J}{E_x} = \text{Current flux per unit field gradient}$$

$$\sigma = n.e.\mu \tag{6.20c}$$

σ_e = Conductivity due to free electrons in the conduction band = $n_e.e.\mu_e$

$\sigma_e = n_e \times e \times \mu_e$

n_e = Number of free electrons per unit volume

e = Electronic charge = 1.602×10^{-19} C.

μ_e = Mobility of electron or electron mobility.

σ_h = Conductivity due to positive holes in the valence band = $n_h.e.\mu_h$

$\sigma_h = n_h \times e \times \mu_h$

n_h = Number of positive holes per unit volume

μ_h = Mobility of positive hole or hole mobility.

Thus the Equation (6.20b) becomes:

$$\sigma = n_e \times e \times \mu_e + n_h \times e \times \mu_h \tag{6.20d}$$

Law of mass action In the case of an intrinsic semiconductor, we have:

Number of free electrons in the conduction band = number of positive holes in the valence band.

Or, $$n_e = n_h = n_i$$

n_i is given by the Equation (6.19):

$$n_i = N e^{-\frac{E_g}{2kT}}$$

Thus the Equation (6.20d) becomes:

$$\sigma = n_i e(\mu_e + \mu_h)$$

$$\sigma = N.e(\mu_e + \mu_h)\, e^{-\frac{E_g}{2kT}} \tag{6.20e}$$

In the case of an extrinsic semiconductor, $n_e \neq n_h$ but *net probability of carrier concentration* of an extrinsic semiconductor is same as that of the corresponding intrinsic semiconductor. In the case of an intrinsic semiconductor: number of free electrons in the conduction band = number of positive holes in the valence band = n_i. On addition of impurities number of free electrons in the conduction band is n_e and number of positive holes in the valence band is n_h. So net probability of finding carrier concentration in the case of an intrinsic semiconductor = $\left(\frac{n_i}{N}\right) \cdot \left(\frac{n_i}{N}\right)$, where '$N$' is the total number of valence electrons, which is almost same in both the intrinsic and extrinsic semiconductor. Net probability of finding carrier concentration in the case of an extrinsic semiconductor = $\left(\frac{n_e}{N}\right) \cdot \left(\frac{n_h}{N}\right)$.

Hence, $$n_i^2 = n_e\, n_h \qquad (6.21)$$

The above equation is the mathematical expression of law of mass action.

Statement Net probability of carrier concentration of an intrinsic or pure semiconductor remains unchanged even after addition of impurities into it.

Conductivity of extrinsic semiconductor In an extrinsic semiconductor, conductivity initially decreases and reaches a minimum, followed by a sharp increase as shown in the following figure:

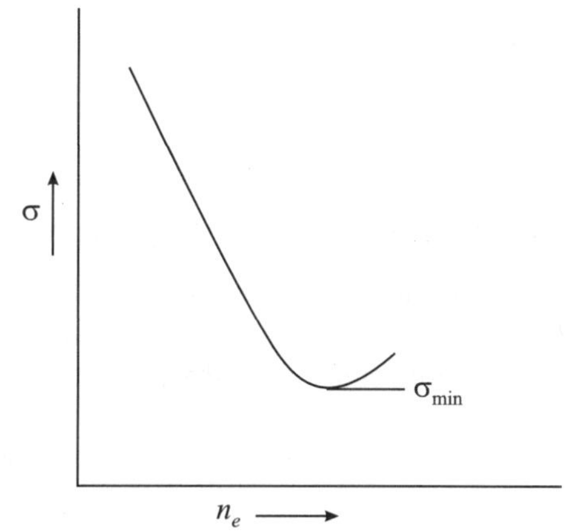

Fig. 6.38 Variation of conductivity with n_e in an extrinsic semiconductor.

To find out the σ_{min}, let us start with the Equation (6.20d).

$$\sigma = n_e \times e \times \mu_e + n_h \times e \times \mu_h$$

Replacing n_h from Equation (6.21), we get:

$$\sigma = n_e\, e\, \mu_e + \left(\frac{n_i^2}{n_e}\right) e\, \mu_h$$

n_i is constant at constant temperature. To find out the σ_{min}, differentiate the above equation with respect to n_e and we get:

$$\frac{d\sigma}{dn_e} = e\mu_e - \frac{n_i^2}{n_e^2} e\mu_h = e\mu_e - \left(\frac{n_i^2}{n_e}\right)\left(\frac{e\mu_h}{n_e}\right)$$

$$= e\mu_e - e\mu_h\left(\frac{n_h}{n_e}\right) \quad [\text{since, } n_i^2 = n_e\, n_h]$$

At the minimum $$\frac{d\sigma}{dn_e} = 0$$

and hence, $$n_e \times e \times \mu_e = n_h \times e \times \mu_h$$

Or, $$\sigma_e = \sigma_h$$

So, $\sigma_{min} = n_e \times e \times \mu_e + n_h \times e \times \mu_h = \sigma_e + \sigma_h$

$= 2\sigma_e = 2n_e \times e \times \mu_e$

Or, $$\sigma_{min} = 2n_e \times e \times \mu_e \qquad (6.22)$$

Conclusion In an extrinsic semiconductor, conductivity initially decreases with increase in n_e and at the minimum, conductivity due to electrons is equal to conductivity due to positive holes.

Effect of temperature on conductivity

From the Equation (6.20e), we have:

$$\sigma = N.e(\mu_e + \mu_h)\, e^{-\frac{E_g}{2kT}}$$

N = Total number of valence electrons per unit volume and $N \propto T^{\frac{3}{2}}$.

Again, mobility decreases with increase in temperature since collisions between lattice points and electrons increase with temperature due to increased lattice vibration. It has been observed that mobility decreases with increase in temperature according to the following equation:

$$\mu \propto \frac{1}{T^{\frac{3}{2}}}$$

However, the above equation is valid at low temperature only. Thus $N(\mu_e + \mu_h)$ is almost independent of temperature, e is the electronic charge and also independent of temperature. So, the Equation (6.20e) can be written as:

$$\sigma = C\,.\,e^{-\frac{E_g}{2kT}}$$

where, $C = N.e(\mu_e + \mu_h)$ = Constant

Taking ln on both sides, we get:

$$\ln \sigma = \ln C - \frac{E_g}{2kT}$$

Thus by plotting $\ln \sigma$ against $\left(\frac{1}{T}\right)$ a straight line of negative slope is obtained.

Again, $\sigma = \dfrac{l}{A.R}$ [Equation (6.20a)]

Putting this value of σ in the above equation we get:

$$\ln\left(\frac{l}{A.R}\right) = \ln C - \frac{E_g}{2kT}$$

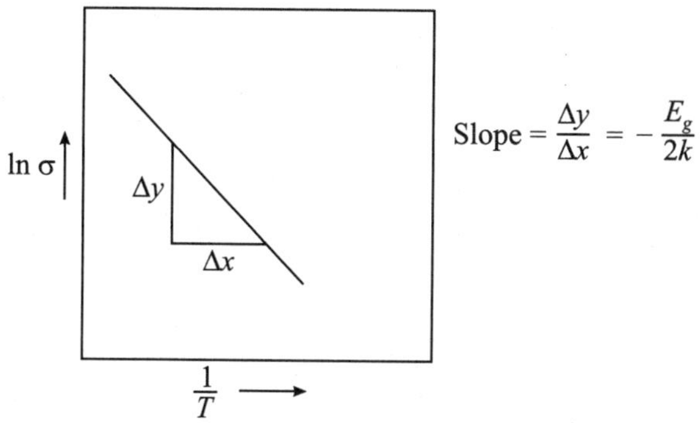

Fig. 6.39(a) Plot of ln σ against $\left(\frac{1}{T}\right)$ for an intrinsic semiconductor.

Or, $$\ln R = \ln \frac{l}{A} - \ln C + \frac{E_g}{2kT}$$

Or, $$\ln R = \ln C' + \frac{E_g}{2kT}$$

where, $$C' = \frac{l}{AC}$$

Simple plotting of (ln R) versus $\left(\frac{1}{T}\right)$ gives a straight line of positive slope as shown in the Figure 6.39(b).

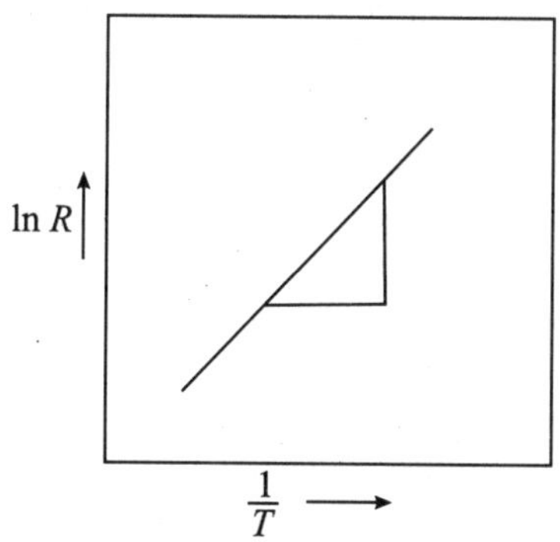

Fig. 6.39(b) Plot of ln R against (1/T) of an intrinsic semiconductor.

From the above figure, we have:

$$\text{slope} = \frac{\Delta y}{\Delta x} = \frac{E_g}{2k} \tag{6.23}$$

Thus by measuring the slope band gap (E_g) can easily be evaluated.

In the case of an extrinsic semiconductor, the plot of ln σ against $\left(\frac{1}{T}\right)$ shows three different zones as shown in the Figure 6.39(c).

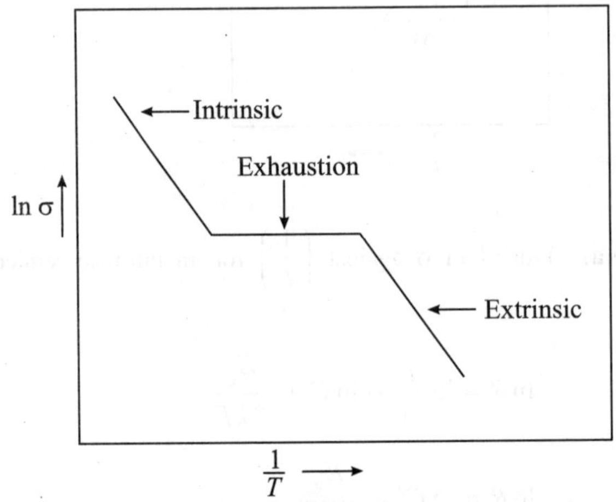

Fig. 6.39(c) Plot of ln σ against $\left(\frac{1}{T}\right)$ for an extrinsic semiconductor.

Extrinsic semiconductors are of two types:

(a) *p*-type semiconductor and (b) *n*-type semiconductor. In the case of a *p*-type semiconductor thermal excitation helps jumping of electrons from valence band to acceptor level. As the energy gap is very low, jumping of electrons starts at a low temperature and due to this jumping, number of conduction holes increases and hence conductivity increases steadily as shown in the Figure 6.39(c). This region is called *extrinsic region*. However, after sometime, acceptor level is totally filled up and it cannot give room to any other foreign electron. Thus, jumping of electrons stop and hence conductivity remains constant over a certain temperature range, which is commonly known as *exhaustion region*, [see the Figure 6.39(c)]. Further increase in temperature results excitation of elections from valence band to conduction band like an intrinsic semiconductor. Thus again, number of conduction electrons increases and consequently conductivity increases. This region is often called *intrinsic region*.

In the case of an *n*-type semiconductor a little energy gap exists between donor level and conduction band. Thus at low temperature electrons in the donor level are able to jump into the conduction band and the process continues with the increase in temperature. Obviously conductivity steadily increases, as shown in the Figure 6.39(c). This region is known as *extrinsic region*.

After sometime all the electrons in the donor level are shifted to the conduction band. Thus further increase in temperature will not increase conductivity of the semiconductor. So conductivity remains constant over a certain temperature range. This region is known as *exhaustion region*, same as obtained in the case of a *p*-type semiconductor. At still higher temperature electrons from the valence band are excited enough to jump to the conduction band just like an intrinsic semiconductor. So again conductivity increases with temperature. This region is known as *intrinsic region*.

6.15 Junction diode and transistor

6.15.1 Junction diode

In *p*-type semiconductor positively charged holes are the majority carriers and in *n*-type semiconductor negatively charged electrons are the majority carrier. If two crystals *p*-type and *n*-type are fused together, due to electrostatic force of attractions and mutual repulsions among electrons in *n*-type crystal, some electrons move into the *p*-type crystal end get arrested by holes to form electric dipoles. These dipoles are formed near the junction of *p* and *n*-crystals, forming a layer called *depletion layer*. This depletion layer actually separates the holes in *p*-type crystal and electrons in *n*-type crystal. The direction of electric field in dipole is such that it opposes the flow of electrons, thereby offering resistance to electrons in entering into the depletion layer [see Fig. 6.40]. Thus to enter into the depletion region, electrons have to cross a potential barrier. With every new entry of electron dipole field strength increases and hence potential barrier increases and ultimately diffusion of electrons across the junction comes to an end. That potential is called *barrier potential*.

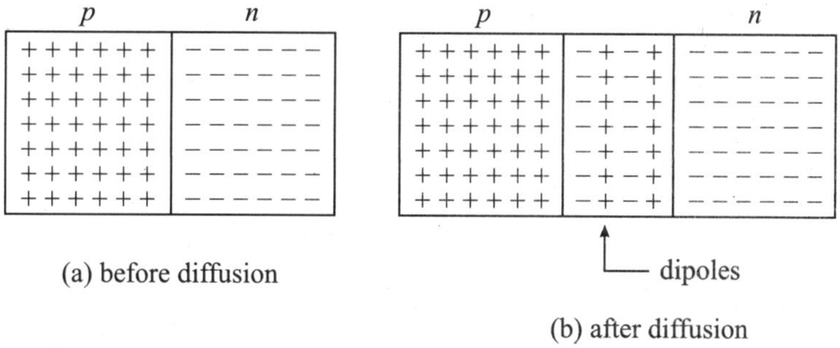

Fig. 6.40 Formation of deletion layer in a *pn*-junction (unbiased diode). The arrow indicates dipole electric field in depletion layer.

This barrier potential is different for different semiconductors, e.g. at 25°C it is 0.3 V for Ge-diode and 0.7 V for Si-diode. With the increase in temperature barrier potential decreases due to increase in field strength. The barrier potential changes with temperature according to the following equation:

$$\Delta V = -0.002 \ \Delta T \tag{6.24}$$

The energy hill We know that acceptor level of *p*-type remains *just above* the valence band and donor level of *n*-type remains *just below* the conduction band. So there is a definite energy gap between acceptor level of *p*-type and valence band of *n*-type. Similar energy gap is also found between donor level of *n*-type and conduction band of *p*-type [see Figure 6.41].

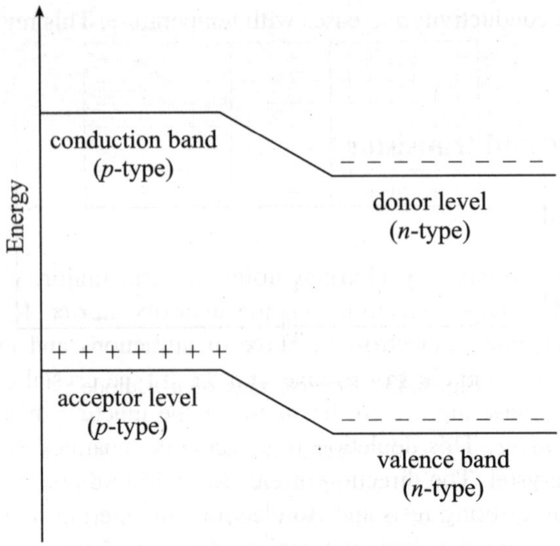

Fig. 6.41 Energy bands after diffusion.

At equilibrium no more electron from the valence band has sufficient energy to get across the junction. Energy must be supplied from outside to help electron climb the energy hill.

Forward bias In the forward bias positive terminal of an external dc-source is attached to the *p*-end and the negative terminal is attached to the *n*-end of diode as shown in the Figure 6.42. Positive holes, reaching to *p*-end, diffuse to depletion layer and recombine with *negative end* of dipoles. Thus number of dipoles decreases and hence depletion layer narrows down. On the other hand, electrons, reaching to *n*-end, diffuse to depletion layer and recombine with *positive end* of dipoles. Again, number of dipoles decreases and hence depletion layer narrows down. As a result of this, height of the energy hill decreases.

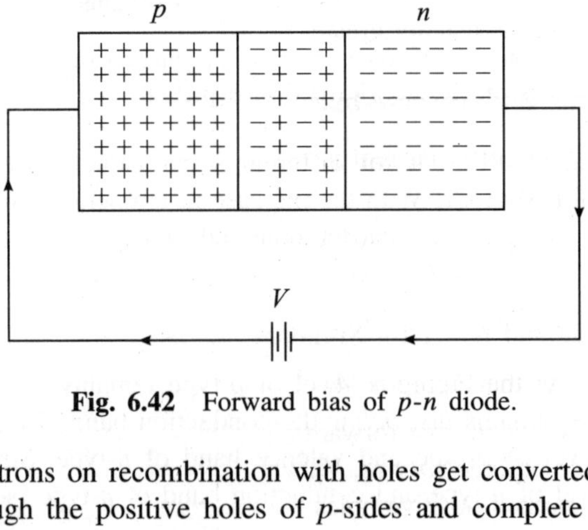

Fig. 6.42 Forward bias of *p-n* diode.

All conduction electrons on recombination with holes get converted into valence electrons, which easily flow through the positive holes of *p*-sides and complete the circuit. Thus heavy forward current is obtained in forward bias.

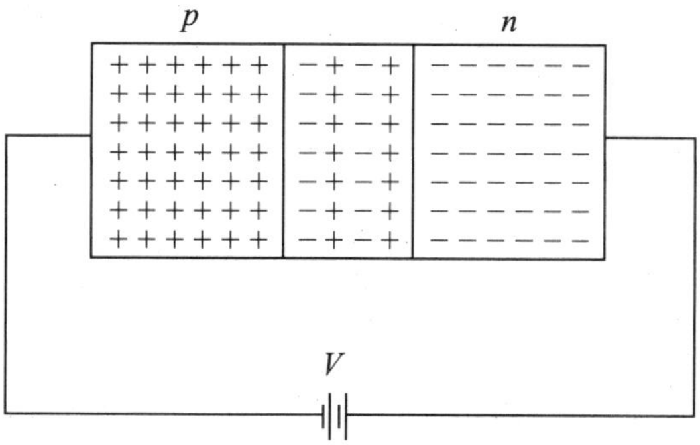

Fig. 6.43 Reverse bias of *p-n* diode

In the reverse-bias condition positive terminal of battery is attached to *n*-end of diode while the negative terminal of battery is attached to *p*-end of diode. So positive end of the diode receives huge electrons from the source and these electrons diffuse to the depletion layer to form dipoles, thereby, increasing the number of dipoles. Similarly negative end of the diode receives huge positive holes from the source and these holes diffuse to the depletion layer to form dipoles, thereby, increasing the number of dipoles. As a result of this, depletion layer widens and hence height of energy hill increases. Thus flow of current is highly restricted. Only a fraction of minority carriers, i.e. electrons in *p*-end is able to cross the junction, as a result of which, a small current flows in the external circuit. This reverse current caused by the minority carriers is called *saturation current* represented by I_S.

Breakdown voltage In the reverse-bias condition electrons are pushed towards the right and holes are pushed towards the left but the dipolar field in the depletion layer resist this process. With the increase in supply voltage pressure on dipolar field increases and ultimately dipoles start breaking into free electrons and positive holes. When all the dipoles break, electrons, as minority carrier, are free to move from *p*-side to *n*-side, as a result of which, heavy current flows through the diode. Voltage, at which this situation prevails, is known as *breakdown voltage*. For rectifier diodes the breakdown voltage is usually greater than 50 V.

Current in forward-bias and reverse-bias conditions

In the forward bias condition heavy current is realized in the external circuit and the expression of forward current (I_F) is given by:

$$I_F = I_S\left[e^{\frac{eV}{\eta kT}} - 1\right] \tag{6.25a}$$

where, $\eta = 1$ for Ge-diode and $\eta = 2$ for Si-diode.

I_F Versus V plot is shown in the Figure 6.44.

In the Figure 6.44 it is evident that forward current rises steeply only after a certain voltage, known as *threshold voltage* or *knee voltage*, V_T, which is also called *barrier potential*. Only after crossing V_T do the electrons and positive holes become free to move across the junction and the current rises steadily.

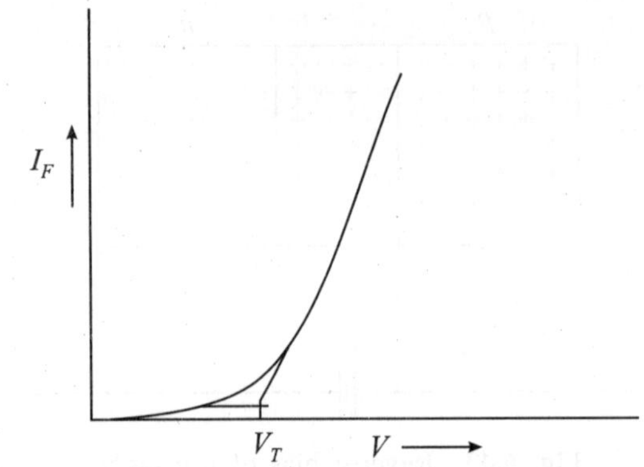

Fig. 6.44 Forward-current-voltage plot of a semiconductor diode.

In the reverse-bias condition only a fraction of minority carrier in *p*-region is able to cross the energy barrier and moves into the *n*-region of the diode. The current established due to jumping of minority carriers in the reverse direction is known as reverse saturation current, represented by I_S. As there is no forward flow of electrons, only reverse saturation current is established in a reverse-bias diode. More accurately the reverse-bias current is given by:

$$I_R = I_S\left[1 - e^{-\frac{eV}{\eta kT}}\right] \tag{6.25b}$$

where, $\eta = 1$ for Ge-diode and $\eta = 2$ for Si-diode.

Bulk resistance While passing current through *p*- and *n*-zones both the latter zones offer resistance, symbolized as r_p and r_n respectively as shown in the Figure 6.45. The sum of r_p and r_n is often called bulk resistance, represented as r_B. So, $r_B = r_p + r_n$.

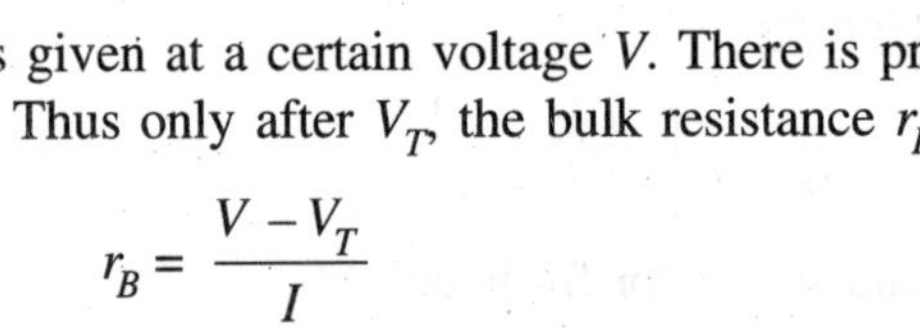

Fig. 6.45 Bulk resistance of a semiconductor diode.

r_B may be evaluated if I_F is given at a certain voltage V. There is practically no current through diode up to the voltage V_T. Thus only after V_T, the bulk resistance r_B is realized. So,

$$r_B = \frac{V - V_T}{I}$$

where, I is the current at the voltage V.

In reverse-bias condition there is no current along forward direction. Only saturation current is realized in the reverse direction [see Figure 6.46a].

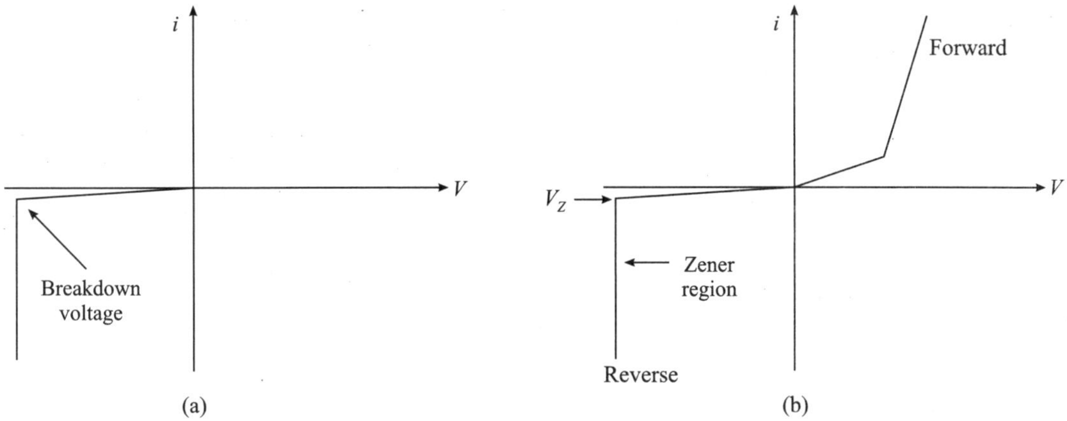

Fig. 6.46 (a) Reverse-bias semiconductor diode curve, (b) Complete diode curve.

At *Zener region* complete breakdown of diode occurs and avalanche of current flows through the circuit. The voltage, at which breakdown of diode occurs, is called *Zener voltage*, represented as V_z [see Figure 6.46b]. Diode, applied in Zener region, is called *Zener diode*.

The ideal diode The ideal diode conducts current in the forward direction at zero voltage but acts like a perfect insulator when reverse-biased. [see Figure 6.47].

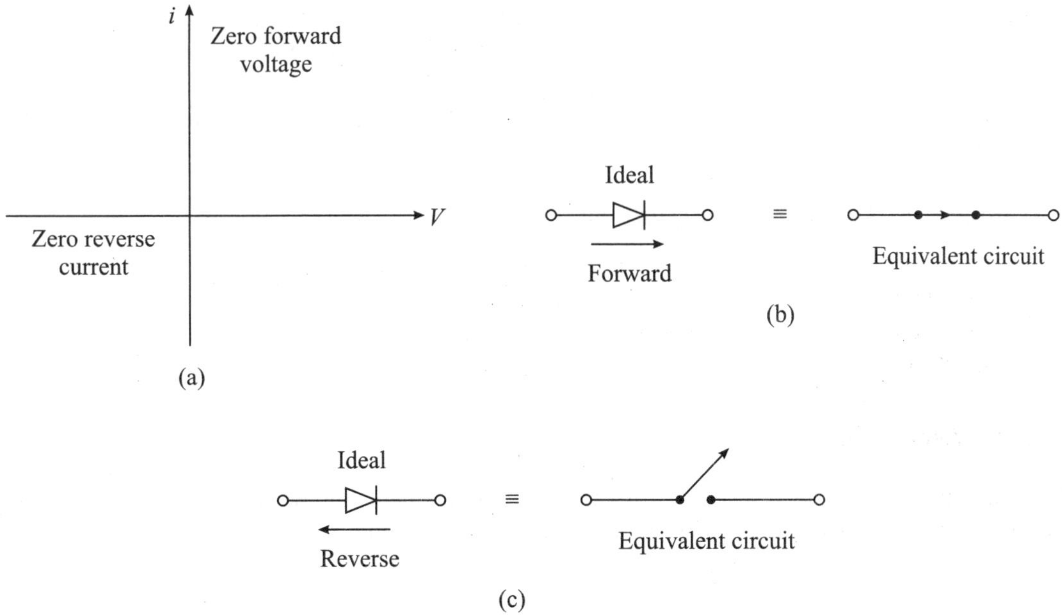

Fig. 6.47 Ideal diode and corresponding forward and reverse current situation.

The actual semiconductor diode contains threshold voltage (V_T), bulk resistance (r_B) and ideal diode. The corresponding equivalent circuit is shown below:

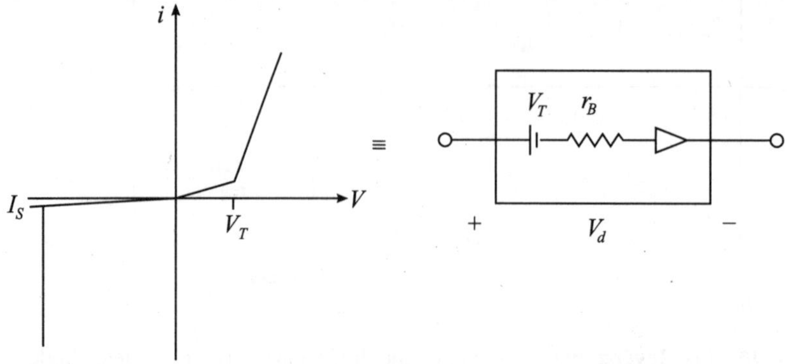

Fig. 6.48 Precise linear equivalent circuit of an actual junction diode.

Peak inverse voltage It is the maximum working voltage across the diode when the latter is reverse-biased. Obviously PIV (Peak Inverse Voltage) is always less than the Zener voltage V_Z.

Advantages of Si-diode over Ge-diode

(a) PIV rating of a Si-diode is around 1000 V but that of a Ge-diode is around 400 V only.
(b) The saturation current (I_S) of Si-diode is very low around 10×10^{-9} Amp. But I_S-value of Ge-diode is around 2×10^{-6} Amp. Furthermore, I_S-value increases with temperature. Increase in I_S-value means deviation from an ideal diode.
(c) Due to low I_S-value working temperature of Si-diode is very high (as high as 200°C) but the same of Ge-diode is only around 100°C.

Advantages of Ge-diode over Si-diode

(a) Threshold voltage (V_T) of Ge-diode is around 0.3 V and that of Si-diode is around 0.7 V. The diode, having lower V_T-value, is closer to the ideal diode.
(b) The factor $\eta = 1$ for Ge-diode but $\eta = 2$ for Si-diode. So the value of I_F for Ge-diode is much higher than that of Si-diode.

6.15.2 Zener diode

Normally junction diode is used in forward-bias arrangement but when the diode is operated under reverse-bias arrangement the diode is called Zener diode, the symbolic representation of which is shown in the Figure 6.49.

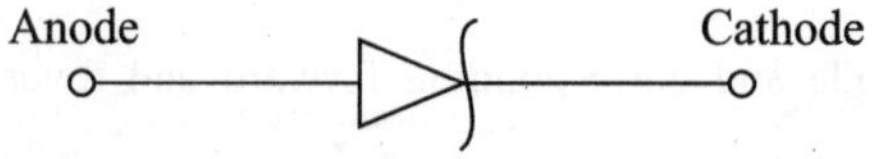

Fig. 6.49 Symbolic representation of Zener diode.

As the diode is connected in reverse-bias condition a small saturation current (I_S) flows through it. With the increase in applied voltage the value of I_S almost remains constant but when the applied voltage reaches a very high value, avalanche of current sets up. The voltage is called *Zener voltage* and the current is called *Zener current.*

Applications Zener diode is mainly used in voltage regulator system to maintain a fairly constant output voltage. One of such regulator circuit is shown in the Figure 6.50.

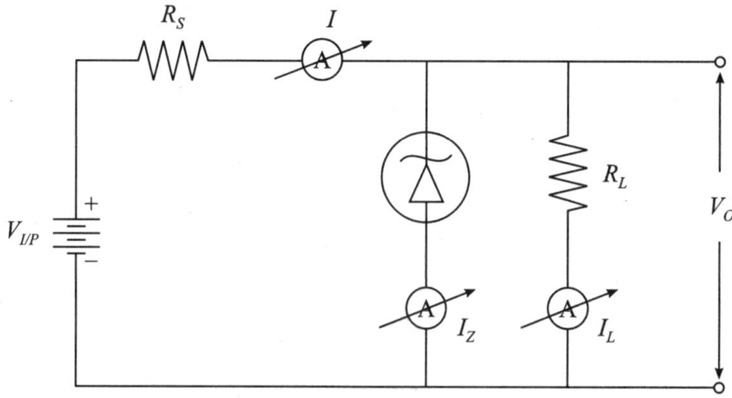

Fig. 6.50 Voltage regulator circuit with Zener diode.

The supply voltage is d.c., represented by $V_{I/P.}$ The output voltage is represented by $V_{O/P}$. The Zener diode is connected in reverse-bias mode. I_L and R_L are the load current and load resistance respectively. R_S is a series resistance I is the total current.

So, voltage drop across R_S is: $V_S = I \times R_S$

$V_{I/p}$ and $V_{O/P}$ are related as: $V_{I/P} = V_S + V_{O/P}$

Again, I is divided into two parts: I_Z and I_L. So, $I = I_Z + I_L$. Any decrease or increase in input voltage ($V_{I/P}$) results in a decrease or increase in I_Z and hence also I. Thus a fluctuation of V_S is observed with respect to fluctuation of supply voltage $V_{I/P}$, keeping $V_{O/P}$ constant.

So, $V_{I/P} = V_S + V_{O/P}$ and $\Delta V_{I/P} = \Delta V_S$, keeping $V_{O/P}$ constant.

This is the principle of voltage regulator with Zener diode.

6.15.3 Tunnel diode

If a semiconducting material is doped with high level of an impurity element, the depletion region in a junction diode, made by that semiconducting material, is very thin so that significant flow of current is established both in forward-bias and reverse-bias conditions, the situation is commonly termed tunneling. The diode is called *tunnel diode*.

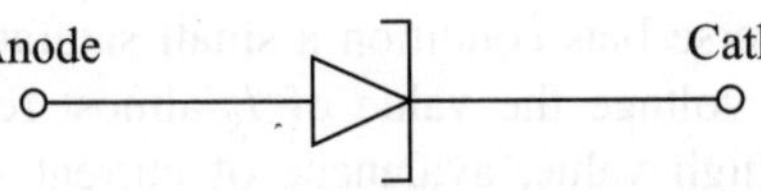

Fig. 6.51 Schematic representation of a tunnel diode.

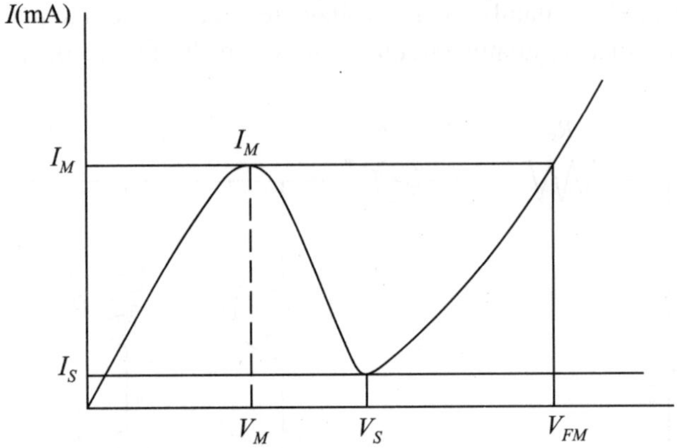

Fig. 6.52 *V-I* characteristic of a tunnel diode under forward-bias condition.

With the increase in forward-bias voltage current increases and reaches a maximum value of I_M. Further increase in forward-bias voltage the current decreases and reaches I_S. The region (V_M to V_S) is known as *negative resistance region* of the diode. However, after V_S, further increase in forwarded-bias voltage results in increase in diode current. This peculiar *V–I* characteristic finds its use as amplifier and oscillator circuits.

6.15.4 Junction transistor

It is a combination of two semiconductor diodes. Its main function is to transfer of resistance and hence the name 'transistor' is derived. Junction transistor is most widely used and was first developed by the Scientist Schockley in 1949. Junction transistor is a combination of two *p-n*-junctions as shown, in the Figures. 6.53(a & b).

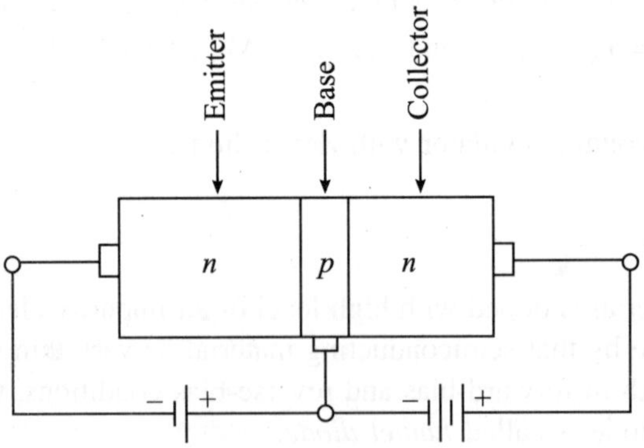

(a) *n-p-n* transistor

Fig. 6.53 (a) Three regions of a *n-p-n*-type transistor.

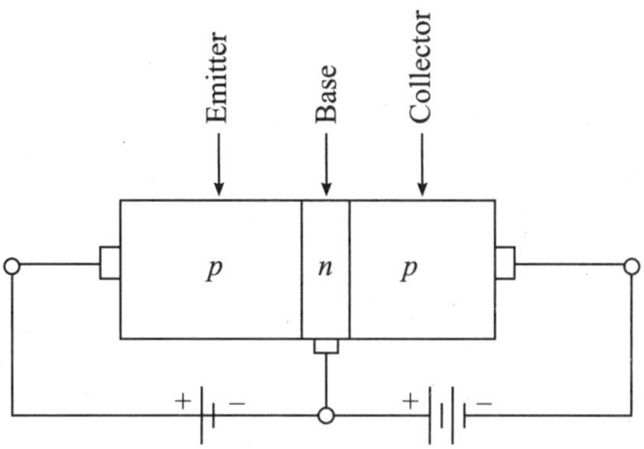

(b) *p-n-p* transistor

Fig. 6.53 (b) Three regions of a *p-n-p*-type transistor.

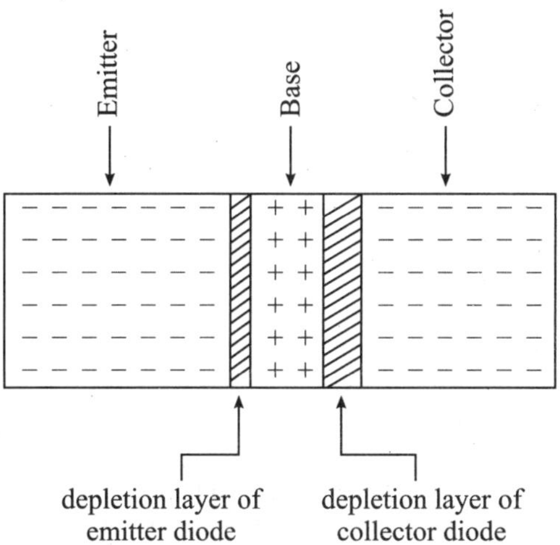

Fig. 6.53 (c) Depletion layers at two junctions of base region for an unbiased transistor.

In *n-p-n*-transistor two *n*-regions are separated by a *p*-region and in *p-n-p*-transistor two *p*-regions are separated by an *n*-region. In *n-p-n*-transistor *n*-region, situated at the left, is connected with a battery in forward-bias condition. In *p-n-p*-transistor also, *p*-region, situated at the left, is connected with a battery in forward-bias condition. Thus heavy current flows from left region to the middle region. In *n-p-n*-transistor heavy flow of electrons occurs from left to right and in *p-n-p*-transistor heavy flow of positive holes occurs from left to right. This left region of the transistor is called *emitter*. Its function is to *pump* carriers to the middle region, which is called *base*.

The function of base is to transfer carriers from emitter region to the right, which is called *collector*, the name implies that it collects carriers from the base region. The collector region is connected with a battery in reverse-bias condition. The *n*-region of the transistor is usually made of *Ge*-crystals, doped with *In*-element or other electron-deficient elements.

The emitter region assumes the *largest area* of the three regions of a transistor and the base region possesses the *smallest* area. The question is why? As the emitter is forward-biased, heavy flow of carrier concentration is developed within this region. To achieve the maximum carrier concentration the area of the emitter region is made the largest. Whenever the carriers pass through the base region, fraction of the carriers is consumed by positive holes to form electron-hole pairs and hence part of the carriers is lost during the passage. In *n-p-n* transistor heavy electrons are flowing through the base *p*-region and during this passage fraction of electrons is consumed in the base to form electron-hole pairs and the remaining electrons enter into the collector region. In the case of a *p-n-p* transistor heavy positive holes are flowing from emitter region to base region and during their passage part of the positive holes is absorbed by the electrons in the base region to form electron-hole pairs. The remaining positive holes enter into the collector region. In order to minimize the loss of carriers during flow of carriers through base region, the area of the latter is set as minimum as possible, since *larger* the area of the bases region *higher* will be the loss of carriers. As some charge carriers are lost in the base region, carrier density and hence current in collector region is always less than that in emitter region. Thus area of collector region is always less than that of emitter region.

Universal representation of junction transistor is shown in the Figure 6.54.

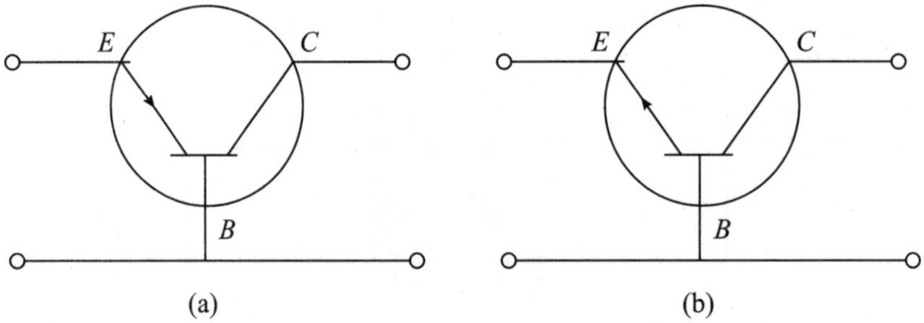

Fig. 6.54 (a) symbolic representation of *p-n-p* transistor
(b) symbolic representation of *n-p-n* transistor.

In the case of an *n-p-n*-transistor heavy electrons flow from emitter to base region but the direction of conventional current is opposite to the direction of flow of electrons. The arrowhead in *n-p-n*-transistor [Figure 6.54(b)] indicates the direction of conventional current. In the case of *p-n-p* transistor heavy positive holes flow from emitter region to base region and the direction of positive hole is same as the direction of current. So the arrowhead in *p-n-p* transistor [Figure 6.54(a)] indicates both the direction of positive hole motion as well as the direction of current.

Configuration of a transistor

As we know transistor has three electrodes, emitter, base and collector and has four terminals. Out of these four terminals, two for the input and two for the output. One electrode is connected to the input and the other is connected to the output. The third electrode must be common. If the common electrode is the emitter, the circuit is called common emitter circuit or *CE* circuit. If the common electrode is the base, it is called common base circuit or *CB* circuit. If the common electrode is the collector the circuit is called common collector circuit or *CC*. circuit. The three different circuit configurations are shown in the following figure.

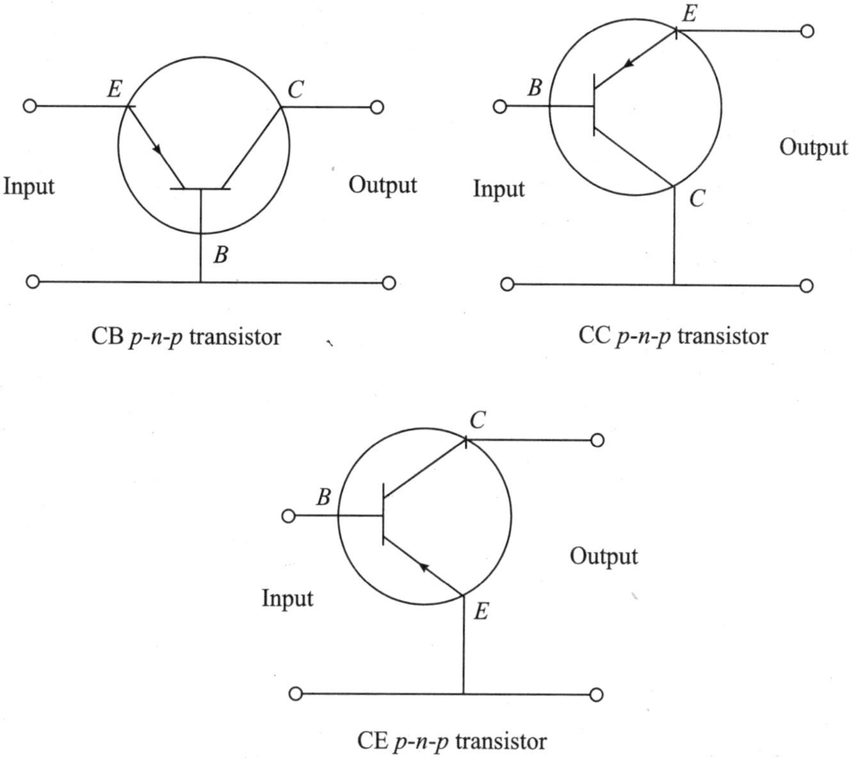

Fig. 6.55 Different transistor circuit configurations.

Current gain In the case of a *CB circuit* the collector voltage V_{CB} is kept constant. The collector current I_C, base current I_B and emitter current I_E are related as follows:

$$I_E = I_B + I_C \tag{6.26a}$$

Differentiating we get:

$$dI_E = dI_B + dI_C \tag{6.26b}$$

The change in collector current dI_C is controlled by the change in emitter current dI_E and the ratio of the two is the control characteristics of *CB* circuit, designated by α.

So,

$$\alpha = \left(\frac{\partial I_C}{\partial I_E}\right)_{V_{CB}}$$

Or, at constant V_{CB},

$$\alpha = \frac{\Delta I_C}{\Delta I_E} \tag{6.27a}$$

During the flow of current from emitter to collector a small fraction of charge carriers is lost in the base region. So, ΔI_C is always less than ΔI_E. Smaller the area of base region, lesser the difference between ΔI_E and ΔI_C. Thus, the value of α is always slightly less than unity.

In the case of a *CE circuit* the output current is I_C and the input current is I_B with the collector voltage V_{CE} kept constant. Thus the change in collector current (dI_C) is controlled by the change in base current (dI_B) and the ratio of this two is the control characteristic of *CE* circuit, designated by β.

So, $$\beta = \left(\frac{\partial I_C}{\partial I_B}\right)_{V_{CE}}$$

Or, at constant V_{CE}, $$\beta = \frac{\Delta I_C}{\Delta I_B} \tag{6.27b}$$

As the collector current is far greater than the base current, the value of β is far greater than unity.

Relation between α and β α is also known as forward-current-transfer ratio and β is often called current amplification factor in a *CE* amplifier. From the Equation (6.26) we have:

$$\Delta I_E = \Delta I_B + \Delta I_C$$

Or, $$\Delta I_B = \Delta I_E - \Delta I_C$$

Dividing both sides by ΔI_B, we get:

$$1 = \frac{\Delta I_E - \Delta I_C}{\Delta I_B} = \frac{\Delta I_C}{\Delta I_B}\left[\frac{\Delta I_E}{\Delta I_C} - 1\right]$$

$$= \frac{\Delta I_C}{\Delta I_B}\left[\frac{1}{\left(\frac{\Delta I_C}{\Delta I_E}\right)} - 1\right] = \beta\left(\frac{1}{\alpha} - 1\right)$$

$$\beta = \left(\frac{\alpha}{1-\alpha}\right) \quad \text{and} \quad \alpha = \frac{\beta}{1+\beta} \tag{6.28}$$

Application The primary application of a transistor is the amplification of current, voltage and power. The amplification can be obtained by using *CB*, *CE* and *CC*-circuits. The gain characteristics of the above transistors are given in the Table 6.5.

Table 6.5 Gain characteristics of transistors.

Characteristics	*CB*	*CE*	*CC*
Voltage gain	Same as in *CE* configuration	Same as in *CB* configuration	< 1
Current gain	$\alpha < 1$	$\beta >> 1$	>> 1
Power gain		Highest	

From the above table it is evident that *CE*-transistor amplifier is the most efficient and widely used while the other two are used in some selective areas.

6.16 Photodiode and photovoltaic cell

Photodiode Sometimes semiconductors are active in presence of light energy or photons. When a semiconductor diode is illuminated in solar flux, valence electrons in the *p*-*n* depletion layer collect ionization energy from solar flux and get converted into conduction electrons, which are free to move. The number of free electrons generated is proportional to the intensity of the incident light. Light intensity is usually expressed as luminous flux.

$$1 \text{ lm/ft}^2 = 1 \text{ foot-candle} = 1 \text{ fc},$$

$$1 \text{ lm} = 1.496 \times 10^{-10} \text{ W}$$

and

$$1 \text{ lm/ft}^2 = 1 \text{ fc} = 1.609 \times 10^{-12} \text{ W/m}^2.$$

In a photodiode the *p*-*n* semiconductor diode is connected to the reverse-bias situation as shown in the following figure.

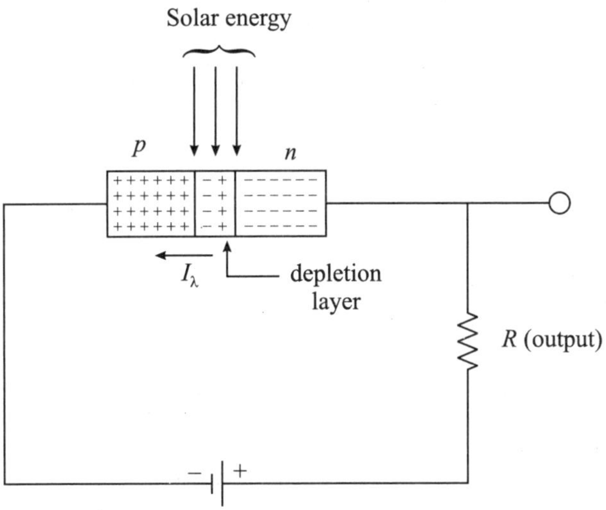

Fig. 6.56 Illumination of depletion layer of a reverse-bias *p*-*n* junction diode by a collimated beam of light.

In the reverse-bias situation, the saturation current I_λ is very small (only a few μA) but when *p*-*n* junction is illuminated by solar energy, the number of minority carriers increases, as a result of which, the value of I_λ increases. Furthermore, I_λ increases with the increase in solar flux intensity, as shown in the Figure 6.57.

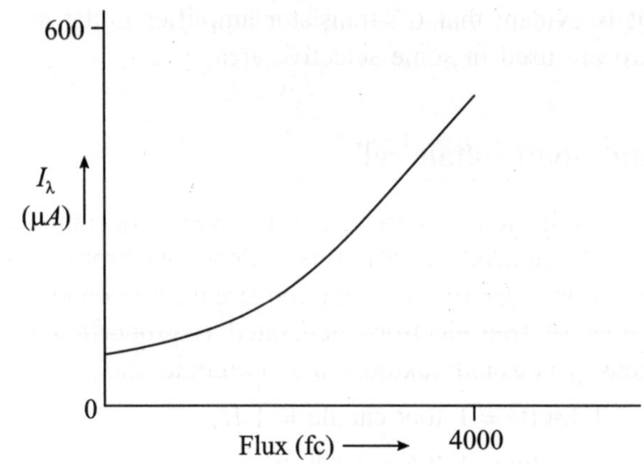

Fig. 6.57 Saturation current (I_λ) versus solar flux plot of a photodiode.

Photovoltaic cell or solar cell

In a photovoltaic cell two metallic conductors are connected with a semiconductor *p-n* junction diode, the junction of which is illuminated by a source to set free the valence electrons. As a result of this, minority carrier density increases, leading to a sharp increase in reverse saturation current and hence a high voltage is established even under open circuit condition. This voltage is known as *photovoltaic voltage*, represented by V_{oc}. V_{oc} increases with the increase in illumination flux density up to a certain level and after that illumination has a little effect on V_{oc} [see Figure 6.58].

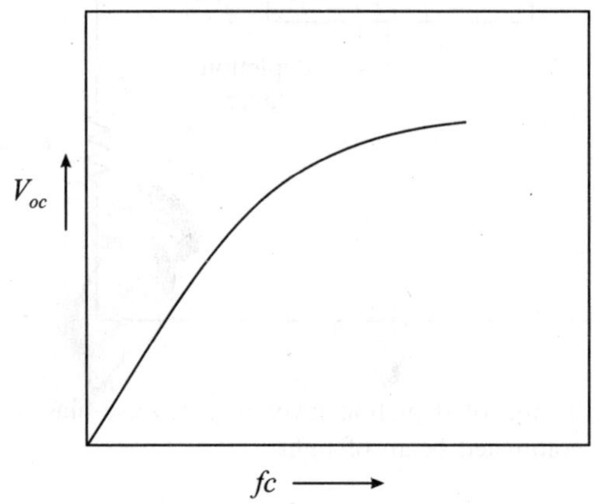

Fig. 6.58 Voltage versus solar flux plot of a photovoltaic cell.

Efficiency (η) of a photovoltaic cell is given by:

$$\eta = \frac{\text{Power output}}{\text{Power input}} \times 100$$

Cadmium Sulphide (CdS) and Cadmium Selenide (CdSe) are the most widely used materials for photovoltaic cell. Their properties are shown in the following table.

Table 6.6 Properties of CdS and CdSe.

Material	E_g*-value* (*in* eV)	*Wavelength* (λ) *at peak spectral response* (Å)	*Response time* (*millisecond*)
CdS	2.42	5100	100
CdSe	1.74	6150	10

Uses Photovoltaic cells are used in burglar alarms, for automatic door opening and switching on streetlights etc. One of the most important applications of photoconductor device is in voltage regulator. The following circuit is maintained in a voltage regulator system.

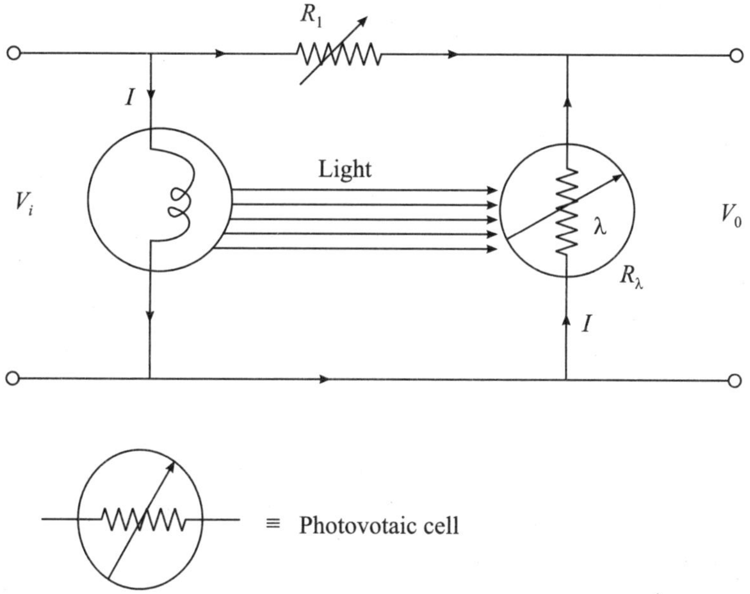

Fig. 6.59 Principle of a voltage regulator circuit.

Working formula

V_i = Input voltage

V_0 = Output voltage

R_1 = Resistance of an on-line resistor

R_λ = Resistance of the photovoltaic cell

I = Current in the circuit

$$I = \frac{V_i}{R_1 + R_\lambda}$$

Again, $$I = \frac{V_0}{R_\lambda}$$

Or, $$V_0 = I\,R_\lambda$$

Or, $$V_0 = \frac{R_\lambda V_i}{R_1 + R_\lambda}$$

Principle When current is passed through the light source the bulb is illuminated generating a beam of light which directly hits the photovoltaic cell. When the latter is active an output current is realized. The working formula is:

$$V_0 = \left(\frac{R_\lambda}{R_1 + R_\lambda}\right)V_i \qquad (6.29)$$

Case-I When the supply voltage V_i drops, brightness and hence illumination of the bulb decreases. So R_λ increases and hence V_0 is maintained constant.

Case-II When the supply voltage V_i increases suddenly brightness and hence illumination of the bulb increases. So, R_λ decreases and hence V_0 is maintained constant.

6.17 Rectifiers

Several electronic devices need d.c. supply. Thus it is required that available a.c. power supply be converted to d.c. supply. The conversion process from a.c. supply to d.c. supply is popularly known as *rectification* and the corresponding devices are called *rectifiers*. Junction diodes are often used as rectifiers.

Classification Depending upon the output response the rectifiers can be classified into two broad categories: (a) Half-wave rectifier and (b) Full-wave rectifier.

Half-wave rectifier The circuit of an ideal half-wave rectifier is shown below:

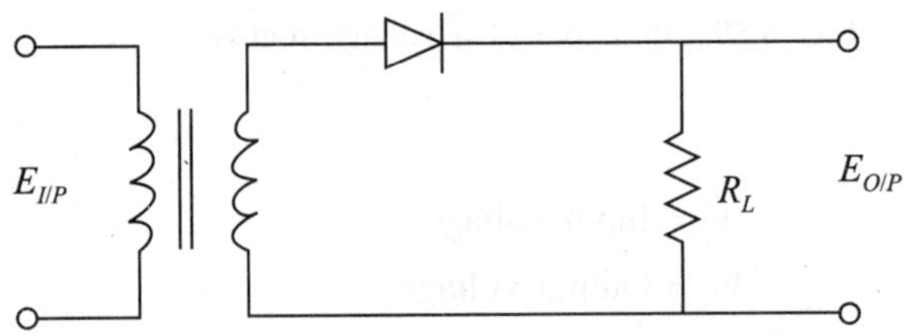

Fig. 6.60 An ideal half-wave rectifier circuit.

In actual practice another resistance R_a is connected in series with the load resistance R_L so that the total impedance of the circuit becomes $(R_a + R_L)$. A practical half wave rectifier circuit is shown in the following figure.

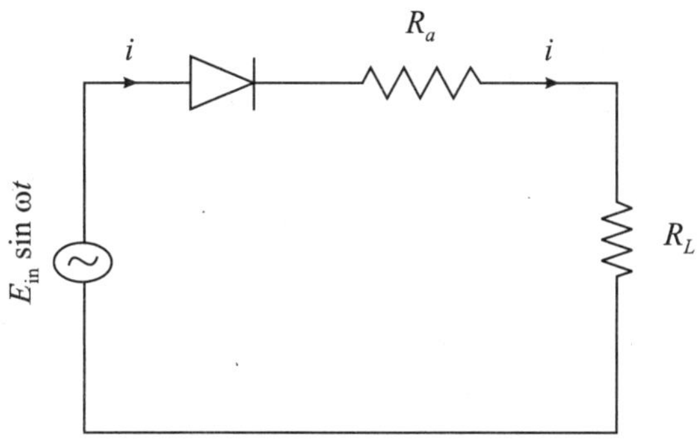

Fig. 6.61 A practical half-wave rectifier circuit.

$$\text{A/C power supply} = E_m \sin(\omega t)$$

$$\text{Output voltage} = E_{O/P} = i.R_L$$

$$\text{Maximum current} = I_m = \frac{E_m}{R_a + R_1}$$

The input voltage and output response are shown in the following figure:

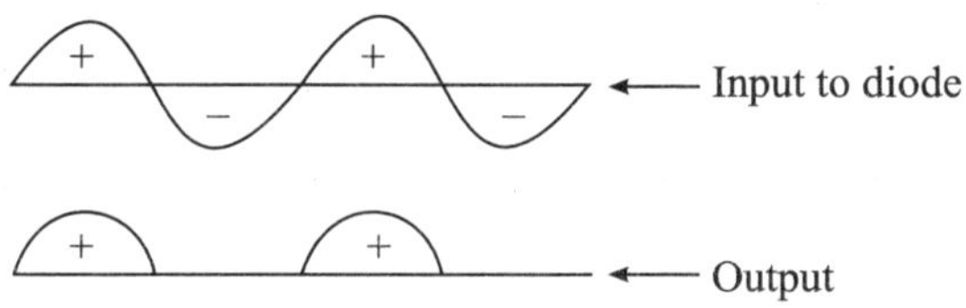

Fig. 6.62 Input voltage and output response of a half-wave rectifier.

It is interesting to note that input voltage is alternating but the output voltage is unidirectional and pulsating, as amplitude changes continuously. Thus the output voltage is not a direct d.c. but a pulsating d.c., containing both a.c. and d.c. components.

The ratio of a.c. component to d.c. component of the output current (or voltage) is known as *ripple factor*, denoted by the symbol 'γ'.

$$\gamma = \frac{\text{effective value of a.c. component of current (or voltage)}}{\text{effective value of d.c. component of current (or voltage)}} = \frac{I_{ac}}{I_{dc}} = \frac{E_{ac}}{E_{dc}}$$

In the half-wave rectifier negative component of input a.c. signal puts the diode in reverse-bias condition and hence is not allowed to pass through the diode. So the negative component of input a.c. signal is not realized at the output.

Thus I_{dc} of a half-wave rectifier is given by:

$$I_{dc} = \frac{1}{2\pi}\int_0^{2\pi} i\,d\theta = \frac{1}{2\pi}\int_0^{\pi} I_m \sin\omega t\,d\theta + \frac{1}{2\pi}\int_{\pi}^{2\pi} 0.\,d\theta$$

$$= \frac{I_m}{2\pi}\int_0^{\pi} \sin \omega t \, d(\omega t) \quad [\text{since, } \theta = \omega t]$$

$$= \frac{I_m}{\pi}$$

$$\boxed{I_{dc} = \frac{I_m}{\pi}}$$

So, $$(I_{dc})_{HW} = \frac{I_m}{\pi}$$

Again, $$I_{rms}^2 = I_{dc}^2 + I_{ac}^2$$

Or, $$\left(\frac{I_{rms}}{I_{dc}}\right)^2 = 1 + \left(\frac{I_{ac}}{I_{dc}}\right)^2$$

Or, $$\frac{I_{ac}}{I_{dc}} = \sqrt{\left(\frac{I_{rms}}{I_{dc}}\right)^2 - 1} = \sqrt{F^2 - 1}$$

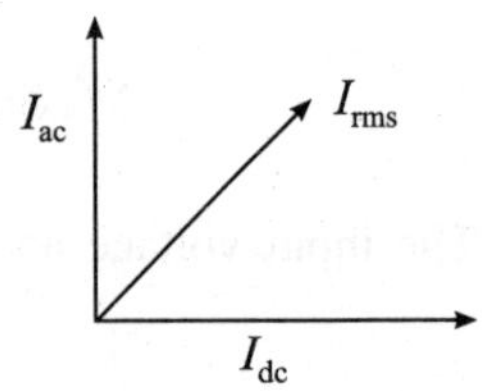

where, $$F = \frac{I_{rms}}{I_{dc}} = \text{Form Factor}$$

Thus ripple factor of a half-wave rectifier is given by:

$$(\gamma)_{HW} = \frac{I_{ac}}{I_{dc}} = \sqrt{F^2 - 1}$$

$$(I_{rms}^2)_{HW} = \frac{1}{2\pi}\int_0^{2\pi} i^2 d(\omega t) = \frac{1}{2\pi}\int_0^{\pi} i^2 d(\omega t) + \frac{1}{2\pi}\int_{\pi}^{2\pi} 0.\, d(\omega t)$$

$$= \frac{1}{2\pi}\int_0^{\pi} I_m^2 \sin^2(\omega t)\cdot d(\omega t) = \left(\frac{I_m^2}{2\pi}\right)\frac{1}{2}\int_0^{\pi}(1 - \cos 2\omega t)\, d(\omega t) = \frac{I_m^2}{4}$$

Or, $$(I_{rms})_{HW} = \frac{I_m}{2}$$

So, $$F = \frac{(I_{rms})_{HW}}{(I_{dc})_{HW}} = \left(\frac{I_m/2}{I_m/\pi}\right) = \frac{\pi}{2} = 1.57$$

and $$(\gamma)_{HW} = \sqrt{F^2 - 1} = \sqrt{1.57^2 - 1} = 1.21$$

So, ripple factor $(\gamma)_{HW} = 1.21$

Efficiency of a half-wave rectifier = $(\eta)_{HW}$

$$(\eta)_{HW} = \frac{\text{d.c. O/P power}}{\text{a.c. I/P power}} = \frac{P_{dc}}{P_{ac}}$$

$$P_{ac} = I_{rms}^2(R_a + R_L) \quad \text{and} \quad P_{dc} = I_{dc}^2 R_L$$

$$(\eta)_{HW} = \frac{I_{dc}^2 R_L}{I_{rms}^2 (R_a + R_L)} = \left(\frac{1}{F^2}\right)\left(\frac{R_L}{R_a + R_L}\right)$$

$$\boxed{(\eta)_{HW} = \left(\frac{2}{\pi}\right)^2 \left(\frac{R_L}{R_a + R_L}\right)}$$

Full-wave rectifier

The circuit of a full wave rectifier is shown below:

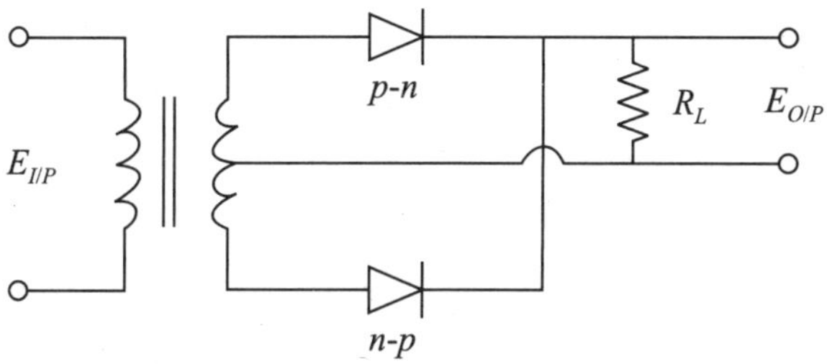

Fig. 6.63 An ideal full-wave rectifier circuit.

The input voltage and output response are shown in the following figure:

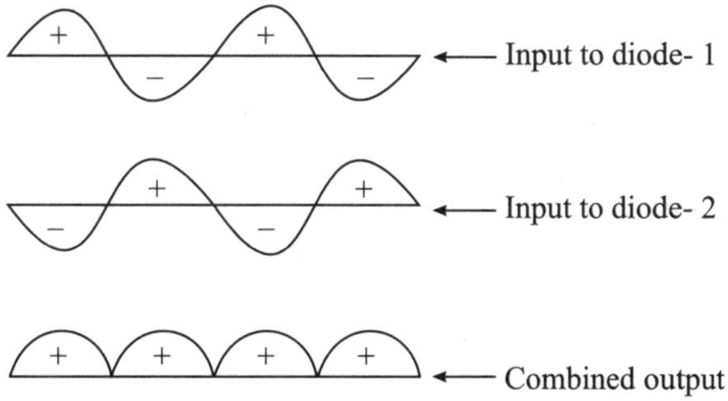

Fig. 6.64 Input voltage and output response of a full-wave rectifier.

In a practical full-wave rectifier a resistor is usually added to each diode.

$$\text{A/C power supply} = E_m \sin \omega t$$

$$\text{Maximum current} = \frac{E_m}{R_a + R_L}$$

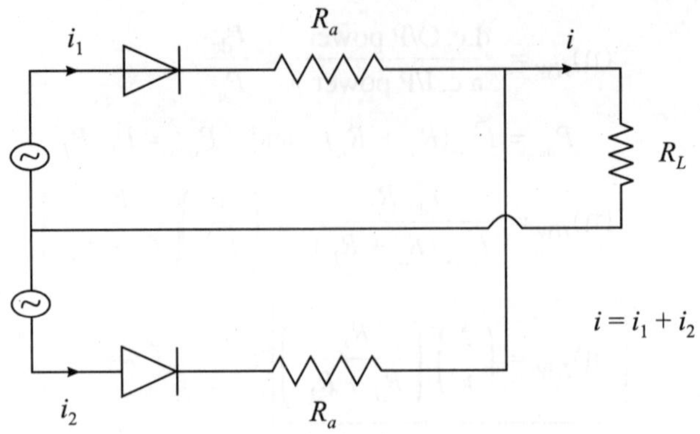

Fig. 6.65 Circuit diagram of a practical full wave rectifier.

Output voltage = iR_L

$$(I_{dc})_{FW} = \frac{1}{2\pi}\int_0^{2\pi} i\,d\theta = \frac{1}{2\pi}\int_0^{2\pi} I_m \sin\theta\,d\theta = \frac{1}{\pi}\int_0^{\pi} I_m \sin\theta\,d\theta$$

$$= \frac{2I_m}{\pi}$$

So,
$$\boxed{(I_{dc})_{FW} = \frac{2I_m}{\pi}}$$

$$(I^2_{rms})_{FW} = \frac{1}{2\pi}\int_0^{2\pi} i^2\,d\theta = \frac{1}{\pi}\int_0^{\pi} I_m^2 \sin^2\theta\,d\theta$$

$$= \frac{I_m^2}{2\pi}\int_0^{\pi}(1-\cos 2\theta)\,d\theta = \frac{I_m^2}{2}$$

So,
$$\boxed{(I_{rms})_{FW} = \frac{I_m}{\sqrt{2}}}$$

Again,
$$I^2_{rms} = I^2_{dc} + I^2_{ac} \quad \text{and} \quad \gamma = \frac{I_{ac}}{I_{dc}} = \sqrt{F^2 - 1}$$

So,
$$F = \frac{(I_{rms})_{FW}}{(I_{dc})_{FW}} = \frac{\left(\dfrac{I_m}{\sqrt{2}}\right)}{\left(\dfrac{2I_m}{\pi}\right)} = \frac{\pi}{2\sqrt{2}} = 1.11$$

$$\text{Ripple factor} = (\gamma)_{FW} = \sqrt{1.11^2 - 1} = 0.48$$

Efficiency of a full wave rectifier = $(\eta)_{FW} = \dfrac{\text{d.c. O/P power}}{\text{a.c. I/P power}} = \dfrac{P_{dc}}{P_{ac}}$

$$P_{ac} = I_{rms}^2 (R_a + R_L) \quad \text{and} \quad P_{dc} = I_{dc}^2 R_L$$

$$(\eta)_{FW} = \frac{I_{dc}^2 R_L}{I_{rms}^2 (R_a + R_L)} = \left(\frac{1}{F^2}\right)\left(\frac{R_L}{R_a + R_L}\right)$$

$$\boxed{(\eta)_{HW} = \left(\frac{2\sqrt{2}}{\pi}\right)^2 \left(\frac{R_L}{R_a + R_L}\right)}$$

7

GENERAL AND CO-ORDINATION CHEMISTRY

SECTION – I CHEMICAL BONDING

7.1 General Chemistry

7.1.1 Atom and molecule

The smallest part of an element is called atom and the smallest part of a compound is called molecule.

Bohr theory of atom

(i) Electrons are moving surrounding the positive nucleus, containing proton and neutron, in different orbits, such that centrifugal force is exactly balanced by the attractive force. Mathematically, we can write that, $\frac{mv^2}{r} = \frac{Ze^2}{r^2}$.

(ii) Electrons, while revolving around the nucleus, do not radiate energy.

(iii) Angular momentum of any electron (mvr) is given by: $mvr = \frac{nh}{2\pi}$, where, n is called principal quantum number, indicating the orbit, e.g., $n = 1$ indicates 1st orbit, $n = 2$ indicates the 2nd orbit etc.

Limitations of Bohr's theory

(i) It is established that a moving charge particle must radiate energy and hence, centrifugal force $\left(\frac{mv^2}{r}\right)$ is less than electrostatic force of attraction $\left(\frac{Ze^2}{r^2}\right)$ and hence electrons are moving towards the nucleus, rendering the atom unstable. The stability of Bohr model is only 10^{-8} sec.

(ii) According to the Bohr model position of an electron can be evaluated by calculating the radius of the orbit and velocity can be also calculated by using the momentum equation: $mvr = \frac{nh}{2\pi}$. This is in contrary to the Heisenberg's uncertainty principle.

7.1.2 Atomic spectra

According to Bohr theory the atomic spectra of a molecule should show only single spectral line for a given energy level since it considers only primary quantum number, '*n*', but in actual practice for a given energy level several spectral lines of different energy levels are obtained. As electron is a charged particle it continuously exerts strong electrostatic field surrounding it and this is believed to be the reason for splitting the spectral lines. To quantify these subsidiary energy levels Orbital Quantum Number or Azimuthal Quantum Number, represented by l, has been introduced. Studies of atomic spectra showed that for a given value of n the values of l ranges from 0 to $(n - 1)$.

In 1896 the scientist Zeeman showed that if the atomic spectra of a molecule is taken under a strong magnetic field, each subsidiary energy level further splits into sub-energy levels of same energy. To quantify these sub-energy levels Magnetic Quantum Number, m, has been introduced. For a given value of l the values of m ranges from $-l$ to $+l$.

As each electron is considered as a tiny magnet, spinning of electrons may be either clockwise or anticlockwise. To quantify the direction of spinning of electron Spin Quantum Number, represented by the symbol s, is introduced. The value of s may be either $+\frac{1}{2}$ or $-\frac{1}{2}$.

7.1.3 Filling up of atomic orbitals by electrons

Pauli's exclusion principle According to this principle no two electrons cannot have same four quantum numbers. Thus for a given orbital, values of n, l and m are fixed but the quantum number s has two options: $+\frac{1}{2}$ or $-\frac{1}{2}$, which signifies that one orbital may contain a maximum of two electrons.

Hund's rule of maximum multiplicity According to this rule, electrons have tendency to remain in unpaired form as long as orbitals of same energy levels (i.e., same values of n and l) are available. Thus if more than one orbital of same energy level is available two electrons will remain in unpaired form.

Aufbau Principle According to this principle electrons have tendency to remain in lowest energy level and hence the lowest energy level is filled up first followed by the next higher energy level and so on. As we know out of the four quantum numbers, n, l, m, s, only n and l represent the energy level, the lowest energy level is that level which has the lowest value of $(n + l)$. However, it may be possible that two energy levels have equal value of $(n + l)$. In that case energy level, having lower value of n, constitutes the lower energy level and is filled up first. $(n + l)$ values of different energy levels are given below:

Table 7.1.1 $(n + l)$ values of different energy levels.

Energy level	$(n + l)$	*Energy level*	$(n + l)$	*Energy level*	$(n + l)$	*Energy level*	$(n + l)$
1s	1 + 0 = 1	4s	4 + 0 = 4	5d	5 + 2 = 7	6f	6 + 3 = 9
2s	2 + 0 = 2	4p	4 + 1 = 5	5f	5 + 3 = 8	6g	6 + 4 = 10
2p	2 + 1 = 3	4d	4 + 2 = 6	5g	5 + 4 = 9	6h	6 + 5 = 11
3s	3 + 0 = 3	4f	4 + 3 = 7	6s	6 + 0 = 6	7s	7 + 0 = 7
3p	3 + 1 = 4	5s	5 + 0 = 5	6p	6 + 1 = 7	7p	7 + 1 = 8
3d	3 + 2 = 5	5p	5 + 1 = 6	6d	6 + 2 = 8		

Thus the energy levels can be arranged as follows:

$1s < 2s < 2p < 3s < 3p < 4s < 3d < 4p < 5s < 4d < 5p < 6s < 4f < 5d < 6p < 7s < 5f < 6d < 7p$... etc.

However, there are some exceptions of this rule, e.g., according to this rule, the electronic structure of Cu: [Ar] $4s^2\ 3d^9$, but actual electronic structure of Cu is: [Ar] $4s^1\ 3d^{10}$. This happens because filled $3d^{10}$-orbital is more stable than $3d^9$-orbital. Similar explanation can be proposed for the electronic structure of Cr: [Ar]$4s^1\ 3d^5$ instead of [Ar]$4s^2\ 3d^4$. In this case half filled $3d^5$-orbital is more stable than $3d^4$-orbital.

7.1.4 Matter waves and Schrödinger wave equation

Heisenberg's uncertainty principle Each and every small particle should be considered as a wave-packet. If the wave-packet is small, position of the particle can be determined very easily but velocity (or momentum) is indeterminate. On the other hand if the size of the wave-packet is large position of the particle is indeterminate but velocity (or momentum) can be determined accurately. Thus it is impossible to determine position and velocity (or momentum) of a particle simultaneously and precisely. If Δp and Δx are the uncertainties of momentum and position of a particle respectively, we can write, according to Heisenberg, that: $\Delta p\,.\,\Delta x \geq \left(\dfrac{h}{4\pi}\right)$.

De Brgolie concept According to this concept each particle possesses both particulate characteristics and wavy characteristics. Following the wavy characteristics we can write that total energy $(T) = h\nu$, where ν is the frequency of the wave and following the particulate characteristics we can write that total energy (T) = mv^2, where, v is the velocity of the particle. Combining the above two concepts, we have: $mv^2 = h\nu = \left(\frac{hv}{\lambda}\right)$, λ is the wavelength of the particle. Hence, $\lambda = \left(\frac{h}{mv}\right) = \left(\frac{h}{p}\right) = \lambda_B$, where λ_B is often called Broglie wavelength.

Time-dependent Schrödinger wave equation According to electromagnetic wave theory we have:

$$\frac{\partial^2 c}{\partial t^2} = c^2 \frac{\partial^2 c}{\partial x^2} \quad \text{[where, 'c' is the velocity of light]}$$

Schrödinger tried to explain electron-wave in the similar way as follows:

$$\frac{\partial^2 \psi}{\partial t^2} = v^2 \frac{\partial^2 \psi}{\partial x^2} \quad \text{[where, } \psi \text{ is the wave function of electron]}$$

At this stage he made some assumptions:

(i) The atomic orbitals are governed by Heisenberg's uncertainty principle and probability distribution function is used to describe the motion of electron.
(ii) The wavelength of electron is calculated by using de Broglie's wave-particle duality concept.
(iii) The electron waves are stationary waves like vibration of a string between two fixed points.

After these assumptions he described the following characteristics of ψ:

(i) It indicates the probability of finding an electron within a specified volume under consideration. It's value may be positive, negative, zero or complex.
(ii) The value of ψ is zero at infinity.
(iii) At any position within the specified volume under consideration ψ has a definite value, since $\psi = \infty$ has no meaning.
(iv) At any position within the specified volume under consideration ψ has only one value because more than one value of ψ at a particular point has no meaning.
(v) Both ψ and $\left(\frac{\partial \psi}{\partial x}\right)$ are continuous within the specified volume under consideration such that $\left(\frac{\partial^2 \psi}{\partial x^2}\right)$ is finite everywhere.
(vi) The point where $\psi = 0$ is called node or nodal point and hence ψ can be described as wave amplitude. As we know in the case of light wave intensity of light is directly proportional to the square of amplitude, the term ψ^2 can be defined as intensity of electron or probability density of electron.
(vii) The function ψ^2 is normalized function and from the knowledge of probability density function we can write that:

$$\int_{-\infty}^{+\infty} \psi^2 \, d\tau = 1 \quad \text{[where, } d\tau \text{ is the volume under consideration]}$$

$d\tau = dx.dy.dz$. For one dimension the above equation becomes:

$$\int_{-\infty}^{+\infty} \psi^2 \, dx = 1$$

The solution of the above equation is given by:

$$\psi = \text{C}e^{-i\omega\left(t - \frac{x}{v}\right)} \tag{7.1.1}$$

where, $\omega = 2\pi\nu$ and ν is the frequency of the wave. v is the velocity of the wave and C is a constant. Again, $v = \nu.\lambda$. Thus, we can write that:

$$\omega = \frac{2\pi v}{\lambda}$$

Or,
$$\frac{\omega}{v} = \frac{2\pi}{\lambda} = k \tag{7.1.2}$$

Differentiating Equation (7.1.1) with respect to t, we have:

$$\frac{\partial\psi}{\partial t} = -i\omega \; \text{C}e^{-i\omega\left(t - \frac{x}{v}\right)} = -i\omega\psi$$

Or,
$$\psi = -\frac{1}{i\omega}\frac{\partial\psi}{\partial t} \tag{7.1.3}$$

Again, we know that:
$$T = E + V \tag{7.1.4}$$

where, T, E and V are the total energy, kinetic energy and potential energy respectively. Multiplying both sides of the Equation (7.1.4) by ψ, we get:

$$T\psi = E\psi + V\psi \tag{7.1.5}$$

Differentiating the Equation (7.1.1) with respect to x we get:

$$\frac{\partial\psi}{\partial x} = \frac{i\omega}{v}\text{C.e}^{-i\omega\left(t - \frac{x}{v}\right)} = \frac{i\omega}{v}\psi \quad \text{and}$$

$$\frac{\partial^2\psi}{\partial x^2} = \frac{i\omega}{v}\frac{\partial\psi}{\partial x} = -\frac{\omega^2}{v^2}\psi = -k^2\psi \quad \text{[from Equation (7.1.2)]}$$

Or,
$$\frac{\partial^2\psi}{\partial x^2} = -k^2\psi \tag{7.1.6}$$

Thus,
$$\psi = -\frac{1}{k^2}\frac{\partial^2\psi}{\partial x^2} \tag{7.1.7}$$

Equation (7.1.3) is the time dependent equation of ψ, Equation (7.1.6) is the one dimensional time independent equation of ψ and Equation (7.1.5) contains both time-dependent and time-independent functions of ψ. Thus combining Equations (7.1.3), (7.1.5) and (7.1.7), we get:

$$-\frac{T}{i\omega}\frac{\partial\psi}{\partial t} = -\frac{E}{k^2}\frac{\partial^2\psi}{\partial x^2} + V\psi \qquad (7.1.8)$$

Again, we know that:

$$E = \frac{1}{2}mv^2 = \frac{1}{2m}(mv)^2 = \frac{p^2}{2m} \qquad (7.1.9)$$

$p = mv$ (momentum) $= \dfrac{h}{\lambda}$ (according to De Broglie's concept).

Thus, $$E = \frac{p^2}{2m} = \frac{h^2}{2m\lambda^2} = \frac{h^2 k^2}{8\pi^2 m}$$

Since, $$\lambda = \frac{2\pi}{k} \quad \text{[from Equation (7.1.2)]}$$

Or, $$E = \frac{h^2 k^2}{8\pi^2 m} \qquad (7.1.10)$$

Again, we know that total energy (T) = $h\nu$.

$$T = h\nu = \frac{h\omega}{2\pi}$$

Or, $$\frac{T}{\omega} = \frac{h}{2\pi} \qquad (7.1.11)$$

Combining Equations (7.1.8), (7.1.10) and (7.1.11), we have:

$$\frac{h}{2i\pi}\frac{\partial\psi}{\partial t} = \frac{h^2}{8\pi^2 m}\frac{\partial^2\psi}{\partial x^2} - V\psi \qquad (7.1.12)$$

The Equation (7.1.12) is the time-dependent one dimensional Schrödinger wave equation. In three dimensions the Equation (7.1.12) becomes:

$$\frac{h}{2i\pi}\frac{\partial\psi}{\partial t} = \frac{h^2}{8\pi^2 m}\left[\frac{\partial^2\psi}{\partial x^2} + \frac{\partial^2\psi}{\partial y^2} + \frac{\partial^2\psi}{\partial z^2}\right] - V\psi$$

Or, $$\frac{h}{2i\pi}\frac{\partial\psi}{\partial t} = \frac{h^2}{8\pi^2 m}\nabla^2\psi - V\psi \qquad (7.1.13)$$

The Equation (7.1.13) is the time-dependent three dimensional Schrödinger wave equation.

Time-independent Schrödinger wave equation From Equation (7.1.1), we have:

$$\psi = Ce^{-i\omega\left(t - \frac{x}{v}\right)}$$

Or, $$\psi = C'e^{i\omega\frac{x}{v}} = C'e^{ikx} \quad (7.1.14)$$

where, $C' = Ce^{-i\omega t}$, since the wave function is assumed to be independent of time and $\frac{\omega}{v} = k$ [Equation (7.1.2)].

Differentiating the Equation (7.1.14) with respect to x we get:

$$\frac{\partial \psi}{\partial x} = ikC'e^{ikx} = ik\psi \quad (7.1.15)$$

Differentiating Equation (7.1.15) with respect to x we get:

$$\frac{\partial^2 \psi}{\partial x^2} = ik\frac{\partial \psi}{\partial x} = -k^2\psi$$

Or, $$\frac{\partial^2 \psi}{\partial x^2} + k^2\psi = 0$$

Replacing k^2 from Equation (7.1.10), we get:

$$\frac{\partial^2 \psi}{\partial x^2} + \frac{8\pi^2 m}{h^2}E\psi = 0$$

Or, $$\frac{h^2}{8\pi^2 m}\frac{\partial^2 \psi}{\partial x^2} + E_x\psi = 0 \quad (7.1.16)$$

The Equation (7.1.16) is the one-dimensional time-independent Schrödinger wave equation, along x-direction.

Along x-direction: $$\frac{h^2}{8\pi^2 m}\frac{\partial^2 \psi}{\partial x^2} + E_x\psi = 0$$

Along y-direction: $$\frac{h^2}{8\pi^2 m}\frac{\partial^2 \psi}{\partial y^2} + E_y\psi = 0$$

Along z-direction: $$\frac{h^2}{8\pi^2 m}\frac{\partial^2 \psi}{\partial z^2} + E_z\psi = 0$$

Thus, in three dimensions, the overall equation becomes:

$$\frac{h^2}{8\pi^2 m}\left[\frac{\partial^2 \psi}{\partial x^2} + \frac{\partial^2 \psi}{\partial y^2} + \frac{\partial^2 \psi}{\partial z^2}\right] + (E_x + E_y + E_z)\psi = 0$$

Or, $$\frac{h^2}{8\pi^2 m}\left[\frac{\partial^2 \psi}{\partial x^2} + \frac{\partial^2 \psi}{\partial y^2} + \frac{\partial^2 \psi}{\partial z^2}\right] + E\psi = 0 \quad [\text{since, } E = E_x + E_y + E_z]$$

Or, $$\frac{h^2}{8\pi^2 m}\left[\frac{\partial^2 \psi}{\partial x^2} + \frac{\partial^2 \psi}{\partial y^2} + \frac{\partial^2 \psi}{\partial z^2}\right] + (T - V)\psi = 0$$

[since, $E = T - V$ from Equation (7.1.4)]

Or, $$\nabla^2\psi + \frac{8\pi^2 m}{h^2}(T - V)\psi = 0 \tag{7.1.17}$$

The Equation (7.1.17) is the three dimensional time-independent Schrödinger wave equation. The solution of time-independent Schödinger wave equation is given by Equation (7.1.14):

$\psi = C'e^{i\omega\frac{x}{v}} = C'e^{ikx}$. Hence, ψ can be written as:

$$\psi = A\,\sin(kx) + B\cos(kx) = \text{A}\,\sin\frac{2\pi}{\lambda}x + \text{B}\cos\frac{2\pi}{\lambda}x \tag{7.1.18}$$

Consider a one dimensional potential well of length L where streams of electrons are continuously vibrating like stationary waves [assumption (iii)] in different energy levels as shown in the Figure 7.1.1.

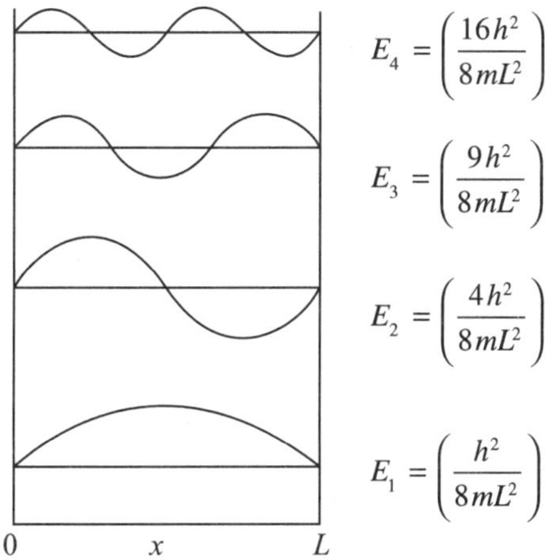

Fig. 7.1.1 Potential well and wave function ψ.

For any energy level two ends of an electron wave are fixed and hence probability of finding electron at both the ends is zero. Thus we can write that:

at $x = 0$, $\psi = 0$ and also at $x = L$, $\psi = 0$.

Applying the first boundary condition in Equation (7.1.18), we have: $0 = 0 + B$.

Or, $$B = 0.$$

Thus Equation (7.1.18) becomes:

$$\psi = A\,\sin\frac{2\pi}{\lambda}x \tag{7.1.19}$$

Applying the second boundary condition in the Equation (7.1.19), we have:

$$\frac{2\pi}{\lambda}L = n\pi$$

Or, $$\lambda = \frac{2L}{n} \tag{7.1.20}$$

This is well illustrated in the figure 7.1.1. As λ has a definite positive value and cannot be ∞, n is always a positive integer and cannot be zero.

Significance of '*n*'

(i) In the potential well n indicates energy level where an electron resides.
(ii) It shows that wavelength of an electron decreases when it resides at higher energy level.
(iii) The pictorial-concept of electrons, revolving round the positive nucleus in an atom (according to Bohr theory), has been changed to well-concept of electron waves, vibrating at different energy levels.

From the Equations (7.1.19) and (7.1.20) we have:

$$\psi = A \sin \frac{n\pi}{L} x \tag{7.1.21}$$

According to the assumption (vii), made by Schrödinger, we have:

$$\int_{-\infty}^{+\infty} \psi^2 \, dx = 1 \quad \text{[for one dimension]}$$

In one dimensional potential well the value of x ranges from 0 to L as shown in the figure. Thus the above integration becomes:

$$\int_0^L \psi^2 \, dx = 1 \tag{7.1.22}$$

From the Equations (7.1.21) and (7.1.22), we have:

$$\int_0^L A^2 \sin^2 \frac{n\pi}{L} x \, dx = 1$$

On integration we get: $A = \sqrt{\frac{2}{L}}$

Putting the value of A in the Equation (7.1.21), we get:

$$\psi = \sqrt{\frac{2}{L}} \sin \frac{n\pi}{L} x$$

Thus, along x-direction: $\psi_x = \sqrt{\frac{2}{L}} \sin \frac{n_x \pi}{L} \cdot x$

Along y-direction: $\psi_y = \sqrt{\frac{2}{L}} \sin \frac{n_y \pi}{L} \cdot y$

Along z-direction: $\psi_z = \sqrt{\frac{2}{L}} \sin \frac{n_z \pi}{L} \cdot z$

As probability has multiplication property, we can write that:

$$\psi = \psi_x \psi_y \psi_z = \sqrt{\frac{8}{L^3}} \sin\frac{n_x \pi}{L} x \,.\, \sin\frac{n_y \pi}{L} y \,.\, \sin\frac{n_z \pi}{L} z \tag{7.1.23}$$

The Equation (7.1.23) is the Schrödinger wave equation in Cartesian coordinate. In spherical coordinate (r, θ, ϕ) the above equation becomes: $x = r \sin\theta \cos\phi$, $y = r \sin\theta \sin\phi$, $z = r \cos\theta$

$$\frac{1}{r^2}\frac{\partial}{\partial r}\left(r^2 \frac{\partial \psi}{\partial r}\right) + \frac{1}{r^2 \sin\theta}\frac{\partial}{\partial \theta}\left(\sin\theta \frac{\partial \psi}{\partial \theta}\right) + \frac{1}{r^2 \sin\theta}\frac{\partial^2 \psi}{\partial \phi^2} \tag{7.1.24}$$

The electron wave function is consisted of radial and polar parts.

$$\psi(r, \theta, \phi) = \psi(r)\ \psi(\theta)\ \psi(\phi) \tag{7.1.25}$$

The radial part the wave functions of electron in different *s*-orbitals are shown in the Figures 7.1.2(a) and (b).

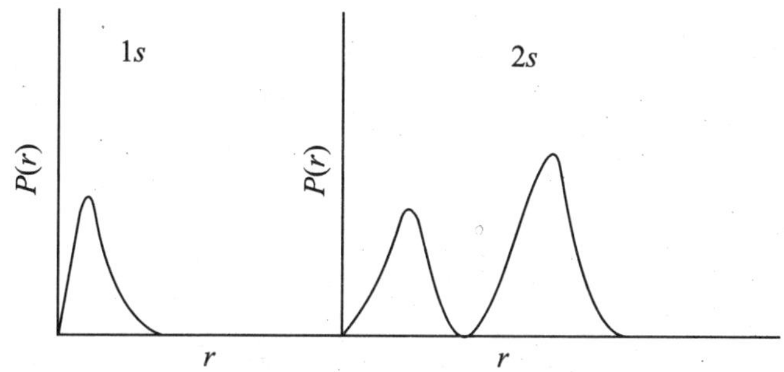

Fig.7.1.2(a) The radial part of electron wave function $[\psi(r)]$ in 1*s*- and 2*s*-orbitals.

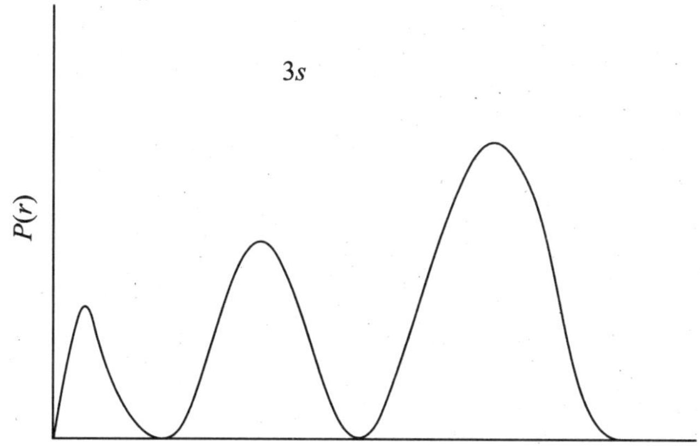

Fig. 7.1.2(b) The radial part of electron wave function $[\psi(r)]$ in 3*s*-orbital.

Node The points, where $\psi = 0$, are called nodes. The number of nodes for a given orbital depends on the primary quantum number (n) as shown in the following table.

Table 7.1.2 Number of nodes of different orbitals.

Orbital	*Number of nodes*
s	$(n - 1)$
p	$(n - 2)$
d	$(n - 3)$
f	$(n - 4)$

Considering the angular part the Schrödinger wave functions of electron, shapes of different orbitals are shown in the Figures 7.1.3(a) and (b).

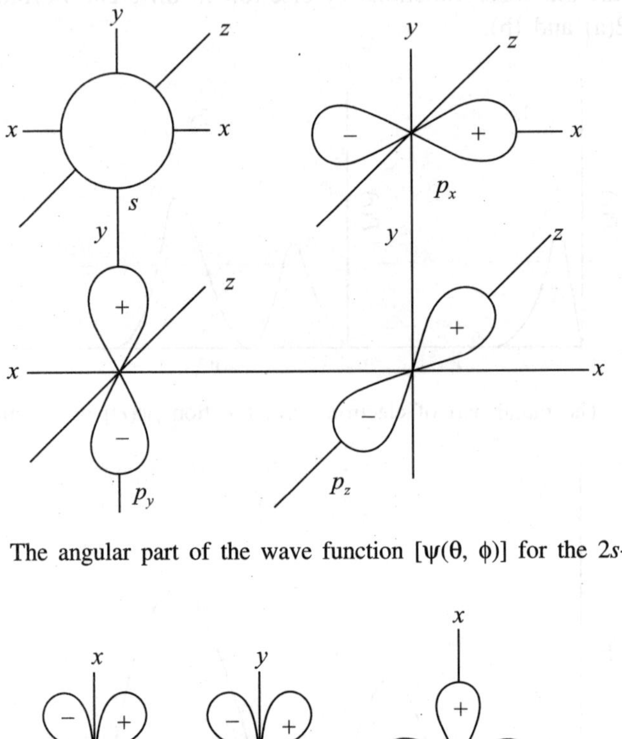

Fig. 7.1.3(a) The angular part of the wave function [$\psi(\theta, \phi)$] for the 2*s*-, 2*p*-orbitals.

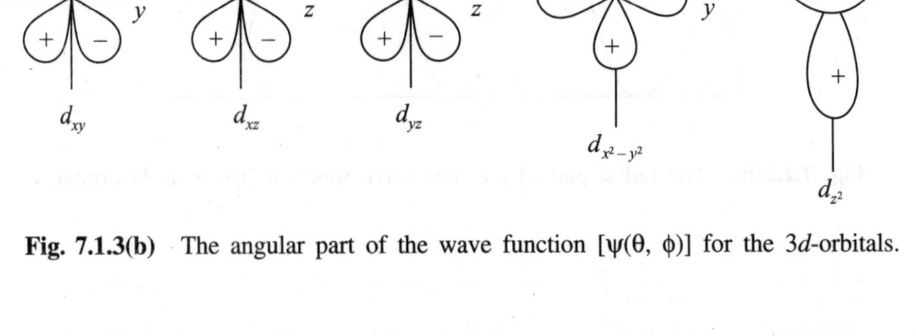

Fig. 7.1.3(b) The angular part of the wave function [$\psi(\theta, \phi)$] for the 3*d*-orbitals.

From the Equation (7.1.10), we have:

$$E = \frac{h^2 k^2}{8\pi^2 m}$$

where, $k = \frac{2\pi}{\lambda}$ and $\lambda = \frac{2L}{n}$ [Equation (7.1.20)].

Putting the values of k and λ we have:

$$E = \frac{n^2 h^2}{8mL^2}$$

Along x-direction: $E_x = \frac{n_x^2 h^2}{8mL^2}$

Along y-direction: $E_y = \frac{n_y^2 h^2}{8mL^2}$

Along z-direction: $E_z = \frac{n_z^2 h^2}{8mL^2}$

Thus the total energy (E) is given by: $E = E_x + E_y + E_z$.

Or, $E = \frac{\left(n_x^2 + n_y^2 + n_z^2\right)h^2}{8mL^2}$

Or, $E_n = \frac{n^2 h^2}{8mL^2}$

where, $n^2 = n_x^2 + n_y^2 + n_z^2$

Significance of 'E'

(i) E_n indicates the energy of electron at a given energy state, n.

(ii) Energy difference between two successive energy states (n) and ($n + 1$) is given by:

$$E_{n+1} - E_n = \Delta E = [(n+1)^2 - n^2]\frac{h^2}{8mL^2} = (2n+1)\frac{h^2}{8mL^2}$$

(iii) $E = \frac{\left(n_x^2 + n_y^2 + n_z^2\right)h^2}{8mL^2}$

Thus for different values of n_x, n_y and n_z, E may have same value. For example, ($n_x = 1$, $n_y = 2$, $n_z = 3$) the value of E is same as for ($n_x = 2$, $n_y = 1$, $n_z = 3$) or ($n_x = 3$, $n_y = 1$, $n_z = 2$) etc. The corresponding energy states, possessing same energy, are called degenerate states.

7.1.5 Uncertainty principle from Schrödinger wave equation

We know from De Broglie's concept that, momentum $(p_n) = \frac{h}{\lambda} = \frac{nh}{(2L)}$, since $\lambda = \frac{2L}{n}$ from Equation (7.1.20). Thus uncertainty in momentum is given by:

$$\Delta p = p_{n+1} - p_n = \frac{(n+1)h}{2L} - \frac{nh}{2L} = \frac{h}{2L}$$

From the Figure 7.1.1 it is evident that, uncertainty in position of an electron is $\Delta x = L$. Thus the uncertainty in momentum is given by: $\Delta p = \frac{h}{(2\Delta x)}$. Or, $\Delta p . \Delta x = \frac{h}{2} > \frac{h}{(4\pi)}$. Thus Schrödinger wave equation agrees with the uncertainty principle.

7.1.6 Validity of Neil Bohr's theory

According to Bohr's theory, the radius of an atomic orbit can be calculated from the following equations:

$$\frac{Ze^2}{r^2} = \frac{mv^2}{r} \quad \text{and} \quad mvr = \frac{nh}{(2\pi)} \tag{7.1.26}$$

where, r is the radius of nth Bohr-orbit. For hydrogen atom atomic number, $Z = 1$.

According to Schrödinger wave equation, we have:

$$E = \frac{1}{2} mv^2 = \frac{(mv)^2}{2m} = \frac{p^2}{2m}$$

Or,

$$\Delta E = \frac{(\Delta p)^2}{2m} \tag{7.1.27}$$

From Heisenberg's uncertainty principle, we have: $\Delta p . \Delta x \geq \frac{h}{(4\pi)}$. However, in most theoretical calculation it has been accepted that:

$$\Delta p \, . \, \Delta x = \frac{h}{2\pi}$$

Or,

$$\Delta p = \frac{h}{2\pi \Delta x}$$

Putting the value of Δp in the Equation (7.1.27), we get:

$$\Delta E = \frac{h^2}{8\pi^2 m (\Delta x)^2} \tag{7.1.28}$$

Again, potential energy (V) is given by:

$$\Delta V = \int_{\infty}^{r} \frac{Ze^2}{4\pi\,\varepsilon_0\, r^2} dr = -\frac{Ze^2}{4\pi\,\varepsilon_0\, r} = -\frac{Ze^2}{4\pi\,\varepsilon_0\,(\Delta x)}$$

since r represents uncertainty in position (Δx). Thus:

$$\Delta V = -\frac{Ze^2}{4\pi\,\varepsilon_0\,(\Delta x)} \quad (7.1.29)$$

So, the total energy (T) is given by: $T = E + V$.

Or, $\Delta T = \Delta E + \Delta V.$

Or, $$\Delta T = \frac{h^2}{8\pi^2 m(\Delta x)^2} - \frac{Ze^2}{4\pi\,\varepsilon_0\,(\Delta x)} \quad (7.1.30)$$

As electron prefers remaining in the lowest energy level corresponding to the minimum value of ΔT. To find out the minimum value of ΔT differentiate the Equation (7.1.30) and set it to zero. Thus, we have:

$$\frac{\partial \Delta T}{\partial \Delta x} = -\frac{h^2}{4\pi^2 m(\Delta x)^3} + \frac{Ze^2}{4\pi\,\varepsilon_0\,(\Delta x)^2} = 0$$

Or, $$\Delta x = \frac{\varepsilon_0\, h^2}{\pi e^2 Z m} \quad (7.1.31)$$

For hydrogen atom $Z = 1$ and the above equation reduces to:

$$\Delta x = \frac{\varepsilon_0\, h^2}{\pi e^2 m} \quad (7.1.32)$$

The value of Δx, calculated from the Equation (7.1.32), is almost same as the Bohr's radius r, calculated from the Equation (7.1.26), although Bohr's theory of atomic structure has been discarded. This has been explained as follows: hydrogen atom has the electronic configuration $1s^1$, i.e., $n = 1$ and $s = 0$. Thus the lone electron of hydrogen atom occupies the $1s$-orbital, which is circular according to Schrödinger wave equation. As the structure of $1s$-orbital is very much similar to Bohr's pictorial model, the radius of Bohr's orbit is very close to the value of Δx. Thus it may be concluded that Bohr's theory of atomic structure is valid for hydrogen-like atoms or ions.

7.1.7 Concept of molecular orbital

According to Schrödinger wave equation electrons are considered as waves as shown in the following figure:

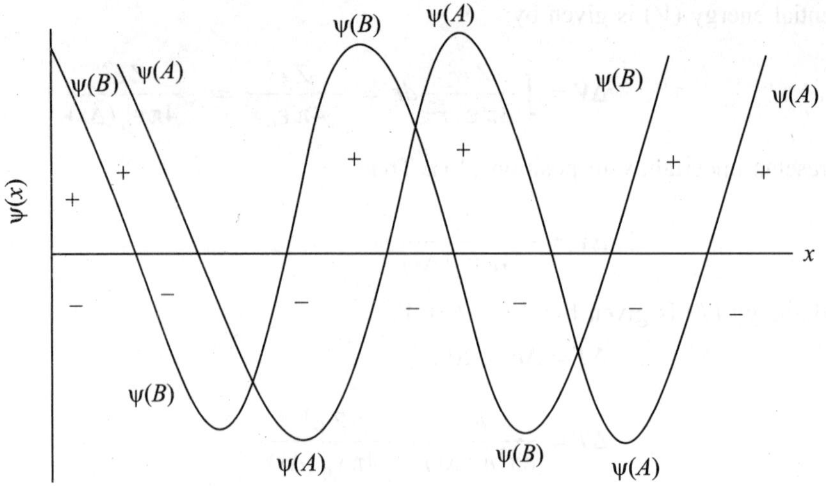

Fig. 7.1.4(a) Overlapping of electron waves and formation of bonding MO.

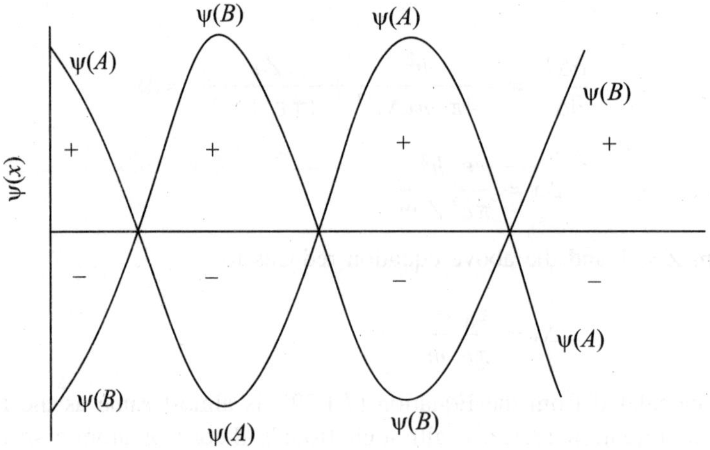

Fig. 7.1.4(b) Overlapping of electron waves and formation of anti-bonding MO.

It is evident from the above figure that when two similar wave functions overlap with each other, constructive interference occurs, giving rise a new resultant wave function. The situation can be considered as overlap of two atomic orbitals, possessing similar wave function, resulting in the formation of a new orbital, known as bonding molecular orbital or bonding MO. Bonding MO is always situated at lower energy level compared to that of atomic orbitals. On the other hand when opposite wave functions overlap with each other, destructive interference occurs and the resultant value of ψ is almost zero. Thus two atomic orbitals, having opposite wave function, always repel each other, producing another molecular orbital, known as anti-bonding molecular orbital or anti-bonding MO, which is always situated at higher energy level compared to that of atomic orbitals. Also it is evident from the above figure that the resultant wave function is totally different from either of the overlapping wave functions. Thus it may be concluded that during formation of a molecular orbital the identities of both the overlapping atomic orbitals are lost. The formation of bonding and anti-bonding molecular orbitals is well illustrated in the following figure:

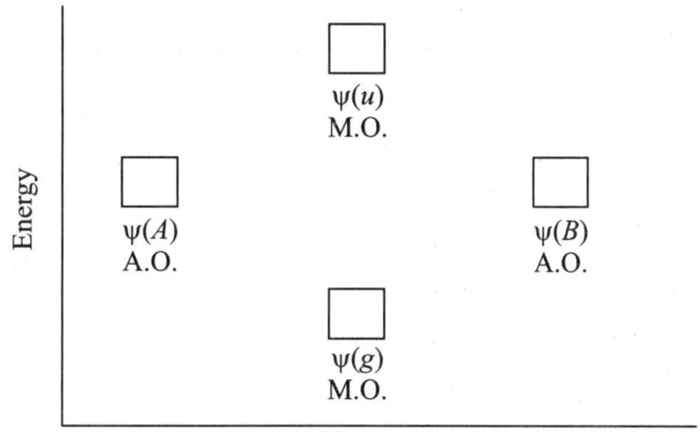

Fig. 7.1.4(c) Formation of bonding and anti-bonding molecular orbitals.

According to Schrödinger wave function concept, during overlapping of two atomic orbitals the superimposing wave functions produce constructive interference and destructive interference with equal probability and hence each overlap of two atomic orbitals always produces two molecular orbitals, one is bonding MO and the other is anti-bonding MO.

Overlapping of orbitals may occur in different ways (a) two atomic orbitals approach each other along an axis and overlap through head-on collision, as a result of which, electrons are distributed along the axis. This type of overlap is called σ-overlap and the corresponding bond is called σ-bond. The bonding MO is denoted by σ and the anti-bonding MO is denoted by σ^*, (b) two atomic orbitals approach each other along an axis and overlap laterally or sidewise, as a result of which electrons are distributed above and below the axis. This type of overlap is called π-overlap and the corresponding bond is called π-bond. Formation of σ- and π-bonds is well illustrated in the following figure.

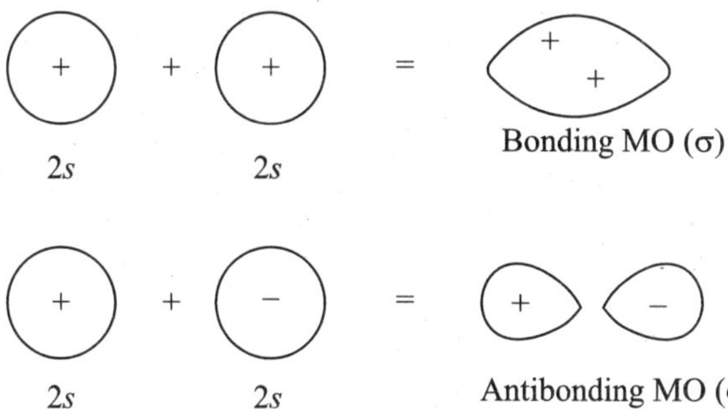

Fig. 7.1.5(a) Schematic representation of formation of σ-bond through overlapping of *s*-orbitals.

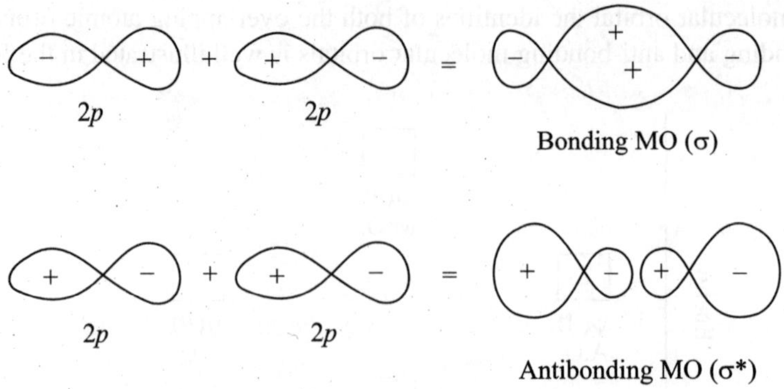

Fig. 7.1.5(b) Schematic representation of formation of σ-bond through overlapping of *p*-orbitals.

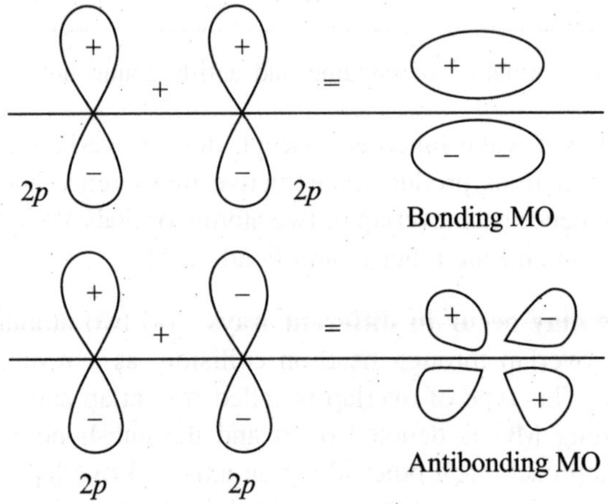

Fig. 7.1.5(c) Schematic representation of formation of π-bond through overlapping of *p*-orbitals.

The filling up of molecular orbitals obeys Pauli's exclusion principle, Hund's rule of maximum multiplicity and Aufbau principle. The different molecular orbitals may be arranged according to their energy levels as shown below:

$$\sigma 1s < \sigma^* 1s < \sigma 2s < \sigma^* 2s < \sigma 2p_x < \pi 2p_y = \pi 2p_z < \pi^* 2p_y = \pi^* 2p_z < \sigma^* 2p_x \ldots$$

For low molecular mass, e.g., H_2, Li_2, Be_2, B_2 and C_2 the above arrangement of molecular orbitals becomes: $\sigma 1s < \sigma^* 1s < \sigma 2s < \sigma^* 2s < \pi 2p_y = \pi 2p_z < \sigma 2p_x < \pi^* 2p_y = \pi^* 2p_z < \sigma^* 2p_x \ldots$

Bond order Bond order of a molecule can be evaluated by using the following formula:

$$\text{Bond order} = \frac{(\text{No. of bonding orbitals} - \text{No. of anti-bonding orbitals})}{2}$$

Examples: N_2: $\sigma 1s^2 < \sigma^* 1s^2 < \sigma 2s^2 < \sigma^* 2s^2 < \sigma 2p_x^2 < \pi 2p_y^2 = \pi 2p_z^2$

$$\text{Bond order} = \frac{(10 - 4)}{2} = 3$$

$\boldsymbol{O_2}$: $\sigma 1s^2 < \sigma^* 1s^2 < \sigma 2s^2 < \sigma^* 2s^2 < \sigma 2p_x^2 < \pi 2p_y^2 = \pi 2p_z^2 < \pi^* 2p_y^1 = \pi^* 2p_z^1$

$$\text{Bond order} = \frac{(10-6)}{2} = 2$$

As two unpaired electrons exist in the electronic structure of oxygen molecule, the latter shows paramagnetic characteristics. Thus paramagnetic behaviour of oxygen molecule can well be explained with the help of molecular orbital theory.

$\boldsymbol{B_2}$: $\sigma 1s^2 < \sigma^* 1s^2 < \sigma 2s^2 < \sigma^* 2s^2 < \pi 2p_y^1 = \pi 2p_z^1$

As two degenerate π-orbitals are available and can accommodate two electrons in unpaired form, $\sigma 2p_x$ is placed at higher energy level with respect to $\pi 2p_y$ and $\pi 2p_z$. Bond order: $\frac{(6-4)}{2} = 1$.

7.1.8 Characteristics features of MO theory

(i) According to this theory two wave functions of two atomic orbitals while interact with each other, two different orbitals are formed according to their mode of interaction. If the interaction is constructive a strong MO is formed, known as bonding MO, lying always at lower energy level than the corresponding atomic orbitals (AO). On the other hand if the interaction is destructive, a weak MO is formed, known as anti-bonding MO, lying always at higher energy level than the corresponding AOs. The formation of bonding and anti-bonding MOs are schematically shown in the Figure 7.1.5.

(ii) According to MO theory two separate molecular orbitals are formed, both of which lose their identity after the formation of MO.

(iii) It cannot explain the phenomenon "Resonance".

(iv) It cannot explain the geometry of a covalent compound.

(v) It can explain the formation of odd electron molecules like H_2^+, NO, O_3 etc.

(vi) It can explain the paramagnetic behaviour of oxygen molecule.

(vii) MO orbitals are filled up by electrons according to Aufbau principle, Hund's rule of maximum multiplicity and Pauli's exclusion principle.

(viii) Total number of bonds or bond order can be evaluated by using MO theory.

(ix) It can successfully explain the metallic bond formation. On the basis of MO theory metals and materials can be classified into three categories: (a) conductors, (b) semiconductors and (c) insulators. In the case of conductor the energy gap between conduction band and valence band is very small or does not exist due to overlapping of the two bands. So easy flow of electrons occur from valence band to conduction band. In the case of semiconductor, the energy gap is small but only a few electrons are able to jump to the conduction band. This jumping of electrons from valence band to conduction band is favoured by thermal means. In the case of insulators electrons are unable to overcome the energy gap and hence they are unable to carry current.

7.1.9 Characteristics features of Valence Bond (VB) theory

(i) According to this theory, atomic orbitals of two elements overlap to form the covalent bond. It considers the electron both as wave and particulate in nature.

(ii) Atomic orbitals (AO) of two elements retain their identity even after overlapping.

(iii) If two atomic orbitals, lobes of which spread along the same axis, approach each other along that axis, head-on overlap occurs to form σ-bond. If two atomic orbitals, lobes of which spread along one axis, approach each other in different axis, perpendicular to that of the former one, lateral overlap occurs to form a weak bond, known as π-bond. The mechanism of formation of σ- and π-bonds are shown in the Figures. 7.1.5(b) and 7.1.5(c).

(iv) Electrons in π-orbital are loosely bound and easily get delocalized under favourable conditions. Thus VB theory well explains the phenomenon "Resonance".

(v) It explains the directional nature of covalent bond and hence geometry of covalent compounds.

(vi) It cannot explain the formation of odd electron molecules like H_2^+, NO, O_3 etc.

(vii) It cannot explain the paramagnetic behaviour of oxygen molecule.

(viii) It cannot explain the metallic bond formation.

7.1.10 Periodic table

In a general sense atoms combine to form molecules but combination of atoms depends on their inherent characteristics. On the basis of inherent characteristics elements, found in the earth's crust, were arranged in a tabular form. Mandeleeve was the first scientist who formed this table on the basis of atomic weight. The table is known as periodic table. Later, Moselay established that atomic number is more fundamental property than atomic weight. Then the periodic table was rearranged on the basis of atomic number. As atomic weight increases steadily with atomic number, only a minor alteration of Mandeleeve's periodic table was enough to develop the modern periodic table.

Periodicity of properties All the properties, exhibited by an element, are governed by electronic configuration of that element. Any property changes continuously along the period starting from Group IA. After zero group the next element again belongs to Group IA and shows similar property with the former element of Group IA. The phenomenon is called periodicity of properties. Some of the properties are described below:

(i) Ionisation potential (I.P.) Amount of energy required to dislodge an electron from an outermost shell or valence shell of an atom is called 1st Ionisation Potential (I.P.) of that atom. I.P. value of an atom depends on electrostatic force of attraction between nucleus and electron, which is inversely proportional to the square of the distance between nucleus and electron (r). Down the group the distance 'r' increases and hence I.P. value decreases down the group. Along the period distance r decreases due to increase in electrostatic force of attraction and hence the I.P. value increases along the period.

Exceptions (a) I.P. value of B should be higher than that of Be but in actual practice the reverse is true. Outermost electronic structure of Be and B are:

	$2s$	
Be:	[↑↓]	→ Fulfilled and stable structure

B: $2s$ [↑↓] $2p$ [↑][ ][ ]

B has a lone-electron in $2p$-orbital, loss of which gives stable configuration in the outermost shell. That is why the I.P. value of B is less than that of Be.

(b) I.P. value of nitrogen is higher than that of oxygen. Outermost shell configuration of N and O are:

N: $2s$ [↑↓] $2p$ [↑][↑][↑] → Half-filled and stable structure

O: $2s$ [↑↓] $2p$ [↑↓][↑][↑]

With the loss of one electron oxygen will get half-filled stable structure like nitrogen but loss of one electron from nitrogen leads to destabilization of stable outermost electronic configuration. Obviously more energy is required in case of nitrogen to dislodge an outermost electron. So I.P. value of nitrogen is higher than that of oxygen.

Ionisation energies of atoms of some selected elements are given in Table 7.1.3.

Table 7.1.3 Ionisation energies of atoms.

Atomic No. (Z)	*Element*	1st I.P. (eV)	*Atomic No. (Z)*	*Element*	1st I.P. (eV)
1	H	13.6	14	Si	8.15
3	Li	5.39	15	P	10.49
11	Na	5.14	16	S	10.36
19	K	4.34	17	Cl	12.97
4	Be	9.32	21	Sc	6.56
12	Mg	7.65	22	Ti	6.82
20	Ca	6.11	23	V	6.74
5	B	8.30	24	Cr	6.77
13	Al	5.99	25	Mn	7.44
31	Ga	6.09	26	Fe	7.89
6	C	11.26	27	Co	7.87
7	N	14.53	28	Ni	7.64
8	O	13.62	29	Cu	7.73
9	F	17.42	30	Zn	9.39

(ii) Electron affinity It is the amount of energy released during the formation of a uninegative ion from its corresponding atom.

$$X(g) + e \rightarrow X^-(g) + E$$

E is the amount of energy released i.e. electron affinity of the atom X. Usually E is expressed as kJ/mole or eV/atom.

The above expression can be written as:

$$X^-(g) \rightarrow X(g) + e - E$$

So, numerically electron affinity is equal to the ionisation potential (but with opposite sign) of an anion.

Electron affinity increases along the period since electrostatic force of attraction increases. The Group-VII A element fluorine (F) is just one electron short of stable Ne-structure and hence possesses very high electron affinity. Furthermore, electron affinity decreases down the group since metallic character increases.

Exception In Group-VII A fluorine should have highest electron affinity but in actual practice it has been observed that electron affinity of chlorine is greater than that of fluorine but after chlorine electron affinity value decreases steadily.

Electronic configuration of F is $1s^2\,2s^2\,2p^5$ and electronic configuration of Cl is $1s^2\,2s^2\,2p^6\,3s^2\,3p^5$. The 2nd shell is far smaller in size than the 3rd shell. So electrons remain congested in the outermost shell of fluorine. Thus any electron, coming from outside, receives repulsion from electron clouds of fluorine, resulting in the decrease in electron affinity. This type of electron repulsion is absent in chlorine due to large size of 3rd shell. The latter has enough space to accommodate an extra electron without repulsion.

Table 7.1.4 Electron affinity values of some selected atoms.

Element	*Electron Affinity* (eV)	*Element*	*Electron Affinity* (eV)
H	0.754	F	3.620
Li	0.590	Cl	3.790
Be	–0.190	Br	3.540
K	0.520	I	3.290
Na	0.340	O	1.467
		N	–0.21

(iii) Electronegativity Electronegativity of an atom is a measure of its ability to attract bond pair of electrons in a compound. Electronegativity depends on the electrostatic force of attraction, experienced by the nucleus. Down the group atomic radius (r) increases and electrostatic force of attraction decreases and hence electronegativity decreases while along the period atomic radius decreases and hence electronegativity increases.

Maximum electronegativity is observed in fluorine and the corresponding electronegativity value has been assigned to be 4, which has been universally accepted as a reference value with the help of which electronegativity of other elements can be easily evaluated by measuring the resonance energy Δ. The electronegativity values of some elements are listed in the Table 7.1.5. As electronegativity depends on electrostatic force of attraction, element, having smallest atomic radius, possesses the highest electronegativity. Thus electronegativity decreases down the group and increases along the period.

Table 7.1.5 Pauling's electronegativity values of some selected elements.

Element	*Electro-negativity*	*Element*	*Electro-negativity*	*Element*	*Electro-negativity*
Li	1.0	Ga	1.6	Bi	1.9
Na	0.9	In	1.7	O	3.5
K	0.8	Ti	1.8	S	2.5
Rb	0.8	C	2.5	Se	2.4
Cs	0.7	Si	1.8	Te	2.1
Be	1.5	Ge	1.8	Po	2.0
Mg	1.2	Sn	1.8	F	4.0
Ca	1.0	Pb	1.8	Cl	2.8
Sr	1.0	N	3.0	Br	2.6
Ba	0.9	P	2.1	I	2.5
B	2.0	As	2.0	At	2.2
Al	1.5	Sb	1.9		

Consider a molecule $A - B$. As the molecule is heterogeneous, it is polar in nature. E_{A-B} is the bond energy, i.e., amount of energy required to break the bond $A - B$. The bonds $A - A$ and $B - B$ are 100% non-polar and E_{A-A} and E_{B-B} are the corresponding bond energies of $A - A$ and $B - B$ respectively. If the bond $A - B$ is 100% non-polar, the bond energy would be $\sqrt{(E_{A-A}\ E_{B-B})}$. Difference between these two energies is known as resonance energy, represented by Δ with unit kcal/mole.

$$\Delta = \left[E_{A-B} - \sqrt{(E_{A-A}\ E_{B-B})}\right] \text{ kcal/mole}$$

According to Pauling, the electronegativity difference = $(\chi_A - \chi_B) = 0.208\sqrt{\Delta}$, where, χ_A and χ_B are the electronegativities of A and B respectively.

Furthermore, polarity increases with the increase in electronegativity difference. For example% ionic character of H – F is 45% but it decreases from HF to HI, as electronegativity decreases from F to I. % Ionic character of HCl, HBr and HI are 19%, 11%, and 4% respectively.

(iv) Atomic radius In true sense atom has no definite boundary and hence it is impossible to determine its absolute dimension. So radius of an atom is the theoretically calculated value, considering Bohr's atomic model. Along the period electrons are added in the same outer shell. Thus electrostatic force of attraction increases and atomic radius decreases but down the group new shell is added. So distance between outermost shell and nucleus increases and hence atomic radius increases.

Shielding or screening effect Electrostatic force of attraction, experienced by nucleus, is partially shielded or screened by inner orbital electrons. *Larger* the screening, less the electrostatic force of attraction, *higher* will be the atomic radius. *s-electrons* show the maximum shielding/screening characteristics which decreases as the size of the orbital increases. Thus *f-electrons* show the minimum screening characteristics. *f-orbital* contains a maximum of 14 electrons but they do not decrease the electrostatic force of attraction sufficiently, as a result of which, atomic radius decreases more than expected. This contraction of atomic radius due to poor shielding effect of *f-electrons* is called *f-contraction.* Similarly contraction of atomic radius due to poor shielding effect of *d-electrons* is called *d-contraction.* Consider the elements of Group-IA. Atomic radius increases steadily from hydrogen to potassium. Rb has filled 3*d-electrons* and hence its atomic radius is close to that of *K*, due to *d-contraction.* Cs has filled 3*d* and 4*d* electrons and hence experiences a further *d-contraction.* Its atomic radius is only slightly higher than that of Rb. Fr has filled 3*d*-. 4*d*-, 5*d*- as well as 4*f-electrons.* So large contraction due to *d*- and *f*-electrons is expected, as a result of which, atomic radius of Fr decreases significantly and falls below Cs.

Table 7.1.6 Atomic radii of some selected atoms.

Element	*Atomic radius (in Å)*	*Element*	*Atomic radius (in Å)*	*Element*	*Atomic radius (in Å)*
H	0.46	Be	1.12	He	1.76
Li	1.52	Mg	1.59	Ne	1.58
Na	1.86	Ca	1.97	Ar	1.88
K	2.31	Sr	2.15	Kr	2.00
Rb	2.44	Ba	2.17	Xe	2.17
Cs	2.62	Ra	2.17	Ru	2.17
Fr	2.50				

(v) Inert pair effect Down the group size of the orbitals of valence shell increases and the electrons have enough space for movement. As a result of this, the paired electrons in the *s*-orbital of valence shell have a little tendency to be unpaired during compound formation unless sufficient amount of energy is supplied. This tendency of electrons to remain in the paired form is called inert pair effect, which increases down the group. Many metals show variable valency due to this inert pair effect, e.g. Ga, In and Tl. They have both +I and +III oxidation states.

ns *np*

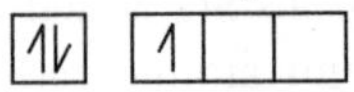 +I state: lower energy

ns *np*

 +III state: higher energy

n = 4 for Ga, n = 5 for In and n = 6 for Tl.

Similarly for Ge, Sn and Pb, we have:

ns *np*

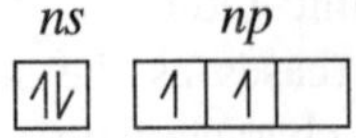 +II state: lower energy

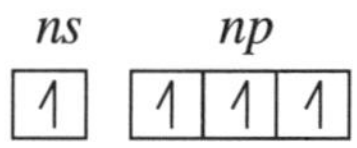

+IV state: higher energy

$n = 4$ for Ge, $n = 5$ for Sn and $n = 6$ for Pb

Among the halides GaCl, InCl and TlCl, TlCl is the most stable and GaCl is the most unstable. GaCl has a strong tendency to form $GaCl_3$. Among the halides $GaCl_3$, $InCl_3$ and $TlCl_3$, $GaCl_3$ is the stablest and $TlCl_3$ is the most unstable. $TlCl_3$ has a strong tendency to form TlCl, since Tl tries to remain in +I oxidation state due to inert pair effect. So $TlCl_3$ is the strongest oxidizing agent and GaCl is the strongest reducing agent. So the order of reducing power of monohalides are:

$$TlCl < InCl < GaCl$$

and the order of oxidizing power of trihalides are:

$$TlCl_3 > InCl_3 > GaCl_3$$

Similarly, in the case of 4th group elements Ge, Sn and Pb the order of reducing power of the halides are:

$$PbCl_2 < SnCl_2 < GeCl_2$$

and the order of oxidizing power of the halides are:

$$PbCl_4 > SnCl_4 > GeCl_4$$

7.1.11 Polarisability and dipole moment

Consider a molecule $A - B$ of which B is more electronegative than A and hence a partial charge separation permanently occurs as $\overset{\delta^+}{A} - \overset{\delta^-}{B}$. If q is the amount of charge developed on each atom and d is the distance between two atoms, the dipole moment (μ) is given by: $\mu = q \times \text{d}$. Usually the unit of d is Å, 1 Å $=10^{-8}$ cm and the unit of q is of the order 10^{-10} esu.

Thus, $\mu = q \times d\ 10^{-10}.\ 10^{-8}$ esu-cm $= q \times \text{d}\ 10^{-18}$ esu-cm $= q \times d\ D$ or Debye, where, $1\ D = 10^{-18}$ esu-cm. If μ and d are given, q can easily be determined from the above equation. If complete charge separation takes place, $A - B$ remains as A^+ and B^-, which means that charge $= 1.602 \times 10^{-19}$ C $= 4.806 \times 10^{-10}$ esu and $A - B$ is considered as 100% ionic compound.

Thus the % ionic character $= \dfrac{q}{e} \times 100.$

Let us consider the following compound:

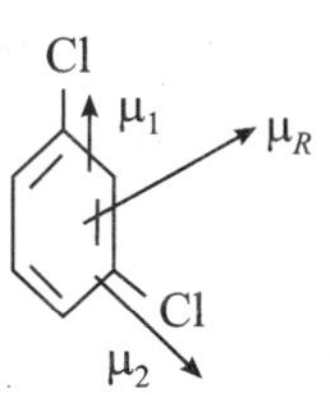

μ_1 and μ_2 are the dipole moments due to two Cl-atoms.

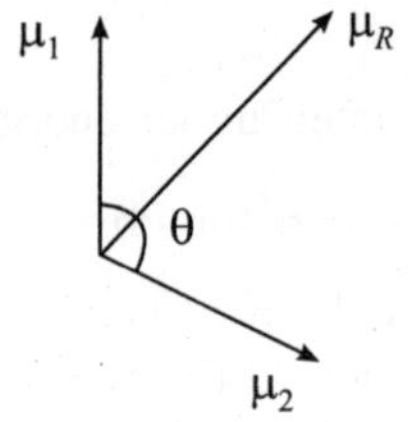

μ_R = Resultant dipole moment and θ = Angle between μ_1 and μ_2

Thus, $$\mu_R = \sqrt{\mu_1^2 + \mu_2^2 + 2\mu_1\mu_2\cos\theta}$$

In this case, $\theta = 120°$ and $\cos 120° = -\left(\frac{1}{2}\right)$.

So, $$\mu_R = \sqrt{\mu_1^2 + \mu_2^2 - \mu_1\mu_2}$$

CO_2: $\overset{\delta-}{O} = \overset{\delta+}{C} = \overset{\delta-}{O}$ $\theta = 180°$ and $\mu_1 = \mu_2 = \mu$

So, $$\mu_R = \sqrt{(\mu_1 - \mu_2)^2} = 0$$

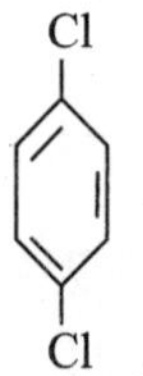

$\mu_R = 0$ (due to same reason)

Table 7.1.7 Dipole moments (μ) of some selected molecules.

Molecule	μ (*D*)	*Molecule*	μ (*D*)	*Molecule*	μ (*D*)
HBr	0.789	SO_2	1.59	CO	0.099
HI	0.381	NH_3	1.482	NO	0.069
H_2O	1.83	SO_3	0	CH_4	0
H_2S	1.02	BF_3	0	H_2	0
HCl	1.041	CO_2	0	N_2	0
HF	1.92	NO_2	0.291		

1C = 3×10^9 esu, 1m = 100 cm and 1 C-m = 3×10^{11} esu-cm.

7.1.12 Chemical bond

According to octet theory atoms of elements do not possess stable electronic configuration and hence they interact among themselves to achieve stable electronic configuration. This interaction between any two atoms leads to the formation of a bond. Any chemical bond arises mostly due to *electronic interaction*. The nature of electronic interaction determines the type of bond. So on the basis of electronic interaction bonds may be classified into *six* different types:

(i) Ionic bond
(ii) Covalent bond
(iii) Co-ordinate or dative bond
(iv) Hydrogen bond
(v) Van der Waals type of bond
(vi) Metallic bond.

(i) Ionic bond When a bond is formed between two atoms by complete transfer of electron(s) from one atom to another, the bond is called ionic bond. Usually strong electropositive element (low I.P. value) and strong electronegative element (high electron affinity) forms this type of bond, e.g., NaCl. Na-atom has one extra electron in the 3rd shell [$1s^2\ 2s^2\ 2p^6\ 3s^1$] and Cl-atom has one electron short from its octet [$1s^2\ 2s^2\ 2p^6\ 3s^2\ 3p^5$]. Complete transfer of electron takes place from Na-atom to Cl-atom to form Na^+ and Cl^-. Then stable Na^+ ion attracts stable Cl^- ions from all directions and vice versa. This electrostatic force of attraction binds both Na^+ and Cl^- ions to form an ionic compound. Hence the bond is called electrovalent or ionic bond.

Formation of an ionic bond depends on two factors

(a) Cationic charge and radius If the cationic charge is high or cationic radius is small, the cation can diffuse the electron cloud of anion, resulting in the formation of covalent bond. Thus large cationic radius and low cationic charge favour the formation of an ionic bond.

(b) Anionic charge and anionic radius If anionic radius is large or charge is high, electron cloud of the anion is easily diffused to form covalent bond. Thus small anionic radius as well as low anionic charge favour the formation of an ionic bond. Strong repulsion exists in high anionic charge, favouring diffusion of anion.
The above two factors is collectively known as Fajan's rule.

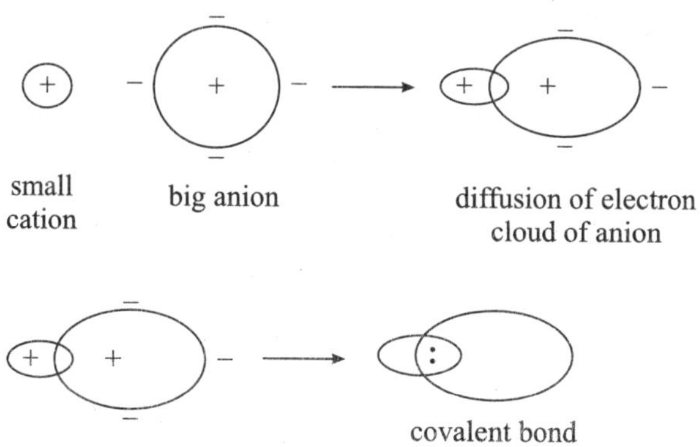

Characteristic features of ionic bonding

(a) ionic bond is formed by complete transfer of electron(s) between the constituent atoms. The ionic species, thus formed, are bound by electrostatic force of attraction.

(b) Ions can be regarded as charged spheres, continuously experience electrostatic forces, which are distributed uniformly in all directions in space. So any ion can attract oppositely charged ion from any direction. In other words, ionic bonds are non-directional.

(c) Each cation is surrounded by a definite number of anions and vice versa, as a result of which, a solid 3-dimensional regular crystal lattice is formed. Thus ionic compounds are usually hard, brittle and high melting solids.

(d) In presence of a polar solvent ionic bond breaks into constituent ions. Thus ionic compounds are usually soluble in polar solvents.

(e) In solution or molten state ions are free and hence any ionic compound appears as a good conductor of electricity in solution or molten state, although it cannot conduct electricity in the solid state.

(f) Ionic compounds undergo ionic reactions in solution. Ionic reactions are fast and instantaneous.

(g) **Molar lattice energy** Amount of energy required to dissociate one mole of an ionic crystal into its constituent ions is called molar lattice energy or in other words amount of energy evolved during formation of one mole of an ionic crystal from its constituent ions is called molar lattice energy. Also the net potential energy, contained in one mole of an ionic solid, is called molar lattice energy of that ionic crystal. Any ionic crystal contains both positive and negative ions, bound by electrostatic force of attraction. Furthermore, each positive ion is surrounded by a definite number of negative ions and vice versa. This is called coordination number. It is the total number of neighbouring anions, surrounding a cation, is called coordination number of that cation. Similarly, the total number of cations, surrounding an anion, is called coordination number of that anion. Thus two types of forces exist in an ionic crystal: (a) electrostatic forces of attraction between two oppositely charged ions and (b) repulsive forces, existing between similar ions. If the cation and anion have the charges Z_+ and Z_- respectively, the electrostatic forces of attraction for one mole of an ionic crystal is given by:

$$F = \frac{(N_A \,.\, A \,.\, Z_+ \,.\, Z_- \,.\, e^2)}{r^2}$$

and the corresponding potential energy is given by:

$$W_{\text{Attraction}} = \int_{\infty}^{r} F.dr = -\frac{(N_A \,.\, A \,.\, Z_+ \,.\, Z_- \,.\, e^2)}{r} \qquad (7.1.33)$$

where, N_A is the Avogadro number, r is the inter-ionic distance and the factor A, known as Madelung constant, is introduced since attractive force is not confined within one positive ion and one negative ion but exists between one positive ion and several negative ions, surrounding it. The above equation shows that energy due to attractive force increases with the increase in inter-ionic distance. The potential energy due to repulsive forces is given by:

$$W_{\text{Repulsive}} = \frac{B}{r^n} \tag{7.1.34}$$

where, B is a constant and the exponent n varies from 8 to 10 depending on the crystal structure. Obviously, the repulsive potential energy decreases with the increase in the inter-ionic distance. Combining the Equations (7.1.33) and (7.1.34) the net potential energy of an ionic crystal is given by:

$$V = W_{\text{Attraction}} + W_{\text{Repulsive}} = -\frac{(N_A \,.\, A \,.\, Z_+ \,.\, Z_- \,.\, e^2)}{r} + \frac{B}{r^n} \tag{7.1.35}$$

The corresponding potential energy diagram is shown in the Figure 7.1.6.

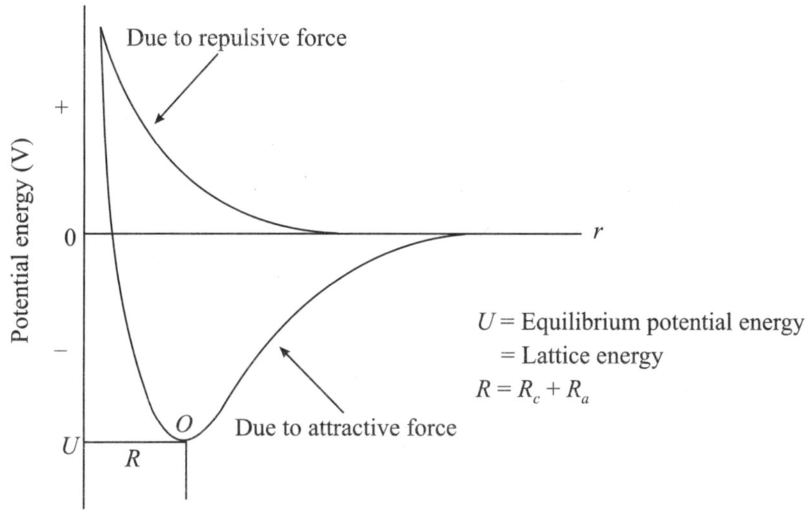

Fig. 7.1.6 Potential energy curve of an ionic compound.

It is evident from the above figure that, there is a minimum where attractive forces are exactly balance by the repulsive forces and the inter-ionic distance, R, at which the minimum occurs, is called equilibrium distance between the cation and anion. Mathematically, R is given by:

$$R = R_c + R_a \tag{7.1.36}$$

where, R_c and R_a are the cationic radius and anionic radius respectively. At the minimum, we have:

$$\left(\frac{\partial V}{\partial r}\right)_{r=R} = 0$$

Hence, we have: $$B = \frac{(N_A \,.\, A \,.\, Z_+ \,.\, Z_- \,.\, e^2 \,.\, R^n)}{(n \,.\, R)}$$

Thus at equilibrium, the net potential energy is given by:

$$U = -\frac{(N_A \, . \, A \, . \, Z_+ \, . \, Z_- \, . \, e^2)}{R} + \frac{B}{R^n} \tag{7.1.37}$$

Putting the value of B in the Equation (7.1.37), we get:

$$U = -\frac{(N_A \, . \, A \, . \, Z_+ \, . \, Z_- \, . \, e^2)}{R}\left(1 - \frac{1}{n}\right) \tag{7.1.38}$$

The value of Madelung constant A, varies from 1.5 to > 5.0, depending on the crystal structure. U, by definition, is called molar lattice energy.

U_{Obs} and U_{Th} The observed value of molar lattice energy can be evaluated experimentally and theoretical value of molar lattice energy can be determined by using Born-Haber cycle, which is based on a cyclic process of formation of an ionic crystal as shown below. Let us consider a general ionic salt, M_aX_b, where, M is usually metal or an electropositive element like Na etc. and X is usually a diatomic electronegative element, like Cl_2 etc. ΔH_f is the molar heat of formation of the ionic solid M_aX_b, ΔH_s is the molar heat of sublimation of the metal M, ΔH_d is the molar heat of dissociation of X_2, I is the molar ionization potential of the metal M, E is the molar electron affinity of the radical X^* and U is the molar lattice energy of the ionic crystal M_aX_b.

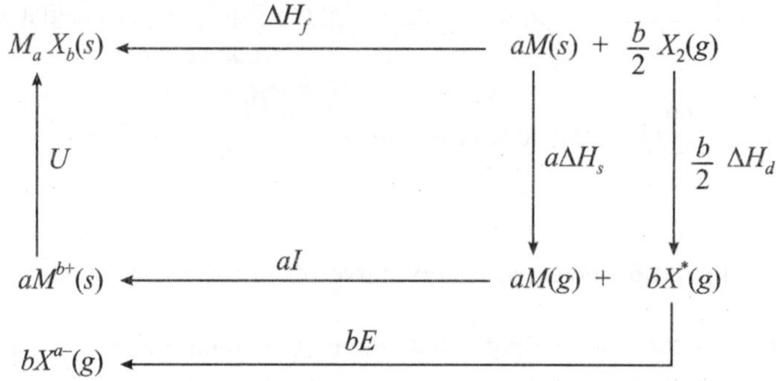

As we know that enthalpy is a state function, we can write that:

$$\Delta H_f = a\Delta H_s + \frac{b}{2}\Delta H_d + aI + bE + U \tag{7.1.39}$$

where, I = amount of energy required to form 1 mole of M^{b+} from 1 mole of M(g).

E = amount of energy released to form 1 mole of X^{a-} from 1 mole of X^*(g).

From the equation the lattice energy (U) can easily be evaluated if other quantities are known. In the above cycle the compound M_aX_b (s) is considered as a perfect ionic compound. Thus the lattice energy is the theoretical lattice energy. The above cycle is known as Born-Haber cycle.

Significances of lattice energy

(i) As evident from the Equation (7.1.38) $|U| \propto (Z_+.Z_-)$. Thus magnitude of lattice energy increases with the increase in charge on ions.

(ii) $|U| \propto \left(\frac{1}{R}\right)$. Thus magnitude of lattice energy decreases with the increase in the inter-ionic distance.

(iii) Percent ionic character of an ionic crystal can be determined by evaluating U_{Obs} and U_{Th}.

$$\% \text{ Ionic character} = \frac{U_{\text{obs}}}{U_{\text{Th}}} \times 100$$

(iv) **Solubility** Solubility of a salt depends on two factors: lattice energy and solvation energy. Solvation energy is the amount of energy released during solvation of an ion, e.g.,

$$Na^+ + n(H_2O) \rightarrow [Na(H_2O)_n]^+ + S \text{ kJ/mole.}$$

This S is known as solvation energy or hydration energy because here water is the solvent. Like lattice energy (U) solvation energy increases with the increase in ionic charge and decreases with the increase in inter-ionic distance. Thus when $U > S$, solvation energy is unable to break the ionic crystal and the salt is insoluble in the solvent. On the other hand, when $U < S$, solvation energy is sufficient to break the crystal and the salt is soluble in solvent. For large anions, like SO_4^{2-}, CO_3^{2-} etc. decrease in lattice energy (U) with the increase in cationic radius is not remarkable whereas decrease in solvation energy with the increase in cationic radius is significant and hence solubility of salt decreases with the increase in cationic radius. One such example is: $BeSO_4$ and $MgSO_4$ are soluble in water but $CaSO_4$, $SrSO_4$ and $BaSO_4$ are totally insoluble in water. Similar observation is also made in the case of carbonate salts. For small anions, like F^-, OH^-, decrease in lattice energy (U) with the increase in cationic radius is significant and hence lattice energy becomes determining factor of a salt to be dissolved into water, e.g., BeF_2 and MgF_2 are insoluble in water since lattice energy is too large to be overcome by hydration energy due to small size of Be^{2+} and Mg^{2+}, whereas CaF_2, SrF_2 and BaF_2 are soluble in water. In the latter case, hydration energy of cation supersedes the corresponding lattice energy of salt.

(h) **Melting point and hardness** Hardness: It is the property of a substance to resist indentation on application of load. Both the melting point and hardness of an ionic solid depend on two factors: (i) charges on the ions: higher the charges on the ions higher will be the melting point and hardness, (ii) inter-ionic distance: lower the inter-ionic distance higher the melting point and hardness. For a given anion inter-ionic distance increases down the group and hence hardness decreases, e.g., hardness of CaO > SrO > BaO. For a given inter-ionic distance hardness increases with the increase in charges on the ions, e.g., in LiCl and SrO, inter-ionic distance in both the cases is 2.57 Å but hardness of SrO is much higher than that of LiCl as SrO has doubly positive and doubly negative charges but LiCl has singly positive and singly negative charges. Due to similar reason NaCl has melting point about 800°C whereas BaO has melting point 1921°C, although inter-ionic distance is nearly same in both the cases (for NaCl inter-ionic distance is 2.81 Å and that for BaO is 2.71 Å). Like hardness for a given anion melting point decreases down the group due to increase in inter-ionic distance, e.g., melting points of NaF, NaCl, NaBr and NaI decrease as follows: 1274°K for NaF, 1073°K for NaCl, 1023°K for NaBr and 925°K for NaI.

(i) **Coordination Number (CN) and Critical Radius Ratio (CRR)** In any ionic crystal each cation is surrounded by a definite number of anions and vice versa. The corresponding number of surrounding anions is called coordination number of that cation and the corresponding number of surrounding cations is called the coordination number of anion. As discussed earlier at equilibrium position attractive force between oppositely charged ions is exactly balanced by repulsive force among anions and the interionic distance (R) is given by: $R = R_c + R_a$, where R_c and R_a are the radii of the cation and anion respectively. At this equilibrium condition the radii ratio of cation to anion $\left(\frac{R_c}{R_a}\right)$ is often called critical radius ratio or CRR.

Significances of CRR

(a) With the increase in CRR coordination number (CN) increases as shown in the following table:

Table 7.1.8 CRR and CN values.

CRR	*CN*
CRR < 0.155	2
0.155 < CRR < 0.225	3
0.225 < CRR < 0.414	4
0.414 < CRR < 0.732	6
0.732 < CRR < 1.000	8

(b) For a given ionic crystal if CN and one of the ionic radii are known, radius of the other ion and the inter-ionic distance may be evaluated.

Limitations of CRR rule

(a) To find out the CRR all ions are considered as spheres, but in actual practice geometry of an ion depends on polarization. Higher is the polarization, more will be the deviation from spherical shape.

(b) To find out CRR of an ionic crystal all ions are assumed to touch each other at equilibrium position and interionic distance 'R' is given by: $R = R_c + R_a$. That limiting situation cannot be achieved in many ionic crystals as the cationic size is either smaller than or greater than the limiting size as shown in the following figure.

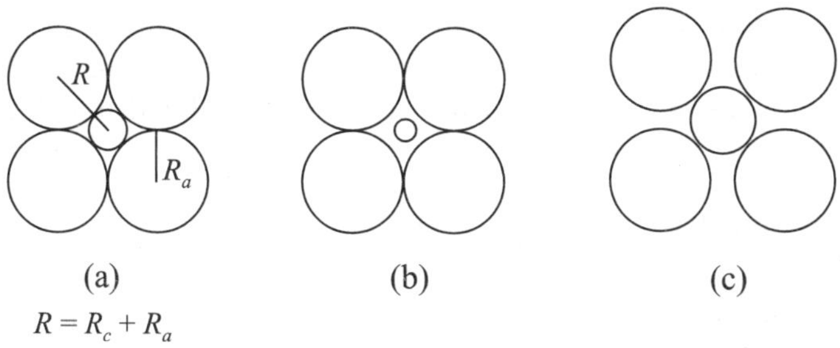

Fig. 7.1.7 (a) Critical Radius Ratio, (b) Deviation of CRR, cationic size is small, (c) Deviation of CRR, cationic size is large.

(c) Due to shielding effect cationic radius decreases significantly but this effect is not considered during calculation of CRR.

(ii) Covalent bond When a bond is formed between two atoms by sharing of electrons, the bond is called covalent bond. Covalent bond is usually formed through overlapping of two atomic orbitals. Shapes of *s*-, *p*- and *d*-orbitals are shown in the Figures. 7.1.3(a) and (b). Electron is believed to have both wavy and particulate characteristics. The wavy characteristics of an electron is assumed to follow sinusoidal curve. If two opposite electronic waves overlap [Figure 7.1.4(b)] the resultant electron density is zero in the overlapping region and the situation is known as antibonding situation. Bonding will occur if and only if two similar electronic waves overlap with each other [Figure 7.1.4(a)]. Furthermore, bonding overlap may be of two types.

(a) Head-on-overlap When both the orbitals are approaching along the axis, head-on overlap takes place and the corresponding bond formed is called σ-bond [Figure 7.1.8(a)].

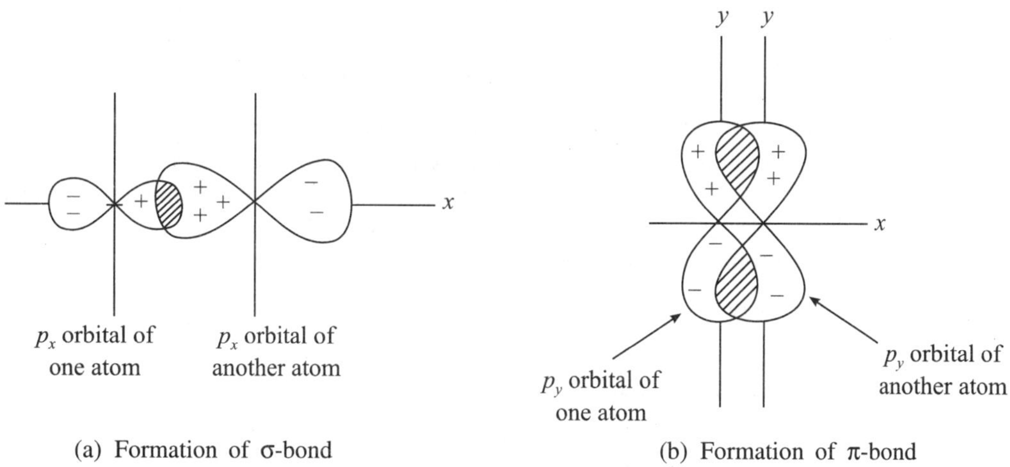

Fig.7.1.8 (a) Head-on overlap: formation of σ-bond
(b) Lateral overlap: formation of π-bond.

(b) **Lateral overlap** When two orbitals overlap perpendicular to the axis of approach, a partial overlap takes place above and below the axis of approach. This type of overlap is known as lateral overlap and the corresponding bond is called. π-bond [Figure 7.1.8(b)].

Characteristic features of covalent bonding:

(a) Covalent bonding is formed by sharing of electrons between constituent atoms.
(b) Covalent bond is formed when two orbitals are approaching along the same axis. Thus covalent bond is highly directional.
(c) Usually molecules of covalent compounds remain in discrete phases and hence covalent compounds exist in gaseous state. There are some covalent compounds where molecules get attached by secondary forces, as a result of which, they exist either in liquid (e.g. water) or in low-melting solid (e.g. iodine) state.
(d) When a covalent bond is formed between two hetero atoms, bond pair of electrons is shifted to more electronegative atom, inducing polarity in the compound. Hence compound achieves some ionic character and the % ionic character induced depends on electronegativity difference between two hetero atoms, as shown in the following chart.

Electronegativity difference	0.1	0.4	0.9	1.4	1.9	3.0	3.3
% ionic character	0.5	4.0	19	39	60	90	100

(e) Non-polar or slightly polar compounds are usually soluble in non-polar solvents like carbon tetra chloride, benzene, ether etc. Highly polar compounds are usually soluble in polar solvents like water, chloroform etc.
(f) Usually covalent compounds are non-conductor of electricity. However, highly polar compounds can conduct electricity but they are not as good conductor as ionic compounds.
(g) In solution covalent compounds undergo molecular reactions, which are slow.

(iii) **Co-ordinate or dative bond** Sometimes it may so happen that lone pair of one atom is equally shared by two atoms. The resultant bond, thus formed, is called coordinate bond. In a coordinate bond one species (atom/molecule/anion) shares its own lone pair of electrons and is called donor species, while the other species accepts the lone pair of electrons of donor to its vacant orbital, and is called acceptor species.

Coordinate bond is of two types:

(a) intermolecular coordinate bond and
(b) intramolecular coordinate bond.

In intermolecular coordinate bond one molecule of a compound shares its lone pair with another molecule of different compound.
Donor molecule is commonly known as Lewis's base and acceptor molecule is known as Lewis's acid. Examples are:

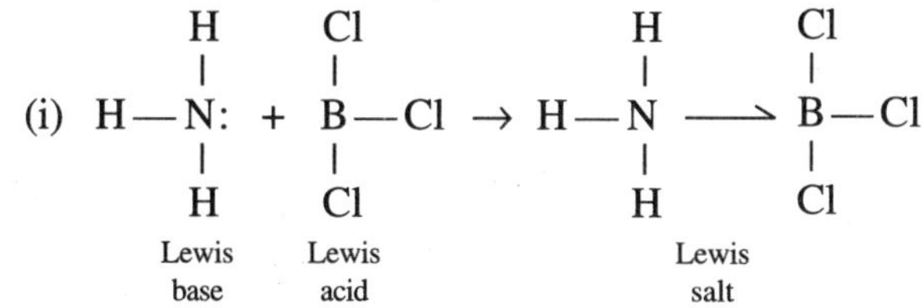

(ii) $Cl^- + FeCl_3 \rightarrow [Cl \longrightarrow FeCl_3]^-$

Lewis base Lewis acid Lewis salt

In intramolelcular coordinate bond, coordinate bond is formed within a molecule. Examples are:

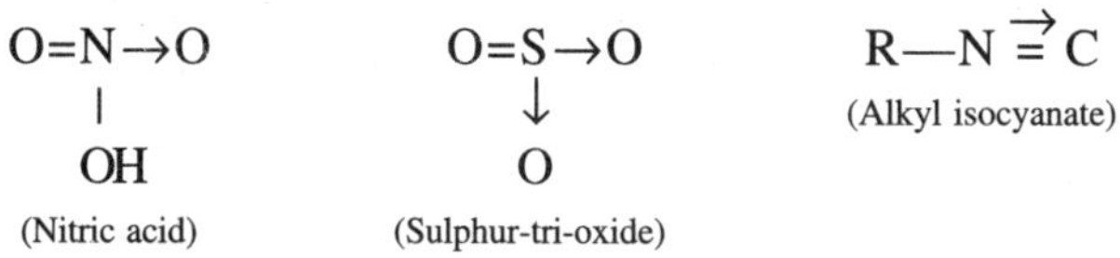

(iv) Hydrogen bond When hydrogen atom is covalently bonded with a strongly electronegative atom such as, F, O, N, Cl etc., polarity or charge separation of the molecule is sufficient to bring positively charged H-atom of one molecule close to an electronegative atom of different molecule of same compound, resulting in the formation of a unique linkage through hydrogen atom. This linkage is called hydrogen bond, e.g., hydrogen bond in HF.

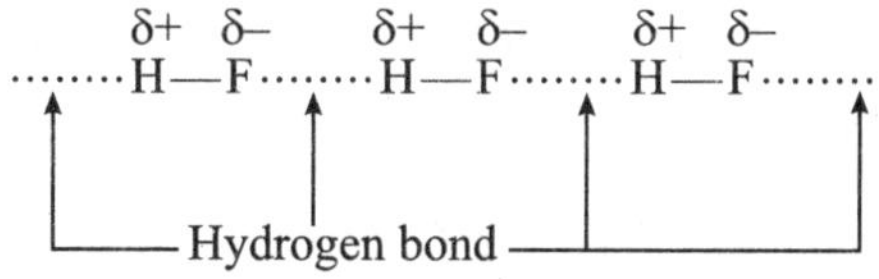

Characteristic features of hydrogen bond:

(a) Strength of hydrogen bond is very less and usually lies within 3 – 10 kcal/mole, whereas the strength of covalent bond is of the order of 100 kcal/mole.
(b) Hydrogen bond is directional and electrostatic in nature.
(c) There are three types of hydrogen bonds usually observed in different compounds, e.g., (i) intermolecular, (ii) intramolecular, (iii) interatomic.

(i) **Intermolecular hydrogen bond** when hydrogen bond is formed between two different molecules, it is called intermolecular hydrogen bond.
One example is *p*-nitrophenol.

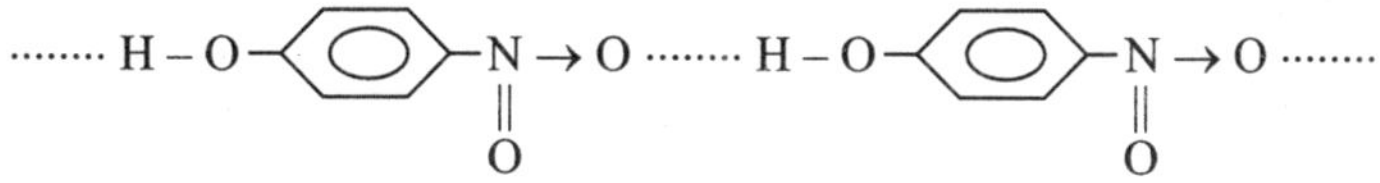

(ii) **Intramolecular hydrogen bond** when hydrogen bond is formed within a single molecule it is called intramolecular hydrogen bond, e.g. *o*-nitrophenol.

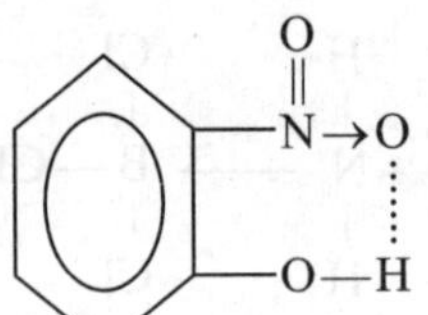

(iii) **Interatomic hydrogen bond** In some cases hydrogen bond is strong enough to abstract the electronegative atom from another molecule and a stable anion is formed. One example is hydrogen fluoride $[(HF)_n]$.

$$\cdots\cdots H{-}F\cdots\cdots H{-}F\cdots\cdots H{-}F\cdots\cdots \rightarrow [F\cdots H\cdots F]^-$$

$[HF_2]^-$ is very much stable, $K[HF_2]$ crystal is known. In $[HF_2]^-$ three atoms are attached through hydrogen bond. This type of hydrogen bond is called interatomic hydrogen bond. Hydrogen bond is the strongest in the case of interatomic hydrogen bond and the corresponding bond energy varies from 20 kcal/mole to 36 kcal/mole.

(d) **Association** There are some compounds where hydrogen bond is confined within two or three molecules. Those compounds exist as dimeric or trimeric form, e.g., benzoic acid exists in dimeric form in benzene.

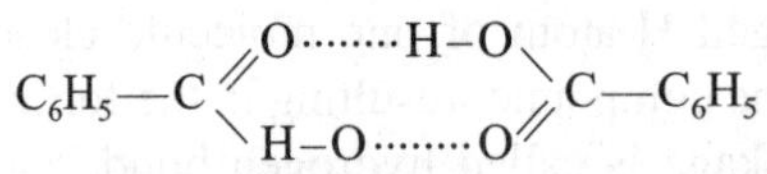

(e) If hydrogen bond is strong enough to associate numerous molecules at a time boiling point or melting point of that compound increases abnormally. Among hydrogen halides hydrogen bond is the strongest in hydrogen fluoride (molecular formula HF) . Numerous molecules of HF are attached through hydrogen bond, as a result of which, hydrogen fluoride appears as liquid of b.p. 19°C. Strength of hydrogen bond is weaker in other hydrogen halids (HCl, HBr, HI) which are all gases. Due to the same reason, water (molecular formula H_2O) is a liquid of b.p. 100°C but hydrogen sulphide (H_2S) is a gas. In case of water, hydrogen bond, under certain condition, may extend tetrahedrally in four directions, resulting in the formation of solid ice crystals of m.p 0°C.

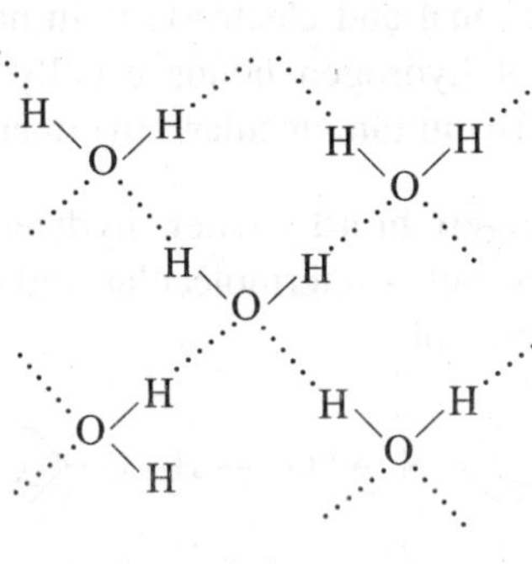

As the hydrogen bonds are longer the ice structure is quite loose and contains sufficient empty space. This is the primary reason for its low density and capacity to form *clathrate compounds.*

(f) New compound may be formed through hydrogen bond. One example is ammonia hydrate. Aqueous ammonia, when cooled, forms the crystals of $H_3N.H_2O$, where ammonia and water are bound by hydrogen bond.
Formation of hydrogen bond and its breakage take place easily at normal temperature and this process is believed to be essential for biological process.

(g) **Solubility** If molecules of a polar solvent are able to form hydrogen bond with solute molecules, the latter partly or totally dissolves into the solvent. Phenol is sufficiently soluble in water but thiophenol is only sparingly soluble in water because water forms strong hydrogen bond with phenol but not with thiophenol due to low electronegativily of S-atom.

(v) **Bond due to intermolecular forces (Van der Waals type of bond):** Apparently dipole moment of a nonpolar molecule (e.g., Br_2) is zero. A molecule is a dynamic system where electrons are continuously revolving around nuclei which are in constant oscillation. Hence distribution of charges surrounding the nuclei may not be uniform, as a result of which, electron cloud shifts towards either atom with equal probability $\left(Br_2 \equiv \overset{\delta+}{Br}—\overset{\delta-}{Br} \text{ and } \overset{\delta-}{Br}—\overset{\delta+}{Br}\right)$, rendering the formation of instantaneous microdipoles. As the probability of shifting electron cloud towards either atom is equal, the resultant dipole moment is zero. The positive charge of one instantaneous microdipole attracts the electron cloud of another molecule, rendering the formation of new instantaneous microdipole. These *fluctuating dipoles* are responsible for developing mutual attraction among the molecules. This is known as intermolecular forces of attraction and the hypothetical bond developed among the molecules is known as Van der Waals type of bond. It is mostly observed in non-polar or slightly polar compounds.

Characteristic features of Van-der-Waals type of bond:

(a) Van-der-Waals type of bond is very weak and energy associated with it only about 1 kcal/mole whereas 100 kcal/mole of energy is required to disrupt covalent forces.
(b) Van-der-Waals type of forces are operative when intermolecular distance is about 3.5Å to 5Å.
(c) The magnitude of Van-der-Waals force of attraction increases with increase in number of electrons per molecule and hence with increase in molecular mass. In the cases of heavy molecules, magnitude of intermolecular forces of attraction is high enough to condense discrete molecules into solids, e.g. iodine and sulphur.

(vi) **Metallic bond** In the case of electronegative elements covalent bonds are formed by sharing of valence electrons but in the case of metals this sharing of electrons is not possible due to their electropositive character. Thus to fulfil their octet numerous metal atoms get attached in a regular manner to form a three dimensional crystal lattice. The lattice energy, thus evolved, is enough to dislodge the valence shell electrons from all metal atoms. These mobile electrons are not free enough to leave the metallic crystal but can move throughout the crystal lattice as gas molecules. This cloud of valence electrons is often called *electron gas* and positive metal ions, thus created, are called *kernels* [Fig.7.1.9].

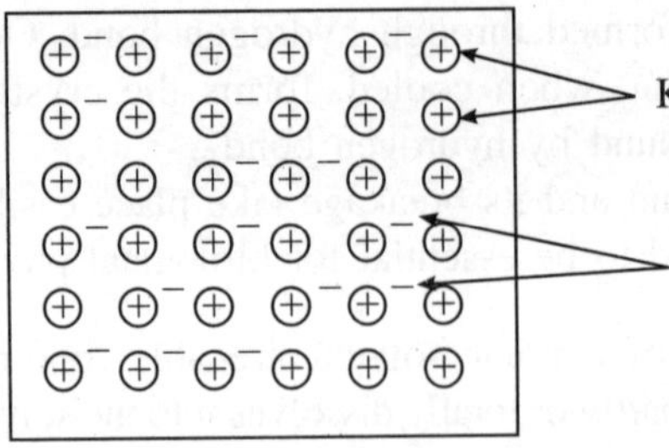

Fig. 7.1.9 Schematic diagram of metal lattice.

All the kernels are linked through electron gas and the bond is called metallic bond.

Merits and demerits of free electron theory

(a) **Metallic lustre** Loosely bound electrons or electron gas readily reflect sunlight, striking on the metal surface. Metals are appearing lusturous due to this reflection of light.
(b) Mobile electrons in a metal are indistinguishable and hence each and every kernel makes an attempt to share all the electrons but not at a time. Thus on application of external stress, layers of kernels slip past one another, as a result of which, metals become malleable and ductile. However, presence of a foreign atom restricts the layer slip and hence brittleness appears. Thus copper is ductile but brass is brittle.
(c) The free moving electron gas is responsible for high electrical and thermal conductivity of metals. Furthermore, thermionic and photoelectric emissions are well explained with the help of electron gas theory.
(d) Electron gas theory fails to explain the specific heat of a metal. The specific heat of an ideal gas, according to kinetic theory, is 3 cal/mole. Specific heat of a metal is expected to be very high due to the presence of free electron gas but in actual practice, specific heat of any metal is observed around the same as that of an ideal gas.

7.1.13 (A) Hybridisation

Electron clouds of different orbitals have different shapes. These electron clouds can overlap in different ways to form σ–, π– and δ– bonds. σ-bond is formed by overlapping of electron clouds along the axis, π-bond is formed by lateral overlap of electron clouds above and below the bond axis, whereas δ-bonds are formed by the overlapping of all four wings of δ-electron clouds arranged in parallel planes. It is well illustrated in the following figure.

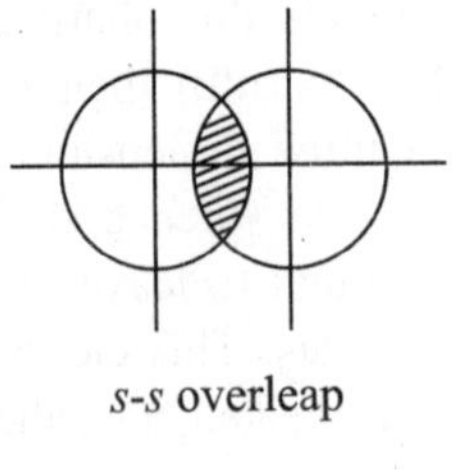

s-s overleap

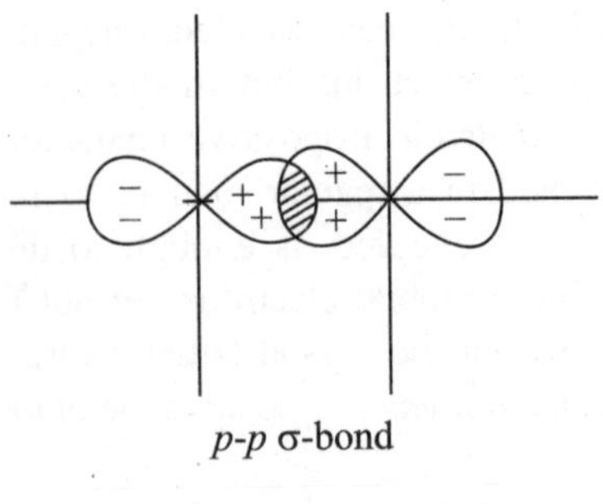

p-p σ-bond

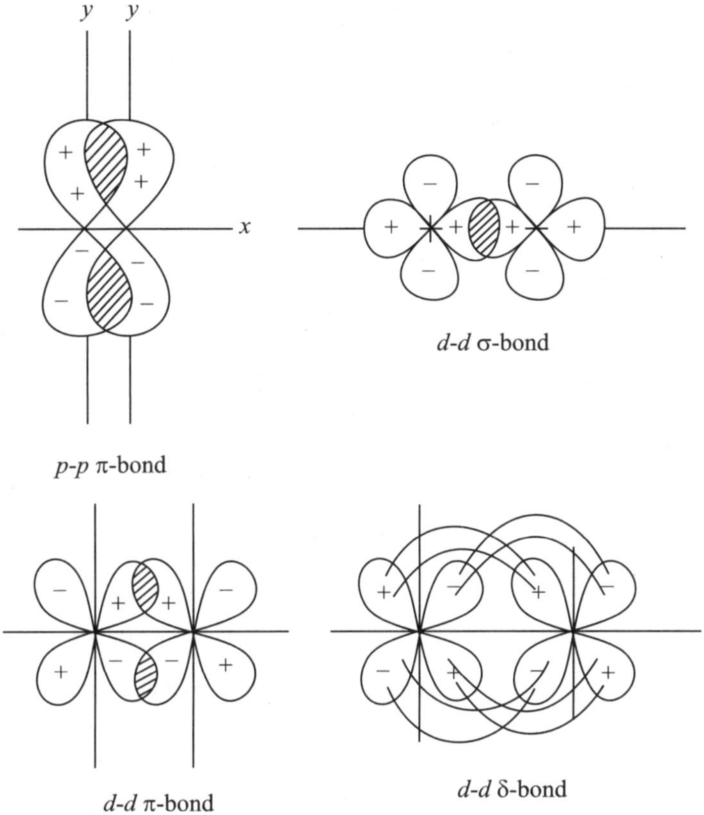

Fig. 7.1.10 Schematic diagram of overlapping of orbitals to form σ–, π– and δ–bonds.

The above figure illustrates the overlapping of similar orbitals (*s–s*, *p–p* or *d–d*) of two atoms but usually atoms form bonds by means of electrons of different energy levels. The question is how electrons with different energy levels form equivalent chemical bonds. Linus Pauling (1931) made an attempt to answer this question with the concept of hybridization of valence orbitals. According to him before formation of any σ-bond mixing or redistribution of valence shell orbitals of an atom takes place to form new orbitals of equivalent energy. These new orbitals are known as *hybrid orbitals*. The process is known as *hybridization* and the amount of energy required for hybridization is called hybridization energy. Only a fraction of energy, evolved during bond formation, is enough for hybridization. Only σ-bonds are resulted from hybrid orbitals. Hybridization of *s*- and *p*-orbitals is well illustrated in the following figure.

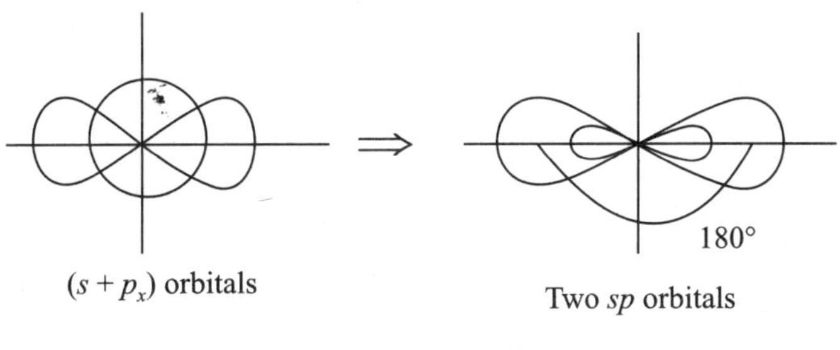

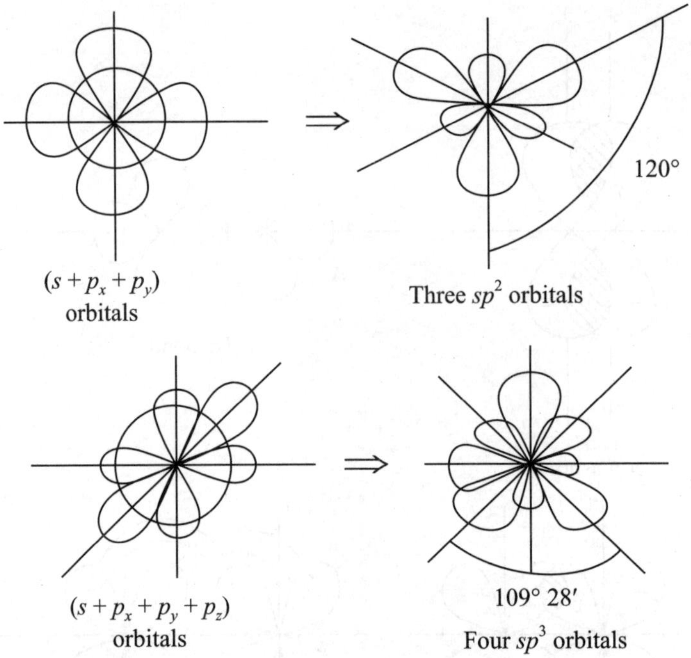

(a) Diagramatic representation of forming hybrid orbitals.

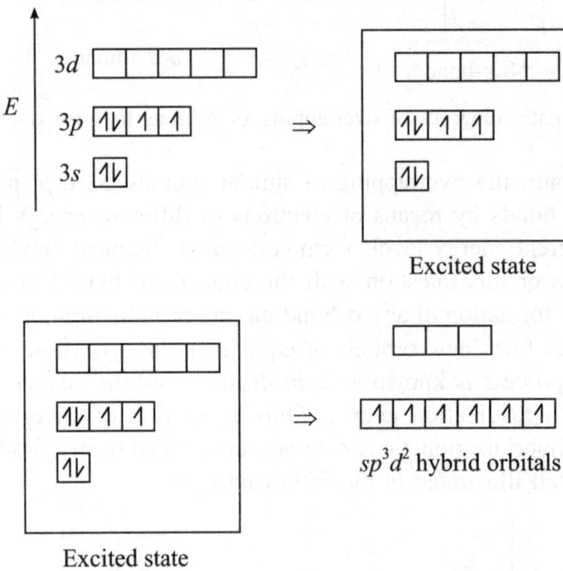

(b) Schematic representation of forming hybrid orbitals.

Fig. 7.1.11 Hybridization of valence orbitals.

Characteristics features of hybridization

(1) The number of hybrid orbitals generated is equal to the number of pure atomic orbitals that participate in hybridization process.
(2) Hybridized orbitals are equivalent in energy and shape.
(3) Only a fraction of energy evolved during bond formation is enough for hybridization of atomic orbitals.
(4) Type of hybridization indicates the geometry of the molecule.
(5) Hund's rule of maximum multiplicity and Pauli's Exclusion Principle are applicable in filling up hybrid orbitals with electrons.
(6) Only σ-bonds are resulted from hybrid orbitals.

(B) The spatial configuration at different states of hybridization

(i) The simplest hybridization is *sp*-hybridization. Example: $BeCl_2$.
Outermost electronic structure of Be:

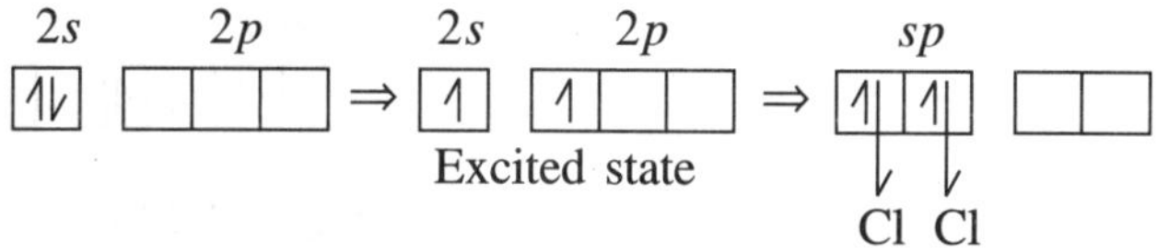

Shape: Cl—Be—Cl Linear Bond angle: 180°

(ii) State of hybridization = sp^2. Example: BF_3.
Outermost electronic configuration of B:

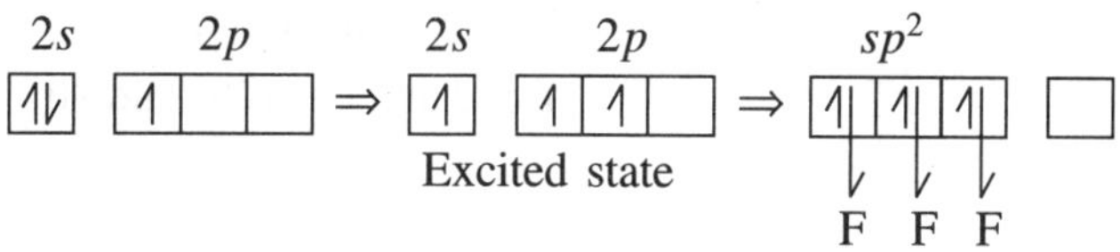

Shape:

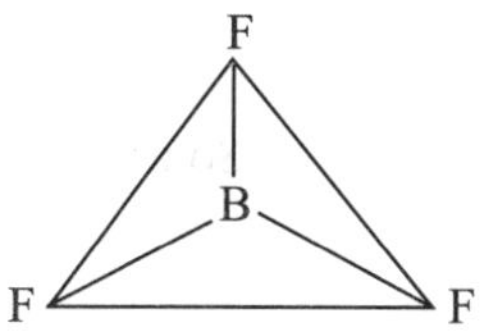

Triangular and planar

Bond angle = $\frac{360°}{3}$ = 120°

(iii) State of hybridization = sp^3. Example: CH_4.
Outermost electronic configuration of C:

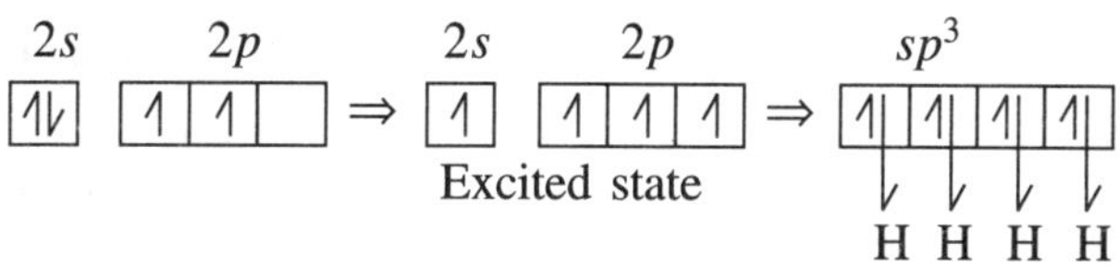

Shape:

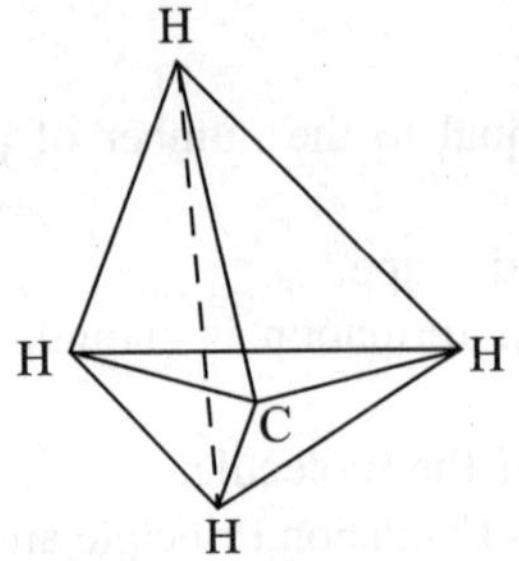

3-dimensional and tetrachedral

Bond angle = 109° 28′

(iv) State of hybridization: sp^3d. Example: PCl_5
Outermost electronic configuration of *P*:

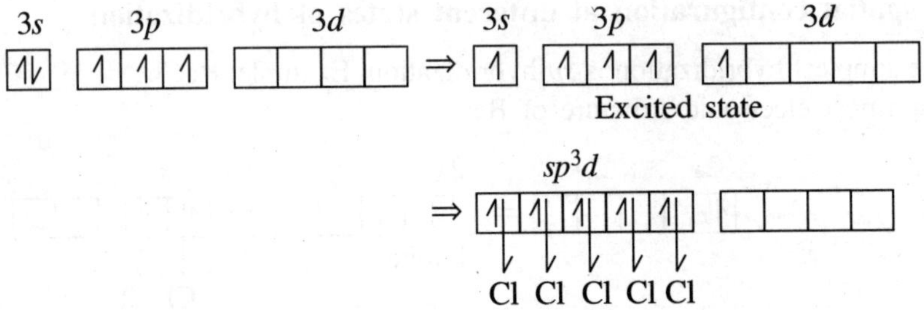

Shape:

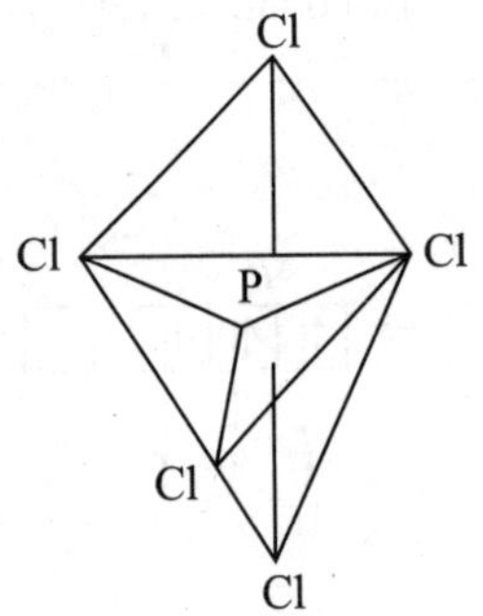

Trigonal bipyranid

Bond angle = 120° and 90°

Two chlorine atoms lie perpendicular to the plane triangle. So angle is 90°. So, bond angle: 120° and 90°.

(v) State of hybridization: sp^3d^2. Example: SF_6.
Outermost electronic configuration of *S*:

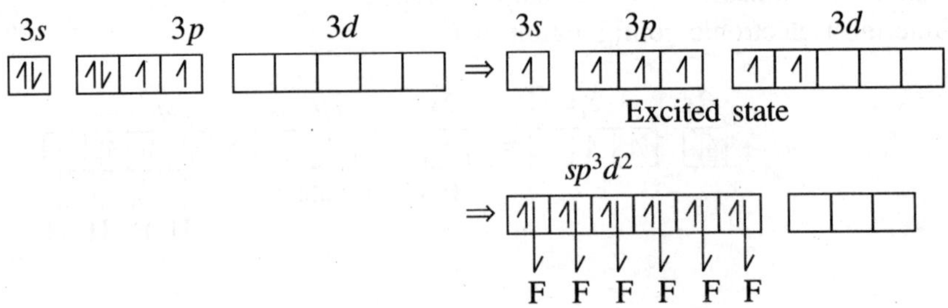

Shape:

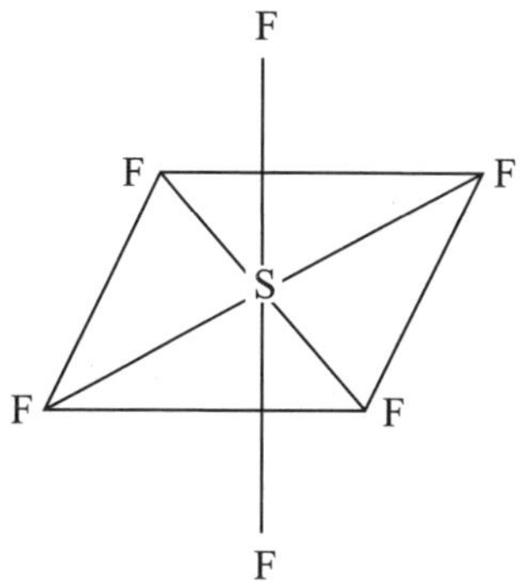

3-dimensional octahedron

Bond angle = 90°

Bond angle Four fluorine atoms forms the square. Bond angle = $\frac{360°}{4}$ = 90°.

Two fluorine atoms lie out of plane so bond angle is again 90°.

So, bond angle: 90°.

(vi) State of hybridization: sp^3d^3. Example: IF_7.

Outermost electronic configuration of I:

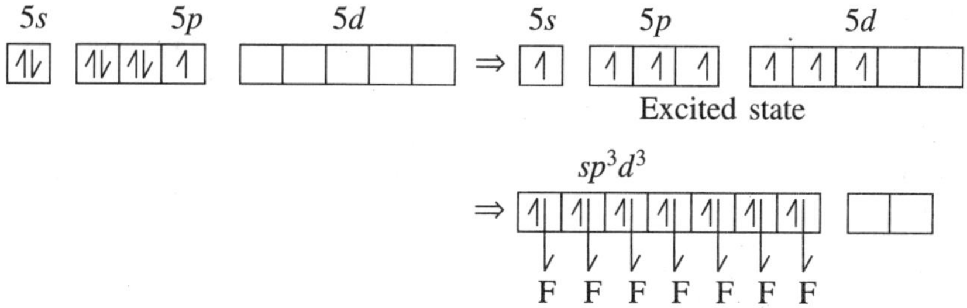

Shape:

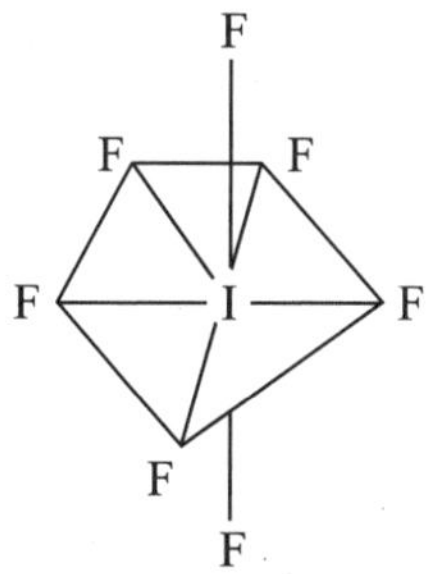

3-dimensional pentagonal bipyramid

Bond angle = $\frac{360°}{5}$ = 72° and 90°

Bond angle Five fluorine atoms lie in one plane. Bond angle = 72°

Two fluorine atoms lie out-of-plane. Bond angle = 90°

(C) VSEPR theory Shape of a covalent compound depends on three factors: (a) hybridization, (b) type of repulsion: lone pair (LP)-lone pair (LP), lone pair (LP)-bond pair (BP) and bond pair (BP)-bond pair (BP), (c) Angle of repulsion: 90°, 120° and 180°. The geometry of many covalent compounds can be well explained with the help of Valence Shell Electron Pair Repulsion theory.

Postulates of VSEPR theory

(1) The electron pairs have a tendency to stay apart as far as possible to achieve maximum stability.

(2) If the central atom of a molecule is surrounded by only bond pair of electrons, shape and geometry of that molecule will be regular.

(3) Geometry of a molecule is distorted if the central atom is surrounded by both lone pairs (LP) and bond pairs (BP). However, distortion of geometry depends on degree of repulsions among electron pairs. The following series is effective in determining degree of repulsion:

LP – LP repulsion > LP – BP repulsion > BP – BP repulsion.

BP-BP repulsion may be ignored with respect to the other two. However, magnitude of BP-BP repulsion depends on the electronegativity difference between the central atom and the other atoms. Furthermore, magnitude of repulsion depends on bond angle and in fact decreases with increase in bond angle according to the following sequence:

90° repulsion > 120° repulsion > 180° repulsion.

(4) With increase in electronegativity of an atom in a molecule bond angle decreases as bond pair of electrons recedes from the central atom, e.g., bond angle in PCl_3 is 100°, that of PBr_3 is 101.5° and that of PI_3 is 102°.

(5) Electronic repulsion is higher in completely filled shells than in incompletely filled shells.

Limitations

(a) This theory fails to explain the shape of the molecules or ions having extensive delocalized π-electron system.

(b) Shape of highly polar molecules and transition metal complexes cannot be explained with the help of this theory.

(c) It is applicable only to simple covalent compounds but it cannot explain ionic bonding. The shape of a covalent compound can be determined by hybridization of the central atom but the actual position of atoms can be determined by considering the type of repulsion and angle of repulsion. However, BP-BP repulsion and 120°, 180° repulsions are very weak and ignored. There are several compounds, where repulsions are needed to be considered. One such example is ClF_3.
Outermost electronic configuration of Cl atom:

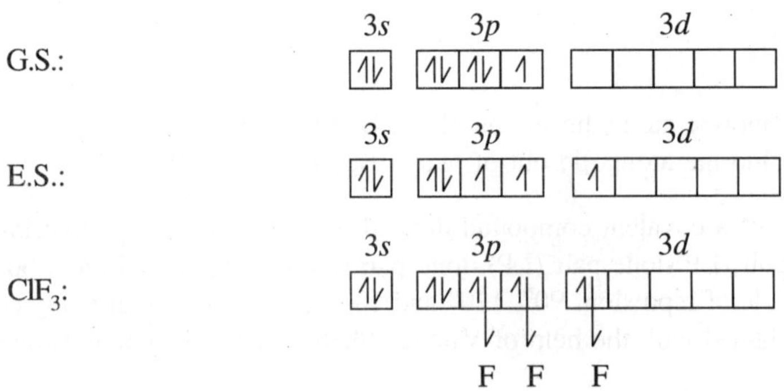

Hybridization of Cl-atom is sp^3d. Shape of ClF_3 is trigonal bi-pyramid. The following three different arrangements are possible for ClF_3:

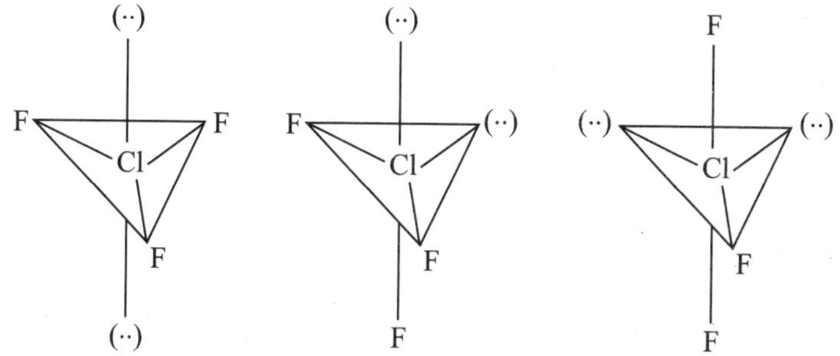

(I) (II) (III)

Structure-I Three F-atoms are in one plane with angle 120° with each other. Two lone pairs are out-of-plane making an angle 90° with each F-atom.
LP-BP repulsions = 6 numbers.

Structure-II One lone pair and one F-atom are out-of-plane.

LP-LP repulsion = 1 number

LP-BP repulsion = 3 numbers.

Structure-III Two F-atoms are out-of-plane.

LP-BP repulsion = 4 numbers.

Maximum number of repulsions exists in structure-I and hence it must not represent the compound ClF_3. In both the structures II and III number of repulsions is 4 but in structure-II there is one LP-LP repulsion, which is absent in structure-III. So, structure-III possesses the least energy and hence represents the compound ClF_3.

7.1.14 Facts, problems and explanation

P. 1 Among boron halides BF_3 should be the strongest Lewis acid followed by BCl_3, BBr_3 etc., as the electronegativity decreases F > Cl > Br, but actually strength of Lewis acid follows the reverse order: $BBr_3 > BCl_3 > BF_3$ – explain.

Ans. The outermost electronic configuration of *B* is:

B: 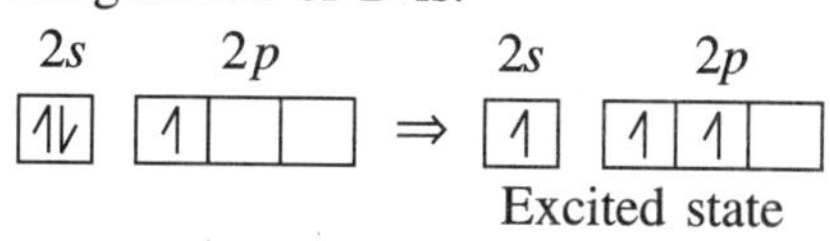

BX_3: 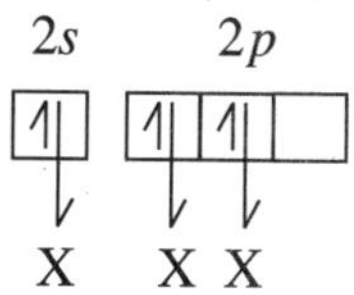

Thus B in BX_3 has vacant p_z orbital and hence has a strong tendency to capture a pair of electrons to fulfil its octet. This vacant p_z orbital is the origin of Lewis acid characteristics of BX_3 compounds. In BF_3 bond pair is shifted to fluorine due to strong electronegativity of fluorine. So nucleus of each fluorine atom is surrounded by a thick electron cloud, developing repulsive forces within it (electron cloud). Thus fluorine has a tendency to diffuse its electron cloud to the vacant p_z orbital to form $p\pi$-$p\pi$ back bonding as follows:

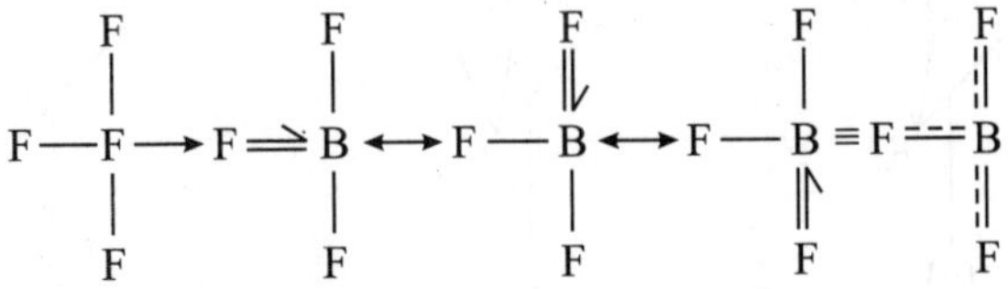

This $p\pi$-$p\pi$ back bonding is the strongest in BF_3, weak in BCl_3 and almost non-existent in BBr_3 due to the following two reasons:

(a) Electronegativity sharply falls from fluorine to bromine.
(b) As the size of halogen atom increases repulsive forces within electron cloud decreases from BF_3 to BCl_3 to BBr_3.

Thus availability of vacant p_z-orbital of boron in BX_3 is maximum in BBr_3 and minimum in BF_3. As we know higher the availability of p_z-orbital of B, higher the acid strength of BX_3, the order of Lewis acid strength is: $BBr_3 > BCl_3 > BF_3$.

P. 2 CCl_4 does not undergo hydrolysis but $SiCl_4$ does – explain.

Ans. In order to hydrolyze CCl_4 carbon atom of CCl_4 has to accept lone pair of oxygen of water molecule, but C-atom in CCl_4 has no spare orbital to accommodate the lone pair of electrons of oxygen of water molecule. So CCl_4 does not undergo hydrolysis.
In the case of $SiCl_4$, Si-atom has available vacant 3*d*-orbital to accept the lone pair of electrons of oxygen of water molecule.

C in CCl_4

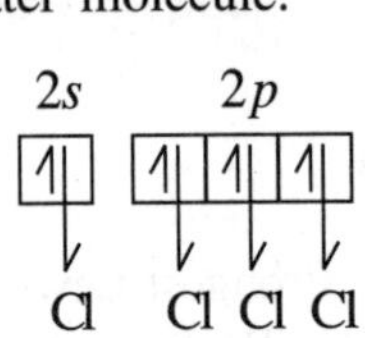

Si in $SiCl_4$

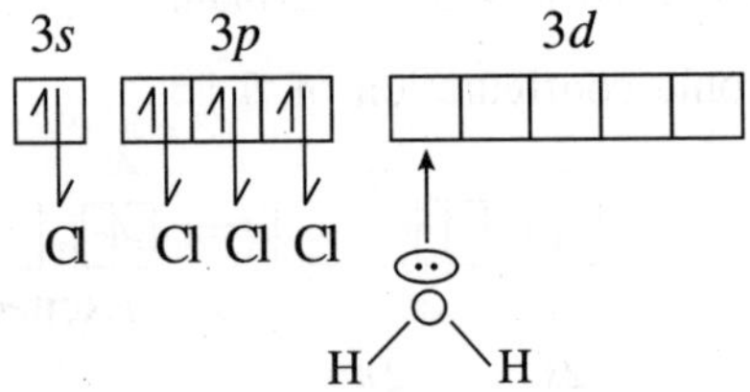

So $SiCl_4$ undergoes hydrolysis according to the following reaction:

$$SiCl_4 + 4H_2O \rightarrow \underset{\text{Orthosilicic acid}}{Si(OH)_4} + 4HCl$$

P. 3 Hydrolysis of NCl_3 gives ammonia (NH_3) but hydrolysis of PCl_3 gives phosphoric acid H_3PO_3 — explain.

Ans. Nitrogen is more electronegative than chlorine. Furthermore, nitrogen in NCl_3 does not have any spare orbital to accept lone pair electrons of oxygen of water molecule.

N in NCl_3

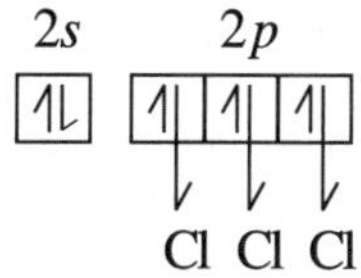

So in presence of water the following reaction may take place.

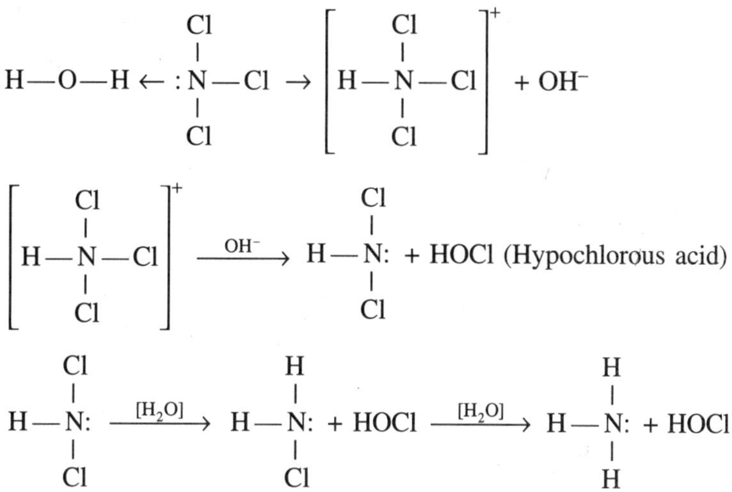

The overall reaction is: $NCl_3 + 3H_2O \rightarrow NH_3 + 3HOCl$

In PCl_3, phosphorous has available vacant 3*d*-orbital to accept lone pair of electrons of oxygen of water. Furthermore, chlorine is far more electronegative than phosphorous.

P in PCl_3

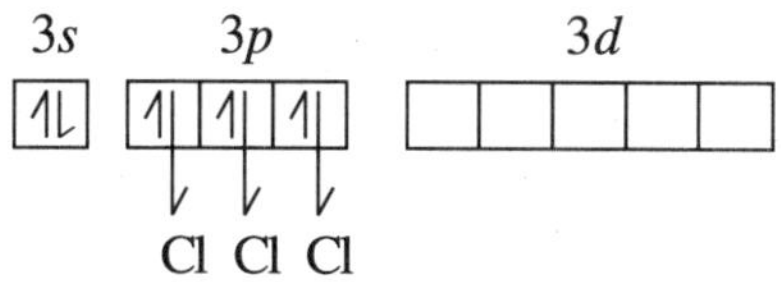

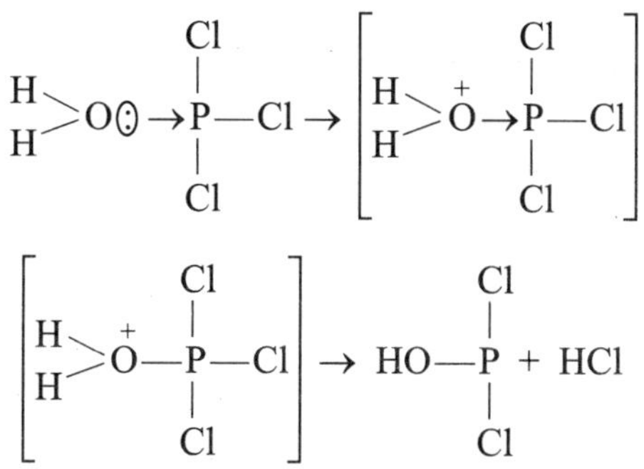

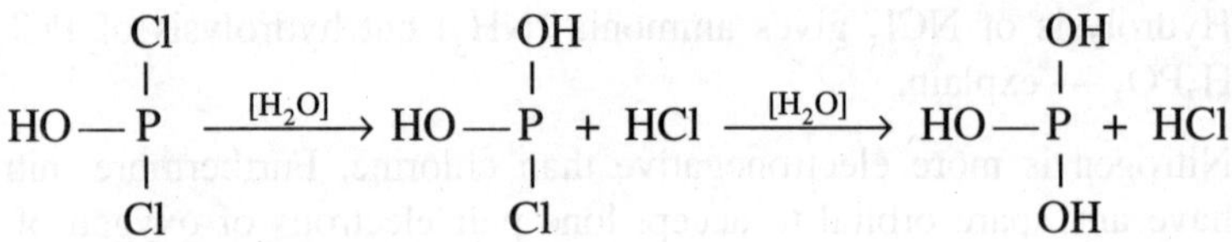

The overall reaction is: $PCl_3 + 3H_2O \rightarrow H_3PO_3 + 3HCl$

P. 4 In CH_4, NH_3, H_2O and HF hybridization of central atom is sp^3 and all compounds have tetrahedral structures, but bond angle in CH_4 is 109°28′, bond angle in NH_3 is 107°, bond angle in H_2O is 104° and bond angle in HF is 102° — explain.

Ans. C in CH_4

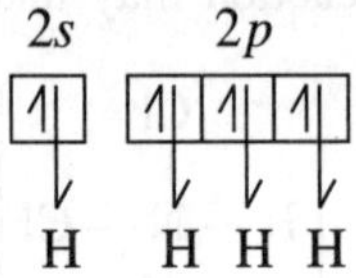

(all four bonds are σ-bonds)

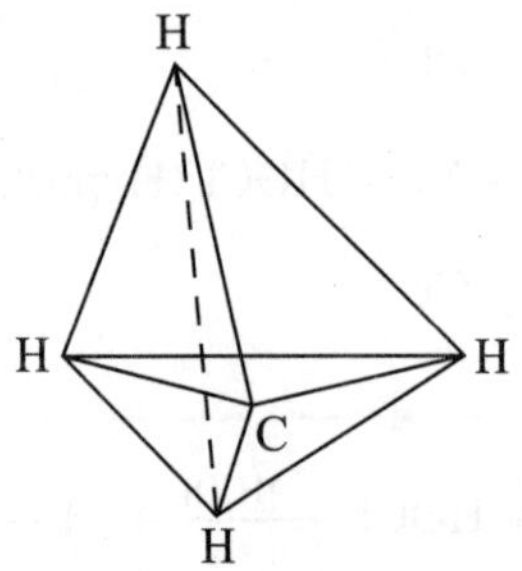

Bond angle = 109°28′

N in NH_3

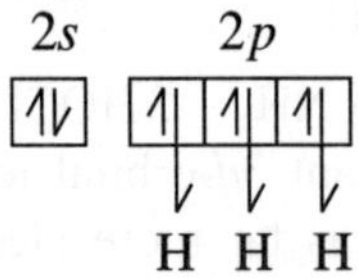

(three σ-bonds and one lone pair)

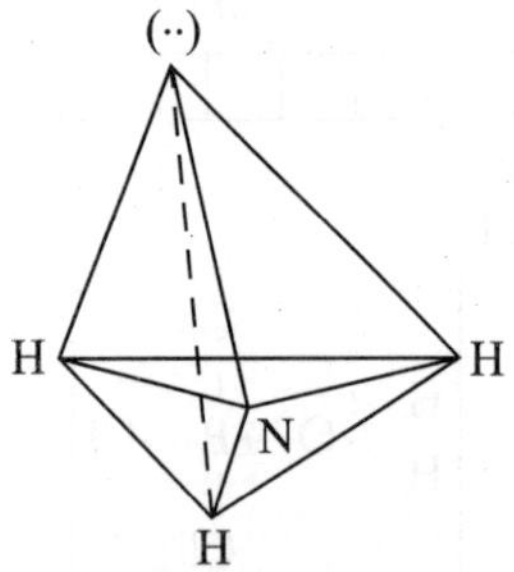

Bond angle = 107°

In ammonia lone pair-bond pair repulsion shortens the tetrahedral bond angle to 107°.

O in H_2O

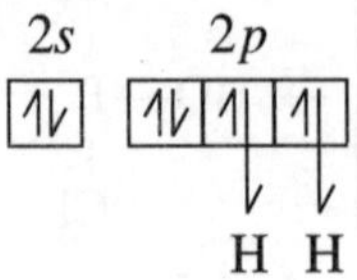

(two σ-bonds and two lone pairs)

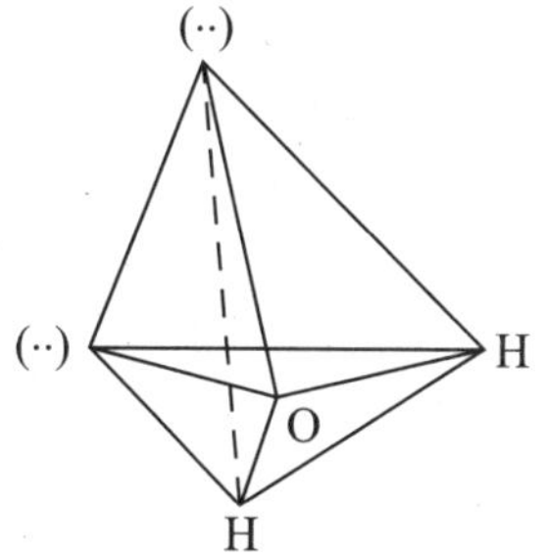

Bond angle = 104°

In water lone pair-lone pair repulsion in addition to lone pair-bond pair repulsion further shortens the tetrahedral bond angle to 104°.

F in HF

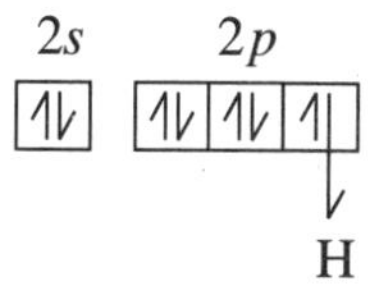

(one σ-bond and three lone pairs)

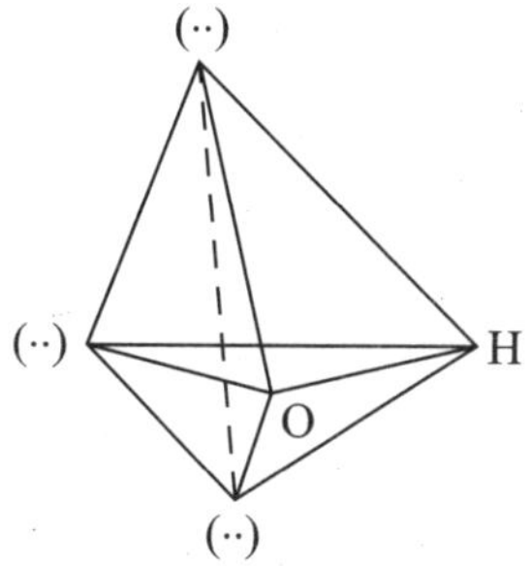

Bond angle = 102°

In hydrofluoric acid intensity of lone pair-lone pair repulsion increases, rendering further shortening of tetrahedral bond angle to 102°.

P. 5 Among the hydrides of Group-V elements bond angle in NH_3 is 107° but bond angle in other hydrides (PH_3, AsH_3, SbH_3 and BiH_3) is 90° — explain.

Ans. In ammonia small 1*s* orbital of hydrogen overlaps with 2*p* orbital of nitrogen. The overlapping area is sufficient to supply necessary hybridization energy and hence NH_3 appears as a tetrahedral structure. So bond angle is 107°. In PH_3, small 1*s* orbital of hydrogen overlaps with large 3*p* orbital of phosphorus and hence overlapping area is not sufficient to supply necessary hybridization energy. As a result of this, p_x, p_y and p_z orbitals of *P* remain perpendicular to each other and obviously, bond angle is around 90°. Overlapping area is further decreased in AsH_3 and SbH_3. It is minimum in BiH_3. That is why, in all three cases bond angle is close to 90°.

P. 6 Ammonia (NH_3) is stronger Lewis base than phosphine (PH_3), although nitrogen is more electronegative than phosphorus.

Ans. Electronic structures of N in NH_3 and P in PH_3 are shown below:

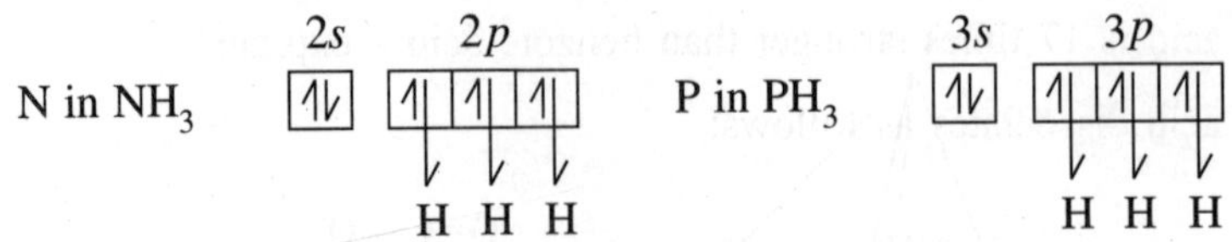

Both NH_3 and PH_3 react with H^+ to form NH_4^+ and PH_4^+ respectively.

$$NH_3 + H^+ \longrightarrow NH_4^+$$

$$PH_3 + H^+ \rightleftharpoons PH_4^+$$

In the case of NH_4^+, small 1*s* orbital of H^+ overlaps with 2*s* oribital of nitrogen. The overlapping area is sufficient to form stable and strong coordinate bond. So, NH_4^+ is stable and hence NH_3 appears as a strong Lewis base.

In the case of PH_4^+, small 1*s* orbital of H^+ overlaps with large 3*s* orbital of phosphorus. The overlapping area is not sufficient and hence the coordinate bond is unstable and weak. Thus PH_4^+ is unstable and readily dissociates into PH_3 and H^+. So PH_3 appears as much weaker Lewis base than NH_3.

P. 7 Electronegativity of halogens decreases in the following order: F > Cl > Br > I. But acid strength of hydrogen halides follows the reverse order:

HF < HCl < HBr< HI — explain.

Ans. To form hydrogen halides small 1*s* orbital of hydrogen overlaps with *p* orbital of halogen. The size of *p* orbital is the smallest in F, then increases down the group and is the largest in I. Thus overlapping area between hydrogen and halogen is maximum in H–F, then decreases to H–Cl to H–Br and is minimum in H–I. Consequently, bond strength is maximum in H–F, decreases to H–Cl to H–Br and is minimum in H–I. As the bond energy in H–I is very low, it readily dissociates in aqueous solution giving H_3O^+.

$$HI + H_2O \rightarrow H_3O^+ + I^-$$

Dissociation of H–Br is difficult since bond energy of H–Br is higher than that of H–I. So, H–Br appears as weaker acid than H–I. Similarly HCl is weaker acid than HBr. Bond energy is the highest in H–F and hence the latter appears as the weakest acid among the hydrogen halides. So the order of acid-strength of hydrogen halides is explained.

P. 8 Explain the following order of acid strength of hypohalous acids:

HOI < HOBr < HOCl

Ans. Hypohalous acid HOX dissociates in aqueous solution as follows:

$$HOX \rightarrow H_3O^+ + OX^-$$

In HOX, the release of H^+ depends on electronegativity of halogen, i.e., X. More the electronegativity of X, thinner the electron cloud on oxygen, faster the release of H-atom as H^+ and hence more will be the acid strength of HOX. This is illustrated as follows:

$$H \rightarrow\!\!- O \rightarrow\!\!- X + H_2O \rightarrow H_3O^+ + OX^-$$

As chlorine is the strongest electronegative atom among chlorine, bromine and iodine, HOCl appears as the strongest acid. Hence the order of acid strength is explained.

P. 9 Salicylic acid is 17 times stronger than benzoic acid – explain.

Ans. Salicylic acid dissociates as follows:

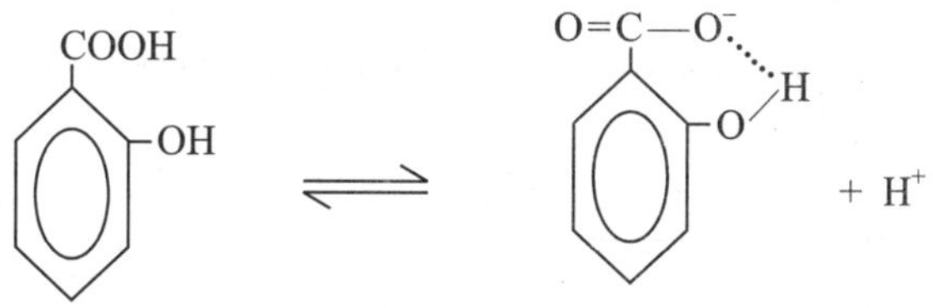

(Salicylic acid) (Salicylate ion)

Salicylate ion is stabilized through hydrogen bonding and hence more stable than salicylic acid itself. Thus the forward reaction is favoured. In the case of benzoic acid benzoate ion is free and not stabilized through hydrogen bonding like salicylate ion.

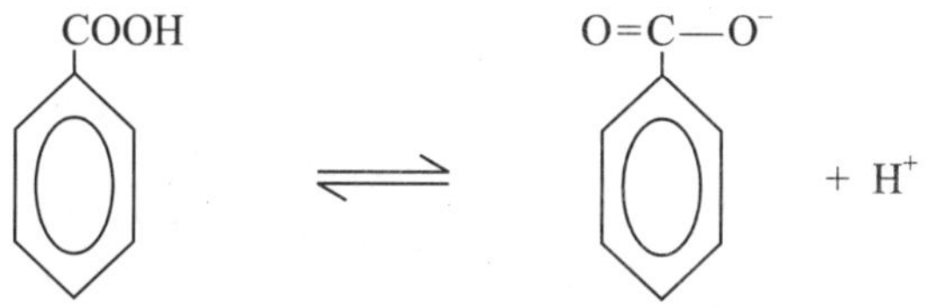

(Benzoic acid) (Benzoiate ion)

Thus forward reaction is not favoured like salicylic acid. That is why, salicylic acid is stronger than benzoic acid.

P. 10 Oxides and oxyacids of fluorine are not known — explain.

Ans. In oxyacids of halogens, halogen remains in *positive* oxidation state which is impossible in the case of fluorine since fluorine is *more* electronegative than oxygen. That is why, oxides and oxyacids of fluorine are not known. Due to the same reason the compounds O_2F_2 and OF_2 are often called *fluorides of oxygen*, where oxygen remains in +I and +II oxidation states respectively.

P. 11 Explain the following order of solubility of ionic compounds in water:

(a) $BeSO_4 > MgSO_4 > CaSO_4 > SrSO_4 > BaSO_4$

(b) $BeF_2 < MgF_2 < CaF_2 < SrF_2 < BaF_2$

Ans. Solubility of an ionic compound in water depends on two factors: (i) Lattice energy of ionic crystal and (ii) Hydration energy of ions. If lattice energy is deciding factor, *higher* the lattice energy, *more stable* the crystal and hence *less* the solubility in water. On the other hand, if hydration energy is the deciding factor, *higher* the hydration energy, *higher* the solubility in water.

If both the cations and anions are small in size lattice energies of those ionic crystals are high enough to dominate over the corresponding hydration energies of the ions. In those cases, solubility of an ionic compound depends on lattice energy of that ionic crystal.

Higher the lattice energy lesser the solubility is. That is the clue to explain the solubility order of fluorides of Group-IIA elements. Fluoride ion is very small in size and common to all fluorides. The size of Be^{2+} cation is the smallest and the size of the cation increases down the group. Hence BeF_2 has the highest lattice energy and hence is the least soluble in water. As the size of the cation increases, lattice energy decreases and hence solubility increases down the group. Maximum solubility is observed in the case of BaF_2.

In the case of sulphates of Group-II elements sulphate ion (SO_4^{2-}) is sufficiently large. Hence hydration energy of a sulphate crystal overcomes the corresponding lattice energy and appears as the deciding factor to determine solubility of sulphates. So smaller the cationic radius, higher the hydration energy, greater the solubility is. So, Be^{2+} has the highest hydration energy. The size of the cation increases down the group and hence corresponding hydration energy decreases. Ba^{2+} has the lowest hydration energy. Hence $BeSO_4$ is readily soluble in water whereas $BaSO_4$ is insoluble in water. Hence the solubility order of sulphates is explained.

P. 12 Anhydrous $AlCl_3$ is covalent but hydrated $AlCl_3$ is ionic — explain.

Ans. In anhydrous $AlCl_3$ the cation is Al^{3+}, which is very small in size and possesses high charge (+3), but the anion is Cl^-, which is comparatively large in size. Thus according to Fajan's rule, the small Al^{3+} cation can easily polarize large electron cloud of Cl^-, leading to the formation of covalent bond and hence anhydrous $AlCl_3$ is covalent in nature.

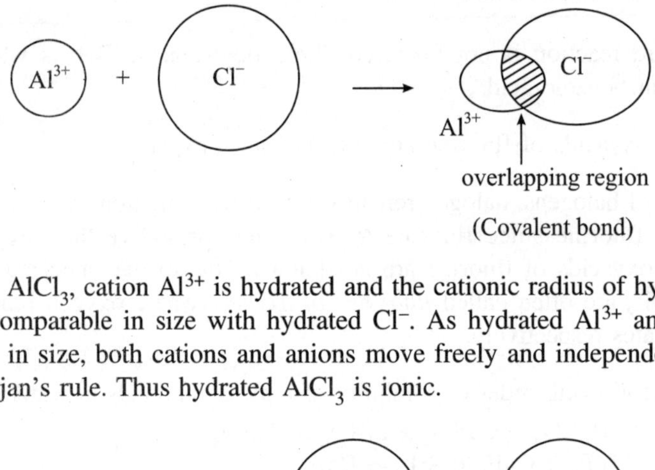

In hydrated $AlCl_3$, cation Al^{3+} is hydrated and the cationic radius of hydrated Al^{3+} is very large and comparable in size with hydrated Cl^-. As hydrated Al^{3+} and hydrated Cl^- are comparable in size, both cations and anions move freely and independently in accordance with the Fajan's rule. Thus hydrated $AlCl_3$ is ionic.

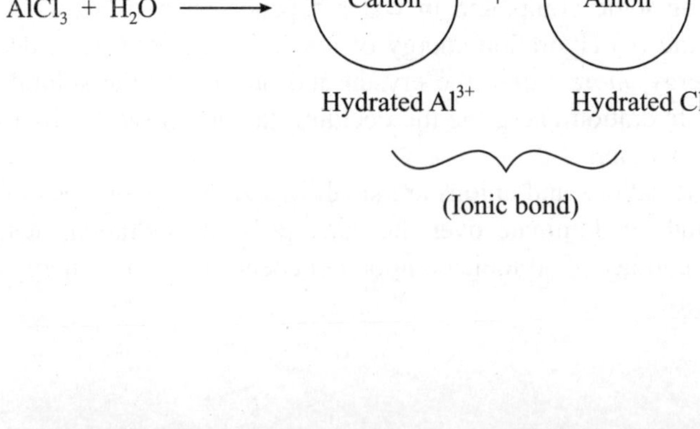

P. 13 Both nitrogen and phosphorus are the elements of Group-V. Phosphorus forms both PCl_3 and PCl_5 but nitrogen forms only NCl_3. Explain.

Ans. The outermost electronic configurations of N-atom and P-atom are given below:

N-atom: 2s [↑↓] 2p [↑ | ↑ | ↑] → G.S. and E.S.

P-atom: 3s [↑↓] 3p [↑ | ↑ | ↑] 3d [| | | |] → G.S.

3s [↑] 3p [↑ | ↑ | ↑] 3d [↑ | | | |] → E.S.

From the above electronic configurations it is evident that N-atom cannot extend its octet due to the absence of any vacant orbital in the outermost shell. So N-atom can accept a maximum of *three* Cl-atoms only to form NCl_3. Whereas due to the presence of vacant 3*d*-orbital P-atom can extend its octet so as to accommodate at most *five* chlorine atoms. Thus P-atom can form both PCl_3 and PCl_5.

P. 14 Bond energy of Cl_2, F_2 and Cl–F are 58, 38 and 61 kcal/mole respectively. Electronegativity of fluorine is 4. Calculate the electronegativity of chlorine.

Ans. $X_F - X_{Cl} = 0.208\sqrt{\Delta}$, where, $\Delta = E_{Cl-F} - \sqrt{(E_{Cl-Cl.} E_{F-F})}$

P. 15 Bond energy of S–S, H–H and H–S are 49, 104.2 and 81 kcal/mole. Electronegativity of hydrogen is 2.1. Calculate the electronegativity of sulphur.

P. 16 Explain the following:

(a) MgO has a higher melting point than NaF.
(b) $SnCl_4$ is a volatile liquid and a poor conductor of electricity.
(c) The melting point of $MgBr_2$ is 700°C while that of $AlBr_3$ is only 97°C.
(d) The melting point of AgCl is only 455°C while that of KCl is 776°C, though the ionic radii of Ag^+ and K^+ ions are almost the same.
(e) HgI_2 is much less soluble in water than $HgCl_2$.
(f) $LiClO_4$ forms a stable trihydrate whereas other alkali metal perchlorates do not.
(g) $NaClO_4$ is about 1000 times more soluble than $KClO_4$ in water.
(h) CsI is nearly 10 times more soluble in water than NaF but it much less soluble than CsF.

Ans. (a) From the knowledge of Born-Lande equation, we have: $|U| \propto (Z_+ . Z_-)$. In the case of MgO: $Z_+ = 2$ and $Z_- = 2$ and in the case of NaF: $Z_+ = 1$ and $Z_- = 1$. Thus lattice energy of MgO is much higher than that of NaF and hence melting point of MgO is higher than that of NaF.

(b) In $SnCl_4$, Sn^{4+} has a very small ionic radius and possesses high charge but Cl^- has comparatively large ionic radius. As a result of this, the electron cloud of Cl^- is easily diffused by Sn^{4+} cation, forming a partial overlap of two orbitals, therby, introducing sufficient covalent character in the compound. Thus, percent ionic character and hence lattice energy of $SnCl_4$ is very low. That is why, $SnCl_4$ a volatile liquid. As the concentration of discrete Sn^{4+} cation decreases significantly due to this partial overlap, $SnCl_4$ appears as a poor conductor of electricity.

(c) In $MgBr_2$ and $AlBr_3$ the common anion is Br^- but the cations are Mg^{2+} and Al^{3+}. Due to possessing higher cationic charge and lower radius, Al^{3+} is far more effective than Mg^{2+} to diffuse the electron cloud of Br^-. Obviously percent ionic character and hence corresponding lattice energy is much less in $AlBr_3$ than in $MgBr_2$. Thus, melting point of $MgBr_2$ is much higher than that of $AlBr_3$.

(d) The electrostatic force of attraction between nucleus and electron is given by:

$F = \left[\dfrac{(Z - sc)e^2}{(4\pi\varepsilon_0 . r^2)}\right]$, where, *r* is the crystal ionic radius, *Z* is the nuclear charge and *sc* is the screening constant.

The value of *sc* is maximum for *s*-orbital and decreases to *p*-orbital, to *d*-orbital and to *f*-orbital. Due to having filled *d*-orbital, the value of *sc* is very low in the case of Ag^+ and corresponding value of '*F*' is also very high. Thus Ag^+ is able to diffuse the electron cloud of Cl^- significantly, whereas K^+ does not have filled *d*-orbital and hence the value of *sc* is higher than that of Ag^+ and hence possesses low value of '*F*'. Thus, polarization is far less in KCl than in AgCl. Consequently, percent ionic character and lattice energy of KCl is higher than that of AgCl. That is why the melting point of KCl is higher than that of AgCl.

(e) In $HgCl_2$ and HgI_2 the cation is same, i.e., Hg^{2+} but the size of anion Cl^- is much lower than that of I^-. Furthermore, Hg^{2+} cation has filled *d*- and *f*-orbitals, thereby, possessing low screening effect. Thus electrostatic force of attraction in Hg^{2+} is remarkably high. Thus, it is obvious that polarization is much higher in HgI_2 than in $HgCl_2$. Thus $HgCl_2$ is far more ionic than HgI_2. As we know, ionic and polar compounds are dissolved in polar solvent like water, $HgCl_2$ is more soluble in water than HgI_2.

(f) In perchlorates, the anion ClO_4^- is large in size. Thus, lower the cationic radius, higher the hydration energy and hence more stable the hydrated cation is. Among the alkali halides Li^+ possesses the smallest cationic radius and hence $LiClO_4$ always forms stable trihydrate, but other hydrated alkali cations are not as stable as hydrated Li^+. That is why $LiClO_4$ forms a stable trihydrate but other alkali metal perchlorates do not.

(g) In both $NaClO_4$ and $KClO_4$, the common anion is ClO_4^-. As the cationic size of Na^+ is much smaller than that of K^+, hydrated Na^+ is much more stable than hydrated K^+. That is why $NaClO_4$ is much more soluble than $KClO_4$.

(h) In CsI and NaF, the cation Cs^+ is much larger than Na^+ and also the size of I^- is much larger than F^-. As we know, $U \propto \left(\dfrac{1}{r}\right)$, where, *r* is the interionic distance, lattice energy of NaF is much higher than that of CsI. So solvation energy is not enough to dissociate the NaF crystal but sufficient to dissociate CsI crystal. That is why, CsI is more soluble in water than NaF. However, in CsF and CsI, the common cation is Cs^+ but the anionic radius of F^- is much less than that of I^-. Solvation power of F^- is much higher than I^- and hence CsI is much less soluble than CsF.

SECTION – II CO-ORDINATION CHEMISTRY

7.2 Co-ordination Chemistry

7.2.1 Complex ion/compound and Werner's theory

A simple ion is a single species and exists as a free ion. If an ion is composed of two or more species the resultant ion is called complex ion and the corresponding compound is called complex compound. For example $[Fe(CN)_6]^{4-}$ is complex anion and $K_4[Fe(CN)_6]$ is called complex compound.

To explain the complex ions Alfred Werner (1892) introduced the concept of *secondary valency*. Transitional elements like others form ionic or covalent bonds to fulfil their octet. This is called primary valency of metals. In addition to that transitional elements have incomplete inner *d*-orbital. Also *d*-electrons have poor screening effect. Both these factors are responsible to develop *residual affinity* in transitional elements, which then attract surrounding anions or species having lone pair of electrons to form coordinate bonds. This residual affinity is often termed secondary valency.

Postulates of Werner's theory

(i) Transitional elements, having incompletely filled *d*-orbital, possess both primary and secondary valencies.

(ii) Primary valency is satisfied by anions while secondary valency is satisfied by either anion or species, having at least one lone pair of electrons.

(iii) Due to secondary valency coordinating groups are attracted towards the central metal cation to form *inner coordination sphere*. The rest of ions, which are not included in the inner coordination sphere, form the *outer coordination sphere*.

(iv) The number of coordinating groups, forming the inner coordination sphere, is called *coordination number*.

One example is $[Cr(NH_3)_6]Cl_3$, where, Cr is the transitional element. Six ammonia molecules form inner coordination sphere, while three chloride ions form outer coordination sphere. Coordination number is 6.

Limitations of Werner's theory

(i) A serious drawback of this theory is that the primary and secondary valencies are indistinguishable in many complexes. One such example is $K_4[Fe(CN)_6]$. Here cyanide ions (CN^-) satisfy both the primary and secondary valencies but it is next to impossible to identify which CN^- satisfies primary valency and which CN^- satisfies secondary valency.

(ii) Warner's theory fails to explain how coordination number is developed in a complex. For a particular coordinating group it (coordination number) may vary metal to metal. Furthermore, a metal may have different coordination number for different coordinating groups.

Sidgwick (1923) proposed the following postulates to explain the limitations of Werner's theory:

(i) Coordinating groups, also called *ligands*, donate their pair of electrons to central metal cation to form inner coordination sphere. So each and every ligand must have at least one pair of electrons.

(ii) Coordination number will be such that central metal cation achieves the electronic configuration of next inert gas. In a complex ion total number of electrons possessed by central metal cation is called the *Effective Atomic Number* (EAN) of that metal. This is also known as EAN-rule of Sidgwick. It is well illustrated in the Table 7.2.1.

Table 7.2.1 EAN of some transitional elements in their complexes.

Complex	*Atomic No.*	*Oxidation state of metal*	*Total No. of electrons possessed by metal cation*	*No. of donated electrons to metal cation*	*EAN– value of metal in the complex*
$[Co(NH_3)_6]^{3+}$	27	+3	24	$6 \times 2 = 12$	$24 + 12 = 36$
$[Fe(CN)_6]^{4-}$	26	+2	24	$6 \times 2 = 12$	$24 + 12 = 36$
$Ni(CO)_4$	28	0	28	$4 \times 2 = 8$	$28 + 8 = 36$
$[NiF_6]^{4-}$	28	+2	26	$6 \times 2 = 12$	$26 + 12 = 38$

7.2.2 Ligand and its classification

For a given complex two important parameters are ligand and coordination number.

Coordination number In a given complex the number of ligands attached to the central metal atom or ion, is called coordination number of that comlplex.

Ligand Any negatively charged ion or neutral species, which can donate its pair of electrons to the central metal atom or cation is called ligand.

Examples are:

Complex	*Ligand*	*Coordination No.*
$Ni(CO)_4$	CO	4
$[Fe(CN)_6]^{3-}$	CN^-	6
$[Cu(NH_3)_4]^{2+}$	NH_3	4

Classification of ligands: Ligands can be classified on three different basis:

(a) On the basis of species.
(b) On the basis of coordinate bonds.
(c) On the basis of coordination power.

(a) On the basis of species ligands can be divided into two categories:

(i) Anionic species, e.g.,

Ligand	*Complex*
CN^-	$[Fe(CN)_6]^{3-}$, $[Ni(CN)_4]^{2-}$ etc.
Cl^-	$[PtCl_4]^{2-}$, $[NiCl_4]^{2-}$, $[CoCl_4]^{2-}$ etc.
SCN^-	$[Fe(SCN)_6]^{3-}$, $[Co(SCN)_4]^{2-}$ etc.
OH^-	$[Zn(OH)_4]^{2-}$ etc.
NO_2^-	$[Co(NO_2)_6]^{3-}$ etc.

(ii) Neutral species, e.g.,

Ligand	*Complex*
NH_3	$[Cu(NH_3)_4]^{2+}$, $[Co(NH_3)_6]^{2+}$ etc.
H_2O	$[Cr(H_2O)_6]^{3+}$ etc.
CO	$[Ni(CO)_4]$, $[Fe(CO)_5]$, $[Cr(CO)_6]$ etc.
NO	$[Co(CO)_3(NO)]$,$[Fe(CO)_2(NO)_2]$ etc.

(b) On the basis of coordinate bonds ligands can be divided into different categories:

(i) **Monodentate ligand** Any ligand, occupying only one coordinating position, is called monodentate ligand e.g., NH_3, H_2O, CN^- etc.

(ii) **Bidentate ligand** Any ligand occupying two coordinating positions at a time, is called bidentate ligand, e.g.,

$$\begin{array}{ccc} \begin{array}{l} O=C-O^- \\ \quad\;\; | \\ O=C-O^- \end{array} & \begin{array}{l} CH_2-NH_2 \\ | \\ CH_2-NH_2 \end{array} & \begin{array}{l} Me-C=NOH \\ \qquad\;\; | \\ Me-C=NOH \end{array} \\ \text{oxalate ion} & \text{en} & \text{DMG} \end{array}$$

en ≡ Ethylene diamine, DMG ≡ Dimethyl glyoxime.

(iii) **Tridentate ligand** Any ligand, occupying three coordinating positions at a time, is called tridentate ligand, e.g.,

$$\begin{array}{l} (CH_2)_2-NH_2 \\ | \\ NH \\ | \\ (CH_2)_2-NH_2 \end{array}$$

dien

dien ≡ Diethylene triamine.

(iv) **Polydentate ligand** Any ligand, occupying more than three coordinating positions at a time, is called polydentate ligands, e.g.,

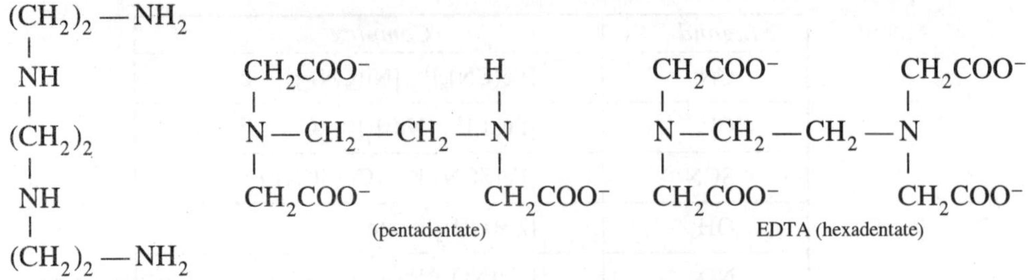

trien (tetradentate) (pentadentate) EDTA (hexadentate)

Trien ≡ Trienthylene tetramine EDTA ≡ Ethylene diamine tetracetic acid

(c) **Basis of coordination power** According to Crystal Field Theory (CFT) while approaching ligands towards central metal cation *d*-orbital of that metal cation splits into two levels. Energy difference between these two levels is called Crystal Field Stabilization Energy (CFSE), represented by Δ. The value of Δ depends on ligand strength or coordination power. Higher the ligand strength (or stronger the ligand), higher the value of Δ and hence more stable the complex is. On the basis of coordination power ligands can be arranged in a series of increasing order of ligand strength as shown below:

$I^- < Br^- < Cl^- < OH^- < F^- < EDTA < H_2O < NH_3 < en < NO_2^- < CN^- < CO, NO$

⟵ weak ligand ⟶ ← moderately strong → ⟵ strong ⟶ ← very strong →

This is known as spectrochemical series.

7.2.3 Isomerism in coordination compounds

Complex compounds often exhibit isomerism, which can be broadly classified into two categories: (a) Structural isomerism, (b) Stereoisomerism

(a) **Structural isomerism** It arises mainly due to chemical structure of a complex compound. On the basis of chemical structure structural isomerism can be subdivided into four categories:

(i) **Ionization isomerism** For a complex compound isomers are produced simply by interchanging of ions between inner and outer coordination spheres, e.g.,

$$[Co(NH_3)_5\,Br]SO_4 \text{ and } [Co(NH_3)_5\,SO_4]Br$$

This type of isomerism is known as ionisation isomerism.

(ii) **Hydrate isomerism** A series of isomers can be produced through exchange of water molecule between inner and outer coordination sphere, e.g.,

$[Cr(H_2O)_6]Cl_3$, $[Cr(H_2O)_5Cl]Cl_2.H_2O$, $[Cr(H_2O)_4Cl_2]Cl.2H_2O$, $[Cr(H_2O)_3Cl_3]3H_2O$

This type of isomerism is known as hydrate isomerism.

(iii) **Linkage isomerism** If a ligand can coordinate with the central metal cation through two different atoms, the isomers, thus produced, are called linkage isomers and the phenomenon is known as linkage isomerism.
Examples are:

$$\begin{array}{c} -N=O \\ \downarrow \\ O \end{array} \quad \text{and} \quad O=N-O-$$

Nitro-linkage   Nitrito-linkage

$$\left[\begin{array}{c} (NH_3)_5\,Co-N=O \\ \downarrow \\ O \end{array}\right]^{2+} \qquad [(NH_3)_5\,Co-O-N=O]^{2+}$$

pentaamine-nitro-cobalt (III) ion   pentaamine-nitrito-cobalt(III) ion

$$S=C=N- \;\leftrightarrow\; -S-C\equiv N$$

(Thiocyanoto)   (Iso-thiocyanato)

$$[Co(SCN)_4]^{2-} \qquad [Co(NCS)_4]^{2-}$$

tetrathiocyanato cobaltate (II) ion   tetraisothiocyanato cobaltate (II) ion

(iv) **Coordination isomerism** If both cation and anion, in an ionic compound, are complex in nature, isomers can be produced just by interchanging coordinating groups between two central metal cations.

One example is: $[Co(NH_3)_6]^{3+}\,[Cr(C_2O_4)_3]^{3-}$ and $[Cr(NH_3)_6]^{3+}\,[Co(C_2O_4)_3]^{3-}$

(v) **Polymerisation isomerism** Some complexes have a tendency to be polymerized through bridge atoms, e.g.,

$[(NH_3)_5\,Co-OH-Co(NH_3)_5]Cl_5$ — one bridge group

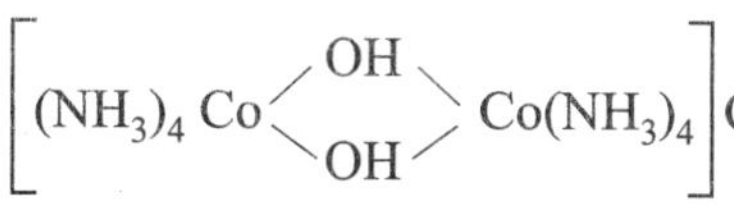

— two bridge groups

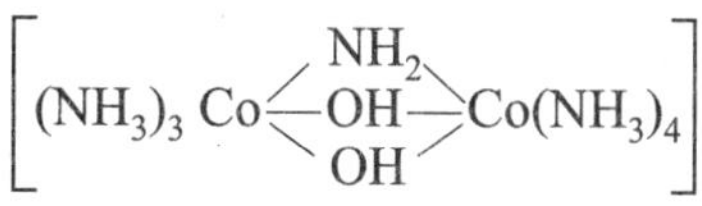

— three bridge groups

These complexes have multiple coordination centres and hence they are called polynuclear complexes. The above are the examples of binuclear complexes.

(b) Steoisomerism For complex compounds two types of stereoisomers are available: (i) Geometrical isomerism and (ii) Optical isomerism.

(i) **Geometrical isomerism** In a planar complex when two same ligands are placed in the same direction the isomer is called cis-isomer but when two same ligands are placed in the opposite direction, the isomer is called trans-isomer. One much example is diamino-dichloro-platinate (II).

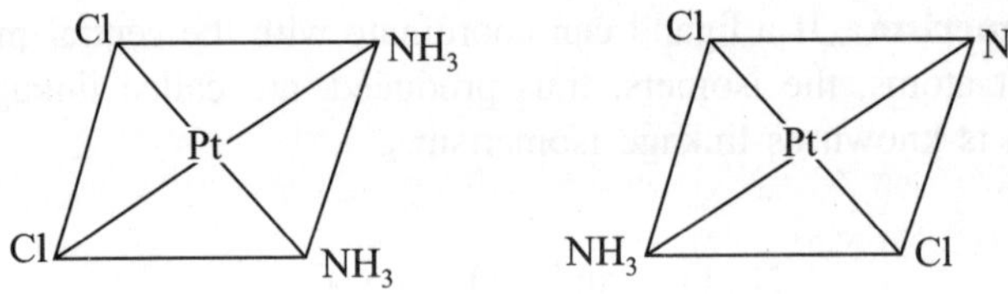

Cis-form Trans-form

(ii) **Optical isomerism** Sometimes position of ligands can make a complex optically active or inactive. The isomers, thus produced, are called optical isomers.

Example: Dichloro-bis-ethylenediamine cobalt (II) optically inactive since both are mirror images with each other.

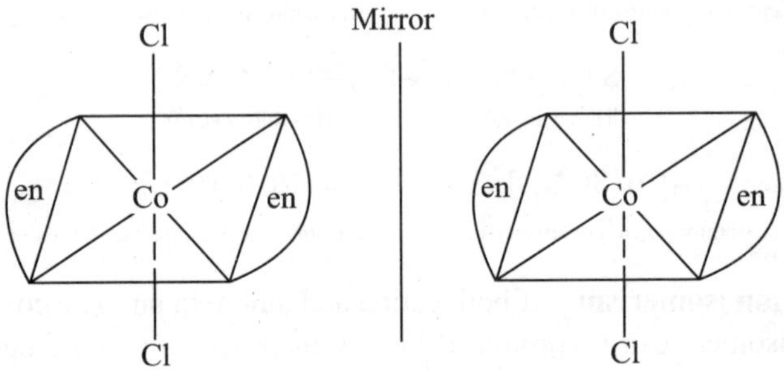

The same compound with different arrangement is optically active since both are not mirror images with each other.

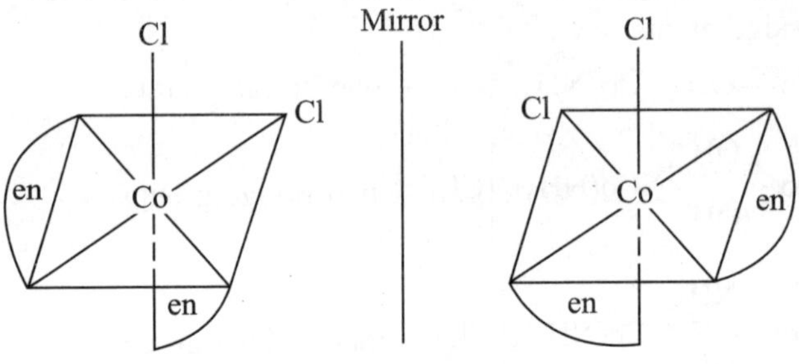

7.2.4 Nomenclature of coordination compounds

According to International Union of Pure and Applied Chemistry (IUPAC) following rules have been adopted for naming of coordination compounds:

(i) Cation should be named first and anion of outer coordination sphere should be named last.
(ii) Complex should be written within third bracket.
(iii) Within a complex ligands should be written alphabetically and the colmplexing cation should be placed last.

(iv) For repeating unidentate ligand prefixes *di-*, *tri-*. *tetra-*, *penta-* etc. are used but for repeating multidentate ligand prefixes *bis-*, *tris-*, *tetrakis-* etc. are used.

(v) For positive complexes complexing agent or central metal cation should be placed after ligand and ended with the name of that metal. Its valency is usually written in Roman letter and within bracket, e.g. Ni (II), Fe (III), Co (III) etc.
For negative complexes the complexing agent is ended with *–ate*, and its valency is usually written in Roman letter and within bracket, e.g. cobaltate (II), ferrate (III) etc.

(vi) Negative ligands, ended with *–ide*, the latter should be replaced with the letter *o*. e.g. chloride to chloro, etc. Negative ligands, ended with *–ite* or *–ate*, the latter should be replaced with *–ito* or *–ato*, e.g, sulphite to sulphito, cyanate to cyanato, etc.

(vii) Complexes with CO and NO are usually known as carbonyls and nitrosyls respectively.

Carbonyls Electronic structure of carbon in CO:

C in CO: 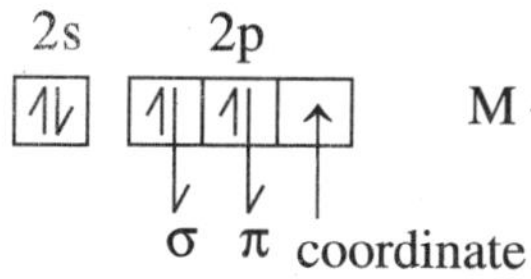 $M \leftarrow :\overset{\leftarrow}{C}=O$

The lone pair of carbon is donated to the central metal cation to form carbonyls.

Nitrosyls Electronic structure of nitrogen in NO:

N in NO: 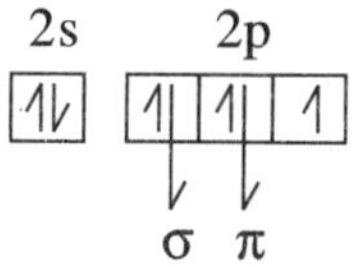 $M \leftarrow :\overset{\leftarrow}{N}=O$

The single electron is donated to the central metal cation and NO becomes NO^+, (Nitrosonium ion), which is isoelectronic with CO, forming complexes (nitrosyls) like CO. As the metal cation gains one electron, per NO molecule, valency of the metal decreases by one unit.
Here are few examples:

$[Cr(NH_3)_6]Cl_3$ ≡ Hexaamine chromium(III) chloride

$[Co(Cl)(CN)_2(NH_3)(H_2O)_2]$ ≡ Amine di-aqua chloro di-cyano cobalt(III)

$K_3[Fe(CN)_4(NO_2)_2]$ ≡ Potassium tetra-cyano di-nitro ferrate(III)

$K[Co(NO_2)_2(SCN)_2(NH_3)_2]$ ≡ Potassium di-amino di-nitro di-thiocyanato cobaltate(III)

$[Cr(NCS)(en)_2(SO_4)]$ ≡ Bis(ethylene diamine) isothiocyanato sulphato chromium(III)

$Na_2[Ni(ONO)Cl_2(NH_3)(OH)(NO)]$ ≡ Sodium amino di-chloro hydroxo nitrito nitrosyl nicklate(II)

$[Pt(NH_3)_4]$ $[PtCl_4]$ ≡ Tetra-amine platinum(II) tetrachloro platinate(II)

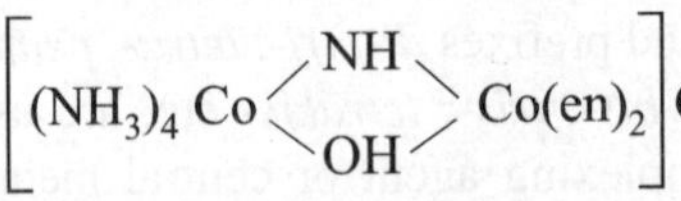

Tetraamine –μ– imido –μ– hydroxo bis(ethylene diamine) dicobalt(III) chloride

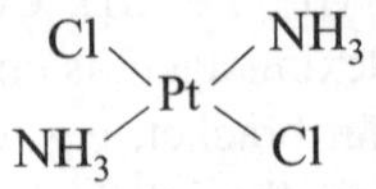

Trans diamine di-chloro platinum(II)

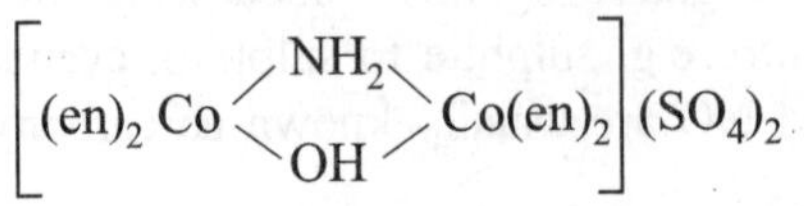

Tetrakis (ethyalne diamine) –μ– amido –μ– hydroxo dicobalt(III) sulphate

Sulphate ion (SO_4^{2-}) and carbonate ion (CO_3^{2-}) may act as both unidentate and bidentate ligand, e.g.,

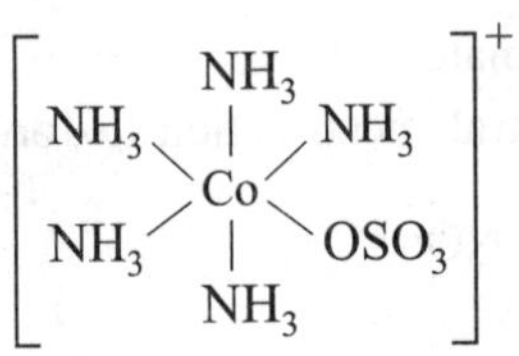

SO_4^{2-} acts as unidentate ligand

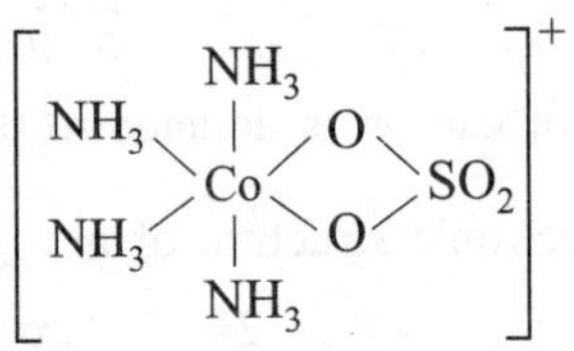

SO_4^{2-} acts as bidentate ligand

7.2.5 Chelates and inner metallic complexes

Complexes, formed by only multidentate ligands, exist in ring structures, which are called chelates. One example is $[Cu(en)_2]^{2+}$:

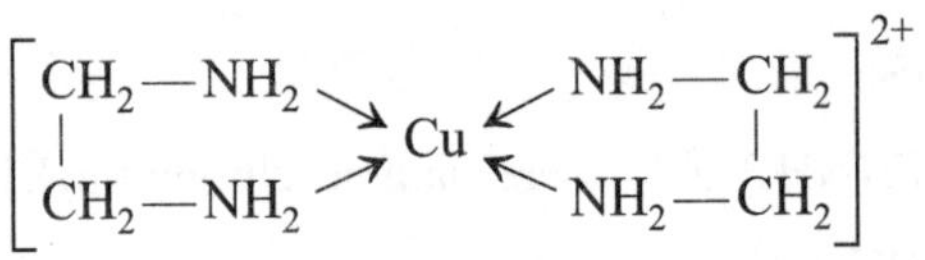

chelation enhances the stability of a complex. Chelates, having 5 or 6 member rings, are very much stable due to reduced strain in such a ring.

Inner metallic complex or inner complex is a special type of chelate, where ligands possess the following characteristics:

(a) Ligand must be bidentate.
(b) Out of two donor groups one should be acidic and another should be neutral.

One of such ligand is acetyl acetone (acac):

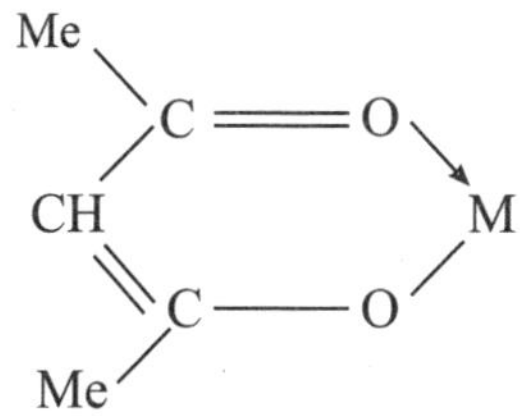

M ≡ Metal cation   acac ≡ $Me-\overset{O}{\overset{\|}{C}}-CH_2-\overset{O}{\overset{\|}{C}}-Me$

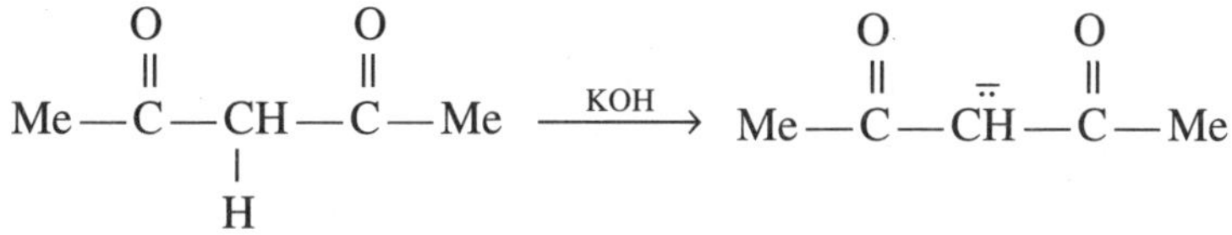

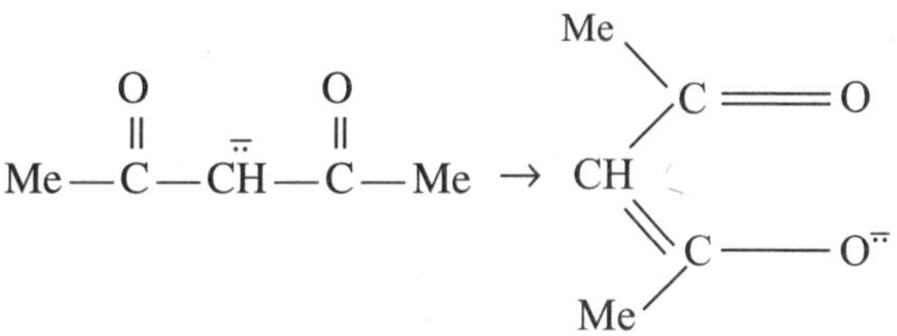

acac has conjugated double bond. So the inner complexes of *acac* are stabilized through resonance, as shown below:

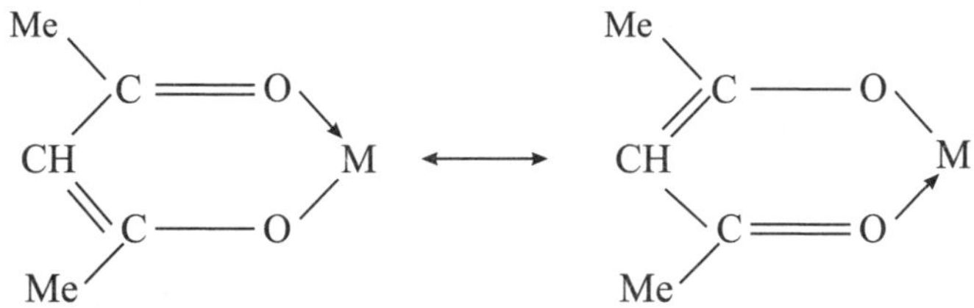

Three different types of inner complexes are known and they are designated as 1st order, 2nd order and 3rd order inner complexes.

1st order inner complex If acidic donor group of bidentate ligand satisfies the valency of central metal cation the complex is called inner complex of 1st order, e.g.,

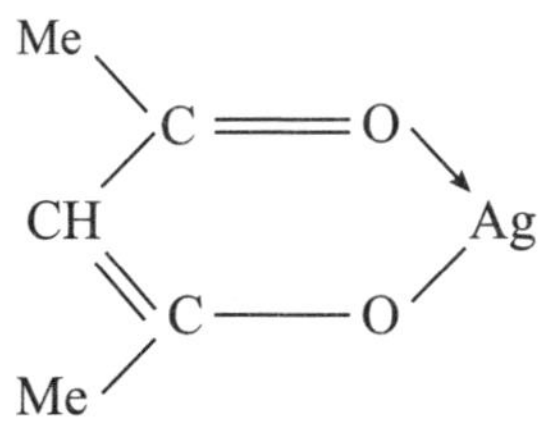

2nd order inner complex If acid group of bidentate ligand does not satisfy the valency of the central metal cation, the inner complex is called 2nd order inner complex, e.g., Na[Co(*acac*)$_3$], [Si(*acac*)$_3$]Cl.

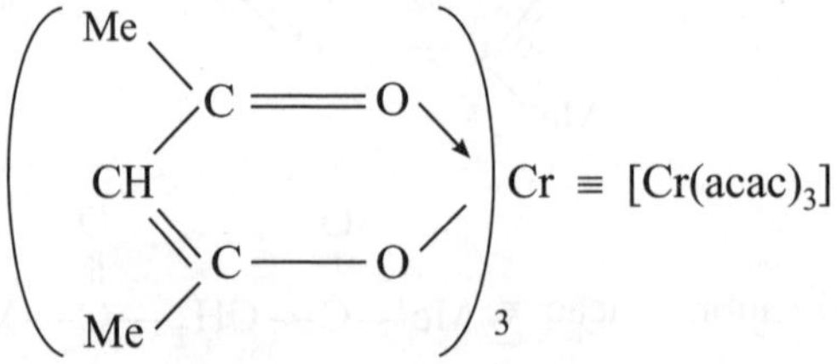

3rd order inner complex There are some inner complexes where bidentate ligands occupy only a few positions of inner coordination sphere, while the remaining positions are occupied by unidentate ligands. Those complexes are called inner complexes of 3rd order, e.g., Na[Co(*acac*)$_2$ (NO$_2$)$_2$], K[Pt(*acac*)Cl$_2$].

Table 7.2.2 Some chelating ligands

Name	*Structure*
Acetyl acetone	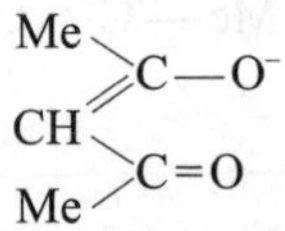
Oxalate ion	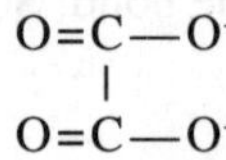
Salicyldehyde ion	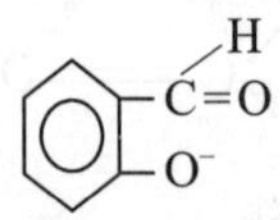
2,2′-Diphenyl	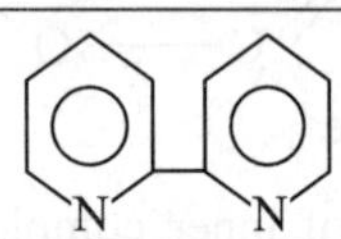
8-hydroxyquinolate ion	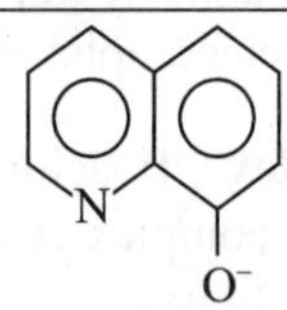
Dimethyl glyoxime	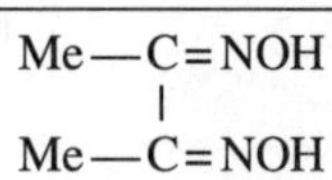
NN′-diethyl-dithio carbamate	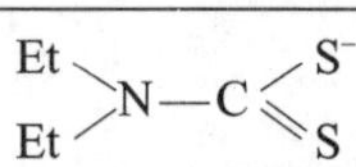

7.2.6 Bonding and characteristic features of complex compounds

Several theories have been proposed to explain bonding and characteristic features of coordination compounds. Only three of them are discussed here in detail:

(a) Valence Bond Theory (VB model).
(b) Crystal Field Theory (CFT model)
(c) Ligand Field Theory (LFT model)

(a) Postulates of VB model

(i) Only empty orbitals of metal cation can take part in bonding with filled orbitals of ligands.
(ii) Before overlapping with ligand orbitals empty orbitals of metal cation undergo hybridization and the geometry of complex compounds depends on the state hybridization.
(iii) Higher the donating power of ligand, larger the overlapping area and more stable the complex is.
(iv) Coordination number of central metal cation is equal to the number of hybridized orbitals.

VB model with the above postulates was proposed by Pauling. Some examples are: Outer shell electronic configuration of metal cation is considered only.

(i) Fe^{2+} :

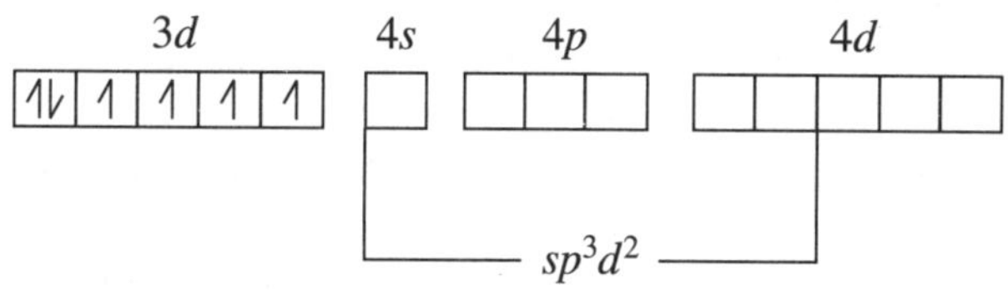

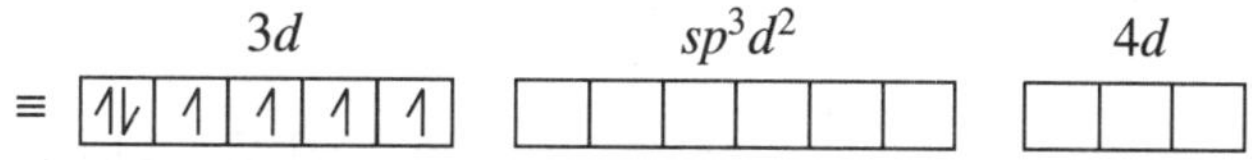

Fe^{2+} in $[Fe(H_2O)_6]^{2+}$ Hybridization = sp^3d^2

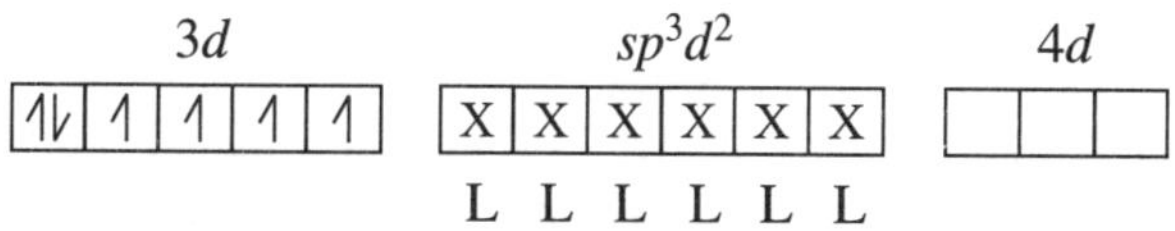

$L = H_2O$ Geometry = Octahedral

(ii) Ni^{2+} :

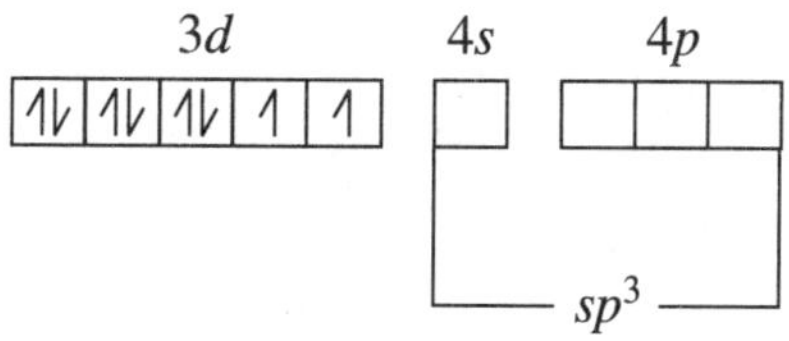

	3*d*	*sp*^{3}
≡	↑↓ ↑↓ ↑↓ ↑ ↑	☐ ☐ ☐ ☐

	3*d*	*sp*^{3}
Ni^{2+} in $[NiCl_4]^{2-}$	↑↓ ↑↓ ↑↓ ↑ ↑	X X X X L L L L

$L \equiv Cl^-$ Hybridization = *sp*^{3} Geometry = Tetrahedral

	3*d*	4*s*	4*p*
(iii) Cu^{2+}	↑↓ ↑↓ ↑↓ ↑↓ ↑	☐	☐ ☐ ☐

	3*d*	*dsp*^{2}	4*p*
≡	↑↓ ↑↓ ↑↓ ↑↓	☐ ☐ ☐ ☐	↑

	3*d*	*dsp*^{2}	4*p*
Cu^{2+} in $[Cu(NH_3)_4]^{2+}$	↑↓ ↑↓ ↑↓ ↑↓	X X X X L L L L	↑

$L \equiv NH_3$ Hybridization = *dsp*^{2} Geometry = Square planar

Limitations of VB model:

(a) It cannot explain magnetic behaviour of complexes.
(b) It cannot explain the colours of complexes.
(c) It cannot explain the distortion of geometry of many complexes such as Cu^{2+}, Tl^{3+} complexes.
(d) Concept of hybrid orbitals is purely hypothetical, since electron has mathematical entity only.

(b) CFT model CFT model was proposed by Bethe and Van Vleck.

Postulates:

(i) Ligands are treated as point charges.
(ii) There is no interaction between metal orbitals and ligand orbitals.
(iii) Electrostatic force of attraction exists between metal cation and ligands. *d*-electrons of metal strongly oppose the field, generating strong repulsion between electrostatic field and *d*-electrons of metal, which eventually brings about splitting of *d*-orbitals.

(c) LFT model LFT model accepts the postulates of CFT model except the 1st one. In addition to that, it considers the shape of *d*-orbitals. So LFT model clearly explains the splitting pattern of *d*-orbital pictorialy. Shapes of five *d*-orbitals are shown in the following figure [Figure 7.2.1.].

The figure shows that electrons are distributed in between the axes in d_{xy}, d_{yz} and d_{xz} orbitals but electrons are distributed along the axes in $d_{x^2-y^2}$ and d_{z^2} orbitals.

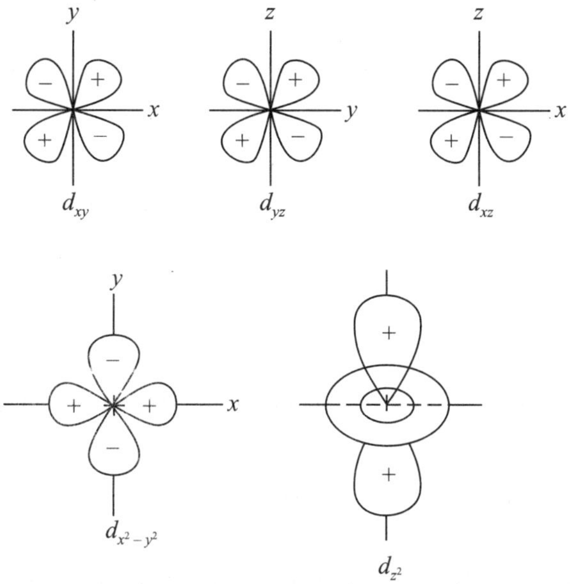

Fig. 7.2.1 Shape of different *d*-oribitals.

Characteristic features of LFT model

(1) To form an octahedral complex filled orbitals of ligands are approaching along the axes. As a result of this, electrons in $d_{z^2-y^2}$ and d_{z^2} exert strong repulsion but much less repulsion is exerienced by electrons in d_{xy}, d_{yz} and d_{xz}.

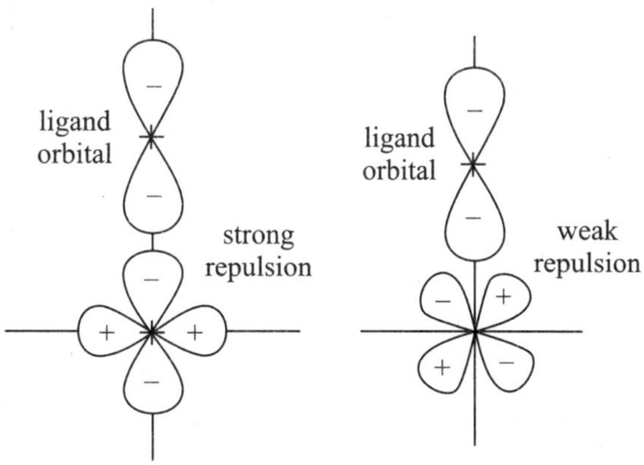

Fig. 7.2.2 Ligand approach in an octahedral complex.

Ultimately all five degenerate *d*-orbitals split into two energy levels, namely t_{2g} and e_{2g} as shown below:

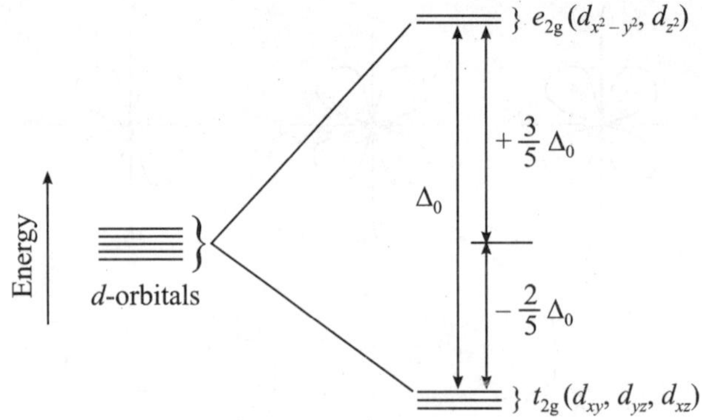

Fig. 7.2.3 Splitting of *d*-orbitals in an octahedral complex.

Due to facing strong repulsion $d_{x^2-y^2}$ and d_{z^2} are placed at higher energy level and the other *d*-orbitals d_{xy}, d_{yz} and d_{xz} are placed at lower energy level. The energy difference between t_{2g} and e_{2g} is called crystal field stabilization energy (CFSE), represented as Δ_0.

(2) To form a tetrahedral complex filled orbitals of ligands are approaching in between the axes. As a result of this, electrons in d_{xy}, d_{yz} and d_{xz} exert strong repulsions whereas electrons in $d_{x^2-y^2}$ and d_{z^2} experience much less repulsion.

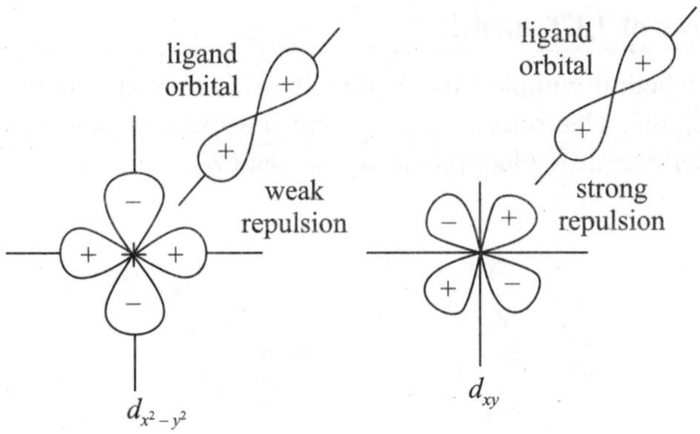

Fig. 7.2.4 Ligand approach in a tetrahedral complex.

So in this case *d*-orbitals split into two energy level, t_{2g} and e_{2g} but t_{2g} is placed at higher energy level, since d_{xy}, d_{yz} and d_{xz} face strong repulsion.
In this case CFSE is represented as Δ_t.
Usually $\Delta_0 >> \Delta_t$, i.e. CFSE in octahedral complex is much higher than that of tetrahedral complex.

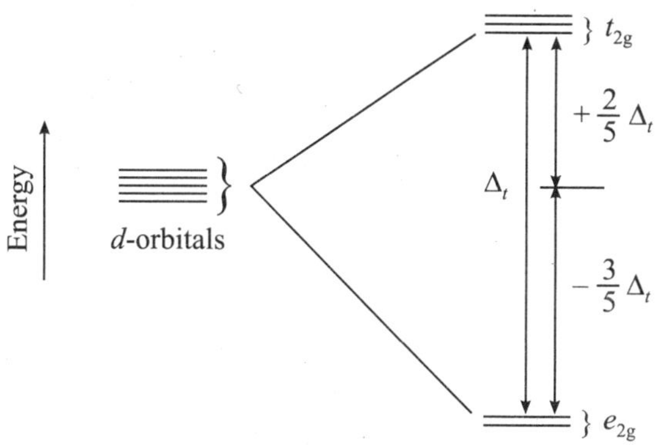

Fig. 7.2.5 Splitting of *d*-orbitals in a tetrahedral complex.

(3) Higher the coordination power of ligand or in other words stronger the ligand (according to spectrochemical series) higher the value of Δ_0 is.

(4) Distribution of *d*-electrons in t_{2g} and e_{2g} largely depends on two factors:

(i) CFSE, i.e. Δ_0 and (ii) pairing energy, i.e. P.

Pairing energy is the amount of energy required to remain two electrons in paired form. In an octahedral complex lower energy level contains three degenerate orbitals. So first three electrons will go to three degenerate orbitals. The 4th electron may either go to higher energy level e_{2g} or remain in paired form in t_{2g}.

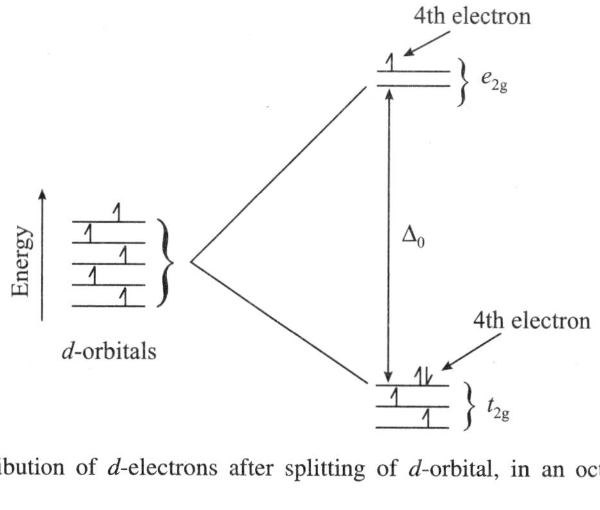

Fig. 7.2.6 Distribution of *d*-electrons after splitting of *d*-orbital, in an octahedral complex.

To go to e_{2g} energy level, the 4th electron has to overcome the CFSE barrier but to remain in t_{2g} energy level the 4th electron must have the pairing energy *P*. In an octahedral complex, if $\Delta_0 > P$, the 4th electron prefers remaining in paired form in t_{2g} level rather to crossing the energy barrier Δ_0 and hence spin multiplicity decreases as is evident from the Figure 7.2.6. In that case t_{2g} level has the privilege to fill up its three orbitals completely and only after that next electron has to cross CFSE barrier and occupies the e_{2g} energy level. As the spin multiplicity decreases, complex is called low spin complex.

If $\Delta_0 < P$ for an octahedral complex, the 4th electron prefers crossing the energy barrier rather to remaining in paired form in t_{2g} level. Thus spin multiplicity of degenerate *d*-orbitals of original metal cation is restored in the octahedral complex, and the complex is often called high spin complex.

Thus for strong ligands like CN^-, NO_2^-, etc. $\Delta_0 > P$ and low spin complexes are formed and for weak ligands like H_2O, F^-, etc., $\Delta_0 < P$ and high spin complexes are formed, e.g. $[FeF_6]^{3-}$ is a high spin complex but $[Fe(CN)_6]^{3-}$ is a low spin complex.

(5) In octahedral complexes, electrons of t_{2g} level acquire CFSE from sunlight and jump to next higher e_{2g} level. This type of transition is known as *d-d transition* of electrons. This type of *d-d transition* is possible in tetrahedral complexes also. To execute *d-d transition* electrons a complex usually absorbs light energy in the visible range and hence it appears coloured. If a complex absorbs colour in the blue-violet range for *d-d transition* of electrons, the complex will appear as orange-red. The absorption and appearance of colour are well illustrated in the Table 7.2.3.

Table 7.2.3 Observed colours with wavelength at different range of colour absorption.

Colour absorbed	*Colour observed*	*Wavelength* (nm)
Yellow-Green	Red-Violet	385 – 417
Yellow	Indigo	417 – 435
Orange	Blue	435 – 476
Red	Blue-Green	476 – 500
Purple	Green	500 – 556
Red-Violet	Yellow-Green	556 – 578
Indigo	Yellow	578 – 610
Blue	Orange	610 – 654
Blue-Green	Red	654 – 781

(6) Hybridization of a complex compound depends on ligand field. In strong ligand field, $\Delta_0 > P$ and electrons remain in paired form in t_{2g} level, as a result of which, one or both the orbitals of e_{2g} level may remain empty. Then e_{2g} orbitals also take part in hybridization and the state of hybridization changes from sp^3d^2 to d^2sp^3. In the case of square planar complexes tetrahedral hybridization (sp^3) changes to square planar (dsp^2) hybridization, in presence of strong ligands.

Example 7.1

$$Fe^{2+} \text{ in } [Fe(H_2O)_6]^{2+} \equiv \text{Hybridization} = sp^3d^2$$

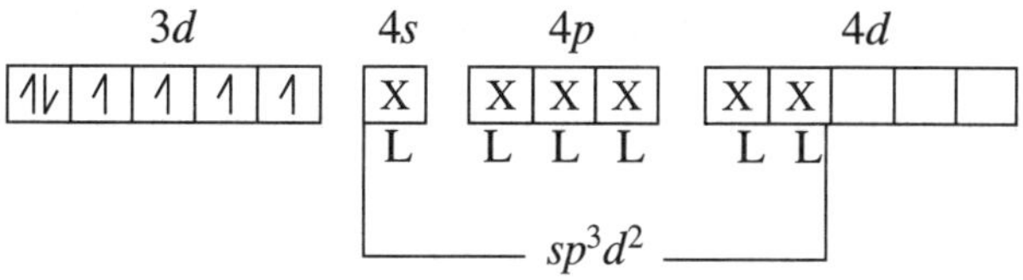

Fe^{2+} in $[Fe(CN)_6]^{4-}$ ≡ Hybridization = d^2sp^3

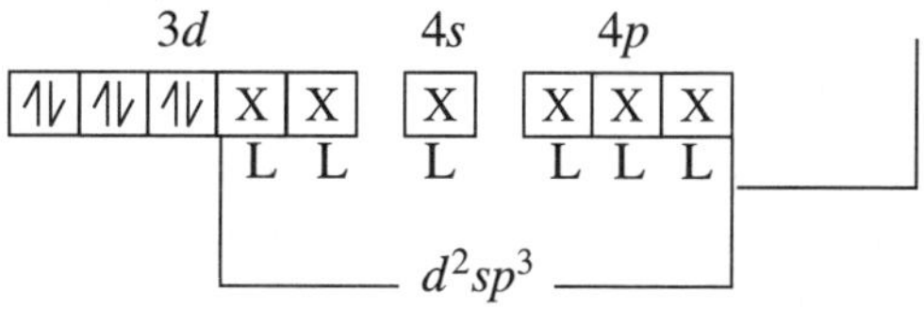

$[Fe(H_2O)_6]^{2+}$ is called high spin complex and $[Fe(CN)_6]^{4-}$ is called low spin complex.

Example 7.2

Ni^{2+} in $[NiCl_4]^{2-}$ ≡ Hybridization = sp^3

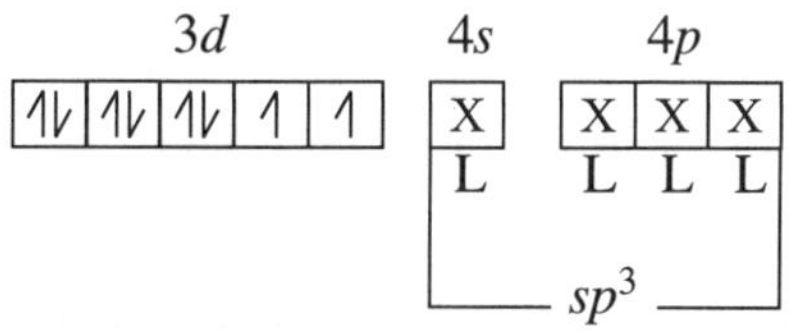

NI^{2+} in $[Ni(CN)_4]^{2-}$ ≡ Hybridization = dsp^2

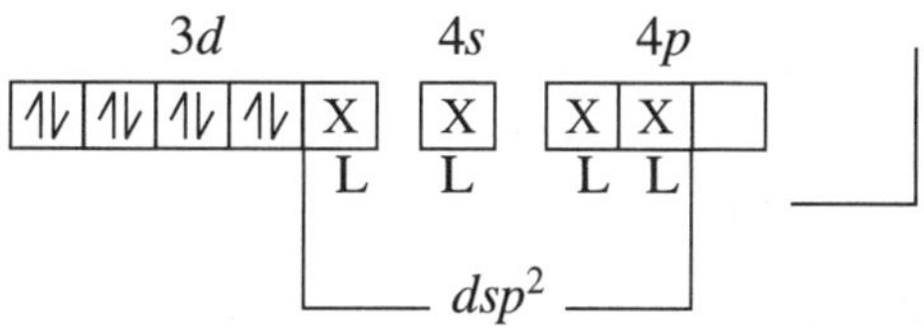

$[Ni(Cl)_4]^{2-}$ is called high spin complex and $[Ni(CN)_4]^{2-}$ is called low spin complex.

In d^2sp^3 and dsp^2 hybridizations inner 3*d*-orbitals take part in hybridization and hence the complexes thus formed, are called *inner orbital complexes*. In sp^3d^2 hybridization outer 4*d*-orbitals take part in hybridization and hence the complexes, thus formed, are called *outer orbital complexes*. It may be noticed that in inner orbital complexes spin multiplicity of 3*d*-electrons of the central metal cation is minimum and in outer orbital complexes spin multiplicity of 3*d*-electrons of the central metal cation remains unaltered or maximum. Thus inner orbital complexes are called low spin complexes and outer orbital complexes are called high spin complexes.

7.2.7 Crystal field stabilization energy (CFSE)

(i) For an octahedral complex the splitting of *d*-orbital is as follows:

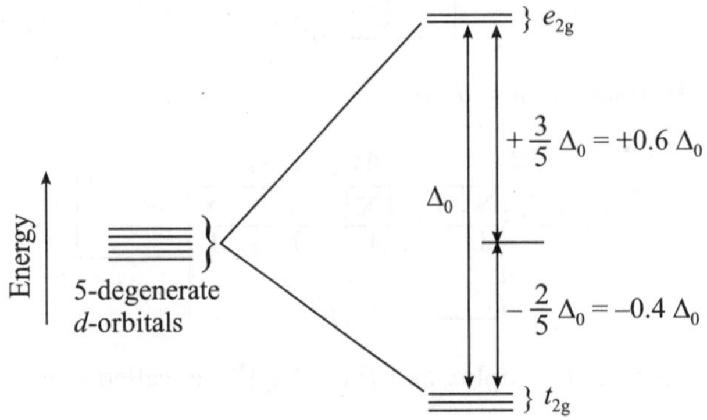

Δ_0 is the CFSE per electron. So for a metal cation of d^n-configuration the total CFSE becomes:

$$\text{CFSE} = \left(-\frac{2}{5}\Delta_0\right)n_t + \left(+\frac{3}{5}\Delta_0\right)n_e = (-0.4\,n_t + 0.6\,n_e)\Delta_0$$

where, n_t = Number of electrons in t_{2g}-level and

n_e = Number of electrons in e_g-level.

The splitting of *d*-orbital in a tetrahedral complex is as follows:

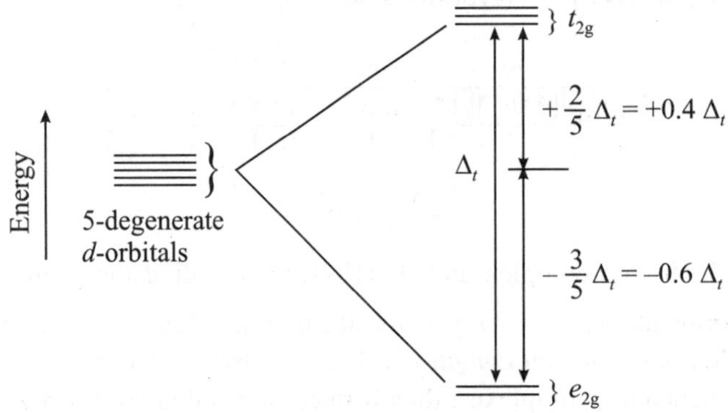

Δ_t is the CFSE per electron. Thus for a metal cation of d^n-configuration the total CFSE is given by:

$$\text{CFSE} = \left(\frac{2}{5}\Delta_t\right)n_t + \left(-\frac{3}{5}\Delta_t\right)n_e = (0.4\,n_t - 0.6\,n_e)\Delta_t$$

where, n_t = Number of electrons in t_{2g}-level and

n_e = Number of electrons in e_g-level.

CFSE values of octahedral complexes and tetrahedral complexes at different d^n-configurations are given in the Table 7.2.4.

Table 7.2.4 CFSE values of octahedral and tetrahedral complexes

d-orbital configuration	*CFSE for octahedral complex*		*CFSE for tetrahedral complex*
	High spin	*Low spin*	
d^1	$-0.4\ \Delta_0$	—	$-0.6\ \Delta_t$
d^2	$-0.4 \times 2\ \Delta_0 = -0.8\ \Delta_0$	—	$-0.6 \times 2\Delta_t$ $= -1.2\ \Delta_t$
d^3	$-0.4 \times 3\ \Delta_0$ $= -1.2\ \Delta_0$	—	$(-0.6 \times 2 + 0.4 \times 1)\Delta_t$ $= -0.8\ \Delta_t$
d^4	$(-0.4 \times 3 + 0.6 \times 1)\Delta_0$ $= -0.6\ \Delta_0$	$-0.4 \times 4\ \Delta_0$ $= -1.6\ \Delta_0$	$(-0.6 \times 2 + 0.4 \times 2)\Delta_t$ $= -0.4\ \Delta_t$
d^5	$(-0.4 \times 3 + 0.6 \times 2)\Delta_0$ $= 0$	$-0.4 \times 5\ \Delta_0$ $= -2.0\ \Delta_0$	$(-0.6 \times 2 + 0.4 \times 3)\Delta_t$ $= 0$
d^6	$(-0.4 \times 4 + 0.6 \times 2)\Delta_0$ $= -0.4\ \Delta_0$	$-0.4 \times 6\ \Delta_0$ $= -2.4\ \Delta_0$	$(-0.6 \times 3 + 0.4 \times 3)\Delta_t$ $= -0.6\ \Delta_t$
d^7	$(-0.4 \times 5 + 0.6 \times 2)\Delta_0$ $= -0.8\ \Delta_0$	$-0.4 \times 6\ \Delta_0$ $= -2.4\ \Delta_0$ 1*e* shifted to 4*d*-orbital	$(-0.6 \times 4 + 0.4 \times 3)\Delta_t$ $= -1.2\ \Delta_t$
d^8	$(-0.4 \times 6 + 0.6 \times 2)\Delta_0$ $= -1.2\ \Delta_0$	—	$(-0.6 \times 4 + 0.4 \times 4)\Delta_t$ $= -0.8\ \Delta_t$
d^9	$(-0.4 \times 6 + 0.6 \times 3)\Delta_0$ $= -0.6\ \Delta_0$	—	$(-0.6 \times 4 + 0.4 \times 5)\Delta_t$ $= -0.4\ \Delta_t$
d^{10}	$(-0.4 \times 6 + 0.6 \times 4)\Delta_0$ $= 0$	—	$(-0.6 \times 4 + 0.4 \times 6)\Delta_t$ $= 0$

(ii) The magnitude of CFSE (Δ_t and Δ_0) depends on three factors:

(a) **Nature of the ligand** Stronger the ligand field, larger the value of CFSE.

(b) **Charge on the metal ion** Higher the charge, greater the electrostatic force of attraction and hence larger the value of CFSE.

(c) **Position of transitional element in the periodic table** Down the group size of *d*-orbital increases. Larger the size of *d*-orbital, higher the repulsion with ligand orbitals and hence higher the value of CFSE. Furthermore, down the group residual affinity of central metal cation increases since poor shielding effect of inner orbitals enhances the electrostatic force of attraction. As a result of this, CFSE of complex increases down the group.

(iii) Δ_t is considerably less than Δ_0 – why ?
Δ_t of a tetrahedral complex is always less than Δ_0 of an octahedral complex due to two reasons:

(a) There are only *four* ligands in a tetrahedral complex instead of six in an octahedral complex. So the tetrahedral ligand field is only two-thirds of that of octahedral ligand field and hence tetrahedral splitting is also-two-thirds of that of octahedral splitting.

(b) In a tetrahedral complex direction of *d*-orbitals does not coincide with the direction of ligands. This further reduces the crystal field splitting by roughly a two-thirds of that of an octahedral splitting.
Thus Δ_t is always less than pairing energy (*P*). So only high spin complex is favoured in any tetrahedral ligand field.

(iv) Conditions favouring to form tetrahedral complexes:

(a) If the ligands are bulky, sufficient steric hindrance, thus generated surrounding the central metal cation, raises the energy barrier high enough to foil chances of forming any octahedral complex. But in a tetrahedral complex steric hindrance is much less pronounced and easily surpassed by CFSE (Δ_t). So tetrahedral complexes and not octahedral complexes, are formed with the bulky ligands.

(b) It is central metal cation remains in low oxidation state electrostatic force of attraction is less and hence the corresponding CFSE is not enough to create octahedral ligand field, although tetrahedral splitting may occur smoothly. Thus at low oxidation state of central metal cation tetrahedral complexes are favoured rather than octahedral complexes.

(c) If ligands are weak, the difference of CFSE between octahedral and tetrahedral splittings is less important than the stability of the complex. In those cases tetrahedral complex is more stable and preferred to octahedral complex.

(d) For certain *d*-orbital configurations, e.g., d^0, d^5 and d^{10} the net CFSE is zero for both the tetrahedral and high-spin octahedral splitting. In that case formation of tetrahedral complex is favoured since less energy is required to form tetrahedral complex than octahedral complex.

7.2.8 John-Teller distortion and square planar complex

In an octahedral complex if the density of electron cloud is symmetrical about the central metal cation the octahedral complex will be a regular octahedron with six equal bonds. However, in some octahedral complexes electronic arrangement is unsymmetrical, e.g. in d^4 configuration only one electron resides in e_{2g} level. In that case two cases are possible: (a) the single electron lies in d_{z^2} or (*b*) the single electron lies in $d_{x^2-y^2}$.

(a) If the single electron belongs to d_{z^2} orbital, two ligands, approaching along *z*-axis, face strong repulsion and remain far away from central metal cation, forming two long coordinate bonds. As $d_{x^2-y^2}$ is empty, remaining four ligands, approaching along *x*- and *y*-axes, do not face much repulsion and eventually remain close to the central metal cation, forming four short coordinate bonds. As a result of this, octahedron gets elongated along *z*-axis and the distortion is called tetrahedral elongation.

(b) If the single electron belongs to $d_{x^2-y^2}$ orbital, two ligands, approaching along the z-axis, face the least repulsion and hence form two short coordinate bonds. The other four ligands, approaching along x- and y-axes, face strong repulsion and stay away from the central metal cation, forming four long coordinate bonds. So the octahedron appears compressed along z-axis and the distortion is called tetrahedral compression. Distortion of type tetrahedral elongation is more common than tetrahedral compression. Furthermore, significant distortion is observed due to unsymmetrical electronic arrangement in e_{2g} state but distortion due to unsymmetrical electronic arrangement in t_{2g} state is insignificant and usually ignored.

Inference If the electronic arrangement in degenerate states of splitted d-orbital is unsymmetrical the octahedral complex will face distortion which is popularly known as John-Teller distortion.

Square planar complex

In an octahedral complex with a strong ligand field unsymmetrical distribution of electrons in e_g-state brings about strong repulsion along z-axis, as a result of which, two ligands along z-axis get detached, leaving a square planar complex. As there is no ligand along z-axis, repulsion along z-axis ceases. So, d_{z^2}, d_{xz} and d_{yz} gain extra stability, leading *further splitting of d-orbital* as shown below:

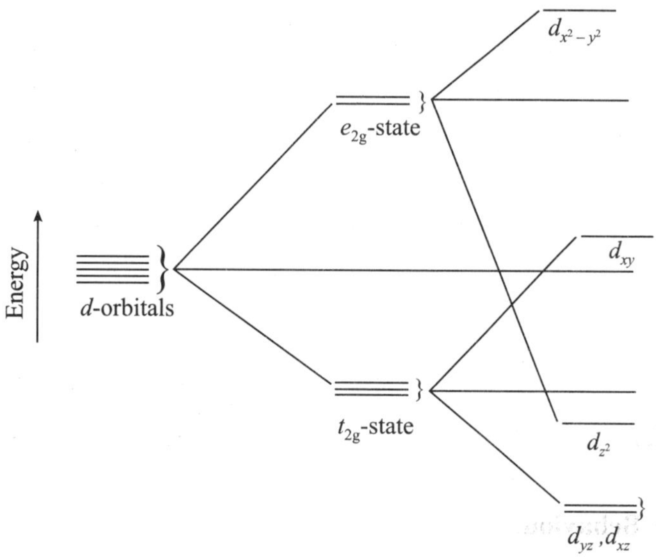

Fig. 7.2.7 Square-planner splitting of d-orbital.

Square planar splitting is very common in Cu^{2+} complexes, where $d_{x^2-y^2}$ possesses two electrons and d_{z^2} contains a lone electron. Thus permanent John-Teller distortion always exists in Cu^{+2} complexes, as a result of which, even in presence of moderately strong ligand like ammonia, d-orbital of Cu^{2+} suffers further splitting, leading to a square planar complex.

In the case of Ni^{2+} octahedral complexes both $d_{x^2-y^2}$ and d_{z^2} possess one electron each but in presence of strong ligand like CN^- pairing of electrons occurs and both the electrons are shifted to $d_{x^2-y^2}$ orbital, leaving d_{z^2} orbital empty. Those Ni^{2+} octahedral complexes suffers from strong John-Teller distortion, leading to a square planar splitting of *d*-orbital. Thus unstable octahedral complexes suffer John-Teller distortion and get converted to stable square planar complexes. The geometry of square planar $[Ni(CN)_4]^{2-}$ and tetrahedral $[Ni(Cl)_4]^{2-}$ are shown in the Figure 7.2.8.

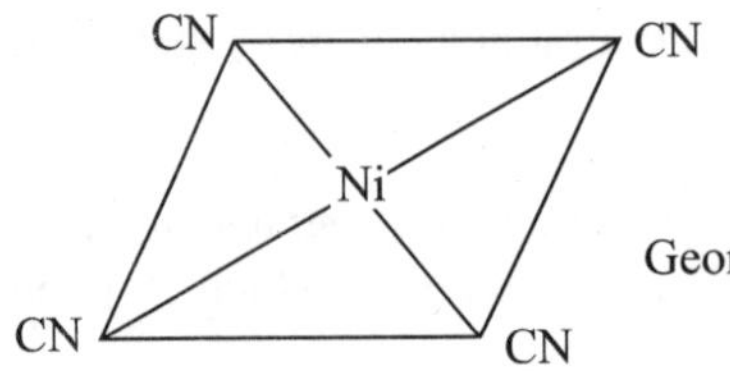

Hybridization of Ni^{2+} – dsp^2
Geometry: Two dimensional square planar
(Low spin complex)

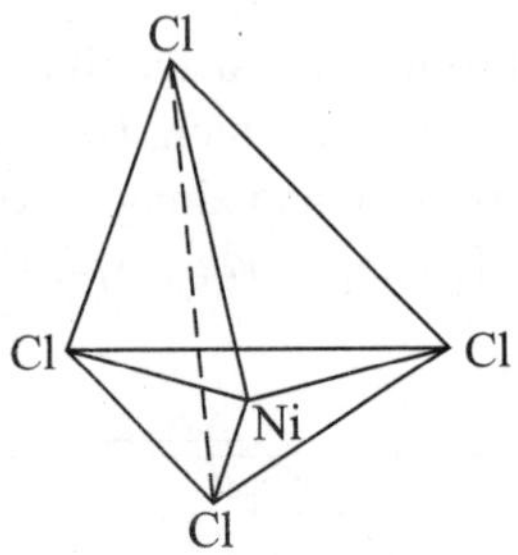

Hybridization of Ni^{2+} – sp^3
Geometry: Three dimensional tetrahedral
(High spin complex)

Fig. 7.2.8 Geometry of square planar and tetrahedral complexes of Ni^{+2}

Table. 7.2.5 Differences between tetrahedral and square planar complexes

Tetrahedral complex	*Square planar complex*
1. Hybridization is sp^3.	1. Hybridization is dsp^2.
2. Geometry is 3-dimensional tetrahedral.	2. Geometry is 2-dimensional square planar.
3. For 1st transitional elements only weak ligands can form tetrahedral complexes, i.e., $[NiCl_4]^{2-}$.	3. For 1st transitional elements only strong ligands can form square planar complexes, i.e., $[Ni(CN)_4]^{2-}$.
4. It is a high spin complex.	4. It is a low spin complex.

7.2.9 Magnetic Behaviour

Magnetic properties of a complex depends on number of unpaired electrons present in the splitted *d*-orbital of the central metal cation. The magnetic moment (μ) of a complex is related to the number of unpaired electrons (*n*) as follows:

$$\mu = \sqrt{n(n+2)} \text{ B.M.}$$

$$1 \text{ B.M. (Bohr Magneton)} = \frac{eh}{4\pi mc} = 5585 \text{ ergs/Gauss}$$

$e = 1.602 \times 10^{-19}$ Coulomb, $h = 6.625 \times 10^{-34}$ J–s, $m = 9.11 \times 10^{-31}$ kg

$c = 3.0 \times 10^8$ m/s

If $\mu = 0$, the complex is said to be *diamagnetic* and if $\mu > 0$, the complex is said to be *paramagnetic*.

s = spin magnetic moment for each electron = $\frac{1}{2}$

n = Total number of unpaired electrons.

S = Total spin magnetic moment = $n\left(\frac{1}{2}\right)$

Or, $S = \frac{n}{2}$

Or, $n = 2S$

Hence, $\mu = \sqrt{n(n+2)} = \sqrt{2S(2S+2)} = \sqrt{4S(S+1)} = 2\sqrt{S(S+1)}$

Or, $\mu = 2\sqrt{S(S+1)}$ B.M.

Magnetic properties of $[FeF_6]^{3-}$ and $[Fe(CN)_6]^{3-}$:

Fe^{3+} in $[FeF_6]^{3-}$

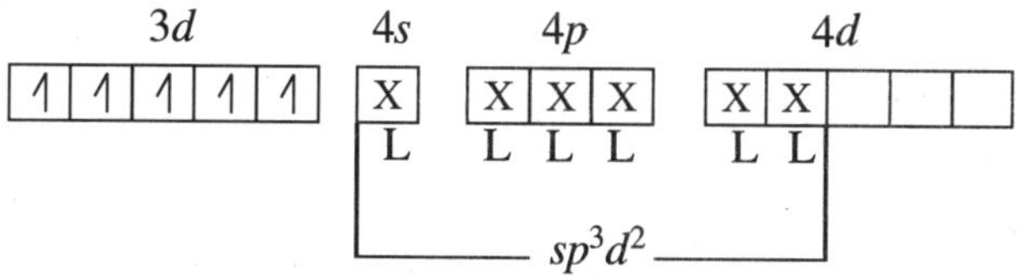

n = Number of unpaired electrons = 5.

$\mu = \sqrt{5(5+2)}$ B.M. = 5.92 B.M.

Fe^{3+} in $[Fe(CN)_6]^{3-}$

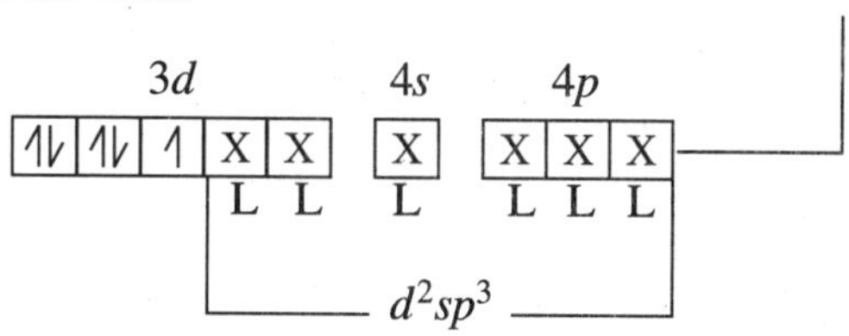

n = Number of unpaired electrons = 1.

$\mu = \sqrt{1(1+2)}$ B.M. = 1.73 B.M.

Thus inner orbital complexes are low spin complexes with low magnetic moment and outer orbital complexes are high spin complexes with high magnetic moment.

7.2.10 π-acid ligand

There are some ligands having vacant orbital within degenerate states, e.g., CO and NO^+.

Outermost electronic configuration of C in CO or N in NO^+ is:

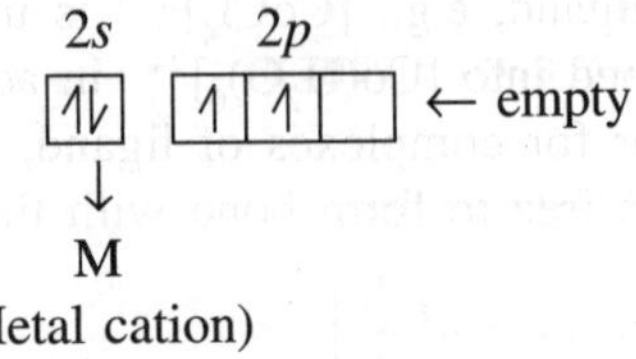

This empty $2p_z$-orbital often overlaps with filled *d*-orbital of central metal cation to form *pπ-dπ* back bonding. As the ligand gains electron pair through back bonding, it is called *π-acid ligand.* Ligand field strength of CO and NO^+ is very high due to the presence of this *pπ-dπ* back bonding. Hence π-acid ligands are categorized as *very strong* ligands and placed at extreme right in the spectrochemical series.

7.2.11 Factors influencing complex formation

(a) **Concentration** The stability of a complex often depends on the concentration of the ligand. In presence of excess of a ligand the nature and characteristics of a complex may change, e.g., tetrahedral cobalt thiocyanate ion changes to octahedral cobalt hexahydrate in presence of excess water.

$$\underset{\text{Tetrahedral and blue}}{[Co(SCN)_4]^{2-}} + 6H_2O \rightarrow \underset{\text{Octahedral and pink}}{[Co(H_2O)_6]^{2+}} + 4SCN^-$$

(b) **Nature of the central metal cation** Electrostatic field, created by the central metal cation, increases significantly as cationic radius decreases or cationic charge increases. So the ratio $\frac{\text{cationic charge}}{\text{cationic radius}}$ has a significant effect in complex formation. Higher the ratio, more stable the complex is.

(c) **Nature of ligand** So far as ligands are concerned, higher the coordination power, larger the CFSE and hence stabler the complex is. Thus, ligands, placed on the right side of the spectrochemical series, form stable complexes.

(d) **Ring formation** Ligands, having multidentate donors, usually form complexes with ring structure, popularly known as chelates. Chelation enhances the stability of a complex, particularly, 5 to 6-member chelates are very much stable due to reduced strain. In some cases, chelates are stabilised through resonance. For example, metal acetyl acetonate.

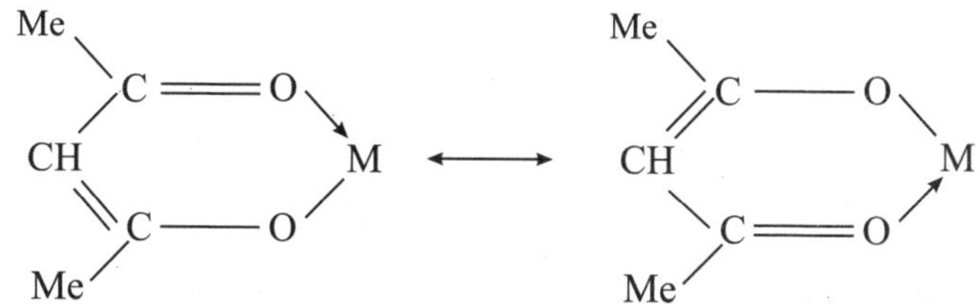

(e) Solvent and pH of the medium Complex with a weak ligand is unstable in a solvent which itself is a strong ligand, e.g., $[CoCl_4]^{2-}$ is unstable since Cl^- is a weak ligand and readily gets transformed into $[Co(H_2O)_6]^{2+}$ in aqueous medium.

pH is an important factor for complexes of ligand, having acid donor group. In those cases acid donor group is free to form bond with the central metal cation only at high pH.

Example 7.1 Ni-dimethyl–glyoxime complex.

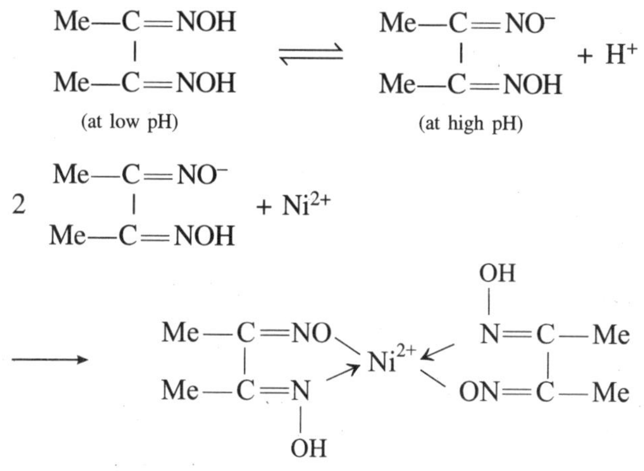

(Red precipitate)

Example 7.2 EDTA complex of metal cations

$$Na_2EDTA \equiv Na_2H_2Y$$

$$Na_2H_2Y \rightarrow H_2Y^{2-} + 2Na^+$$

$$M^{2+} + H_2Y^{2-} \rightarrow MY^{2-} + 2H^+ \quad \text{(at high pH)}$$

The reaction is favoured only in alkaline medium. The pH is maintained around 10 by ammonium hydroxide-ammonium chloride buffer solution.

(f) In aqueous solution complex ion dissociates like a weak electrolyte, e.g., $[Ag(CN)_2]^-$ dissociates in aqueous medium as follows:

$$[Ag(CN)_2]^- \rightleftharpoons Ag^+ + 2CN^-$$

At equilibrium, we can write that:

$$K_{\text{inst}} = \frac{[Ag^+][CN^-]^2}{\left[\{Ag(CN)_2\}^-\right]}$$

The dissociation constant K_{inst} is often called instability constant. Stronger the ligand field, lower the value of instability constant and hence more stable the complex is.

Table 7.2.6 Instability Constant (K_{inst}) values of some complex ions in aqueous solution.

Commplex	K_{inst}
$[Ni(NH_3)_6]^{2+}$	9.80×10^{-9}
$[Ag(NO_2)_2]^-$	1.48×10^{-3}
$[Ag(NH_3)_6]^+$	6.80×10^{-8}
$[Ag(S_2O_3)_2]^{3-}$	3.47×10^{-14}
$[Ag(CN)_2]^-$	1.41×10^{-20}

SOLVED PROBLEMS

P. 1 Explain why complexes of $[Co(NH_3)_6]^{3+}$, $[Co(en)_3]^{3+}$ and $[Co(NO_2)_6]^{3-}$ orange yellow in colour while $[CoF_6]^{3-}$ and $[Co(H_2O)_6]^{3+}$ are blue in colour.

Ans. According to spectrochemical series (NH_3) moderately strong ligand while (en) and (NO_2^-) are strong ligands. Thus the CFSE or Δ_0 of the corresponding complexes, i.e., $[Co(NH_3)_6]^{3+}$, $[Co(en)_3]^{3+}$ and $[Co(NO_2)_6]^{3-}$ is so high that, the above complexes absorb colour in indigo blue region for *d-d* transition of electron and hence they themselves appear as orange-yellow in colour. Whereas F^- and H_2O are considered as very weak ligands and hence Δ_0 of the corresponding complexes, i.e. $[CoF_6]^{3-}$ and $[Co(H_2O)_6]^{3+}$ is very low. So they absorb colour in orange-red region for *d-d* transition of electron and themselves appear as blue.

P. 2 Explain why $[Co(NH_3)_6]^{2+}$ is easily oxidized to $[Co(NH_3)_6]^{3+}$ even in presence of mild oxidizing agent.

Ans. Outershell electronic configuration of Co^{2+} is given by:

	3*d*	4*s*	4*p*	4*d*
Co^{2+}	↑↓ ↑↓ ↑ ↑ ↑	☐	☐☐☐	☐☐☐☐☐

In presence of moderately strong ligand, like ammonia, Co^{2+} forms inner orbital octahedral complex with hybridization d^2sp^3. To accommodate ligands in 3*d*-orbital one electron is shifted to higher 4*d*-orbital. Electronic configuration of Co^{2+} in $[Co(NH_3)_6]^{2+}$ is given by:

Co^{2+} in $[Co(NH_3)_6]^{2+}$

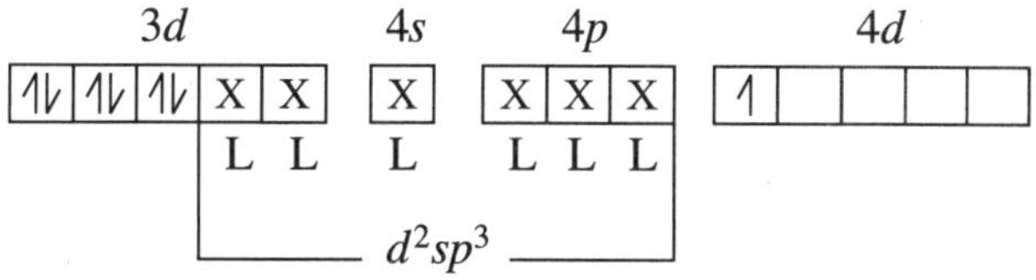

The lone electron in high energy 4*d*-orbital is very loosely bound and hence easily gets detached even in presence of mild oxidizing agent, like air. Thus Co^{2+} is oxidized to Co^{3+} and the complex $[Co(NH_3)_6]^{2+}$ changes to $[Co(NH_3)_6]^{3+}$.

P. 3 Write the EAN-value of central metal cation of the following:

$$[Ni(NH_3)_6]^{2+},\ [Zn(NH_3)_4]^{2+},\ [Ni(CN)_4]^{2-},\ [CoF_6]^{3-}.$$

Ans. $[Ni(NH_3)_6]^{2+}$: Ni^0: $[Ar]3d^84s^2$, Ni^{2+} : $[Ar]3d^8$.

Total number of electrons = 26. Each ligand (NH_3) coordinates with Ni^{2+} through its pair of electrons. Thus Ni^{2+} gains two electrons from each ligand.
So Ni^{2+} in $Ni(NH_3)_6]^{2+}$ gain $2 \times 6 = 12$ electrons from six ligands.
So, total number of electrons surrounding the nucleus of $Ni^{2+} = 26 + 6 \times 2 = 26 + 12 = 38$.
So, EAN – value of $Ni^{2+} = 38$.

$[Zn(NH_3)_4]^{2+}$: Zn^0: $[Ar]3d^{10}4s^2$, Zn^{2+} : $[Ar]3d^{10}$.

Total number of electrons = 28.
Gain of electrons from four monodentate ligands = $4 \times 2 = 8$.
Total number of electrons surrounding the nucleus of $Zn^{+2} = 28 + 8 = 36$.
EAN – value of $Zn^{+2} = 36$.

$[Ni(CN)_4]^{2-}$: Ni^0: $[Ar]3d^84s^2$, Ni^{2+} : $[Ar]3d^8$.

Total number of electrons = 26.
Gain of electrons from four monodentate ligands = $4 \times 2 = 8$.
Total number of electrons surrounding the nucleus of $Ni^{+2} = 26 + 8 = 34$.
EAN – value of $Ni^{+2} = 34$.

$[CoF_6]^{3-}$: Co^0: $[Ar]3d^74s^2$, Co^{3+} : $[Ar]3d^6$.

Total number of electrons = 24.
Gain of electrons from four monodentate ligands = $6 \times 2 = 12$.
Total number of electrons surrounding the nucleus of $Co^{3+} = 24 + 12 = 36$.
EAN – value of $Co^{3+} = 36$.

P. 4 Write the IUPAC name of:

(i) $[Co(H_2O)_2(NH_3)_4]Cl_3$ ⇒ Tetra-amine di-aqua cobalt(III) chloride.

(ii) $[Co(CN)(CO)_2(NO)]^+$ ⇒ Di-carbonyl cyano nitrosyl cobalt(II) ion.

(iii) $[Co(en)_2(Cl)(ONO)]^+$ ⇒ Chloro bis(ethylene diamine) nitrito cobalt(III) ion.

(iv) $[Pt(NH_3)_3(NO)Cl_2]Br_2$ ⇒ Tri-amino di-chloro nitrosyl platinum(IV) bromide.

P. 5 Write the formula of:

(i) Triamine chloro cyano nitro cobalt (III) ≡ $[Co(NH_3)_3(Cl)(CN)(NO_2)]$
(ii) Tetra-amine diaqua nickel (II) sulphate ≡ $[Ni(NH_3)_4(H_2O)_2]SO_4$
(iii) Dichloro-bis (ethylene diamine) Cobalt (III) chloride ≡ $[CoCl_2(en)_2]Cl$
(iv) Penta-amine-isothiocyanato chromium (III) ion ≡ $[Cr(NH_3)_5(NCS)]^{2+}$

P. 6 Calculate the magnetic moment of the following complexes:

$$[Co(NH_3)_6]^{3+},\ [Fe(H_2O)_6]^{3+},\ [Fe(CN)_6]^{4-}$$

Co^{3+} in $[Co(NH_3)_6]^{3+}$

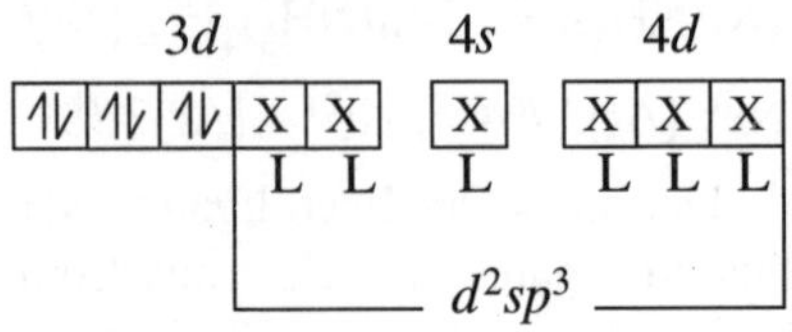

X indicates pair of electrons donated by each ligand to the central metal cation. Magnetic moment = μ = 0 B.M., since no unpaired electron is available in the complex.

$[Fe(H_2O)_6]^{3+}$

Fe^{3+} in $[Fe(H_2O)_6]^{3+}$

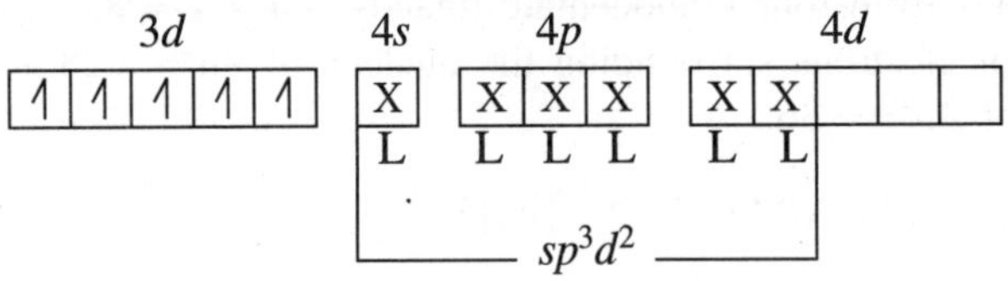

Number of unpaired electrons = n = 5.

Magnetic moment = $\mu = \sqrt{n(n+2)}$ B.M. = $\sqrt{5(5+2)}$ B.M. = 5.916 B.M.

$[Fe(CN)_6]^{4-}$

Fe^{2+} in $[Fe(CN)_6]^{4-}$

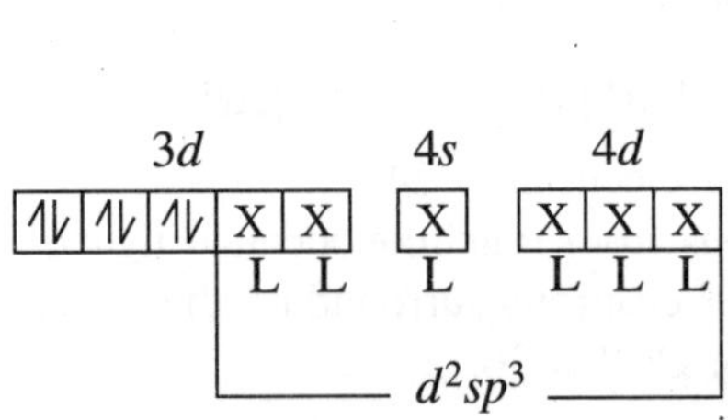

Number of unpaired electrons = n = 0. Magnetic moment = μ = 0 B.M.

P. 7 Explain the geometry and magnetic behaviour of the following:

$$[NiCl_4]^{2-},\ [Ni(CN)_4]^{2-},\ [Ni(CO)_4],\ [CoF_6]^{3-},\ [Co(NO_2)_6]^{3-}.$$

(i) Ni^{2+} in $[NiCl_4]^{2-} \equiv$ Hybridization = sp^3

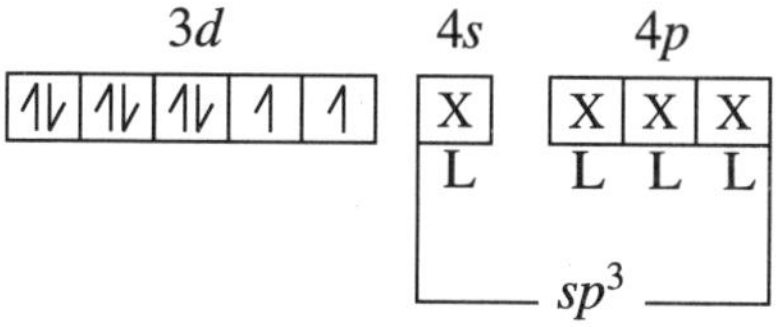

No. of unpaired electrons = 2

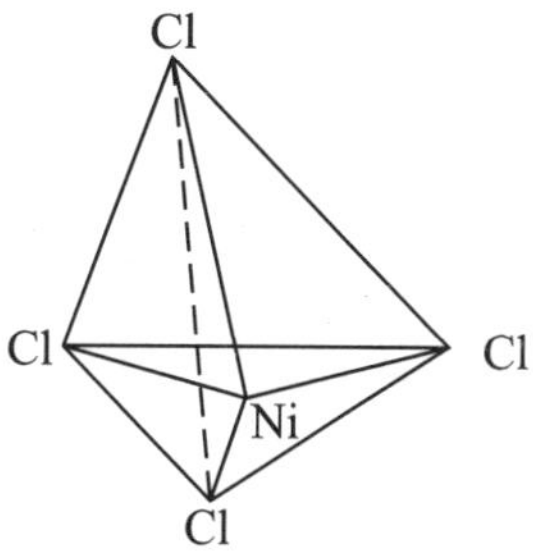

Geometry: Tetrahedral

Magnetic moment = $\mu = \sqrt{n(n+2)}$ B.M.

$$\mu = \sqrt{2(2+2)}\ \text{B.M.} = \sqrt{8} = 2.828\ \text{B.M.}$$

(ii) Ni^{2+} in $[Ni(CN)_4]^{2-} \equiv$ Hybridization = dsp^2

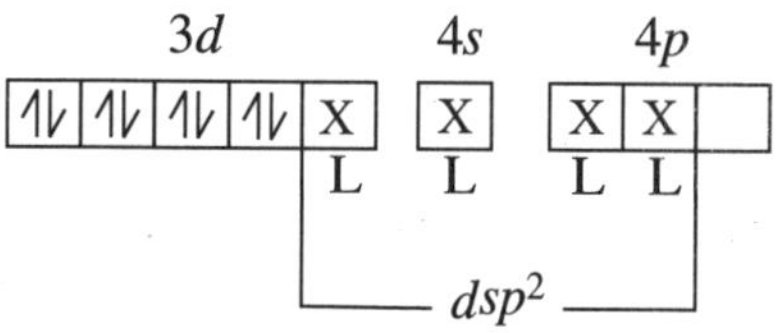

No. of unpaired electrons = 0

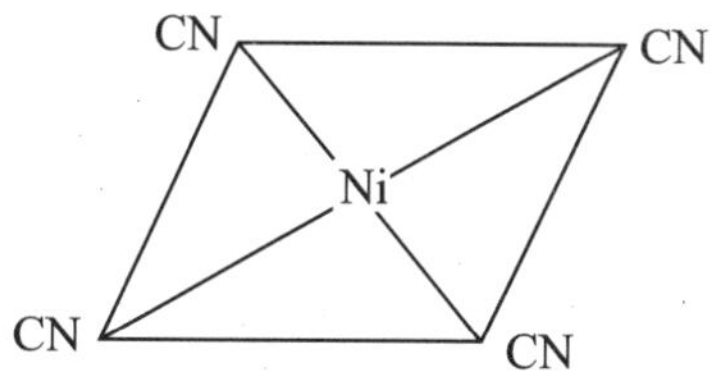

Geometry: Square planar

Magnetic moment = $\mu = \sqrt{n(n+2)}$ B.M. = 0 B.M.

(iii) Ni^0 in $[Ni(CO)_4]$

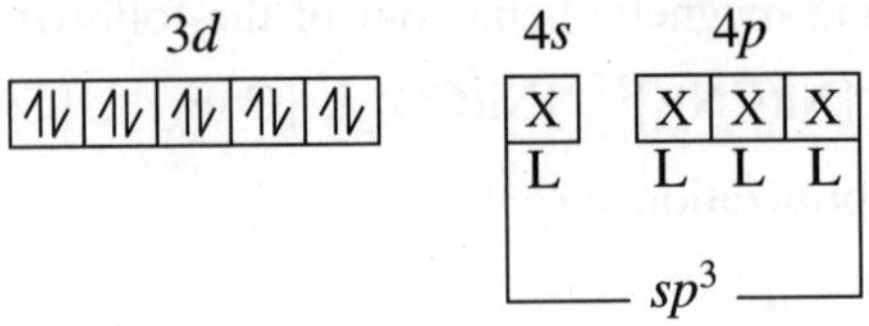

Hybridization = sp^3 Geometry = Tetrahedral

Number of unpaired electrons = 0

Magnetic moment = $\mu = \sqrt{n(n+2)}$ B.M. = 0 B.M.

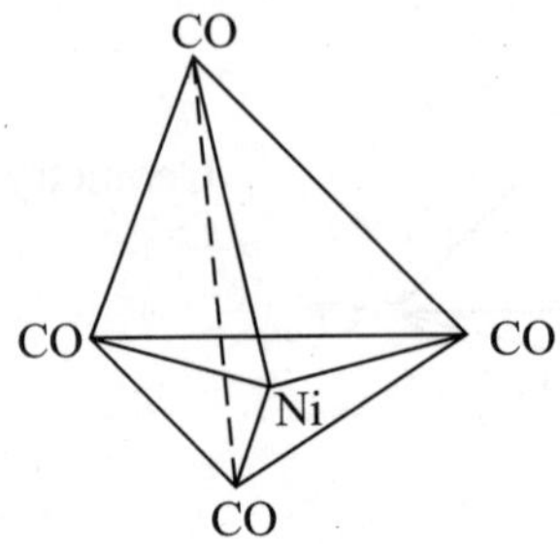

(iv) Co^{3+} in $[CoF_6]^{3-}$

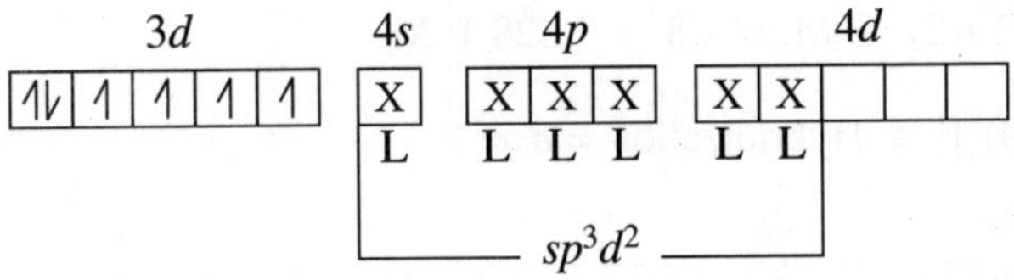

Hybridization = sp^3d^2 Geometry = Octahedral

Number of unpaired electrons = 4

$$\text{Magnetic moment} = \mu = \sqrt{4(4+2)} \text{ B.M.} = \sqrt{24} \text{ B.M.}$$

$$= 4.899 \text{ B.M.}$$

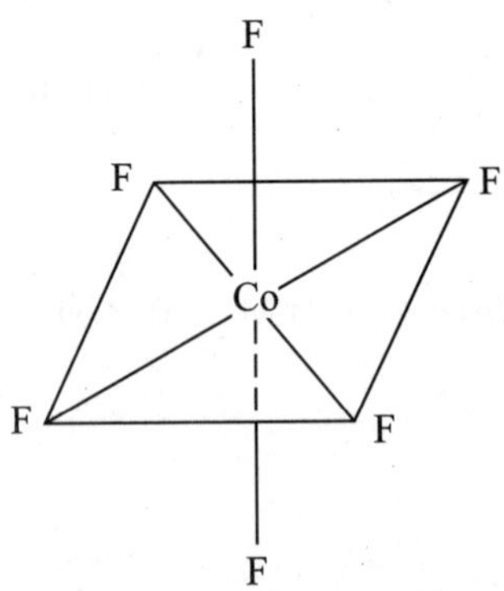

(iv) Co^{3+} in $[Co(NO_2)_6]^{3-}$

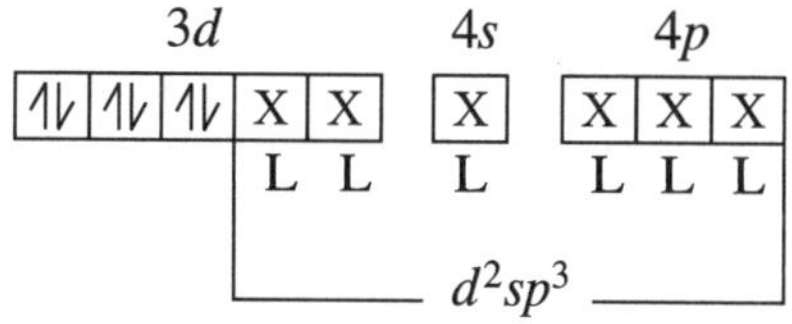

Hybridization = d^2sp^3 Geometry = Octahedral

Number of unpaired electrons = 0

Magnetic moment = $\mu = \sqrt{n(n+2)}$ B.M. = 0 B.M.

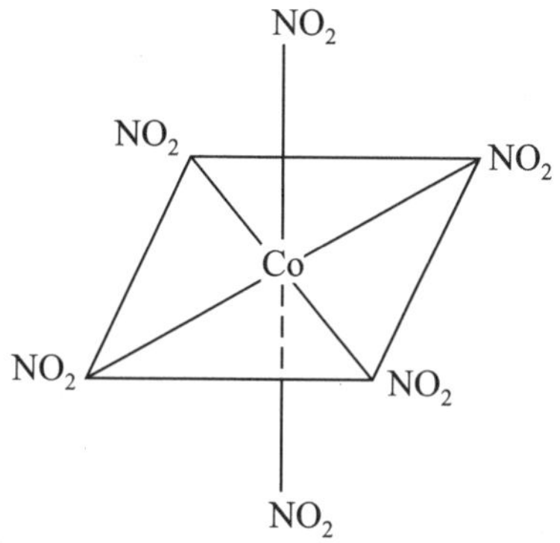

P. 8 What is instability constant? What is its significance?

Ans. In aqueous solution complex ion reversibly dissociates into its constituents. The corresponding equilibrium constant is known as instability constant, e.g. $[Ag(CN)_2]^-$ dissociates as:

$$[Ag(CN)_2]^- \rightleftharpoons Ag^+ + 2CN^-$$

$$K_i = \frac{[Ag^+][CN^-]^2}{\left[\{Ag(CN)_2\}^-\right]} = \frac{C_{Ag^+} \cdot C^2_{CN^-}}{C_{[Ag(CN)_2]^-}}$$

K_i is called instability constant of the complex $[Ag(CN)_2]^-$.

Significance *Higher* the value of K_i, *greater* the tendency of complex to dissociate or in other words, lesser the stability of the complex is. Obviously *weaker* the ligand field, *higher* the value of K_i is.

P. 9 Which of the following should possess higher conductivity in 1 molar solution?

(a) $[Co(NH_3)_5SO_4]Cl$ and $[Co(NH_3)_4(SO_4)Cl]$

(b) $[Cr(NH_3)_5NO_2]Br_2$ and $[Cr(NH_3)_3(NO_2)_3]$

Ans. (a) $[Co(NH_3)_5SO_4]Cl$ ionizes in aqueous medium as follows:

$$\underset{(C)}{[Co(NH_3)_5SO_4]Cl} \rightleftharpoons \underset{(\alpha C)}{[Co(NH_3)_5SO_4]^+} + \underset{(\alpha C)}{Cl^-}$$

but $[Co(NH_3)_4(SO_4)Cl]$ itself is a complex and does not ionize at all in aqueous medium. So in aqueous medium C moles of $[Co(NH_3)_4(SO_4)Cl]$gives $2\alpha C$ moles of ions but $[Co(NH_3)_4(SO_4)Cl]$ does not give any ion, where C = concentration of the complex = 1 M and α is the degree of dissociation. Thus $[Co(NH_3)_5SO_4]Cl$ should possess higher conductivity than $[Co(NH_3)_4(SO_4)Cl]$ in 1 molar solution.

(b) $[Cr(NH_3)_5NO_2]Br_2$ ionizes in aqueous solution as follows:

$$\underset{(C)}{[Cr(NH_3)_5NO_2]Br_2} \rightleftharpoons \underset{(\alpha C)}{[Cr(NH_3)_5NO_2]^{2+}} + \underset{(2\alpha C)}{2Br^-}$$

Total moles of ions available = $\alpha C + 2\alpha C = 3\alpha C$, where C = concentration of the complex = 1 M and α is the degree of dissociation.
Being itself a complex $[Cr(NH_3)_3(NO_2)_3]$ does not ionize at all in aqueous solution. So obviously in 1 M solution, $[Cr(NH_3)_5NO_2]Br_2$ should possess higher conductivity than that of $[Cr(NH_3)_3(NO_2)_3]$.

P. 10 $[Cr(NH_3)_4Cl_2]^+$ ion absorbs light in the visible region. The wavelength corresponding to maximum absorption is 750 nm. Calculate CFSE for the complex.

$$h = 6.625 \times 10^{-34} \text{ J–s}, \ c = 3.0 \times 10^8 \text{ cm/s}, \ N = 6.023 \times 10^{23}/\text{g.mole}.$$

Ans. Light energy absorbed by single molecule of the complex corresponding to maximum absorption $= \dfrac{hc}{\lambda}$

So, $\dfrac{hc}{\lambda}$ is equivalent to CFSE of a single complex ion or mathematically, we can write that:

$$\Delta_0 = \frac{hc}{\lambda}$$

where Δ_0 = CFSE of the octahedral complex ion.

$$\text{So CFSE for 1 mole of ions} = N\Delta_0 = \frac{Nhc}{\lambda} \text{ J/g mole}$$

$$= \frac{6.023\times10^{23}\times6.625\times10^{-34}\times3.0\times10^8}{750\times10^{-9}}$$

$$= \frac{6.023\times6.625\times3.0}{750}\times10^6 = 15.96\times10^4 \text{ J/g mole}$$

$$\approx 16\times10^4 \text{ J/g mole}$$

$$\text{CFSE} = 16\times10^4 \text{ J/g mole.}$$

8

GENERAL ORGANIC AND POLYMER CHEMISTRY

SECTION – I ORGANIC CHEMISTRY

8.1.1 Introduction

Organic Chemistry is also known as carbon chemistry. Two main constituents of any organic compound are carbon and hydrogen. Many organic compounds contain some other elements, e.g., oxygen, nitrogen, halogen etc.

Classification All organic compounds are divided into two main categories:

(a) Aliphatic compounds
(b) Aromatic compounds

Aliphatic compounds can be further subdivided into two classes, namely

(i) Open chain compounds e.g. C_3H_8 (propane), represented by $CH_3—CH_2—CH_3$
(ii) Cyclic compounds, e.g. C_6H_{12} (Cyclohexane), represented by:

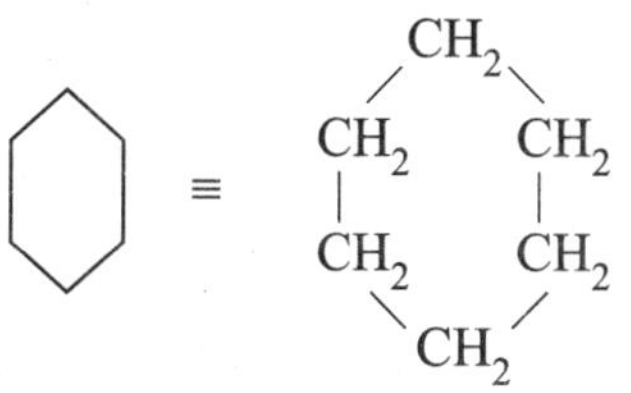

(Cyclohexane)

Aromatic compounds Compounds having planar ring structure and containing delocalized $(4n + 2)$ π-electrons, are called aromatic compounds. The π-electrons form clouds above and below the ring. One example is benzene, having chemical formula C_6H_6. No. of π-electrons = $4 \times 1 + 2 = 6$.

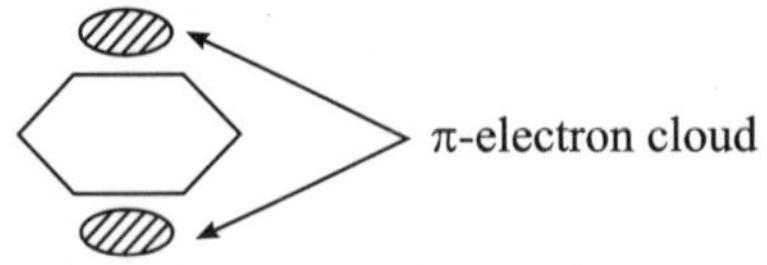

Functional Groups Any group, which is reactive to foreign agent, added from outside, is called functional group. Each functional group is a part of an organic compound and usually contains a reactive atom like oxygen, nitrogen, halogen etc. Examples are:

—OH ≡ Alcoholic group, —CHO ≡ Aldehyde group, —CO— ≡ Ketone group or Acyl group, —COOH ≡ Carboxylic acid group, —COOR ≡ Carboxylic ester group, —NO_2 ≡ Nitro group, —NH_2 ≡ Amino group, —X ≡ Halogen group, —SH ≡ Thiol group, —O— ≡ Ether group, —S— ≡ Thio-ether group.

8.1.2 Reaction mechanism

Organic compounds undergo chemical reactions through functional groups. The principle of understanding a reaction process is called reaction mechanism. Reaction mechanism is usually divided into three main categories:

(a) **One-step reaction**, where reactants are converted into products without forming any intermediate.
(b) **Two-step reaction**, where reactants are converted into products through one intermediate.
(c) **Three-step reaction**, where reactants are converted into products through two intermediates.

The potential energy diagrams of one- and two-step processes are shown in the Figure 8.1.1.

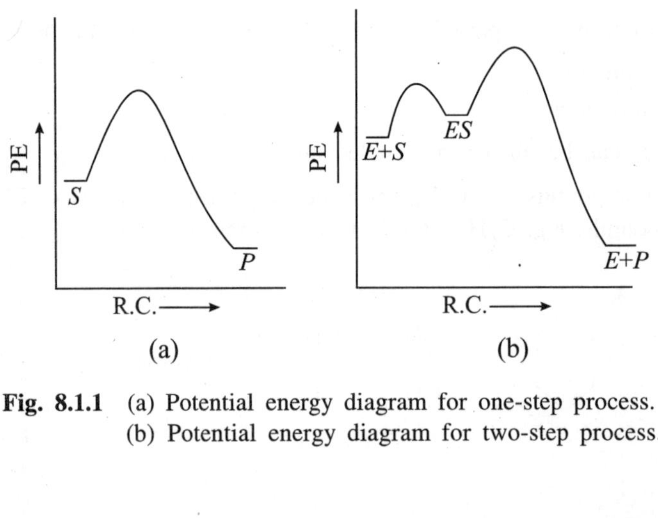

Fig. 8.1.1 (a) Potential energy diagram for one-step process.
(b) Potential energy diagram for two-step process.

Organic reactions often proceed through multi-steps. In any multi-step reaction the rate of the slowest step determines the rate of the whole reaction.

8.1.3 Controlling factors

Followings are the controlling factors to determine the mechanism of an organic reaction:

(a) Inductive effect
(b) Electromeric effect
(c) Resonance effect
(d) Hyperconjugative effect.

(a) Inductive effect Presence of a strong electronegative hetero-atom like oxygen brings about a permanent charge separation in an organic molecule, e.g., $^{\delta+}CH_3 \rightarrow OH^{\delta-}$.
Due to electronegativity difference oxygen attracts the C—O bond pair towards itself and gain some –ve charge. Consequently some +ve charge is developed on the neighbouring carbon atom. The phenomenon is called negative inductive effect (–I effect) and it arises due to the presence of an electronegative atom or electron-withdrawing group. The following series shows the decreasing order of –I effect of different groups/atoms.

$$—NO_2 > —CN > —COOH > —F > —Cl > —Br > —I$$

There are other groups, which collectively push the electron-cloud towards the neighbouring carbon atom. The phenomenon is called positive inductive effect (+I effect). Usually alkyl groups show the +I effect. In tertiary carbon atom three alkyl groups are attached to the central carbon atom, which ultimately shows *very strong* +I effect, whereas methyl group shows the *least* +I effect since only hydrogen atoms are attached to the central carbon atom. The following series shows the decreasing order of +I effect of different alkyl groups:

$$(Me)_3C \rightarrow\!\!- \; > \; (Me)_2CH \rightarrow\!\!- \; > \; Me—CH_2 \rightarrow\!\!- \; > \; CH_3 \rightarrow\!\!-$$

(b) Electrometric effect In an unsaturated organic molecule transfer of π-electron pair takes place in presence of an attacking agent. The phenomenon is called electrometric effect. Attacking agents are of two types.

(i) Electron-seeking, also known as electrophiles. There are two types of electrophiles:

(1) positively charged electrophiles, e.g., H^+, Cl^+, $[C{=}O]^+$, $[SO_3H]^+$ etc.
(2) neutral electrophiles, e.g., BF_3, $AlCl_3$, $FeCl_3$ etc.

(ii) Electron-donating, also known as nucleophiles. There are three types of nucleophiles:

(1) negatively charged nucleophiles, e.g. F^-, Cl^-, OH^-, CN^-, NH_2^-, NO_2^- etc.
(2) neutral nucleophiles, e.g., H_2O, NH_3, ROH etc.
(3) ambident nucleophile, e.g., CN^- can donate its electron pair from both sides to form either

$$N \overset{\rightarrow}{\equiv} C^- \quad \text{or} \quad C \overset{\rightarrow}{\equiv} N^-$$

(Cyanate) (Isocyanate)

NO_2^-, can donate its electron pair either through nitrogen or through oxygen

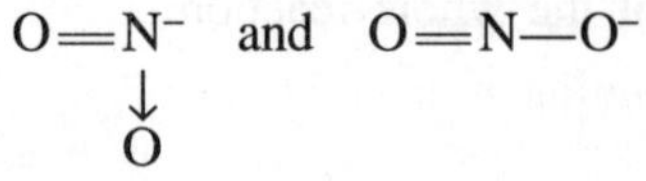

(Nitro) (Nitrito)

Positive electromeric effect (+*E* effect) If the attacking agent is an electrophile the organic molecule donates its π-electron pair to the attacking agent and the phenomenon is called +*E* effect, e.g.,

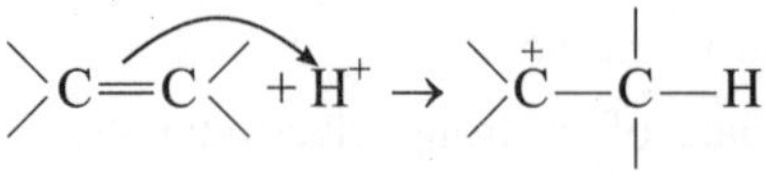

Negative electromeric effect (–*E* effect) If the attacking agent is a nucleophile the π-electron transfer occurs within the molecule and the phenomenon is called –*E* effect, e.g.,

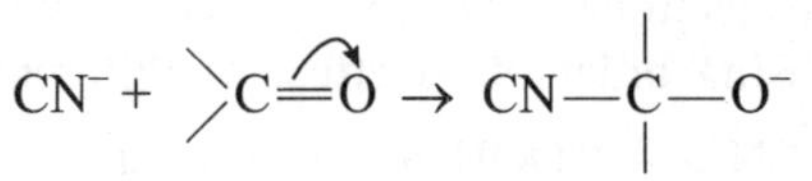

(c) Resonance effect If a molecule or ion can be represented by two or more Lewis structures the phenomenon is called resonance and each of these structures is called resonance structure or canonical form. Resonance structures are connected by double headed arrow ↔, e.g.,

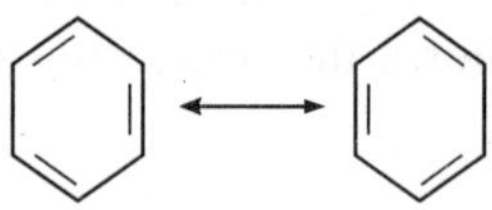

Characteristic features of resonance:

(a) In a molecule any one of the resonating structures does not represent the actual molecule nor does it explain all physical and chemical properties of the molecule.
(b) Position of each atom remains same in all resonating structures.
(c) Energy content of all the canonical forms is nearly the same.
(d) Only shifting of electron pair takes place, resulting in delocalization of charges. Hence energy of actual molecule is less than that of any canonical structure. This energy difference is known as resonance stabilization energy.
(e) Resonance structure, having less energy content, contributes more in the actual molecule, Equivalent or identical canonical structures make equal contribution to the actual molecule. For example, acetate ion:

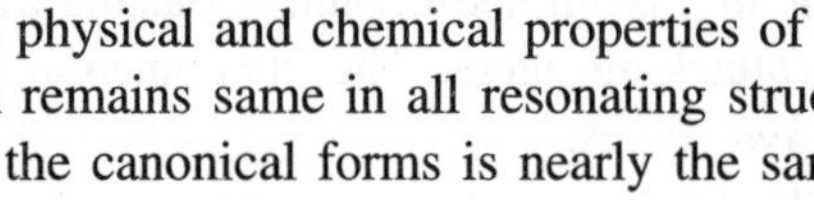

(I) (II)

(I) and (II) are identical or equivalent resonating structures.

(f) Resonating structures cannot be separated. They have only hypothetical existence.

(g) Resonating structure, containing greater number of covalent bonds, contributes more in the actual molecule, e.g.,

$$\underset{\text{(I)}}{CH_2{=}CH{-}CH{=}CH_2} \longleftrightarrow \underset{\text{(II)}}{\overset{-}{C}H_2{-}CH{=}CH{-}\overset{+}{C}H_2} \leftrightarrow \underset{\text{(III)}}{\overset{+}{C}H_2{-}CH{=}CH{-}\overset{-}{C}H_2}$$

Structure (I) is the most stable since it contains maximum number of covalent bonds.

(h) Resonating structure, in which each atom has complete valence shell, is more stable and makes larger contribution to the actual molecule. For example:

$$\underset{\text{(I)}}{\overset{+}{C}H_2{-}\ddot{\underset{..}{O}}{-}H} \longleftrightarrow \underset{\text{(II)}}{CH_2{=}\overset{+}{\underset{..}{O}}{-}H}$$

Carbon atom in structure – (I) has 6 electrons in the valence shell and has 8 electrons in structure – (II) in the valence shell. Thus population of structure – (II) is more than that of structure – (I) in actual molecule.

(i) Resonating structure, possessing no charge separation, is more stable than that with charge separation.

$$\underset{\text{(I)}}{CH_2{=}CH{-}\ddot{\underset{..}{Cl}}{:}} \longleftrightarrow \underset{\text{(II)}}{\overset{-}{C}H_2{-}CH{=}\overset{+}{\underset{..}{Cl}}{:}}$$

The population of structure – (I) is more than that of structure – (II) in actual molecule.

+R (+M effect) In a molecule or an ion if canonical structures are formed by donation of a pair of electrons from an atom or a group, the latter is said to have a +R or +M effect. The following groups show the +R (+M) effect:
–OH, –OR, $-NH_2$, –Cl, ...etc., e.g.,

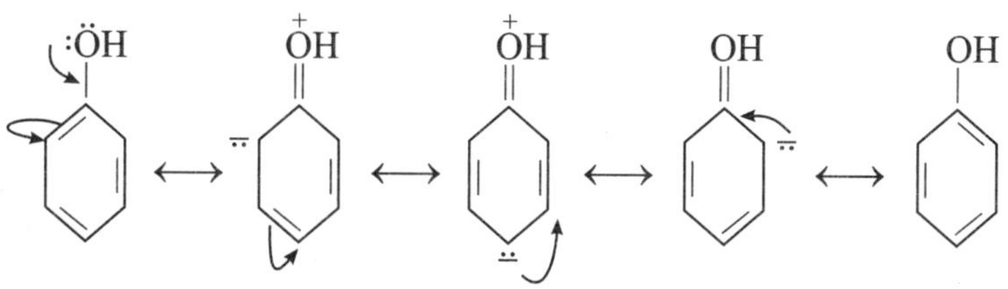

–R (–M) effect In some molecules/ions canonical structures are formed through withdrawal of a pair of electrons by an atom/or a group, then the latter is said to have –R(–M) effect. Examples are: $-NO_2$, –CN, > C = O^-, –COOR etc.

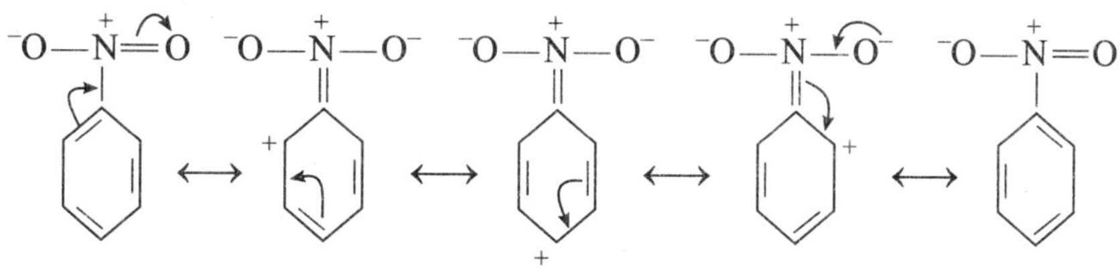

(d) Hyperconjugative effect (Baker-Nathan effect) In case of resonance canonical structures are formed through delocalization of π-electrons but when canonical structures are formed through delocalization of σ-electrons the effect is called hyperconjugative effect, first established by Baker and Nathan.

One example is delocalization of σ-electrons in propylene.

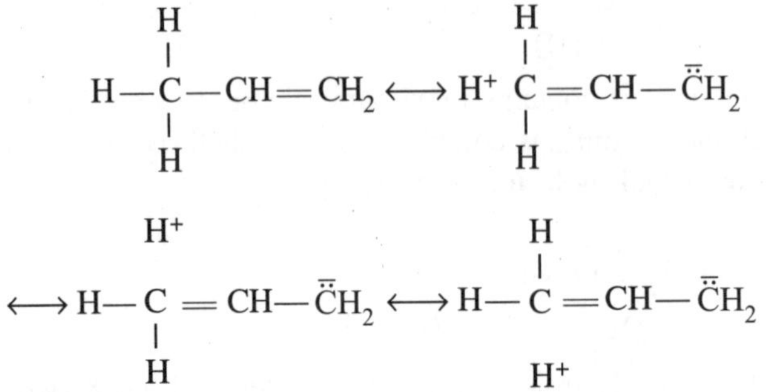

Characteristic features of hyperconjugative effect:

(i) H^+ ions in canonical structures are not free but are bound by the neighbouring π-cloud.

(ii) No bond exists between H^+ ion and C-atom. Hence hyperconjugative effect is also known as no bond resonance.

(iii) As only α-hydrogen atoms are responsible for hyperconjucative effect the effect is more and more pronounced with the increase in the number of α-hydrogen atoms. So methyl group has the maximum capability to slow hyperconjugative effect. The order of decrease in hyperconjugative effect is shown below:

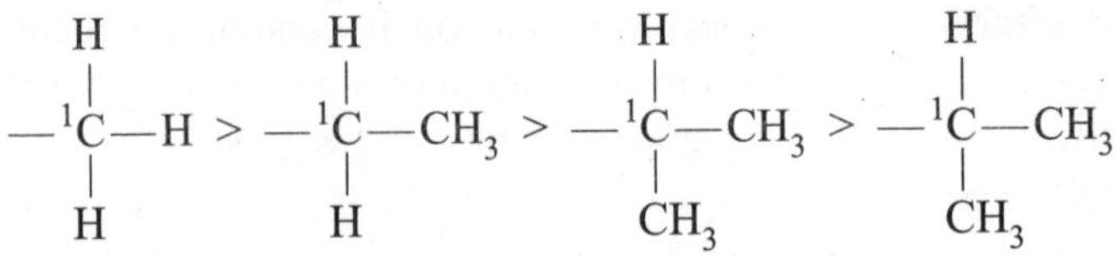

3 α–H atoms	2 α–H atoms	1 α–H atoms	No α–H atoms

1: α—carbon atom. The H-atom, attached to the α—C atom is called α—H atom.

(iv) Sometimes nature of product obtained in an organic reaction can be explained with the help of hyperconjugative effect, e.g. bromination of toluene.

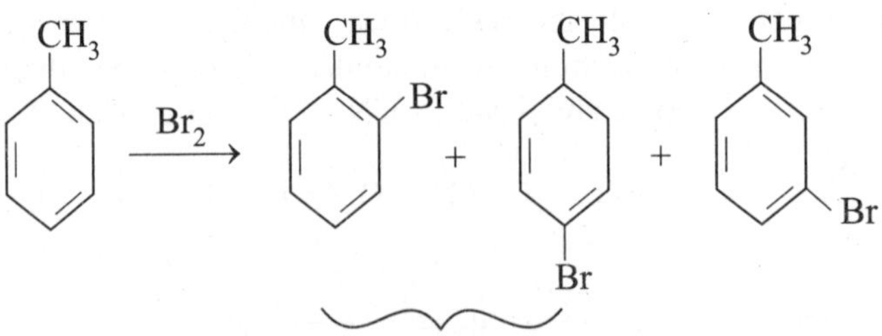

(Major or exclusive product)

The electrophile Br^+ attacks the benzene ring either at, *o*- or *p*- or *m*-position but the attack occurs only at that position where electron concentration is the highest.

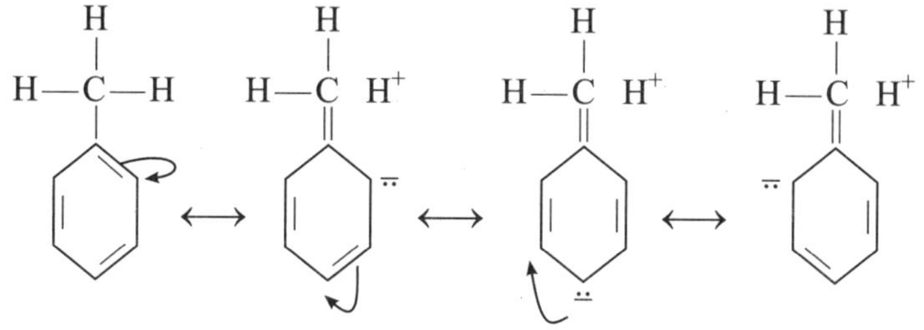

Similar resonating structures are also possible with the other two hydrogen atoms. It is evident that –ve charge or electron cloud is concentrated at *o*- and *p*-positions and hence ortho – and para-bromo toluenes are the exclusive products of the above reaction.

(v) Shortening of C—C single bond adjacent to double/triple bond can be explained by hyperconjugative effect.

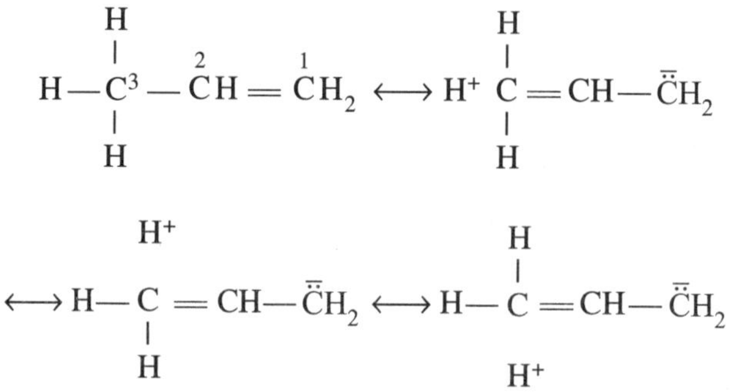

C^2—C^3 bond length is shortened due to partial double bond character.

(vi) Stability of different alkenes can be well explained with the help of hyperconjugative effect. In any alkene higher the number of α—H atoms, higher the number of hyperconjugative structures and consequently greater is its stability.

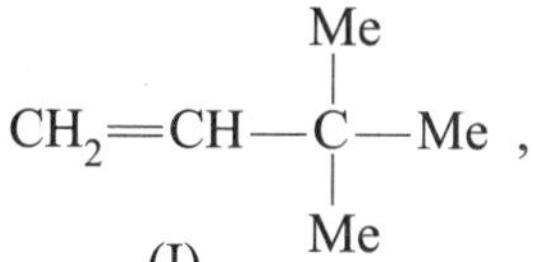
(I)

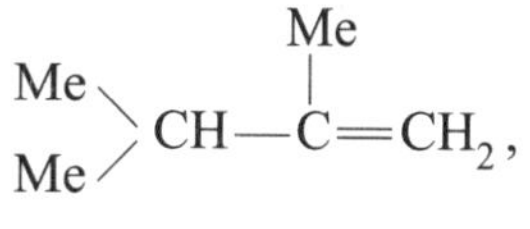
(II)

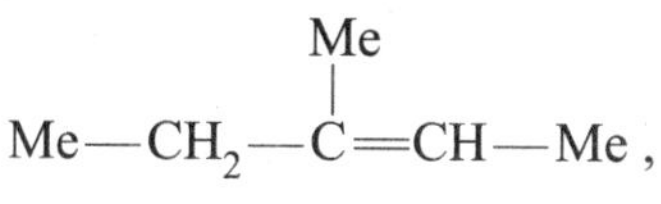
(III)

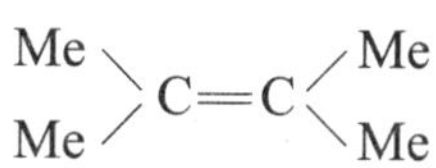
(IV)

Structure (I) has no α—H atom, structure (II) has 4 number of α—H atoms, structure (III) has 8 number of α—H atoms, structure (IV) has 12 number of α—H atoms. Thus the stability of alkanes is of the following order:

alkane (IV) > alkane (III) > alkane (II) > alkane (I)

The stability order indicates that the most substituted alkene is the most stable.

(vii) Stability of carbonium ion can be explained in the similar way. Tertiary carbonium ion possesses maximum number of α—H atoms and hence is the most stable.

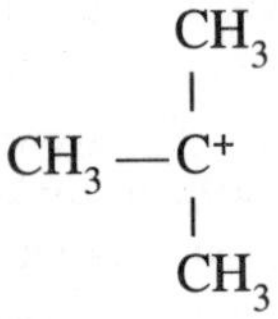

Tertiary carbonium ion, containing 9 α—H atoms

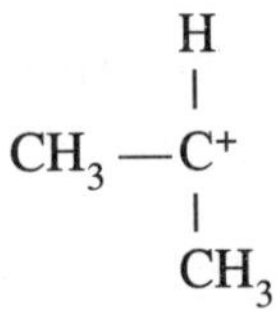

Secondary carbonium ion, containing 6 α—H atoms

$CH_3—CH_2^+$ Primary carbonium ion, containing 3 α—H atoms

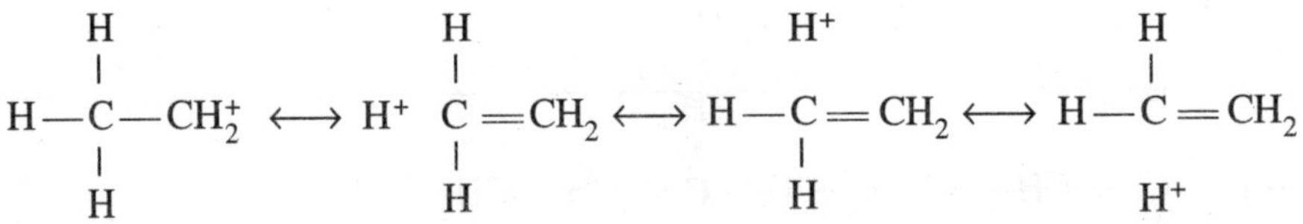

CH_3^+ Methyl carbonium ion, having no α-hydrogen atom.

(viii) Free radicals also get stabilized through α—H atoms. Hence tertiary free radical is the stablest free radical and methyl free radical is the most unstable free radical. The stability order is as follows:

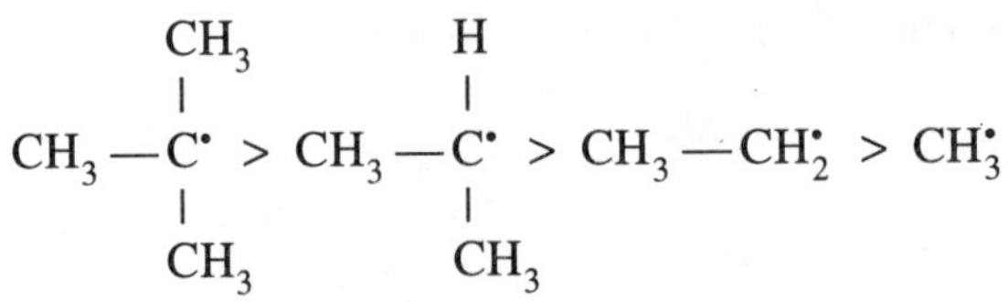

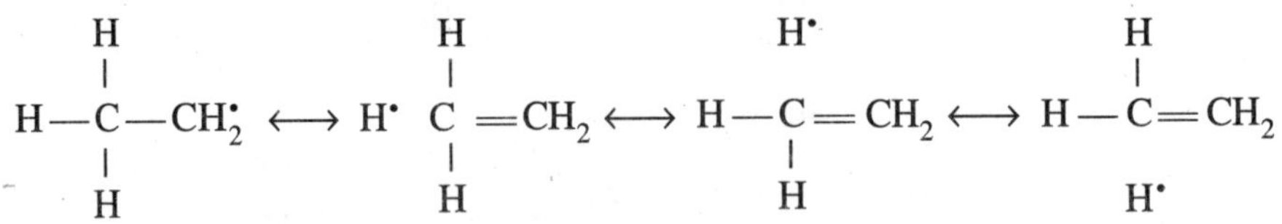

8.1.4 Effect of hybridization on acidity and electronegativity

Carbon remains in different states of hybridization in its compounds, e.g.,

Compound	*Formula*	*Hybridization of C-atom*	*% s-character*	pK_a
Acetylene	HC≡CH	*sp*	50	25
Ethylene	$CH_2{=}CH_2$	sp^2	33.34	44
Ethane	$CH_3{-}CH_3$	sp^3	25	50

PK_a values of the above compounds show that higher the *s*-character higher the acidity. So acetylene is the strongest acid among them. Outermost shell of carbon atom has *s*- and *p*-orbitals. 2*s*-orbital is much closer to the nucleus than 2*p*-orbital. Thus electron in *s*-orbital is strongly attracted towards the nucleus. Hence carbon atom, having higher % *s*-character, experiences stronger electrostatic force of attraction and hence shows higher electronegativity. Obviously carbon atom of acetylene is the strongest electronegative carbon atom among the three compounds and hence acetylene appears as the strongest acid among them.

8.1.5 Classification of organic reactions

Organic reactions can broadly be classified into three categories,

(i) Substitution reactions, (ii) Elimination reactions, (iii) Addition reactions.

8.1.5.1 Substitution reactions

If an atom or group, attached to carbon atom, is replaced by another atom or group, the reaction is said to be substitution reaction.

Substitution reactions may be classified into three categories:

(a) Free-radical substitution reaction
(b) Electrophilic substitution reaction
(c) Nucleophilic substitution reaction.

(a) Free-radical substitution When substitution reaction takes place through the formation of free radical the reaction is called free-radical substitution. Usually it takes place in presence of sun-light or photoradiation. One example is chlorination of methane:

$$Cl_2 \xrightarrow{h\nu} 2Cl^{\bullet} \quad \text{(Chain initiation)}$$

$$\left.\begin{aligned} CH_4 + Cl^{\bullet} &\longrightarrow CH_3^{\bullet} + HCl \\ CH_3^{\bullet} + Cl_2 &\longrightarrow CH_3Cl + HCl \end{aligned}\right\} \text{(Chain propagation)}$$

$$\left.\begin{aligned} Cl^{\bullet} + Cl^{\bullet} &\longrightarrow Cl_2 \\ CH_3^{\bullet} + Cl^{\bullet} &\longrightarrow CH_3Cl \\ CH_3^{\bullet} + CH_3^{\bullet} &\longrightarrow C_2H_6 \end{aligned}\right\} \text{(Chain termination)}$$

- **(b) Electrophilic substitution** If the substitution takes place through electrophilic attack the reaction is said to be electrophilic substitution reaction. It usually takes place in three steps:

Step-I: Formation of an electrophile

(1) $R{-}Cl + AlCl_3 \xrightarrow{\text{Fast}} R^+ + AlCl_4^-$

(2) $RCOCl + FeCl_3 \xrightarrow{\text{Fast}} RCO^+ + FeCl_4^-$

(3) $HO{-}NO_2 + H^+ \xrightarrow{\text{Fast}} H_2O^+ - NO_2 \xrightarrow{\text{Fast}} NO_2^+ + H_2O$

(4) $H_2SO_4 + H^+ \xrightarrow{\text{Fast}} {}^+SO_3H + H_2O$

(5) $Cl{-}Cl + AlCl_3 \xrightarrow{\text{Fast}} Cl^+ + AlCl_4^-$

Step-II: Electrolphilic attack

(a) 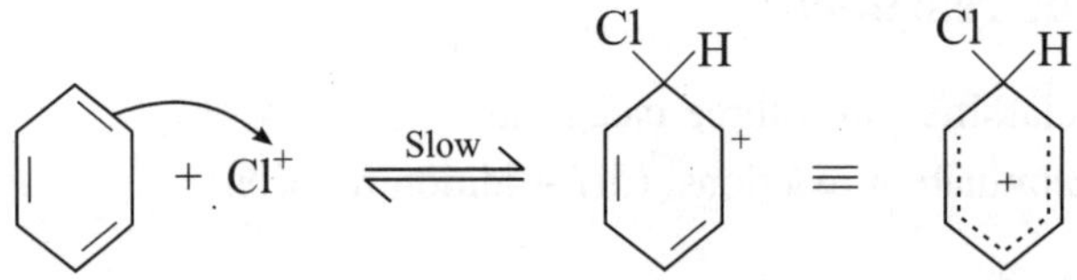

(b) 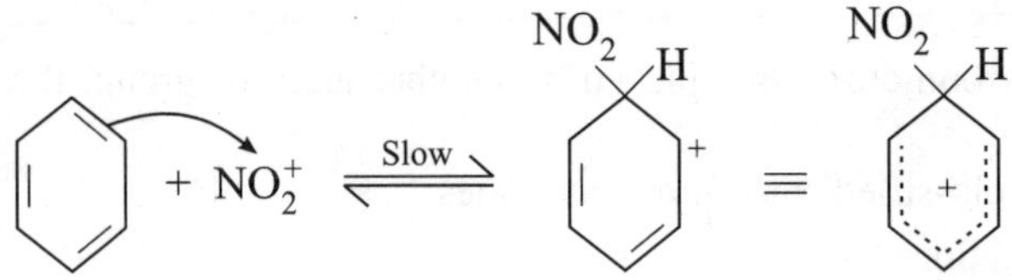

(c) 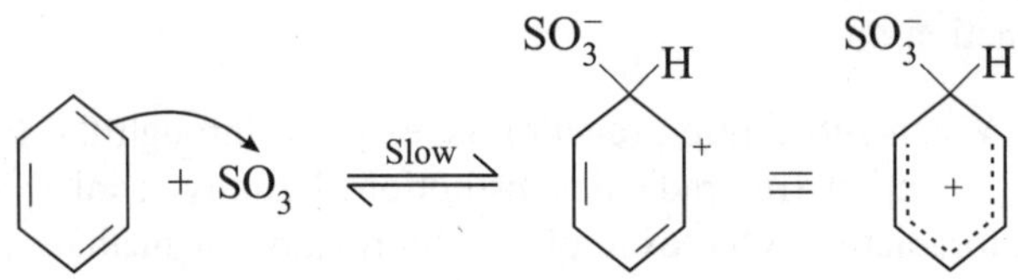

(d) 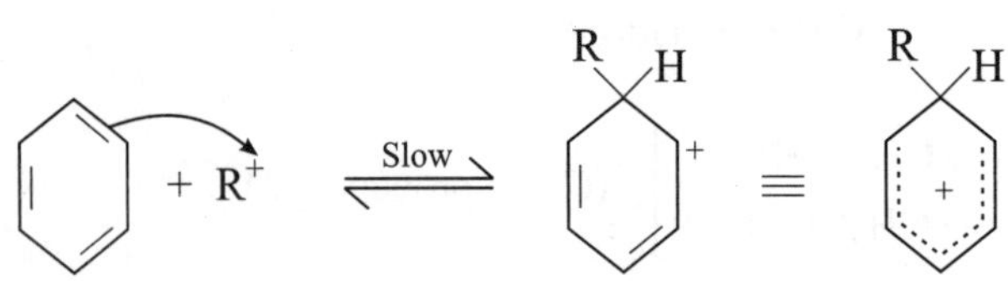

(e) 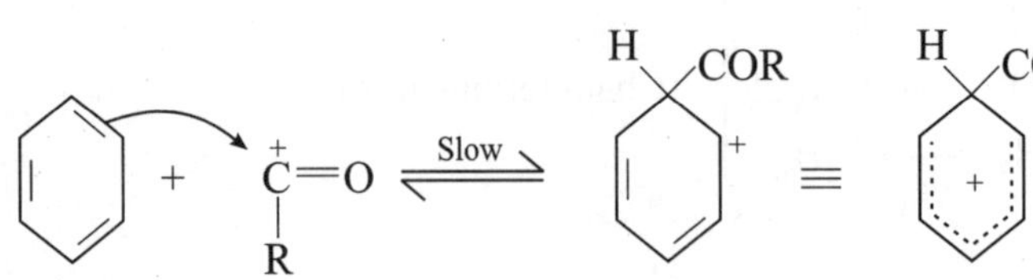

Step-III: Elimination of proton

(a) 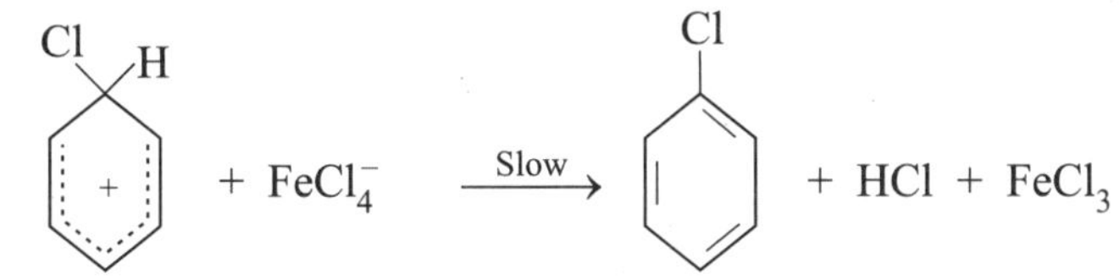

(b) 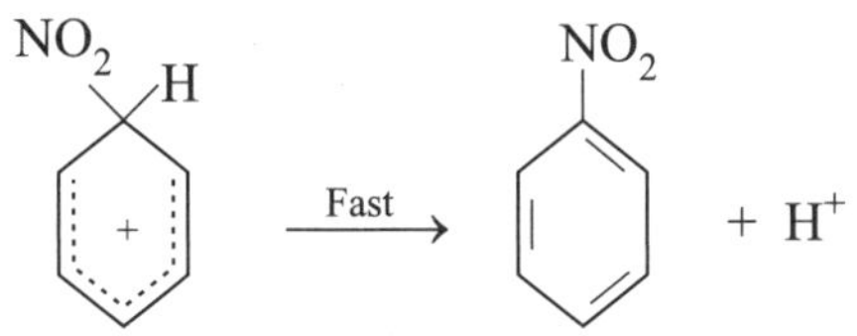

(c) 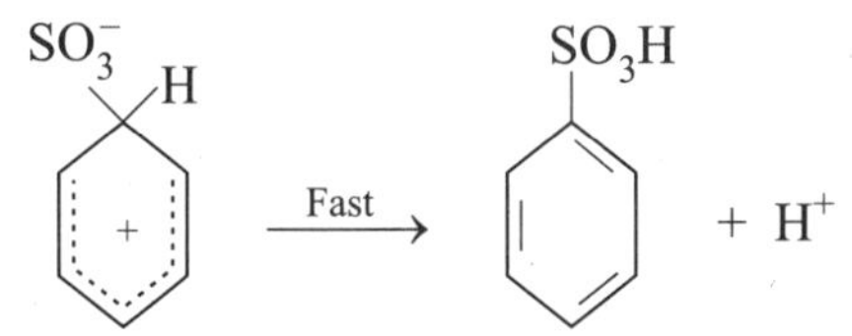

(d) 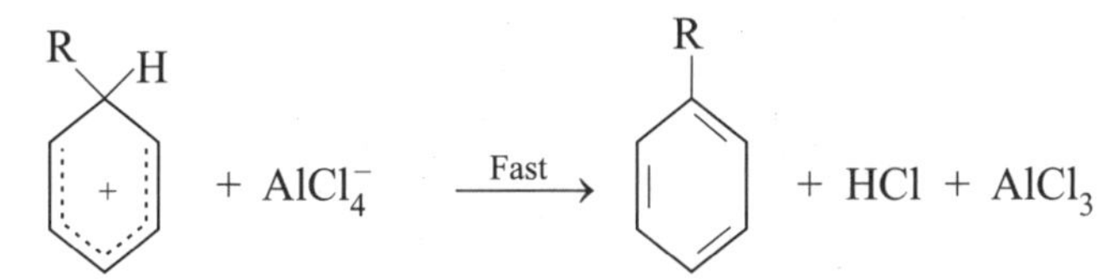

(e) 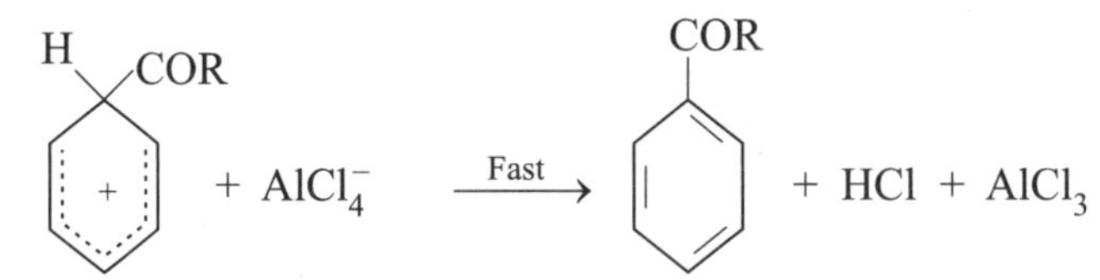

Experimental results show that step–II is the slowest step and hence it is the rate-determining step. Thus, the overall rate of the above reaction (J) can be written as:

$$J = k[C_6H_6]\ [Y^+]$$

Where benzene (C_6H_6) is the substrate and Y^+ is the electrophile. The corresponding potential energy diagram is shown in the following figure.

In this type of reaction electrophile plays the key role in substitution reaction and the latter is often called electrophilic substitution reaction, represented as S_E2, because two species are involved in the rate determining step.

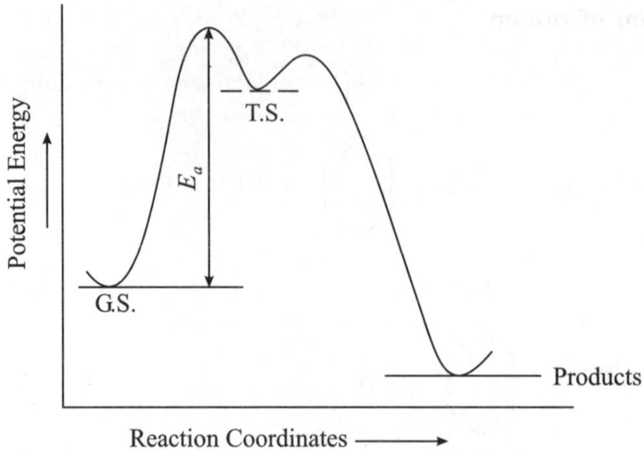

Fig. 8.1.2 Potential energy curve of an electrophilic aromatic substitution reaction.

Nucleophilic substitution reactions or S_N reactions

In these reactions nucleophiles play the major role. The main components are substrate, nucleophile and solvent. The substrate mainly consists of an alkyl group, on which nucleophilic attack occurs and a leaving group. Leaving group has a strong tendency to retain its pair of electrons and hence is considered as a weak base. On the other hand, nucleophile has a strong tendency to donate its pair of electrons and hence is considered to be a strong base. The overall substitution reaction may proceed in two different ways. In one reaction mechanism only one species is involved in the rate determining step. That particular substitution reaction is designated as S_N1.

Example Overall reaction

$$\text{RBr} \xrightarrow[\text{Solvent}]{\text{KOH}} \text{ROH} + \text{KBr}$$

Step–I: $$\text{R—Br} \xrightarrow[\text{Solvent}]{\text{Slow}} \text{R}^+ + \boxed{\text{Br}^-} \leftarrow \text{leaving group}$$

The carbonium ion R^+ is stabilized through solvation.

Step–II: $$\text{R}^+ + \boxed{\text{OH}^-} \xrightarrow[\text{Solvent}]{\text{Fast}} \text{R—OH}$$

Step–I is the slowest step, which involves only one reactant species. Hence the substitution reaction is called S_N1 reaction.

In another type of substitution reaction mechanism, two species are involved in the rate determining step. That type of *substitution* reaction is designated as S_N2.

Example Overall reaction

$$\begin{matrix}\text{R}\\ \text{R}\end{matrix}\!\!>\text{CH—Br} \xrightarrow[\text{Solvent}]{\text{KOH}} \begin{matrix}\text{R}\\ \text{R}\end{matrix}\!\!>\text{CH—OH} + \boxed{\text{Br}^-} \leftarrow \text{leaving group}$$

Step-I: 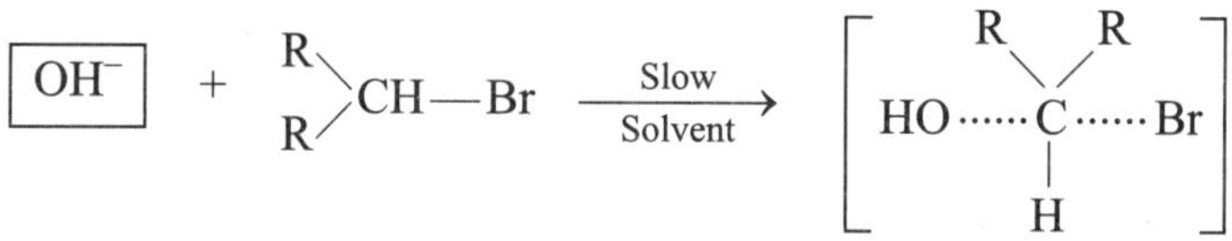

Transition state (T.S.)

Step-II: 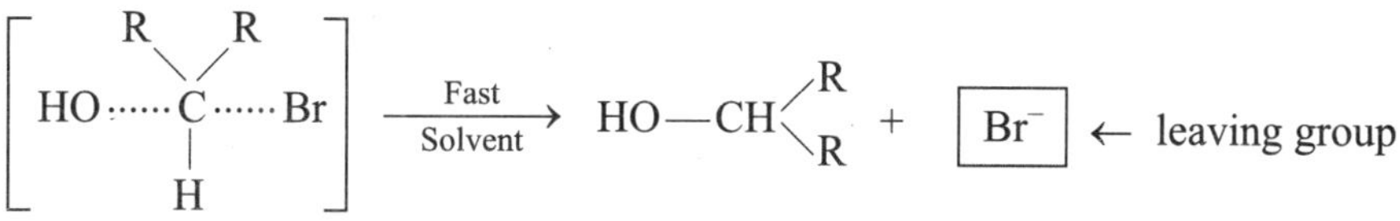

Transition state (T.S.)

Step–I is the slowest step, which involves two reactant species. Hence the substitution reaction is called S_N2 reaction. Moreover, T.S. complex is destabilized through steric hindrance as five groups are attached to the central carbon atom.

Characteristic features of S_N1 reaction

(a) The rate of the S_N1 reaction is given by: rate (J) = k[RBr]
Thus an S_N1 reaction is independent of the concentration of base.

(b) The rate of an S_N1 reaction largely depends on the stability of the cation formed in the rate determining step. Furthermore, stability of the carbocation depends on inherent stability mechanism of the carbocation as well as on the polarity of the solvent.

(c) Stability of a carbocation increases as the number of alkyl groups in the carbocation increases, since each alkyl group stabilizes the positive charge on the central carbon atom through hyper-conjugative effect or inductive effect.

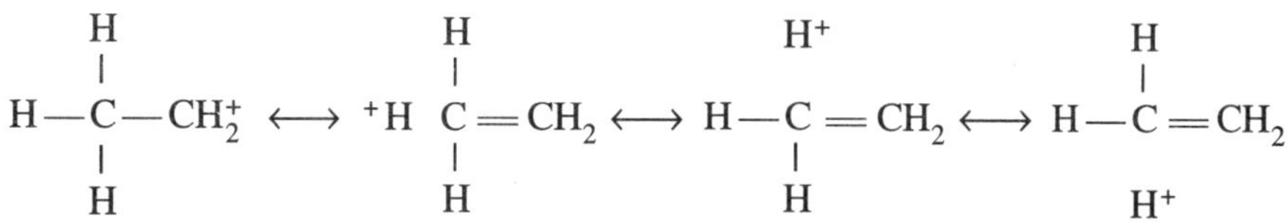

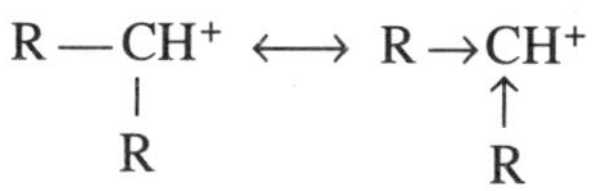

So it is evident that, tertiary carbocation, i.e., R_3C^+ is the stablest carbocation followed by secondary carbocation, i.e., R_2CH^+ and primary carbocation, i.e., RCH_2^+. Methyl carbonium ion, i.e., CH_3^+ is the least stable carbonium ion. Thus tertiary halide quickly responds to S_N1 mechanism but methyl halide hardly follows S_N1 mechanism, due to its high activation energy as reflected in the following figure.

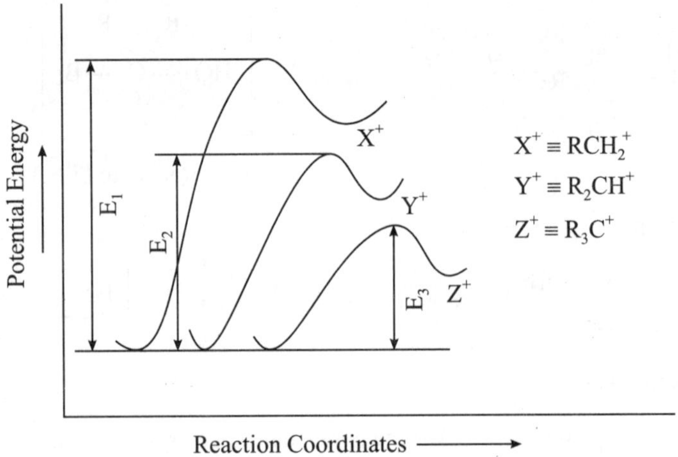

Fig. 8.1.3 Potential energy curves for S_N1 reactions. E_1, E_2 and E_3 are the activation energies for the formation of primary, secondary and tertiary carbonium ions.

It is evident from the above figure that activation energy for primary carbonium ion is very high and in actual practice primary carbonium ion (RCH_2^+) do not follow S_N1 reaction path. However, there are some exceptions:

(i) In presence of an alkali solution, benzyl chloride ($C_6H_5CH_2Cl$) follows S_N1 mechanism through the formation of primary carbonium ion $C_6H_5\ CH_2^+$ as follows:

$$C_6H_5 - CH_2Cl \underset{\text{Solvent}}{\overset{\text{Slow}}{\rightleftharpoons}} C_6H_5CH_2^+ + Cl^-$$

$$C_6H_5CH_2^+ + OH^- \xrightarrow[\text{Solvent}]{\text{Fast}} C_6H_5 - CH_2OH$$

This anomalous behaviour of benzyl chloride is believed to be due to pronounced stability of benzylium cation ($C_6H_5\ CH_2^+$) through resonance as shown below:

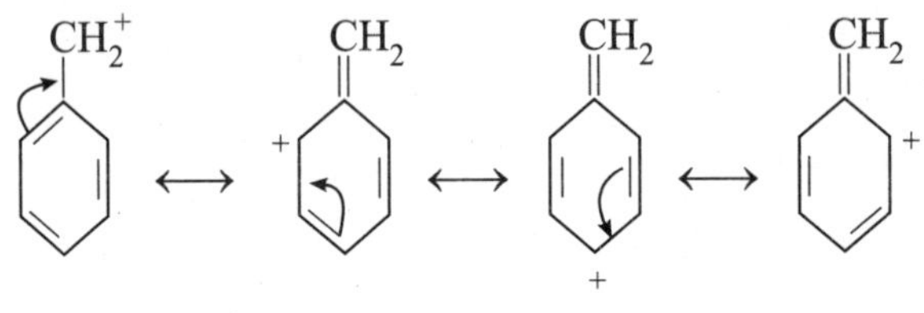

(ii) Allyl carbonium ion shows similar resonance stabilization and hence allyl halide follows S_N1 reaction path.

$$CH_2{=}CH{-}CH{-}Cl \underset{\text{Solvent}}{\overset{\text{Slow}}{\rightleftharpoons}} CH_2{=}CH{-}CH_2^+ + Cl^-$$

$$CH_2{=}CH{-}CH_2^+ + OH^- \xrightarrow[\text{Solvent}]{\text{Fast}} CH_2{=}CH{-}CH_2OH$$

Allylium ion stabilizes through resonance as follows:

$$CH_2{=}CH{-}\overset{+}{C}H_2 \longleftrightarrow {}^{+}CH_2{-}CH{=}CH_2$$

(d) Nature of solvent has pronounced effect on S_N1 reaction mechanism. It has been observed that higher the polarity (or higher the dielectric constant ε) faster is the reaction. One example is solvolysis of tertiary halide Me_3CBr. The reaction rate in aqueous ethanol system is 3×10^4 times faster than that in pure ethanol system. The pronounced stabilization of T.S. compound or carbonium ion is believed to be the prime reason for this rate difference. The reactant species is neutral and hence stabilization through solvation is not important but in the transition state separation of charges occurs leading to the formation of two ions. Both the dipolar T.S. compound and charged species are highly stabilized through solvation and hence forward reacation path of the following equilibrium is favoured in presence of a polar solvent.

$$R{-}X \rightleftharpoons \underset{\text{(T.S.)}}{\left[\overset{\delta+}{R}\cdots\cdots\overset{\delta-}{X}\right]} \rightarrow R^+ + X^-$$

As the dielectric constant of aqueous ethanol is much higher than that of pure ethanol, charged species are more stable in aqueous ethanol rather than in pure ethanol itself. Obviously S_N1 reaction rate is much higher in aqueous ethanol than that in pure ethanol. In fact the charged species in S_N1 reaction need to be stabilized through solvation or otherwise S_N1 reaction cannot take place. That is why, S_N1 reaction is almost uncertain in gas phase reactions.

(e) If substitution reaction proceeds through S_N1 mechanism, more than one product is often obtained. This happens due to rearrangement of carbonium ions, formed in the transition state.

Example I

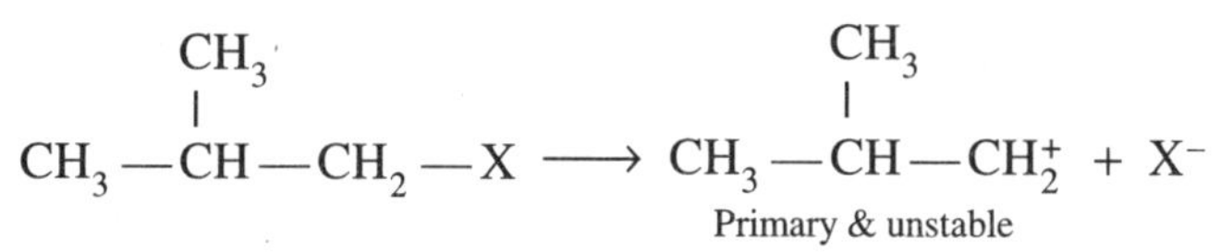

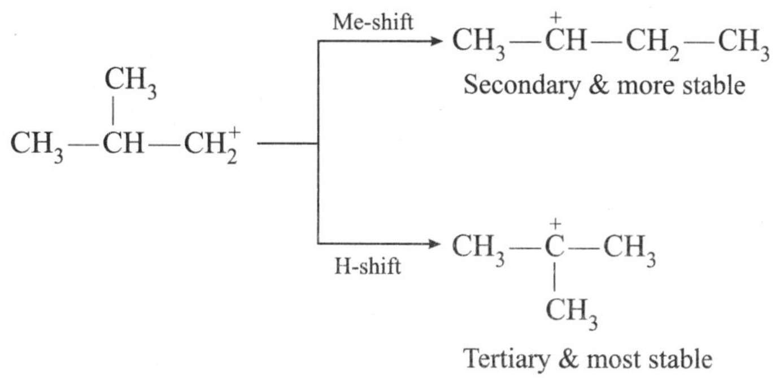

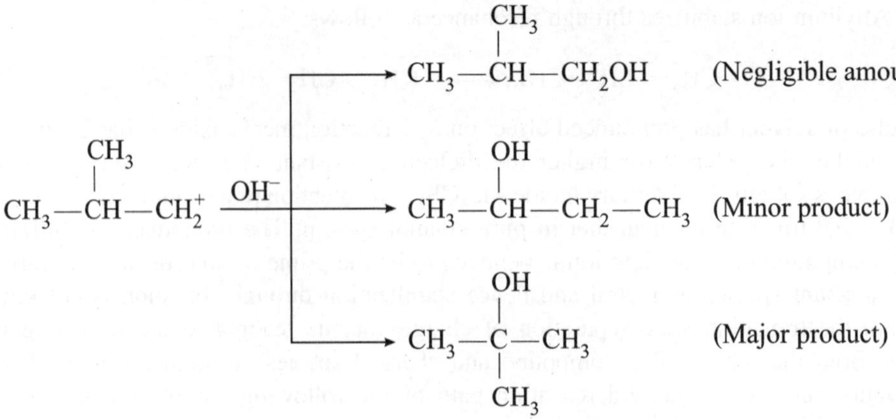

Rearrangement of carbonium ions in the T.S. of S_N1 reaction may also occur through resonance, thereby producing a mixture of products. One such example is the reaction between benzyl chloride with hydroxyl ion.

$$C_6H_5—CH_2Cl \underset{\text{Solvent}}{\overset{\text{Slow}}{\rightleftharpoons}} C_6H_5\ CH_2^+ + Cl^-$$

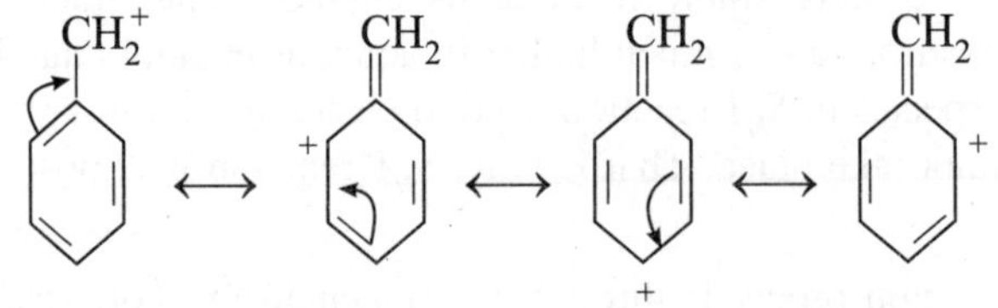

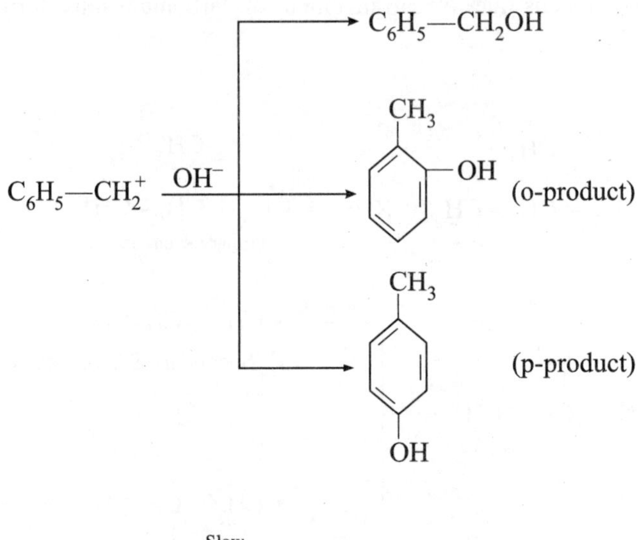

Example II

$$RCH═CH—CH_2—X \underset{\text{Solvent}}{\overset{\text{Slow}}{\rightleftharpoons}} RCH═CH—CH_2^+ + X^-$$

$$RCH═CH—CH_2^+ \longleftrightarrow R\overset{+}{C}H—CH═CH_2$$

$$\mathrm{RCH{=}CH{-}CH_2^+ + OH^- \xrightarrow[\text{Solvent}]{\text{Fast}} RCH{=}CH{-}CH_2{-}OH}$$

$$\mathrm{R\overset{+}{C}H{-}CH{=}CH_2 + OH^- \xrightarrow[\text{Solvent}]{\text{Fast}} \underset{\displaystyle OH}{\underset{|}{RCH}}{-}CH{=}CH_2}$$

(f) Racimization As we have discussed, a carbonium ion is formed during S_N1 mechanism. The carbon atom in the carbomium ion is sp^2 – hybridized and planar. So steric hindrance has no role to play due to wide bond angle (120°). So three groups, attached to the central carbon atom, can rearrange among themselves freely. As a result of this, nucleophile may attack the central carbon atom from both sides (front and rear positions) with equal probability. Thus an optically active compound, when undergoes S_N1 reaction, always produces a racemic mixture with complete loss of optical activity.

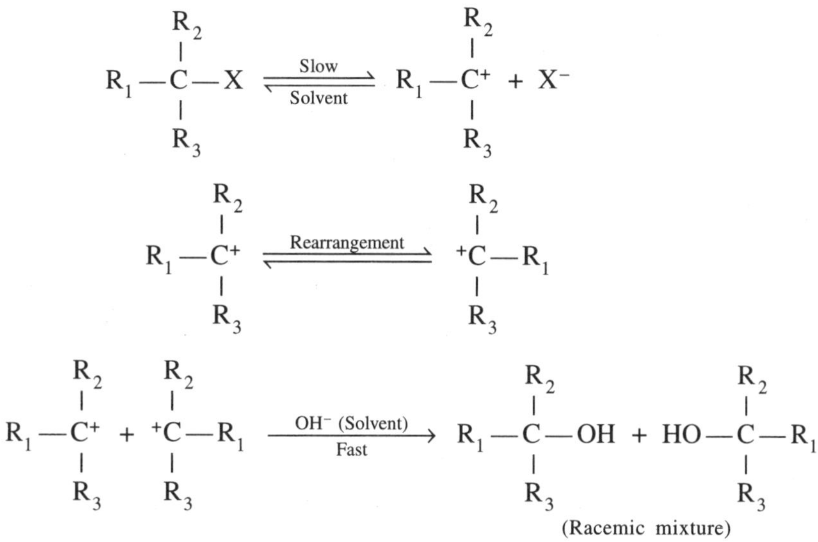

Characteristic features of S_N2 reaction

(a) The rate of the S_N2 reaction is given by:

$$\text{Rate (J)} = k[\text{RBr}]\,[\text{OH}^-]$$

$$k = \text{Reaction rate constant.}$$

Thus an S_N2 reaction depends on the concentration of base.

(b) In the T.S. of an S_N2 reaction five groups are attached to the central carbon atom. So steric hindrance plays an important role in S_N2 reaction.

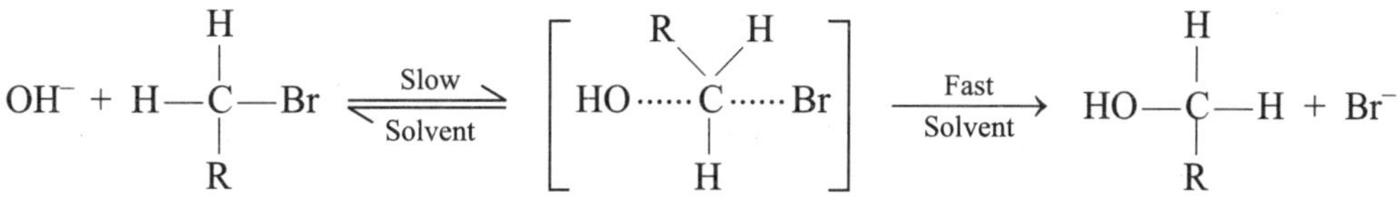

(T.S.)

Besides the nucleaphile and leaving group, three more groups are attached to the central Carbon atom. If all the three groups are H-atoms, T.S. is stable as steric hindrance is minimum. As the number of R-group increases or bulky group is attached, the transition state compound suffers from significant steric hindrance. If the steric hindrance is sufficiently high, the T.S. compound is unstable and the reaction mechanism follows S_N1 mechanism rather than S_N2 mechanism. One of such example is the hydrolysis of neopentyl bromide.

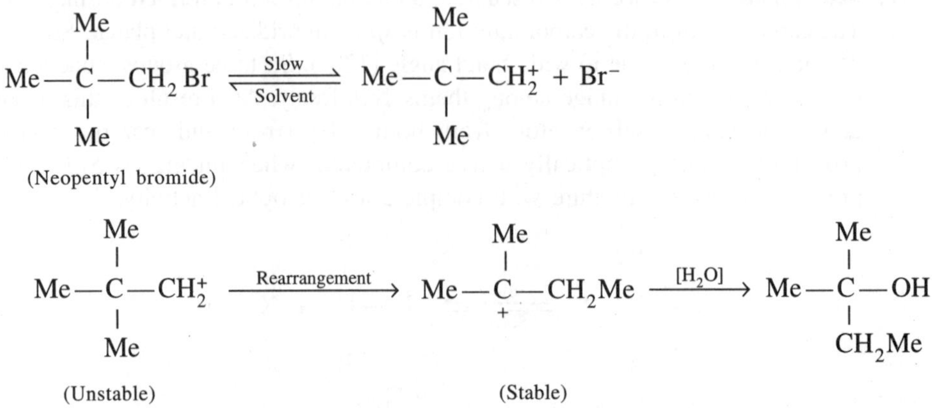

Neopentyl bromide is a primary halide. Thus the hydrolysis reaction should follow S_N2 reaction path but the T.S. is highly unstable due to bulky neopentyl group. So the hydrolysis prefers following S_N1 mechanism to S_N2 mechanism.
Similar observation is also made in the case of hydrolysis of isobutyl bromide, where large Isobutyl group destabilizes the T.S. compound of S_N2 reaction, as a result of which, hydrolysis reaction follows S_N1 reaction path.

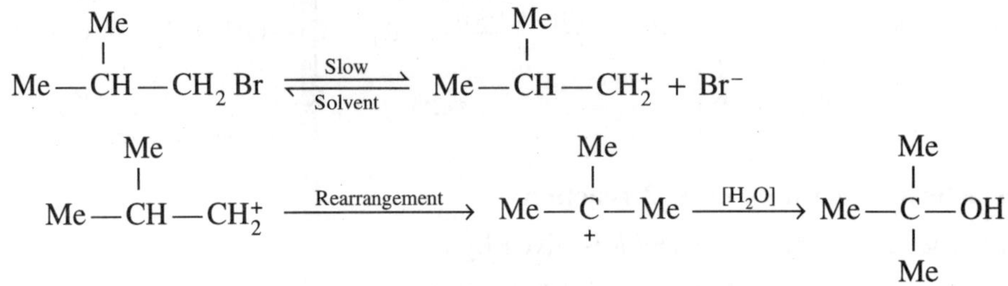

So the reactivity of S_N2 substitution decreases from primary substrate to tertiary substrate. S_N2 reactivity order is:

$$CH_3X > RCH_2X > R_2CHX > R_3CX$$

It has been observed that in an aprotic solvent like DMF, S_N2 reaction rate of ethyl bromide (C_2H_5Br) is 10^4 times higher than that of tert-butyl bromide [$(CH_3)_3CBr$] and S_N2 reaction rate of methyl bromide (CH_3Br) is around 40 times higher than that of ethyl bromide (C_2H_5Br).

(c) S_N2 reaction mechanism depends on nature of nucleophile and leaving group. In S_N2 reaction mechanism nucleophilic attack occurs before departing the leaving group. Thus strong nucleophile and weak leaving group favours S_N2 reaction. Conversely, weak nucleophile and strong leaving group combination usually favours S_N1 mechanism. One example is substitution reaction of neopentyl bromide in presence of C_2H_5OK and C_2H_5OH. As ethoxide ion ($C_2H_5O^-$) is much stronger nucleophile than ethanol (C_2H_5OH), substitution reaction follows S_N2 mechanism in presence of C_2H_5OK but the same reaction follows S_N1 mechanism in presence of C_2H_5OH.

S_N2 reaction without rearrangement:

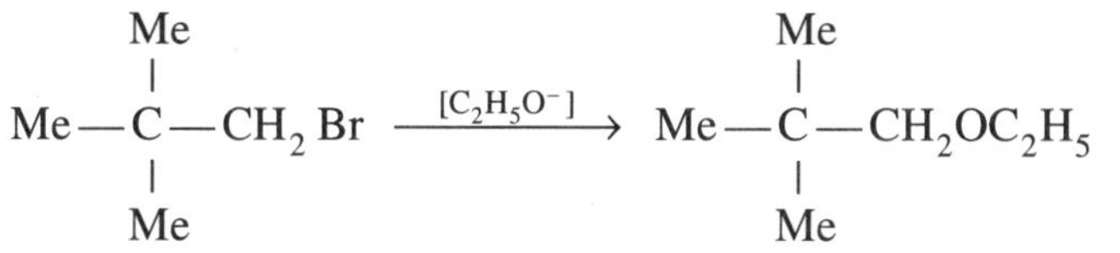

S_N1 reaction with rearrangement:

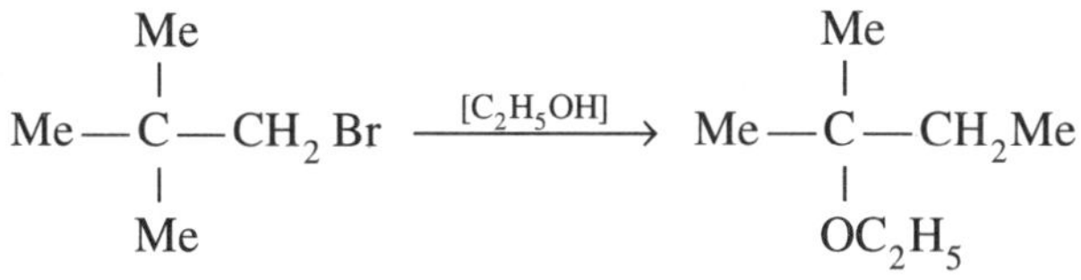

(d) Reaction rate in an S_N2 mechanism largely depends on the effectiveness of nucleophile. In presence of a protic solvent nucleophile easily gets solvated through the formation of hydrogen bond. As a result of this, electron cloud of the nucleophile is bound and not easily available for nucleophilic attack. However, in presence of a aprotic or non-protic solvent, electron cloud of the nucleophile remains comparatively free and hence available for nucleophilic attack due to the absence of any hydrogen bonding. Here are two examples in support of this discussion.

Example 1 The nucleophile [N_3^-] reacts with CH_3I in presence of a solvent through S_N2 mechanism. Two solvents of similar polarity were chosen. One is methanol (CH_3OH), which is a protic solvent and the other is dimethyl formamide (DMF), which is non-protic but has similar polarity to methanol. It has been observed that reaction rate in DMF is 10^4 times higher than that in methanol.

Example 2 If the above substitution reaction is carried out in methanol and DMSO (dimethyl sulphoxide), it will be observed that reaction rate in DMSO is 10^9 times higher than that in methanol, although DMSO is more polar than methanol. Methanol (CH_3OH) is a protic solvent and shields the *electron* pair of nucleophile while $HCONMe_2$ (DMF) and Me_2SO (DMSO) are aprotic solvents and hence make the electron pair of nucleophile free.

High polarity of the solvent often changes the reaction mechanism. In S_N2 reaction, charge in T.S. is delocalized and hence stabilization through *solvation* is less but in *nucleophile* charge density is high and hence sufficiently stabilized through solvation. So T.S. compound is more and more unstable compared to nucleophile as polarity of the solvent increases thereby, decreasing the S_N2 reaction rate. In fact the reaction, which follows S_N2 mechanism in non-polar or less polar solvent, proceeds through S_N1 reaction path in highly polar solvent.

(f) Inversion of configuration In S_N2 mechanism leaving group is attached to the carbon atom through a weak bond in the transition state. As one side of the carbon atom is blocked by the leaving group, nucleophile has to attack the carbon from the opposite side to complete the T.S. compound. Five groups are attached to the central carbon atom, restricting the rotation of groups. However, positions of nucleophile and leaving group are in opposite directions and hence inversion of configuration occurs in the final product. If the substrate is an optically active compound, the sign of direction of rotation of plane polarized light reverses after substitution, maintaining cent percent optical purity. For example 2-bromooctane when undergoes S_N2 reaction in presence of sodium hydroxide, 2-octanol is produced with inversion of configuration.

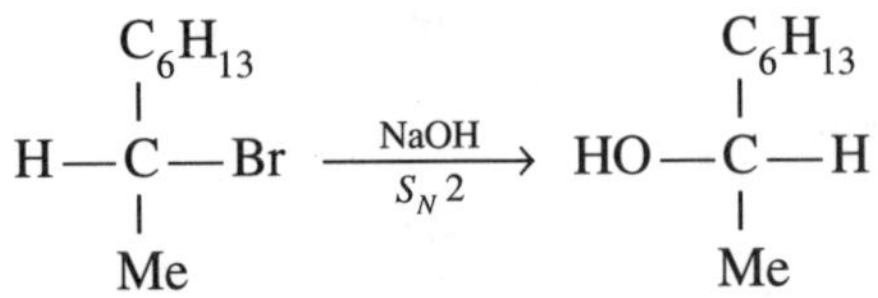

2-bromooctane $\theta = -39.6°$ 2-octanol $\theta = +10.3°$

S_Ni mechanism As we have seen earlier that S_N2 reaction mechanism always proceeds inversion of configuration with maintaining optical purity. However, there are some reactions where two species are involved in the rate determining step but retention of configuration is resulted in the final product. This type of reaction is known as intermediate nucleophilic substitution reaction, denoted by the symbol S_Ni.

Example 1 1-phenyl ethanol undergoes substitution reaction in presence of thionyl chloride ($SOCl_2$).

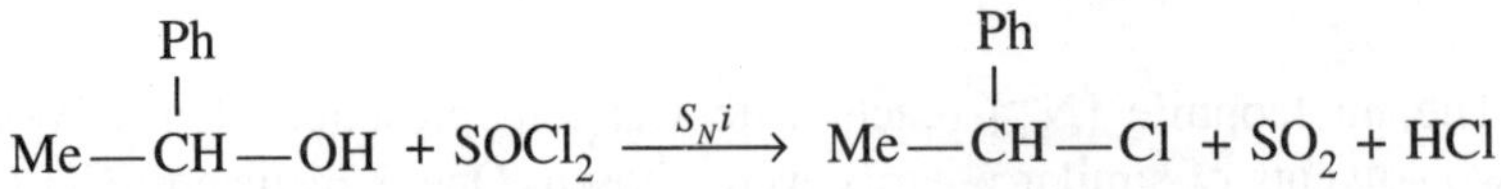

The rate of the reaction is given by:

$$J = k[(Me)(Ph)CHOH]\,[SOCl_2]$$

The following mechanism illustrates the above substitution reaction with retention of configuration:

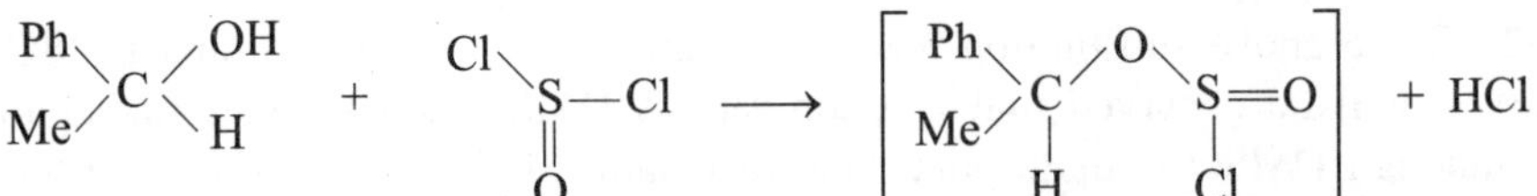

T.S.

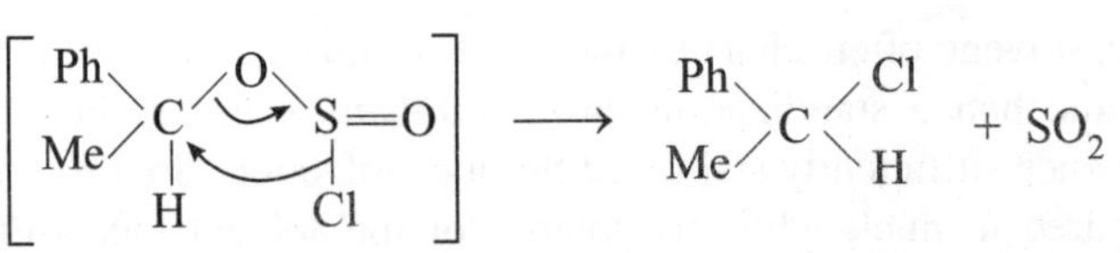

However, if the same reaction is carried out in presence of a base, e.g., pyridine (C_6H_5N), inversion of configuration occurs, presumably because of formation of the salt $C_6H_5NH^+Cl^-$.

Pyridine readily reacts with HCl, released in the first step of S_Ni reaction, to set free the Cl^- ion, which then attacks the T.S. compound from backside, leading to inversion of configuration.

$$\underset{\text{(Pyridine)}}{C_6H_5N} + HCl \longrightarrow [C_6H_5NH]^+ + Cl^-$$

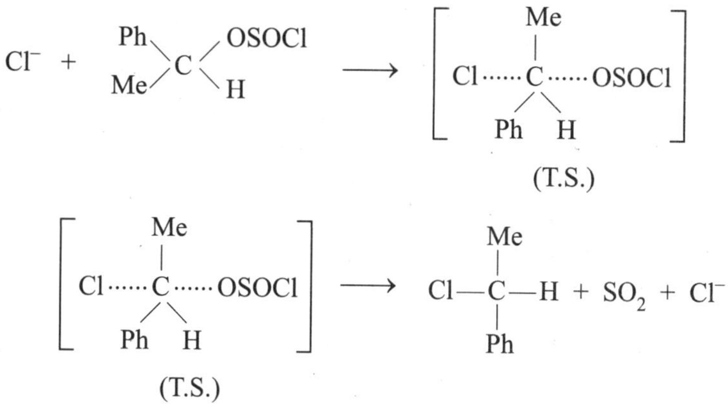

Example 2 In presence of a suitable neighbouring group retention of configuration occurs through participation of neighbouring group. Ethyl chloride thio ether undergoes hydrolysis very fast with retention of configuration according to the following mechanism.

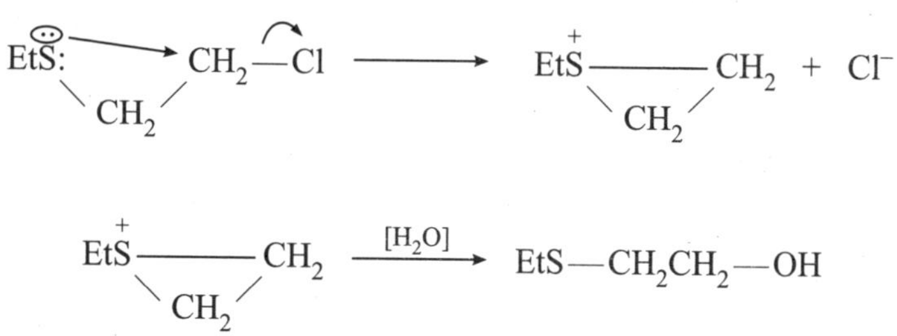

However, ethyl ethyl chloride ether [$EtOCH_2CH_2Cl$] undergoes hydrolysis very slowly due to strong electronegativity and poor donor characteristics of oxygen with respect to sulphur.

Another important example of this class is the substitution reaction of 2-bromopropanoate ion with hydroxyl ion (OH^-).

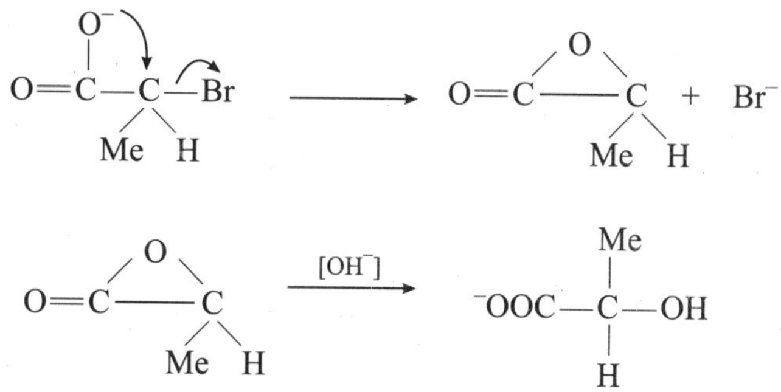

8.1.5.2 Elimination reactions

The reaction, in which atoms or groups are detached from an organic compound to from an unsaturated organic compound along with an inorganic byproduct, is known as elimination reaction. In most elimination reactions inorganic by-product is HY, where, Y may be Cl, Br, CN, OH, MeO, EtO, PhO, PhSO etc. Elimination always takes place in presence of a strong base, which abstracts H-atom from β-position of the substrate. The leaving group occupies α-position of the substrate. As α- and β-positions of a substrate are involved in the elimination reaction, the latter is often called αβ-elimination or β-elimination or 1, 2-elimination reaction. Two different mechanisms have been established so far to explain different elimination reactions.

(i) In one mechanism only one species is involved in the rate determining step and the reaction proceeds through the formation of carbonium ion.

$$\underset{\displaystyle \overset{|}{Y}}{CH_3-CHR} \underset{\text{Solvent}}{\overset{\text{Slow}}{\rightleftharpoons}} [CH_3-\overset{+}{C}HR] + Y^-$$

$$B^{\ominus} \longrightarrow H$$

$$[\overset{|}{C}H_3-\overset{+}{C}HR] \xrightarrow[\text{Solvent}]{\text{Fast}} CH_2{=}CHR + [BH]^+$$

$$[BH]^+ + Y^- \xrightarrow{\text{Fast}} HY + B$$

This type of multi-step elimination reaction is known as *E*l reaction.
Reaction rate (J) is given by:

$$J = k[CH_3CHRY]$$

Like S_N1 reaction *E*l reaction is independent of base concentration and depends on the stability of carbocation. Thus elimination reaction is most favoured when leaving group is attached to tertiary carbon atom and least favoured when leaving group is attached to primary carbon atom.

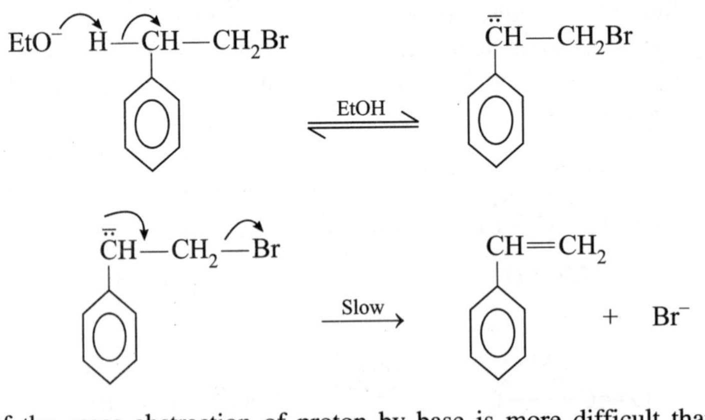

In most of the cases abstraction of proton by base is more difficult than detachment of leaving group and hence carbonium ion is formed in the intermediate. However, in some cases carbanion is formed in the intermediate. This carbanion is the conjugate base of the substrate and hence the mechanism is often called El_{CB} mechanism. One example is the formation of styrene from bromoethyl benzene in presence of sodium ethoxide and ethanol. Poor leaving group and stability of carbanion are the two main criteria to be fulfilled to execute El_{CB} mechanism.

(ii) In other type of elimination reactions no intermediate ions are formed. Attack of base on β-hydrogen atom and detachment of leaving group are taking place simultaneously, giving rise a transition state (T.S.) compound, which is ultimately converted into products.

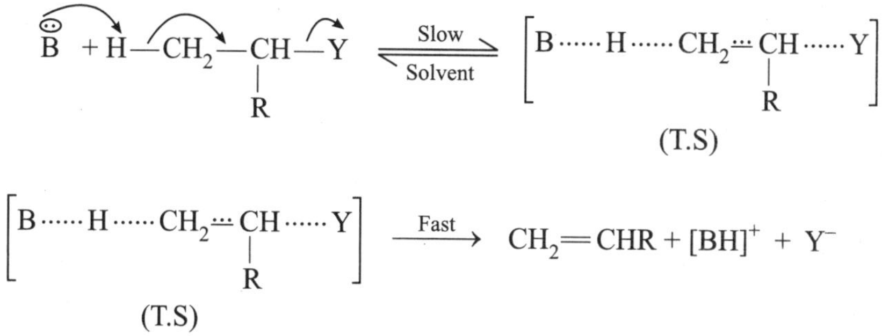

$$[BH]^+ + Y^- \xrightarrow{\text{Fast}} HY + B$$

The reaction rate (J) is given by:

$$J = k[B]\,[\,CH_3CHRY]$$

As two species are involved in the rate determining step, the elimination reaction is often called *E*2 reaction. Two factors are very important in executing *E*2 mechanism.

(1) The base must be soft enough or in other words, the base must be very good nucleophile. So, *E*2 reaction rate is much higher in presence of NH_2^- rather than in presence of OH^- or even in presence of OR^-.

(2) *E*2 reaction rate is significantly high in presence of an aprotic polar solvent like DMSO but is rather slow in presence of a protic polar solvent like water or alcohol, since the base easily gets solvated in a protic solvent through hydrogen bonding. Thus the effectiveness of the base is reduced significantly.

*E*2 reaction is highly stereospecific. In the T.S. compound both base and leaving group are attached to two successive carbon atoms. If base and leaving group remain on the same side with respect to the main carbon chain, strong steric hindrance prevails, thereby, destabilizing the T.S. compound. Thus to form stable T.S. compound base always attacks from opposite side with respect to the leaving group. Thus elimination of two groups always occurs from two opposite sides of carbon main chain. This type of elimination reaction is known as trans-elimination.

The above example illustrates that two different conformations of the same compound give rise two different geometrical isomers on trans-elimination.

trans-elimination:

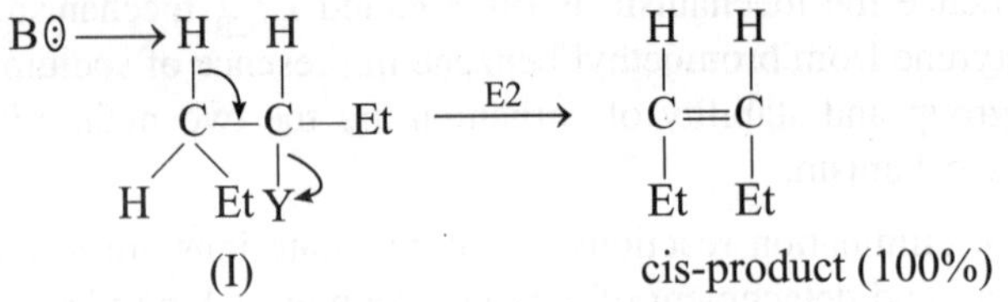

trans-elimination:

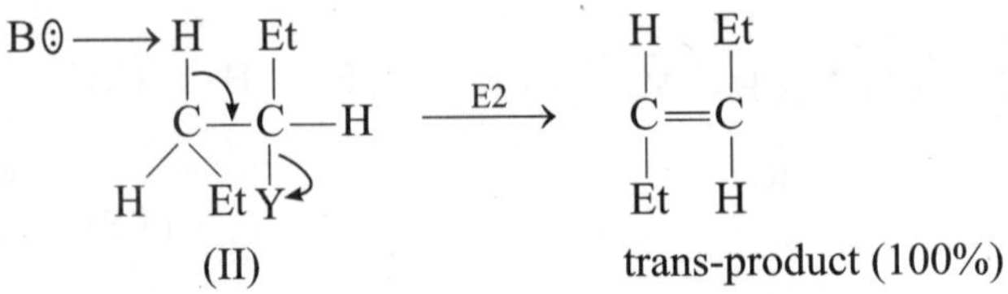

Differences between *E*1 and *E*2

(i) In the case of *E*1, the reaction rate increases with the increase in radical sequence 1° → 2° → 3° as the stability of carbonium ion increases from 1° (primary) to 2°(secondary) to 3° (tertiary).
The same trend of reaction rate with radical sequence is also observed in the case of *E*2 also since with the increase in radical sequence stability of the product, i.e. alkene increases due to branching.

(ii) In the case of *E*1 only one species is involved in the rate determining step but in the case of *E*2 two species are involved in the rate determining step.

(iii) In the case of *E*1 mechanism the reaction rate does not depend on the concentration of base but in the case of *E*2 mechanism the reaction rate depends on the concentration of base.

(iv) In the case of *E*1 mechanism the rate of reaction is independent of the nature of the base. In the case of *E*2 mechanism the nature of the base plays a vital role in determining the rate of reaction.

Stronger is the base *faster* the elimination of proton is, and hence the rate of reaction *increases*.
In general *E*1 mechanism is favoured with secondary and tertiary substrates with low base concentration or in presence of weak base, particularly, where the solvent, acts as the base.

- Dehydrohalogenation of isopropyl bromide is very slow in alcoholic KOH but the same is very fast by t-BuO$^-$K$^+$ in DMSO – explain.

Ans. Usually isopropyl bromide follows *E*1 mechanism during elimination reaction but as KOH is a very strong base, elimination reaction occurs very fast. Thus the reaction follows *E*2 mechanism and hence the reaction rate is slow.
In presence of t-BuO$^-$K$^+$ elimination reaction follows *E*1 mechanism since t-BuO$^-$K$^+$ is a very weak base compared to KOH. Thus the reaction occurs at a very fast rate.

In alcoholic KOH:

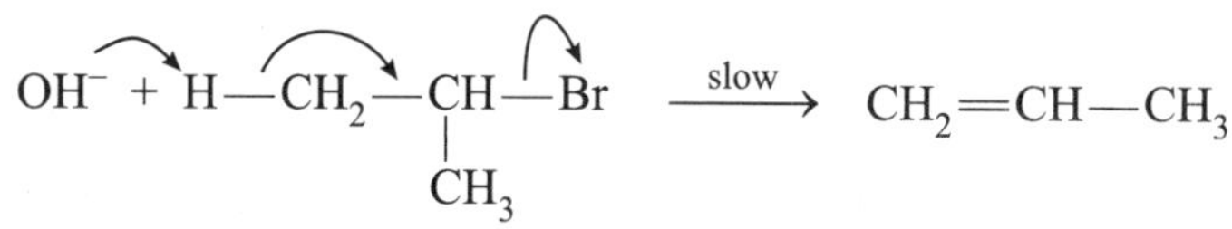

In DMSO:

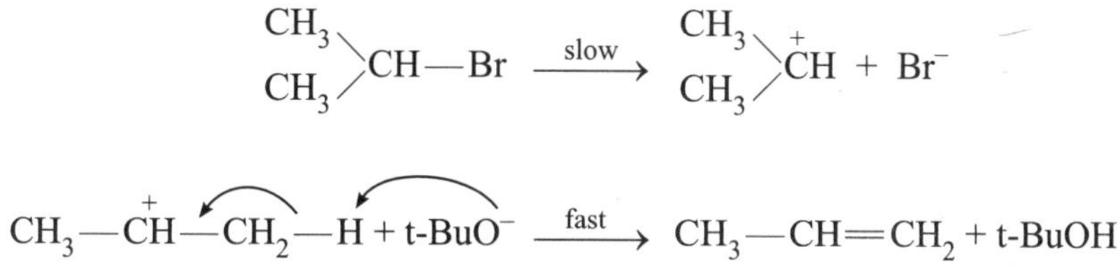

Orientation in elimination reactions It has been established that elimination always occurs through two successive carbon atoms forming a double bond. In most of the elimination reactions, H-atom is one of the eliminating groups. However, if more than one β-hydrogen atom is available in a compound, two or more products are formed after elimination and one of them appears as a major product. Formation of major product depends on leaving group Y. Let us take an example to illustrate this.

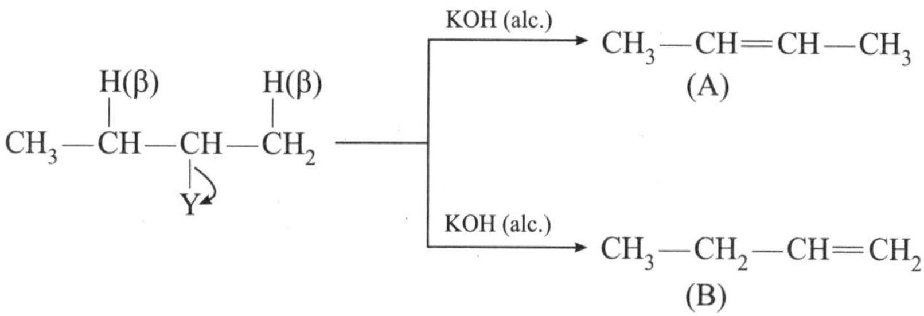

In product *A* two H-atoms of double bonded carbon atoms are substituted by alkyl groups but in product *B* one H-atom of double bonded carbon atoms is substituted by alkyl group. So, product *A* is often called the most substituted product and product *B* is called the least substituted product.

Saytzeff's rule This rule was first developed by Alexander Saytzeff in 1875.

Statement If the leaving group (Y) is a neutral and weak electronegative group like, –Br, –I, –OH etc. the most substituted product will be the major product.

So, in the previous example *A* will appear as the major product and *B* will appear as the minor product if the leaving group is Br etc. Product *A* is known as 2-butene whereas product *B* is called 1-butene.

As we have discussed earlier, stability of an alkene depends on number of α—H atoms available. Higher the number of α—H atoms, more stable the alkene is. So, it is evident that 2-butene is more stable than 1-butene and hence has higher probability to appear as the major product. Thus the concept of stability of alkene further justifies the Saytzeff's rule. In actual practice almost 80% 2-butene is obtained after elimination reaction and only 20% 1-butene is formed.

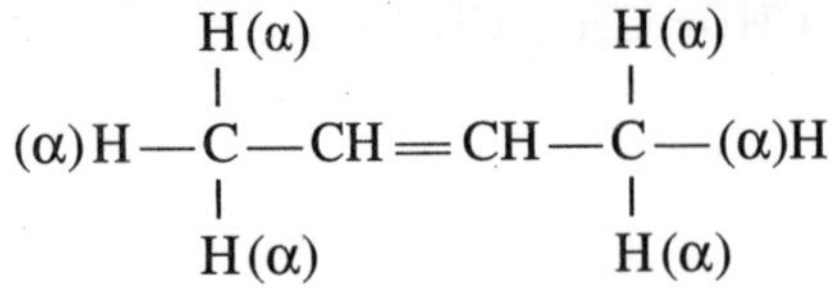

2-butene

No. of α—H atoms = 6

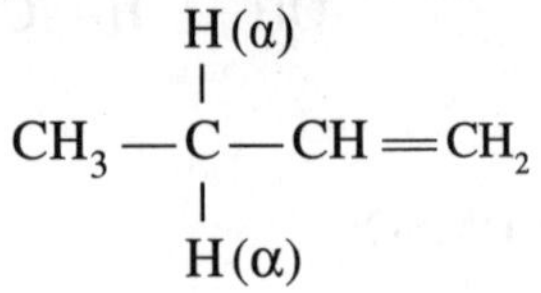

1-butene

No. of α—H atoms = 2

In the case of *E*1 reactions alkenes are formed through the formation of carbonium ion. Here also, abstraction of H-atom by base always leads to the formation of more stable alkene, i.e., most substituted alkene.

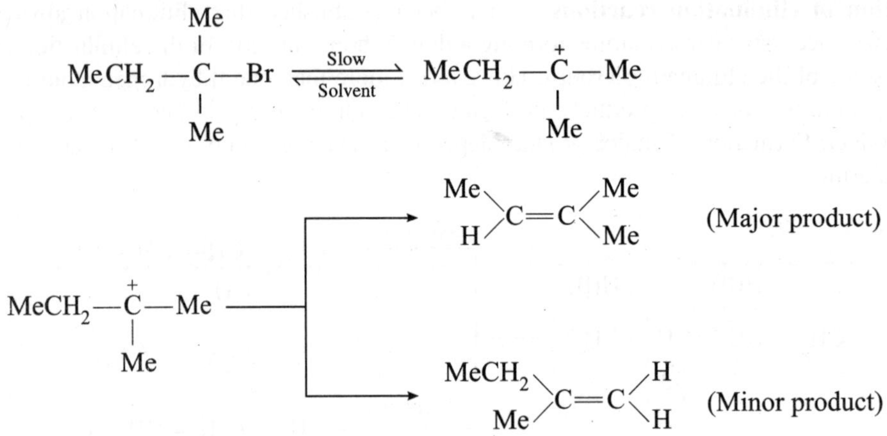

So, *E*l mechanism also follows Saytzeff's rule. There are some *E*l reactions where β—H atom is not available in T.S. carbonium ion. In that case carbonium ion undergoes rearrangement followed by β-elimination. One example is elimination of HBr in neopentyl bromide.

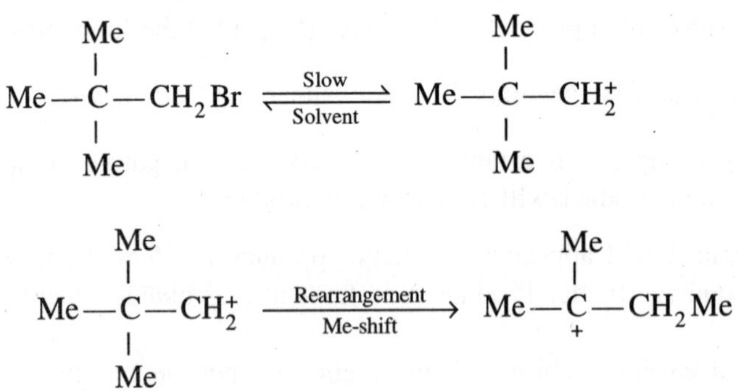

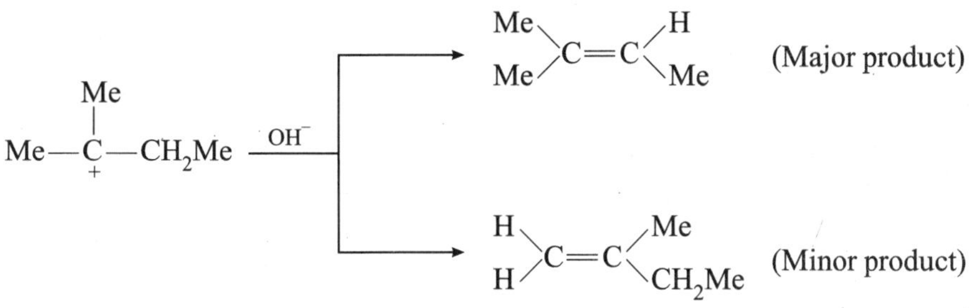

Hofmann rule This rule of elimination was developed by Hofmann in 1851.

Statement If the leaving group (Y) is strong electronegative group the least substituted product is obtained after elimination.

Thus in the previous example 1-butene appears as a major product and 2-butene appears as a minor product if Y is -SMe2+ or -NMe3+.

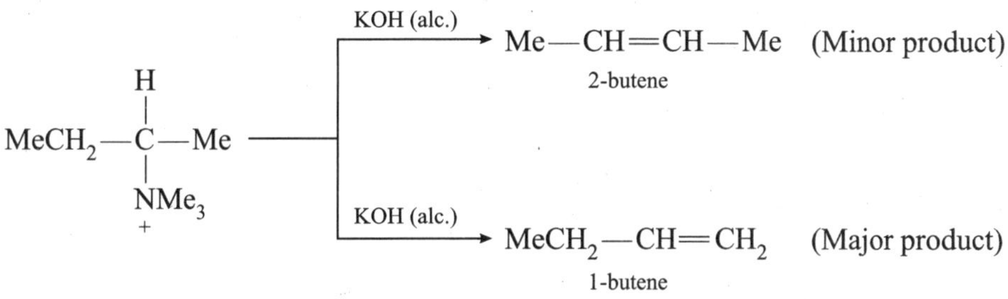

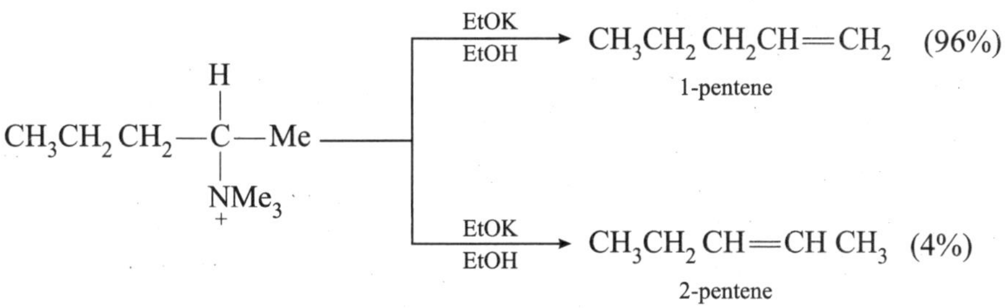

If the leaving group is sufficiently strong electron withdrawing group, the corresponding β-H-atoms are significantly acidic.

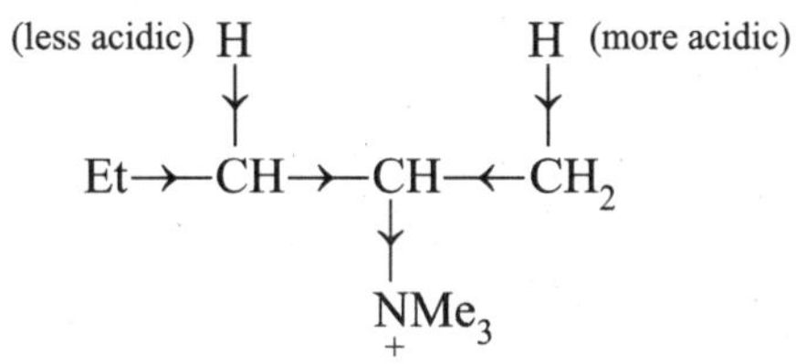

Abstraction of β—H atom depends on acidic nature of the latter. More acidic the β—H atom, the latter is more easily detached in presence of a base. Furthermore, acidity of β—H atom decreases due to +I effect of alkyl group. Thus it is obvious that β—H atom, attached to the least substituted carbon atom, is more acidic and hence will react quickly in presence of a base.

Saytzeff rule versus Hofmann rule Usually Saytzeff rule is followed when the leaving group is weak electronegative like, –I, –Br, –OH, –OR etc. The same elimination reaction follows Hofmann rule if the leaving group is sufficiently electronegative like, –Cl, –F, $-SMe_2^+$, $-NMe_3^+$ etc. For example:

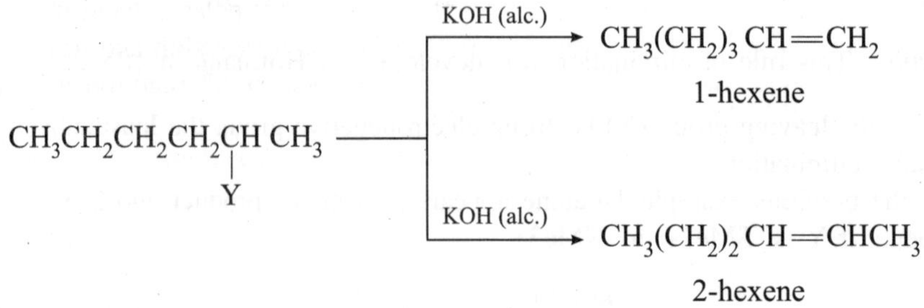

If Y = I, % 2-hexene = 81 and % 1-hexene = 19.

If Y = F, % 2-hexene = 30 and % 1-hexene = 70.

If different alkyl groups are attached to a quaternary compound, several β—H atoms are available but the most acidic β—H atom reacts with the base to form the major alkene.

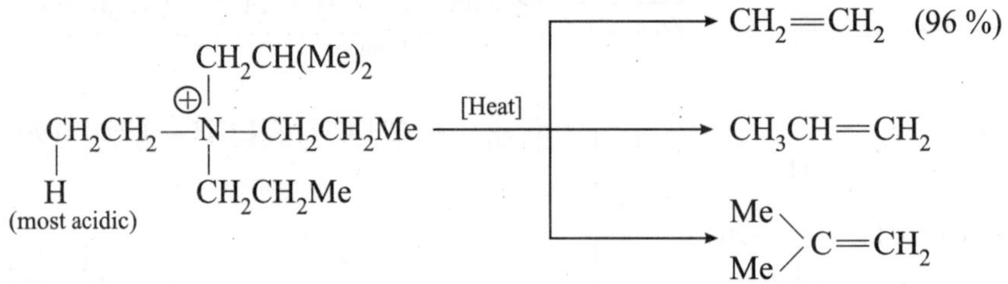

Higher alkyls have strong +I effect and hence provide more effective support to the electropositive nitrogen than lower alkyls. Obviously lower alkyls are more susceptible to be detached from positively charged nitrogen.

Elimination Vs. Substitution:

(i) In the case of S_N2 mechanism the rate of reaction increases as the nature of the substrate changes from tertiary to primary.

In the case of *E*2 mechanism the rate of reaction increases as the radical sequence increases from 1° to 3°.

So when both S_N2 and *E*2 mechanism are favourable the proportion of elimination product increases as the nature of substrate changes from primary to tertiary. In fact tertiary alkyl halide prefers following *E*2 mechanism exclusively rather to S_N2 mechanism.

(ii) **Nature of base/nucleophile** Usually strong bases are very good acceptor of protons but have poor donating characteristics to other atoms and hence they are considered as

weak nucleophiles. Nucleophile has a tendency to donate its pair of electrons to any atom but has a little tendency to abstract proton from a compound. Thus strong bases are weak nucleophiles and strong nucleophiles are weak bases.
In presence of a strong bases abstraction of proton is much easier and elimination reaction is favoured. On the other hand in presence of a strong nucleophile substitution reaction is preferred.

(iii) **Nature of solvent** In presence of a non-polar or slightly polar solvent the intermediate compound or ion does not get chance to stabilize through solvation and hence elimination reaction particularly *E*2 mechanism is favoured, since in *E*2 reaction no intermediate compound is formed.
In presence of a polar solvent, especially protic solvent, substitution reaction is favoured due to enhanced stability of intermediate compound through solvation, particularly through hydrogen bonding.

(iv) Usually high temperature favours elimination reaction. The bonds formed during substitution reaction, are stable at low temperature. Thus substitution reactions are favoured at low temperature.

(v) In both S_N1 and *E*1 reactions the intermediate is the same. i.e., carbocation but the products are different.

$$R{-}X \xrightarrow[\text{or } S_N1]{E1} R^+ + X^-$$

(a) $$R^+ \xrightarrow[\text{elimination}]{\text{In case of}} -\overset{|}{C}=\overset{|}{C}-$$

(b) $$R^+ + Z^- \xrightarrow[\text{substitution}]{\text{In case of}} R{-}Z$$

8.1.5.3 Addition reactions

Addition reactions may broadly be categorized in two different classes:
(a) Electrophilic addition and (b) Nucleophilic addition.

(a) Electrophilic addition In these type of addition reactions, electrophile is the attacking agent. Two different electrophiles are considered here. One is halogen ion and the other is hydrogen ion.

Case – I Halogen ion acts as an electrophile: In the halogen group Br_2 (Bromine) is the most reactive to carbon–carbon double bond.

Step-I: $$Br_2 \longrightarrow Br^+ + Br^-$$

Step-II:

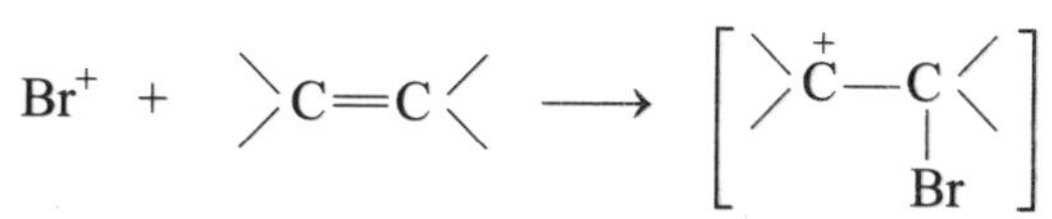

(T.S.)

Step-III:

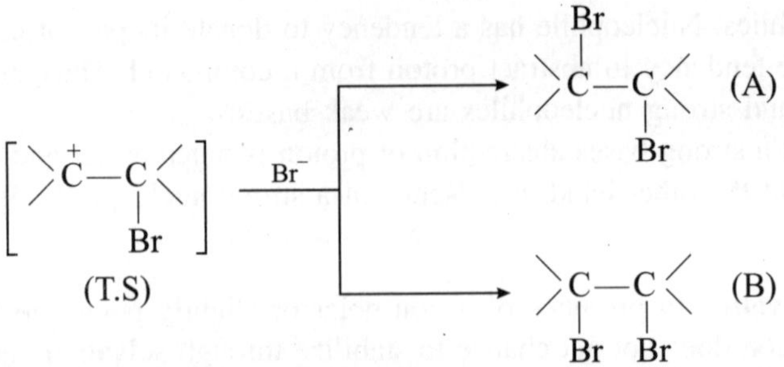

If free carbonium ion is formed in the T.S., Br^- may attack the carbonium ion from both side with equal probability giving rise two isomers (A and B) with equal concentration. The product (A) is formed by trans-addition whereas the product (B) is formed by cis-addition. However, the nature of product depends on stability of carbonium ion. In presence of alkyl substituents, usually carbonium ion is stabilized through bridge formation as follows:

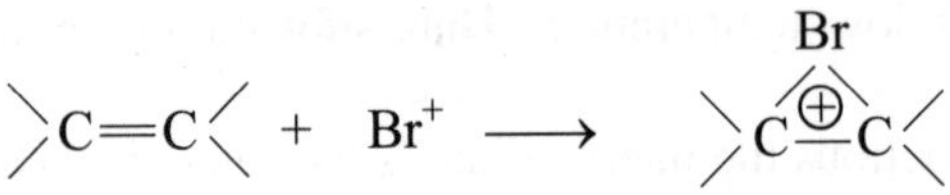

Bromonium ion

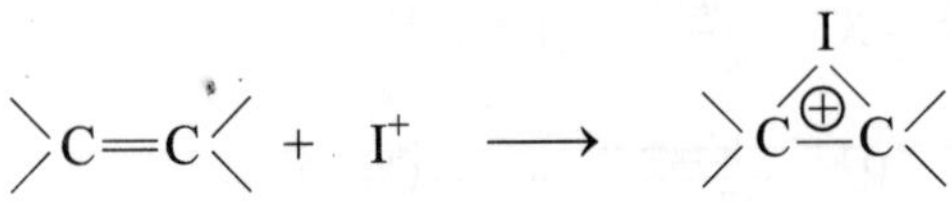

Iodonium ion

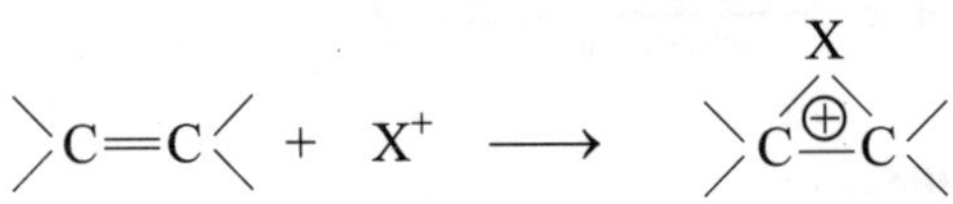

Halonium ion

The positively charged cyclic halonium ion is stable and prevents the attack of nucleophile from halogen side. So, almost 100 % trans-addition product is obtained.

If carbonium ion is stabilized by other means, cyclisation occurs partly, allowing both cis- and trans-additions. It is well illustrated in the following example.

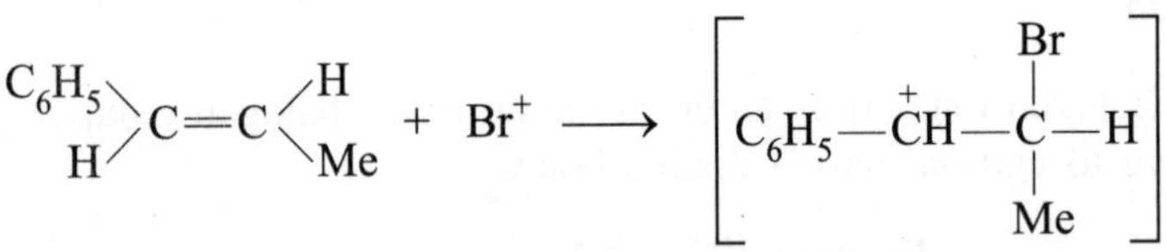

The T.S. carbonium ion is highly stabilized through phenyl group and hence bridge formation through Br-atom is prevented. The analysis of final product shows that both cis- and trans-additions take place.

Addition of hypohalous acid to carbon-carbon double bond also takes place through the formation of halonium ion:

$$HOCl \longrightarrow Cl^{+} + OH^{-}$$

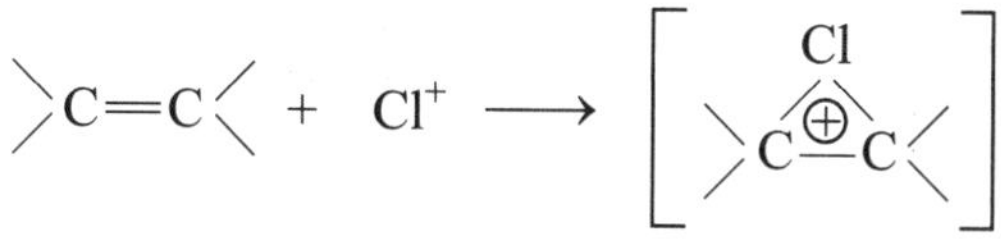

(T.S.)
Chloronium ion

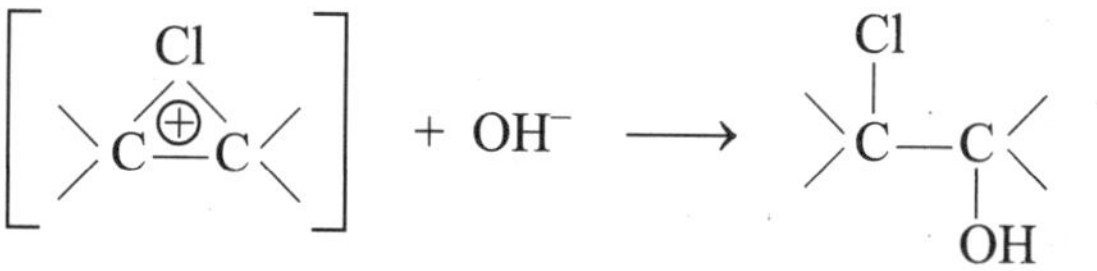

The overall rate of reaction largely counts upon the substituents, attached to the carbon atoms, taking part in the addition reaction. If the substituents are electron donating groups, like alkyl groups, the T.S. carbonium ion is stabilized through +I effect and hence rate of reaction increases significantly. On the other hand, presence of any electron withdrawing group destabilizes the T.S. carbonium ion, thereby, decreasing the overall reaction rate. For example 1-butene ($CH_3CH_2 \rightarrow CH = CH_2$) undergoes addition reaction around 100 times faster than ethylene does but if the same addition reaction is performed on bromo-ethylene, the reaction rate is 100 times ($Br \leftarrow CH = CH_2$) shower than that of ethylene.

Case – II Hydrogen ion as an electrophile. Addition of halogen acid (HX) to carbon-carbon double bond gives some interesting results.

$$HX \longrightarrow H^{+} + X^{-}$$

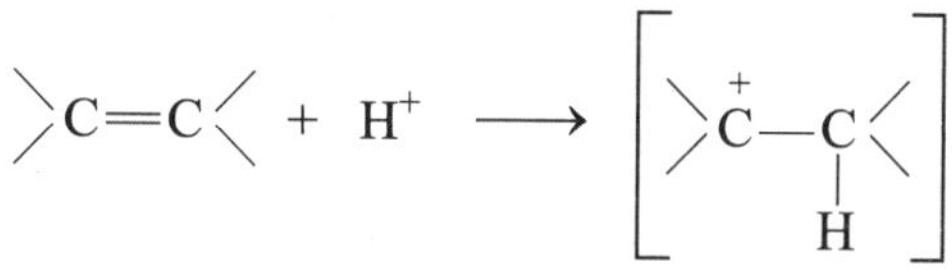

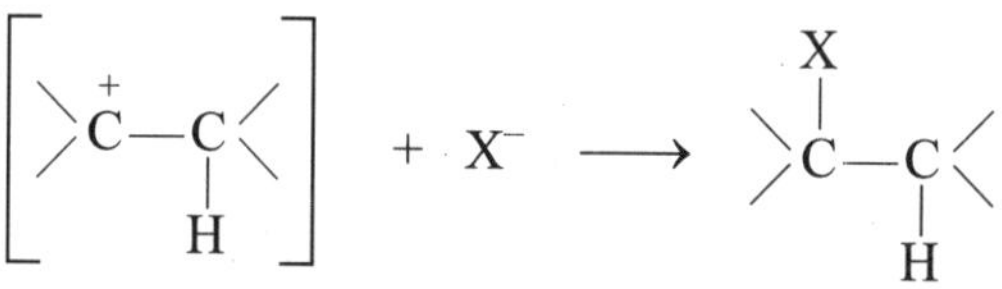

Addition of H^+ to carbon-carbon double bond occurs in such a way that the most stable carbonium ion is formed, which means that, H-atom prefers joining to the least substituted carbon atom, facilitating the formation of stable carbonium ion. Thus halogen atom, in the next step, has no other option but to join to the most substituted carbon atom. In this case formation of bridged carbocation is not possible due to lack of lone pair of electrons of H^+ ion. This observation was closely studied by the scientist Markownikov and gave the following rule:

Markownikov's rule In addition reaction, where hydrogen ion is an electrophile, the latter always prefers attaching to the least substituted carbon atom.

(i) 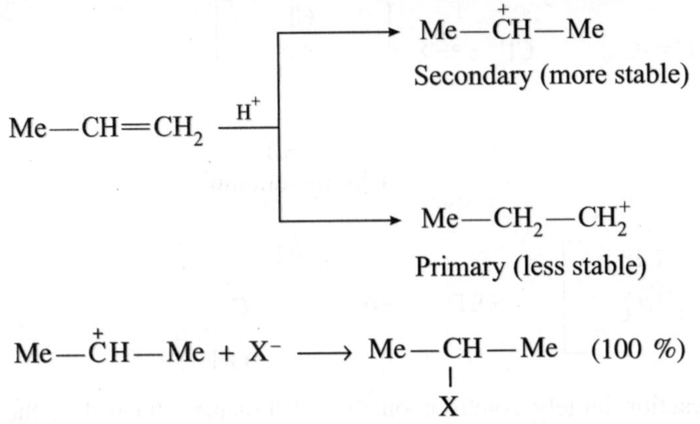

(ii) 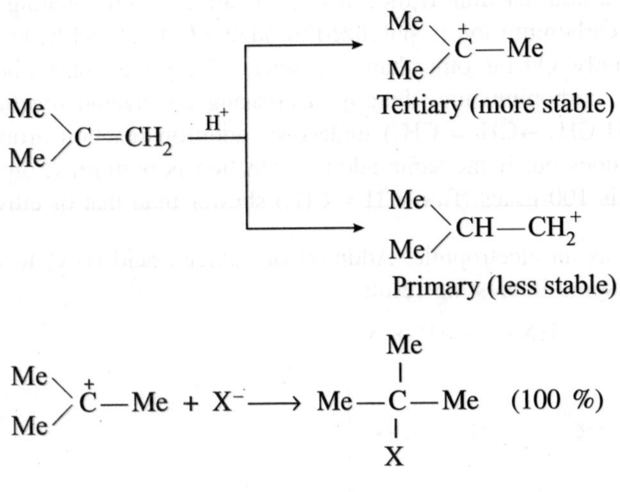

(iii) 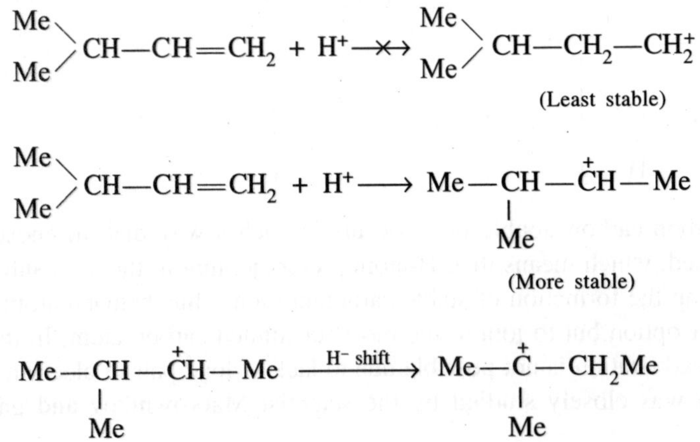

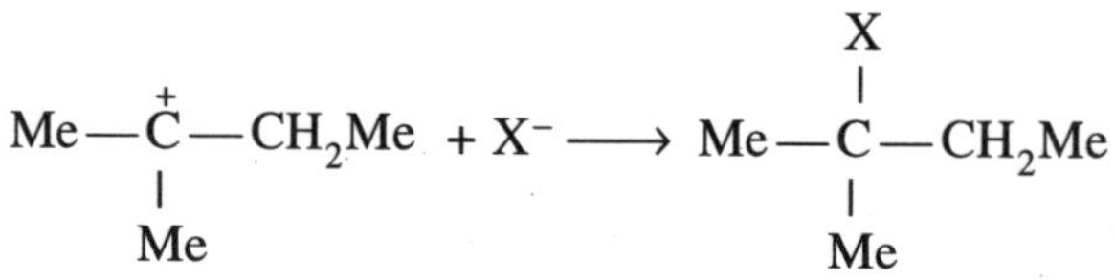

(iv)

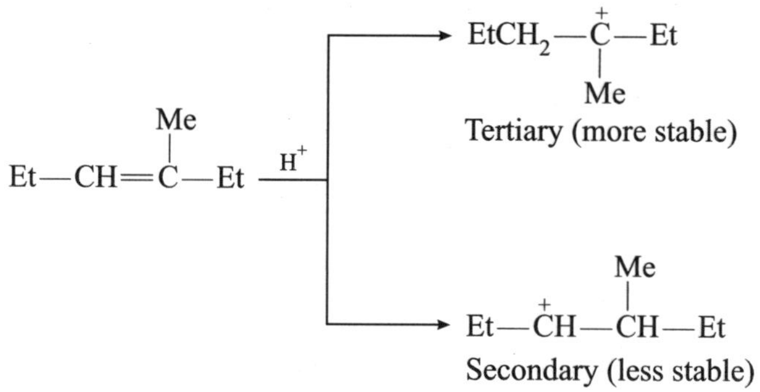

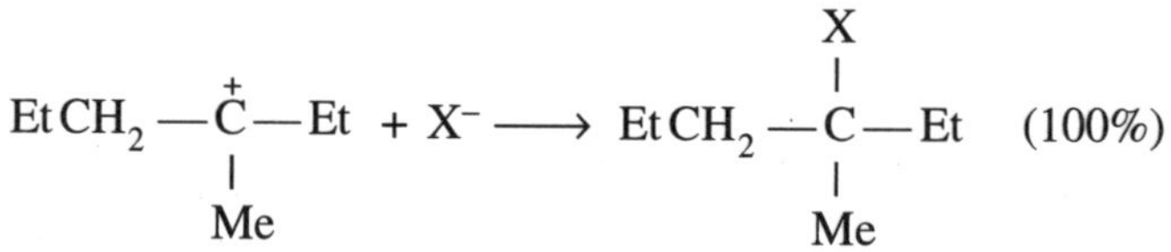

Detailed study of addition reactions shows the following order of reactivity of alkenes towards the addition of acids:

$$Me_2C{=}CH_2 > MeCH{=}CHMe;\ MeCH_2CH{=}CH_2;\ MeCH{=}CH_2 > CH_2{=}CH_2 > CH_2{=}CH{-}Cl$$

The potential energy diagram of addition reaction is shown in the following figure:

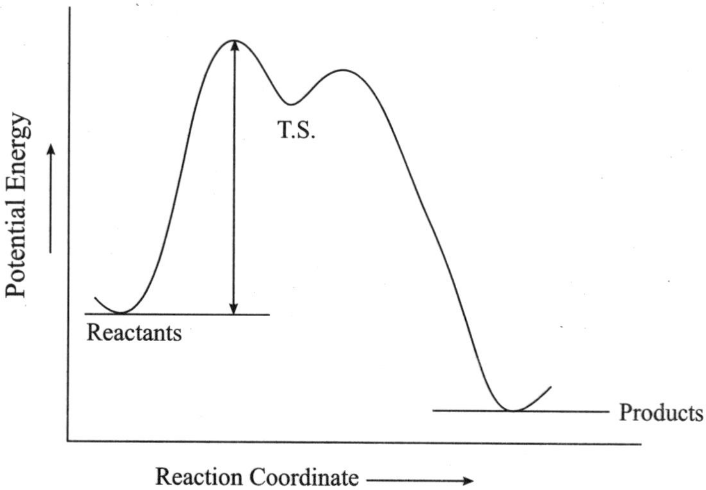

Fig. 8.1.4 Potential energy diagram of an addition reaction.

Anti-Markownikov's rule It is well established that Markownikov's rule is valid if the addition reaction proceeds through ionic mechanism. However, the same addition reaction does not follow Markownikov's rule if the reaction proceeds through free radical mechanism. Peroxides are very good generators of free radicals. Thus in presence of peroxides addition reaction proceeds through free radical mechanism and the rule of addition is:

Hydrogen atom of the attacking agent joins to the most substituted carbon atom, while the other part of the attacking agent attacks to the least substituted carbon atom.

Anti-Markownikov's rule is also known as peroxide effect. Detailed research works reveal that only Hydrogen Bromide (HBr) strongly responds to the peroxide effect.

Probable mechanism:

(i)

$$\text{HO—OH} \xrightarrow{h\nu} 2\text{OH}^*$$

$$\text{OH}^* + \text{H—Br} \longrightarrow \text{H}_2\text{O} + \text{Br}^*$$

$$\text{Br}^\bullet + \text{Me—CH=CH}_2 \longrightarrow \text{Me—}\dot{\text{C}}\text{H—CH}_2\text{—Br} \quad \text{(A)}$$

Secondary (more stable)

$$\text{Br}^\bullet + \text{Me—CH=CH}_2 \longrightarrow \text{Me—CH(Br)—CH}_2^\bullet \quad \text{(B)}$$

Primary (less stable)

$$\underset{\text{(A)}}{\text{Me—}\dot{\text{C}}\text{H—CH}_2\text{—Br}} + \text{HBr} \longrightarrow \underset{(100\%)}{\text{MeCH}_2\text{CH}_2\text{Br}} + \text{Br}^*$$

In the transition state two free radicals *A* and *B* are formed but *A* is far more stable than *B*. Thus only one product is obtained.

Stability of free radicals follows the following order:

$$\underset{\text{(allylic)}}{\text{—}\dot{\text{C}}\text{H—CH=CH—}} \longleftrightarrow \underset{\text{(allylic)}}{\text{—CH=CH—}\dot{\text{C}}\text{H—}} > \text{tertiary} > \text{secondary} > \text{primary} > \text{CH}_3^\bullet > \underset{\text{(vinylic)}}{\text{—}\dot{\text{C}}\text{=CH—}}$$

(ii)

$$\text{Me}_2\text{CH—CH=CH}_2 + \text{Br}^\bullet \longrightarrow \text{Me}_2\text{CH—}\dot{\text{C}}\text{H—CH}_2\text{—Br}$$

$$\text{Me}_2\text{CH—}\dot{\text{C}}\text{H—CH}_2\text{—Br} \xrightarrow{\text{H}^\bullet\ \text{shift}} \underset{\text{(Stable)}}{\text{Me}_2\dot{\text{C}}\text{—CH}_2\text{CH}_2\text{Br}}$$

$$\text{Me}_2\dot{\text{C}}\text{—CH}_2\text{CH}_2\text{Br} + \text{HBr} \longrightarrow \text{Me}_2\text{CH—CH}_2\text{CH}_2\text{Br}$$

Note Free radical addition of HI or HCl to an alkene does not follow anti-Markownikov's rule, since I* is not reactive enough to add on to a double bond and in the case of HCl, bond is too strong to be broken homolytically.

Addition to conjugated diene

(i) An electrophilic addition to a conjugated diene results a carbocation which is stabilized through resonance. As more than one resonating structure is available, more than one product is obtained. One example is addition of Br_2 to butadiene.

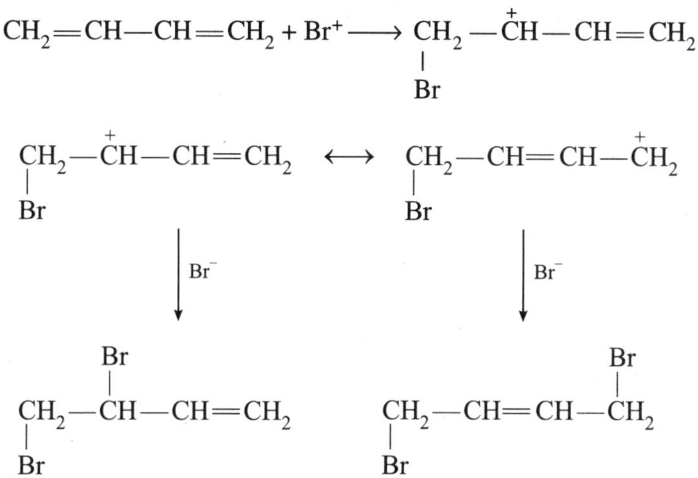

(1, 2-dibromo-3-butene) (1, 4-dibromo-2-butene)

(ii) In the case of unsymmetrical conjugated diene only stable resonating structures are believed to represent the T.S. compound. For example after electrophilic addition 2, 4-pentadiene gives rise two sets of resonating structures, as shown below:

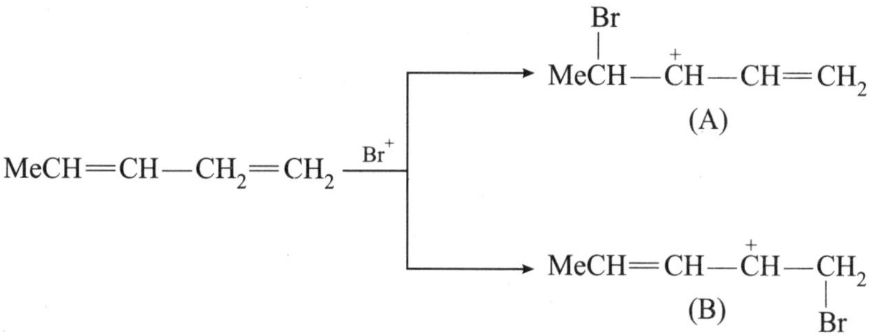

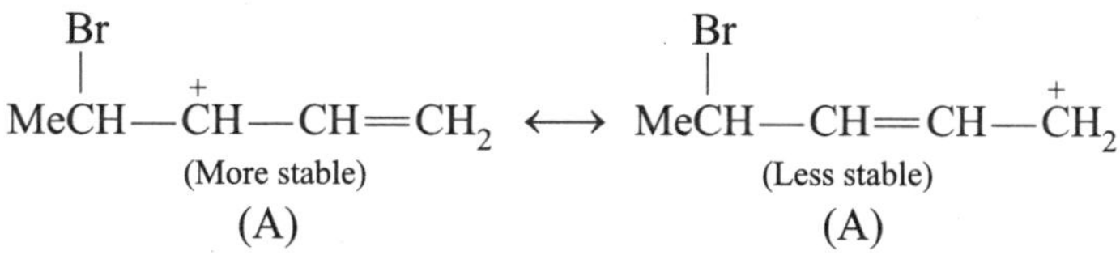

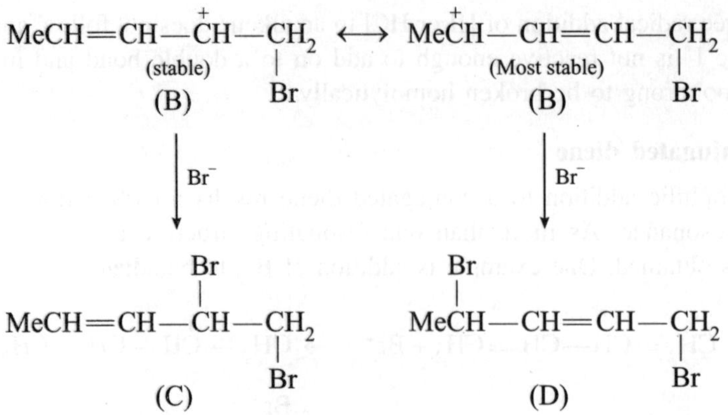

It is evident that the carbocation (B) represents the T.S. compound and hence the products are (C) and (D).

(iii) Electrophilic addition to an α, β-unsaturated ketone gives some interesting results.

For example, addition of HX to 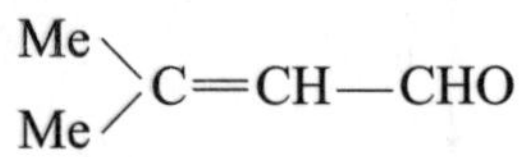

Step–I: 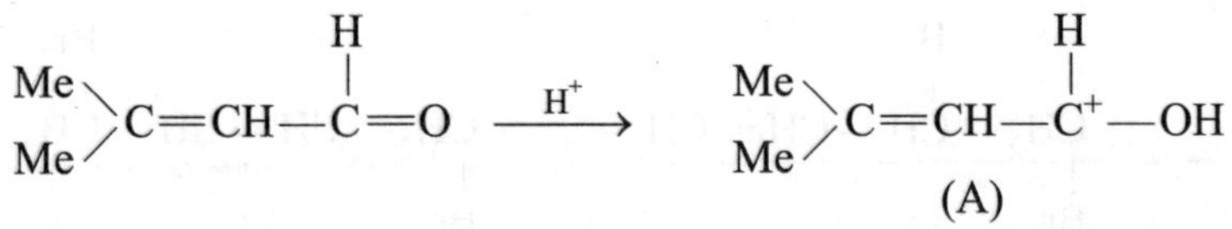

Step–II: 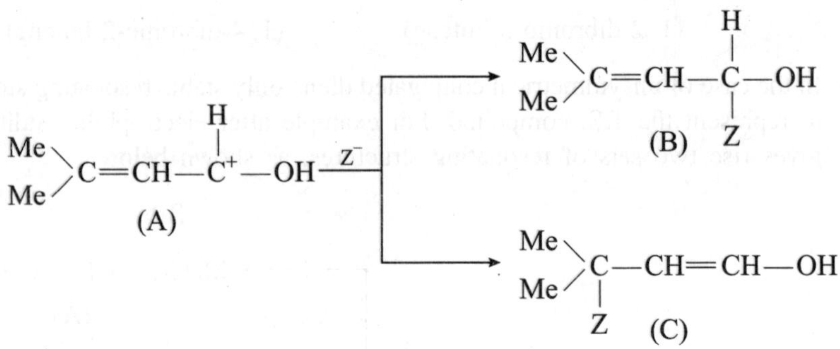

The product (C) is stabilized through tantomerism:

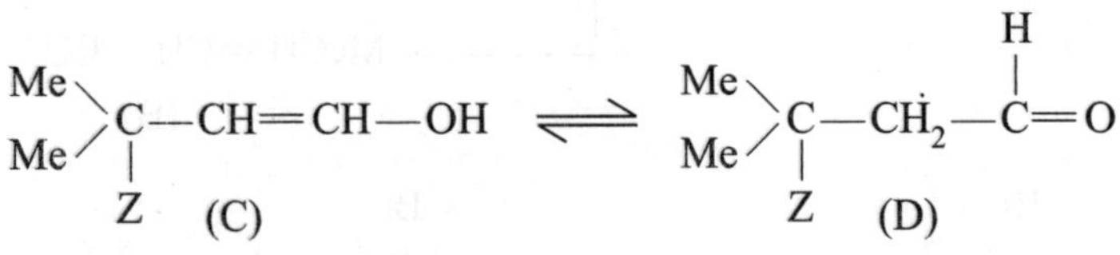

This kind of tantomerism is absent in the case of (B) and hence only (C) along with its tantomer (D) is obtained in the final product.

(b) Nucleophilic addition

(i) If the alkene contains strong electron withdrawing group, opening up of the double bond is facilitated through nucleophilic addition. Nucleophilic addition always proceeds through the carbanion formation and obviously the most stable carbanion constitutes the transition state.

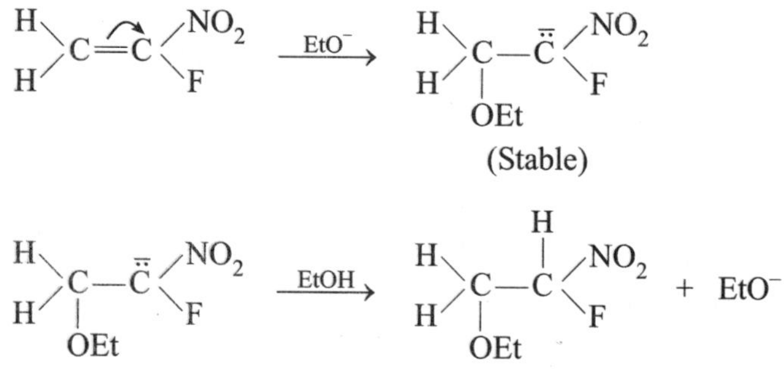

(ii) In the case of Nucleophilic addition to α, β-unsaturated compound, carbanion in the transition state undergoes resonance.

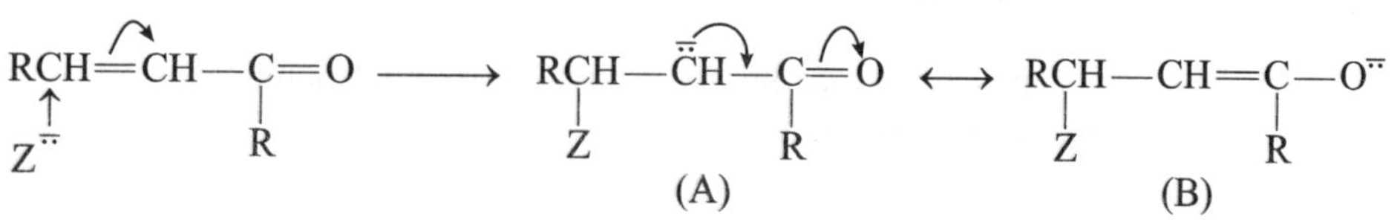

(B) is more stable than (A) because the negative charge is placed on the more electronegative oxygen atom.

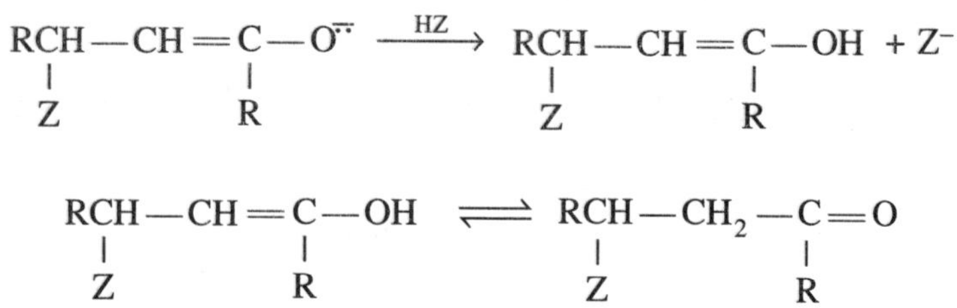

(iii) Some fatty acid compounds generate stable carbanion in presence of a base. That carbanion then acts as a nucleophile and attacks α, β-unsaturated compound.

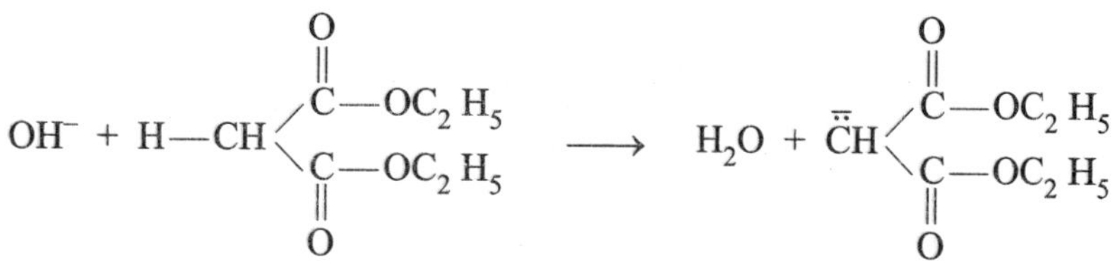

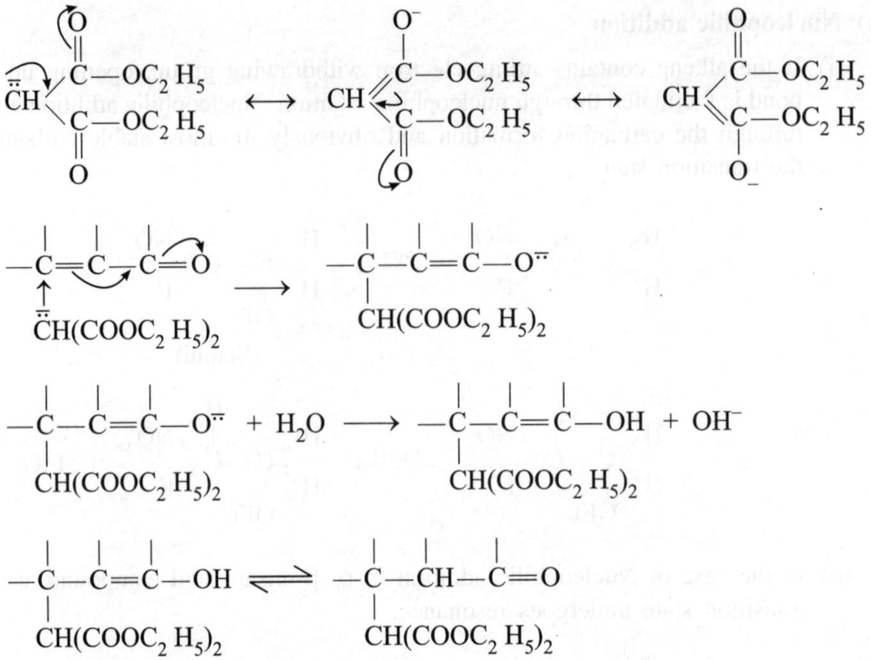

This type of nucleophilic addition reaction, where nucleophile is a carbanion, is known as *Michael addition reaction.*

8.1.6 (i) Aromatic substitution reactions

(a) Nitration of benzene:

Step-I:

$$HO{-}NO_2 + H^+ \xrightarrow{\text{Fast}} H_2O^+{-}NO_2 \xrightarrow{\text{Fast}} NO_2^+ + H_2O$$

Step-II:

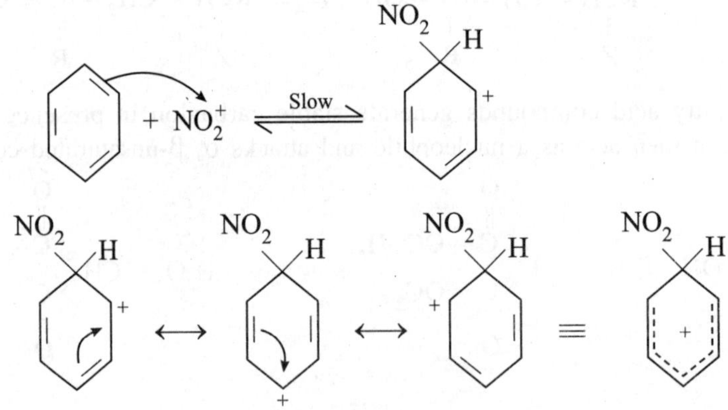

Step-III:

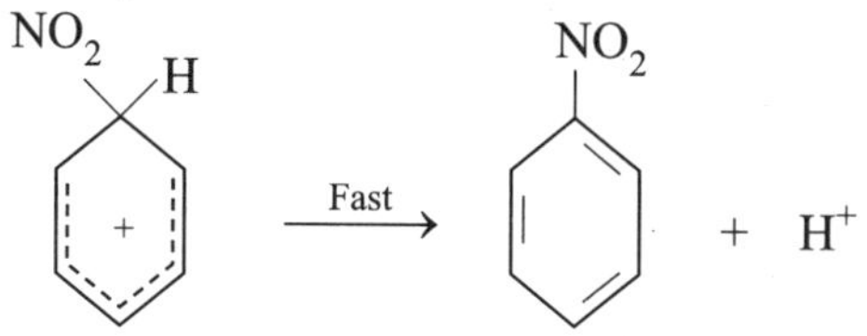

(Nitrobenzene)

(b) Sulphonation of benzene:

Step-I:

$$2H_2SO_4 \rightleftharpoons H_3O^+ + HSO_4^- + SO_3$$

Step-II:

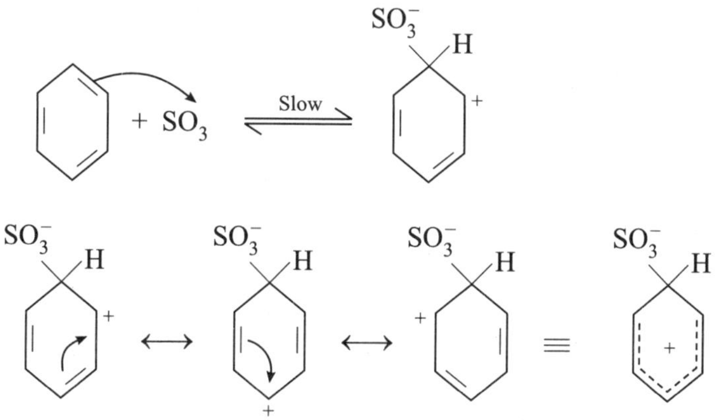

Step-III:

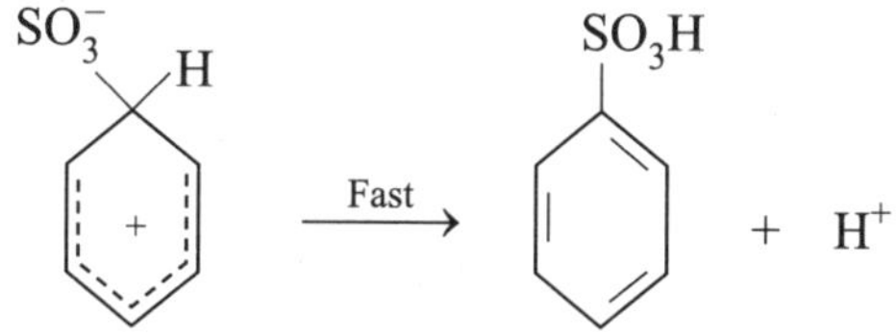

(Benzene sulphonic acid)

(c) Alkylation of benzene [Friedel-Crafts reaction]:

Step–I:

$$R{-}Cl + AlCl_3 \xrightarrow{\text{Fast}} R^+ + AlCl_4^-$$

Step–II:

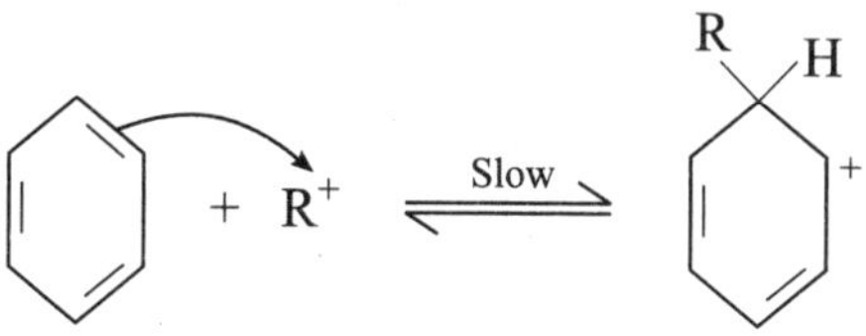

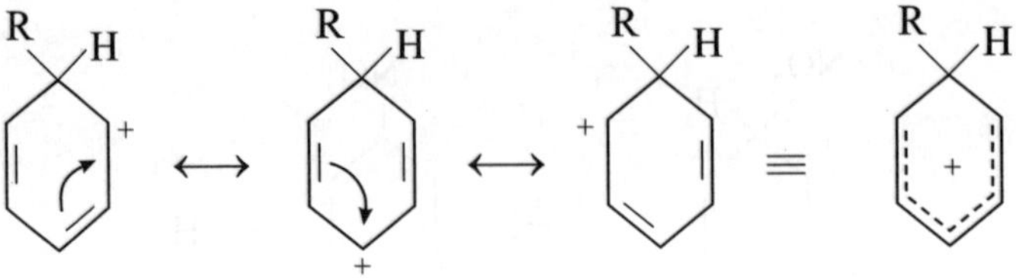

Step-III:

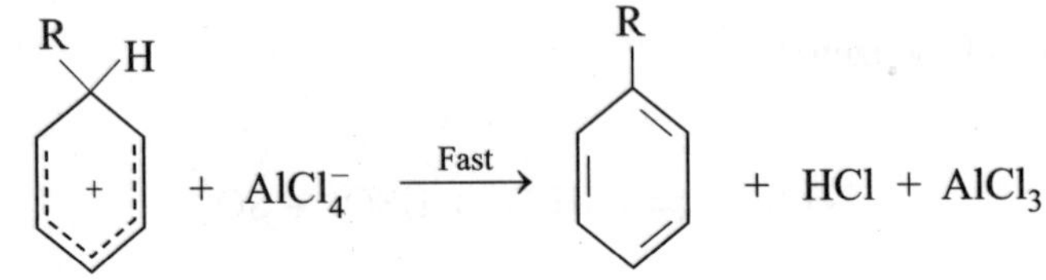

Aliphatic – Aromatic hydrocarbons are often called Arenes.

(d) Halogenetion of benzene:

Step-I:

$$Cl—Cl + FeCl_3 \xrightarrow{\text{Fast}} Cl^+ + FeCl_4^-$$

Step-II:

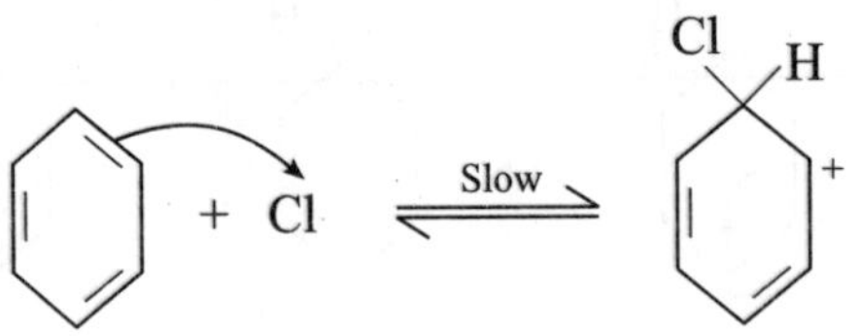

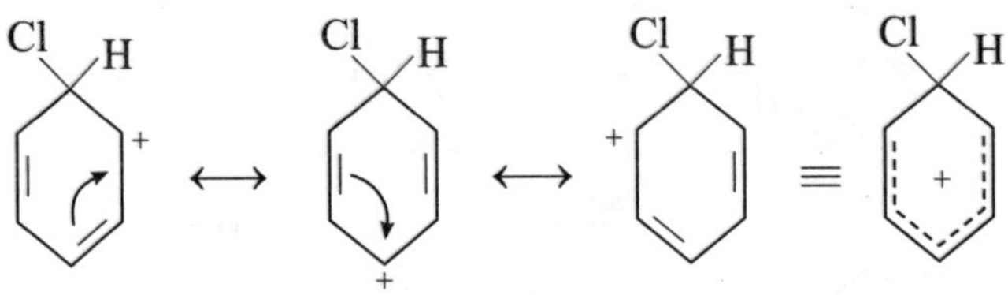

Step-III:

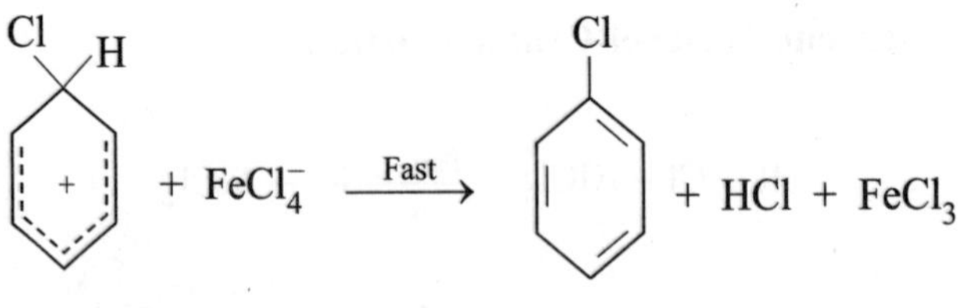

(ii) Disubstitution in benzene ring

The major product depends on –Su group. If the –Su group supports the substitution by donating electrons, –Su group is called activating group. If –Su group delays or prohibits the substitution by *withdrawing* electrons, –Su group is called deactivating group.

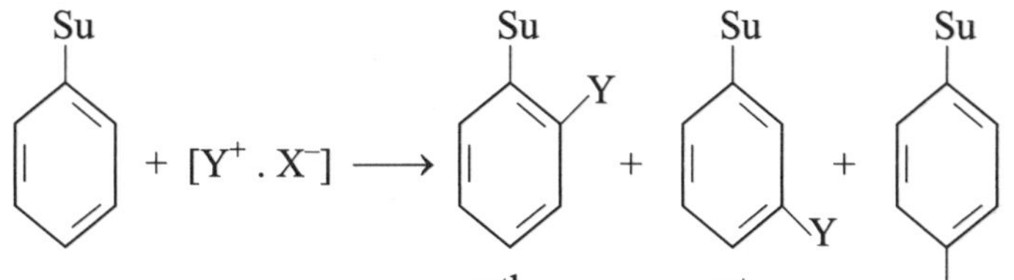

ortho- (o-) meta- (m-) para- (p-)

- **Activation of an aromatic ring may be of two types:**
  - (a) Inductive activation and hyperconjugative activation
  - (b) Resonance activation

(a) Inductive activation If a group attached to benzene ring, supports an electrophilic substitution through inductive effect the activation is called inductive activation.

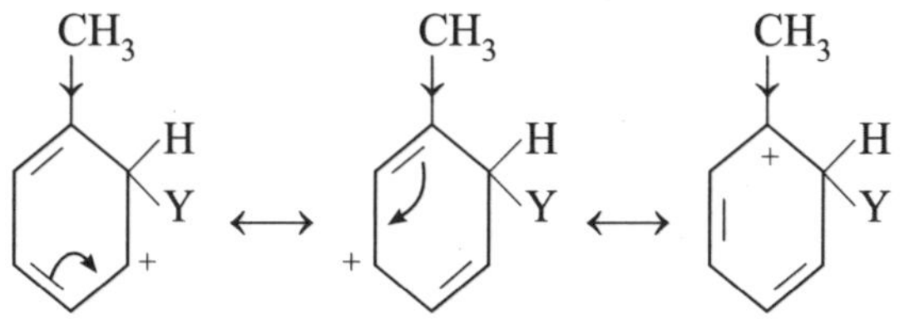

(I) (o-attact)

Structure (I) is stable since electron deficiency of the electropositive carbon atom is by and large nullified by +I effect of $-CH_3$ group or in other words positive charge of the carbon atom is delocalized through $-CH_3$ group by hyperconjugative effect:

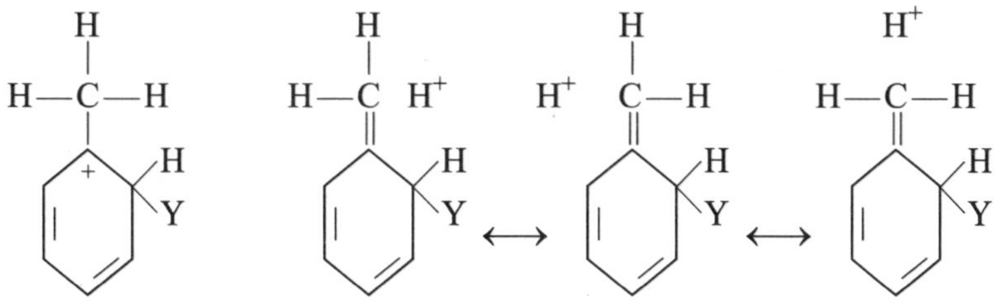

(I)

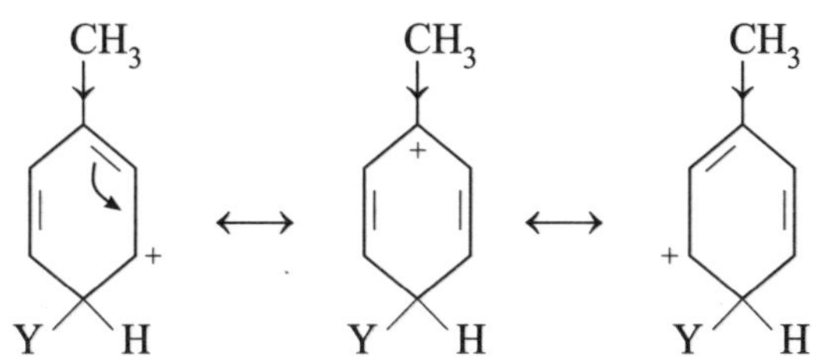

(II) (p-attack)

Structure (II) is stable due to the same reason as mentioned above.

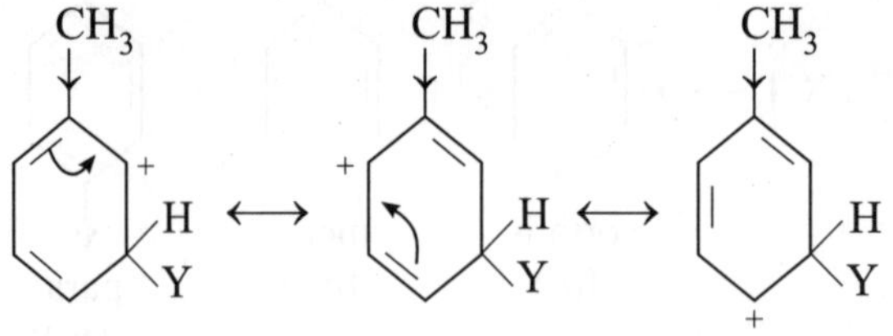

(*m*-attact)

No such stable structure [structure (I) or structure (II)] is available in the case of *m*-substitution. Hence *o*- and *p*-products appear as major products.

(b) Resonance activation When a group, attached to benzene ring or other aromatic ring, supports an electrophilic substitution through resonance effect the activation is called resonance activation.

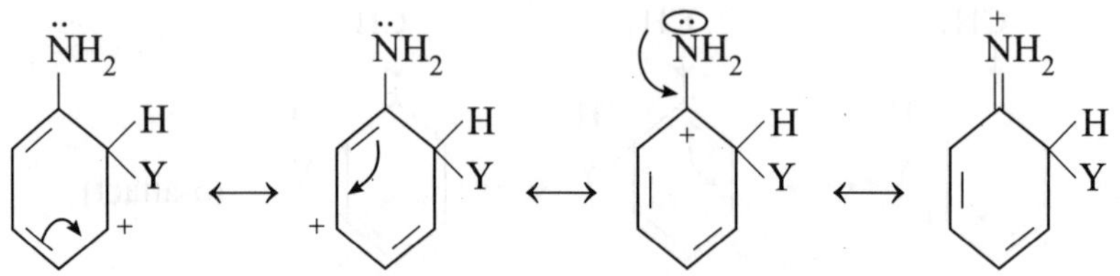

(*o*-attack) (4 resonating structures)

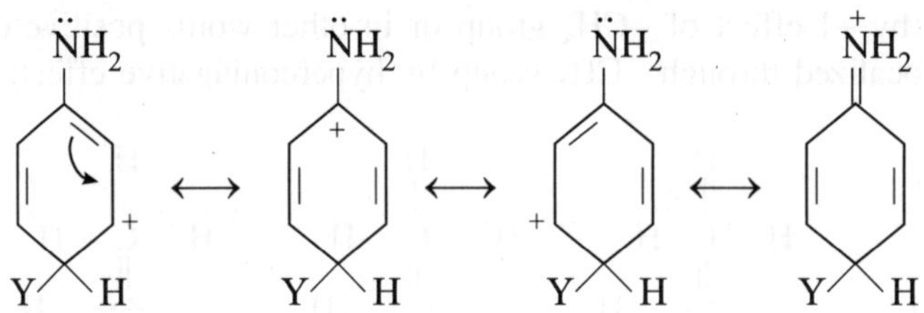

(*p*-attack) (4 resonating structures)

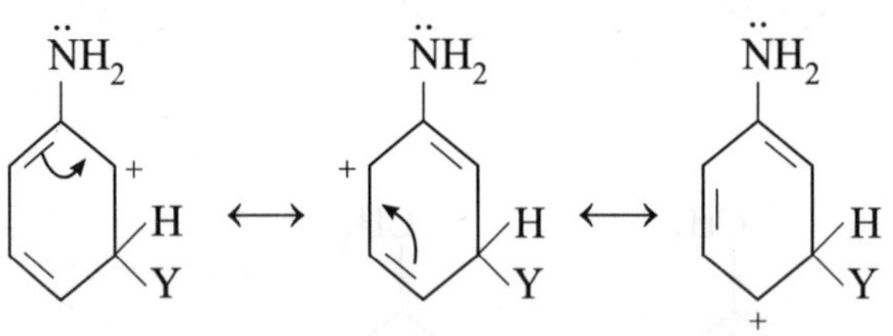

(*m*-attack) (3 resonating structures)

In the cases of o- and p-substitution reactions, lone pair of nitrogen takes part in resonance and the intermediate is stabilized through four resonating structures whereas the intermediate of *m*-substitution is stabilized through three resonating structures only. Obviously *o*- and *p*-products appear as major products.

Thus resonance activation of $-NH_2$ group is illustrated as:

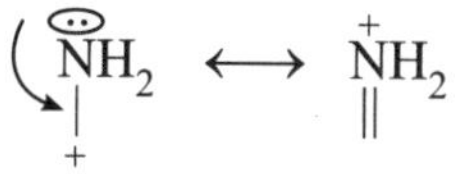

Similarly resonance activation of $-OR$ group can be written as:

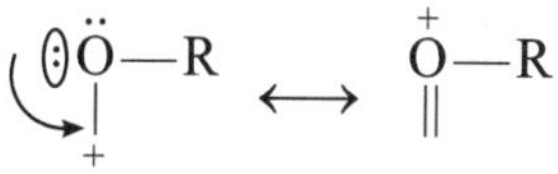

Resonance activation is far stronger than inductive activation.

All known activating groups may be divided into three categories:

(i) Strongly activating: $-NH_2$, $-NHR$, $-NR_2$, $-OH$
(ii) Moderately activating: $-OR$, $-NHCOR$
(iii) Weakly activating: $-R$, $-Ph$.

• In the case of *electrophilic substitution* deactivation of an aromatic ring may take place either (a) through inductive effect or (b) through resonance effect.

(a) Inductive deactivation If a group, attached to an aromatic ring, delays or prohibits electrophilic substitution through $-I$ effect, the deactivation is called inductive deactivation, e.g. substitution in chlorobenzene.

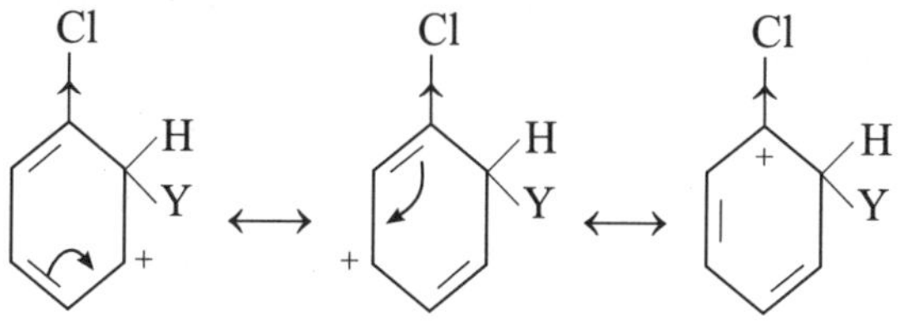

(*o*-attact)

(I)

Structure (I) is highly unstable since +ve charge is located on electron-difficient carbon atom.

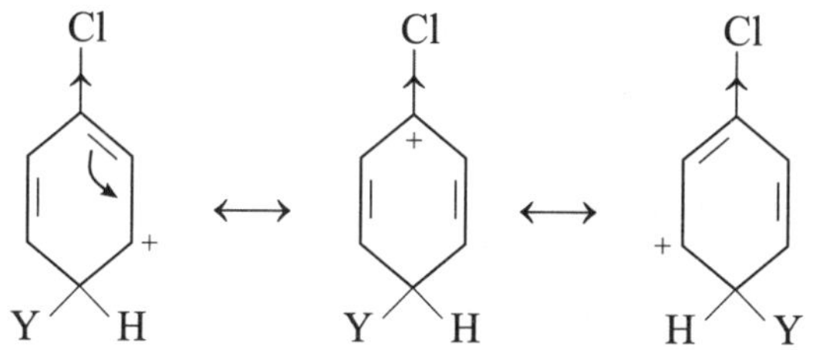

(*p*-attack)

(II)

Structure (II) is also unstable due to the same reason as mentioned above.

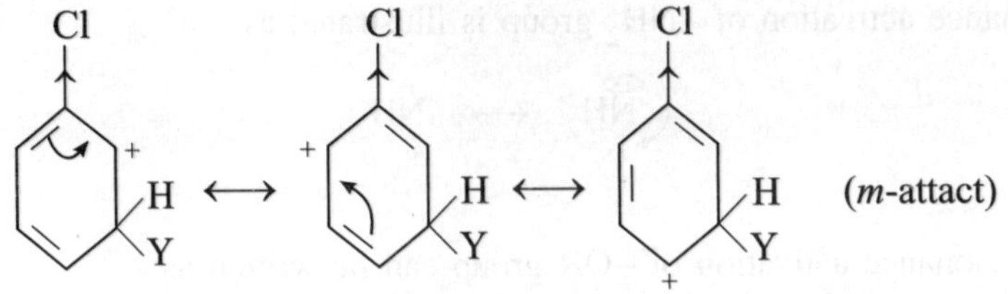

In the case of *m*-substitution there is no such canonical form which is as unstable as resonating structure (I) or (II). Obviously, *m*-attack is far more favourable than the corresponding *o*- or *p*-attack.

(b) Resonance deactivation There are some groups, which deactivate aromatic ring through resonance, e.g. $-NO_2$ group.

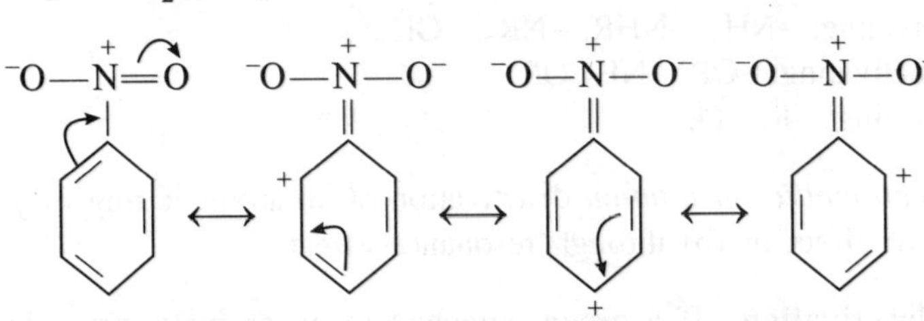

It is interesting to note that +ve charge is located on *o*- and *p*-positions but not on *m*-position. Thus obviously electron density is minimum at *o*- and *p*-positions. Hence *m*-product appears as the major product. Examples of some deactivating groups are given below according to the order of decreasing deactivation.

$$-NO_2 > -CN > -COOH > -SO_3H > -CHO > -COR > [-N(CH_3)_3]^+$$

It is interesting to note that chlorine promotes deactivation of aromatic ring through –I effect but at the same time supports electrophilic substitution of aromatic ring through resonance.

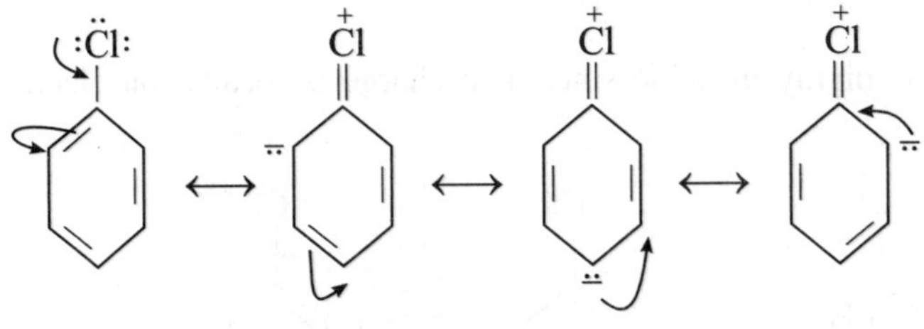

(iii) Orientation in disubstituted benzene ring

Case – I: 1, 3-dimethyl benzene

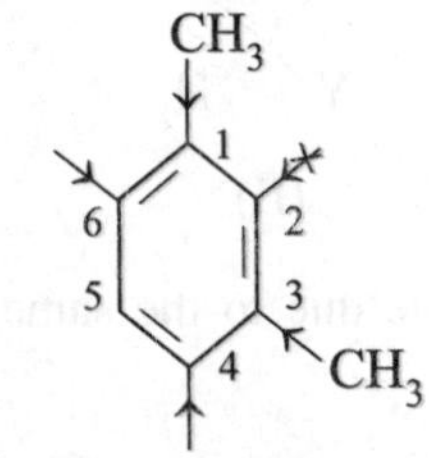

Here both the groups are reinforcing through +I effect. So obviously 4th and 6th positions are favourable for 3rd substitution. Probability of 3rd substitution at the 2nd position is very low since the 2nd position is sterically hindered by two bulky methyl groups.

Case – II: 1, nitro-3, methyl benzene

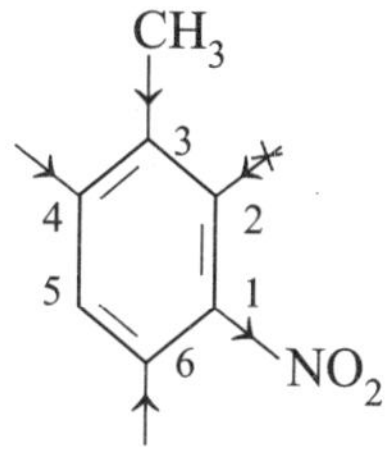

In this case, $-CH_3$ group has + I effect but $-NO_2$ group is strongly deactivating group. The rule is that reinforcing group i.e. here $-CH_3$ group controls the orientation of the 3rd substitution. So the 3rd substitution is mostly oriented at 4th and 6th positions. Substitution at the 2nd position is not favoured due to steric hindrance.

Case – III: 2-methyl phenol

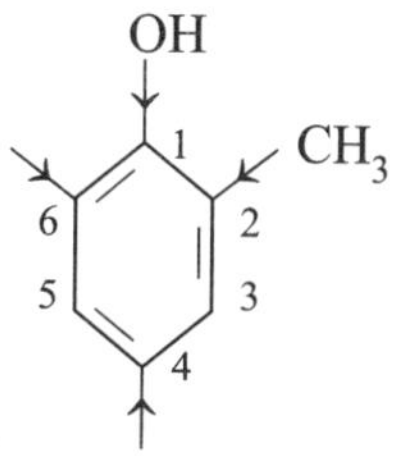

Here both the groups are reinforcing. The rule is that stronger activating group controls the 3rd substitution. Thus 4th and 6th positions are favourable for the 3rd substitution, since $-OH$ group is stronger activating group than $-CH_3$ group.

Case – IV:

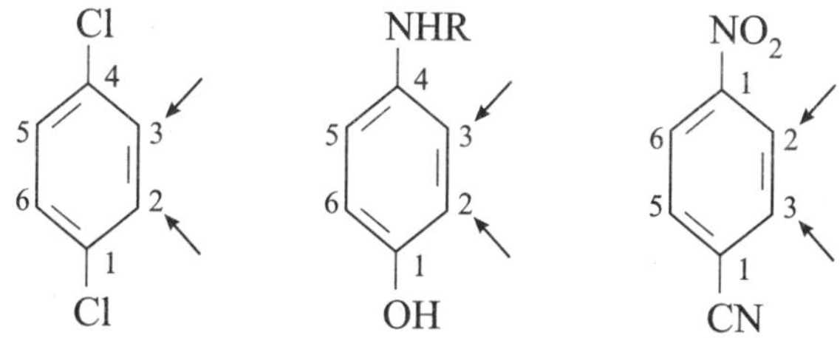

When two competitive groups are present at para-positions the 2nd and 3rd positions are equally favourable for the 3rd substitution.

8.1.7 Some rearrangement reactions with mechanism

(a) Beckmann rearrangement (1886) In Beckmann rearrangement oximes are converted into corresponding amides in presence of a catalyst.

$$\underset{\text{(Ketoxime)}}{R-\overset{\overset{\displaystyle R'}{|}}{C}=N-OH} \xrightarrow{\text{Catalyst}} \underset{\text{(Amide)}}{RCONHR'}$$

A Lewis acid or protonic acid is usually preferred as catalyst e.g. $PCl_5/SOCl_2/SbCl_3$ in a weak acid like acetic acid.

Mechanism:

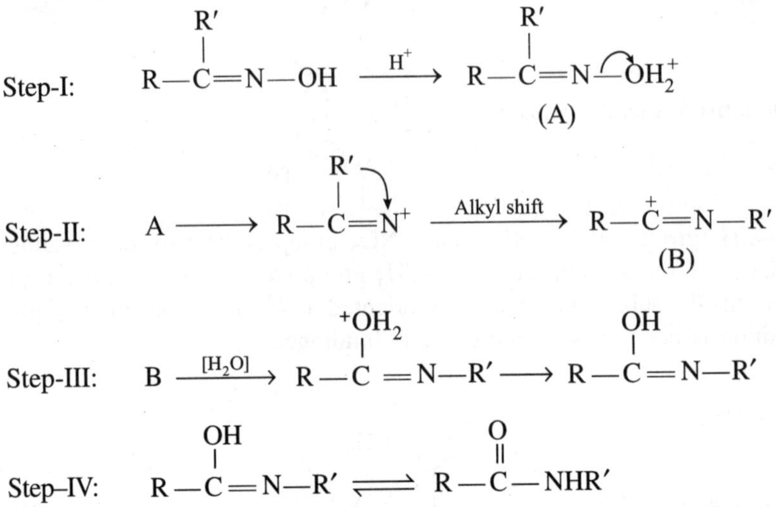

This is an example of intramolecular rearrangement. An asymmetric oxime may have two different structures.

(I) (II)

The group, lying in the trans-position with respect to –OH group, migrates to the neighbouring nitrogen atom. So, RCONHR′ is the exclusive product when structure (I) undergoes Beckmann rearrangement, but in the case of structure (II) R′CONHR appears as the exclusive product.

One of the important applications of Beckmann rearrangement is ring expansion, which is illustrated below:

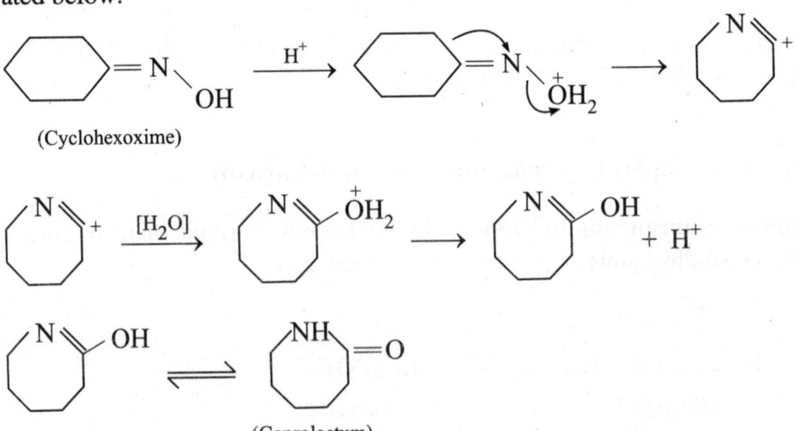

If the migrating group of an oxime is stable the rearrangement follows intermolecular mechanism rather than intramolecular mechanism. If two different oximes are used instead of one, a mixture of amides is resulted.

Step-I:

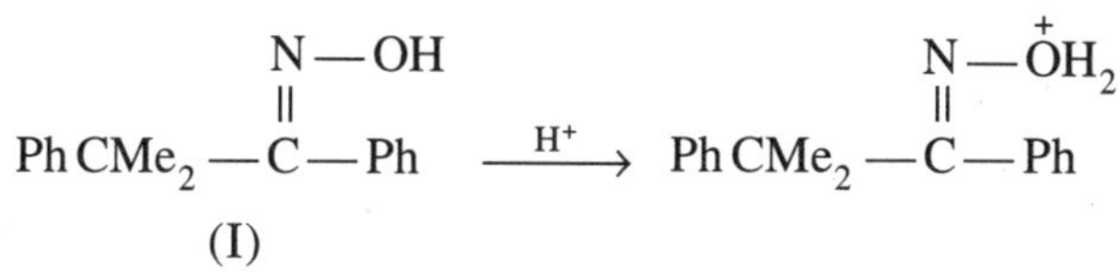

Step-II:

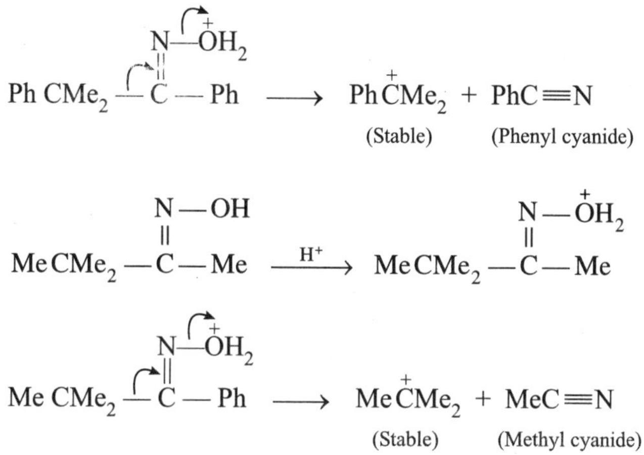

Step-III:

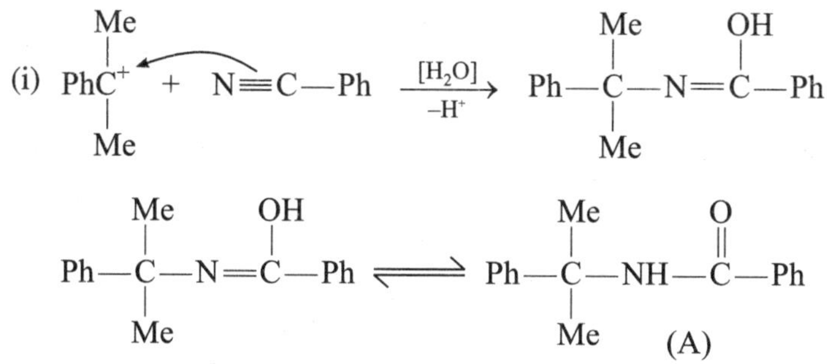

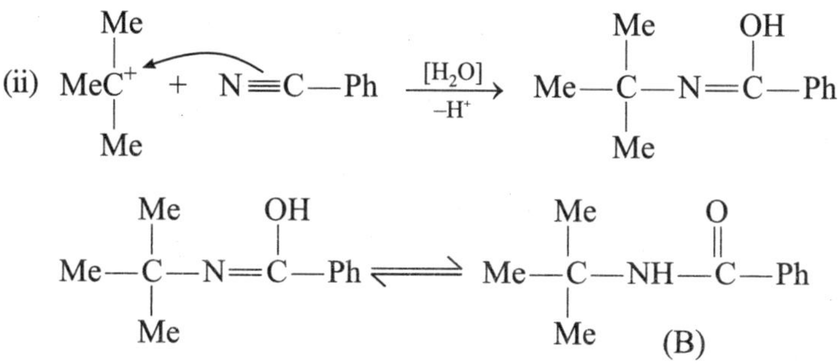

(iii) 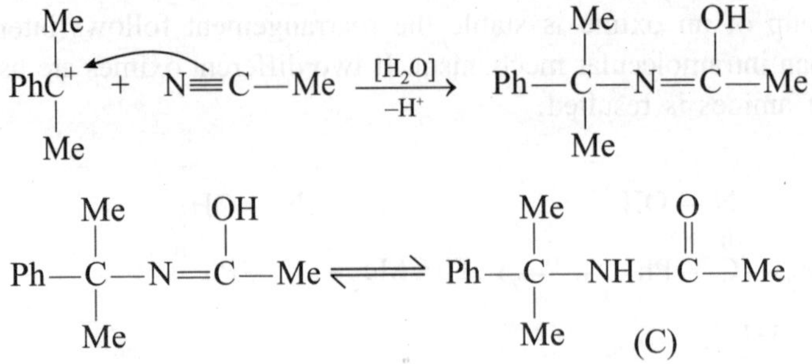

(iv) 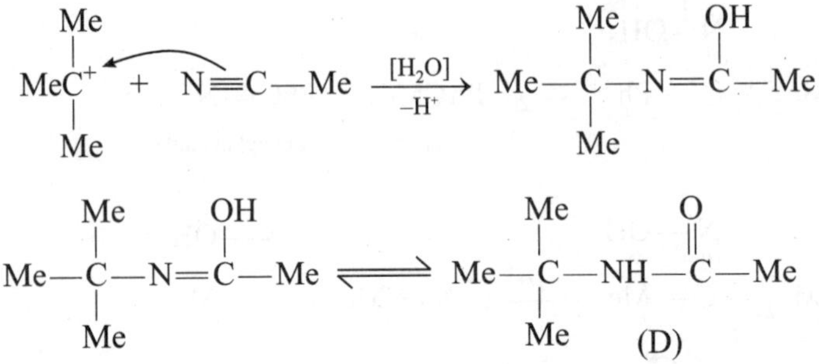

Production of mixture of oximes indicates that the reaction follows intermolecular rearrangement. In the 2nd step the migrating group, instead of migrating, forms a stable carbonium ion and a cyanide, which are converted to oxime in presence of water in the 3rd step.

In intramolecular rearrangement migrating group prefers migrating rather to forming carbonium ion since the latter is unstable.

(b) Hoffmann reaction In this rearrangement reaction amides are converted to amines with lesser number of carbon atoms in presence of sodium hypobromite. The overall reaction is given by:

$$RCONH_2 + (Br_2 + NaOH) \longrightarrow RNH_2 + 2NaBr + Na_2CO_3 + 2H_2O$$

Mechanism:

Step-I: 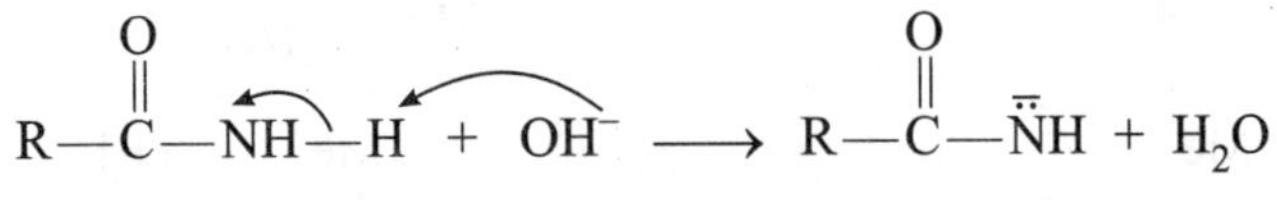

Step-II: 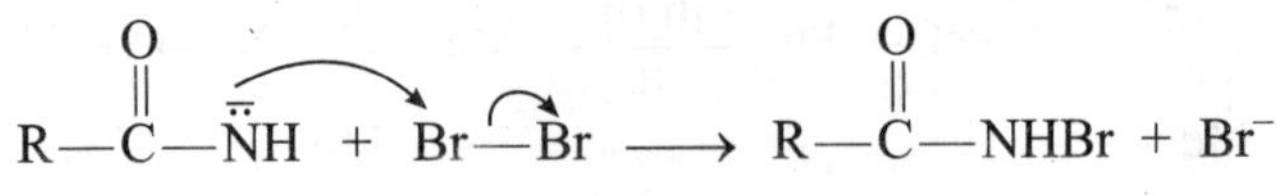

Step-III: 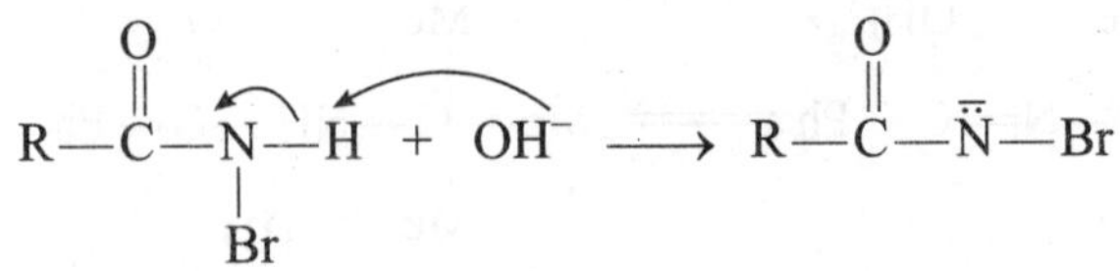

Step-IV:

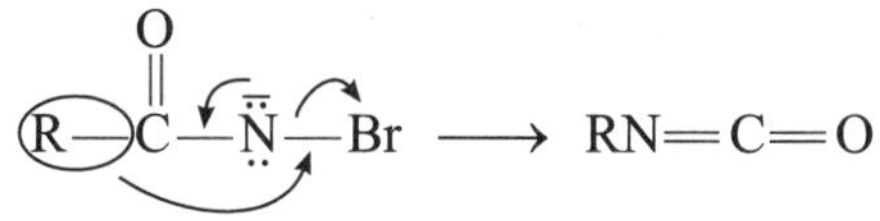

Step-V:

$$RN{=}C{=}O \xrightarrow{[H_2O]} R\bar{\ddot{N}}{-}\underset{}{\overset{\overset{+}{O}H_2}{\overset{|}{C}}}{=}O \longrightarrow RNH{-}\overset{OH}{\overset{|}{C}}{=}O$$

$$RNH{-}\overset{OH}{\overset{|}{C}}{=}O \longrightarrow \underset{\text{(amine)}}{RNH_2} + CO_2$$

(c) Cannizaro reaction Aldehydes, containing no α-hydrogen atoms, undergo a rearrangement in presence of an alkali to form both the corresponding acid and alcohol. The mechanism proceeds with a hydride ion shift. The overall reaction is given by:

$$2HCHO \xrightarrow{OH^-} CH_3OH + HCOO^-$$

$$2PhCHO \xrightarrow{OH^-} PhCH_2OH + PhCOO^-$$

Mechanism:

Step-I:

Step-II:

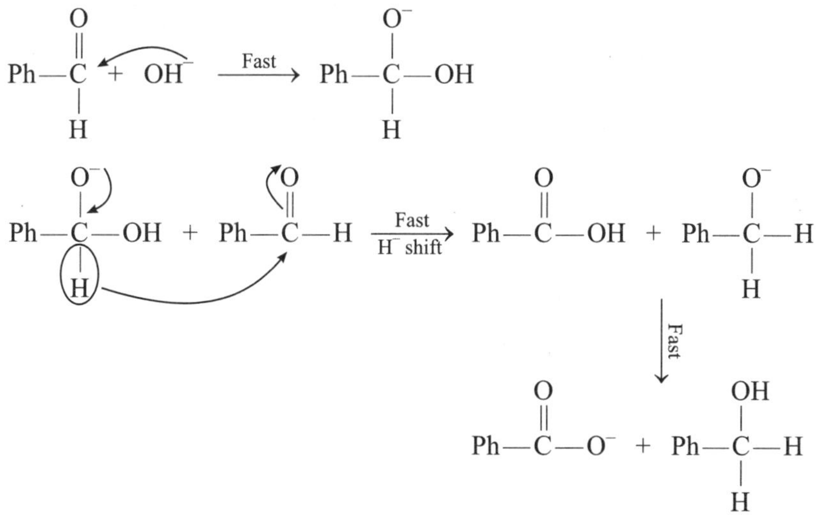

If the two aldehydes are different, more reducing or stronger aldehyde gets converted into corresponding acid, e.g. formaldehyde (HCHO) is much stronger reducing agent than 2, 2-dimethyl propanal $\left(\begin{array}{c} CH_3 \\ \downarrow \\ CH_3 \rightarrow C{-}CHO \\ \uparrow \\ CH_3 \end{array}\right)$ since in the latter +I effect of 3 methyl groups decreases the reducing property. So when mixture of these two aldehydes undergoes Cannizaro reaction formaldehyde gets oxidized to formic acid salt. The overall reaction is:

$$CMe_3CHO + HCHO \xrightarrow{OH^-} CMe_3CH_2OH + HCOO^-$$

Intramolecular Cannizaro reaction When cannizaro reaction takes place within a molecule the reaction is called intramolecular Cannizaro reaction, e.g. 1-oxo-1-phenyl-ethanal undergoes intramolecular Cannizaro reaction according to the following mechanism:

Step-I:

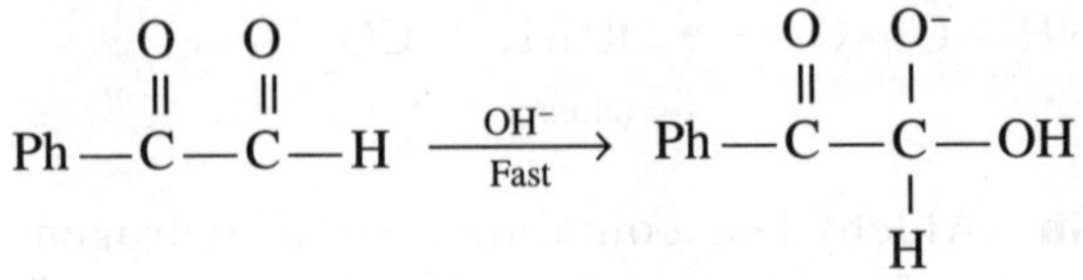

Step-II:

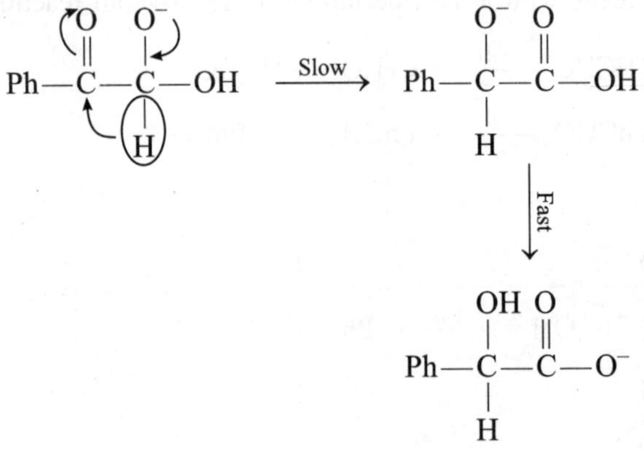

Another example is bezil-benzilic acid rearrangement:

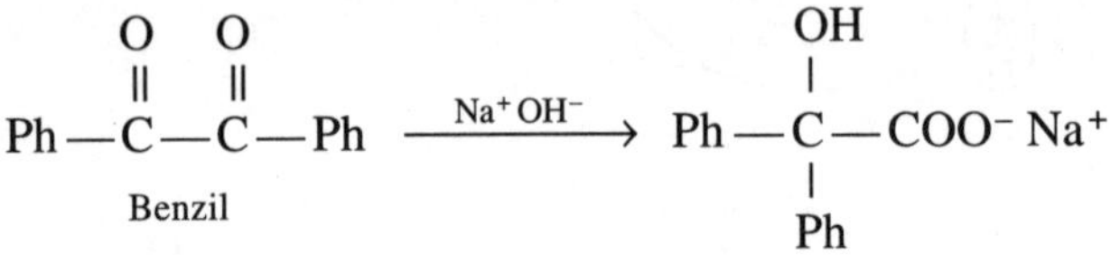

Benzil

Na-salt of benzilic acid

Aldol condensation Aldehydes, containing α-hydrogen atoms, undergo aldol condensation, instead of Cannizaro reaction, since hydroxyl ion, instead of attacking carbonyl group, attacks α-hydrogen atom. One such example is condensation of acetaldehyde.

$$2CH_3CHO \xrightarrow{OH^-} CH_3-\underset{}{\overset{OH}{\overset{|}{C}H}}-CH_2CHO$$

$$\text{Rate} = k.[CH_3CHO]^2\ [OH]^-$$

Mechanism:

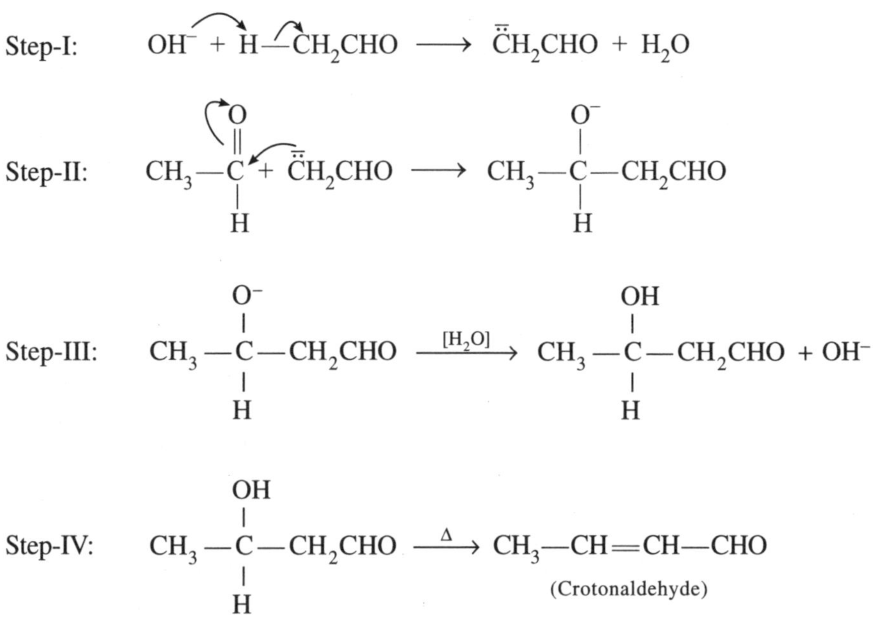

(d) Halogenation of ketones Halogenation of ketone can be carried out either in presence of acid catalyst or in presence of base catalyst.

(i) Acid catalyzed halogenation In presence of acid ketone is converted into its corresponding enol-form, which undergoes halogenation reaction. This is well illustrated in the following example:

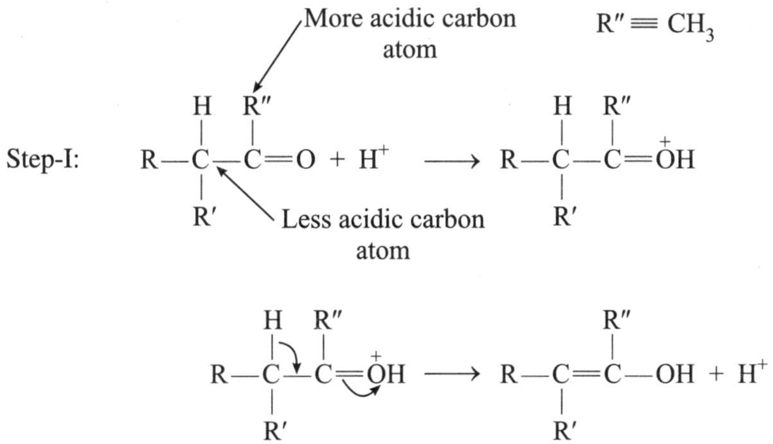

Step-II:

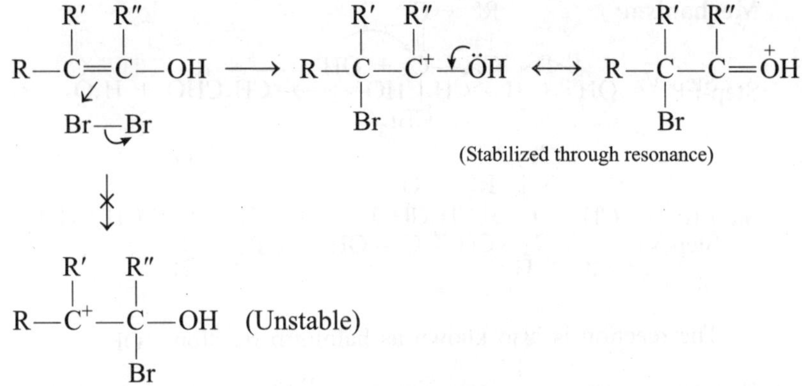

Step-III:

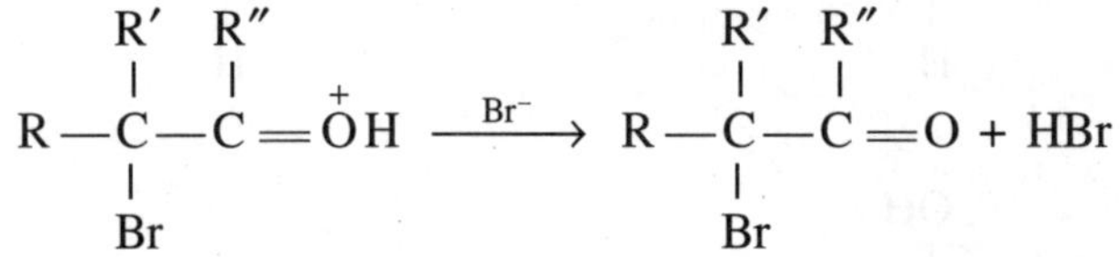

Here it is interesting to note that bromination occurs at less acidic carbon atom.

(ii) Base-catalyzed halogenation In presence of a base halogenation takes place at more acidic carbon atom.

Step-I:

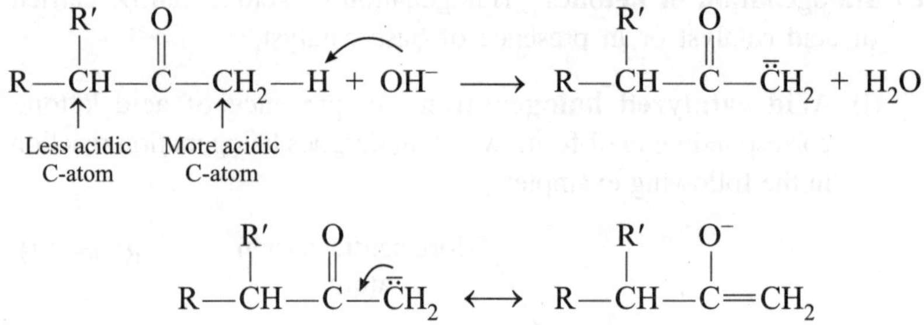

Step-II:

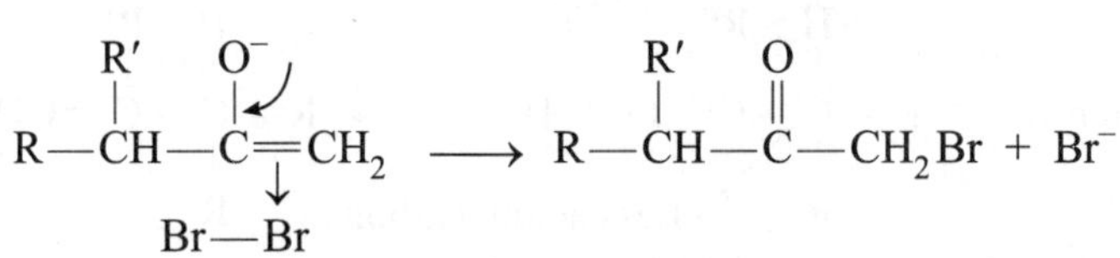

Step-III:

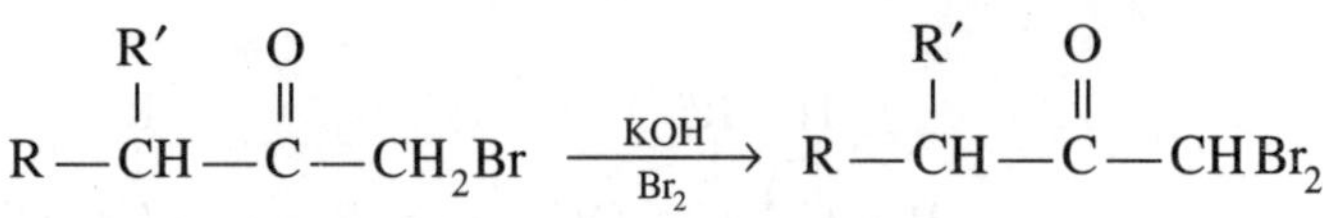

Step–IV:

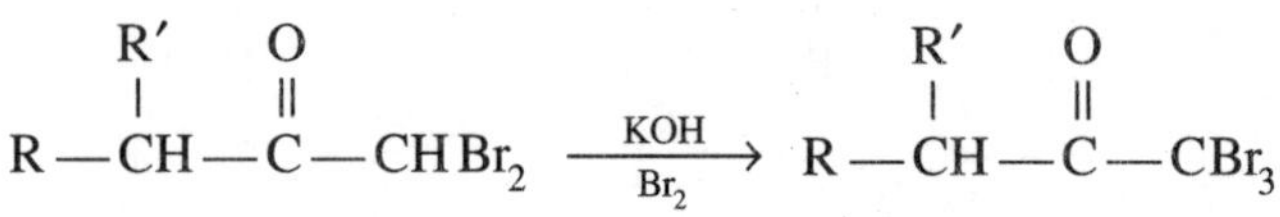

Step-V: $$R{-}\underset{\large R'}{CH}{-}\underset{\large CBr_3}{\overset{\large O}{\overset{\|}{C}}} + OH^- \longrightarrow R{-}\underset{\large R'}{CH}{-}\overset{\large O}{\overset{\|}{C}}{-}OH + CBr_3^-$$

Step-VI: $$R{-}\underset{\large R'}{CH}{-}\overset{\large O}{\overset{\|}{C}}{-}OH + CBr_3^- \longrightarrow R{-}\underset{\large R'}{CH}{-}\overset{\large O}{\overset{\|}{C}}{-}O^- + \underset{\text{(Bromoform)}}{CHBr_3} \downarrow$$

The reaction is also known as haloform reaction.

(e) Diel's Alder reaction If a conjugated diene *A* is heated with an unsaturated compound *B*, called dienophile, a cyclo-addition product *C*, called adduct, is formed according to the following reaction:

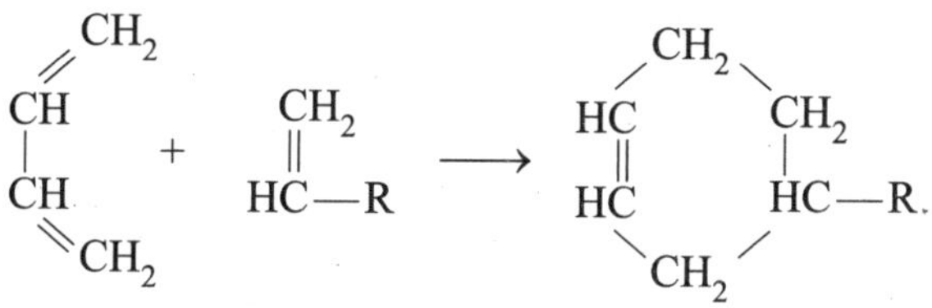

(A) (B) (C)

Reaction is usually carried out in an inert solvent (e.g., benzene, toluene etc.) and in absence of any catalyst. Diene *A* may be of different types:

(a) acyclic, e.g., ⟶ 

(b) cyclic, e.g., ⟶ 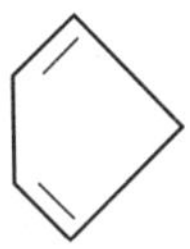

(c) heterocyclic, e.g., ⟶ 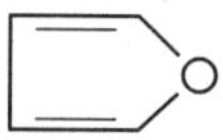

(d) aromatic hydrocarbons containing at least three benzene rings,

e.g., ⟶ 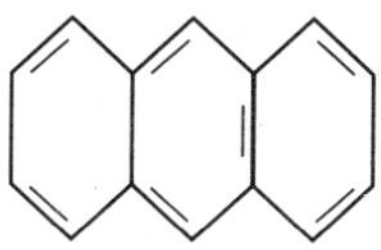

Dienophile *B* may be of different types:

(a) simple alkene, e.g., $CH_2{=}CH{-}R$

(b) alkylene, e.g., $CH\equiv C—R$

(c) α, β-unsaturated carbonyl compound, e.g., $CH_2=CH—CHO$

Mechanism Mechanism of the reaction may proceed in three different ways: (i) Ionic, (ii) free radical and (iii) one-step concerted mechanism.

(i) Ionic mechanism Both diene *A* and dienophile *B* form Zwitterions. In the next step two Zwitterions combine to form the adduct.

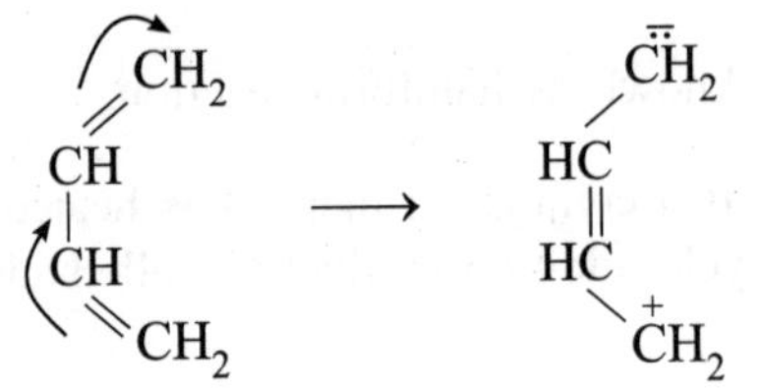

butadiene

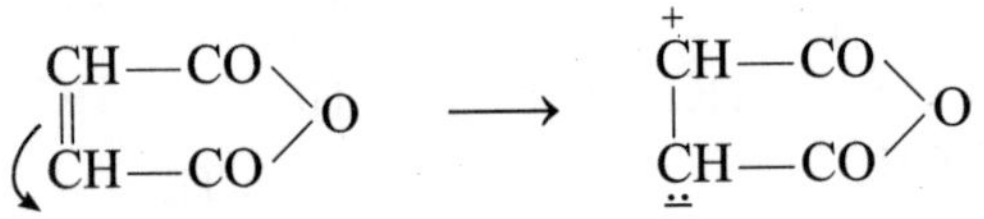

maleic anhydride

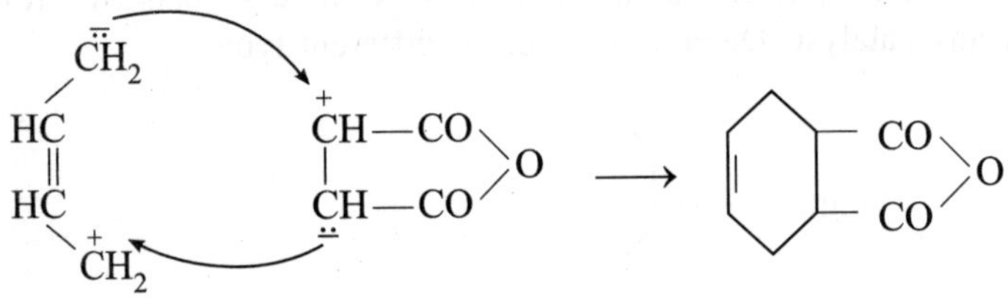

(ii) free radical mechanism

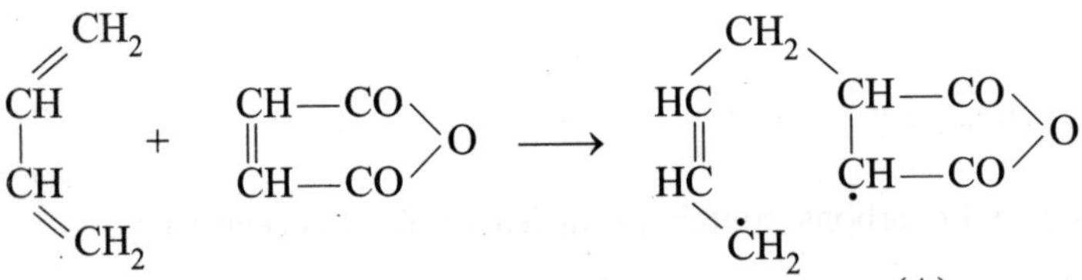

(A)

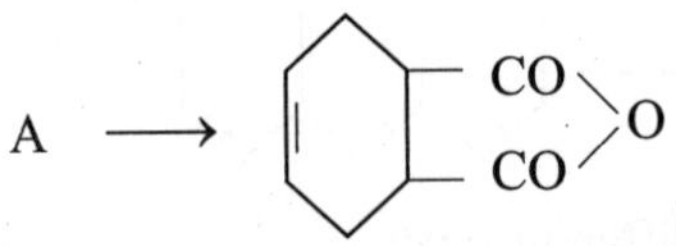

(iii) One-step concerted mechanism

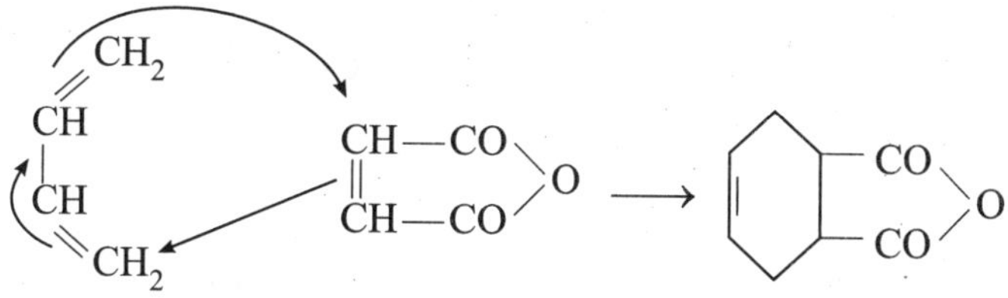

(f) Reimer–Tiemann reaction Aromatic compounds having phenolic group and some heterocyclic compounds undergo Reimer–Tiemann reaction, where, aldehyde group (–CHO) is included to the aromatic ring during the course of reaction.

Reagents: (i) Chloroform ($CHCl_3$), (ii) Potassium hydroxide (KOH),

(ii)

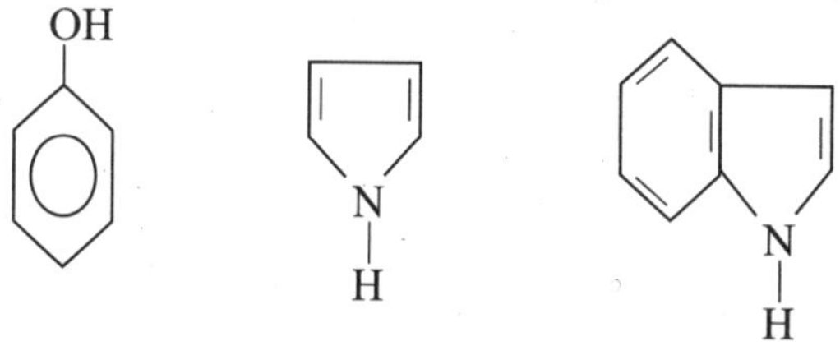

Phenol Pyrrole Indole

Temperature: 340°K

Overall reaction:

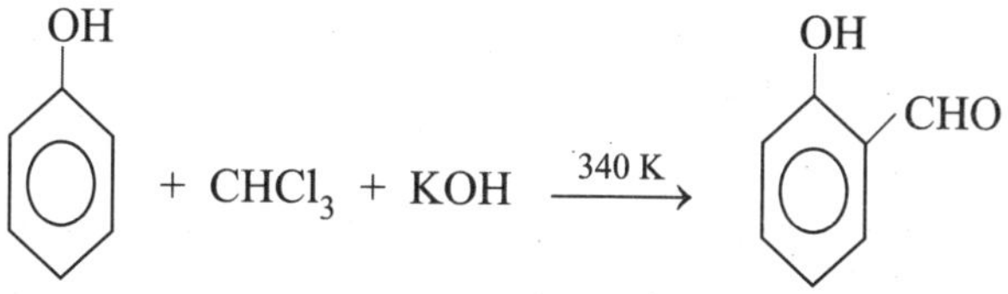

Mechanism:

Step-I: $CHCl_3 + OH^- \rightleftharpoons \ddot{\bar{C}}Cl_3 + H_2O$

$\ddot{\bar{C}}Cl_3 \longrightarrow :CCl_2 + Cl^-$

$:CCl_2 \equiv$ Dichlorocarbene

C in $:CCl_2 \equiv$ 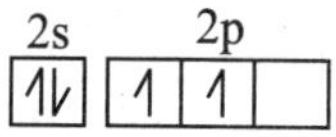

Step-II:

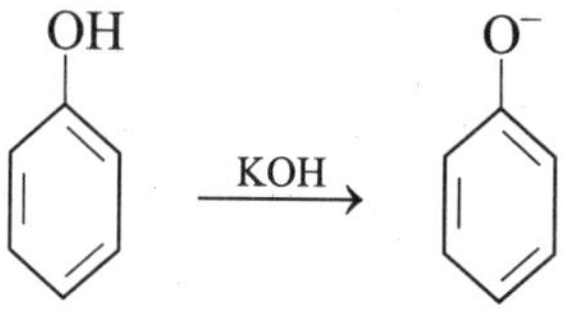

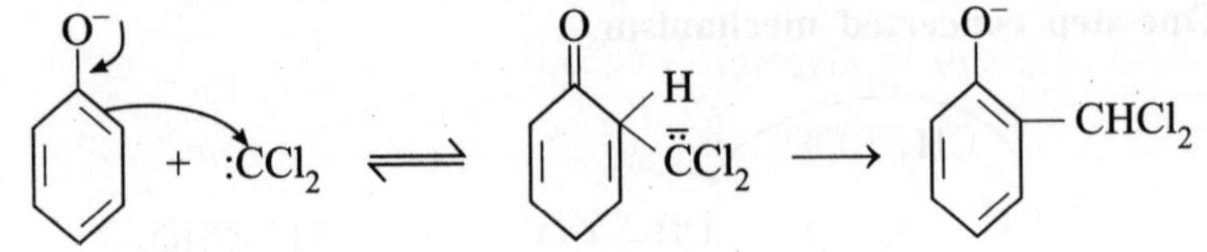

Step-III:

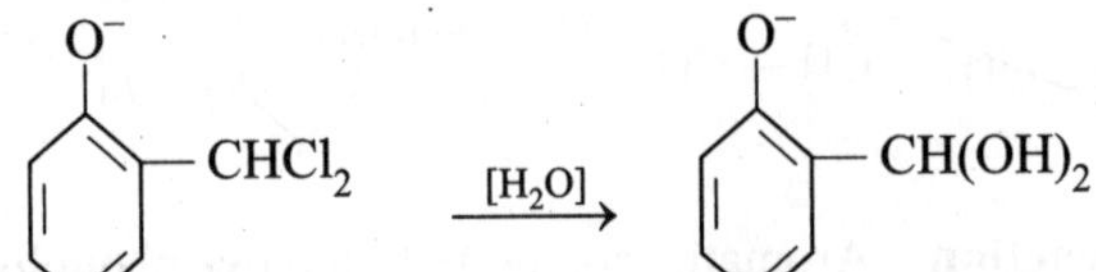

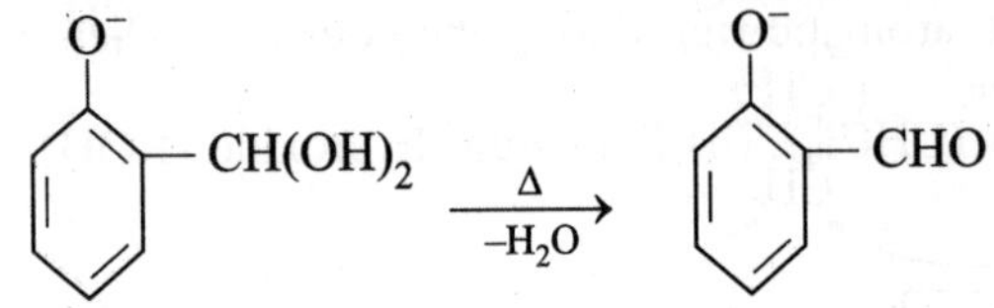

Step-IV:

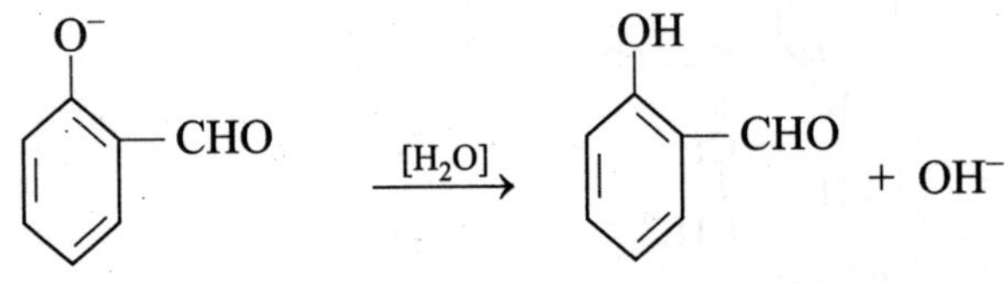

Salicyldehyde

(g) Skraup synthesis It is a method of preparation of quinoline from aniline.

Reagents:

Aniline:  Glycerol: CH_2OH–$CHOH$–CH_2OH Nitrobenzene: 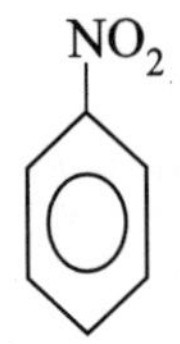

Ferrous sulphate ($FeSO_4$) and concentrated sulphuric acid (H_2SO_4).

Overall reaction:

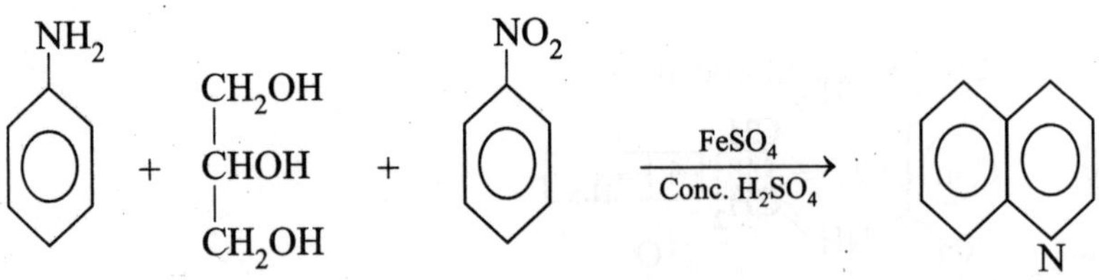

Quinoline

Step-I: Dehydration of glycerol:

$$\begin{matrix} CH_2OH \\ | \\ CHOH \\ | \\ CH_2OH \end{matrix} \xrightarrow{\text{Conc. } H_2SO_4} \underset{\text{Acrolein}}{CH_2{=}CH{-}CHO}$$

Step-II:

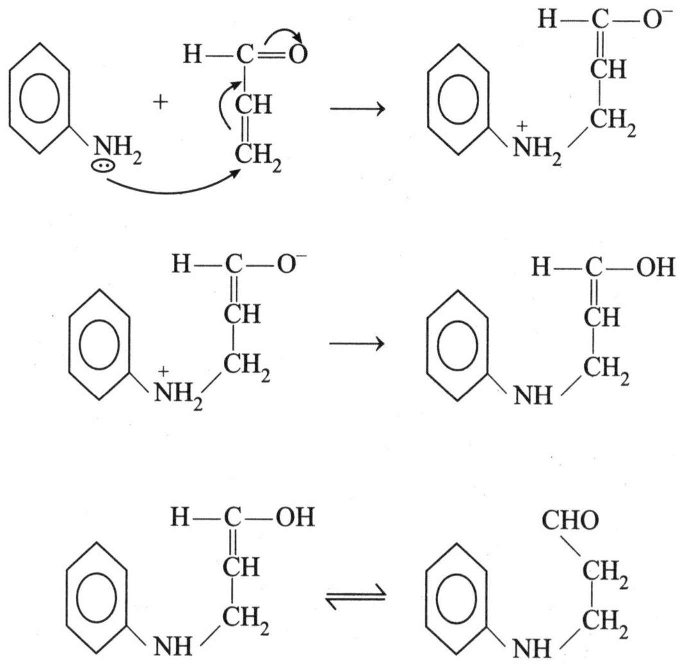

Step-III:

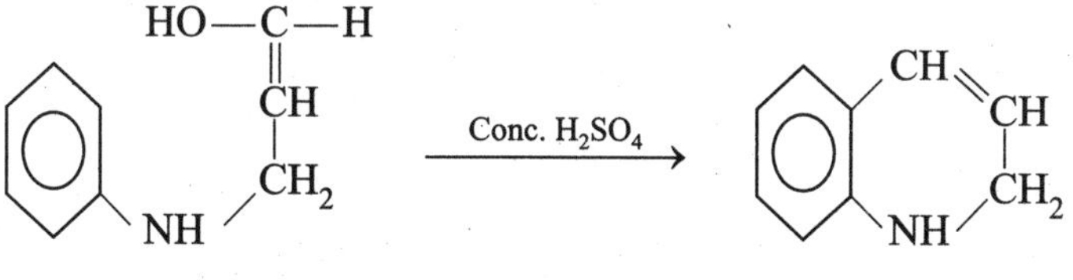

Step-IV:

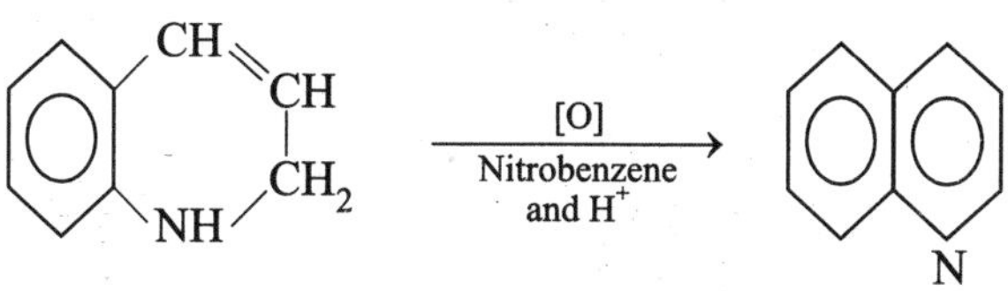

Quinoline

The function of ferrous sulphate ($FeSO_4$) is to retard the reacton rate considerably to avoid explosion.

SECTION – II POLYMER CHEMISTRY

8.2.1 Introduction

What is a polymer? The answer is:

(i) Geometrically it is a long chain coil, often represented as:

(ii) Chemically it is a macromolecule represented as •, which is actually a very long chain macromolecule, constituted by several numbers of a particular chemical unit often called repeat unit.

(iii) Structurally it is represented as —[—MU—]$_n$—, where MU is known as repeat unit and n is known as degree of polymerization.

Degree of polymerization (DP) The overage number of repeat units attached in a single polymer chain is known as degree of polymerization (DP), represented either by $\bar{X}_n$ or by $\bar{X}_w$. $\bar{X}_n$ is the number average degree of polymerization and is given by:

$$\bar{X}_n = \frac{\sum_{i=1}^{\infty} x_i N_i}{\sum_{i=1}^{\infty} N_i}$$

where, N_i is the number of ith species, having DP x_i.

$\bar{X}_w$ is the weight average degree of polymerization and is given by:

$$\bar{X}_w = \frac{\sum_{i=1}^{\infty} x_i N_i M}{\sum_{i=1}^{\infty} N_i M}$$

where, M is molecular weight of monomer unit and N_i is the number of ith species, having DP x_i.

Molecular weight and Molecular weight distribution As we have seen earlier, degree of polymerization (DP) varies chain to chain or molecule to molecule. So, molecular weight of one polymer molecule differs from that of another polymer molecule. Thus molecular weight of a polymer should be expressed in average molecular weight, quite distinguished from monomer or low molecular weight species where molecular weight of each species remains same.

The average molecular weight of a polymer can be expressed either as number average $\bar{M}_n$ or as weight average $\bar{M}_w$.

$$\bar{M}_n = \frac{\text{Sum of (No. of species} \times \text{Mol.wt. of } i\text{th species)}}{\text{Total no. of species}}$$

If N_i and M_i are the number of ith species and molecular weight of ith species respectively, we have:

$$\bar{M}_n = \frac{\sum_{i=1}^{\infty} N_i M_i}{\sum_{i=1}^{\infty} N_i}$$

$$\bar{M}_w = \frac{\text{Sum of (Total wt. of all } i\text{th species} \times \text{Mol.wt. of } i\text{th species)}}{\text{Total wt. of all species}}$$

If N_i, M_i and w_i are the number of ith species, molecular weight of ith species and total weight of all ith species respectively, we have:

$$w_i = N_i M_i \quad \text{and} \quad \bar{M}_w = \frac{\sum_{i=1}^{\infty} w_i M_i}{\sum_{i=1}^{\infty} w_i} = \frac{\sum_{i=1}^{\infty} N_i M_i^2}{\sum_{i=1}^{\infty} N_i M_i}$$

So molecular weight distribution curve can be obtained by plotting weight of each species against the corresponding molecular weight (Figure 8.2.1).

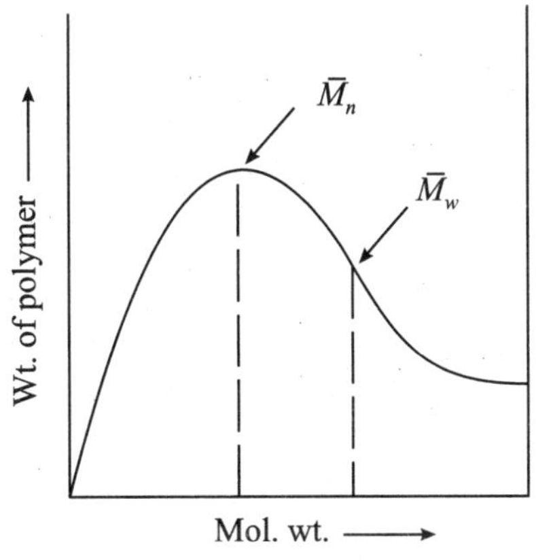

Fig. 8.2.1 Distribution of molecular weights in a typical polymer.

Polydispersity index (PI) The ratio of two molecular weight averages, i.e., $\left[\frac{\bar{M}_w}{\bar{M}_n}\right]$ is known as polydispersity index (PI). Lower the value of $\left[\frac{\bar{M}_w}{\bar{M}_n}\right]$ means narrower the distribution, that means number of species, having different molecular weight, is less, i.e., maximum number of molecular species possess almost same molecular weight. Obviously polymer, having low polydispersity index, possesses superior physical characteristics to others. The lowest possible value of PI is 1.

8.2.2 Classification of polymers

Polymers can be classified on different basis.

(i) On the basis of origin
(ii) On the basis of physical appearance
(iii) On the basis of polymerization processes.

(i) On the basis of origin polymers can be divided into three categories:
(a) Natural, (b) Semi-synthetic, (c) Synthetic.

(a) Natural polymer Polymeric materials, which are occurring naturally, are known as natural polymers. Examples are: Cotton, Silk, Wool, Rubber, Cellulose etc.

(b) Semi-synthetic polymer Polymeric materials, derived from natural polymers, are called semi-synthetic polymers. Examples are: Cellulose acetate, cellulose nitrate, ethyl cellulose, methyl cellulose etc.

(c) Synthetic polymer Polymers, which are manufactured from the corresponding monomers, are called synthetic polymers. Examples are: Polyethylene, polypropylene, polyvinyl chloride etc.

(ii) On the basis of physical appearance polymers can be divided into three categories:
(a) Elastomer, (b) Plastic, and (c) Fibre.

(a) Elastomers The materials, which show long range elasticity on the application of stress, are known as elastomers. Elastomers can be further divided into three broad categories, namely, (1) natural elastomer or natural rubber, (2) synthetic elastomer or synthetic rubber and (3) thermoplastic elastomer.
Naturally occurring elastomer is known as natural rubber, represented as NR.
Elastomers, manufactured artificially are known as synthetic rubber, e.g., Styrene Butadiene Rubber, represented as SBR, Acrylonitrile Butadiene Rubber represented as NBR, Poly Butadiene Rubber, represented as PBR, etc.
Usually elastomers are easily deformable semi-solid materials at room temperature. But there are some elastomers which are not deformable at room temperature but at high temperature they show usual elastic characteristics. Those materials are known as

thermoplastic elastomers. Examples are: Styrene- Butadiene-Styrene, represented as SBS, Polyurethane elastomers, represented as PU, etc.

(b) Plastics The polymeric materials, which are rigid, dimensionally stable and usually brittle, are known as plastics. Plastics can be divided into two broad categories: (1) Thermoplastics and (2) Thermosets.

The materials, which have high molecular weight and a definite softening point range, are known as thermoplastic materials. Examples are: Polyvinyl chloride, represented as PVC, Polystyrene, represented as PS, Polymethyl methacrytale, represented as PMMA, etc.

Thermosets are very hard, brittle, infusible and insoluble in any solvent. They are low molecular weight material and do not have any softening point range. Thermosetting materials usually exist in liquid, which on crosslinking at high temperature gets converted into a insoluble and infusible three-dimensional network. This low molecular weight high viscous liquid polymer is often called resin. Examples are: Phenolic resins, also known as Bakelite, amino resins, polyester resins etc.

(c) Fibres Materials, which exist as a very long continuous filament or staple and possess following characteristics, are known as fibres.

(1) length(l) to diameter(d) ratio must be over 100. The $\frac{l}{d}$ ratio exceeds 1000 for almost all naturally occurring fibres.
(2) Generally fibres possess high softening point.
(3) They have high mechanical strength and stiffness close to that of metals.
(4) They are light weight high molecular weight material.

Examples are: nylon, polyester, aramid etc.

(iii) On the basis of polymerization processes polymers can be classified in two different categories: (a) Condensation polymers and (b) Addition polymers.

(a) Condensation polymers Polymers, which are produced by condensation mechanism, are known as condensation polymers. In this mechanism low molecular weight material is produced as by-product along with the polymer.

Examples are:

(1) Preparation of polyester through condensation of a diol and a dibasic acid.

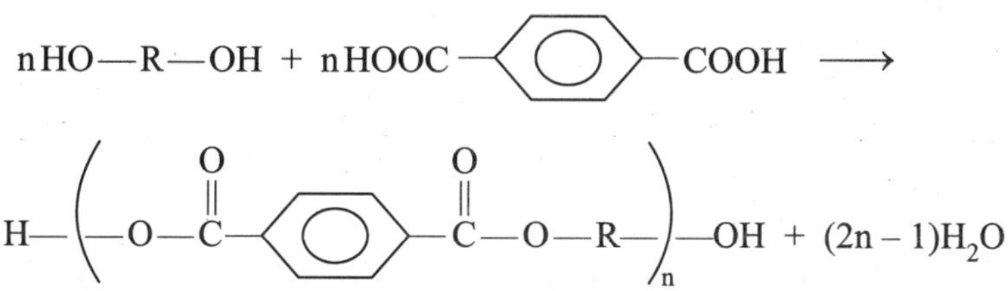

Polyester resin

By-product is water.

(2) Preparation of nylon 6, 6:

$$NH_2—(CH_2)_6—NH_2 + HOOC—(CH_2)_4—COOH \longrightarrow$$

$$NH_2 \left[(CH_2)_6—NHCO—(CH_2)_4—CO \right] OH + (2n-1)H_2O$$

By-product is water.

It is to be noted that to execute condensation polymerization, reacting species must have more than one functional groups. Furthermore, rate of condensation decreases as reaction proceeds due to accumulation of by-product. Continuous removal of byproduct somewhat increases the reaction rate. Molecular weight of condensation polymer usually lies within 10^4.

(b) Addition polymers Polymers, which are produced through addition mechanism without giving off any by-product, are known as addition polymers. Usually unsaturated compounds undergo addition polymerization. Mechanisms of addition polymerization may be divided into three categories:

(1) free radical mechanism,
(2) ionic mechanism,
(3) coordination mechanism.

Examples of addition polymerization:

$$n\,CH_2{=}CH_2 \longrightarrow \left(CH_2—CH_2 \right)_n$$

Ethylene — Polyethylene (PE)

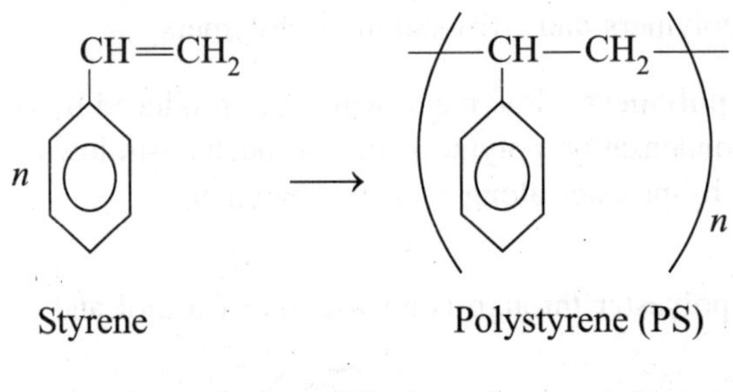

Styrene — Polystyrene (PS)

(1) Free radical mechanism In free radical mechanism initiators are employed to generate free radicals in the reaction mixture. These free radicals then attack monomer molecules to form monomer free radicals, which eat up monomer molecules continuously to form macroradicals. The mechanism proceeds through following steps:

Initiation step: $I \longrightarrow 2R^*$

Initiator — Initiator free radical

$$\underset{}{R^*} + \underset{\text{Monomer}}{M} \longrightarrow \underset{\substack{\text{Monomer}\\ \text{free radical}}}{M_1^*}$$

Peroxides are the most common initiators.

Propagation step:

$$M_1^* + M \longrightarrow M_2^*$$
$$M_2^* + M \longrightarrow M_3^*$$

– – – – – – – – – – –

– – – – – – – – – – –

$$M_n^* + M \longrightarrow M_{n+1}^*$$

Termination step:

$$M_x^* + M_y^* \longrightarrow M_{x+y} \quad \text{(Polymer)}$$

(Termination by combination)

$$M_x^* + M_y^* \longrightarrow \underset{\text{(Polymer)}}{M_x} + \underset{\text{(Polymer)}}{M_y}$$

(Termination by disproportionation)

Free radical reaction proceeds very fast and usually occurs at high temperature around 100°C. Reaction is near to complete within a few minutes and hence molecular weight of the resulting polymer is very high, in the range of $10^5 – 10^7$. In order to control the reaction rate inhibitors are added to the reaction mixture. These inhibitors act as scavengers of free radicals to form stable products. Some common inhibitors are nitrobenzene, thiophenol etc.

(2) Ionic mechanism Ionic mechanism of a polymerization reaction may proceed either by cationic process or by anionic process. Ionic mechanism is much faster than free radical mechanism. Ionic reaction usually occurs at very low temperature at 0°C or lower but take only a few seconds to complete. Molecular weight of an ionic polymer is usually higher than that of free radial polymer. Furthermore, narrow molecular weight distribution is usually observed in ionic polymers.

In cationic polymerization process, Lewis acids, namely, $AlCl_3$, BF_3, $SnCl_2$, $TiCl_4$ etc. act as catalyst. The catalyst attacks monomer in presence of a co-catalyst to form carbonium ion, which then attacks another monomer. The process continues and the chain grows on, leading to a macrocarbonium ion, which is then terminated by an anion. Cationic polymerization is favoured if R-group has *+I* effect.

Initiation step:

$$\underset{\text{Catalyst}}{BF_3} + \underset{\text{Co-catalyst}}{H_2O} \longrightarrow [BF_3OH]^- H^+$$

$$[BF_3OH]^- H^+ + CH_2{=}\underset{\substack{|\\ R}}{CH}$$

$$\longrightarrow [BF_3OH]^- \; CH_3-\underset{\substack{|\\R}}{\overset{+}{C}H} \equiv [BF_3OH]^- \; HM_1^+$$

Propagation step:

$$[BF_3OH]^- \; HM_1^+ + CH_2=\underset{\substack{|\\R}}{CH} \longrightarrow [BF_3OH]^- \; HM_2^+$$

$$[BF_3OH]^- \; HM_2^+ + CH_2=\underset{\substack{|\\R}}{CH} \longrightarrow [BF_3OH]^- \; HM_3^+$$

- -

- -

$$[BF_3OH]^- \; HM_n^+ + CH_2=\underset{\substack{|\\R}}{CH} \longrightarrow [BF_3OH]^- \; HM_{n+1}^+$$

Termination step:

$$[BF_3OH]^- \; HM_x^+ + M \longrightarrow M_x + [BF_3OH]^- \; HM_1^+$$

In anionic polymerization process, catalyst generates nucleophile, which then attacks monomer to form carbanion. This carbanion acts as nucleophile and attacks another monomer. The process continues and ultimately a macroanion is formed. Termination of macroanion is not an easy process. In some cases solvent molecules interact with macroanion to form a dead polymer. In some cases negative end of the macroanion is entrapped within the chain. Those polymers are known as *living polymers*. In presence of another monomer the *living polymer* attacks that new monomer and the chain continues to grow. Anionic polymerization is favoured if *R*-group has *–I* effect.

Initiation step: Some common catalysts used for anionic polymerization are: potassium amide (KNH_2), lithium amide ($LiNH_2$), *n*-butyl lithium (Li-*n*-Bu), sodium naphthenate etc.

$$KNH_2 \rightarrow K^+ + NH_2^-$$

$$NH_2^- + CH_2=\underset{\substack{|\\R}}{CH} \longrightarrow NH_2-CH_2-\underset{\substack{|\\R}}{\overset{..}{C}H} \equiv NH_2M_1^-$$

Propagation step:

$$NH_2M_1^- + M \longrightarrow NH_2M_2^-$$

$$NH_2M_2^- + M \longrightarrow NH_2M_3^-$$

- - - - - - - - - - - - - - -

- - - - - - - - - - - - - - -

$$NH_2M_x^- + M \longrightarrow NH_2M_{x+1}^-$$

Termination step:

$$NH_2M_x^- + \underset{\text{(Solvent)}}{NH_3} \longrightarrow NH_2M_xH + NH_2^-$$

Anionic polymerization is usually carried out at –75°C or below. Polymers with very high molecular weight and significantly narrow molecular weight distribution can be obtained by this technique. Thermoplastic elastomers can be prepared only by this technique. Polymerization process is very fast even faster than cationic polymerization process.

(3) Coordination mechanism Coordination polymerization technique was first discovered in early 1950 and a consolidated and valuable works on coordination polymerization was first published by Ziegler and Natta. The catalyst, used in this technique, is called Ziegler-Natta catalyst, which consists of metal alkyls of groups I–III with halides and other derivatives of transition metals of groups IV–VIII.
A list of Ziegler-Natta catalysts is given in the Table 8.2.1.

Table 8.2.1 Components of Ziegler-Natta catalysts.

Component – I *Organometallic Compound*	*Component – II* *Transition metal salt*
Triethyl aluminium	Titanium tetra chloride
Diethyl aluminium chloride	Vanadium trichloride
Diethyl aluminium chloride	Triacetyl acetone vanadium
Butyl lithium	Titanium tetrachloride
Butyl magnesium iodide	Titanium trichloride
Ethyl aluminium dichloride	Dichlorodicyclopentadienyl titanium

At first two components of Ziegler-Natta catalyst system combines to form a complex, leaving behind an empty site to accommodate the growing polymer chain. Monomer is adsorbed on the metal surface and transferred to the growing chain through a particular direction, leading to form stereospecific polymer.

Initiation step:

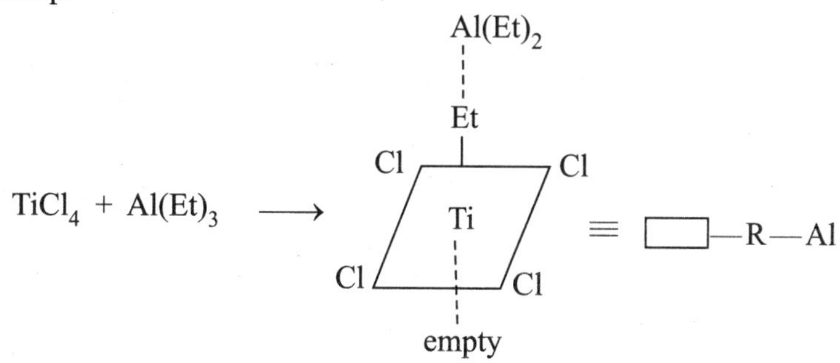

where, *R* indicates Et and □ indicates metal bar. Monomer *M* is adsorbed on the empty side of the metal bar:

Propagation step:

$$M + \square\text{—R—Al} \longrightarrow M\text{—}\square\text{—R—Al}$$

$$M\text{—}\square\text{—R—Al} \longrightarrow \square\text{—MR—Al}$$

$$M + \square\text{—MR—Al} \longrightarrow \square\text{—M}_2\text{R—Al}$$

- -

- -

$$M + \square\text{—M}_x\text{R—Al} \longrightarrow \square\text{—M}_{x+1}\text{R—Al}$$

Termination step:

$$\square\text{—M}_{x+1}\text{R—Al} \longrightarrow \square\text{—MR—Al} + M_x$$

Table 8.2.2 Differences between condensation polymerization and addition polymerization

Sl. No.	*Condensation polymerization*	*Addition polymerization*
1	Low molecular weight polymer of the order of $10^4 - 10^5$ is produced.	High molecular weight polymer of the order of $10^6 - 10^9$ is produced.
2	Monomer with multiple functionality is the basic requirement to execute condensation polymerization.	Unsaturation in monomer is the basic requirement to execute addition polymerization.
3	Two different types of monomers are required to execute condensation polymerization.	Only one type of monomer is enough to execute addition polymerization.
4	Byproduct is always formed in condensation polymerization.	No byproduct is formed in addition polymerization.
5	Reaction rate is slow.	Reaction rate is fast.
6	Both thermoplasts and thermosts can be produced by condensation method.	Only thermoplasts can be produced by addition method.

8.2.3 Tacticity

Configuration of a molecule indicates arrangements of constituent atoms in space. Arrangements of atoms may change due to rotation of single bonds and the structures, thus produced, are known as conformations. In the case of a polymer molecule several repeating units are attached through single bonds. If the repeating unit contains one asymmetric carbon atom the corresponding polymer molecule contains a series of asymmetric carbon atoms. In that case numerous stereoregular arrangements are possible through C—C single bond rotation. Such polymers exist in several stereoregular forms. The phenomenon was first established by Staudinger (1932) and is known as tacticity. Three types of tacticity are very common in asymmetric polymers, namely, isotactic, syndiotactic and atactic. Consider a monomer, substituted olefin, namely, vinyl chloride

$CH_2{=}CH{-}Cl$. On polymerization vinyl chloride forms polyvinyl chloride (PVC). The molecule of PVC may have three different stereoregular forms, as shown below:

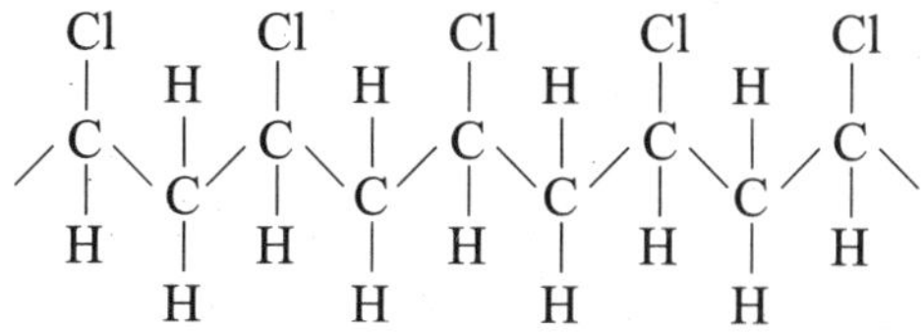

(a) All Cl-atoms remain above the plane (isotactic)

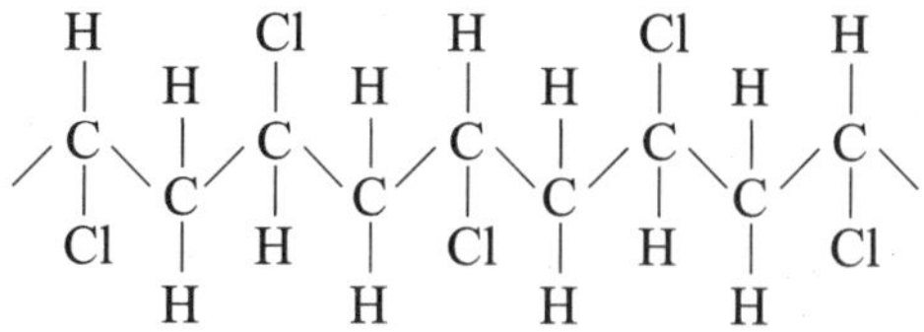

(b) Cl-atoms are placed alternately above and below the plane (syndiotactic)

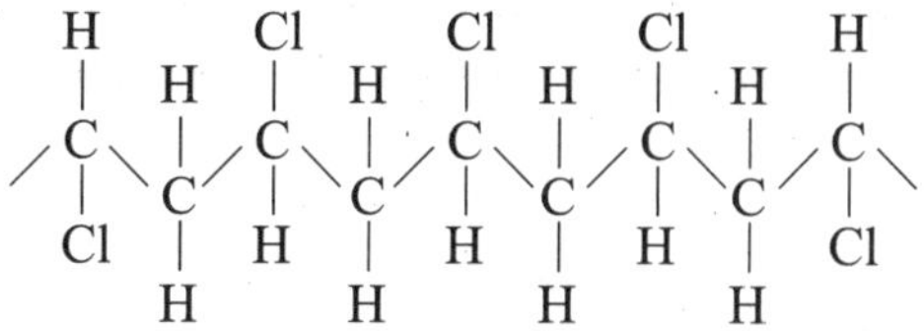

(c) Cl-atoms lie above and below the plane irregularly (atactic)

Fig. 8.2.2 Diagrams of different stereoregular arrangements of PVC polymer molecule.

The above figure illustrates different stereoregular arrangements of PVC polymer chain. In one arrangement all the chlorine atoms are placed above the plane of the main chain and the corresponding sequence is known as *isotactic*. The polymer is represented as *iso-PVC*. In another arrangement the chlorine atoms lie alternately above and below the main polymer chain and the corresponding sequence is termed *syndiotactic*. The polymer is represented as *syn-PVC*. In the third arrangement chlorine atoms are placed irregularly above and below the main polymer chain and the corresponding random sequence is known as *atactic*. The polymer is represented as *PVC only*.

Geometrical isomerism If the repeating unit contains one double bond the resulting polymer may have geometrical isomerism, i.e., chloroprene on polymerization forms polychloroprene. The structure of chloroprene is:

$$CH_2{=}CH{-}\underset{\displaystyle Cl}{\underset{|}{C}}{=}CH_2$$

Polymerization of chloroprene yields the following isomers of repeating units:

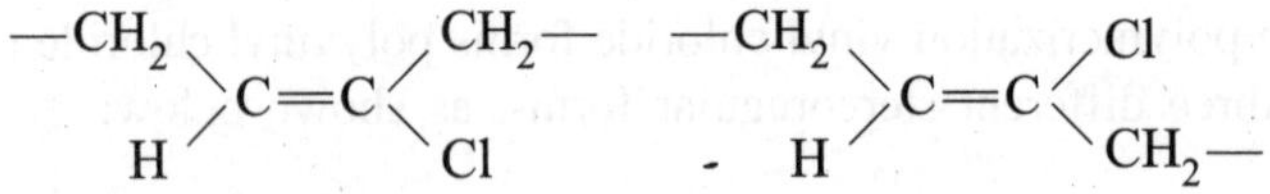

cis-1, 4 trans-1, 4

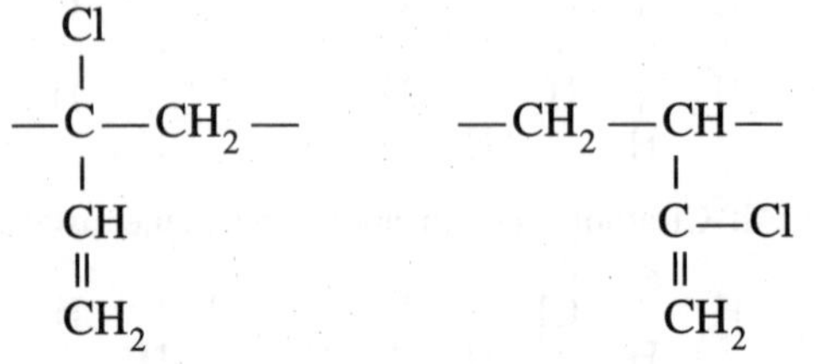

1, 2-isomer 3, 4-isomer

8.2.4 Copolymer

Definition If an addition polymerization reaction is carried out using two different types of monomers, the resultant polymer chain consists of both the monomer units. The polymer is called copolymer.

Kinetics of chain propagation in copolymerization

Let M_1 and M_2 represent two different type of monomers. During initiation both $M_1^\bullet$ and $M_2^\bullet$ are produced. Four different propagation steps can be distinguished:

1. $M_1^\bullet + M_1 \xrightarrow{k_{11}} M_1^\bullet$ (8.2.1)
2. $M_1^\bullet + M_2 \xrightarrow{k_{12}} M_2^\bullet$ (8.2.2)
3. $M_2^\bullet + M_2 \xrightarrow{k_{22}} M_2^\bullet$ (8.2.3)
4. $M_2^\bullet + M_1 \xrightarrow{k_{21}} M_1^\bullet$ (8.2.4)

At the steady state condition it may be approximated that:

$$k_{21}[M_2^\bullet][M_1] = k_{12}[M_1^\bullet][M_2] \qquad (8.2.5)$$

The rate of consumption of 1st monomer M_1 is given by:

$$-\frac{d[M_1]}{dt} = k_{11}[M_1^\bullet][M_1] + k_{21}[M_2^\bullet][M_1] \qquad (8.2.6)$$

The rate of consumption of 2nd monomer M_2 is given by:

$$-\frac{d[M_2]}{dt} = k_{22}[M_2^\bullet][M_2] + k_{12}[M_1^\bullet][M_2] \qquad (8.2.7)$$

The monomer reactivity ratios r_1 and r_2 are defined as:

$$r_1 = \frac{k_{11}}{k_{12}} \text{ and } r_2 = \frac{k_{22}}{k_{21}} \tag{8.2.8}$$

Dividing (8.2.6) by (8.2.7), we get:

$$\frac{d(M_1)}{d(M_2)} = \frac{k_{11}[M_1^\bullet][M_1]+k_{21}[M_2^\bullet][M_1]}{k_{22}[M_2^\bullet][M_2]+k_{12}[M_1^\bullet][M_2]} = \frac{[M_1](k_{11}[M_1^\bullet]+k_{21}[M_2^\bullet])}{[M_2](k_{22}[M_2^\bullet]+k_{12}[M_1^\bullet]}$$

$$= \left(\frac{[M_1]}{[M_2]}\right)\frac{\left(\frac{k_{11}}{k_{21}}\frac{[M_1^\bullet]}{[M_2^\bullet]}+1\right)}{\left(\frac{k_{22}}{k_{21}}\frac{[M_2^\bullet]}{[M_2^\bullet]}+\frac{k_{12}}{k_{21}}\frac{[M_1^\bullet]}{[M_2^\bullet]}\right)}$$

Putting the value of $[M_1^\bullet]/[M_2^\bullet]$ from equation (8.2.5), we have:

$$\frac{d(M_1)}{d[M_2]} = \left(\frac{[M_1]}{[M_2]}\right)\frac{\left(r_1\frac{[M_1]}{[M_2]}+1\right)}{\left(\frac{[M_1]}{[M_2]}+r_2\right)} \tag{8.2.9}$$

F_1 = Fraction of monomer M_1, added to the main chain of the polymer.

So, $$F_1 = \frac{d[M_1]}{d[M_1]+d[M_2]}$$

F_2 = Fraction of monomer M_2, added to the main chain of the polymer.

So, $$F_2 = \frac{d[M_2]}{d[M_1]+d[M_2]}$$

f_1 = Fraction of unreacted monomer M_1 and f_2 = Fraction of unreacted monomer M_2.

So, $$f_1 = \frac{[M_1]}{[M_1]+[M_2]} \text{ and } f_2 = \frac{[M_2]}{[M_1]+[M_2]}$$

$$F_1 + F_2 = 1 \text{ and } f_1 + f_2 = 1$$

Using the values of f_1, f_2, F_1 and F_2, equation (8.2.9) takes the form

$$F_1 = \frac{r_1 f_1^2 + f_1 f_2}{r_1 f_1^2 + 2 f_1 f_2 + r_2 f_2^2} \tag{8.2.10}$$

Case – 1: Ideal copolymerization

In some copolymerization reactions two radicals $M_1^\bullet$ and $M_2^\bullet$ display equal preferences to react with either monomer. In those cases we can write that $r_1r_2 = 1$. The copolymerization is called ideal copolymerization. So the equation (8.2.9) reduces to:

$$\frac{d(M_1)}{d[M_2]} = r_1\frac{[M_1]}{[M_2]}.$$

Or, $$\frac{d\ln[M_1]}{d\ln[M_2]} = r_1 \tag{8.2.11}$$

The equation (8.2.10) reduces to:

$$F_1 = \frac{r_1 f_1}{r_1 f_1 + f_2} \tag{8.2.12}$$

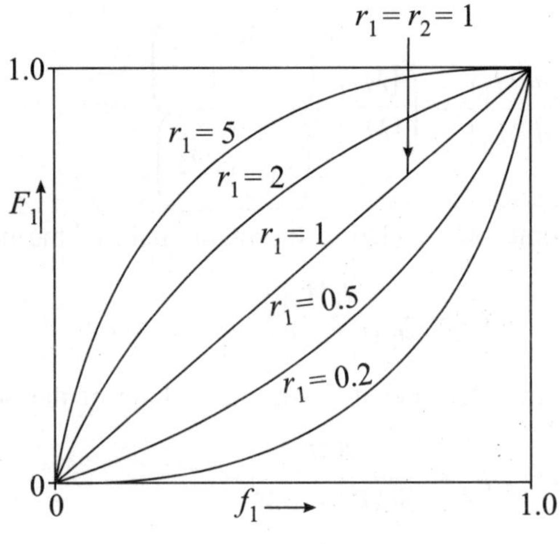

Fig. 8.2.3 Polymer fraction versus unreacted monomer fraction curves for different values of r_1 and $r_1r_2 = 1$.

Case – 2: Alternating copolymerization

Alternating coplolymer can be obtained only when $M_1^\bullet$ reacts with only M_2 and $M_2^\bullet$ reacts with only M_1. So, $k_{11} = k_{22} = 0$ and $r_1 = r_2 = 0$. So, the equations (8.2.9) and (8.2.10) becomes:

$$\frac{d[M_1]}{d[M_2]} = 1$$

and $$F_1 = \frac{f_1 f_2}{2 f_1 f_2} = \frac{1}{2} \tag{8.2.13}$$

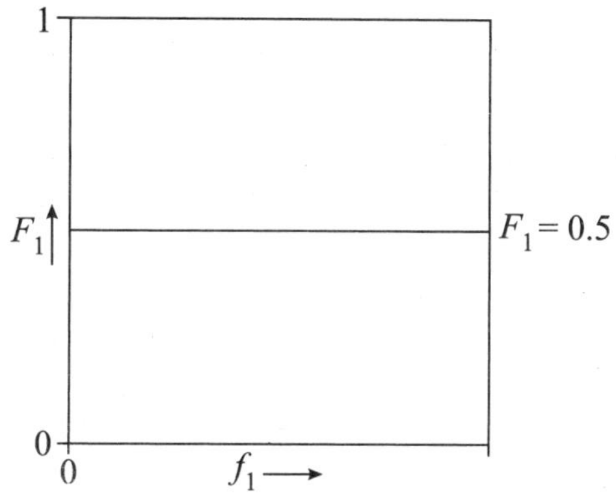

Fig. 8.2.4 Polymer fraction versus unreacted monomer fraction curve for alternating copolymer.

Case – 3: Non-ideal copolymerization

The following copolymerization curves are obtained for non-ideal copolymerization:

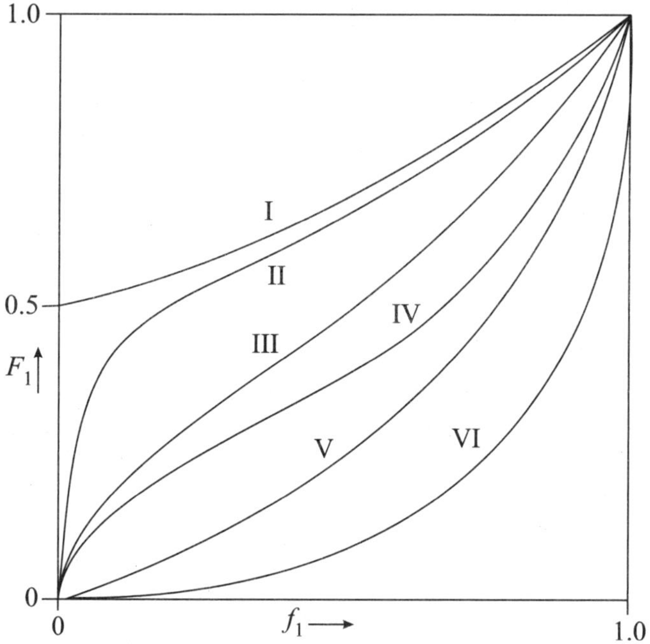

Curve–I: $r_1 = r_2 = 0.5/0.0$ Curve–II: $r_1 = r_2 = 0.5/0.1$
Curve–III: $r_1 = r_2 = 0.5/0.5$ Curve–IV: $r_1 = r_2 = 0.5/1.0$
Curve–V: $r_1 = r_2 = 0.5/2.0$ Curve–VI: $r_1 = r_2 = 0.5/4.0$

Fig. 8.2.5 Polymer fraction versus unreacted monomer fraction curves for non-ideal copolymerization.

The following observations can be made from the above figures:

(a) For a given value of r_1, as r_2 decreases, tendency to form alternating copolymers increases at a given value of f_1 (the line oo′).

(b) For a given value of r_1, as r_2 increases the alternating copolymer is formed at higher value of f_1.

Two other important copolymers are (a) block copolymer and (b) graft copolymer.

(i) Block copolymer Two different polymer chains, constituted by two different repeat units, may be joined to form a single polymer chain. The resultant polymer is called block copolymer. It is represented as following

$—[(— M_1 —)_m — (— M_2 —)_n] —$, where, m and n are different. M_1 and M_2 are two different monomer repeat units. Two examples are:

(a) Styrene-butadiene rubber: The repeat unit of the copolymer is represented as

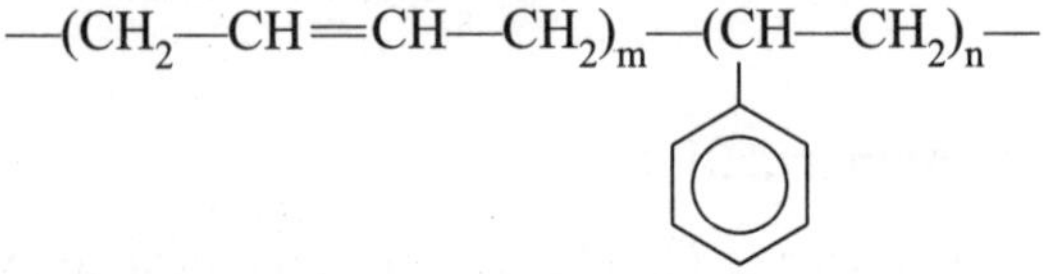

(b) Nitrile rubber: The repeat unit of the copolymer is represented as

$$—(CH_2—CH{=}CH—CH_2)_m—(\underset{\displaystyle CN}{\underset{|}{CH}}—CH_2)_n—$$

Usually block copolymer is prepared by anionic mechanism because living macroradical is available in anionic mechanism. At first one monomer is fed into the reaction vessel. After complete consumption of that monomer, 2nd monomer is fed into the reaction vessel and the polymerization proceeds through living macroradical.

(ii) Graft copolymer There are some polymers, which act as very good free radical acceptor. Those polymers can act as chain transfer agents. Thus during free radical polymerization of a given monomer those polymers are added as chain transfer agents and macroradical is terminated, forming a pendent polymer chain, suspended from the backbone of the other polymer. Thus the resultant polymer contains two different monomers. The main chain is constituted by one monomer while the pendent chain is constituted by other monomer. The process is called grafting and the polymer is called graft copolymer. It is represented as follows

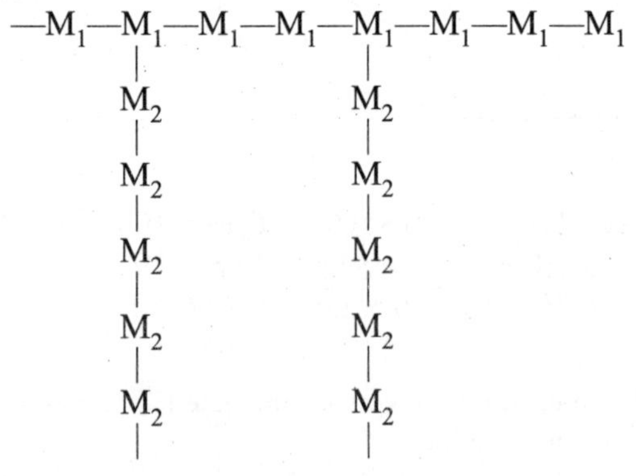

One very good example of graft copolymer is grafted NR. It has been found that NR acts as very good free radical acceptor. Thus when NR is used as chain transfer agent in a given free radical polymerization, grafting occurs as follows

Step-1

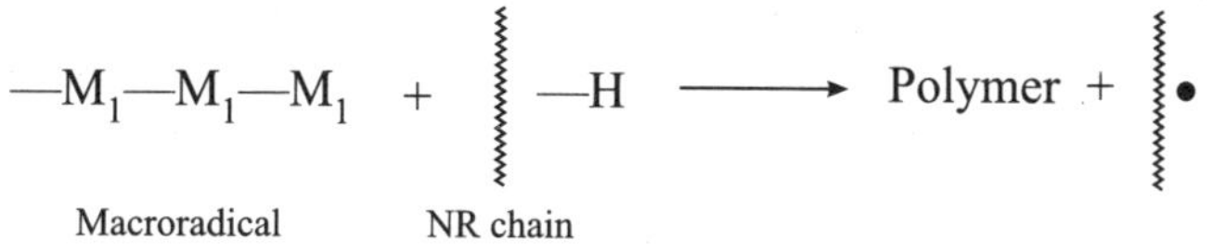

Step-1

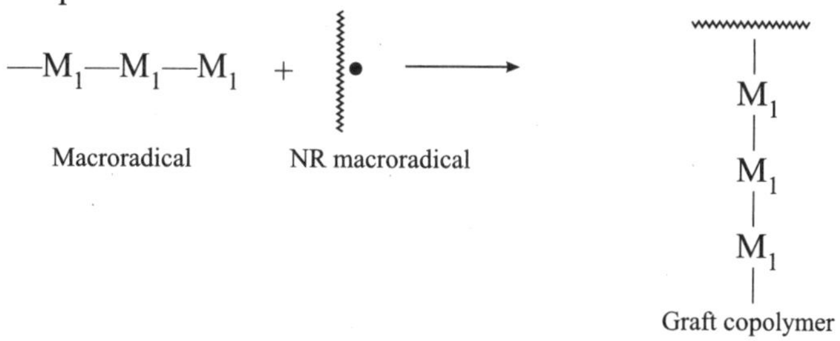

8.2.5 Intermolecular forces in polymer

Each polymer molecule is a long chain and usually exists in coil/spiral form as shown below:

The coil is flexible and often changes its dimension depending on the stress received. For simplicity a polymer chain is divided into several segments such that each segment is rigid and does not change its dimension, but can move in various directions. This type of motion of segment is popularly known as *segmental motion.*

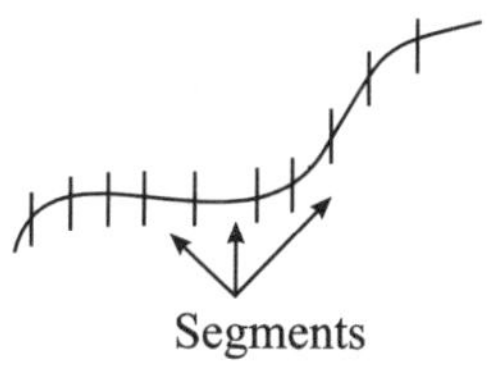

In each polymer there exists several types of secondary forces which bind the polymer chains together. These secondary forces are popularly known as intermolecular forces or Van-der Waals forces. The intermolecular forces are of the following types:

(a) **Dipole forces** If a monomer has a dipole moment, corresponding polymer molecule possesses a large dipole moment, as a result of which, strong intermolecular or interchain forces of attraction develops among the polymer chains. This type of forces is known as dipole forces. This dipole forces is responsible for molecular orientation and chain alignment. However, magnitude of dipole forces decreases with temperature since thermal energy opposes the molecular orientation and destroys the chain alignment.

(b) **Induction forces** In presence of polar molecules, non-polar molecules sufficiently get induced to form dipoles, known as induced dipoles. A strong intermolecular force is developed among permanent and induced dipoles, what is known as induction forces. The induction forces is usually small in magnitude and independent of temperature.

(c) **Dispersion forces** As electrons are considered as moving charge particles as well as possess wave characteristics, fluctuation of electron clouds often occurs in non-polar molecules, due to which, positive and negative centres are seperated. As a result of this, some polarity is developed within the non-polar molecules. However, this polarity vanishes after some time. Thus, the corresponding dipole moment is called instantaneous dipole moment. These instantaneous dipoles induce polarity to other non-polar molecules, as a result of which, a strong attractive force builds up among the molecules. This is known as dispersion forces. Obviously dispersion forces exist in non-polar molecules only and are independent of temperature.

(d) **Hydrogen bond** If the monomer unit contains polar group, such as carboxyl, amine or amide group, the corresponding polymer molecule contains numerous such polar groups. These polar groups are responsible for forming strong hydrogen bonds among the polymer chains. Strength of hydrogen bonds is enough to bring the polymer chains closer and get themselves aligned. However, with the increase in temperature thermal energy increases and ultimately it is enough to break hydrogen bonds.

Due to the presence of these strong secondary forces polymer chains are tightly bound and polymer appears rigid and brittle. However, with the increase in tempeature magnitude of secondary forces decreases sufficiently, encouraging segmental motion.

8.2.6 Glass transition temperature (Tg)

The intensity of intermolecular forces is very high in polymers due to large size of polymer molecules. Furthermore polymer molecules, due to having large size, have a tendency to exhibit segmental motion. Thus two opposing factors are always operating in all poymers:

(a) Intermolecular forces of attraction, which has a tendency to bind all polymer chains tightly to freeze any primary or secondary motion.
(b) Macromolecules have a natural tendency to execute segmental motion, which is considered as a secondary motion.

The long range elasticity, the important characteristic of any polymer, solely depends on segmental motion. In order to execute segmental motion each molecule must have a definite *free volume* within which segmental motion exists. With the increase in temperature the amount of *free volume* increases and hence the intensity of segmental motion also increases. Usually at low temperature intermolecular forces of attraction dominates over segmental motion and the polymer appears as a rigid and dimensionally stable form, what is known as *plastic state* or *glassy state*. The glassy state is defined as the state where viscosity is more than 10^{13} Pa-s. At higher temperature segmental motion dominates over the intermolecular forces of attraction and the polymer appears as a highly elastic form, what is known as *elastic state*. The transition temperature, above which, plastic material gains the property of elasticity, is known as *glass transition temperature*, represented as T_g. After T_g specific volume and hence *free volume* increases sharply with temperature as shown in the Figure 8.2.6.

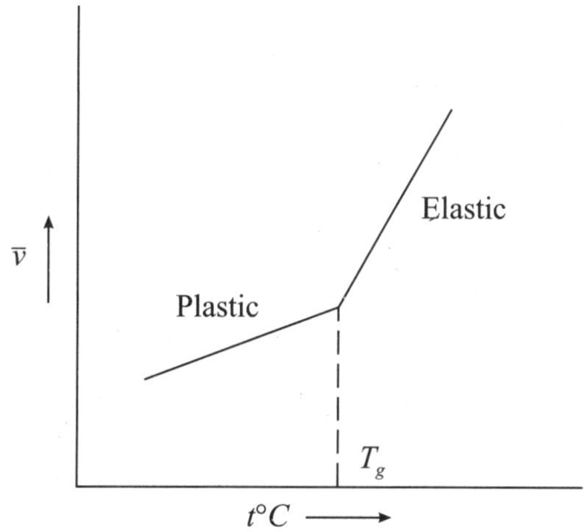

Fig. 8.2.6 Specific volume Vs. temperature plot of a polymer sample.

A noticeable change of slope has been observed around T_g and an obvious phase change takes place. If segmental motion persists at room temperature, only elastic phase will be observed at room temperature. Existence of plastic phase is solely due to secondary binding forces and hence the transition is often termed secondary phase transition. There are some polymers for which T_g is well above the room temperature and they appear as plastics at room temperature. Examples are: Polystyrene (PS), Polymethylmethacrylate (PMMA), Polyvinyl chloride (PVC) etc. There are plenty of polymers, for which T_g is well below the room temperature and they appear as rubbers at room temperature. Examples are:

Natural Rubber or NR $T_g = -73$°C,

Polychloroprene Rubber or CR $T_g = -71$°C,

Butyl Rubber $T_g = -88$°C etc.

It is to be noted that T_g-values of the above polymers are measured by Differential Scanning Caloriemetry (DSC) method.

8.2.7 Crystalline and amorphous polymers

As we have discussed earlier, at room temperature, when cohesive forces or intermolecular forces of attraction is sufficiently higher than segmental motion force, material becomes dimensionally stable, rigid and brittle. If the monomer unit possesses symmetry, high degree of regularity may be achieved in all polymer chains. Highest degree of regularity is found in PE, followed by PP, Nylon, PET etc. Those polymers appear crystalline like metals and ionic solids. As the dimension and nature of polymer chain are widely different from that of metal atom or ion, the nature of crystallinity is also different.

8.2.7.1 Crystalline polymers

Highly crystalline polymers are rigid, high melting, and less affected by solvent penetration. Crystallinity makes polymers strong, but also lowers their impact resistance. As an example, samples of polyethylene prepared under high pressure (5000 atm.) have high crystallinity (95 – 99%) but are extremely brittle.

In case of ionic solids, like NaCl the atoms are arranged in a regular sequence to form a 3D-crystal but in case of a polymer, chains are large enough so that they cannot be arranged in such a sequence like an ionic solid. A polymer chain remains folded regularly up and down as shown below. This is called lamellar.

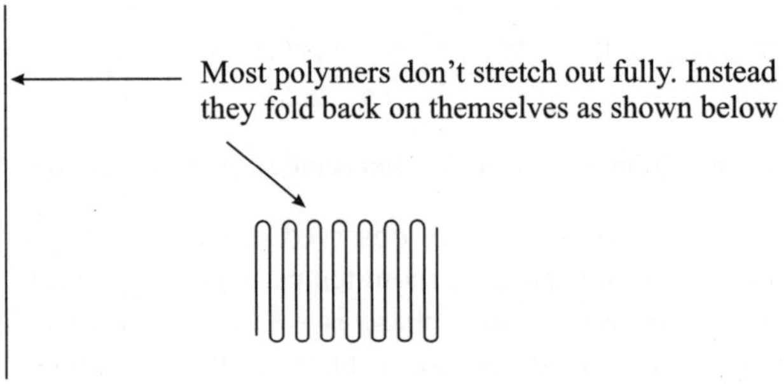

For polyethylene the length the stretched chain before folding is about 100 Å.

Several of such chains remain folded, as shown above, and stacked together to form a plate-like crystal, which is known as lamella. Thickness of the crystal plate is about 10 to 20 nm.

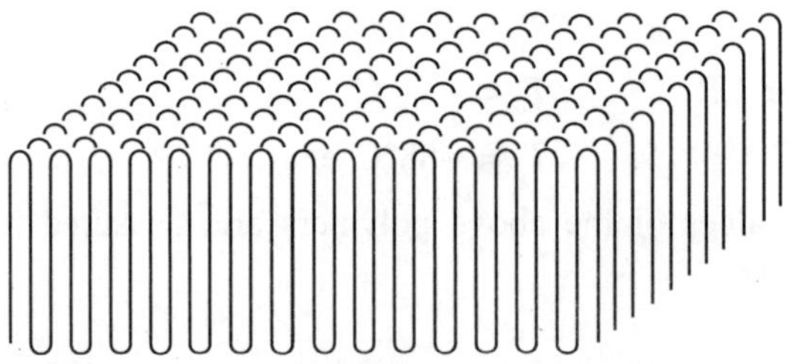

They can fold back and stack. A stack of polymer chains, folded back on themselves, is called lamella.

These lamella are oriented in perpendicular to the face of the crystal. As the polymer chains are close-packed in lamella, the specific volume of the crystalline zone is low.

8.2.7.2 Amorphous polymers

Polymer chains with branches or irregular pendant groups cannot pack together regularly enough to form crystals. The polymer chains are placed wide apart without any sequence. The specific volume of the amorphous zone is large. These polymers are said to be **amorphous**. Amorphous polymers are softer, have lower melting points, and are penetrated more by solvents than their crystalline counterparts.

8.2.7.3 Semi-crystalline polymers

Semi-crystalline polymers have both crystalline and amorphous regions. Semi-crystallinity is a desirable property for most plastics because they combine the strength of crystalline polymers along with the flexibility of amorphous. Semi-crystalline polymers can be tough with an ability to bend without breaking. A semi-crystalline polymer has two components, namely, (i) Crystalline region and (ii) Amorphous region.

Several crystalline regions remain dispersed in the amorphous matrix. Each lamellar propagates along the radius of a sphere as shown below. That crystalline polymer is called ***spherulite***.

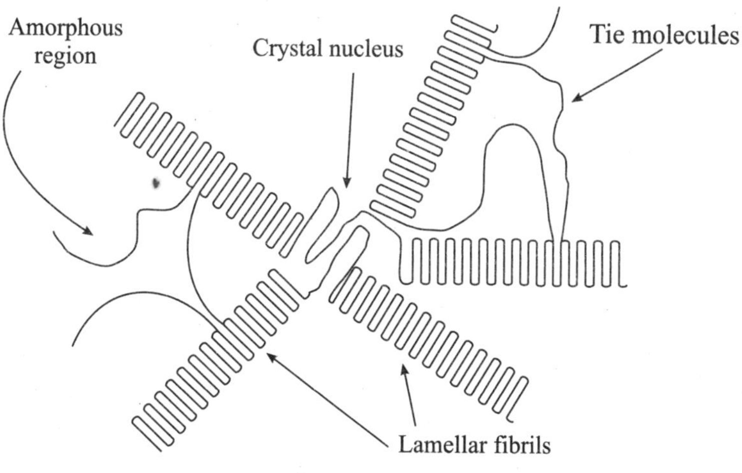

A polymer crystalline spherulite

8.2.7.4 Percent crystallinity in a semi-crystalline polymer

Usually most of the plastics are considered semi-crystalline. So, the total volume of the polymer is the sum of the volume of crystalline region and that of amorphous region. If V, V_a and V_c are the total volume, amorphous volume and crystalline volume respectively, we can write that

$$V = V_a + V_c$$

Again, volume is the ratio of mass to density. So, we have

$$\frac{m_s}{\rho_s} = \frac{m_a}{\rho_a} + \frac{m_c}{\rho_c}$$

where, m_s, m_a and m_c are the total mass of the sample, mass of amorphous region and mass of crystalline region respectively. ρ_s, ρ_a and ρ_c are total density of the sample, density of the amorphous region and density of the crystalline region respectively.

Dividing both sides by m_s, we have

$$\frac{1}{\rho_s} = \frac{m_a/m_s}{\rho_a} + \frac{m_c/m_s}{\rho_c} = \frac{f_a}{\rho_a} + \frac{f_c}{\rho_c} = \frac{(1-f_c)}{\rho_a} + \frac{f_c}{\rho_c} = \frac{1}{\rho_a} + f_c\left(\frac{1}{\rho_c} - \frac{1}{\rho_a}\right)$$

Or,
$$\frac{1}{\rho_s} = \frac{1}{\rho_a} + f_c\left(\frac{1}{\rho_c} - \frac{1}{\rho_a}\right)$$

where, f_a and f_c are the mass reactions of amorphous region and crystalline region respectively. Hence, we have

$$f_c = \frac{\left(\frac{1}{\rho_s} - \frac{1}{\rho_a}\right)}{\left(\frac{1}{\rho_c} - \frac{1}{\rho_a}\right)}$$

If P_c is the percent crystallinity of the polymer, we can write that

$$P_c = f_c \times 100 = \frac{\left(\frac{1}{\rho_s} - \frac{1}{\rho_a}\right)}{\left(\frac{1}{\rho_c} - \frac{1}{\rho_a}\right)} \times 100$$

ρ_a and ρ_c values of some polymers are given below

Polymer	ρ_a *(gm/cm3)*	ρ_c *(gm/cm^3)*
Polyethylene	0.853	1.004
Poly(vinyl alcohol)	1.291	1.350
Poly(vinyl chloride)	1.412	1.477
Poly(vinylidene chloride)	1.775	1.957
Poly(ethylene terephthalate)	1.336	1.514

This refers to the overall level of crystalline component in relationship to its amorphous component. The percent crystallinity is directly related to many of the key properties exhibited by a semi-crystalline polymer. With increase in crystallinity the following observations are made:

(a) Tensile strength, modulus, tear strength, stiffness and hardness increase.

(b) Toughness and impact strength decrease.

(c) Elongation at break decreases.
(d) Solubility in solvent and permeability decrease.
(e) Optical clarity decreases.
(f) Creep resistance increases.

Applications, where the percent crystallinity plays critical role, include:

- Injection moulded parts and components
- Plastic food and beverage containers
- Fibres
- Films
- Automotive components

8.2.7.5 Factors influencing the crystallinity

(a) ***Molecular weight*** Higher the molecular weight of a polymer, more irregular will be the chain and hence entropy increases or compactness decreases. So, percent crystallinity decreases.

(b) ***Chain branching*** Chain branching imparts irregularity in the chain, increasing the distance between the chains. So, entropy increases or compactness decreases and hence percent crystallinity decreases.

(c) ***Symmetry in repeat unit*** This property of repeat unit plays an important role in determining percent crystallinity in a given polymer. Maximum symmetry is observed in case of ethylene, where only carbon and hydrogen are present in the repeat unit. In case of other alkenes, maximum symmetry is observed in case of isotactic sequence, followed by syndiotactic sequence, whereas atactic polymer does not possess any symmetry. Entropy decreases or compactness increases with increase in symmetry.

8.2.8 Inorganic Polymers

There are many polymers where the repeat unit does not contain carbon atom in the main chain. Examples are: polydichlorophosphazene, silicone rubber, boron nitride fibres etc.

Polydichlorophosphazene:

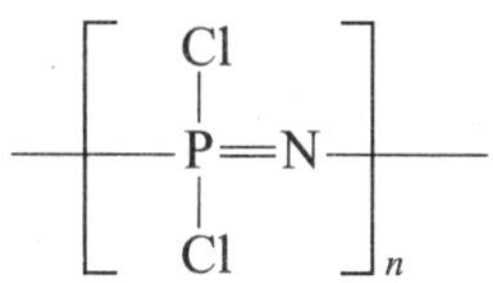

(i) It is highly elastic rubber like material.
(ii) It can be crosslinked with ammonia. Crosslinking reaction is shown below:

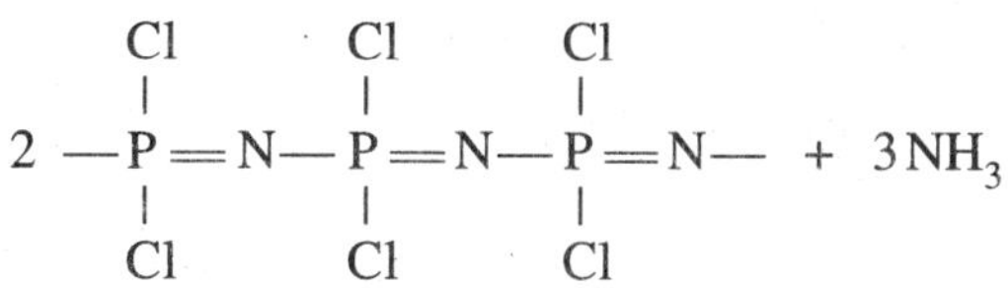

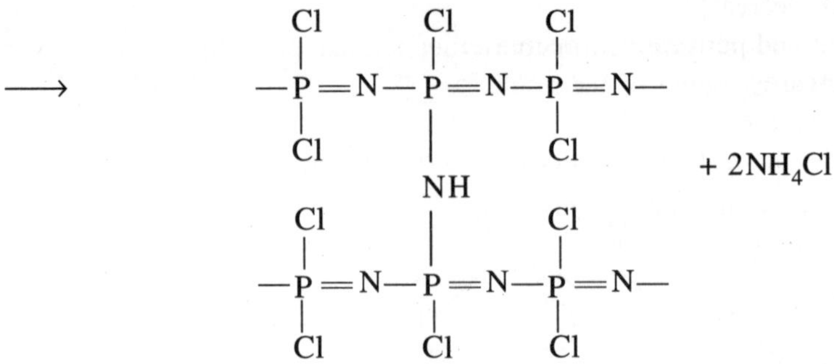

Polydietherphosphazene:

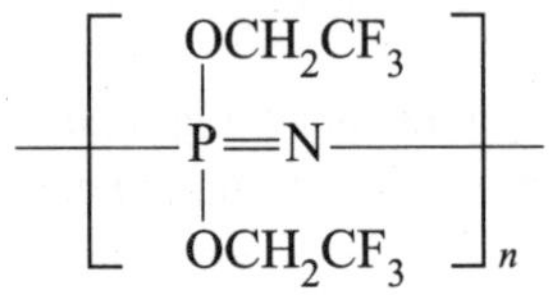

This is a stable fluorinated inorganic thermoplastics, having melting point, (T_m) = 242°C and glass transition temperature (T_g) = –66°C.

Silicon polymer:

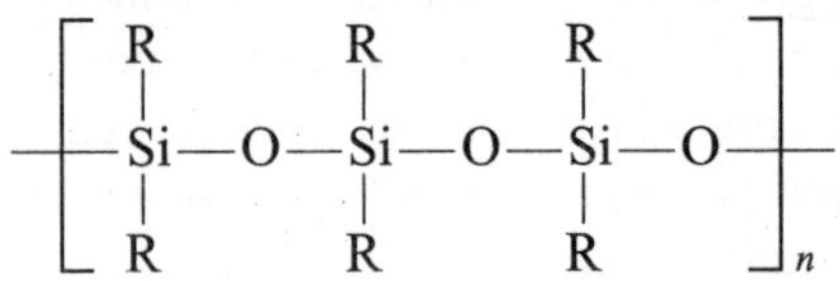

This is the chemical structure of silicone resin, an inorganic resin. On the basis of —R-group different types of silicone resins are known to us. Furthermore, silicone resins of various molecular weights are available. Low molecular weight silicone resins appear as viscous liquids. High molecular weight silicone resin appears as semi-solid grease like material. Silicone polymer with very high molecular weight, behaves rubbery and is known as silicone rubber.

Silicone rubber has excellent oxidation and ozone resistance as well as good water resistance. Its permeability to gases is very high. Strong Si—O—Si bond is responsible for its high temperature resistance characteristics. It maintains elastic characteristics even at 250°C. So it is a primary constituent of those products, used at high temperatures. It is widely used in manufacturing oil seals, lubricants, protective coatings, adhesives etc. The most important application area of silicone rubber is in the manufacturing of medical equipment since silicone rubber is totally inert and non-toxic to blood, serum and enzymes. Because of its high electrical insulation characteristics silicone rubber is used in wire and cable insulations also.

Boron Nitride polymer Boron nitride fibre (BN) is an inorganic fibre. It is manufactured from boron oxide B_2O_3, treated with ammonia. BN fibre is highly stable even at temperatures of 1800°C – 2000°C.

8.2.9 High temperature polymers

There are many polymers which can retain their mechanical properties even at very high temperatures, as high as 500°C. Some of them are:

Polyimides:

(i) 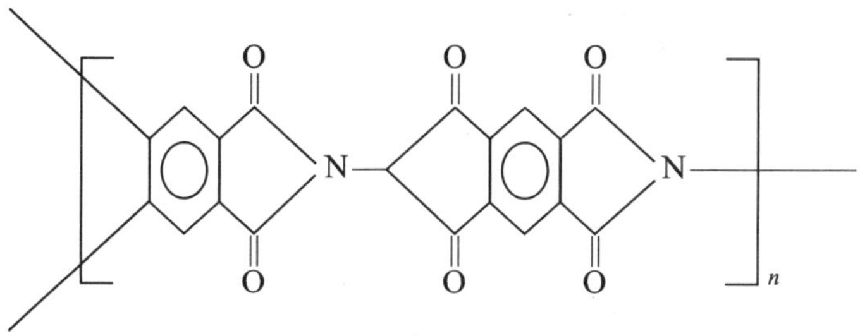

(ii) Polycondensation of aromatic dianhydrides and aromatic amines will lead to the following high temperature polymer:

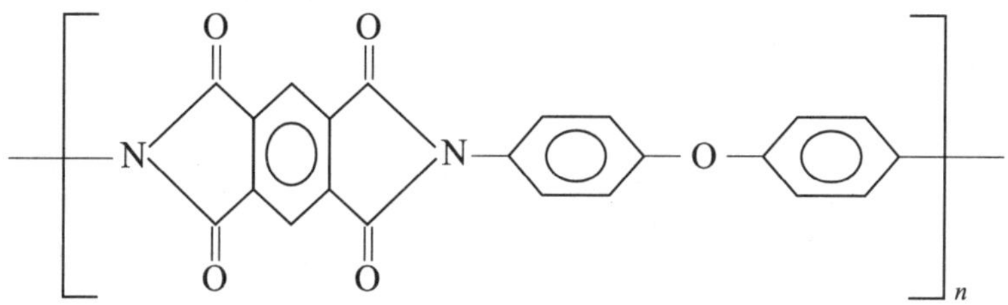

Poly(phenylene oxide) (PPO): 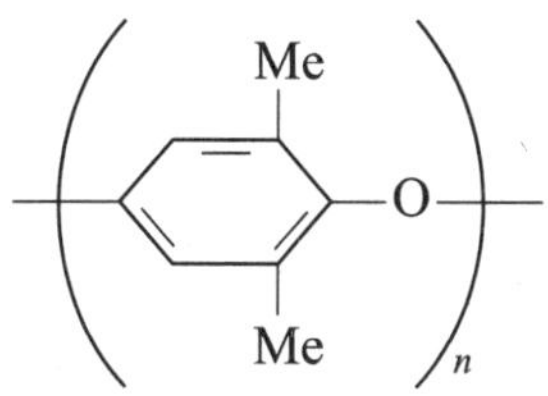

All these polymers are highly stable even at 300°C for months. They have good electrical properties and resistance to water.

8.2.10 Conducting polymers

In general polymeric materials are insulators but there are few polymers having the specific resistance range: 10^{-4} to 10^{-6} Ω-cm. Conductivity of polymers, having specific resistance of the order of 10^{-6} Ω-cm, can be compared with that of copper. Polymers, having conjugated dienes, have enough labile electrons, which can easily flow to the empty conduction band. Some examples of conductive polymers are:

Polyacetylene: $-\!\left(CH{=}CH{-}CH{=}CH\right)_n\!-$

Poly-p-phenylene: 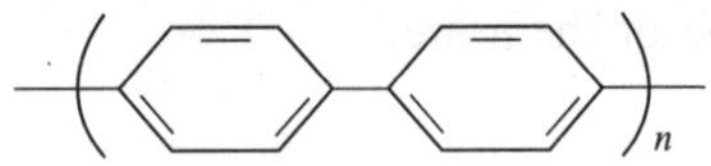

Polypyrrole: 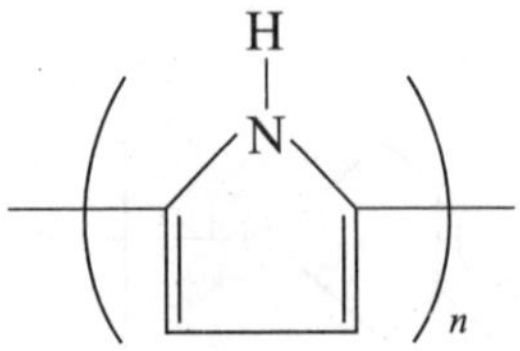

Conductivity of polymers may further be improved by appropriate doping. *n*-type conductive polymer can be made by doping with metals such as Na, Li etc. whereas *p*-type conductive polymer can be made by doping with halogens.

n-doping: $P + nM \longrightarrow [P]^{n-} + nM^{+}$ where, M = Na, Li etc.

p-doping: $P + \frac{1}{2}mX_2 \longrightarrow [P]^{m+} + mX^{-}$ where, $X_2 = Br_2$, I_2 etc.

These dopant atoms do not replace any polymer molecule from the crystal like semiconductors, but form a new energy band very close to valance band or conduction band of polymer molecules. As a result of this overlapping occurs among valence band, conduction band of polymers and new energy band of dopant atoms, leading to a sharp increase in concentration of free electrons and holes. This increase in free electrons and holes increases the conductivity of polymers several times.

Advantages of conducting polymers The main advantages of conductive polymers are their light weight, flexibility and high conductivity.

Applications Conducting polymers are mostly used in manufacturing electrical and electronic appliances like solar cells, photovoltaic devices, diodes and transistors, telecommunication system etc.

8.2.11 Biopolymers

There are some polymers, which are the main constituents of plant and animal cells. These polymers are called biopolymers. Examples are carbohydrates, (i.e., polysaccharides) and proteins.

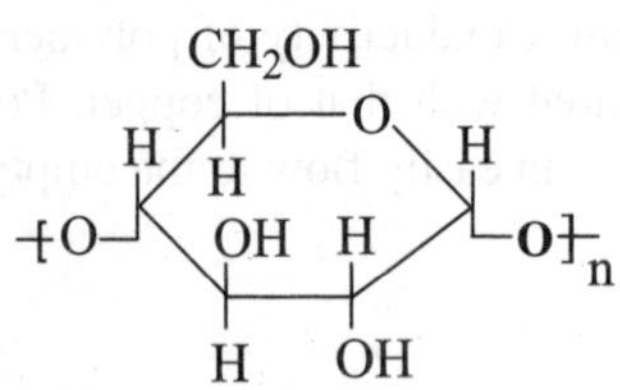

Structure of polysaccharide

Peptides are low molecular weight polymers, derived from amino acids. The simplest form of peptide is dipeptide, the formula of which is $H_2N — CH_2 — CONH — CH_2 — COOH$. Peptide often undergoes resonance, contributing some double bond character in C — N bond, therby restricting the rotation of molecule about C — N axis.

A polypeptide always contains $—NH_2$ group at one end and —COOH group at the other end. Protein is a polypeptide, having a very high molecular weight. Any biological cell consists of proteins, where amino groups and carboxyl groups are arranged in a definite sequence, defined by the sequence of gene, which is encoded in the genetic code.

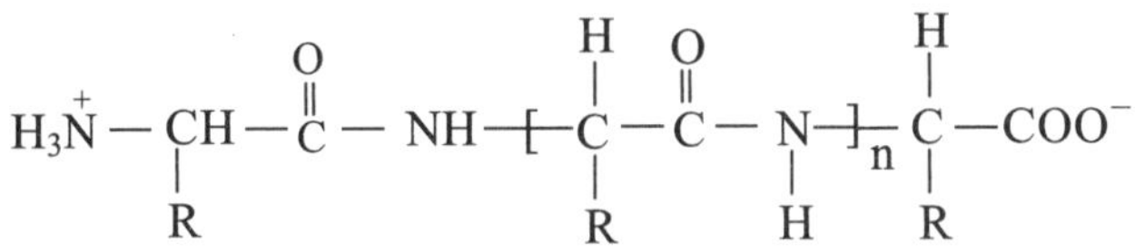

Structure of protein

8.2.12 Vulcanization

Due to segmental motion polymer chains of a rubber slip past one another on application of stress. Thus on release of stress only a fraction of chains is able to return to its original state and hence the rubber material gets deformed permanently. This chain slippage phenomenon is the prime reason for rubber deformation. Somehow if the polymer chains are interlinked, chain slippage will remarkably be restricted and hence rubber material shows long-range elasticity without permanent deformation even on the application of high stress. This interlinking of polymer chains is usually carried out by chemical means and the process is known as crosslinking or vulcanization. The vulcanization leads to the formation of a 3-dimesional network, which is responsible for enhancement of following properties of rubber: (a) Tensile strength, (b) Modulus, (c) Hardness, (d) Tear resistance, (e) Solvent resistance.

Vulcanizing agents (i) The first and most important vulcanizing agent is sulphur (S), discovered by Charls Goodyear in 1839. Besides sulphur some other agents are also used for vulcanization, e.g. selenium (Se), tellurium (Te), organic peroxides, phenolic resins etc.

(ii) Inorganic accelerator, namely, zinc oxide (ZnO), which is also a part of activator.

(iii) **Activator** The function of activator is to promote accelerator to react with sulphur, to form a complex, which ultimately undergoes crosslinking reaction with rubber. The common activator is a mixture of metal oxides and stearic acid. At the outset metal oxide reacts with stearic acid to form metal stearate, which then reacts with organic accelerator. Zinc oxide (ZnO) is widely used for this purpose since among the metal stearates, Zinc stearate is the most friendly and widely accetable to rubber matrix.

(iv) **Organic accelerators** Some of common organic accelerators are:
Mercapto-benz-thiazole (MBT),
Cyclohexyl-benz-thiazole (CBS),
Tetramethyl-Thiuram-disulfide (TMTD) etc.

Vulcanization process At first rubber is mixed with vulcanizing agent, accelerators and activators in a mixing mill. Then it is placed in a mould which is inserted into a press, where high temperature and pressure is applied for a certain time in order to execute the necessary chemical reactions. After that the press is released and the vulcanized material is collected from the mould followed by cooling under water.

Structure In any vulcanized rubber long chain polymers are interlinked by a flexible sulphur chain as shown below:

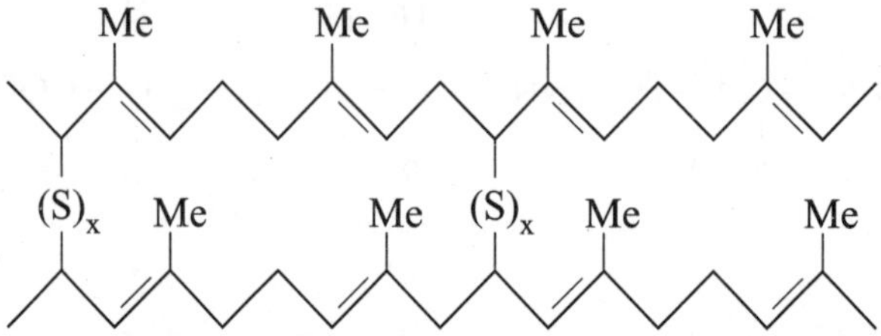

In absence of accelerators and activators the value of x is large and varies point to point. Hence the nature of crosslinking is erratic, giving rise poor quality product. In presence of accelerators and activators the value of x is usually restricted to 2 at almost all points and hence regular and high quality crosslinked product is obtained.

8.2.13 Preparation, structure and uses of some common polymers

(i) Natural rubber The rubber which is obtained from the natural source is known as natural rubber, represented as NR. It is obtained as a latex from Hevea tree. Hence it is also called Hevea rubber. Latex is an emulsion of rubber containing 25–40% rubber hydrocarbon, solvent, emulsifier and stabilizer, i.e. protein material or fatty acids. To recover rubber, latex is first coagulated by 1% acetic acid solution, washed with solvent and dried. Two different forms of rubber may be obtained.

(a) Crepe rubber After coagulation rubber is bleached by small amount of sodium bisulphate. The coagulum, thus obtained, is about 1 mm thickness and finally dried at about 50 °C. This rubber is known as crepe rubber.

(b) Smoked sheet Thicker sheets, obtained after coagulation, are dried in smokehouses at about 50°C within the smoke-atmosphere, generated from burning wood or coconut shells.

Structure The main constituent of NR is cis-polyisoprene. The chemical structure of isoprene is:

$$CH_2{=}CH{-}\underset{}{\overset{CH_3}{\overset{|}{C}}}{=}CH_2$$

Isoprene on polymerization gives either cis-polymer or trans-polymer as shown below:

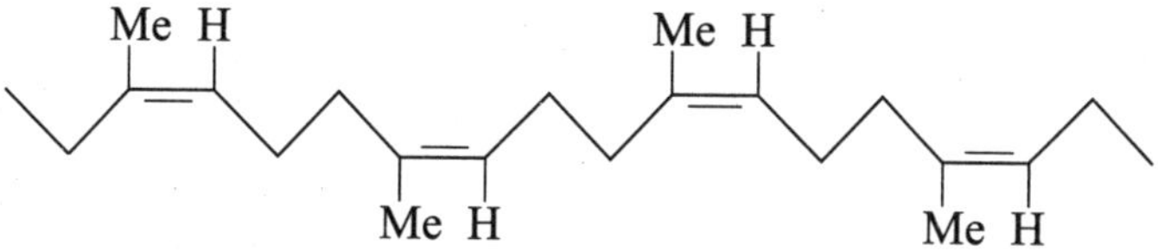

Cis, 1, 4-polyisoprene

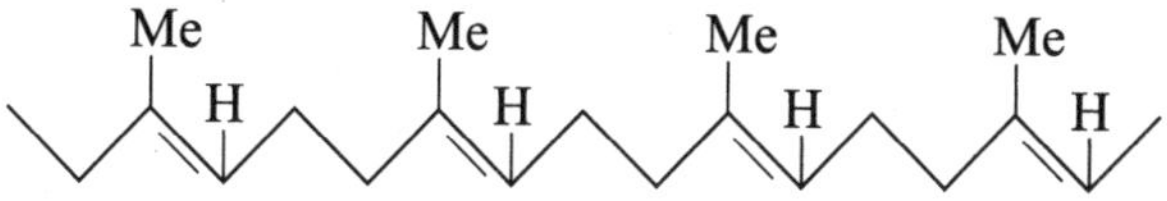

Trans, 1, 4-polyisoprene

NR contains about 97% cis-polyisoprene. It is soft and highly elastic material, whereas Gatta parcha containing about 97% trans-polyisoprene, is a very hard thermoplastic solid. Cis-polyisoprene can be synthesized from isoprene either by using butyl lithium as catalyst, or by coordination polymerization by using titanium tetrachloride and triisobutyl aluminium.

Applications NR has wide application area. It is used in the manufacture of tyres, cables, conveyor belts, V-belts, hoses, shoe-soles, moulded items and so many.

(ii) SBR It is an abbreviation of styrene butadiene rubber. Obviously the main constituents are styrene and butadiene. The chemical structure of SBR is shown below:

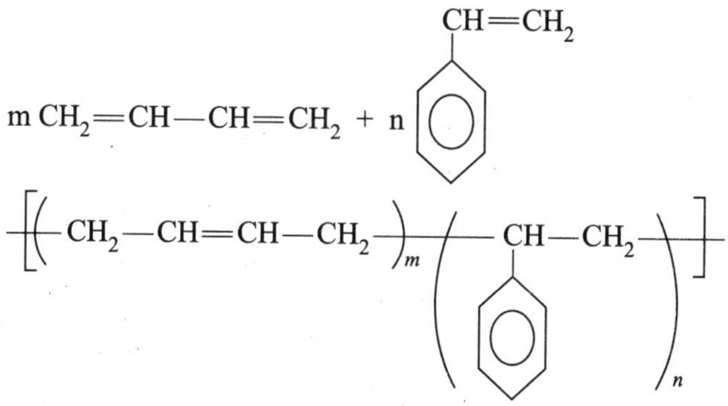

The weight ratio of butadiene unit to styrene unit is maintained around 3 : 1. The butadiene unit contains three different forms:

(a) $-\underset{\underset{CH_2}{\overset{||}{\underset{}{CH}}}}{\overset{}{\underset{|}{CH}}}-CH_2-$ — 20%

1, 2-butadiene unit

(b) 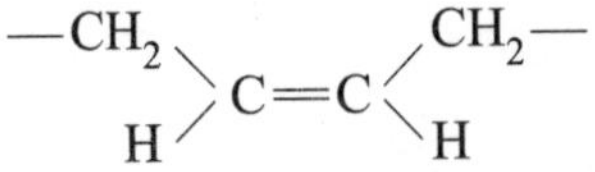 — 20%

cis, 1,4-butadiene unit

(c) 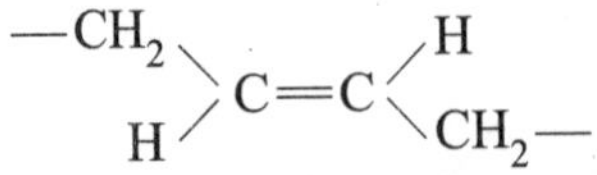 — 60%

trans, 1,4-butadiene unit

Preparation SBR rubber is usually prepared in two different techniques:

(a) emulsion polymerization technique
(b) solution polymerization technique.

Emulsion polymerization is carried out in presence of acative initiators, such as cumene hydroperoxide, *p*-menthane hydroperoxide and antifreeze components. The product is known as cold rubber, because temperature of reaction is maintained at around 5°C.
In solution polymerization technique the components (butadiene and styrene) undergo copolymerization through anionic mechanism in presence of alkyllithium catalyst. The product is called solution SBR rubber.

Properties So far as mechanical strength is concerned, SBR is comparable to NR. However, SBR loses its mechanical strength fast at elevated temperatures. Ozone resistance and abrasion resistance of SBR are superior to NR but poor resilience, cut growth resistance and high heat build-up are the major demerits of SBR compared to NR.

Applications SBR is widely used in the manufacture of several rubber products such as, tyre components, footwear components, electrical insulation, belting, hoses, moulded and extruded items, adhesives etc.

(iii) Nitrile rubber The repeat unit of nitrile rubber (NBR) contains two constituents: acrylonitrile and butadiene.

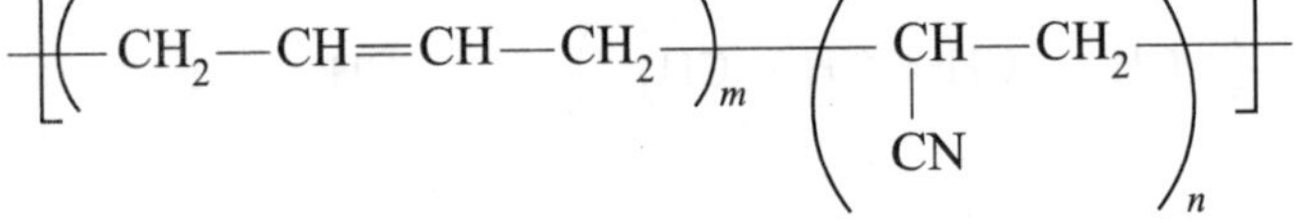

NBR is available in different grades depending on acrylonitrile content. NBR is particularly important due to its oil resistant characteristics, which mostly depends on acrylonitrile content. In NBR acrylonitrile content varies between 18% on mole basis (only fair oil resistant) and 40% on mole basis (extremely oil resistant). NBR is manufactured in emulsion polymerization technique.

Properties The most important characteristic of NBR is its oil resistance. Besides that it has good mechanical strength, abrasion resistance, heat and oxidation resistance. It can be used in contact with water and antifreeze solutions.

Applications The main applications of NBR is in the manufacture of following products: Oil seals and gaskets, tank liners, gasoline hoses, adhesives etc.

(iv) Polyethylene (PE) The chemical structure of polyethylene is given below:

$$-\!\!\left(CH_2-CH_2\right)_n\!\!-$$

It is usually manufactured by solution polymerization technique, where benzene or chlorobenzene is used as solvent. Monomer, i.e., ethylene is subjected to high temperature (around 250°C) and high pressure (between 1000 and 3000 atm.) in presence of a free radical initiator, such as traces of oxygen, peroxides, hydroperoxides etc. Under this condition ethylene molecules get polymerized to form polyethylene.

Properties Polythylene is characterized by high crystallinity and hence has high melting point, low specific gravity but at the same time it is highly flexible and tough material. Melting point of PE varies between 115°C – 145°C depending on the grades of PE. Wide range of crystallinity 75% – 85% is found in PE. Generally crystallinity in PE varies with side chain branching. Specific gravity of PE ranges between 0.88 – 0.99. Different types of PE are available depending on molecular weight and crystallinity. Some are described below.

Low density polyethylene (LDPE)

It has higher branching than HDPE. Molecular weight is low with specific gravity lies within 0.91 to 0.94. It can withstand temperature up to 90°C. Intermolecular forces in LDPE are weak. Hence it is less crystalline.

Properties

- It has moderate tensile strength but high resilience.
- Excellent resistance to dilute and concentrated acids and bases.
- Good resistance to aldehydes, ketones and oils.
- Poor resistance to aliphatic hydrocarbons, aromatic hydrocarbons, oxidizing agents and halogenated hydrocarbons.

Applications

It is widely used to manufacture:

- Containers, different types of bottles, tubes.
- Plastic bags for electronics components and different moulded items.
- Trays and other containers.
- Food containers, laboratory apparatus and containers.
- Plastic wraps.
- Different parts of computers, e.g., hard disc drives, data cards, optical disc drives etc.
- Milk and juice containers.

Very low density polyethylene (VLDPE)

It is a linear polyethylene with high fraction of short chain branches. Due to this high level of short chain branches, chains are less compact, making the polymer light weight. Specific gravity of the polymer lies within 0.880 – 0.915. It is usually made by copolymerization of ethylene and short-chain α-olefin, e.g., 1-butene or 1-hexene or 1-octene using metallocene catalysts.

Applications

It is widely used to manufacture:

- Hoses and tubes, ice and frozen food bags.
- Food packaging.
- Impact modifiers

Linear low density polyethylene (LLDPE)

Molecular weight of this type of polymer is far less than HDPE but possesses high crystallinity due to having low fraction of short chain branches. It is usually made by copolymerization of ethylene and long-chain α-olefin using metallocene catalysts. LLDPE, manufactured by using metallocene catalysts is represented as mLLDPE.

Properties

- The polymer has high tensile strength and impact strength compared to that of low density polyethylene (LDPE). It is very flexible and has high elongation at break.
- High environmental stress cracking resistance.
- Significant resistance to chemicals and UV radiations.
- Good electrical properties.

Applications

It is widely used to manufacture:

- Plastic bags and sheets of very low thickness.
- Toys, lids, buckets and containers.
- Flexible tubes.
- Cable insulation.

High density polyethylene (HDPE)

It is manufactured by using Ziegler-Natta catalyst. Chains are mostly linear with a few branching and closely packed. Specific gravity is around 0.97. Very strong intermolecular forces exists in polymer chains. Molecular weight is higher than MDPE or LDPE.

Properties

- It has high tensile strength.
- It is hard, opaque and less flexible.
- It has high temperature withstanding capacity up to 120°C.
- It is resistant to many inorganic chemicals and organic solvents.

Applications

It is widely used to manufacture:

- Plastic containers, such as plastic bags, milk containers, oil containers, detergent bottles, chemical containers etc.
- Cell liners in sanitary landfills.
- Cable insulation.

Very high molecular weight polyethylene (UHMWPE)

It is a thermoplastic. Polymer chains are very long and hence molecular weight is very high, of the order of 3×10^6 to 6×10^6. Intermolecular forces are very strong and hence packing is extremely good. Chain branching is minimum. Specific gravity of this grade of PE is close to 1.0.

Properties

- Melting point is very high, close to 145°C.
- It is odourless, tasteless and nontoxic.
- The material is highly crystalline.
- High modulus and tensile strength.
- Very tough and very high impact strength.
- Excellent chemical resistance. It fails only in presence of oxidizing acids.
- Excellent abrasion resistance. It is 10 times more abrasion resistant than carbon steel.
- Very low coefficient of friction

Applications

It is widely used to manufacture:

- Artificial limbs.
- High modulus fibres for bulletproof jackets.
- Plastic containers, such as plastic bags, milk containers, oil containers, detergent bottles, chemical containers etc.

(v) Polypropylene (PP) The chemical structure of polypropylene is:

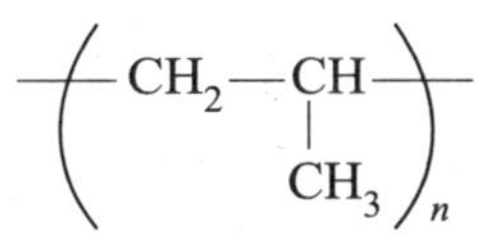

Different types of polypropylene (isotactic, syndiotactic or atactic) can be manufactured by using Ziegler-Natta catalyst.

Properties Polypropylene is the lightest plastic known, with specific gravity of 0.85. It is a highly crystalline solid (around 60% Crystallinity) with a melting point of 170°C. Due to high crystallinity PP possesses high mechanical strength, stiffness and hardness. However, it becomes brittle at low temperature.

Applications Polypropylene is widely used in manufacturing numerous extruded and injection moulded items. It is also used in manufacturing machine parts, household items etc.

(vi) Polyvinyl chloride (PVC) The chemical structure of PVC is:

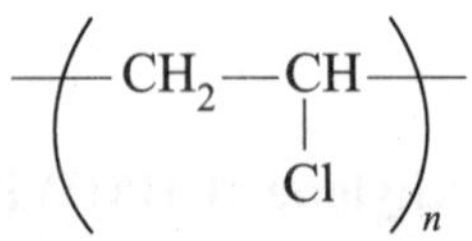

Preparation of monomer Ethylene and chlorine are allowed to react in liquid phase at 30 to 50°C in presence of iron chloride catalyst. Ethylene dichloride $[CH_2(Cl)CH_2(Cl)]$ is formed. This ethylene dichloride, on heating in tubes around 500°C, gives vinyl chloride with the release of hydrochloric acid.

$$Cl{-}CH_2{-}CH_2{-}Cl \xrightarrow{500°C} CH_2{=}CH{-}Cl + HCl$$

The resulting vapour is a mixture of vinyl chloride, ethylene dichloride and hydrogen chloride. The vapour mixture is then quenched in the downstream and finally vinyl chloride is separated by distillation.

Manufacture of PVC:

Polymerisation of vinyl chloride is carried out in suspension technique. The following are the ingredients used for polymerization:

Monomer	:	Vinyl chloride	–	30 – 50 parts
Dispersing agent	:	Gelatine	–	0.001 parts
Modifier	:	Trichloroethylene	–	0.1 parts
Initiator	:	Caproyl peroxide	–	0.001 parts
Demineralised water	:	–	–	90 parts

Disodium hydrogen phosphate is used as a buffer to maintain pH around 10. The function of trichloroethylene is to control molecular weight of polymer. Particle shape, size and size distribution may be controlled by varying the dispersing systems and the rate of stirring. Typical polymerization vessel is shown in the Figure 8.2.7. Reaction time is 1 to 2 hours. Reaction temperature is 50°C and pressure is 0.7 MPa when pressure falls to about 0.07 MPa, the excess monomer is removed and the polymer is collected.

Structure and properties of PVC The structure of PVC is shown below:

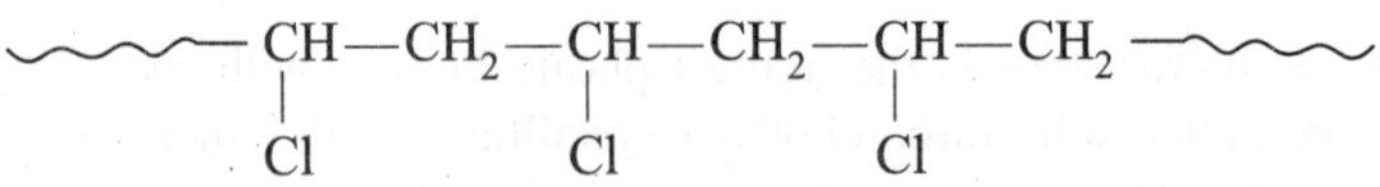

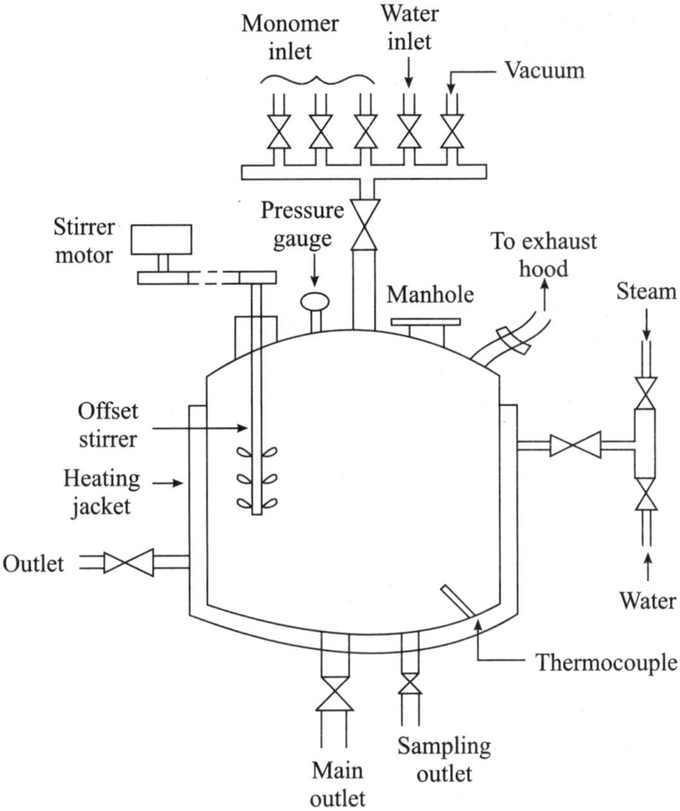

Fig. 8.2.7 Typical polymerization vessel for the manufacture of PVC.

Molecular weight: $\bar{M}_w = 1 \times 10^6$ to 3.5×10^6

Glass transition temperature: $T_g = +\ 80°C$

PVC has high electrical resistance, high ozone and weather resistance properties. Rigid PVC is dimensionally stable. It has better clarity, oil resistance and barrier properties than polyolefines. PVC can be made flexible and elastic like rubber by using plasticizers.

Applications of PVC:

Rigid PVC : Pipes and fittings
Profiles
Film and sheet
Blow moulded products
Records.

Plasticized PVC : Film and sheet
Wire and cable insulation
Shoe soles
Injection moulded products.

8.2.14 Thermosetting resins

Thermosetting resins are usually low molecular weight polymeric material. Two important thermosetting resins are:

(a) phenolic resins or bakelite
(b) amino resins.

Bakelite Bakelite is formed either in two-stage process or in three-stage process.

(i) **Reactions, taking place during resin formation** Resin may be manufactured either by acid-catalyst or by alkali-catalyst. In acidic medium phenol:formaldehyde ratio is maintained 1:0.8. An addition reaction takes place between phenol and formaldehyde to form monomethylol, dimethylol or trimethylol derivatives as shown below:

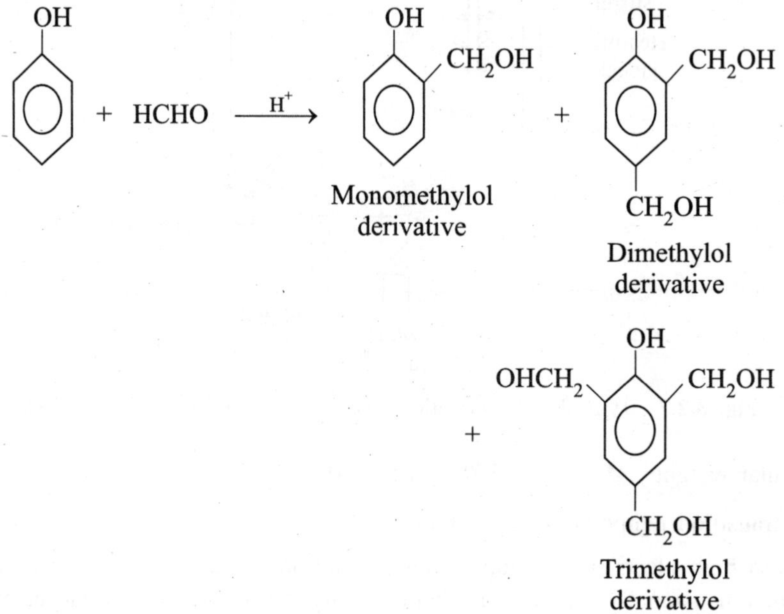

Since formaldehyde content is less in the mixture, monomethylol derivative is the main product. These monomethylol derivatives condense with phenols to form low molecular weight soluble linear polymer, known as Novalacs.

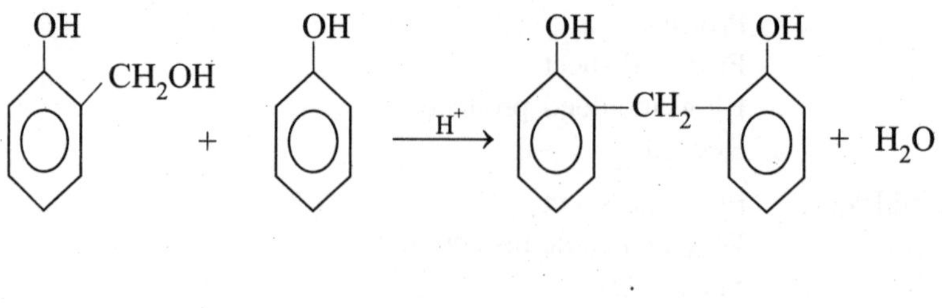

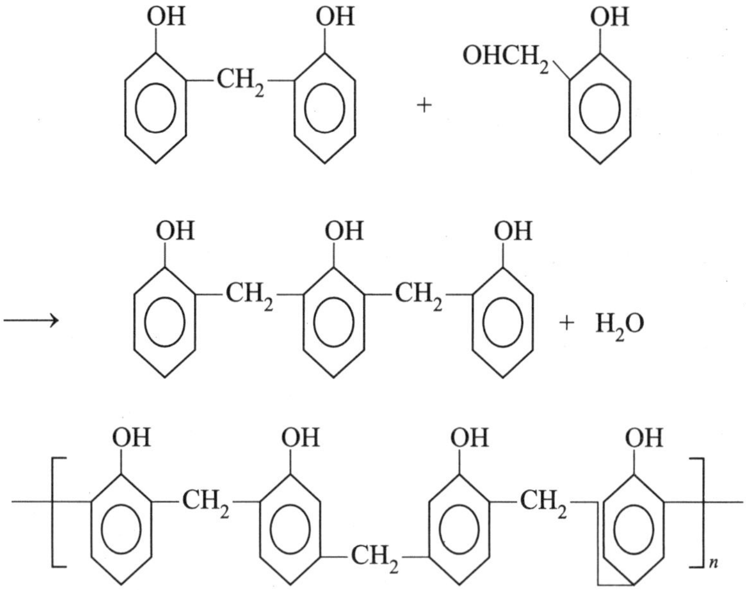

Novolac (Molecular weight ~ 1000)

In alkaline medium phenol:formaldehyde ratio is maintained 1:1.25. As formaldehyde content is more the resin contains free alcohol groups and the resin is called Resols.

Step-I:

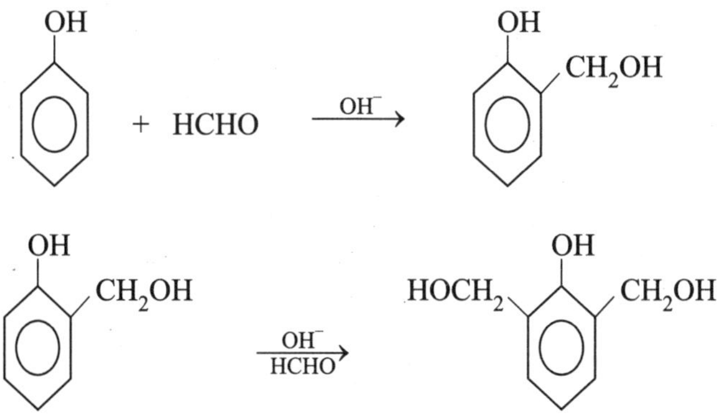

Step-II:

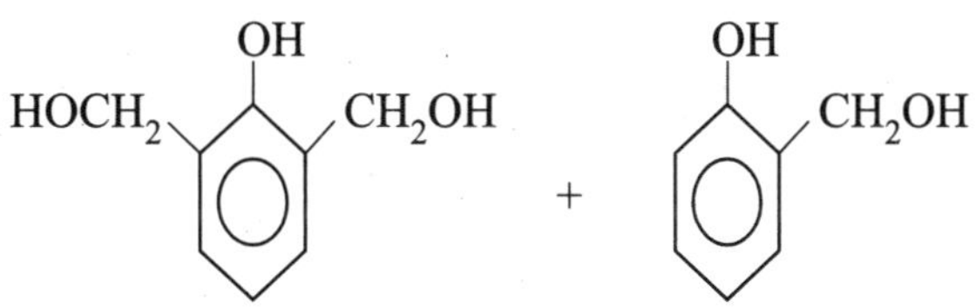

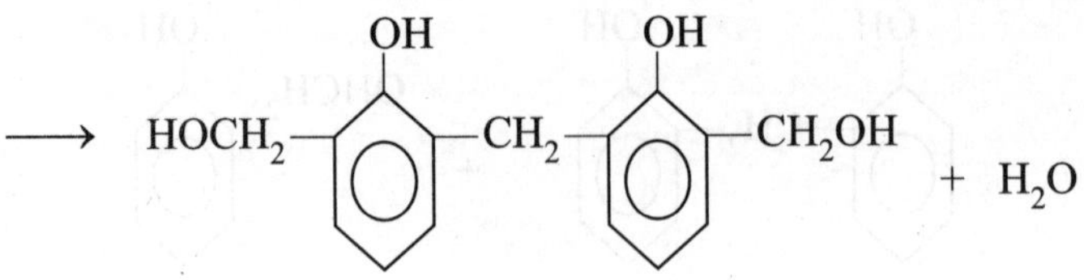

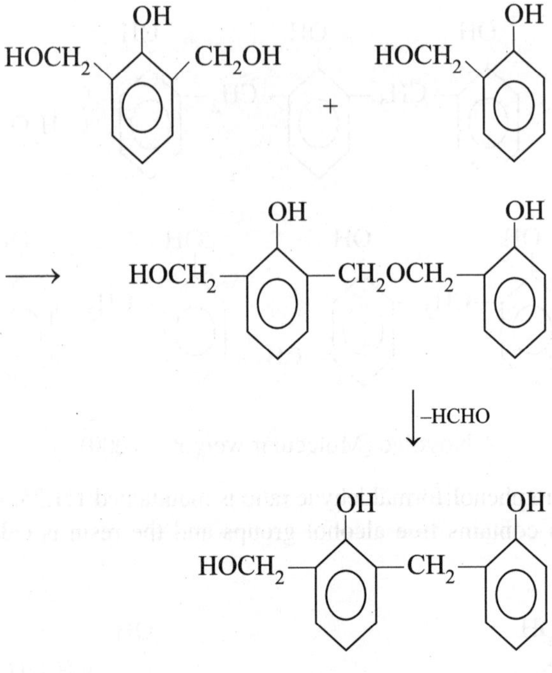

Step-III:

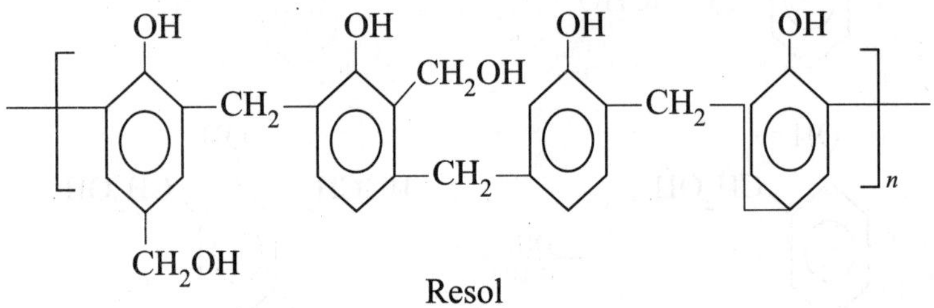

Resol

Step-IV: Resoles are soluble in organic solvents and contains alcohol groups.

(ii) Manufacturing process

Step-I Phenol and formaldehyde are allowed to react in presence of acid or base catalyst. In acid catalytic process phenol:formaldehyde ratio is maintained around 1:0.8 and the product is called *Novolac*. In base catalytic process phenol:formaldehyde ratio is maintained around 1:1.25 and the product is called *Resol*. Both Novalac and Resol are fusible and soluble in organic solvents and this soluble resin is known as *A-Stage resin*.

Step-II In a blender the resin is mixed with other additives like fillers, colourants, lubricants etc. as well as hexamethylenetetraamine, popularly known as *hexa*. Hexa provides enough methylene groups (—CH_2—) to increase phenol:formaldehyde mole ratio to 1:1.5. The mixture is then allowed to roll on a heated mixing mill to carry out condensation polymerization, leading to formation of B-stage resin. B-stage resin is nearly insoluble in organic solvents but fusible under heat and pressure.

Step-III The B-stage resin under hot condition is poured into the desired mould. Where resin takes the shape of the mould. Finally it is heated at high temperature and pressure. Enough crosslinking takes place at this stage to give insoluble, infusible resin, which is known as *C-stage resin.*

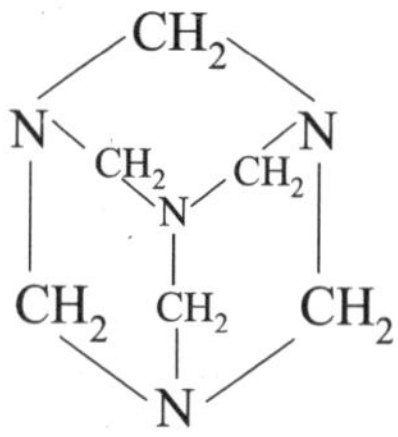

Hexamethylene tetraamine

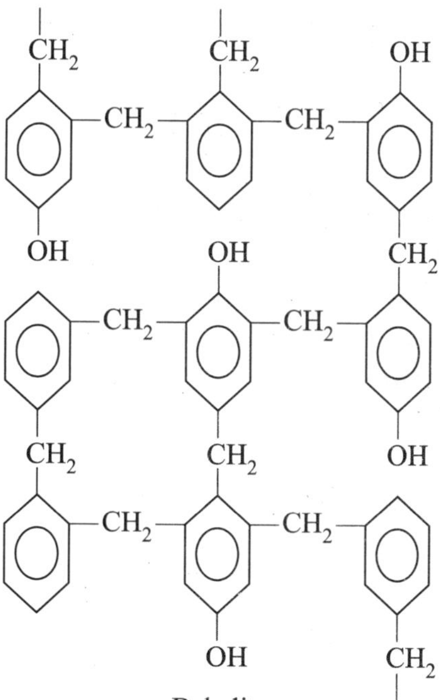

Bakelite

C-stage resin, insoluble and infusable structure

(iii) Properties Phenolic resins have good electrical insulation characteristics and chemical resistance. They have excellent dimensional stability and absorb little or no moisture.

(iv) Applications Phenolic resins are mainly used in manufacturing electrical appliances such as switches, plugs etc. Besides that it finds uses in heater and kettle handles, telephone parts, radio and television cabinet etc. Laminated products is one of the major outlets of phenolic resin. It is mainly used in decorative laminates, wall coverings, industrial laminates etc. Some percentage of phenolics is used to manufacture injection-moulded products. Phenolic resins are also used in the manufacture of ion-exchange resins.

Amino resins Amino resins are made either from urea or from melamine. Resins, derived from urea and formaldehyde is called Urea-Formaldehyde (UF) resins and resins, derived from melamine and formaldehyde, is called Melamine Formaldehyde (MF) resins. Following chemical reactions take place during formation of UF resin.

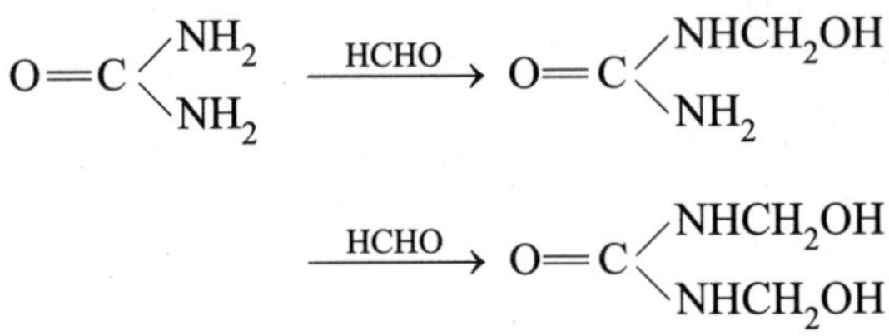

(i)

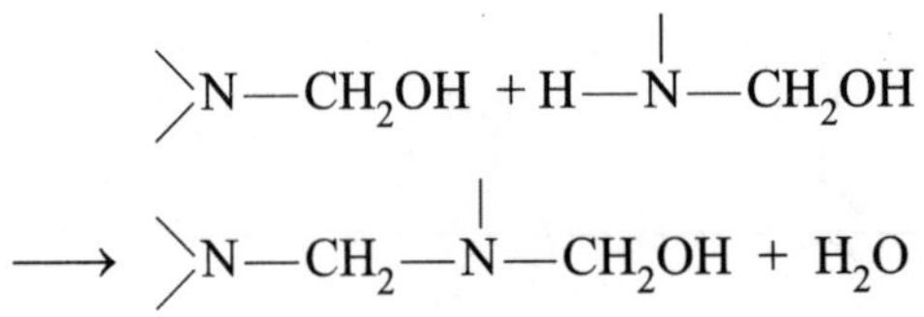

(ii)

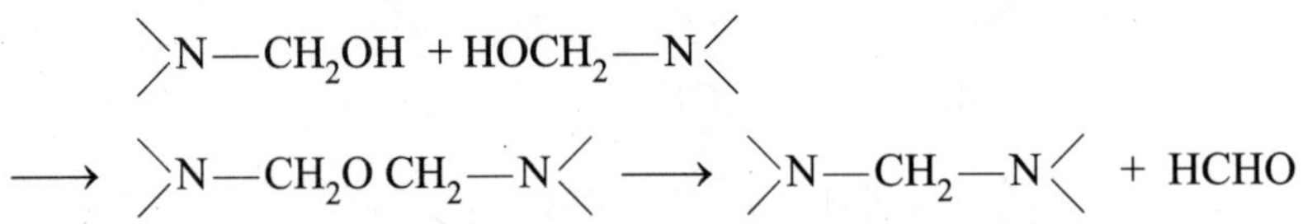

(iii)

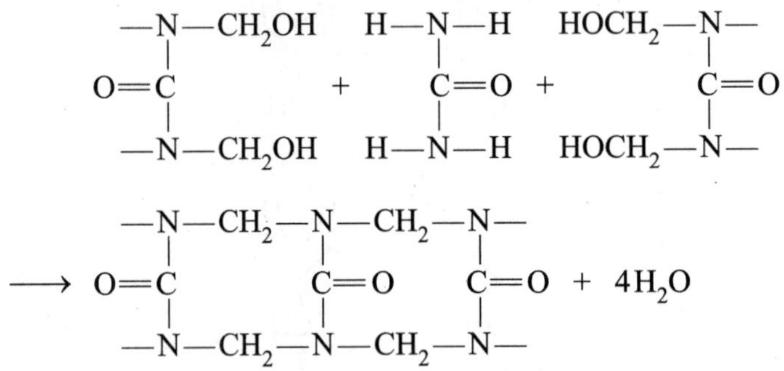

Urea:formaldehyde ratio is maintained 2:1 and the condensation reaction is carried out in presence of a base catalyst at about 50°C. The resin, thus produced is *A-stage resin* and water soluble. This is then mixed with other ingredients and dried to form *B-stage resin.* The *B-stage resin,* in presence of excess formaldehyde, gets converted into infusible and insoluble 3-dimensional thermosets.

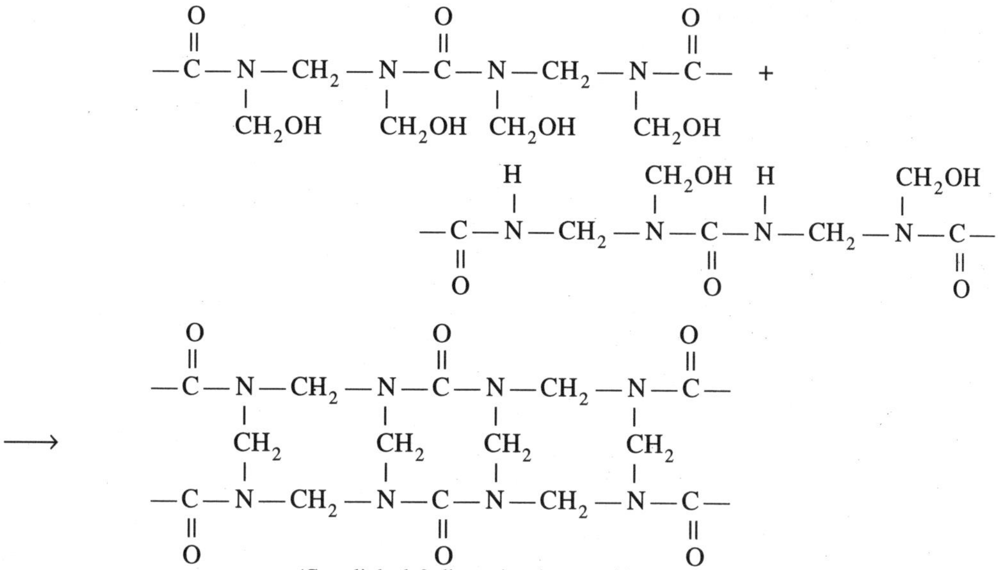

(Crosslinked 3-dimensional network)

Following reactions take place during formation of MF resin.

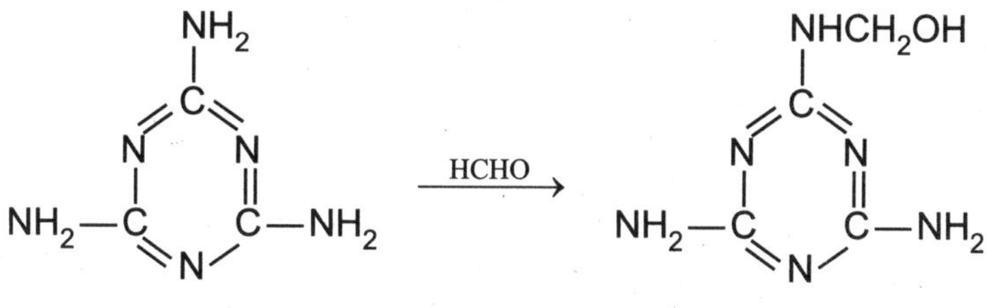

Melamine

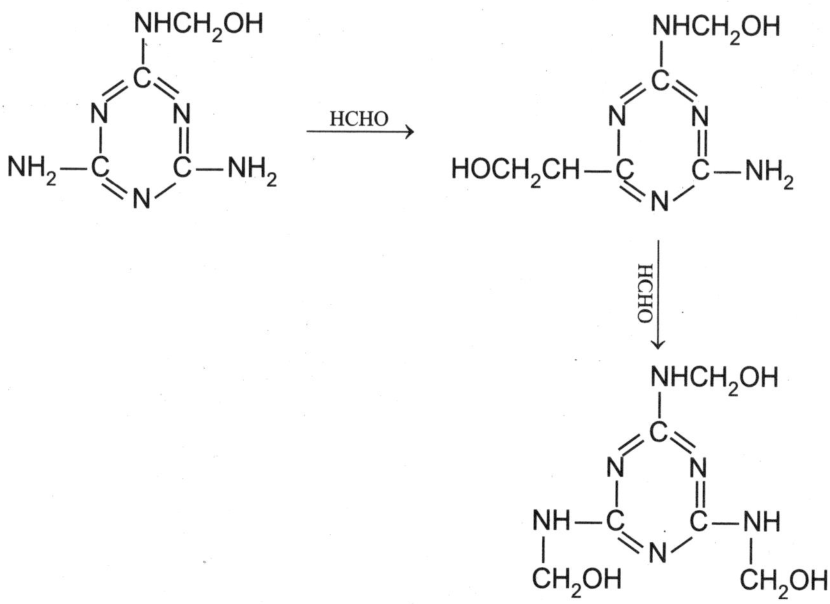

Trimethylol melamine

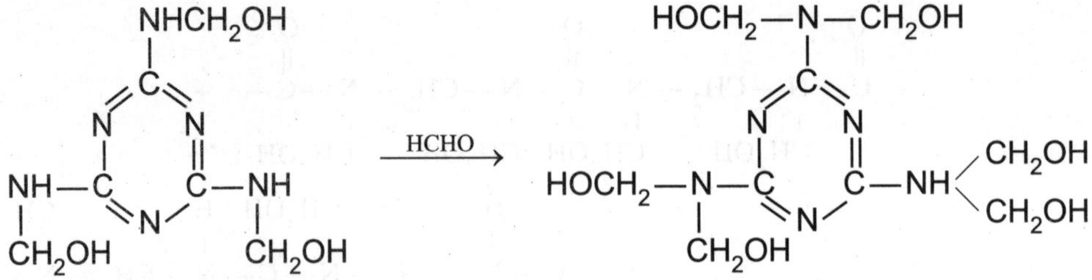

Trimethylol melamine Hexamethylol melamine

Properties and applications Amino resins are clear and colourless as well as possess high tensile strength and hardness but their impact strength, heat and moisture resistance are worse than those of phenolic resins.

Extensive use of MF-resin is in the area of adhesives for plywood and furniture. They are also used in protective coatings, textile treatment, paper treatment etc. High equality dinnerware is manufactured by cellulose filled MF resins.

Table 8.2.3 Differences between thermoplasts and thermosets

Sl. No.	*Thermoplasts*	*Thermosets*
1.	Generally thermoplasts are of high molecular-weight of the order of $10^6 - 10^9$.	Generally thermosets are of low molecular weight of the order of 10^4.
2.	Thermoplasts are soluble in suitable organic solvents.	Thermosets are insoluble in any organic solvent.
3.	Each thermoplast has a definite softening point, above which it melts and below which it solidifies.	Thermosets have no softening point. It is infusible. However, it burns on heating.
4.	It is two dimensional structure and polymer chains are bound by strong intermolecular forces.	It is a three dimensional structure and polymer chains are interconnected through crosslinking.
5.	It softens on heating and hardens on cooling. So it can be remoulded.	It hardens on heating to produce 3D-network and never softens. So it cannot be remoulded.

8.2.15 Fibres

Polyester fibre Polyester is manufactured through condensation reactions between dimethyl terephthalate and ethylene glycol.

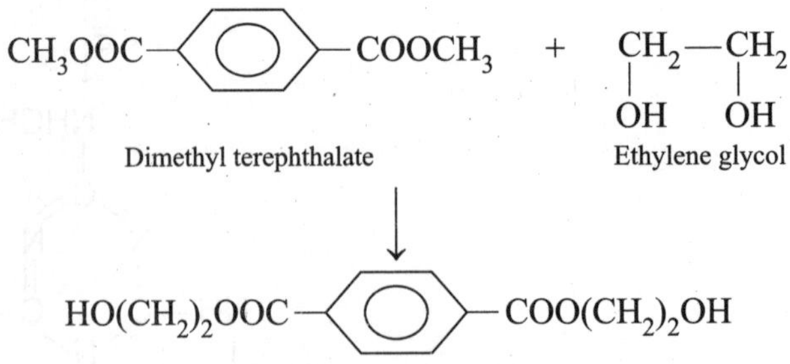

Diethylene glycol terephthalate

The ester undergoes condensation reactions at low pressure and high temperature to form the polymer polyethylene terephthalate (PET).

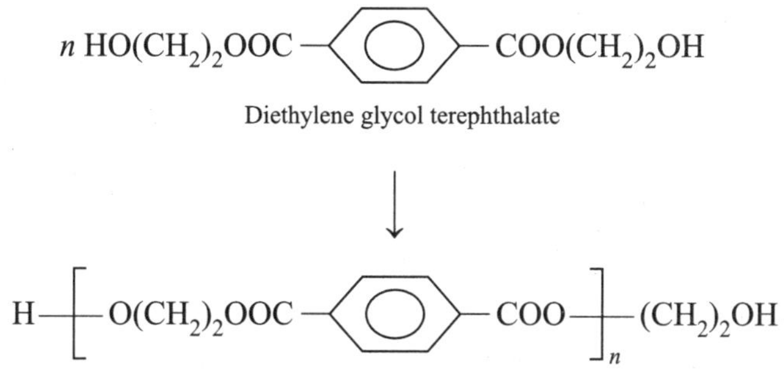

Diethylene glycol terephthalate

PET

This bulk polymer is then converted to fibre by spinning. Spinning is usually performed in different techniques, e.g., dry spinning, wet spinning and melt spinning. Out of them the most important is melt spinning. In melt spinning technique molten polymer is passed through a spinneret, containing two guide rolls, rotating at different speeds. As a result of this, stretching and molecular orientation take place and molten polymer takes the form of fibre filament. This is illustrated in the Figure 8.2.8.

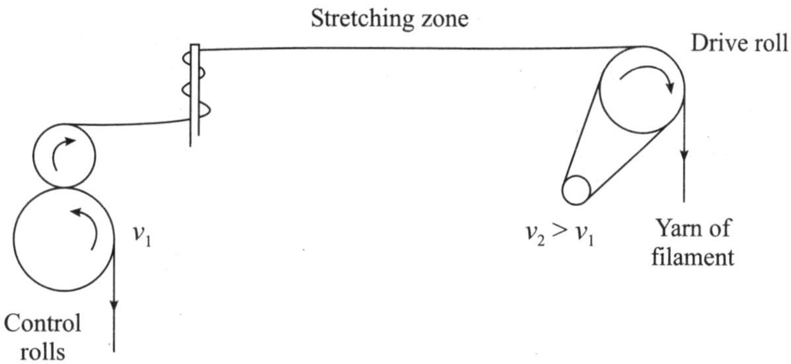

Fig. 8.2.8 Drawing process of a fibre.

Properties PET fibre is highly crystalline and has a sharp melting point of 265°C. It has good mechanical properties and can retain the mechanical properties even at high temperature up to 150°C to 175°C. It has good chemical, solvent and moisture resistance.

Applications PET fibre is widely used in textile industry in making synthetic garments. It is usually blended with wood to have better crease and wrinkle resistance. Currently it is used in manufacturing of tyre cords.

Nylon fibre Nylon is a condensation product of amino group and carboxyl group. Nylon has different grades depending on the number of carbon atoms in the repeating unit. One of such important grade is Nylon 6, 6. It is a condensation product of hexamethylenediamine and adipic acid. Each of the two constituents provide 6 carbon atoms to the repeating unit and hence the product is called nylon 6, 6. The condensation reaction is given below:

$$n\,NH_2—(CH_2)_6—NH_2 + n\,HOOC—(CH_2)_4—COOH$$

Hexamethylene tetraamine   Adipic acid

$$\downarrow$$

$$H\!\left[HN—(CH_2)_6—NHOC—(CH_2)_4—CO\right]_n\!OH$$

Nylon 6,6

The linkage —CONH— is known as polyamide linkage.

Preparation At first hexamethylene diamine and adipic acid (1:1) are allowed to react to form salt, which is then transferred to an autoclave where polymerization is carried out at 250 psi and 270°C to 280°C for 3 – 4 hrs. Around 1 mole % acetic acid is added to maintain the viscosity.

Properties Nylon possesses high mechanical strength, elasticity, toughness and abrasion resistance. It is highly crystalline and crystalline melting point is 265°C. It maintains mechanical properties even at 150°C. Solvent resistance of nylon is very high, although it is dissolved in some polar solvents, such as, phenol, cresols, formic acid etc. However its electrical resistance is not good due to having polar groups in repeating units.

Applications Nylons are mostly used in the manufacture of apparels, carpets, ropes etc. Nylon fibre cord is extensively used in the manufacture of tyres.

Besides Nylon-6,6, other important grades of Nylon are Nylon-6 and Nylon-11.

Nylon-6

The raw material for manufacturing nylon-6 is caprolactum, a cyclic amide, as shown below:

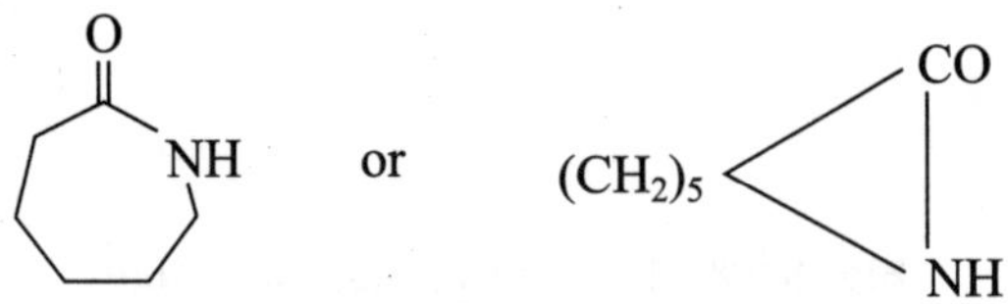

The caprolactum is heated in a closed chamber at about 260°C for about 4 to 5 hours under nitrogen atmosphere to avoid any oxidation. Under this condition, C – N bond of caprolactum breaks, forming free radicals, which undergo rapid polymerization, producing Nylon-6. The polymerization process is called ring opening polymerization. The product is obtained as a molten mass, which is then drawn through spinnerets to form fibres. The reaction is shown below:

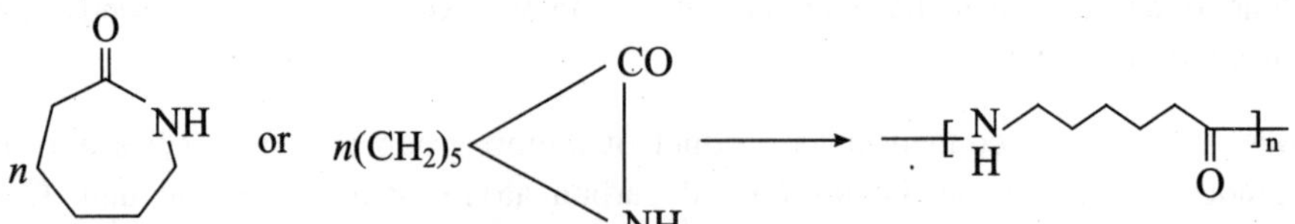

Caprolactum   Nylon-6

As the monomer, caprolactum, contains 6 carbon atoms, the repeat unit also contains 6 carbon atoms and hence the name, nylon-6.

Properties

1. Nylon-6 fibres appears as lustrous. It is sufficiently elastic and possesses high tensile strength.
2. It has high abrasion resistance as well as high acid and alkali resistance.
3. It is wrinkle-proof and can absorb moisture up to 2.4%.

Nylon-11

The raw material for manufacturing Nylon-11 is decamethylene amino acid. It undergoes self condensation polymerization to produce the polymer, Nylon-11. As the repeat units contains 11 carbon atoms and amide linkage, the polymer is commercially known as PA11. The following reaction takes place during polymerization process.

$$\underset{\text{Decamethylene amino acid}}{2n\,NH_2\text{—}(CH_2)_{10}\text{—}COOH} \longrightarrow \underset{\text{Nylon-11}}{H\text{—}[NH\text{—}(CH_2)_{10}\text{—}CO]_n\text{—}OH} + (2n\text{-}1)\,H_2O$$

Properties

1. Melting point	185°C
2. Specific gravity	1.04
3. Maximum water absorption	0.4%
4. Tensile strength	55 MPa
5. Elongation at break	300%
6. U.V. and Chemical resistance	Good

Explain the following

Q. 1 In S_N1 mechanism, 3-Cl-1 butene is more than 5000 times as reactive as allyl chloride but in S_N2 mechanism with sodium ethoxide in ethanol, the reactivity of allyl chloride is more than 20 times as reactive as 3-Cl-1 butene – explain.

Ans. S_N1 mechanism proceeds with the formation of carbocation. Carbocation, obtained from 3-Cl-1-butene is far more stable than that obtained from allyl chloride due to the strong + I effect of methyl group.

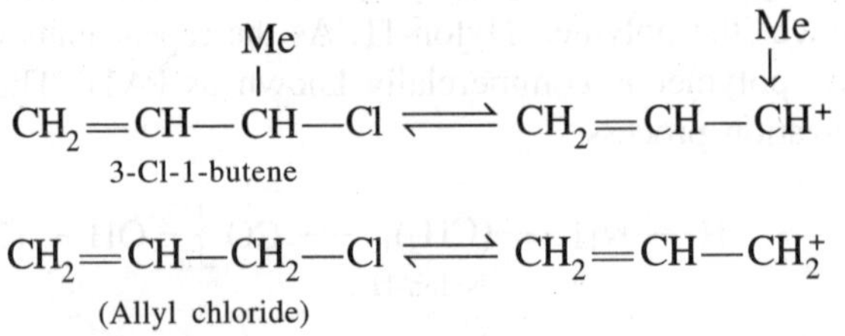

As solvolysis reaction proceeds through S_N1 mechanism, 3-Cl-1-butene shows much higher reactivity than allyl chloride.
However, in presence of strong base like C_2H_5ONa, the substitution reaction follows S_N2 mechanism. In the T.S. of S_N2 mechanism five groups are attached to the central carbon atom. In the case of 3-Cl-1-butene, presence of Me-group introduces significant steric hindrance, rendering the T.S. compound unstable. This type of steric hindrance is absent in the case of allyl chloride.

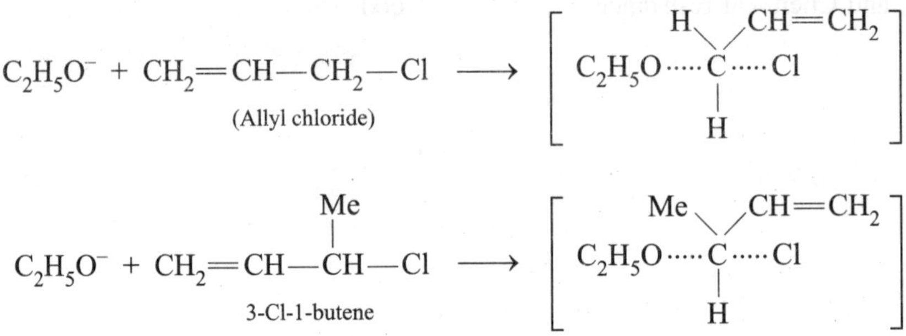

(Less stable)

This stability criterion of T.S. compound makes the allyl chloride more reactive than 3-Cl-1-butene.

Q. 2 Addition of $BrCCl_3$ to 1, 3-butadiene in presence of H_2O_2 gives rise two products $Cl_3C{-}CH_2{-}CH(Br){-}CH{=}CH_2$ and $Cl_3C{-}CH_2{-}CH{=}CH{-}CH_2(Br)$ – explain.

Ans. In presence of peroxide the addition reaction proceeds according to the following mechanism:

$$HO{-}OH \longrightarrow 2HO^{\bullet},$$

$$HO^{\bullet} + BrCCl_3 \longrightarrow {}^{\bullet}CCl_3 + HO{-}Br$$

$$^{\bullet}CCl_3 + CH_2{=}CH{-}CH{=}CH_2 \longrightarrow CCl_3{-}CH_2{-}\dot{C}H{-}CH{=}CH_2$$

$$\underset{(A)}{CCl_3{-}CH_2{-}\dot{C}H{-}CH{=}CH_2} \longleftrightarrow \underset{(B)}{CCl_3{-}CH_2{-}CH{=}CH{-}\dot{C}H_2}$$

$$A + BrCCl_3 \longrightarrow CCl_3{-}CH_2{-}\underset{\displaystyle Br}{\underset{|}{CH}}{-}CH{=}CH_2 + {}^{\bullet}CCl_3$$

$$B + BrCCl_3 \longrightarrow CCl_3{-}CH_2{-}CH{=}CH{-}CH_2{-}Br + {}^{\bullet}CCl_3$$

Q. 3 State with explanation, the major product obtained after dehydrohalogenetion of 4-Br-1-hexene.

Ans.

$$MeCH_2{-}\underset{\displaystyle Br}{\underset{|}{CH}}{-}CH_2{-}CH{=}CH_2$$

(4-bromo-1-hexene)

$$MeCH_2{-}\underset{\displaystyle Br}{\underset{|}{CH}}{-}CH_2{-}CH{=}CH_2 \xrightarrow{HBr} \begin{cases} \underset{(A)}{MeCH_2{-}CH{=}CH{-}CH{=}CH_2} \\ \underset{(B)}{Me{-}CH{=}CH{-}CH_2{-}CH{=}CH_2} \end{cases}$$

As conjugated diene is far more stable than unconjugated diene, the product (A) appears as the major product.

Q. 4 t-BuCl on hydrolysis yields chiefly t-BuOH and the said reaction is not appreciably affected by dissolved sodium fluoride. However, instead of water if DMSO is used as a solvent, t-BuCl yields isobutylene in presence of same sodium fluoride – explain.

Ans. t-BuCl produces stable teriary carbonium ion in the 1st step of reaction.

$$Me{-}\underset{\displaystyle Me}{\underset{|}{\overset{\displaystyle Me}{\overset{|}{C}}}}{-}Cl \;\underset{Solvent}{\overset{Slow}{\rightleftharpoons}}\; Me{-}\underset{\displaystyle Me}{\underset{|}{\overset{+}{C}}}{-}Me$$

This reaction occurs spontaneously and is not influenced by any solvent. In presence of a protic solvent like water, the carbonium ion is further stabilized through hydrogen bonding and hence denies the access of other attacking agent into its domain. Furthermore, F^- of NaF is a strong base but in presence of water, it is solvated through hydrogen bonding and hence the effectiveness of F^- as a base is significantly reduced. As a result of this,

NaF is unable to attack the carbonium ion in presence of water and hence carbonium ion undergoes hydrolysis to produce the corresponding alcohol:

$$\mathrm{Me{-}\overset{+}{C}(Me){-}Me + H_2O \xrightarrow[\text{Fast}]{Cl^-} t\text{-}BuOH + HCl}$$

However, the scenario in presence of an aprotic polar solvent, like DMSO, is different. In DMSO, carbonium ion is free due to absence of any hydrogen bonding. Due to the same reason F^- of NaF is also free and appears as a strong base. Thus F^- readily attacks the carbonium ion, facilitating elimination reaction.

$$\mathrm{F^- + H{-}CH_2{-}\overset{+}{C}(Me){-}Me \xrightarrow[\text{Fast}]{DMSO} CH_2{=}C(Me){-}Me}$$

(iso-butylene)

Q. 5 Explain the following reaction with mechanism:

(i) $\mathrm{MeCH(OH)CH{=}CH_2 \xrightarrow[{[H_2O]}]{HBr}}$ $\mathrm{MeCH(Br)CH{=}CH_2}$ + $\mathrm{MeCH{=}CHCH_2Br}$

(ii) $\mathrm{MeCH{=}CHCH_2OH \xrightarrow[{[H_2O]}]{HBr} ?}$

Ans. (i) $\mathrm{MeCH(OH){-}CH{=}CH_2 \xrightarrow{+H^+} MeCH(\overset{+}{O}H_2){-}CH{=}CH_2}$

$$\mathrm{MeCH(\overset{+}{O}H_2){-}CH{=}CH_2 \longrightarrow Me\overset{+}{C}H{-}CH{=}CH_2 \longleftrightarrow MeCH{=}CH{-}CH_2^+}$$

$$\mathrm{Me\overset{+}{C}H{-}CH{=}CH_2 \xrightarrow[-H^+]{HBr} MeCH(Br)CH{=}CH_2 \qquad MeCH{=}CH{-}CH_2^+ \xrightarrow[-H^+]{HBr} MeCH{=}CHCH_2Br}$$

(ii) $\mathrm{MeCH{=}CH{-}CH_2{-}OH \xrightarrow{+H^+} MeCH{=}CH{-}CH_2{-}OH_2^+}$

$$\mathrm{MeCH{=}CH{-}CH_2{-}OH_2^+ \xrightarrow{-H_2O} MeCH{=}CH{-}CH_2^+ \longleftrightarrow Me\overset{+}{C}H{-}CH{=}CH_2}$$

$$\mathrm{MeCH{=}CH{-}CH_2^+ \xrightarrow[-H^+]{HBr} MeCH{=}CH{-}CH_2Br \qquad Me\overset{+}{C}H{-}CH{=}CH_2 \xrightarrow[-H^+]{HBr} MeCH(Br){-}CH{=}CH_2}$$

Q. 6 When 1, 4-hexadien-1-ol is dissolved in sulphuric acid, only 3, 5-hexadiene-2-ol is obtained – explain.

Ans.

$$CH_2{=}CH{-}\underset{\displaystyle OH}{\underset{|}{CH}}{-}CH{=}CHMe \xrightarrow{[H_2SO_4]} CH_2{=}CH{-}\underset{\displaystyle ^{+}OH_2}{\underset{|}{CH}}{-}CH{=}CHMe$$

$$CH_2{=}CH{-}\underset{\displaystyle ^{+}OH_2}{\underset{|}{CH}}{-}CH{=}CHMe \longrightarrow CH_2{=}CH{-}\overset{+}{C}H{-}CH{=}CHMe + H_2O$$

$$\underset{(A)}{CH_2{=}CH{-}\overset{+}{C}H{-}CH{=}CHMe} \longleftrightarrow \underset{(B)}{CH_2{=}CH{-}CH{=}CH{-}\overset{+}{C}HMe}$$

The carbonium ion [B] is far more stable than [A] due to the following reasons:

1. Strong +I effect of methyl group stabilizes the carbonium ion [B] but this type of stabilization is absent in [A].
2. A stable conjugated diene group, attached to the carbocation [B], stabilizes the latter through resonance. This type of stabilization is absent in [A].

Thus the T.S. mainly contains the carbocation [B] and hence only one product is obtained in the next step.

$$\underset{(B)}{CH_2{=}CH{-}CH{=}CH{-}\underset{\displaystyle Me}{\underset{|}{\overset{+}{C}H}}} + H_2O \longrightarrow \underset{(3,5\text{-hexadiene-2-ol})}{CH_2{=}CH{-}CH{=}CH{-}\underset{\displaystyle OH}{\underset{|}{CH}}Me} + H^+$$

Q. 7 Addition of Br_2 or HBr to 1, 3-butadiene gives rise chiefly 1, 2-product at low temperature but chiefly 1, 4-product is obtained at high temperature – explain.

Ans.

$$CH_2{=}CH{-}CH{=}CH_2 + Br_2 \longrightarrow \underset{(\text{Chief product at } -80°C)}{CH_2(Br)CH(Br)CH{=}CH_2}$$

$$CH_2{=}CH{-}CH{=}CH_2 + Br_2 \longrightarrow \underset{(\text{Chief product at } +40°C)}{CH_2(Br){-}CH{=}CH{-}CH_2(Br)}$$

In the first step electrophilic addition occurs:

$$CH_2{=}CH{-}CH{=}CH_2 + Br_2 \longrightarrow \underset{\displaystyle Br}{\underset{|}{CH_2}}{-}\overset{+}{C}H{-}CH{=}CH_2$$

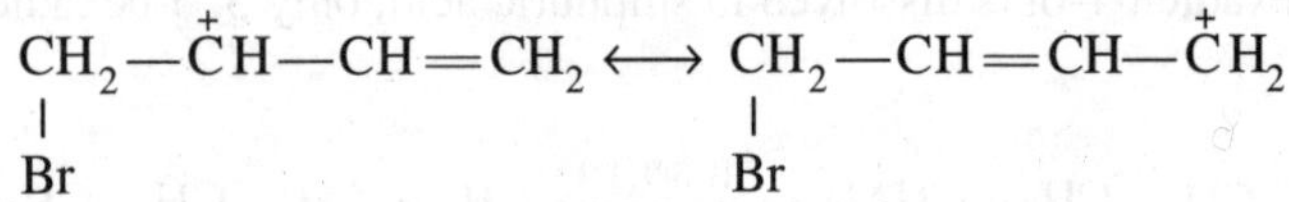

(A) (B)

The nucleophilic addition to [A] occurs directly:

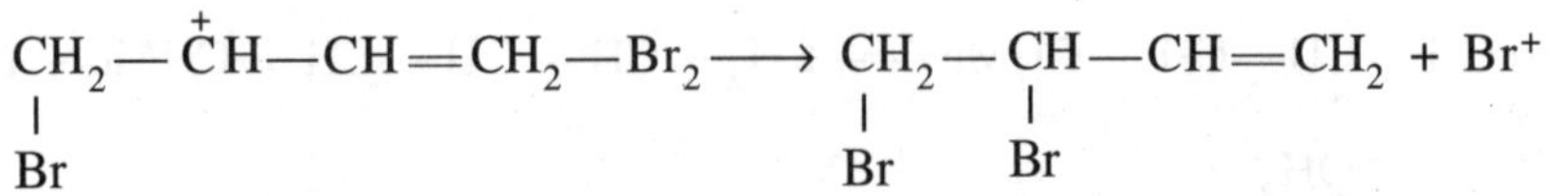

However, the nucleophilic addition to [B] is believed to take place through cyclisation:

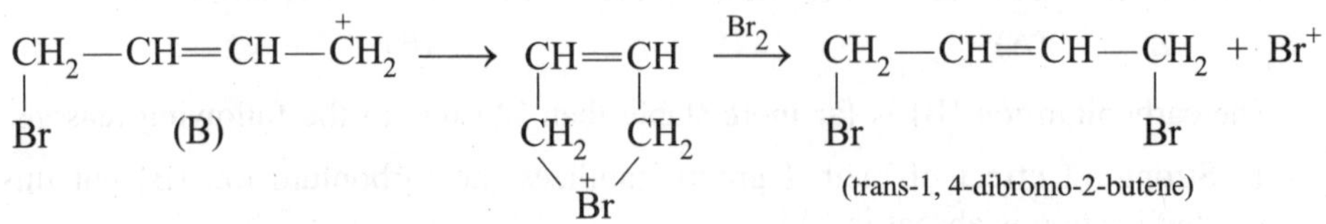

The cyclic bromonium ion is very much stable and more stable than [A], but cyclisation takes place only at high temperature. So 1, 4-product is favourable at high temperature only. Furthermore, 1, 4-product is usually obtained in the trans-form. As nucleophilic addition to [A] occurs directly, 1, 2-product appears as the chief product at low temperature.

Q. 8 Explain the ozonolysis reaction of 2-methyl propene with mechanism.

Ans.

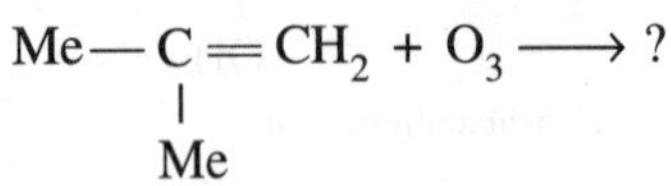

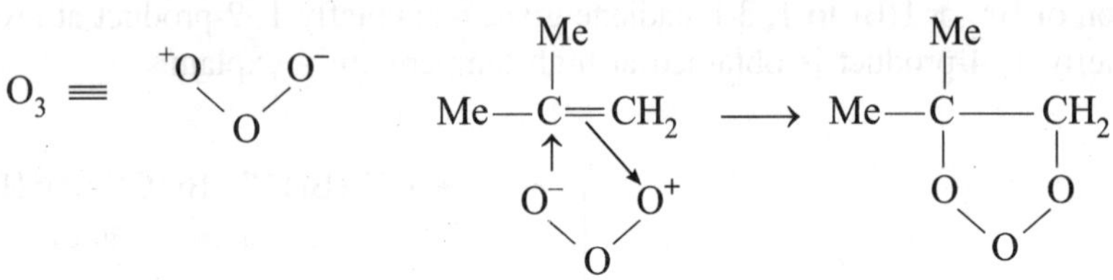

(Molozonide)

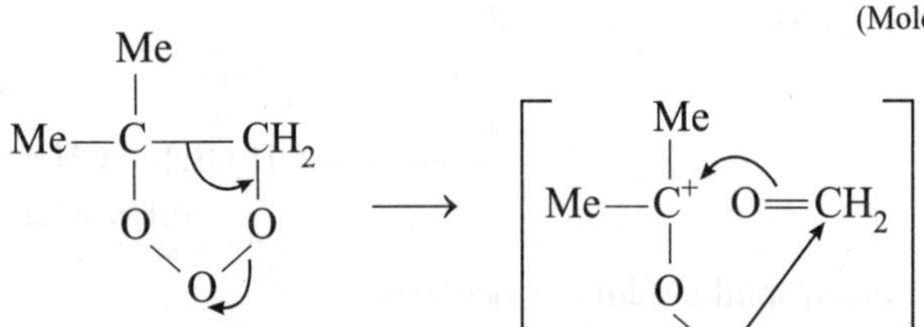

(Molozonide)

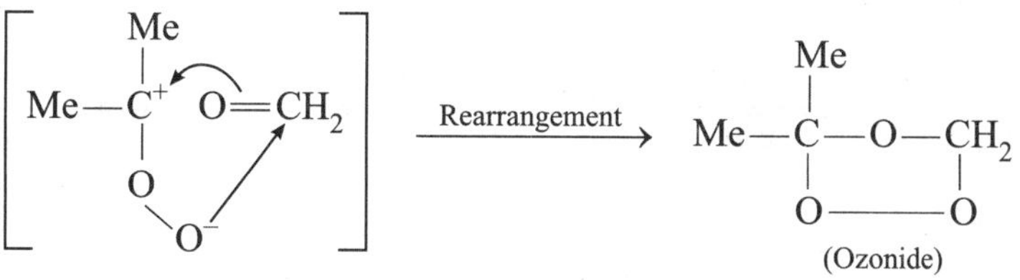

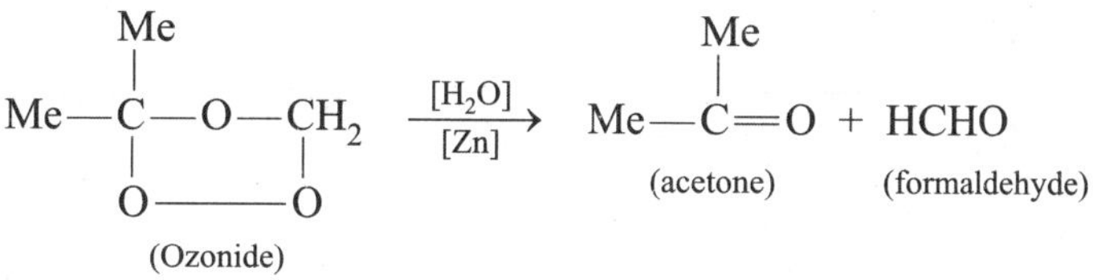

Generally the reaction is carried out in an inert solvent like carbon-tetra-chloride (CCl_4). Ozonide is an explosive compound. To hydrolyze the ozonide safely, a reducing agent (e.g. Zn-metal) is often used. Otherwise ozonide will produce hydrogen peroxide [H_2O_2]. This hydrogen peroxide then further oxidizes the products ketones and aldehydes to the corresponding acids.

Q. 9 Predict the mechanism of the following reaction:

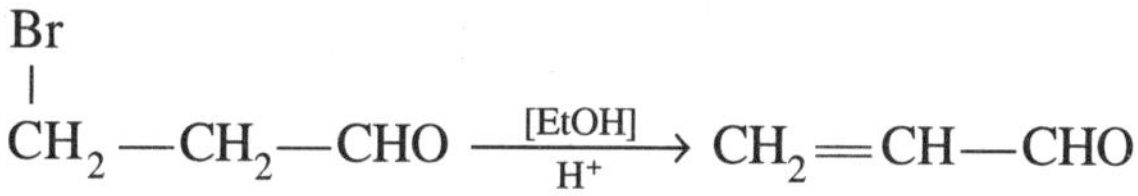

Ans. EtOH protects the aldehyde group with the formation of acetals.

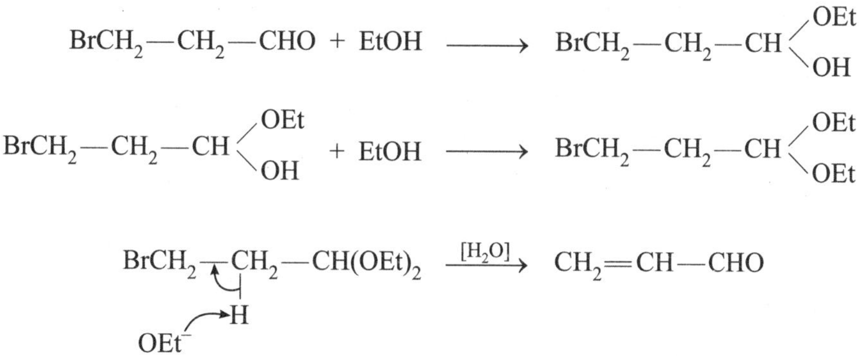

Q. 10 Explain acyloin condensation with mechanism.

Ans. Diesters like $(CH_2)_n(COOEt)_2$ undergoes acyloin condensation in presence of sodium and ethanol:

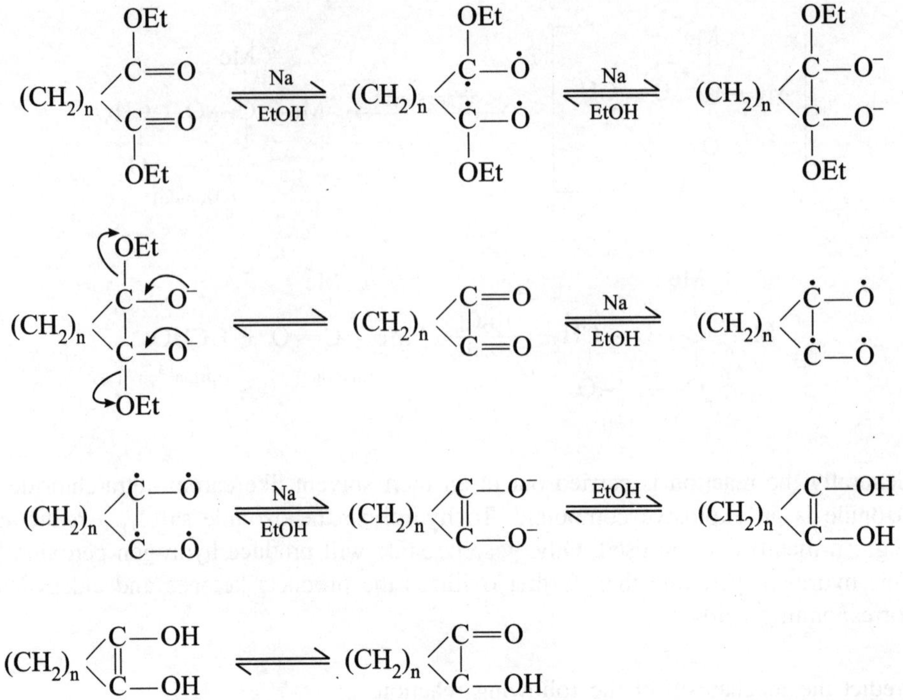

9

FUELS AND COMBUSTION

9.1 Definition

Any substance, which gives off heat on combustion, is called fuel. The main constituent of any fuel is carbon. Fuels, obtained from earth's crust are known as fossil-fuels.

9.2 Classification

Fuels may be classified into two main categories: (i) Primary and (ii) Secondary. Fuels, obtained from natural source, are known as primary fuels and fuels, derived from primary fuels, are known as secondary fuel.

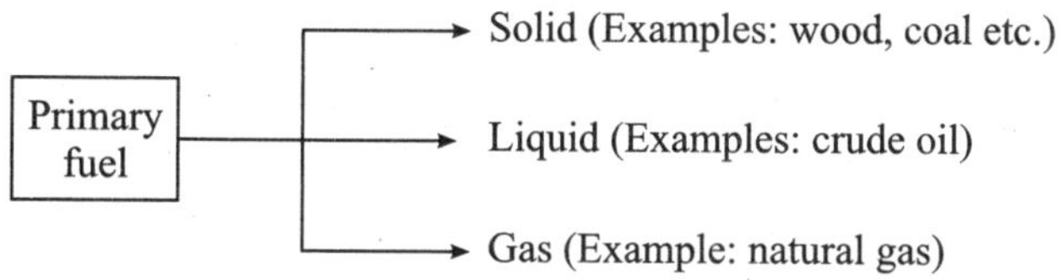

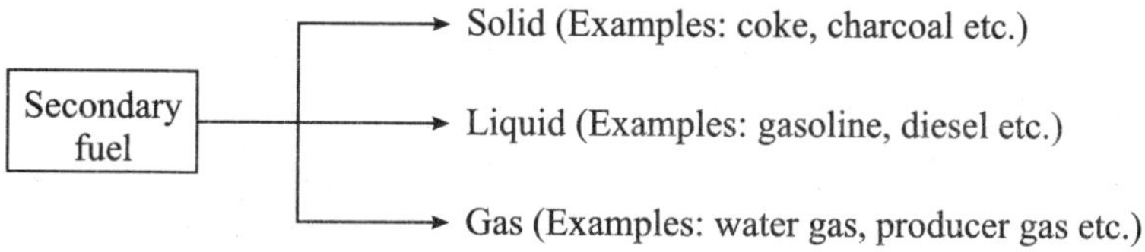

All primary and secondary fuels can be divided into three categories depending on their physical appearance.

9.3 Some definitions

(a) **Calorific value** Amount of heat liberated after burning unit mass of fuel completely is known as calorific value of that fuel. It is usually expressed in any of the three units:

(i) **Calorie** Amount of heat energy required to raise the temperature of one gm of water from 15°C to 16°C is known as 1 calorie.

(ii) **B.T.U.** Amount of heat energy required to raise the temperature of one pound of water from 60°F to 61°F is known as British Thermal Unit or B.T.U.

(iii) Amount of heat energy required to raise the temperature of one pound of water from 15°C to 16°C is known as Centigrade Heat Unit or C.H.U.

1 B.T.U. = 252 Cal = 0.252 kcal and 1 kcal = 3.968 B.T.U. = 2.2 C.H.U.

So unit of calorific value may be expressed as any of the following units.

(i) Cal/gm or kcal/kg, (ii) B.T.U./lb, (iii) kcal/m^3, (iv) B.T.U./ft^3

$$1 \text{ kcal/kg} = 1.8 \text{ B.T.U./lb and } 1 \text{ B.T.U./ft}^3 = 9.3 \text{ kcal/m}^3$$

Calorific value of a fuel may be of two types.

(i) **Gross Calorific Value (GCV)** If a unit mass or unit volume of a fuel is burnt completely in a closed chamber and the products of combustion are allowed to cool to room temperature or 15°C, the total amount of heat received is known as gross calorific value or GCV. In this process water vapour, thus produced during combustion, gets condensed to release the latent heat which is included in GCV.

(ii) **Net Calorific Value (NCV)** If a unit mass or unit volume of a fuel is burnt completely in an open chamber and the products of combustion are allowed to escape, the amount of heat received is known as Net Calorific Value or NCV. Obviously latent heat of water vapour is not included in NCV.
So, we can write that: GCV = NCV + Latent heat of water vapour.
% hydrogen in a given fuel sample is assumed H.

Latent heat of steam = 587 kcal/kg.

Or, $$\text{NCV} = \text{GCV} - 0.09\,\text{H} \times 587 \text{ kcal/kg, since}$$

$$1 \text{ kg of } H_2 \equiv 9 \text{ kg of steam and hence } \left(\frac{\text{H}}{100}\right) \text{ kg of } H_2 \equiv 0.09\,\text{H kg of steam}$$

(b) **Ignition temperature** Ignition temperature of a fuel is that temperature at which the fuel starts burning. Lower the ignition temperature more danger the fuel is. Conversely, higher the ignition temperature more difficult to ignite the fuel is.

(c) Flame temperature The maximum temperature, attained during burning of coal in open air, is known as flame temperature. In the determination of flame temperature it is assumed that heat, liberated during combustion, does not heat up the surrounding air. In actual practice this assumption has no meaning and hence flame temperature is never attained in actual combustion.

9.3.1 Calculation of calorific value

$$\text{NCV} = (\text{GCV} - 0.09\ \text{H} \times 587)\ \text{kcal/kg}$$

where,
H = Weight of hydrogen per 100 kg fuel.
O = Weight of oxygen per 100 kg fuel.
C = Weight of carbon per 100 kg fuel.
S = Weight of sulphur per 100 kg fuel.

(i) HCV calculation by Bomb calorimeter:

$$\text{GCV} = \frac{(W_{\text{H}_2\text{O}} + W_E)(T_2 - T_1)}{F}$$

$W_{\text{H}_2\text{O}}$ = Mass of water in calorimeter in kg
W_E = Water equivalent of calorimeter in kg
= (Weight of apparatus) × (specific heat)
= (Calorific value of equivalent water)
= (Weight of equivalent water) × (specific heat of water)
= Weight of equivalent water [since sp. heat of water = 1kcal/kg°C]
T_1, T_2 = Initial and final temperatures of water in the calorimeter respectively.
F = Mass in kg of fuel sample burnt in the calorimeter.

(ii) Theoretical GCV calculation by Dulong's formula:

$$\text{GCV} = \left[8080\text{C} + 34500\left(\text{H} - \frac{\text{O}}{8}\right) + 2250\text{S}\right]\ \text{kcal per 100 kg coal}$$

9.4 Solid fuel

The most important solid fuel is coal. Coal is a highly carbonaceous matter comprising of carbon (C), hydrogen (H), nitrogen (N) and oxygen (O) and a trace amount of non-combustible inorganic matter. The quality of coal mostly depends on the colorific value, which further depends on carbon content of coal. On this basis the *best* quality coal is *anthracite*, having C-content around 93% and calorific value around 8700 kcal/kg. The *worst* quality coal is *peat*, having C-content 57% and calorific value around 5000 kcal/kg.

9.4.1 Analysis of coal

Coal sample is analyzed in two different aspects, (i) to determine ingredient composition e.g., moisture content etc. This type of analysis is known as proximate analysis, (ii) to determine elemental composition, i.e. carbon content, hydrogen content etc. This type of analysis is known as ultimate analysis.

Proximate analysis In this method of analysis the following components are determined.

(i) **Moisture** W_1 gm, of coal sample is taken in a silica crucible, which is then kept in a hot air oven at around 105°C for an hour. Sample is taken out and kept in a desiccator for cooling and finally the weight of the sample is taken. If the final weight is W_2 gm, $(W_1 - W_2)$ gm is the loss in weight due to evaporation of moisture.

$$\% \text{ Moisture content} = \left(\frac{W_1 - W_2}{W_1}\right) \times 100$$

Significance Lesser the moisture content, better the quality of coal.

(ii) **Volatile matter** A W gm of coal sample is taken and moisture is driven off by following the above procedure. The dried sample (W_1 gm) is taken in a silica crucible with lid, which is then kept in a muffle furnace at around 950°C for 7 minutes. The sample is taken out and kept in a desiccator until it attains the room temperature. Weight of that sample is taken. Suppose it is W_2 gm. Then $(W_1 - W_2)$ gm indicates the loss in weight due to evaporation of volatile matter other than moisture, where, W_1 = Weight of dried coal sample, i.e., coal sample free from moisture.

$$\% \text{ Volatile matter content} = \left(\frac{W_1 - W_2}{W}\right) \times 100$$

Significance Lesser the volatile matter content better the quality of coal.

(iii) **Ash-content** W_1 gm of coal sample, free from moisture and volatile matter, is taken in a silica crucible without lid, which is then kept in a muffle furnace at around 750°C for 1 hour. It is then taken out and allowed to cool down in air. After cooling weight of the sample is taken. That sample is again kept in the muffle furnace at 750°C for 30 mins. followed by cooling and weighing. If the two weights are nearly equal there is no need to repeat the process. Suppose, W_2 is the final weight of the sample in gm. The treated coal sample is nothing but the ash or inorganic matter.

$$\% \text{ Ash content} = \frac{W_2}{W} \times 100$$

where, W = Amount of coal sample taken initially and W_1 = weight of that coal sample free from moisture and volatile matter.

Significance Presence of ash increases the storage cost, opposes free flow of air and heat. So, lower the ash content better the quality of coal.

(iv) Fixed carbon content

Fixed carbon content = 100 – %(moisture + volatile matter + ash).

Significance Higher the percentage of carbon, less the amount of volatile matter and other impurities, higher the calorific value and hence better the quality of coal.

Ultimate analysis In ultimate analysis elementwise coal composition is determined.

(a) Carbon content A small quantity of coal sample (W_1 gm) is burnt in presence of enough oxygen. The flue gas, thus obtained, is passed through a tube, filled with known weight of potassium hydroxide, which absorbs carbon-di-oxide from flue gas. The increase in weight of potassium hydroxide is calculated and hence wt % of carbon in coal sample is calculated from the following equation:

$$\text{wt \% of C} = \left(\frac{\Delta W \times 12}{W_1 \times 44}\right) \times 100$$

ΔW = Increase in weight of KOH-bed = Weight of CO_2

44 gms of CO_2 contains 12 gms of C.

(ΔW) gms of CO_2 contains $\dfrac{12\,\Delta W}{44}$ gms of C

W_1 = weight of coal sample taken.

Significance Higher the % of carbon, better the quality of coal.

(b) Hydrogen content After absorbing carbon-di-oxide, the flue gas is then passed through a tube, containing dry calcium chloride, which absorbs water vapour of the flue gas. Again increase in weight is determined and hydrogen content can be evaluated by using the following formula.

$$\text{wt \% of H} = \left(\frac{\Delta W \times 2}{W_1 \times 18}\right) \times 100$$

ΔW = Increase in weight of $CaCl_2$-bed = weight of water vapour

W_1 = weight of coal sample taken.

18 gms of H_2O contains 2 gms of H (Hydrogen)

ΔW gm of H_2O contains $\dfrac{2 \times \Delta W}{18}$ gm of H.

Significance Hydrogen-content is desired to be high in a coal sample since hydrogen has high colorific value.

(c) Nitrogen content A certain amount of coal (say W_1 gm) is powdered and heated with a mixture of concentrated sulfuric acid (H_2SO_4) and potassium sulphate (K_2SO_4) in a long-necked flask. The solution is allowed to stand until a clear solution is obtained. That clear solution is then treated with excess potassium hydroxide (KOH). The ammonia is liberated from the top and allowed to get absorbed in a known volume of a standard acid solution. Volume of acid consumed by ammonia can be easily evaluated by back-titration with standard sodium hydroxide (NaOH) solution.

V = volume of acid consumed by ammonia in ml.

N = Strength of acid in Normality (gm-eq./litre).

$N \times \left(\dfrac{V}{1000}\right)$ = Amount of acid in gm-equivalent in volume V ml.

≡ Amount of ammonium hydroxide (NH_4OH) in gm-equivalent or gm.mole

$= \dfrac{N \times V \times 14}{1000}$ gm of nitrogen, since 1 gm-equivalent or 1 gm-mole of NH_4OH contains 14 gm of nitrogen.

W_1 = Amount of coal sample taken in gm.

$$\text{wt \% of } N_2 = \frac{N \times V \times 14}{W_1 \times 1000} \times 100 = \frac{N \times V \times 1.4}{W_1}$$

Significance Nitrogen has no calorific value and hence nitrogen-content of coal is desired to be low.

(d) Sulphur content A known weight of coal sample (say, W_1 gm) is heated under high temperature and pressure (as is performed in Bomb Calorimeter). All sulphur gets converted to H_2SO_4, which belongs to washings and collected after heat treatment. The washings is then treated with barium chloride ($BaCl_2$) solution, which reacts with H_2SO_4 to form barium sulphate ($BaSO_4$). $BaSO_4$ is precipitated out of the solution. The precipitate is filtered, washed and dried. Finally the dried sample is weighed.

Molecular weight of $BaSO_4$ = 233, Molecular weight of S = 32

W_1 = weight of coal sample taken. W_2 = weight of $BaSO_4$

$$\text{wt \% of } S = \frac{W_2}{W_1} \times \frac{32}{233} \times 100 = \frac{W_2 \times 32 \times 100}{W_1 \times 233}$$

since, 233 gm $BaSO_4$ contains 32 gm S,

so, W_2 gm $BaSO_4$ contains $\left[\left(\dfrac{32}{233}\right) \times W_2\right]$ gm S

Significance Sulphur-content increases the calorific value of coal but deleterous effects of sulphur oxides (SO_2 and SO_3) suggest that sulphur-content of coal be low.

(e) Oxygen content

Weight % of O = 100 – weight % (C + H + N + S + ash)

Significance Presence of oxygen decreases the calorific value as well as cooking power of coal.1% increase in oxygen-content decreases the calorific value by about 1.7%.

9.4.2 Carbonization of coal

If a coal sample is heated at high temperature in absence of air large quantity of volatile matter is removed, creating pores in the coal. The process is known as carbonization and the product of carbonization is called coke.

Differences between coal and coke

Coal	*Coke*
(1) It burns with heavy smoke.	(1) It burns easily with little smoke.
(2) It has low calorific value.	(2) It has high calorific value due to negligible amount of volatile matter and ash.
(3) It possesses no porosity but has low mechanical strength.	(3) It possesses huge porosity and has high mechanical strength.
(4) It burns with long flame which is desired for reverberatory furnaces.	(4) It burns with short flame which is particularly preferred in blast furnaces.

Carbonization of coal can be carried out either at low temperature or at high temperature and the processes are known as low temperature carbonization (LTC) and high temperature carbonization (HTC) respectively.

(i) Low temperature carbonization Characteristic features of LTC are:

(a) Carbonization temperature is about 500 – 700°C.
(b) Volatile matter content of LTC-Coke is about 5 – 15%.
(c) As LTC-Coke is not mechanically strong it is particularly useful for domestic purpose.
(d) Calorific value of LTC-Coke varies from 6500 to 9500 kcal/m^3
(e) LTC-Coke is soft and burns without smoke.

(ii) High temperature carbonization Characteristic features of HTC are:

(a) Carbonization temperature is about 900 – 1200°C.
(b) Maximum volatile matter content of HTC-Coke is 3%.
(c) As HTC-Coke is mechanically strong it is particularly useful for metallurgical purposes.
(d) Calorific value of HTC-Coke varies from 5400 to 6000 kcal/m^3.
(e) HTC-Coke is hard and burns with a shooty flame.

Byproducts of carbonization After carbonization huge gas is released from the oven as a byproduct, which is known as coke oven gas. The following are the main constituents of coke oven gas: Ammonia, hydrogen sulphide, naphthalene, benzene, tar and moisture.

9.5 Liquid Fuel

Huge quantities of viscous oil, obtained from the earth's crust, is known as petroleum or crude oil.

Elemental composition of crude petroleum

Carbon	—	80 – 87%
Hydrogen	—	11 – 15%
Sulphur	—	around 3.0%
Nitrogen and Oxygen	—	Rest

9.5.1 Classification

Crude petroleum can be classified into three broad categories:

(i) **Paraffin-based crude oil** The main composition of this crude oil is paraffins, ranging from CH_4 to $C_{35}H_{72}$. The higher hydrocarbons like $C_{18}H_{38}$ are known as waxes, which appear as semi-solid masses.

(ii) **Asphalt-based crude oil** The main composition of this oil is naphthalene and naphthalene-derivatives.

(iii) **Mixed crude oil** Crude oil, containing competitive amount of paraffins and naphthalenes, is known as mixed crude oil. The mixed crude oil is usually enriched with waxes.

9.5.2 Fractional distillation of crude petroleum

As we have seen earlier, crude petroleum contains several fractions of different boiling point range. In order to separate these fractions crude petroleum is passed into a fractionating column, where different temperatures have been set at different stages, such that different fractions of crude oil are evaporated at different stages and hence separated. The following fractions are separated:

(a) **Low boiling fraction** Hydrocarbons, containing a maximum of 4 carbon atoms, is separated below 30°C and is known as liquefied petroleum gas or LPG. The main composition are liquid propane and liquid butane. It is mainly used as a domestic fuel.

(b) **Petroleum ether** That fraction of petroleum, containing C_5 to C_7 hydrocarbons and boiling point range 30°C – 70°C, is known as petroleum ether. It is mainly used as a solvent.

(c) **Gasoline/Petrol/Motor spirit** That fraction of petroleum, containing C_5 to C_9 hydrocarbons and boiling point range 40°C – 120°C, is known as gasoline. Carbon and hydrogen contents of gasoline are 84% and 15% respectively. The rest is nitrogen, sulphur and oxygen. It has a high calorific value (around 11,250 kcal/kg.) and is highly volatile and inflammable. It is used as a primary fuel in internal combustion engines. Pentane and octane are the two main constituents of gasoline. Its specific gravity varies from 0.70 to 0.74.

(d) **Naphtha/Solvent spirit** That fraction of petroleum, containing C_9 to C_{10} hydrocarbons and boiling point range 120°C – 180°C, is known as naphtha or solvent spirit. It is mainly used as a solvent of organic matters and also as a cleaning agent.

(e) **Kerosene oil** That fraction of petroleum, containing C_{10} to C_{16} hydrocarbons and boiling point range 180°C – 250°C, is known as kerosene oil. It has a calorific value of approximately 11,000 kcal/kg. It is namely used as a domestic fuel such as in kerosene stoves, lanterns etc. Its specific gravity varies from 0.80 to 0.85.

(f) **Diesel oil/Fuel oil/Gas oil** That fraction of petroleum, containing C_{12} to C_{18} hydrocarbons and boiling point range 250°C – 320°C, is known as diesel oil. Its specific gravity varies from 0.86 to 0.95. It has a calorific value of around 11,000 kcal/kg. It is mainly used as a diesel engine fuel.

(g) **Heavy oil** That fraction of petroleum, containing C_{17} to C_{30} hydrocarbons and boiling point range 320°C – 400°C, is known as heavy oil. It may be further fractionated into four constituents:

(i) The lightest fraction is the lubricating oil.
(ii) Next higher fraction is the petroleum jelly or Vaseline.
(iii) Next higher fraction is the grease.
(iv) The heaviest fraction is the paraffin wax.

These are mainly used as lubricants, in the manufacture of boot polishes, candles, etc.

(h) **Residue** After completion of fractionation of petroleum, heavy sludge, collected from the bottom, is known as residue. It has very high boiling point (above 400°C). Long chain hydrocarbon with carbon content more than C_{30}, is the main constituent of this residue. Residue may be of two types:

(i) Asphalt It is mainly used as a water proofing component and road making component.

(ii) Petroleum coke It is mainly used as a fuel.

9.5.3 Gasoline/Petrol

Gasoline can be classified into three categories on the basis of origin:
(a) Natural, (b) Semi-synthetic, (c) Synthetic.

(a) **Natural** Gasoline, obtained by fractional distillation of crude petroleum, is known as *natural gasoline or straight-run gasoline*. It has some demerits:

(i) Yield is very low. Only 20% of the crude petroleum is the natural gasoline.
(ii) It contains some undesirable straight chain unsaturated hydrocarbons and sulphur compounds.
(iii) Storage stability and anti-knocking characteristics are not good enough for practical purposes.

To improve the quality of straight run gasoline, refining and reforming processes are carried out.

Refining: Step-I In this process, straight run gasoline is treated with an alkaline solution of sodium plumbite with a small amount of sulphur. This reagent converts sulphur compounds of gasoline into di-sulphides, which is then extracted by a suitable solvent. The preocess is known as Sweetening process or Doctor's process. The following reactions may occur:

$$2RSH + Na_2PbO_2 \rightarrow Pb(SR)_2 + 2NaOH$$

$$Pb(SR)_2 + S \rightarrow PbS\downarrow + R{-}S{-}S{-}R$$

Lead sulphide (PbS) is removed by filtration and disulphide is removed by solvent extraction technique.

Step-II The treated gasoline, thus obtained, is passed through a porous bed of fuller's earth, which absorbs colouring matters and olefins.

Step-III Refined gasoline is blended with a pinch of inhibitors (to improve storage stability) and heavier fraction of petroleum (to meet the desired specification).

Reforming Straight run gasoline contains a considerable amount of alkanes and cyclic alkanes, which reduces anti-knock characteristics of gasoline. To improve the anti-knock characteristics, gasoline is subjected to reforming process where *n*-alkanes are converted to aromatics, lower alkanes and branched alkanes. The reforming process is usually carried out in presence of a catalyst (platinum supported alumina). When gasoline is allowed to pass through the catalyst bed at around 500°C and 40 atmospheric pressure, the following reactions are believed to take place:

Degradation:

$$CH_3(CH_2)_8CH_3 \xrightarrow{\text{calalyst}} CH_3(CH_2)_5CH_3$$

Isomerisation:

$$\underset{n\text{-hexane}}{CH_3(CH_2)_4CH_3} \xrightarrow{\text{calalyst}} CH_3{-}\underset{\underset{CH_3}{|}}{CH}{-}CH_2{-}CH_2{-}CH_3$$

2-methyl pentane

Aromatization:

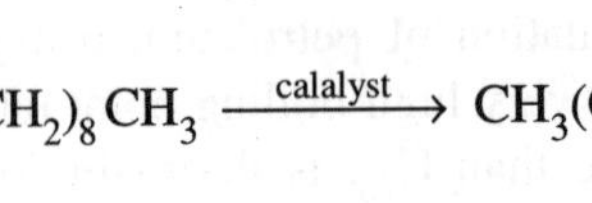

Cyclohexane benzene

$CH_3(CH_2)_5CH_3$
n-heptane

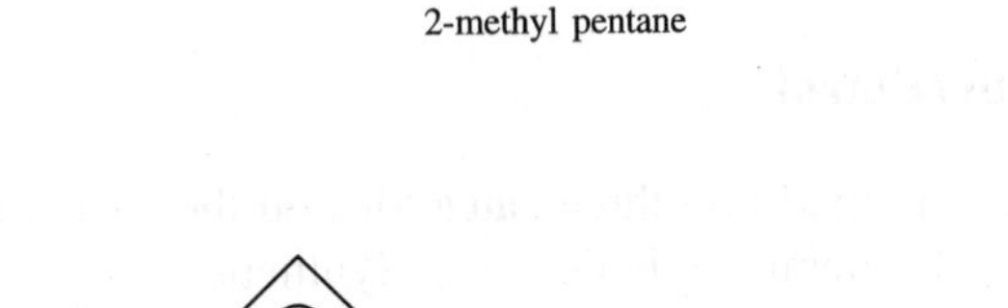

methyl cyclohexane toluene

(b) Semi-synthetic During fractionation of crude petroleum large quantity of heavier fraction, when subjected to high temperature and pressure (around 500°C and 100 atmospheric pressure), gets converted to lower alkanes and hydrogen. This decomposition of higher hydrocarbons to lower ones is known as cracking. The cracked products are seperated by fractional distillation method. Large quantity of gasoline is obtained by cracking process. This gasoline is known as semi-synthetic petrol.

Cracking process can also be carried out in presence of catalyst [aluminium silicate $Al_2(SiO_3)_3$ or alumina Al_2O_3]. In that case very low pressure (1 – 5 atm.) is enough to execute the cracking process. This semi-synthetic petrol has some merits over natural petrol:

(i) Yield of gasoline is very high, (ii) It possesses better anti-knock charactreristics, (iii) It contains a very little amount of sulphur bearing compounds.

(c) Synthetic petrol Semi-synthetic petrol is obtained by cracking of higher fractions of crude petroleum, whereas synthetic petrol is obtained by polymerisation of gases, a by-product of craking. Polymerisation is usually carried out in presence of a catalyst (e.g., phosphoric acid) at around 200°C.

$$2CH_3\text{—}CH{=}CH_2 \xrightarrow{\text{calalyst}} CH_2{=}CH\text{—}CH_2\text{—}\underset{\displaystyle CH_3}{\underset{|}{CH}}\text{—}CH_3$$

$$CH_3\text{—}CH{=}CH_2 + CH_3CH_2\text{—}CH = CH_2 \xrightarrow{\text{calalyst}}$$

$$CH_2{=}CH\text{—}CH_2\text{—}CH_2\text{—}\underset{\displaystyle CH_3}{\underset{|}{CH}}\text{—}CH_3$$

Constituent of synthetic petrol

Synthetic petrol is known as *polymer gasoline*.

9.5.4 Knocking in petrol engine

Petrol tank contains liquid fuel/air mixture. After ignition liquid fuel readily forms vapour and the resultant volume of fuel vapour/air mixture is much higher than that of liquid fuel/air mixture. This sudden expansion of volume drives the gas mixture towards the piston which is then moving forward. The design is such that the cylinder volume changes to execute the process smoothly. The ratio of cylinder volume at the end of down stroke to the volume at the end of up stroke of the piston is known as compression ratio. Higher the compression ratio higher the efficiency of engine is. The value of compression ratio primarily depends on the constituents of gasoline.

For an ideal fuel after ignition the flame proceeds smoothly towards the other end. As gasoline is highly inflammable, after ignition of fuel at one end last part of fuel may burn spontaneously. The phenomenon is known as *knocking*, which may cause severe damage to the cylinder.

Another problem of gasoline is *preignition*. Temperature of fuel tank usually remains at temperature higher than room temperature. So there is a possibility of spontaneous ignition even before sparking. The phenomenon is known as *preignition*. Sparking is necessary to ignite the fuel when it is desired. The preignition poses a threat on gasoline tankers. The parameter, indicating the ignition quality of a gasoline, is called *Octane number*. Knocking tendency of a fuel exclusively depends on the chemical structure of fuel constituents. It has been established that knocking tendency is maximum for straight chain alkanes followed by olefins, naphthalenes and aromatics. Aromatics have the least knocking tendency. For practical purpose *n*-heptane (straight chain alkane) and iso-octane (branched chain alkane) are taken as standards. The *n*-heptane knocks very quickly and hence its anti-knock value is taken as *zero*. On the other hand iso-octane knocks with much difficulty and its anti-knock value is taken as *hundred*.

$$CH_3—(CH_2)_5—CH_3$$

n-heptane
Octane Number = 0

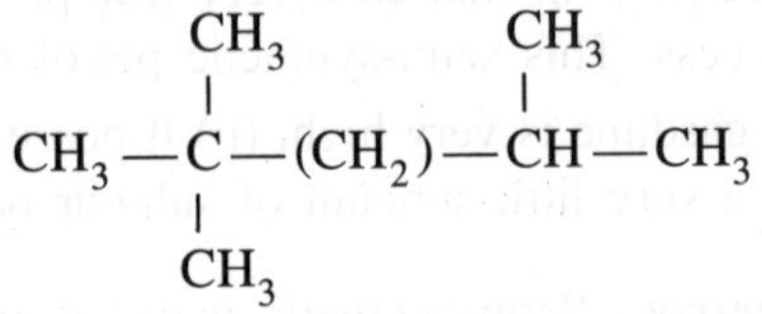

2:2:4-trimethyl pentane
(iso-octane)
Octane Number = 100

Octane number or rating Octane number is the % of iso-octane present in a mixture of iso-octane and *n*-heptane. If the combustion characteristics of the mixture matches with that particular gasoline, the latter is said to have the same octane number. Thus a fuel of octane number 70 has the same combustion characteristics with that of the mixture containing 70% iso-octane and 30% *n*-heptane.

In many cases it has been observed that the fuel has low octane number. To improve the anti-knocking characteristics various chemical reagents are added to the fuel. The common reagent used for this purpose is tetraethyl lead with a trace amount of ethylene dibromide or 1, 2-dibromo ethane.

Tetraethyl lead or TEL ≡ $(C_2H_5)_4Pb$. Ethylene dibromide ≡ $CH_2Br\text{-}CH_2Br$.

Tetraethyl telluride $[(C_2H_5)_4Te]$ is also used istead of TEL.

Flame propagates through chain oxidation process by free radical mechanism. TEL produces sufficient lead oxide particles which inhibit the chain oxidation process and hence the knocking characteristics of fuel is sufficiently reduced. However, presence of sulphur compounds in fuel reduces the effectiveness of TEL. Deposition of lead oxide in engine is not desired as it pollutes the environment significantly. To remove the lead oxide particles ethylene dibromide is added. Ethylene dibromide reacts with lead oxide to form lead halide, which is carried away to atmosphere by exhaust gases.

$$PbO + CH_2Br—CH_2Br \rightarrow PbBr_2 + CH_2{=}CH_2$$

Dosage of TEL

(i) In motor fuel the required amount of TEL is 0.5 ml per litre of petrol.
(ii) In aviation fuel the required amount of TEL is 1.5 ml per litre of petrol.

TEL is extremely poisonous and use of TEL threatens the environment with the release of poisonous lead in the environment.

9.5.5 Knocking in diesel fuel

Diesel fuel contains longer chain hydrocarbons than petrol or gasoline. The ignition characteristics of diesel fuel is far worse than that of petrol. In the latter case problem arises due to spontaneous combustion of the last part of fuel, i.e., knocking but the same knocking, in the case of diesel fuel, arises due to *delay in combustion* in the first part of fuel. Knocking in the diesel fuel is characterized by *cetane number.*

Cetane number It is the % of *n*-hexadecane present in a mixture of *n*-hexadecane and α-methyl naphthalene.

$CH_3—(CH_2)_{14}—CH_3$ [Cetane Number = 100]

(*n*-hexadecane)

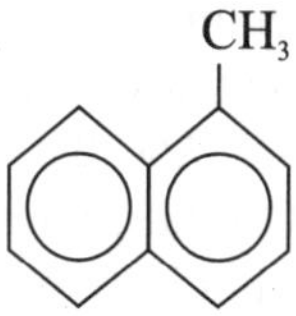

[Cetane Number = 0]

Time lag between sparking and ignition is the least in the case of *n*-hexadecane and hence its cetane number is taken as 100 whereas the time lag is long in the case of α-methyl naphthalene and hence the corresponding cetane number is taken as zero.

Decreasing order of *ignition quality* of hydrocarbon constituents of a *diesel fuel*:

n-alkanes > naphthalenes > alkenes > branched alkenes > aromatics.

Decreasing order of knocking quality of hydrocarbon constituents of a gasoline.

Straight-chain paraffins > branched-chain paraffins >
olefins (alkanes) > cycloparaffins (naphthalenes)> aromatics

It is interesting to note that napthalenes, which have *good* ignition quality in diesel fuel, have *poor* knocking quality in gasoline. Similarly, branched-chain paraffins or isoparaffins have *good* knocking quality in gasoline but have *poor* ignition quality in diesel fuel.

High speed/Medium speed/Low speed diesel Speed of a diesel fuel is classified on the basis of time lag in ignition. The time lag is very short in the cases of *high speed* diesel oil (cetane number lies between 45 and 60). In the cases of *medium speed* diesel oil and *low speed* diesel oil the corresponding cetane numbers are around 35 and 25 respectively.

9.6 Aviation fuel

The fuel, which is mainly used in aero planes, is called aviation fuel. There are two types of aviation fuel

(a) **Aviation gasoline** It is basically a mixture of different octanes such that the mixture has octane number close to 100. It is mainly used in passenger aero planes and racing cars.

(b) Jet fuel It is mainly used in jet aero planes, which are moving with velocity, more than 3 times the sonic velocity. The temperature of fuel tank is very high, as high as 325°C. So the jet fuel must have sufficient thermal stability. Usually, jet fuel is a mixture of gasoline and kerosene such that aromatic content lies within 25%.

9.7 Biodiesel

Biodiesel is a mixture of long-chain alkyl esters, preferably methyl, ethyl or propyl esters. It is produced through reaction between lipids (vegetable oil or animal fat) and alcohol.

The important characteristics of biodiesel are pour point and cloud point as discussed below:

Pour point and Cloud point As the temperature of biodiesel decreases fluidity of biodiesel molecules decreases. When the temperature drops below a certain temperature, molecules aggregate together forming crystals and fluidity of biodiesel is lost. That temperature is called ***pour point*** of biodiesel. Further decrease in temperature encourages crystal growth. When crystal diameter becomes greater than quarter of the wavelengths of visible light, i.e., $\lambda/4$, the biodiesel appears cloudy and the temperature is called ***cloud point*** of the biodiesel. Crystal growth continues with decrease in temperature and ultimately the fuel does not pass through 45μ filter. The temperature at which this happens is called ***cold filter plugging point***.

The main disadvantage of biodiesel is its moisture contamination. As alcohol is hygroscopic in nature, the biodiesel may contain small amount of moisture, which may create several problems

(i) Moisture reduces the calorific value of the fuel.
(ii) Moisture causes corrosion in fuel tank as well as pitting in the piston.
(iii) Moisture encourages microbe generation and growth, leading to damage of fuel tank.

Manufacturing of biodiesel

A variety of oils can be used to produce biodiesel. These include:

1. Virgin oil feedstocksoyabean oils, jatropha oils, sunflower oils, rapeseed oils, coconut oils, etc.
2. Animal fats can also be used.

Mechanism of trans-esterification

Trans-esterification process is used to reduce the viscosity of triglycerides. Trans-esterification of triglycerides produces fatty acid alkyl esters and glycerol. The process is carried out in a reaction vessel. Continuous stirring is needed to expedite the reaction. The heavier fraction, i.e., glycerol layer settles down at the bottom of the reaction vessel. It is collected from the bottom of the reaction vessel as by-product. Diglycerides and monoglycerides are the intermediates formed in this process. The mechanism of trans-esterification is described below:

Triglycerides + R′OH → Diglycerides + RCOOR′
Diglycerides + R′OH → Monoglycerides + RCOOR′
Monoglycerides + R′OH → Glycerol + RCOOR′

RCOOR′ is the constituent of biodiesel.

When methanol is used for trans-esterification, R is CH_3 and the process is called methanolysis. Methanolysis of jatropha oil requires three moles of methanol as shown below:

$$\underset{\text{Triglycerides}}{\begin{array}{l} CH_2COOR_1 \\ | \\ CHCOOR_1 \\ | \\ CH_2COOR_1 \end{array}} + \underset{\text{Methanol}}{3\ CH_3OH} \longrightarrow \underset{\text{Glycerol}}{\begin{array}{l} CH_2OH \\ | \\ CHOH \\ | \\ CH_2OH \end{array}} + \underset{\text{Methyl esters}}{3\ R_1COOCH_3}$$

R_1COOCH_3 is the constituent of biodiesel. Biodisel is the lighter fraction and collected from the top of the reaction vessel.

Table 9.1: Fatty acids composition of jatropha oil

Fatty acid contribution of triglycerides	*Composition (wt. %)*
Palmitic acid	18.22
Stearic acid	5.14
Oleic acid	28.46
Linoleic acid	48.18
Average Molecular Weight	874

Advantages of biodiesel

(i) Biodisel is non-flammable, non-toxic and non-explosive.
(ii) It has higher flash point (150°C) compared to petroleum diesel (64°C).
(iii) It is biodegradable and eco-friendly.
(iv) Sulphur content of biodiesel is close to zero.
(v) It reduces the emission of CO_x since the carbon in biodiesel emission is recycled from the carbon in the atmosphere.
(vi) Lubricity of biodiesel is significantly high, ensuring longer service life of engine.
(vii) Biodiesel can be blended with petroleum diesel in any proportion.
In many occasions biodiesel is mixed with petroleum diesel and the mixture is used as fuel. The amount of biodiesel in a given fuel mix is expressed as B factor as described below:
(a) Fuel mix containing 100% biodiesel is referred to as B100
(b) Fuel mix containing 30% biodiesel is referred to as B30.
(c) Fuel mix containing 10% biodiesel is referred to as B10.
(d) Fuel mix containing 5% biodiesel is referred to as B5.

The fuel mix, having high percentage of biodiesel, is considered to be eco-friendly.

9.8 Gaseous fuel

Gaseous fuel can provide high calorific value which can be used in several purposes. On the basis of origin gaseous fuels can be classified into following categories:

(a) Natural gas, (b) Coal gas, (c) Oil gas, (d) Producer gas, (e) Water gas (or blue gas).

(a) Natural gas Gas associated with crude petroleum is known as natural gas. The gas is first treated to remove moisture, hydrogen sulphide, nitrogen, dust etc. since the above mentioned components do not add any calorific value to natural gas.

Composition:

CH_4 — Around 80 to 90%
C_2H_6 — Around 8 to 10%
H_2 — Around 3%
$(CO + CO_2)$ — The rest

Calorific value: 12,000 to 14,000 kcal/m^3.

Uses It is (i) used as domestic fuel and (ii) used as raw materials for the manufacture of chemicals and carbon black.

(b) Coal gas During carbonization of coal large amount of gas is produced. The gas contains solvent vapours, hydrogen sulphide etc. Hydrogen sulphide is removed by moist ferric oxide and solvent vapour is removed by creosote oil. The resultant gas is known as coal gas.

Composition:

CH_4 — 32%, C_2H_4 — 3%
C_2H_2 — 2% H_2 — 40%
$(CO + CO_2 + N_2)$ — The rest

Calorific value: 4900 kcal/m^3.

(c) Oil gas When kerosene oil is subjected to cracking a mixture of gas and liquid tar products are obtained. The gas, thus collected, is known as oil gas. It is stored over water.

Composition:

CH_4 — Average 30%
H_2 — Average 50%
CO — Average 10%
CO_2 — Average 3%

Calorific value: 4500 to 5400 kcal/m^3.
It is used as laboratory gas.

(d) Producer gas If a stream of air-steam mixture is allowed to pass over a red hot coal or coke bed (temperature around 1100°C), a mixture of combustible and non-combustible gases is obtained. This mixture of gases is called producer gas.

Composition:

Combustible gases: CO — Around 28%, H_2 — Around 10%
Non-combustible gases: N_2 — Around 54%, CO_2 — Around 3%

Calorific value: 1300 kcal/m^3

Producer gas of high calorific value can be obtained by carrying out the reaction using oxygen and super heated steam on coal/coke bed at high pressure (about 30 atm.). Under this reaction condition some carbon monoxide (CO) and carbon-di-oxide (CO_2) are converted to methane (CH_4). Calorific value of the resultant gas mixture is calculated about 4050 kcal/m^3.

(e) Water gas (or blue gas) If steam with a little amount of air is allowed to pass over red hot coal/coke bed (temperature around 1000°C), a mixture of carbon monoxide and hydrogen with a little amount of non-combustible gases (carbon-di-oxide and nitrogen) is obtained according to the following reaction:

$$C + H_2O \rightarrow CO + H_2 \quad \text{(endothermic reaction)}$$
$$C + O_2 \rightarrow CO_2 \quad \text{(exothermic reaction)}$$
$$2C + O_2 \rightarrow 2CO \quad \text{(exothermic reaction)}$$

The resultant gas mixture is known as water gas or blue gas.

Composition: H_2 — 51%, CO — 41%, N_2 — 4%, CO_2 — 4%.

Calorific value: 2800 kcal /m^3.

Semi-water gas If air-steam mixture is taken in such a proportion that heat absorbed in endothermic reaction is balanced by heat evolved in exothermic reaction, the resultant gas mixture is called semi-water gas. The contains around 53% nitrogen and around 3% hydrocarbon.

Composition: H_2 — 12%, CO — 30%, N_2 — 53%,
CO_2 — 2%, CH_4 — 3%,

Calorific value: 1700 kcal/m^3.

Carburetted water gas If some gaseous hydrocarbons are mixed with water gas the resultant gas mixture is called carburetted water gas.

Composition: H_2 — 35%, N_2 + CO_2 — 5%
CO — 25%, Hydrocarbons — 35%
(saturated and unsaturated)

Calorific value: 4500 kcal/m^3.

9.9 Combustion of fuel

The main composition of any fuel is carbon and hydrogen. It also contains a trace amount of sulphur. During combustion of fuel by air, carbon is oxidized to carbon-di-oxide or a mixture of carbon-di-oxide and carbon monoxide if the air supply is inadequate, whereas hydrogen is oxidized to water and sulphur is oxidized to sulphur-di-oxide. Measurement of CO_2-content, CO-content and O_2-content in the outgoing gas-mixture is usually carried out in *Orsat apparatus* and the process is known as *flue gas analysis*. The Orsat apparatus mainly contains three tubes T_1, T_2 and T_3, filled with potassium hydroxide solution, alkaline pyrogallic acid solution and ammoniacal cuprous chloride solution respectively. The flue gas mixture is allowed to pass through the tubes T_1, T_2 and T_3 consecutively such that all carbon-di-oxide is absorbed in the tube T_1, all oxygen is absorbed in the tube T_2 and all carbon monoxide is absorbed in the tube T_3. Hence CO_2-content, CO-content and O_2-cotent in the flue gas can easily be determined.

In air oxygen content is 23% (by wt.) or 21% (by volume or by mole) and nitrogen is assumed to occupy the rest. As nitrogen is an inert gas the amount of nitrogen remains intact before and after combustion. Thus if nitrogen content in the flue gas mixture is known amount of air supplied to the fuel can easily be determined.

For example 5 m^3 of N_2 gas $\equiv \left(\frac{5}{0.79}\right) = 6.33\ m^3$ of air and the amount of O_2-gas supplied to the fuel for combustion = $6.33 \times 0.21 = 1.33\ m^3$. If the CO_2-content in the flue gas mixture is known, amount of carbon present in the fuel can easily be determined since we know that: 44 gms of CO_2 contains 12 gms of carbon.

If the water vapour content of the flue gas mixture is known amount of hydrogen present in the fuel can be determined since we know that 18 gms of H_2O contains 2 gms of hydrogen.

If the SO_2-content of the flue gas mixture is known the amount of sulphur present in the fuel can be determined since we know that 64 gms of SO_2 contains 32 gms of sulphur.

Weight-basis calculations can easily be converted to volume-basis calculations by using the following conversions.

1 gm mole of any gas occupies 22.4 litres at N.T.P. i.e. at 273°K and 1 atm. pressure.

1 lb-mole of any gas occupies 359 ft^3 at N.T.P.

$$\text{Molecular weight of air} = 0.79 \times 28 + 0.21 \times 32 = 28.84$$

$$1\text{ lb of air occupies } \frac{359}{28.84} = 12.45\ ft^3 \text{ at N.T.P.}$$

Example 9.1 A coal sample contains C = 62.4%, H = 4.1%, O = 6.9%, N = 1.2%, S = 0.8%, moisture = 15% and ash = 9.6%, on weight- basis. The coal sample was burnt in such a way that the flue gas contains:

CO_2 — 12.9%, CO — 0.2%, O_2 — 6.1%, N_2 — 80.8% (on weight basis).

Calculate (i) the weight of air theoretically required per kg of coal, (ii) the weight of dry flue gas obtained per kg of coal, (iii) the weight of air actually used.

Sol. 100 kg of coal contains:

C — 62.4 kg
H — 4.1 kg
O — 6.9 kg
N — 1.2 kg
S — 0.8 kg

$$\text{So theoretical air required is} = \frac{\left[\frac{32}{12}C + 8\left(H - \frac{O}{8}\right) + S\right]}{0.23} \text{ kg per 100 kg of coal}$$

since oxygen content in air is 23% (by wt.).

Example 9.2 A sample of liquid fuel was fired in a bomb calorimeter for the determination of its calorific value and the following observations were made:

Weight of fuel burnt = 0.78 gm

Water equivalent of the calorimeter = 680 gm

Volume of water in calorimeter = 2.5 lit.

Initial temperature of water = 36.4°C.

Final temperature of water = 39°C.

Calculate the calorific value of fuel in kcal/kg

Sol.

$$\text{Calorific value} = \frac{(W_{H_2O} + W_E)(T_2 - T_1)}{F}$$

$$(T_2 - T_1) = (39 - 36.4) = 2.6°C$$

$$W_{H_2O} = 2500 \text{ cc} \times 1 \text{ gm/cc} = 2500 \text{ gm}$$

$$W_E = 680 \text{ gm. } F = 0.78 \text{ gm}$$

$$GCV = \frac{(2500 + 680) \times 2.6}{0.78} = 10{,}600 \text{ cal/gm}$$

Example 9.3 Calculate the volume of air under normal condition of temperature and pressure theoretically required for the combustion of one kg of the coal having following composition:

Moisture – 2%	Oxygen – 8%
Carbon – 60%	Nitrogen – 1%
Hydrogen – 5%	Ash – 24%

What will be the percentage composition of the dry products of combustion if 50% excess air is used?

Sol.

Moisture – 2% = 0.02 kg

Carbon – 60% = 0.60 kg

Hydrogen – 5% = 0.05 kg

Oxygen – 8% = 0.08 kg

Nitrogen – 1% = 0.01 kg

Ash – 24% = 0.24 kg

Combustion equations are:

$$C + O_2 = CO_2 \quad \text{and} \quad 2H_2 + O_2 = 2H_2O$$

So, 12 kg carbon requires 32 kg oxygen
0.6 kg carbon requires 1.6 kg oxygen.

Similarly, 0.05 kg hydrogen requires $\frac{32}{4} \times 0.05 = 0.4$ kg of oxygen.

Total oxygen required = (1.6 + 0.4) kg = 2.0 kg

Oxygen present in the coal = 0.08 kg

Net oxygen required = (2 – 0.08) kg = 1.92 kg

Assuming ideal behaviour of gases we have:

$$PV = nRT \quad \text{and} \quad n = \frac{1.92}{32} = 0.06 \text{ kg-mole.}$$

Volume of oxygen at N.T.P. = 22.4 litre/g-mole

= 22.4 × 1000 litre/kg-mole

= 22400 litre/kg-mole.

So, volume occupied by 0.06 kg-mole of oxygen at NTP = 0.06 × 22,400 lit.

= 1344 lit.

Theoretical volume of oxygen required = 1344 lit.

$$\text{Theoretical volume of air} = 1344 \times \frac{100}{21} = 6400 \text{ lit.}$$

Air – supply is 50% in excess.

So, oxygen supplied = 1.92 × 1.5 = 2.88 kg

Excess oxygen = 2.88 – 1.92 = 0.96 kg

Total nitrogen = weight of air – weight of O_2

$$\text{Weight of air} = \frac{2.88 \times 100}{23} = 12.52 \text{ kg}$$

Weight of N_2 = (12.52 – 2.88) = 9.64 kg

Nitrogen present in the coal = 0.01 kg

Total weight of N_2 = (9.64 + 0.01) = 9.65 kg

12 kg carbon produces 44 kg CO_2

$$0.6 \text{ kg carbon produces } \frac{44 \times 0.6}{12} = 2.2 \text{ kg } CO_2$$

Constituents of dry flue gas:

Gas		*Wt. in (kg)*		*M.W.*		*No. of moles.*		*Mole (%) or volume (%)*
CO_2	–	2.20	–	44	–	0.05	–	11.9
O_2	–	0.96	–	32	–	0.03	–	7.14
N_2	–	9.65	–	23	–	0.34	–	80.96
						Total = 0.42		

Problem 9.1 Analysis of a coal sample gave the following result:

C = 69%, H = 5.4%, S = 1.0%,

N = 1.6% O = 15%, Ash = 8%

Calculate the theoretical quantity of air for the complete combustion of 1.0 kg of this coal.

Problem 9.2 The percentage composition of coal is:

C = 90.0%, H = 3.5%, O = 3.0%,

N = 1.0%, S = 0.5%

Calculate the theoretical volume of oxygen and air required at 27°C and 1 atm. for the combustion of 100 kg of coal.

Example 9.4 The percentage composition by weight of a sample of coal is as follows.

C = 81.0%, H = 5.0%, O = 8.5%,

S = 1.0%, N = 1.0%, Ash = 3.5%

(a) Find the amount of air required for the complete combustion of 1 kg of this coal.
(b) Calculate the Gross and Net calorific values of this coal.

Given: C = 8060 kcal/kg

H = 34000 kcal/kg

S = 22000 kcal/kg

Sol. (a) weight of constituents/kg of coal

C = 81% = 0.81 kg

H = 5% = 0.05 kg

O = 8.5% = 0.085 kg

S = 1.0% = 0.01 kg

N = 1.0% = 0.01 kg

Ash = 3.5% = 0.035 kg

Oxygen required for combustion of carbon in the coal

$$= \frac{32}{12} \times 0.81 = 2.16 \text{ kg}$$

Oxygen required for combustion of hydrogen in the coal

$$= \frac{32}{4} \times 0.05 = 0.40 \text{ kg}$$

Oxygen required for combustion of sulphur in the coal

$$= \frac{32}{32} \times 0.01 = 0.01 \text{ kg}$$

Total oxygen required = 2.16 + 0.40 + 0.01 = 2.57 kg
Oxygen present in the coal = 0.085 kg
Required oxygen = (2.57 – 0.085) = 2.485 kg

$$\text{Required air} = 2.485 \times \frac{100}{23} = 10.8 \text{ kg}$$

(b) Calorific value:

$$\text{HCV} = \left[8080\text{C} + 34500\left(\text{H} - \frac{\text{O}}{8}\right) + 2250\text{S}\right] \text{kcal/kg}$$

$$\text{HCV} = \left[8080 \times 0.81 + 34500\left(0.05 - \frac{0.085}{8}\right) + 2250 \times 0.01\right] \text{kcal/kg}$$

$$= 7925.7 \text{ kcal/kg}$$

Weight of steam formed by the combustion:

$$2H_2 + O_2 = 2H_2O$$

4 kg of hydrogen yields 36 kg of steam

0.05 kg of hydrogen yields $\frac{36}{4} \times 0.05$ kg of steam = 0.45 kg of steam

Latent heat of steam at 100°C = 540 kcal/kg

NCV = HCV – 0.45 × 540 = (7925.7 – 243) = 7682.7 kcal/kg.

Problem 9.3 The percentage composition of a sample of anthracite coal is:

C = 90%, H = 3.5%,
O = 3%, N = 1%,
S = 0.5%, Ash = Remaining.

If 50% excess air is supplied for the combustion, estimate the volumetric analysis of dry flue gases.

Example 9.5 In a boiler trial analysis, done on the flue gas, shows the following result (by volume).

CO_2 = 10%, CO = 1.5%,
O_2 = 8% N_2 = 80.5%

The coal used has the following analysis by weight:

C = 82%, H_2 = 4.2%, O_2 = 4.8%

The temperature of the flue gases is 310°C and that of air in boiler house is 16°C. Calculate the heat carried away by the excess of air/kg of coal, taking the average specific heat of air as 0.238 kcal/(kg.°C).

Sol.

Flue gas		*Volume* (m³)		*Weight* (kg) *or mole*
CO_2	–	0.1	–	0.1 × 44 = 4.4 kg
CO	–	0.015	–	0.015 × 28 = 0.42 kg
O_2	–	0.08	–	0.08 × 32 = 2.56 kg
N_2	–	0.805	–	0.805 × 28 = 22.54 kg
				Total = 29.92 kg

$$\text{Weight of } CO_2 \text{ per kg of flue gas} = \frac{4.4}{29.92} = 0.147 \text{ kg}$$

$$\text{Weight of CO per kg of flue gas} = \frac{0.42}{29.92} = 0.014 \text{ kg}$$

$$\text{Weight of } O_2 \text{ per kg of flue gas} = \frac{2.56}{29.92} = 0.085 \text{ kg}$$

$$\text{Weight of } N_2 \text{ per kg of flue gas} = \frac{22.54}{29.92} = 0.747 \text{ kg}$$

$$\text{Weight of carbon in } CO_2 = \frac{12}{44} \times 0.147 = 0.04 \text{ kg}$$

$$\text{Weight of carbon in CO} = \frac{12}{28} \times 0.014 = 0.006 \text{ kg}$$

$$\text{Total} = 0.046 \text{ kg}$$

Total weight of carbon/kg of flue gas = 0.046 kg
Weight of carbon per kg of fuel = 0.82 kg

$$\text{Weight of dry flue gas per kg of fuel burnt} = \frac{0.82}{0.046} = 17.82 \text{ kg}$$

Oxygen (O_2) required for the combustion of CO per kg of flue gas

$$= \frac{16}{28} \times 0.014 = 0.008 \text{ kg}$$

Excess O_2 = (0.085 – 0.008) = 0.077 kg since weight of oxygen per kg flue gas = 0.085 kg
Excess oxygen per kg fuel burnt = 0.077 × 17.82 = 1.37 kg

$$\text{Excess air per kg of fuel burnt} = \frac{1.37 \times 100}{23} = 6.0 \text{ kg}$$

Heat carried away by the excess air = m × s × $(t_2 - t_1)$.

m = 6.0 kg, s = 0.238 kcal/kg°C and $(t_2 - t_1)$ = (310 – 16) = 294°C.

m × s × $(t_2 - t_1)$ = 6 × 0.238 × 294 kcal = 419.832 kcal.

Example 9.6 A gaseous fuel contains the following by volume:

CO_2 = 2.6%	CO = 6.0%
O_2 = 0.6%	N_2 = 3.6%
H_2 = 48%	CH_4 = 34%
C_2H_4 = 5.2%	

Calculate the minimum volume of air required for the complete combustion of 100 m^3 of the gas. If 50% excess air is supplied calculate the volume of each of the dry constituents of flue gas.

Sol. The combustion reactions are:

$$CO + \frac{1}{2}O_2 = CO_2$$

$$H_2 + \frac{1}{2}O_2 = H_2O$$

$$CH_4 + 2O_2 = CO_2 + 2H_2O$$

$$C_2H_4 + 3O_2 = 2CO_2 + 2H_2O$$

Volume of O_2 required for the combustion of CO = $\frac{6 \times 1}{2}$ = 3 m^3

[since 1 volume of CO requires $\frac{1}{2}$ volume of O_2].

Volume of O_2 required for the combustion of H_2 = $\frac{48 \times 1}{2}$ = 24.0 m^3

[since 1 volume of H_2 required $\frac{1}{2}$ volume of O_2]

Volume of O_2 required for the combustion of CH_4 = $\frac{34 \times 2}{1}$ = 68.0 m^3

[since 1 volume of CH_4 requires 2 volume of O_2]

Volume of O_2 required for the combustion of C_2H_4 = $\frac{5.2 \times 3}{1}$ = 15.6 m^3

[since 1 volume of C_2H_4 requires 3 volume of O_2]

Total volume of O_2 required = 3 + 24 + 68 + 15.6 = 110.6 m^3

Volume of O_2 present in the fuel gas = 0.6 m^3

Volume of minimum O_2 required = 110.6 – 0.6 = 110 m^3

As air is supplied in 50% excess, volume of O_2 supplied = 110 × 1.5 = 165 m^3.

Volume of air supplied = $\frac{165 \times 100}{21}$ = 786 m^3

Volume of minimum air required = $\frac{110 \times 100}{21}$ = 524 m^3

The products of the dry flue gas are = CO_2, O_2 and N_2

1 volume of CO gives 1 volume of CO_2.

6 volume of CO gives 6 volume of CO_2 = 6 m^3 of CO_2

1 volume of CH_4 gives 1 volume of CO_2

34 volume of CH_4 gives 34 volume of CO_2 = 34 m^3 of CO_2

1 volume of C_2H_4 gives 2 volume of CO_2

5.2 volume of C_2H_4 gives 5.2 × 2 = 10.4 volume of CO_2 = 10.4 m^3 of CO_2

Total volume of CO_2 in the flue gas = (6 + 34 + 10.4) m^3 = 50.4 m^3

Volume of CO_2 present in fuel gas = 2.6 m^3

Total volume of CO_2 in flue gas = (2.6 + 50.4) m^3 = 53.0 m^3

Volume of O_2 supplied from air = 165 m^3

Volume of O_2 required for combustion = 110 m^3

Excess O_2 present in the flue gas = (165 – 110) m^3 = 55 m^3

Volume of N_2 = (volume of air supplied – volume of O_2 in air)

= (786 – 165) m^3 = 621 m^3

Volume of N_2 present in the fuel gas = 3.6 m^3

Volume of N_2 in flue gas = (621 + 3.6) m^3 = 624.6 m^3

In the flue gas: CO_2 = 53.0 m^3, O_2 = 55.0 m^3, N_2 = 624.6 m^3

Problem 9.4 A coke-oven gas of the following percentage composition by volume was used for firing gas retorts:

$$CO_2 = 1.4;\ CO = 5.1;\ O_2 = 0.5;\ N_2 = 4.2;\ H_2 = 57.4;\ CH_4 = 28.5;$$
$$C_2H_4 = 2.9.$$

The dry flue gases on analysis by Orsat apparatus, gave the following results by volume:

$$CO_2 = 15\%,\ CO = \text{NIL},\ O_2 = 2.5\%,\ N_2 = 82.5\%$$

Calculate the excess air supplied for the burning of 100 m^3 of coke-oven gas.

Problem 9.5 A fuel oil containing C = 85.1% and H = 14.9% was fired in an internal combustion engine and the volumetric analysis of dry flue gas reported to be as follows: CO_2 = 11%, CO = NIL, O_2 = 9%, N_2 = 80%

Calculate: (i) Weight of dry flue gas/kg of fuel oil burnt.
(ii) Weight of excess air/kg of fuel oil burnt.

(i) Theoretically it is assumed that all carbon and hydrogen are oxidized to CO_2 and H_2O respectively.

C	+	O_2	=	CO_2
12 kg		32 kg		44 kg
62.4 kg		$\frac{32}{12} \times 62.4$ kg		$\frac{44}{12} \times 62.4$ kg

Or, 62.4 kg carbon requires 166.34 kg oxygen to form 228.8 kg carbon-di-oxide.

$2H_2$	+	O_2	=	$2H_2O$
4 kg		32 kg		36 kg
4.1 kg		$\frac{32}{4} \times 4.1$ kg		$\frac{44}{4} \times 4.1$ kg

Or, 4.1 kg hydrogen requires 32.8 kg oxygen to form 36.9 kg water vapour.

S	+	O_2	=	SO_2
32 kg		32 kg		64 kg
0.8 kg		0.8 kg		1.6 kg

Or, 0.8 kg sulphur requires 0.8 kg oxygen to form 1.6 kg sulphur-di-oxide.
So, total oxygen required to burn 100 kg of coal
= (166.34 + 32.8 + 0.8) kg = 199.94 kg
Coal already contains 6.9 kg of oxygen.

Thus net oxygen required = (199.94 – 6.9) = 193.04 kg

$$\text{Hence, theoretical air required} = \frac{193.04}{0.23}$$
$$= 839.30/\text{kg } 100 \text{ kg of coal}$$
$$= 8.3930 \text{ kg/kg of coal}$$

(ii) 100 kg of flue gas contains:
12.9 kg CO_2, 0.2 kg CO, 6.1 kg O_2, 80.8 kg N_2

44 kg CO_2 contains 12 kg carbon

12.9 kg CO_2 contains $\frac{12}{44} \times 12.9$ kg carbon = 3.52 kg carbon

28 kg CO contains 12 kg carbon

0.2 kg CO contains $\frac{12}{28} \times 0.2$ kg carbon = 0.086 kg carbon

So total carbon content in 100 kg of flue gas
= (3.52 + 0.086) = 3.606 kg of carbon.

Again carbon content in 100 kg of coal = 62.4 kg

$$\text{Fraction of carbon converted of } CO_2 = \frac{3.52}{3.606} = 0.976$$

Amount of CO_2 produced theoretically = 228.8 kg.

Actual amount of CO_2 produced = 228.8 × 0.976 kg = 223.31 kg

% CO_2 in flue gas = 12.9%

$$\text{Weight of flue gas} = \frac{223.31}{0.129} = 1731.1 \text{ kg}$$

$$\text{Weight of dry flue gas/kg of fuel} = \frac{1731.1}{100} = 17.311 \text{ kg}$$

(iii) Air required theoretically to convert all carbon to carbon-di-oxide is 839.30 kg per 100 kg of fuel.
Actually, some carbon monoxide (CO) are also produced along with CO_2.
Amount of CO in flue gas = 1731.1 × 0.002 kg = 3.4622 kg

CO	+	$\frac{1}{2}O_2$	=	CO_2
28 kg		16 kg		44 kg
3.4622 kg		$\frac{16}{28} \times 3.4622$ kg		$\frac{44}{28} \times 3.4622$ kg

So, 3.4622 kg CO requires 1.981 kg O_2 to get converted to CO_2.

Amount of O_2 in the flue gas = 1731.1 × 0.061 = 105.6 kg

Excess oxygen = (105.6 – 1.981) kg = 103.619 kg

Hence, excess air = $\frac{103.619}{0.23}$ = 450.52 kg

So, air supplied for burning the coal sample = (839.30 + 450.52) = 1289.82 kg

Weight of air supplied per kg of fuel burnt = $\frac{1289.82}{100}$ = 12.8982 kg/kg of fuel burnt.

Calculation of oxygen required to burn a fuel:

1. Minimum oxygen required to burn a fuel
   = (Theoretical oxygen required) – (Oxygen present in the fuel)
2. Theoretical oxygen required:
   A solid or liquid fuel mainly contains carbon (C) hydrogen (H), sulphur (S) and water vapour (H_2O)

$$C + O_2 = CO_2 \quad \text{12 kg C needs 32 kg oxygen}$$
$$2H_2 + O_2 = 2H_2O \quad \text{4 kg } H_2 \text{ needs 32 kg oxygen}$$
$$S + O_2 = SO_2 \quad \text{32 kg S needs 32 kg oxygen}$$

Consider 1 kg of a fuel contains C kg of carbon, H kg of hydrogen, S kg of sulphur and O kg of oxygen. Some hydrogen is bound as water vapour and not available for burning. In 1 mole of water vapour, 2 kg of H_2 is bound with 16 kg of O_2.

So, O kg of oxygen is linked with $\frac{2}{16} \times O = \frac{O}{8}$ kg of hydrogen

Thus in 1 kg fuel, out of H kg of hydrogen, $\left(\frac{O}{8}\right)$ kg is not available for burning.

So, available hydrogen = $\left(H - \frac{O}{8}\right)$ kg

For C kg carbon oxygen required = $\frac{32}{12} \times C$ kg

For $\left(H - \frac{O}{8}\right)$ kg hydrogen oxygen required = $\frac{32}{4} \times \left(H - \frac{O}{8}\right)$ kg

For S kg sulphur oxygen required = S kg

So, theoretical oxygen required is = $\left[\frac{32}{12}C + 8\left(H - \frac{O}{8}\right) + S\right]$ kg per 100 kg of fuel

10

CHARACTERIZATION OF ORGANIC COMPOUNDS

10.1 The ultraviolet spectrum

10.1.1 Introduction

What is spectroscopy?

Ans: It is the analysis of the electromagnetic radiations, scattered or emitted by molecules.

Absorption spectrum It is observed due to transition of electrons from lower energy level to higher energy level.

Emission spectrum It is observed due to transition of electrons from higher energy level to lower energy level.

Atomic spectra It is observed due to transition of electrons between atomic energy levels.

Molecular spectra It is observed due to transition of electrons between molecular energy levels.

Applications of molecular spectroscopy

(i) Study of molecular structure, type of primary and secondary bonds, bond energy, bond length and bond angle.
(ii) To monitor change in concentrations of reactants and products in kinetic studies.
(iii) To understand the colour of different compounds.

Spectrophotometry When radiations are passed through a transparent layer of a sample, a fraction of radiation is absorbed and hence intensity of the emitted radiations decreases. This decrease in intensity of radiation is the origin of spectrophotometry.

The electromagnetic spectrum can be broadly classified into different regions depending on the wavelength-range, as shown in the Table 10.1.

Table 10.1 Classification of electromagnetic spectrum

Spectral region	*Wavelength range*	*Energy change involved*
γ-rays	100 pm → 1 pm	Nuclear
X-rays	10 nm → 100 pm	*K*- and *L*-shell electrons
Ultraviolet or U.V.	100 nm → 400 nm	Middle and valence shell electrons in Molecular Orbital
Visible	400 nm → 780 nm	Valence shell electrons in Atomic Orbitals
Infrared or IR	0.750 μm → 1000 μm	Molecular vibration and rotation
Microwave	0.100 cm → 50 cm	Molecular rotation, spin orientation (ESR)
Radiowave	1.0 m → 30,000 m	Spin orientation in magnetic field (NMR)

N.B. 1 μm = 10^{-6} m, 1 nm = 10^{-9} m, 1 pm = 10^{-12} m.

Table 10.2 Sources of different rays

Ray	*Wavelength range*	*Source*
Ultraviolet	200 nm to 400 nm	Discharge through deuterium gas or Xenon in quartz.
Visible	400 nm to 780 nm	Sunlight
Near Infrared	800 nm to 30 μm	Nernst filament, which is a heated ceramic filament containing rare-earth oxides.
Far Infrared	30 μm to 3 mm	Mercury arc inside a quartz envelop.
Nuclear Magnetic Resonance	3 m to 3 km	These radiations are generated due to oscillation of electric currents in coils of wire.

So, light, having wavelength range 100 nm → 400 nm is called ultraviolet light (U.V.) and U.V. light region, having wavelength close to 400 nm, is usually called near U.V. region. With the absorption of U.V. rays, electrons get energized and undergo shifting from stable electronic levels to a number of sub-levels, as a result of which, a broad U.V. spectrum band is obtained. U.V. spectrum of a compound usually shows a few broad humps. Thus a U.V. spectrum is characterized by position of the hump and intensity of absorption, i.e. extinction coefficient (ε_{max}).

If U.V. ray is passed through a sample, the latter absorbs energy for excitation of electrons. So, intensity of emergent ray is different from that of incident ray. If I_0 and I are the intensities of incident and emergent rays, absorbance is given by **Lambert-Beer law**.

Lambert's law When a monochromatic beam of light is passed through a pure homogeneous compound or solution, fraction of radiation is absorbed and the decrease in intensity of light with respect to thickness of the compound/solution is proportional to the intensity of the emergent light. If I_0 and I are the intensity of incident and emergent monochromatic light respectively, we have:

$$-\frac{dI}{dx} \propto I$$

Beer's law When a monochromatic beam of light is passed through a pure homogeneous compound or solution, fraction of radiation is absorbed and the decrease in intensity of light with respect to thickness of the compound/solution is proportional to the intensity of the emergent light as well as the concentration of the solution too. Thus mathematically, we have

$$-\frac{dI}{dx} \propto I.c.$$

Or,
$$-\frac{dI}{dx} = \varepsilon.I.c$$

$$\int_{I_0}^{I} -\frac{dI}{I} = \varepsilon.c\int_0^l dx.$$

On integration, we get:

$$\ln\frac{I_0}{I} = \varepsilon'.c.l.$$

Or,
$$\log\frac{I_0}{I} = \varepsilon.c.l$$

Where, I = Intensity of the emergent light and
c = Concentration of the solution in mole/l.
ε = Molar absorption coefficient and
l = Length of the solution in cm.
ε = 2.303ε'.

For a given solution if c = 1 unit and $I_0 = 10I$, $\varepsilon' = 1/l$. So molar extinction coefficient or molar absorptivity is the inverse of the length of the solution.

$-\log\frac{I}{I_0}$ is called optical density or absorbance (A) of the sample and $\frac{I}{I_0}$ is called transmittance (T) of the sample.

$$A = \log\frac{I_0}{I} = -\log\frac{I}{I_0} = \log T$$

The SI unit of ε is m^2/mole but in general the unit is expressed as litre $mole^{-1}cm^{-1}$.

Selection of solvent During U.V. spectroscopic analysis, it is desired that the solvent not absorb any U.V. ray and not interact with solute molecules. Usually less polar solvents, such as, hexane, cyclohexane, diethyl ether etc. are used for this purpose. Solvents, like water, ethanol etc. are transparent in U.V. region but usually they are not preferred due to their high polarity. Solvents, like benzene, chloroform, carbon tetrachloride etc. are avoided as they show strong absorbance in U.V. region.

A molecule of a compound possesses several bonding and antibonding orbitals. Bonding orbitals are of three types: σ-type, π-type and *n*-type. Whereas antibonding orbitals are designated by the symbol σ^*–, π^*–respectively.

Bonding pair of electrons are usually located in σ- and π-orbitals, whereas non-bonding electrons e.g. lone pair of electrons are located in an *n-orbital.*

Antibonding orbitals are associated with high energy and unstable whereas bonding orbitals are associated with low energy and stable.

10.1.2 Electronic transitions

Following electronic transitions are usually observed due to U.V. absorption:

(a) $\sigma \rightarrow \sigma^*$ transition,
(b) $n \rightarrow \sigma^*$ transition.
(c) $\pi \rightarrow \pi^*$ transition,
(d) $n \rightarrow \pi^*$ transition.

(a) $\sigma \rightarrow \sigma^*$: This type of transition is usually observed in saturated compounds and requires high energy. Usually short length (120 – 125 nm) is required. So to avail $\sigma \rightarrow \sigma^*$ transition evacuated spectrophotometer is necessary, rendering the method very costly.

(b) $n \rightarrow \sigma^*$: This type of transition is usually observed in saturated compounds. Unpaired and unshared electrons are shifted to higher energy level. Energy required is much less than $\sigma \rightarrow \sigma^*$ and hence comparatively longer wavelength (190 – 380 nm) is enough to execute this type of transition.

(c) $\pi \rightarrow \pi^*$: This type of transition is usually observed in organic compounds containing double bonds. This is the most useful transition to detect the structure of different organic compounds. Energy required is less than $n \rightarrow \sigma^*$ and longer wavelength (190 – 380 nm) is enough to execute this type of transition. The value of molar extinction coefficient (ε) is very high, of the order of 10^4 *l* mole^{-1}cm^{-1} or greater. So the strongest signal strength is observed for this type of transition. $\pi \rightarrow \pi^*$ is very common and known as allowed transitions.

(d) $n \rightarrow \pi^*$: This type of transition is usually observed in unsaturated organic compounds containing hetero atom with lone pair of electrons. Energy required is much less than $\pi \rightarrow \pi^*$ and occurs at longer wavelength region. In most cases this type transitions remains forbidden. The value of molar extinction coefficient (ε) is low, within 10^2 *l* mole^{-1}cm^{-1}.

Organic compounds, having one or more double bonds, usually show $\pi \rightarrow \pi^*$ transition and the corresponding absorption bands are usually observed in far U.V. region. However, for conjugated systems less energy is required for transition and hence absorption bands are shifted towards the near U.V. region. Higher the degree of conjugation, larger the shifting of absorption bands towards the visible region. If there is enough number of double bonds in conjugation, the compound appears coloured, e.g. β-carotin, having 11 C–C double bonds in conjugation, appears yellow.

In the case of conjugated systems if a π-electron is shifted to higher energy level, the excited state undergoes resonance through conjugated double bonds and becomes more stable than ground state and hence less energy is required to bring the molecule in the excited state.

$$-C{=}C{-}C{=}C{-}C{=}C- \xrightarrow{\text{U.V. rays}} -C-\dot{C}-C{=}C{-}C{=}C-$$

(ground state) (excited state)

$$-C-\dot{C}-C{=}C{-}C{=}C- \leftrightarrow -C-C{=}C-\dot{C}-C{=}C-$$

$$\leftrightarrow -C-C{=}C{-}C{=}C-\dot{C}- \text{ etc.}$$

So higher the number of double bonds greater is the resonance and hence less and less energy is required to bring the molecule to the excited state.

10.1.3 Identification of organic compounds

U.V. spectrum is not used to direct identification of functional groups. It is mainly used to identity (i) double bond, (ii) conjugation system, (iii) aromatic ring and (iv) location of *substituents*.

Chromophores and auxochromes

As we know higher the number of C=C bonds U.V. absorption takes place at longer wavelength. There are some groups where small change in structure leads to significant change in wavelength of U.V. radiation absorbed. These structural units are known as *chromophores*. Examples are:

C=C, C≡C, C≡N, N=N, C=O etc.

There are polar groups, where U.V. absorption promote $n \rightarrow \sigma^*$ and $n \rightarrow \pi^*$ transitions, which usually take place at longer wavelengths. These polar groups, when conjugated with chromophores, the U.V. absorption bands are further modified. These polar groups, usually possessing lone pair of electrons, are known as *auxochromes*. Examples are:

–OH, –OR, –SH, –X groups etc.

Table 10.3 Absorption wavelengths of some chromophores and auxochormes

Chromophore	*Example*	*Auxochromes*	*Example*	*Type of excitation*	λ_{max} (*nm*)
C=C	$CH_2{=}CH_2$			$\pi \rightarrow \pi^*$	171
C≡C	HC≡CH			$\pi \rightarrow \pi^*$	173
C = O	$(CH_3)_2CO$			$\pi \rightarrow \pi^*$ $n \rightarrow \pi^*$	189 279
O=N → O	CH_3NO_2			$\pi \rightarrow \pi^*$ $n \rightarrow \pi^*$	201 274
		$-\ddot{\underset{..}{S}}-$	$C_6H_{13}SH$	$n \rightarrow \sigma^*$	224
		$-\ddot{\underset{..}{I}}:$	CH_3I	$n \rightarrow \sigma^*$	258
		$-\underset{..}{\bar{\ddot{N}}}-$	$(CH_3)_3N$	$n \rightarrow \sigma^*$	227

Inference In an organic compound conjugation of double bonds shifts the absorption bands to longer wavelength but at the same time intensity decreases, e.g., $(CH_3)_2CO$ shows an absorption maximum at 189 nm due to $\pi \rightarrow \pi^*$ transition and at 279 nm due to $n \rightarrow \pi^*$ transition. In the case of $CH_2=CH-COOH$ λ_{max} of $n \rightarrow \pi^*$ transition is observed at 324 nm and that of $\pi \rightarrow \pi^*$ transition is observed at 219 nm.

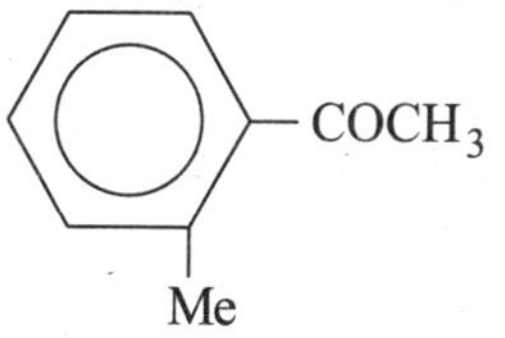

o-isomer: λ_{max} = *243 nm*

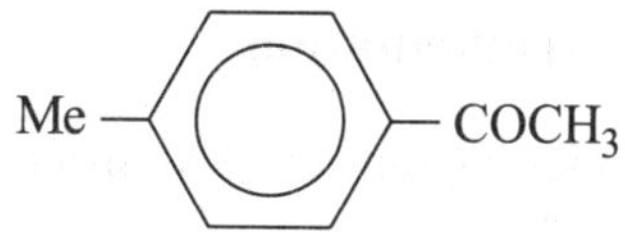

p-isomer: λ_{max} = *252 nm*

In case of o-isomer conjugation systen is disturbed due to steric hindrance of ortho methyl group. So the absorption occurs at shorter wavelength. This type of steric hindrance is absent in the case of p-isomer and hence absorption occurs at longer wavelength.

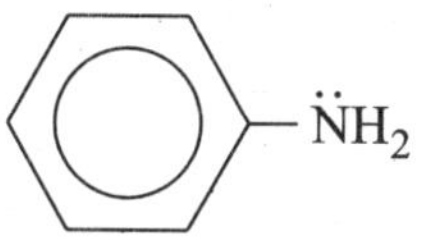

Lone pair of nitrogen takes part in resonance, λ_{max} = 230 nm

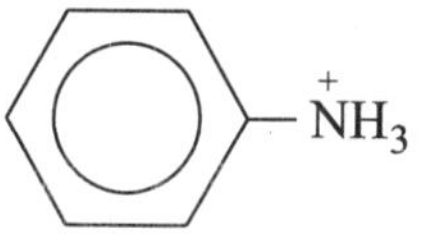

Lone pair of nitrogen is not available for resonance, λ_{max} = 203 nm

As the number of double bonds increases, λ_{max} value also increases.

$CH_2 = CH_2$

λ_{max} = 170 nm

$CH_2 = CH - CH = CH_2$

λ_{max} = 215 nm

λ_{max} = 255 nm

λ_{max} = 255 nm

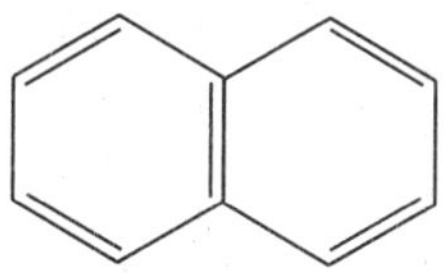

λ_{max} = 314 nm

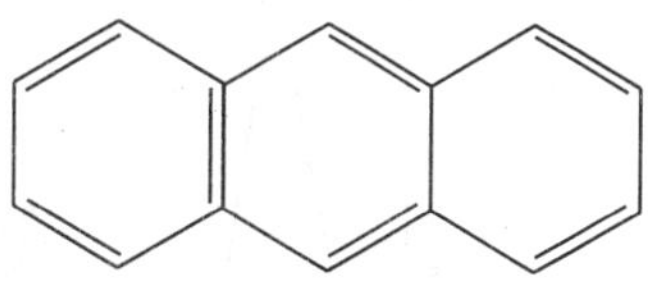

λ_{max} = 380 nm

As the conjugation system extends, absorption spectrum is shifted towards the longer wavelength.

Bathochromic shift or red shift If λ_{max} value of a molecule increases or absorption maximum is shifted to longer wavelength due to the presence of functional groups, the shift is called bathochromic shift or red shift.

Hypsochromic shift or blue shift If λ_{max} value of a molecule decreases or absorption maximum is shifted to shorter wavelength due to the presence of functional groups, the shift is called hypsochromic shift or blue shift.

Hyperchromic shift When intensity of absorption maximum increases, the shift is called hyperchromic shift.

Hypochromic shift When intensity of absorption maximum decreases, the shift is called hypochromic shift.

10.1.4 Instrumentation

The schematic diagram of U.V. spectrophotometer is shown in the Figure 10.1.

Description of the process A beam of U.V. ray is allowed to pass through the sample as well as through the reference sample, called, solvent. Both the sample and solvent absorb characteristic wavelengths. Transmitted beam from sample and reference are passed to the detection unit where intensity of the transmitted beam is determined for both the cases of sample and solvent. If I and I_0 are the intensities of transmitted beam from sample and solvent respectively, the absorbance A is given by:

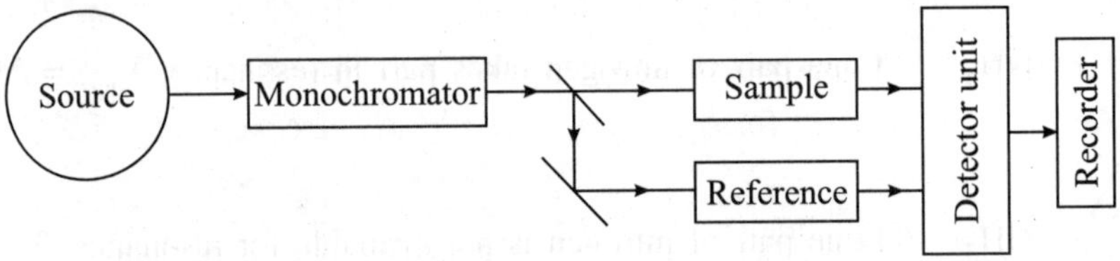

Fig. 10.1 Schematic diagram of a U.V. spectrophotometer.

$$A = \log \frac{I_0}{I} = \varepsilon . t \qquad (10.1)$$

where, ε = Molar extinction coefficient and t = Sample thickness.

Absorbance versus wavelength plot is recorded in the recorder. U.V. absorption spectra of benzene and naphthalene are shown in the Figure 10.2.

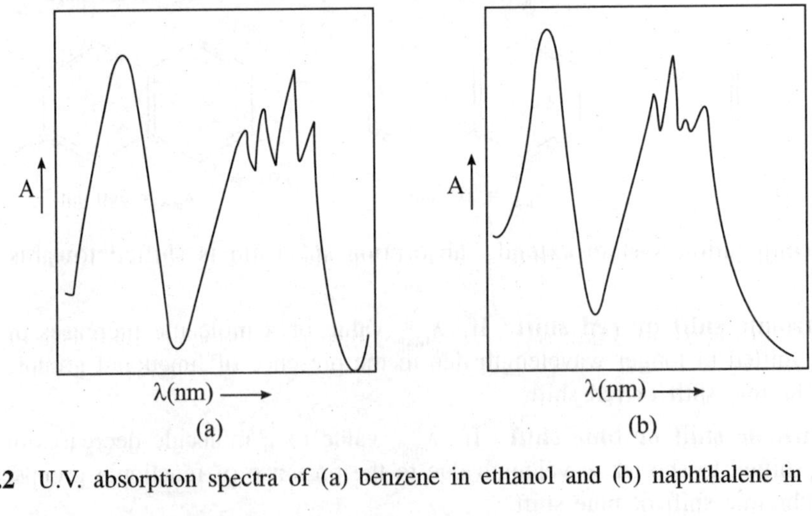

Fig. 10.2 U.V. absorption spectra of (a) benzene in ethanol and (b) naphthalene in methanol.

10.1.5 Applications

(i) It can be used to measure the extent of conjugation within molecules.
(ii) It can be used to differentiate conjugated dienes from non-conjugated dienes.
(iii) It can be used to reveal the number and location of substituent attached to the carbon atoms of the conjugated systems.
(iv) It can be used to detect aromatic rings within a molecule.
(v) It can be used to detect cis- and trans- isomers.
(vi) Qualitative and quantitative analysis of compounds.
(vii) Detection of impurities in organic compounds.
(viii) Determination of molecular weight.
(ix) Determination of dissociation constant of weak acid and base.

10.2 The infrared spectroscopy

In any compound, atoms and molecules are continuously vibrating, as a result of which, stretching and bending of bonds are continuously occurring. Let us consider a molecule ABC. Stretching vibration results elongation and contraction of bonds B–A and B–C. There are different types of bonding vibrations, which are well illustrated in the following figures.

(a) Scissoring Bond angle α increases or decreases as atoms A and C are moving apart or closer (Fig. 10.3a).

(b) Rocking Bond angle α remains constant but the positions of two atoms A and C change (Fig. 10.3b).

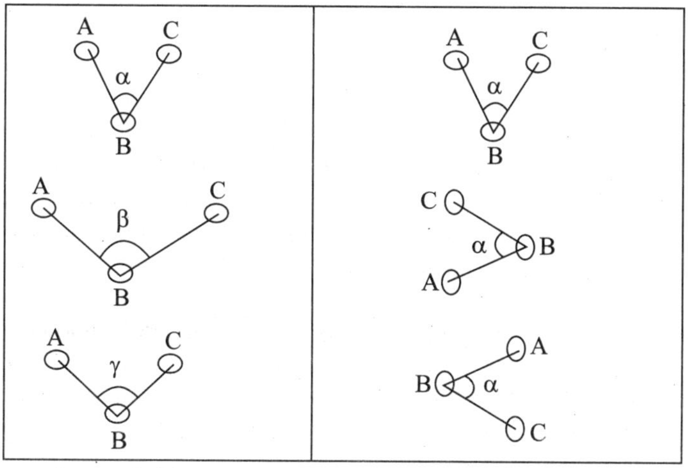

(a) (b)

Fig. 10.3 In-plane deformation, that means, bonds get deformed in the same plane.

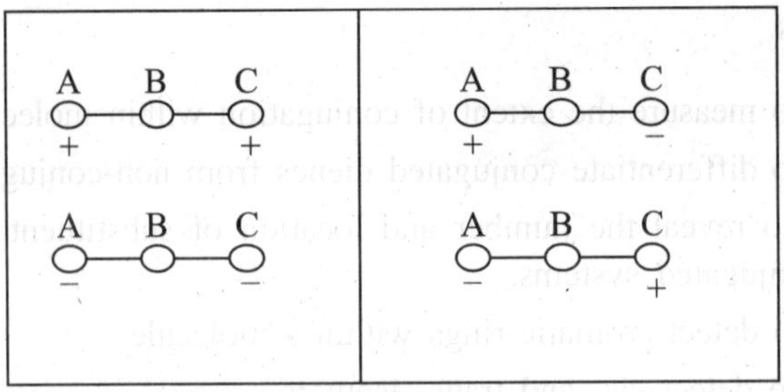

(a) (b)

Fig. 10.4 Out of *phase* deformation, that means, atoms are vibrating at different planes.

(a) Wagging Both the atoms A and C are vibrating simultaneously either in upward direction (represented by the symbol '+') or in downward direction (represented by the symbol '–') (Fig. 10.4a).

(b) Twisting During vibration one atom (A or C) moves upward and the other atom (C or A) moves downward. To execute these vibrations (stretching and bending) almost all organic molecules absorb light energy in the intermediate infrared region of the electromagnetic spectrum (λ = 1500 nm to 20,000 nm) (Fig. 10.4b).

Functional groups quickly respond to the IR spectra. Also absorption in IR region is highly selective to each functional group.

It has been observed that all the functional groups absorb light energy within 1000 cm^{-1} to 3000 cm^{-1}. Thus the structure of any organic compound could be identified by observing its absorption in the aforesaid region. This region of absorption (1000 cm^{-1} to 3000 cm^{-1}) is called *fingerprint region*.

In IR spectroscopy we often use *wave number* instead of frequency. Wave number is the number of waves per cm. and expressed as cm^{-1} and its value is $\frac{1}{\lambda}$. So each and every functional group has a characteristic IR frequency. Any noticeable change in characteristic IR frequency indicates the structural change of functional group. For example, presence of hydrogen bonding offers hindrance in atomic vibration and it is reflected in IR-absorption. Not only hydrogen bonding but also conjugation, neighbouring substituent, angle strain etc. are also responsible for noticeable shifting of characteristic frequency. Absorption due to C–H stretching vibration reflects the hybridization characteristics of carbon, e.g., *sp^3-hybridized* carbon (usually alkanes) absorbs within 2800 – 3000 cm^{-1}, *sp^2-hybridized* carbon (alkenes and aromatic rings) absorbs within 3000 – 3100 cm^{-1}, whereas *sp-hybridized* carbon (alkynes) absorbs around 3300 cm^{-1}.

Absorption due to C–H out-of plane bending vibration is an important criteria to identify functional group, e.g., alkenes absorb within 800 – 1000 cm^{-1} for C–H out-or plane bending vibration but aromatic rings absorb within 675 – 870 cm^{-1} for C–H out-of-plane bending vibration. Some characteristic IR-absorption frequencies of different groups are shown in the Table 10.4.

Table 10.4 Characteristic infrared absorption frequencies

Bond	*Compound Type*	*Frequency Range* (cm^{-1})
C–H	Alkanes	2850 – 2960 1350 – 1470
C–H	Alkenes	3020 – 3080 675 – 1000
C–H	Aromatic rings	3000 – 3100 675 – 870
C–H	Alkynes	3300
C=C	Alkenes	1640 – 1680
C≡C	Alkynes	2100 – 2260
C=C	Aromatic rings	1500 – 1600
C–O	Alcohols, ethers, carboxylic acids, esters	1080 – 1300
O–H	Monomeric alcohols, phenols, hydrogen bonded alcohols, carboxylic acids.	3610 – 3640 3200 – 3600 2500 – 3000
N–H	Amines	3300 – 3500
C–N	Amines	1180 – 1360
C≡N	Nitriles	2210 – 2260
$-NO_2$	Nitro compounds	1515 – 1560 1345 – 1385

10.2.1 Applications

1. Infrared spectroscopy is very common and widely used in industry, research and forensic laboratory.
2. It is used to identify functional groups and structures of organic substances. Most of the functional groups respond within finger print region (1500 cm^{-1} – 400 cm^{-1}). The peaks corresponding to different functional groups are usually found in group frequency region (4000 cm^{-1} – 1500 cm^{-1}). Thus characteristic peaks due to different functional groups are observed in both group frequency region and finger print region.
3. **Identification of compound:** Each and every compound shows several absorption bands in both group frequency and finger print regions. Two different compounds cannot have all absorption peaks at the same positions. As IR spectrum of each compound is known, any unknown compound can be identified by comparing the corresponding IR spectra. However, IR spectroscopy method fails to distinguish optical isomers, i.e., geometrical isomers, diastereomers and enantiomers since they have identical structures.
4. Kinetics of a chemical reaction can be studied with the help of IR spectroscopy. During the course of reaction characteristic absorption bands due to reactants disappear,

corresponding peak heights decrease and at the same time characteristic absorption bands due to products appear and corresponding peak heights increase. Thus by studying IR spectra of a reaction mixture in regular interval one can study rate of reaction. Corresponding change in concentration and rate constant can also be determined.

5. IR spectroscopy is very much useful to determine the presence of impurities in a compound. Any impurity, present in the compound will give additional peak. This can be done by comparing IR spectra of the pure compound and the impure one.
6. Quantitative analysis can be performed by using IR spectroscopy. A given peak height of a pure compound is significantly reduced when it is mixed with other compounds. Thus by measuring and comparing peak heights quantity of a substance can be determined in a given mixture.
7. It is a useful technique to detect the presence of hydrogen bonding in an organic compound.

10.3 Nuclear magnetic resonance (NMR) spectroscopy

10.3.1 Introduction

If a nucleus contains odd number of protons or neutrons, each single proton or neutron spins about its own axis, as a result of which, a magnetic moment is developed. Thus nucleus, having odd mass number, behaves as a tiny bar magnet. The fundamental nucleus of this kind is proton (^1H). Other common nuclei are ^{13}C and ^{19}F. In presence of an external magnetic field two different orientations of this tiny bar magnet are possible.

(i) Low energy orientation where the tiny bar magnet is aligned in parallel with the external field.

(ii) High energy orientation where the tiny bar magnet is aligned against the external field as shown in the Figure 10.5.

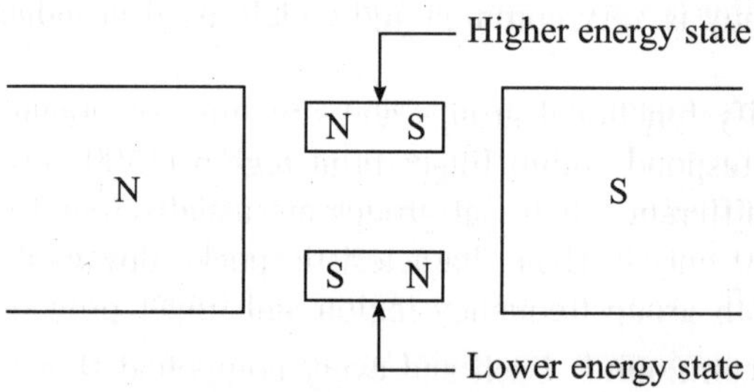

Fig. 10.5 Orientations of tiny bar magnet in an external magnetic field. ΔE = Energy difference between two states.

Usually the nucleus belongs to lower energy state. If sufficient energy is supplied in the form of electromagnetic radiation, the tiny magnet flips over to the less stable alignment, which is against the magnetic field, i.e., the nucleus is shifted to higher energy state.

The frequency of electromagnetic (EM) radiation needed for this shifting is given by:

$$\nu = \frac{g \, . \, H_0}{2\pi} \tag{10.2}$$

where, ν = frequency in Hz.

H_0 = strength of the magnetic field in gauss.

g = magnetogyric ratio.

The value of g is 26750 for the proton. In a magnetic field of 14,086 gauss, i.e. at H_0 = 14,086 Gauss, the corresponding frequency is given by:

$$\nu = \frac{26750 \times 14086}{2 \times 3.14} = 60 \text{ MHz}$$

Thus EM-radiation, required to produce NMR-spectra, lies in the radio frequency range, much lower than infrared frequency range.

Thus in presence of an external magnetic field the magnetic moment of the nucleus is solely responsible for giving signal in this radio frequency region. That is why, the spectrum is often called Nuclear Magnetic Resonance (NMR) spectrum.

Although frequency of absorption mostly depends on nuclear magnetic moment, surrounding environment of the nucleus has noticeable influence on NMR-spectra. Thus the *effective field* strength is different from the *applied field* strength due to environmental factor. The effective field strength of all protons is same but they absorb at different applied field strength which is actually measured and plotted against the absorption of radiation as illustrated in the following figure (Figure 10.6).

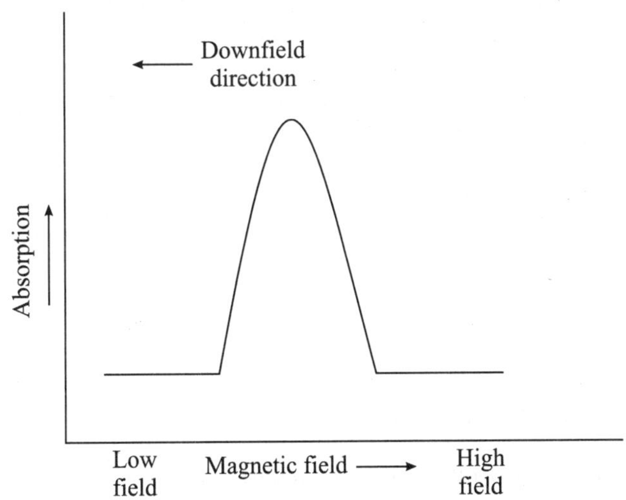

Fig. 10.6 NMR-signal of a proton.

10.3.2 Various aspects of NMR-spectrum

(a) Number of signals NMR-signal depends on the type of proton. So the number of NMR-signals of a compound depends on the number of different types of protons available in the compound.

(b) Position of a signal Position of proton NMR-signal depends on electronic environment of the corresponding proton.

(c) Intensity of a signal Intensity of a proton NMR-signal depends on the number of similar protons available corresponding to that particular NMR-signal.

(d) Splitting of a signal A proton NMR signal is largely influenced by neighbouring atoms and ultimately peak of the signal splits.

(a) Signal in a proton NMR-spectrum Consider the following three organic compounds:

$$\overset{2(a)}{CH_3}-\overset{1(b)}{CH_2}-Cl \qquad\qquad \overset{3(a)}{CH_3}-\overset{2(b)}{CH_2}-\overset{1(c)}{CH_2}-Cl$$

(A) (B)

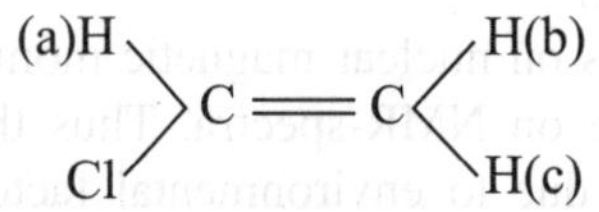

(C)

(i) In compound *A*, electronic environment of hydrogen atom of C^1-carbon atom is different from that of C^2-carbon atom. So two different signals (*a* and *b*) are obtained in the corresponding NMR-spectrum. In compound *B*, three signals (*a*, *b* and *c*) are obtained due to same reason. In compound *C*, three signals are obtained. One H-atom is in the cis-position but the other H-atom is in the trans-position with respect to Cl-atom. That is why two H-atoms (*b* and *c*) of the same carbon atom give two different signals.

(ii) In compound *A* two H-atoms attached to C^1-atom are of same kind and hence they are called equivalent protons. Similarly, C^2-atom of compound A possesses three equivalent protons. H-atom, attached to C^1-atom is environmentally different from H-atom, attached to C^2-atom. So protons of

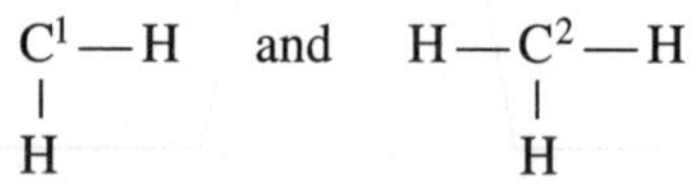

are called non-equivalent protons.

(b) Position of a signal: Shielding and deshielding effects When a molecule is placed in a magnetic field electrons are circulating to generate a secondary magnetic field, known as induced magnetic field. If the induced magnetic field, created by surrounding electrons of a H-atom, opposes the applied field, the resultant field strength decreases and the proton is said to be *shielded*. Conversely, if the same induced magnetic field reinforces the applied field the resultant field strength increases and the proton is said to be *deshielded*.

A normal proton, without any shielding or deshielding effects absorbs EM-radiation at a certain magnetic field strength. In the case of a shielded proton, required magnetic field strength is higher (than normal proton) and hence EM-radiation takes place at higher field, known as upfield. In the case of a deshielding proton, required magnetic field strength is lower than that of normal proton and hence EM-radiation takes place at lower field, known as *downfield*. This shifting of position in NMR spectra is called *chemical shift*, represented by δ. The chemical shift (δ) is given by the following expression:

$$\delta = \frac{\text{Observed shift (Hz)}}{60 \times 10^6 \text{ (Hz)}} \times 10^6 \text{ ppm}$$

The compound tetra-methyl silane (TMS) has 12 protons. As electronic and neighbouring environment of all protons is same, all 12 protons are called equivalent protons. That is why TMS is used as reference sample with respect to which chemical shift of an unknown sample is calculated. δ-value of TMS is taken as zero universally, chemical shift is expressed in terms of τ which is related to δ according to the following equation: τ = 10 – δ. Obviously, τ value of TMS is 10.

(c) Intensity of a signal Intensity of a signal or peak height largely depends upon the number of equivalent protons. Higher the number of equivalent protons larger the peak height or intensity. Area under an NMR signal is directly proportional to number of protons giving rise to the signal. For example, in the case of the following organic compound,

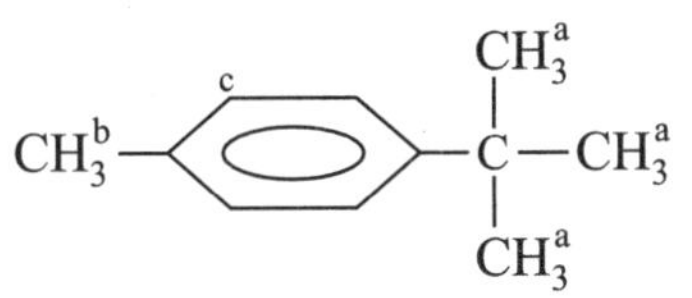

the three methyl groups, attached to the tertiary carbon atom, are identical in nature. The *nine* hydrogen atoms of three methyl groups are equivalent and collectively gives one signal (a). The aromatic ring contains *four* equivalent hydrogen atoms and collectively gives one signal (c). The lone methyl group, attached to the aromatic ring has *three* equivalent hydrogen atoms and collectively gives one signal (b). So, the peak area is maximum for the signal (a), followed by the signal (c) and signal (b). The number of protons associated with each signal can be computed by measuring the peak area.

The compound $C_{11}H_{16}$ gives three signals *a*, *b* and *c* in NMR-spectra.

The ratio of peak height = 8.8 : 2.9 : 3.8

Total peak height = 8.8 + 2.9 + 3.8 = 15.5

Total number of hydrogen atoms = 16.

Number of protons per unit = $\frac{16}{15.5}$ = 1.03

Number of protons corresponding to signal a = 8.8 × 1.03 = 9.1 ≈ 9.0.

Number of protons corresponding to signal b = 2.9 × 1.03 = 3.0.

Number of protons corresponding to signal c = 3.8 × 1.03 = 3.9 ≈ 4.0.

(d) Splitting of a NMR-signal In many NMR-spectra it has been observed that peak of a signal splits into a group of peaks. The environment of adjoining protons is the primary reason for splitting of a peak.

(i) If only one proton exists in the neighbouring carbon atom that proton may align itself with respect to the applied field (H_0) in two different ways:

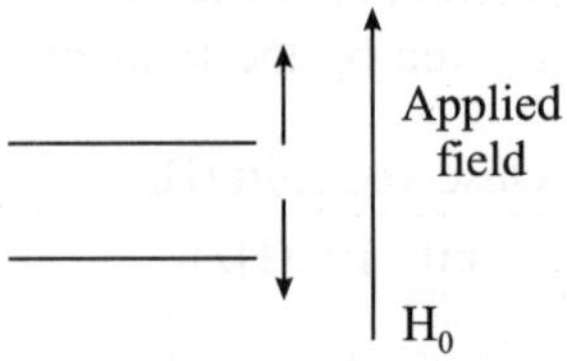

Adjacent carbon atom —C—H

The proton may remain in two possible alignments as shown above and population of two alignments is same. So the peak splits into two equal intensities (1:1).

(ii) If two protons are present in the neighbouring carbon atom they align themselves in the following ways with respect to the applied field.

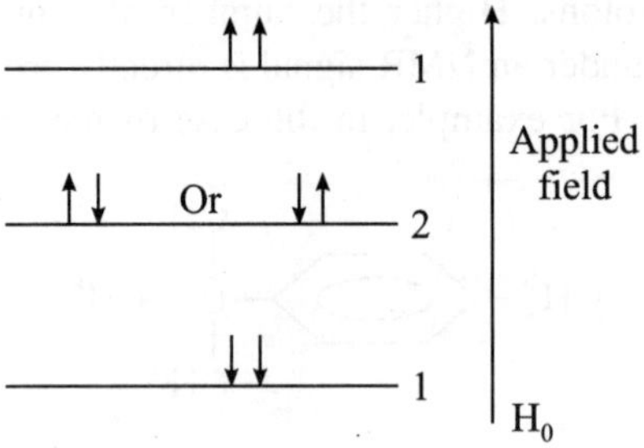

Spin-spin coupling for adjacent protons $-CH_2-$. The peak splits into three with the intensity ratio 1:2:1.

(iii) If three protons are present in the neighbouring carbon atom the possible alignments are:

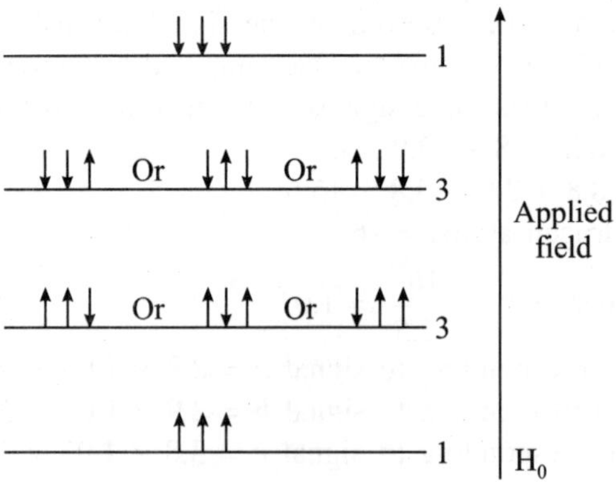

Spin-spin coupling for adjacent three protons $-CH_3-$. The peak splits into four with intensity ratio: 1:3:3:1.

If there are n number of protons in the neighbouring carbon atom the peak splits into $(n + 1)$ fragments. Intensity ratio can be obtained from the following formula:

$$(1+ x)^n = 1 + {}^nC_1.x + {}^nC_2.x^2 + {}^nC_3.x^3 + \ldots\ldots.. + \ldots\ldots.. x^n$$

The peak intensity ratio = $1 : {}^nC_1 : {}^nC_2 : {}^nC_3 : \ldots\ldots.. : 1$

Example 10.1 NMR-spectrum of

$$\overset{2(a)}{CH_3}-\overset{1(b)}{CHBr_2}$$

In the signal (a), 3 equivalent protons are present whereas in the signal (b) only one equivalent proton is present. Again the presence of Br-atom in C^1-atom has deshielding effect and hence the signal (b) is observed in downfield. Presence of any *electron withdrawing group* has *deshielding effect*, and presence of *electron donating group* has *shielding effect*.

Signal (a)	Signal (b)
(i) up-field	(i) Down-field
(ii) Area ≡ 3	(ii) Area ≡ 1
(iii) Doublet, since number of neighbouring H-atoms is 1	(iii) Quartet, since number of neighbouring H-atoms is 3

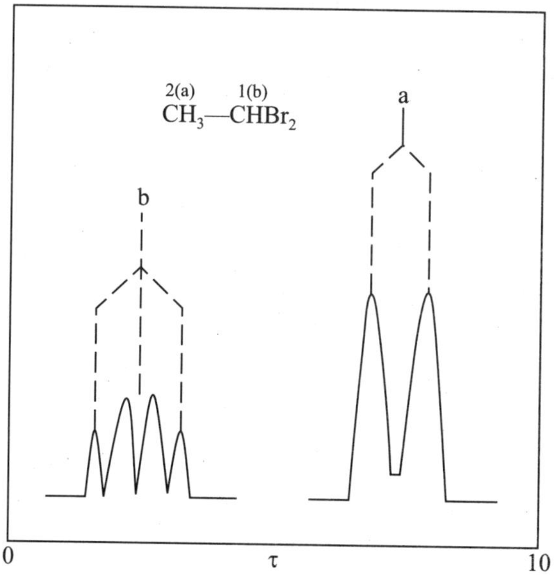

Fig. 10.7 NMR-spectrum of CH_3-CHBr_2.

Downfield shift is noticeably observed in halogenated compounds, e.g.,

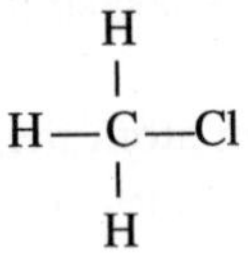 $\delta = 3.0$ $\tau = 7.0$	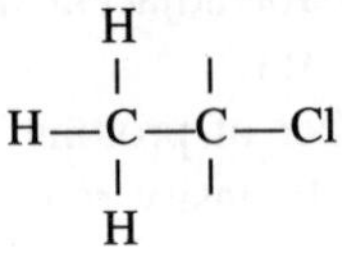 $\delta = 1.5$ $\tau = 8.5$
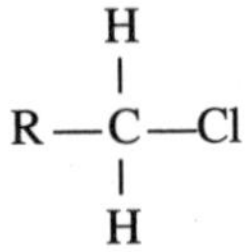 $\delta = 3.4$ $\tau = 6.6$	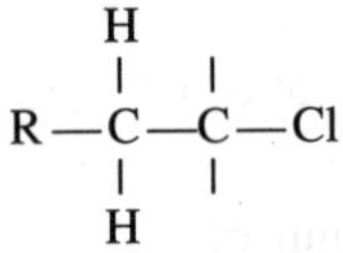 $\delta = 1.7$ $\tau = 8.3$
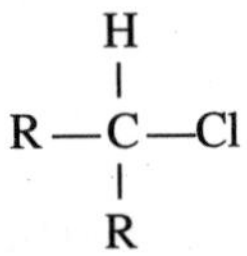 $\delta = 4.0$ $\tau = 6.0$	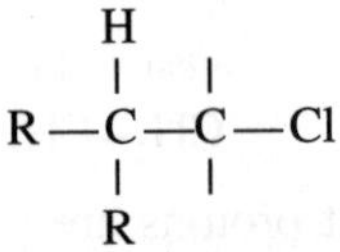 $\delta = 1.6$ $\tau = 8.4$
Direct attachment to Cl-atom shifts the signal to the downfield, i.e., high δ and low τ	Indirect attachment to Cl-atom shows an upfield signal, i.e., low δ and high τ

Example 10.2

$$\underset{a}{CH_3}—\underset{b}{CH}\,Cl—\underset{c}{CH_3}$$

Two Me-groups are chemically equivalent and hence give signal at the same position. The proton signal of *b* lies in the downfield due to the presence of Cl-atom.

Example 10.3

$$\underset{a}{CH_3}—\underset{b}{CH_2}—\underset{c}{CH_2}—Cl$$

Proton signal of *a* is different from that of *b* due to the presence of C—Cl attachment in *b*.

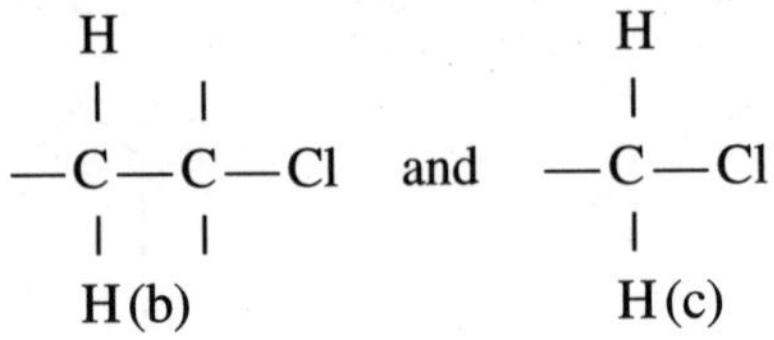

The signal *b* lies in the downfield and the signal *c* lies further downfield due to direct attachment of Cl-atom.

Example 10.4

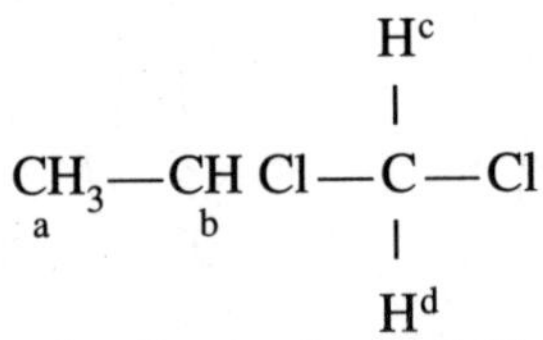

Four signals are obtained. The signal *a* lies in the upfield direction.

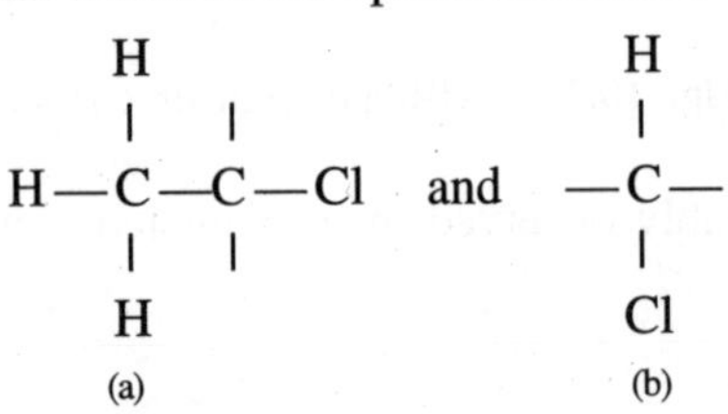

The signal *b* lies in the downfield. The two protons of end carbon atom are chemically equivalent but stereochemically different as shown below:

$CH_3CH(Cl)CH_2Cl \rightarrow$

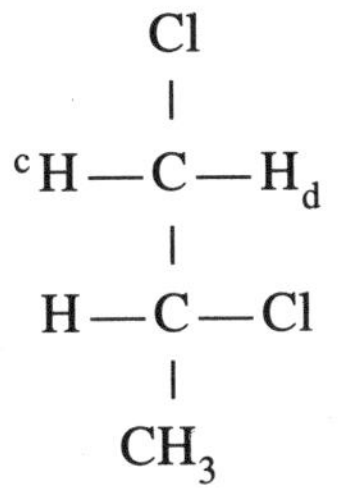

The proton single *d* lies in the downfield due to presence of neighbouring Cl-atom but proton signal *c* lies in the upfield compared to proton signal *d* due to the absence of any such Cl-atom in the neighbourhood.

Table 10.5 Characteristic proton chemical shifts

	Chemical shift (ppm)	
	δ	τ
Cyclopropane	0.2	9.8
Primary: RCH_3	0.9	9.1
Secondary: R_2CH_2	1.3	8.7
Tertiary: R_3CH	1.5	8.5
Vinylic: C=C–H	4.6 → 5.9	5.4 → 4.1
Acetylenic: C≡C–H	2 → 3	8 → 7
Aromatic: Ar–H	6 → 8.5	4 → 1.5
Benzylic: Ar–C–H	2.2 → 3	7.8 → 7
Allylic: C=C–C–H	1.7	8.3
Fluorides: H–C–F	4 → 4.5	6 → 5.5
Chlorides: H–C–Cl	3 → 4	7 → 6
Bromides: H–C–Br	2.5 → 4	7.5 → 6
Iodides: H–C–I	2 → 4	8 → 6
Alcohols: H–C–OH	3.4 → 4	6.6 → 6
Ethers: H–C–OR	3.3 → 4	6.7 → 6
Esters: RCOO–C–H	3.7 → 4.1	6.3 → 5.9
Esters: H–C–COOR	2 → 2.2	8 → 7.8
Acids: H–C–COOH	2 → 2.6	8 → 7.4
H–C–C=O	2 → 2.7	8 → 7.3
Aldehyde: R–CH=O	9 → 10	1 → 0
Hydroxylic: R–OH	1 → 5.5	9 → 4.5
Phenolic Ar–OH	4 → 12	6 → –2
Enolic C=C–OH	15 → 17	–5 → –7
Carboxylic RCOO–H	10.5 → 12	–0.5 → –2
Amino $R–NH_2$	1 → 5	9 → 5

10.3.3 Applications

1. It is the most powerful, sophisticated and versatile tool to determine structures of simple organic compounds, polymers and biopolymers. It also provides different structure related information of a given substance. It is basically used in organic chemistry, polymer chemistry and biochemistry.
2. It is the only tool to determine atomic-resolution structure of biomacromolecules.
3. It is used to quantify different properties of biomacromolecules.
4. It is a useful technique for determination of residual structures of unfolded proteins and the structures of folding intermediates. It also provides information about chemical properties of functional groups in biomacromolecules.
5. It is a useful tool for accurate and direct detection of hydrogen bonding interactions and other weak intermolecular interactions.
6. It is used to determine structure of protein and enzyme as well as to detect drugs. Metabolic analysis can also be done by using this technique.
7. It is a very useful technique to detect impurity as well as to quantify percent impurity.

11

PHASE EQUILIBRIUM

11.1 Phase rule

(i) Phase It is a physically distinct, chemically homogeneous and mechanically separable region, separated by other parts of the system by a definite boundary. The number of phases is represented by *P*. Examples: (a) gas phase: mixture of two or more gases always constitutes a single gas phase since gases are miscible in all proportions, (b) mixture of two immiscible liquids (e.g., benzene and water) will form two distinct phases.

(ii) Component In an equilibrium system the minimum number of independent variables with the help of which all phases at equilibrium can be described successfully, are called components. The number of components is represented by *C*. Examples:

(a) Ice(s) $\rightleftharpoons$ Water(l) $\rightleftharpoons$ Water vapour(g)

The component H_2O successfully describes all the three phases. So the above system is a single component system.

(b) $CaCO_3(s) \rightleftharpoons CaO(s) + Co_2(g)$

The number of components is apparently 3 and the number of phases is 3 but 3 components are linked by a chemical reaction. Thus at least one component is dependent variable, since $\sum x_i = 1$. In this case $x_1 + x_2 + x_3 = 1$. If x_1 and x_2 are known x_3 can easily be evaluated $[x_3 = 1 - (x_1 + x_2)]$. Thus the third component is dependent variable. So the total number of components is $(3 - 1) = 2$. Restriction: Any chemical reaction among the components restricts the number of independent variables.

(iii) Degrees of freedom It is the minimum number of independent variables, including pressure and temperature, which are required to describe an equilibrium system perfectly. The number of degrees of freedom is represented by *F*.

Phase Rule For an equilibrium system, uninfluenced by external forces like gravitational force, electrical or magnetic forces etc, the total number of degrees of freedom is related to the number of components (C) and the number of phases (P) according to the following equation:

$$F = C - P + 2 \tag{11.1}$$

where the factor 2 on the right hand side indicates the two independent variables, pressure and temperature. It was first proposed by Willard Gibbs in 1874.

Proof As we know for any equilibrium system the chemical potential of ith component is equal in all phases at equilibrium. If there are P number of phases, represented by α_1, α_2, α_3 ... α_p etc, in equilibrium, for the 1st component, we can write that: $(\mu_1)^{\alpha_1} = (\mu_1)^{\alpha_2} = (\mu_1)^{\alpha_3} = (\mu_1)^{\alpha_4} = \ldots = (\mu_1)^{\alpha_P}$, where, $(\mu_1)^{\alpha_1}$ is the chemical potential of the 1st component in α_1 phase. Again $\Sigma\alpha_1 = P$. Similarly, we can write that:

$$(\mu_2)^{\alpha_1} = (\mu_2)^{\alpha_2} = (\mu_2)^{\alpha_3} = (\mu_2)^{\alpha_4} = \ldots = (\mu_2)^{\alpha_P} \quad \text{...for the 2nd component}$$

$$(\mu_3)^{\alpha_1} = (\mu_3)^{\alpha_2} = (\mu_3)^{\alpha_3} = (\mu_3)^{\alpha_4} = \ldots = (\mu_3)^{\alpha_P} \quad \text{...for the 3rd component}$$

....

....

$$(\mu_C)^{\alpha_1} = (\mu_C)^{\alpha_2} = (\mu_C)^{\alpha_3} = (\mu_C)^{\alpha_4} = \ldots = (\mu_C)^{\alpha_P}$$

where C is the total number of components. Thus for phase α_1, the total number of independent components is $(C - 1)$, since $\Sigma x_i = 1$,
where x_i is the mole fraction of the ith component. As there are P number of phases, the total number of independent variables is $P(C - 1)$. Considering pressure and temperature are the two independent variables, the total number of independent variables becomes $[P(C - 1) + 2]$. Each component is distributed in P number of phases in equilibrium. So the number of restrictions for each component is $(P - 1)$. As there are C number of components, the total number of restrictions or chemical equations is $C(P - 1)$. Thus the total number of independent variables = degrees of freedom (DOF) = F.

$$F = [P(C - 1) + 2] - C(P - 1) = C - P + 2$$

Or,
$$F = C - P + 2 \tag{11.2}$$

For $F = 0$, the system is said to be invariant, which means that all the independent variables are fixed at that point. For $F = 1$, the system is said to be univariant, which means that one independent variable may be varied. For $F = 2$, the system is said to be bivariant, which means that two independent variables may be varied.

11.2 Single component system

Water system is the simplest example of a single component system.

$$\text{Ice(s)} \rightleftharpoons \text{Water(l)} \rightleftharpoons \text{Water vapour(g)} \quad \text{(shown in the Figure 11.1)}$$

$C = 1$ for water system and hence, $F = C - P + 2 = 1 - P + 2 = 3 - P$. Within AOB or BOC or COA only single phase exists. So $P = 1$ and hence, $F = 3 - 1 = 2$. The system is bivariant. Both pressure and temperature vary independently. Along the curve OA both liquid and vapour phases coexist. So, $P = 2$ and hence, $F = 3 - P = 3 - 2 = 1$. So if pressure varies independently, temperature variation depends on pressure variation. So temperature is a dependent variable. Similarly if temperature varies independently, pressure will be dependent variable. The system is called univariant system and the curve is called *boiling point curve*. Along the curve OC both liquid and solid phases coexist and the curve is called *melting point curve*. Along the curve OB both solid and vapour coexist and the curve is called *sublimation curve*. At the point O, three phases solid, liquid and vapour coexist and hence, $F = 3 - P = 3 - 3 = 0$. The system is called invariant system. All the variables are constant at O and cannot vary independently. The point O is known as *triple point*. Pressure and temperature at O are 0.00603 atm (4.58 mm of Hg) and 0.0075°C respectively. The point D in the Figure 11.1 indicates the boiling point of water at normal pressure, i.e., 1atm. The point A indicates the highest point where two phases (liquid and vapour) coexist and is known as critical point above which the liquid and vapour phases are indistinguishable from each other.

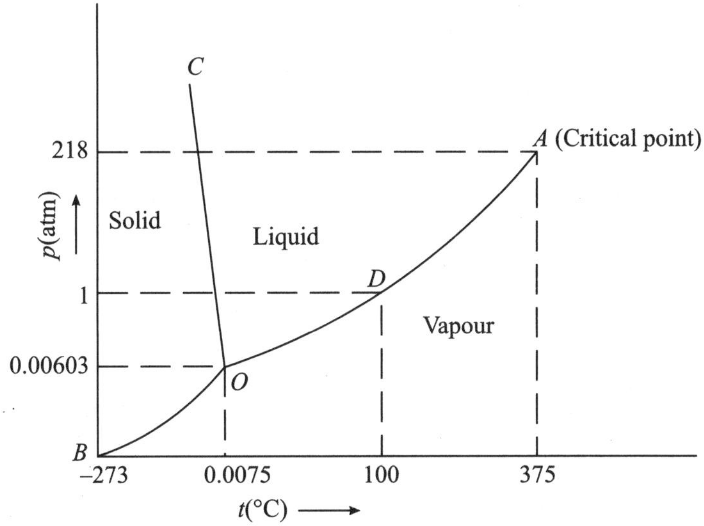

Fig. 11.1 Phase diagram (pressure vs. temperature) of water system (not to scale).

11.3 Two-component liquid-vapour system

11.3.1 Homogeneous mixture

Consider an ideal liquid mixture, e.g., water and ammonia. Boiling point of water is 100°C while that of ammonia is –33.3°C. Thus with the addition of ammonia boiling point of water decreases as shown in the Figure 11.2. Within the envelop *ocado*, heterogeneous mixture of two phases (liquid and vapour) exist. Only vapour phase exists above the curve *oca* and liquid phase exists below the curve *oda*. At any particular composition x_2 the point d indicates the temperature at which the liquid starts boiling. The point d is called *bubble point* and the locus of the point d, i.e., the curve *oda* is known as bubble point curve. At the same composition x_2, the point *c* indicates the temperature at which the vapour starts condensing. The point *c* is called *dew point* and the locus of the point *c*, i.e., the curve *oca* is known as the dew point curve. The temperature difference $(t_c - t_d)$ is known as boiling range or condensing range. The isothermal curve *LV* is known as tie line. Along the tie line the liquid and vapour phases are in equilibrium. At *L* mole fraction of *B* in the liquid phase is x_2 and at *V* mole fraction of *B* in the gas phase is y_2.

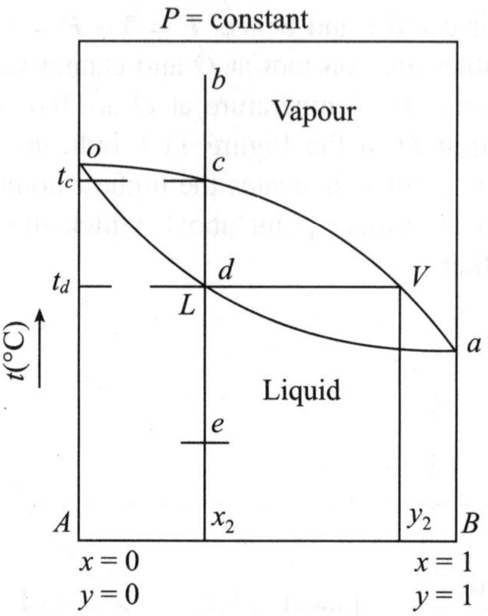

Fig. 11.2 Binary liquid-liquid phase diagram of completely miscible water/ammonia system. Component *A*: higher boiling liquid (water) and component *B*: lower boiling liquid (ammonia).

11.3.2 Azeotropic mixture

Ideal solution Ideal solution is the solution where enthalpy of mixing (ΔH^M) of all the components is zero. Mathematically, we can write: $\Delta H^M = 0$.

Raoult's law Vapor pressure of an ideal solution depends on vapor pressure of each component as well as mole fraction of each component present in the solution. Thus, at equilibrium, the total vapor pressure (*p*) of an ideal solution is given by:

$$p = x_A p_A^* + x_B p_B^* + x_C p_C^* + \ldots\ldots\ldots = \sum x_i p_i^* = p_1 + p_2 + p_3 + \ldots\ldots = \sum p_i, \text{ where } p_i = x_i p_i^*$$

p_i^* = Vapor pressure of the pure component

x_i = Mole fraction of that component in the solution

In the case of a two-component ideal solution, we have: $p = x_A p_A^* + x_B p_B^*$.

The corresponding pressure versus concentration plot is shown in Fig. 11.3.

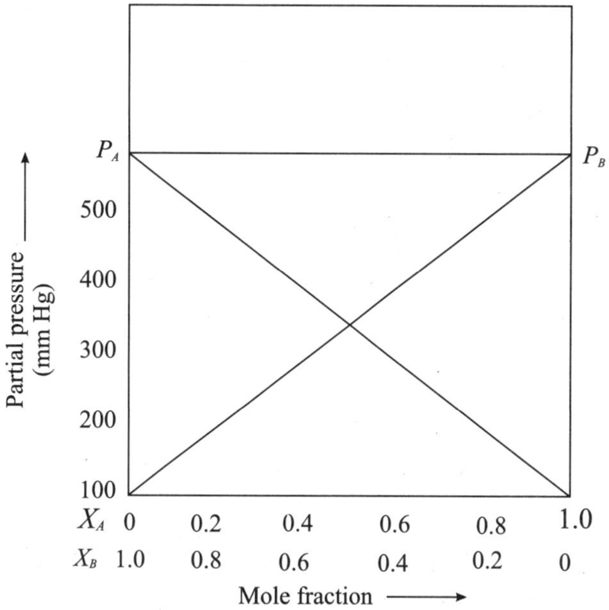

Fig. 11.3 Phase diagram of a two-component ideal solution, illustrating Raoult's law.

In the case of an ideal liquid mixture, e.g., in liquid water/ammonia system Raoult's law holds good but in the case of real mixture deviation of Raoult's law is a common phenomenon. The phase diagrams of positive and negative deviation of Raoult's law are shown in the Figures 11.4 and 11.5 respectively. In the case of positive deviation of Raoult's law, vapour pressure is higher than the corresponding ideal vapour pressure and hence boiling point usually decreases with mole fraction giving rise a minimum (Figure 11.4). At the minimum (the point *A* in the Figure 11.4) bubble point curve and the dew point curve merge and the mixture, like a pure substance, boils or condenses at a constant pressure with a constant composition. The mixture is called *azeotrope*. As the boiling point of azeotrope is lower than that of either component, it is called *minimum boiling azeotrope*. In the case of negative deviation of Raoult's law vapour pressure is lower than the corresponding ideal mixture and hence boiling point usually increases with the mole fraction, giving rise a maximum (Figure 11.4). At the maximum (the point *B* in the Figure 11.5) buble point curve and dew point curve merge and the corresponding mixture is called *maximum boiling azeotrope*.

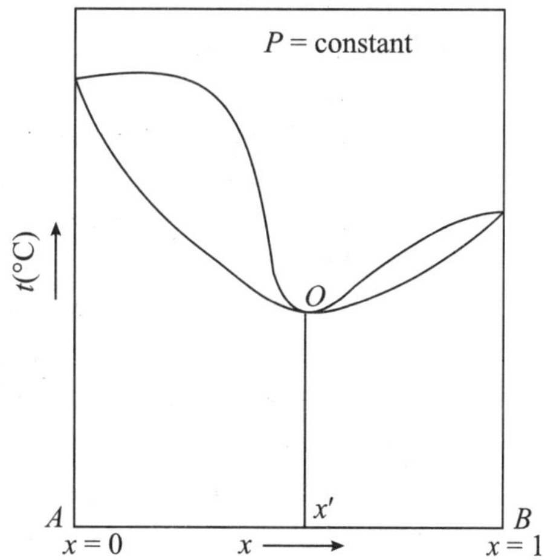

Fig. 11.4 Binary liquid-liquid phase diagram, showing positive deviation of Raoult's law. System: $CHClF_2$ and CCl_2F_2 mixture.

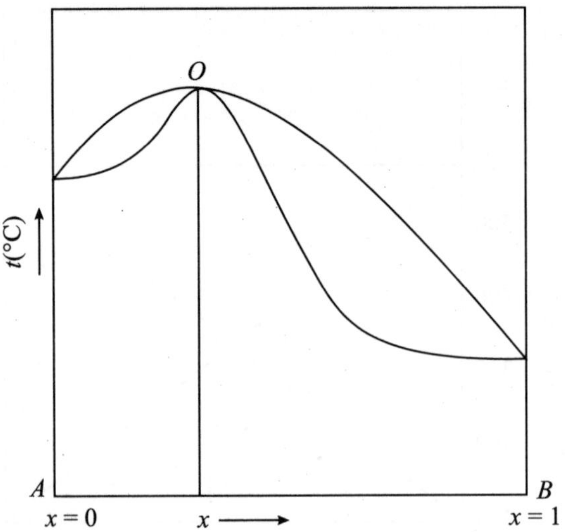

Fig. 11.5 Binary liquid-liquid phase diagram, showing negative deviation of Raoult's law. System: acetone/chloroform mixture.

11.4 Binary phase diagram for solid solutions

11.4.1 Solid solution

If an alloy contains more than one component but exists as a homogeneous single phase, it is called a solid solution. Solid solution can be divided into two categories

(a) Substitutional solid solution There are some solid solutions where some lattice sites of the parent crystal are substituted by atoms of another metal, giving rise to a homogeneous alloy. This homogeneous alloy is called substitutional solid solution. One such example is Cu/Ni alloy, where the type of crystal lattice remains unchanged but some Cu atoms are replaced by Ni atoms. However, formation of the solid solution depends on the following conditions:

(i) Size difference between atoms of the two elements must be less than 15%.

(ii) Valency of two elements must be the same.

(iii) Electronegativity difference between the two elements should be close to zero.

(iv) Crystal structures of the two elements must be same.

The above four conditions are popularly known as ***Hume-Rothery conditions***. Solid solution, Cu/Ni alloy is shown in Fig. 11.6.

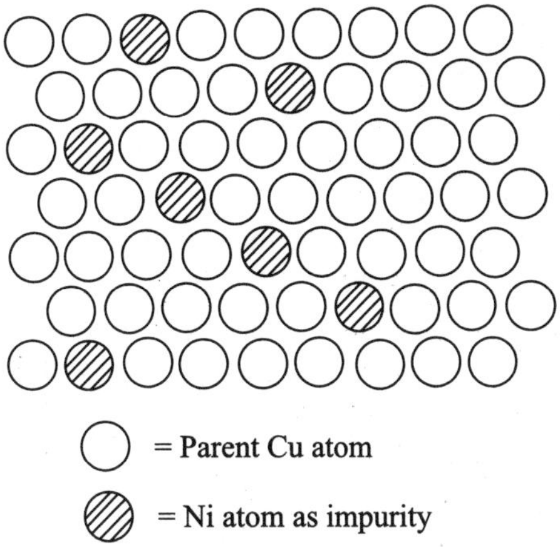

Fig. 11.6 Substitutional solid solution: Cu/Ni alloy.

(b) Interstitial solid solution There are some solid solutions where foreign atoms slowly diffuse into the voids of another metallic crystal, giving rise a homogeneous alloy. This type of solid solution is called interstitial solid solution. One such example is austenitic steel, where C atoms diffuse into the voids of Fe crystal. Figure 11.7 illustrates the Fe/C interstitial solid solution.

Fig. 11.7 Interstitial solid solution: austenitic steel.

11.4.2 Completely miscible solid solutions

In the case of solid solution no gaseous phase is present and hence effect of pressure may be neglected. Hence the degrees of freedom is given by: $F = C - P + 2 - 1 = C - P + 1$. The Figure 11.8 shows the binary phase diagram of a completely miscible solid solution (Cu/Ni) system. From the figure it is evident that freezing point of Cu increases with the addition of Ni or conversely freezing point of Ni decreases with the addition of Cu. The upper curve (*lheak*) is known as *liquidus or freezing curve* and the lower curve (*lcfgk*) is known as *solidus or melting curve*. In between the solidus and liquidus curves two phases coexist and $F = 1$ but above the liquidus curve and below the solidus curve one phase exists with $F = 2$. (*oabcd*) is the cooling curve of an alloy solution of a particular composition. Along *oa* the liquid is homogeneous. At the point *a* the solid phase starts separating out and the composition of solid and liquid phases can easily be evaluated by using the tie line *ag*. If the solution is cooled further the point *b* is reached. The corresponding compositions of liquid and solid phases can be evaluated by using the tie line *ef*. At the point *e* the phase is liquid, enriched with copper and at the point *f* the phase is solid, enriched with nickel. C_l and C_s represent % Ni in the liquid and solid phases respectively. The point *c* on the solidus curve is the last point of heterogeneity, Further cooling reaches the point *d* where homogeneous solid phase results. Other examples of this class of solid solution are Ag/Cu, Ge/Si etc.

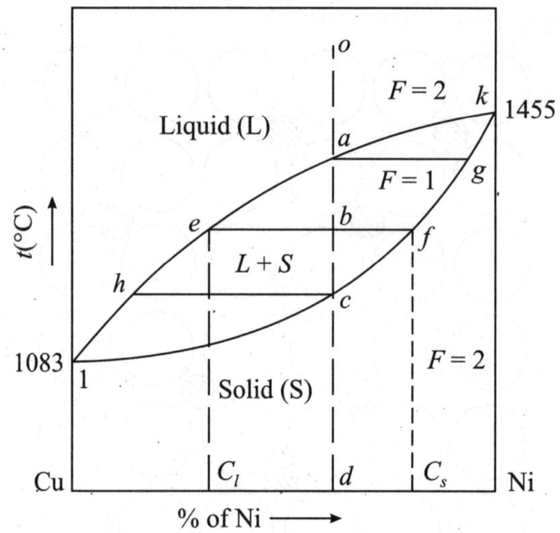

Fig. 11.8 Binary phase diagram (temperature vs. composition) of Cu/Ni system.

11.4.3 Metals are miscible in liquid phase but immiscible in solid phase

One of the good examples of this type of system is Cd/Bi system, the phase diagram of which is shown in the Figure 11.9.

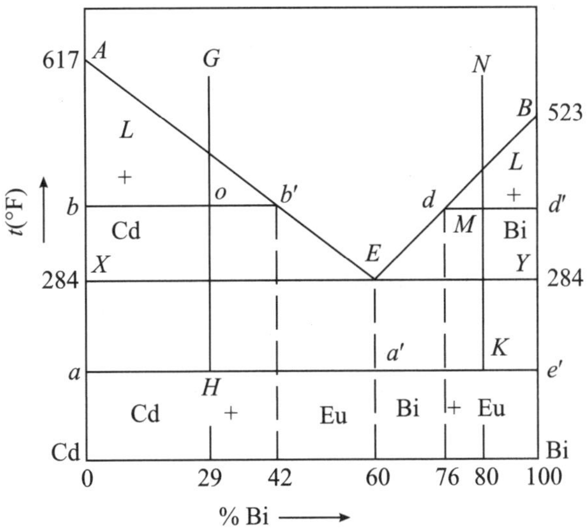

Fig. 11.9 Binary phase diagram (temperature vs. composition) of Cd/Bi system.

The line *AE* represents freezing point curve of Cd. According to colligative properties freezing point of Cd decreases with addition of Bi. Both liquid phase and solid Cd phase coexist within the triangle *AXE*. Similarly the line BE represents the freezing point of Bi, which shows that with the increase in Cd content the freezing point of Bi decreases. Within the triangle *BEY* liquid phase and solid Bi phase coexist. At the point *E* two freezing point curves AE and BE meet and the following reaction takes place at *E*:

$$\text{Liquid} \underset{\text{Heating}}{\overset{\text{Cooling}}{\rightleftharpoons}} \text{Cd(s)} + \text{Bi(s)}$$

Thus three phases coexist at *E* and hence $F = C - P + 1 = 2 - 3 + 1 = 0$. The point *E* is known as eutectic point and the line *XEY* is known as *eutectic line*. At the eutectic point eutectic temperature is 284°F and composition 60% Bi/40% Cd. Consider the point *G*, where the molten alloy contains 29% Bi. If the alloy is cooled, the point *o* is reached. At *o* both melt and solid Cd(s) coexist. Further cooling touches the eutectic line *XEY*, where both Cd(s) and eutectic mixture (Eu) start separating out. The cooling ends at the point *H*, which is a mixture of Cd(s) and eutectic mixture (Eu) only.

Structural composition/lever rule Assume that *w* be the net amount of molten alloy at *G*. Bi-content at *G* is 29% , i.e., 0.29 w ...(i).

At the point *o* molten phase appears at *b′* and the amount of molten phase is *y*. The solid Cd phase appears at *b* and its amount is *x*. At *b*, Bi-content is zero (0) and at *b′*, Bi-content is 42%. Thus net Bi-content = $(0x + 0.42y)$ ...(ii)

Equating (i) and (ii) we get, $0x + 0.42y = 0.29w = 0.29(x + y)$ [since, $w = x + y$]

Or, $(0.42 - 0.29)y = (0.29 - 0)x$

Or, $$y = \frac{(0.29-0)x}{(0.42-0.29)} \quad \text{(iii)}$$

and $$(x + y) = \frac{(0.42-0)x}{(0.42-0.29)} \quad \text{(iv)}$$

From (iv) we get, $\frac{x}{(x+y)} = \frac{(0.42-0.29)}{(0.42-0)}$

Or, $$\% \text{ Cd-phase} = \frac{(100x)}{(x+y)} = \frac{100(0.42-0.29)}{(0.42-0)} = 30.95\%$$

Again from (iii) and (iv), we have:

$$\frac{y}{(x+y)} = \frac{(0.29-0)}{(0.42-0)}$$

Or, $$\% \text{ melt phase} = \frac{100\,y}{(x+y)} = \frac{100(0.29-0)}{(0.42-0)} = 69.05\%$$

To find out the composition at *o* draw a line parallel to composition axis through the point *o*. The line *bob*' is called *lever*.

$$\text{So } \% \text{ melt phase} = \frac{100(ob)}{(bb')} = \frac{100(0.29-0)}{(0.42-0)} = 69.05\% \text{ and}$$

$$\% \text{ Cd(s) phase} = \frac{100(ob')}{(bb')} = \frac{100(0.42-0.29)}{(0.42-0)} = 30.95\%$$

To find out the composition at *H* draw a lever *aHa′*. At *a′* 100% Eu and at the point *a* 100% Cd(s) phase exist. Thus at *H*:

$$\% \text{ Eu} = \frac{100(\text{aH})}{(aa')} = \frac{100(0.29-0)}{(0.60-0)} = 48.33\% \text{ and}$$

$$\% \text{ Cd(s) phase} = \frac{100(\text{Ha}')}{(aa')} = \frac{100(0.60-0.29)}{(0.60-0)} = 51.67\%$$

Similarly to find out the composition at *M* draw a lever *dMd′*. At the point *d′* 100% Bi(s) phase and at the point *d* 100% melt exist. Thus at *M*:

$$\% \text{ melt} = \frac{100(\text{Md}')}{(dd')} = \frac{100(100-80)}{(100-76)} = 83.33\% \text{ and}$$

$$\% \text{ Bi(s) phase} = \frac{100(\text{dM})}{(dd')} = \frac{100(80-76)}{(100-76)} = 16.67\%$$

To find out the composition at *K* draw a lever *a*′Ke′. At the point *e*′ 100% Bi(s) phase and at the point *a*′ 100% Eu phase exist. Thus at *K*:

$$\% \text{ Eu} = \frac{100(\text{Ke}')}{(a'e')} = \frac{100(80-80)}{(100-60)} = 50\% \text{ and}$$

$$\% \text{ Bi(s) phase} = \frac{100(\text{a}'\text{K})}{(a'e')} = \frac{100(80-60)}{(100-60)} = 50\%$$

Now consider the point *N*, where Bi-content is around 80%. If the molten alloy is cooled from the point *N*, the point *M* is reached. At *M* both melt and Bi(s) phases coexist. On further cooling to 284°F liquid phase disappears and eutectic mixture (Eu) and Bi(s) phases are separated out at constant composition. At the point *K* both Eu-phase and Bi(s)-phase coexist.

Another example of this type of system is Ag/Pb system, binary phase diagram of which is shown in the Figure 11.10.

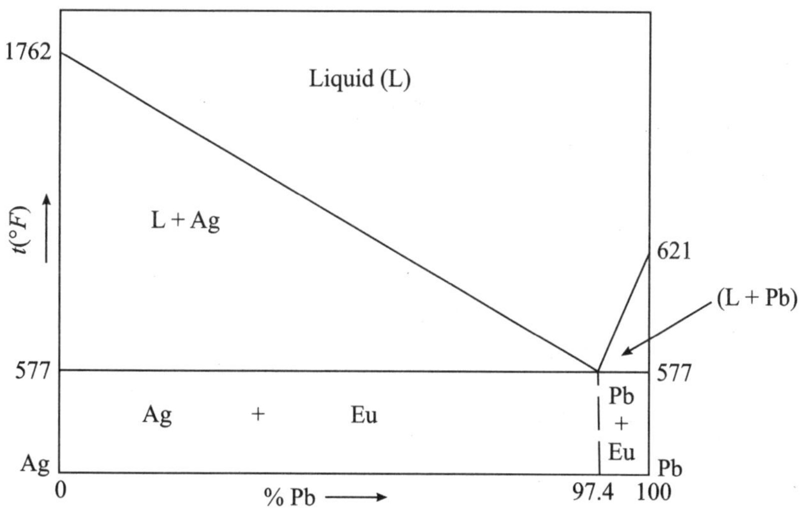

Fig. 11.10 Binary phase diagram (temperature vs. composition) of Ag/Pb system.

11.4.4 Metals are miscible in liquid phase but only partially miscible in solid phase

One of the examples of this type of system is Pb/Sn system. In this system Sn is partially miscible in Pb-phase to form homogeneous α-phase, containing a maximum of 18% Sn. Similarly Pb is partially miscible in Sn-phase to form β-phase, containing a maximum of 3% Pb. The binary phase diagram of Pb/Sn system is shown in the Figure 11.11.

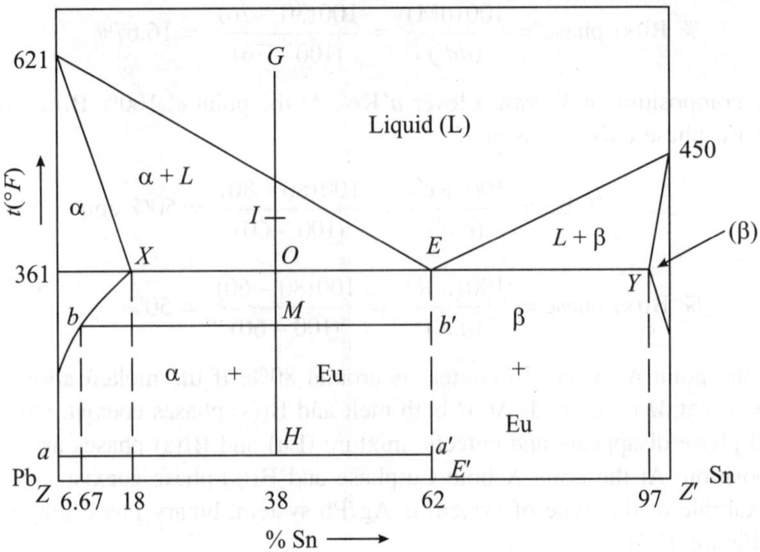

Fig. 11.11 Binary phase diagram (temperature vs. composition) of Pb/Sn system

In the Figure 11.11, *XEY* represents the eutectic line and *E* is the eutectic point, having coordinate (361°F, 62% Sn). The molten alloy having composition 38% Sn is cooled continuously. At the point *G* the molten alloy is homogeneous. At the point *O* melt phase disappears, giving rise two solid phases α- and β-phases. At the point *M* draw the lever *bMb′*.

$$\% \text{ Eu} = \frac{100(\text{bM})}{(bb')} = \frac{100(38-6.67)}{(62-6.67)} = 56.62\% \text{ and}$$

$$\% \ \alpha\text{-phase} = \frac{100(\text{Mb}')}{(bb')} = \frac{100(62-38)}{(62-6.67)} = 43.38\%$$

The cooling line ends at the point *H* where another lever *aHa′* is drawn.

$$\% \text{ Eu} = \frac{100(\text{aH})}{(aa')} = \frac{100(38-0)}{(62-0)} = 61.29\% \text{ and}$$

$$\% \text{ Pb(s)} = \frac{100(\text{Ha}')}{(aa')} = \frac{100(62-38)}{(62-0)} = 38.71\%$$

Microstructures at different points of cooling line *GIOMH* are shown in the Figure 11.12. At the point *G* in the Figure 11.11 homogeneous liquid phase exists. The line *EE′* in the same figure represents eutectic composition. Alloys belonging to the region *EXZE′* are known as *hypoeutectic alloys* and alloys belonging to the region *EYZ′E′* are known as *hypereutectic alloys*.

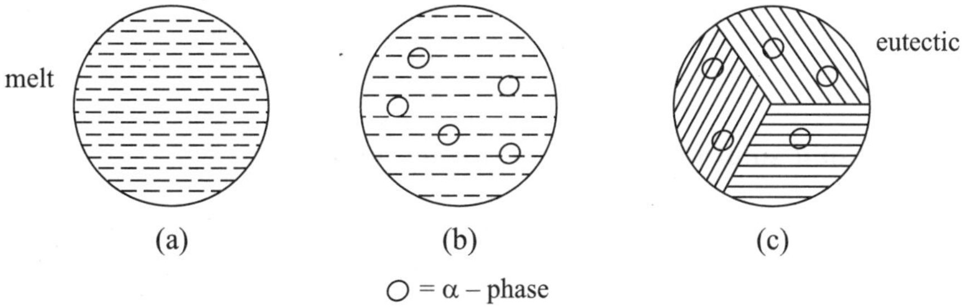

Fig. 11.12 Microstructures of Pb/Sn system at different temperatures of cooling line *GIOMH*, (a) homogeneous liquid phase (melt) at the point *G* in the Figure 11.11, (b) heterogeneous phase (melt and α-phase) at the point *I* in the Figure 11.11, (c) heterogeneous phase, consisting of eutectic and α-phase, at the point *M* in the Figure 11.11.

Another system of this type is Ag/Cu system shown in the Figure 11.13.

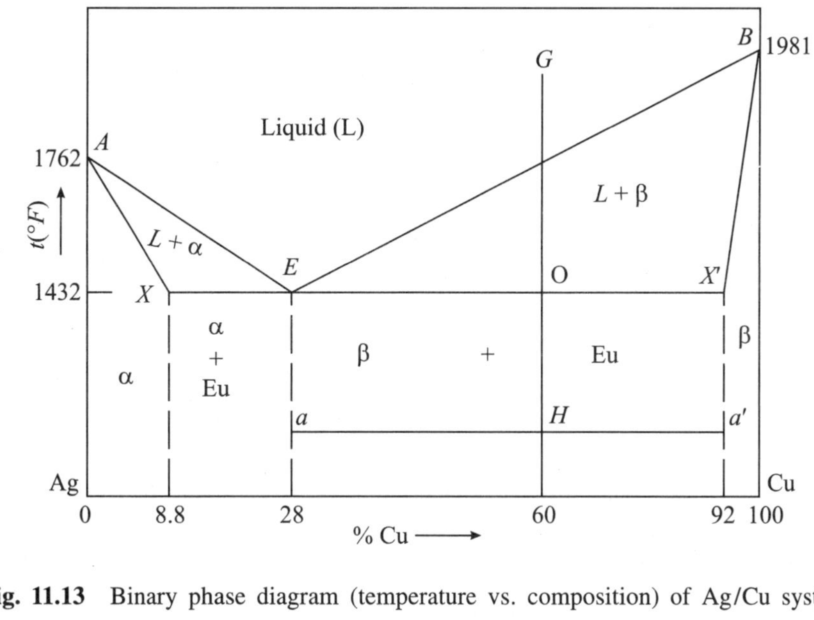

Fig. 11.13 Binary phase diagram (temperature vs. composition) of Ag/Cu system.

Eutectic point is at 1432°F with 28% Cu. Maximum concentration of Cu in α-solid solution is 8.8%. Maximum concentration of Ag in β-solid solution is 8%. The line GOH represents cooling line of an alloy having composition of 60% Cu and 40% Ag. At the point *H* draw the lever *aHa′*.

$$\% \text{ Eu} = \frac{100(\text{Ha}')}{(aa')} = \frac{100(92-60)}{(92-28)} = 50\% \text{ and}$$

$$\% \ \beta\text{-phase in the alloy} = \frac{100(\text{aH})}{(aa')} = \frac{100(60. - 28)}{(92 - 28)} = 50\%$$

11.4.5 Binary solid solutions having peritectic point

Consider the case of an alloy, where the melting points of the components differ widely. In that case, on addition of low melting component, melting point of the alloy decreases sharply and the solidus line, thus formed, overshadows the solidus line of low melting component. As a result of this, two solidus lines do not intersect but there exists a point where three phases are in equilibrium. The point is known as *peritectic point* and the horizontal line passing through this point is known as peritectic line. One of such system is Ag/Pt system. The binary phase diagram of Ag/Pt system is shown in the Figure 11.14.

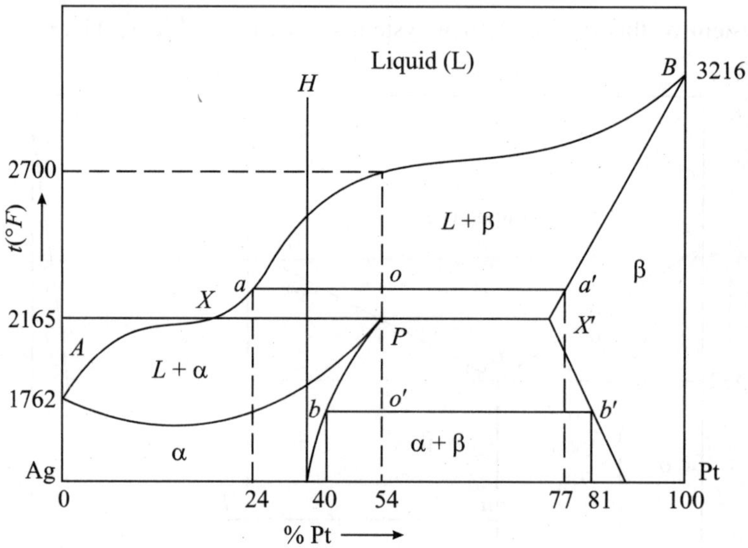

Fig. 11.14 Binary phase diagram (temperature vs. composition) of Ag/Pt system

Draw a lever through the point *o*, i.e., *aoa′*.

$$\% \ \beta\text{-phase in the alloy} = \frac{100(ao)}{(aa')} = \frac{100(54 - 24)}{(77 - 24)} = 56.6\% \text{ and}$$

$$\% \text{ melt} = \frac{100(oa')}{(aa')} = \frac{100(77 - 54)}{(77 - 24)} = 43.4\%$$

Draw another lever through the point *o′*, i.e., *bo′b′*.

$$\% \ \beta\text{-phase} = \frac{100(bo')}{(bb')} = \frac{100(54 - 40)}{(81 - 40)} = 34.1\% \text{ and}$$

$$\% \ \alpha\text{-phase} = \frac{100(o'b')}{(bb')} = \frac{100(81-54)}{(81-40)} = 65.9\%$$

P is the peritectic point where two phases disappear giving rise a new phase. The overall reaction is given by:

$$\text{Liquid} + \beta(s) \underset{\text{Heating}}{\overset{\text{Cooling}}{\rightleftharpoons}} \alpha(s)$$

Thus three phases coexist at P and hence the point P is invariant. The line XPX' is known as peritectic line. It is to be noted that at eutectic point one phase, on cooling, gives two new phases but at peritectic point two phases disappear on cooling giving rise a new phase. Consider the alloy containing 40% Pt at the point H, where only homogeneous liquid phase exists. On cooling to around 2500°F β-particles begin to form and the β-particles grow in size as the cooling proceeds. The heterogeneous phase is cooled further to peritectic temperature where α-phase crystals begin to form surrounding the β-phase particles. On further cooling α-phase particles grow in size through diffusion of Pt from β-phase and diffusion of Ag from α-phase. If sufficient time is allowed for diffusion the formation of homogeneous α-phase particles will be completed. The following figure shows the change in microstructure taking place during cooling. It is evident from the Figure 11.15 that at room temperature only α-phase exists below 40% Pt composition but the microstructure shows that *dendritic* β-phases are entrapped within the α-phase, as shown in the Figure 11.15(e).

Reason During solidification α-phase begins to form surrounding the β-phases. As the α-grains grow in size β-grains are entrapped and are unable to diffuse out of the α-phase. As a result of this, β-grains are entrapped within the α-phase. If the alloy is subjected to hot working, shielding of α-phase disappears, rendering to a complete diffusion of β-grains and equilibrium microstructure is reached.

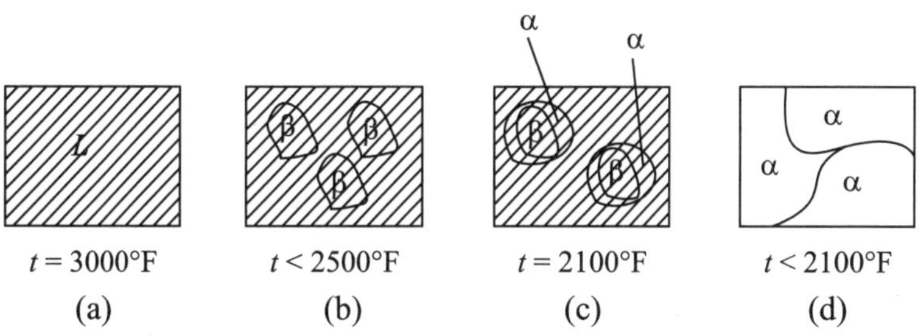

Fig. 11.15 Microstructures of Ag/Pt system at different temperatures: **(a)** homogeneous liquid phase at 3000°F, **(b)** heterogeneous phase containing liquid and β-phases at below 2500°F, **(c)** α-phase, β-phase and liquid phase at around peritectic temperature, **(d)** only α-phase just below peritectic temperature, i.e., 2100°F.

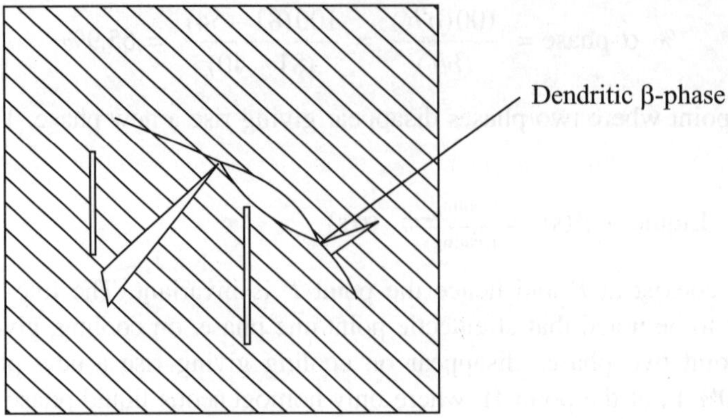

Fig. 11.15(e) Microstructure of Ag/Pt alloy containing less than 40 % Pt. Dentridic β-grains are encased within the α-phase.

Another example of peritectic system is Fe/Ni system, the binary phase diagram of which is shown in the Figure 11.16. The point *O* is called peritectic point, where the following reaction takes place:

$$\text{Liquid} + \delta(s) \underset{\text{Heating}}{\overset{\text{Cooling}}{\rightleftharpoons}} \gamma(s)$$

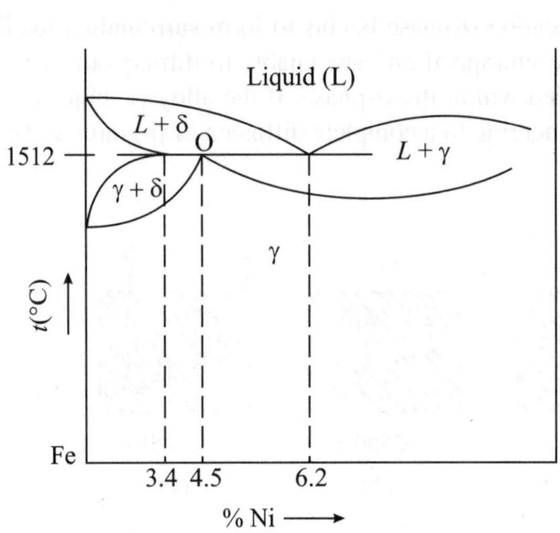

Fig. 11.16 Binary phase diagram (temperature vs. composition) of Fe/Ni system.

11.4.6 Compound forming alloys

In some cases during formation of alloys two components react to form compounds, known as intermetallic compounds, which are similar to common compounds like, water, sodium chloride etc. These intermetallic compounds are usually very hard, brittle with little or no strength. They are primarily used as hardeners. Examples of intermetallic compounds are Fe_3C, $CuAl_2$, $NiCd_7$, CuSn, SbSn etc. Some of the hardest intermetallic compounds are tungsten carbide (WC), boron carbide (B_4C) etc. Both WC and B_4C are as hard as diamond. Binary phase diagram of compound forming alloys may be of three types:

(i) In some cases intermetallic compound has a definite melting point and retains its composition in the molten state. This type of intermetallic compound is called compound with *congruent melting point*. One example of this type of alloy is Zn/Mg alloy. The Figure 11.17 shows the binary phase diagram of Zn/Mg alloy. Melting points of Zn and Mg are 788°F (420°C) and 1202°F (590°C) respectively. The components Zn and Mg form an intermetallic compound $MgZn_2$ which has a sharp melting point at 1094°F. The line CF represents the compound line. The mole percent of Mg in $MgZn_2$ is 33.33%.

Eutectic points

(1) Point B temperature is 716°F (380°C), composition is 6.67% Mg (or, 79% Zn and 21% $MgZn_2$), reaction is given by:

$$\text{Liquid} \underset{\text{Heating}}{\overset{\text{Cooling}}{\rightleftharpoons}} MgZn_2(s) + Zn(s)$$

(2) Point D temperature is 656°F (347°C), composition is 74% Mg (or, 39% $MgZn_2$ and 61% Mg), reaction is given by:

$$\text{Liquid} \underset{\text{Heating}}{\overset{\text{Cooling}}{\rightleftharpoons}} MgZn_2(s) + Mg(s)$$

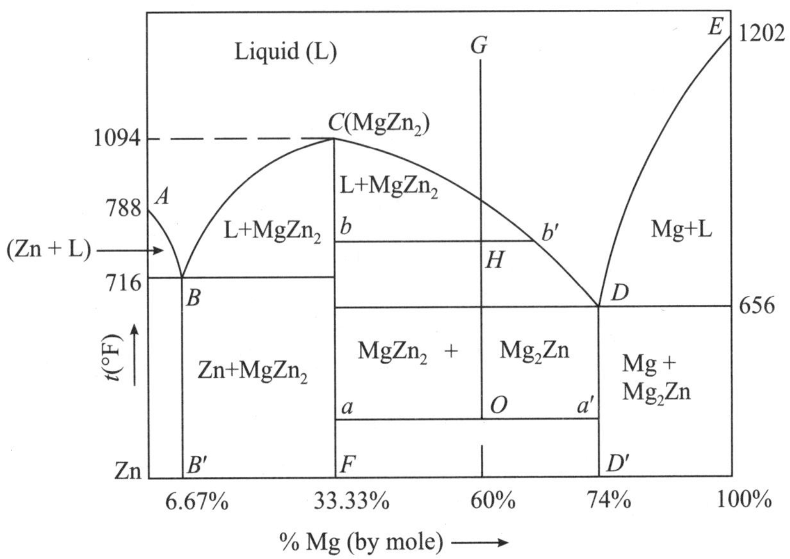

Fig. 11.17 Binary phase diagram (temperature vs. composition) of Zn/Mg system.

At the point F, $MgZn_2$ is 100%. If an alloy, having 60% Mg is cooled from the point G, the cooling line GHO is formed. At the point O, $MgZn_2$-content is 60.5% and Mg-content is 39.5%. At the point D, $MgZn_2$-content is 39% and Mg-content is 61%. A lever aoa' is drawn through the point O. the line DD' represents eutectic composition.

$$\text{Thus \% eutectic} = \frac{100(ao)}{(aa')} = \frac{100(39.5-0)}{(61-0)} = 64.75\% \text{ and}$$

$$\%\ MgZn_2 = \frac{100(oa')}{(aa')} = \frac{100(61-39.5)}{(61-0)} = 35.25\%$$

(ii) In some cases melting point of the intermetallic compound is far lower than that of pure metal. In that case the intermetallic compound is formed under peritectic reaction. The melting point of the intermetallic compound is not at all realized practically and the intermetallic compound is said to have an *incongruent melting point*. It is well illustrated in the Figure 11.18. An alloy contains two components A and B, having melting points marked in the Figure 11.18, G and K respectively. The metal B is low melting point and forms an eutectic point E at 80% of B. The metal A has high melting point and GS represents melting point curve of the alloy AB. At 20% B, a compound A_mB_n is formed but the compound has far lower melting point than the metal A. HS represents the melting point line of the compound A_mB_n. Thus above the line CHS heterogenous phase exists but below the line solid phase exists. At the point H three phases coexist through a reaction equilibrium as follows:

$$\text{A(s) + Liquid} \underset{\text{Heating}}{\overset{\text{Cooling}}{\rightleftharpoons}} A_mB_n\text{(s)}$$

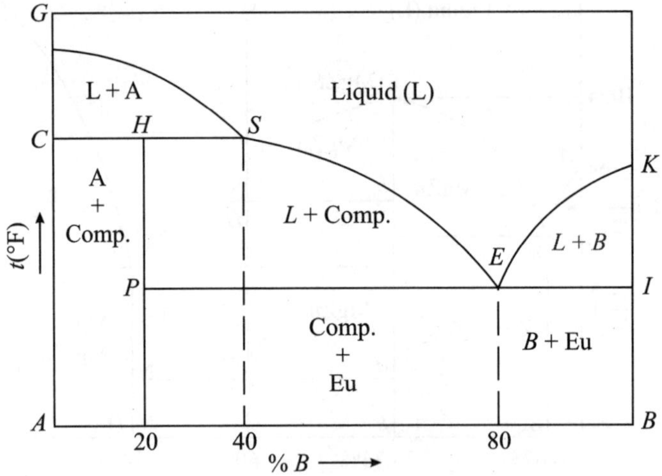

Fig. 11.18 Phase diagram of a compound-forming alloy, having peritectic reaction.

Thus the point *H* is called peritectic point. The curve *SE* represents melting point curve of the compound. The line *CHS* indicates that at the point *S* liquid melt, having 40% *B*, reacts with pure metal *A*, represented by the point *C*, to form the compound A_mB_n of composition 20% *B*. The point *H* represents the incongruent melting point of the compound A_mB_n.

(iii) In some cases components of an alloy form different compounds at different compositions such that one compound has a *congruent melting point* and the other compound has an *incongruent melting point.* One such example is Ni/Mg alloy. The binary phase diagram of Ni/Mg alloy is shown in the Figure 11.16. Two eutectic points are evident, one is at E_1, where E_1E_1' represents corresponding eutectic composition and the other is at E_2, where E_2E_2' represents corresponding eutectic composition.

(1) Eutectic at E_1 temperature is 2012°F, overall Mg-content is 23%, and reaction is given by:

$$\text{Melt} \underset{\text{Heating}}{\overset{\text{Cooling}}{\rightleftharpoons}} MgNi_2(s) + Ni(s) \quad \text{[composition: 69\% } MgNi_2 \text{ and 31\% Ni]}$$

Melting point of the compound $MgNi_2$ is 2102°F.

(2) Eutectic at E_2 temperature is 945°F, overall Mg-content is 89%, and reaction is given by:

$$\text{Melt} \underset{\text{Heating}}{\overset{\text{Cooling}}{\rightleftharpoons}} Mg_2Ni(s) + Mg(s) \quad \text{[composition: 33.33\% } Mg_2Ni \text{ and 66.67\% Mg]}$$

(3) Peritectic point *P* Temperature is 1400°F, overall Mg-content is 66.67%, and reaction is given by:

$$\text{Melt} + MgNi_2(s) \underset{\text{Heating}}{\overset{\text{Cooling}}{\rightleftharpoons}} Mg_2Ni(s) \quad \text{[composition: 40\% } MgNi_2 \text{ and 60\% Eu(2)]}$$

Consider an alloy melt, having composition of 48% Mg. The location of alloy melt is at *G* in the Figure 11.19. *Goo′* is the cooling line of the alloy melt. Draw a lever *aoa′* through the point *o*, where heterogeneous mixture of two phases melt and $MgNi_2$ is present. At the point *C*, 100% $MgNi_2(s)$ is present, % Mg in the melt at *a′* is 50%, % Mg in the heterogeneous mixture at *o* is 22.22% and % Mg at *a* is 0%. At the point *o*:

$$\%\ MgNi_2(s) = \frac{100(oa')}{(aa')} = \frac{100(50-22.22)}{(50-0)} = 55.56\% \text{ and}$$

$$\%\ \text{Melt} = \frac{100(ao)}{(aa')} = \frac{100(22.22-0)}{(50-0)} = 44.44\%$$

Draw another lever *bo′b′* through the point *o′*.

Point *C*: 100% $MgNi_2(s)$, point *P′*: 100% $Mg_2Ni(s)$, point *o′*: 44.44% $Mg_2Ni(s)$.

$$\text{Thus \% MgNi}_2\text{(s)} = \frac{100(o'b')}{(bb')} = \frac{100(100-44.44)}{(100-0)} = 55.56\%$$

It is interesting to note that, at o, % melt is 44.44% and the same melt is totally converted to the compound Mg_2Ni(s) on cooling to o'.

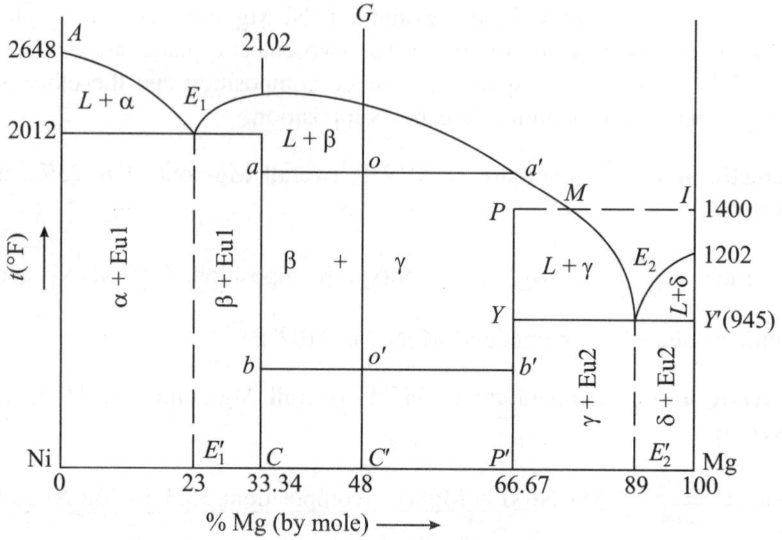

Fig. 11.19 Binary phase diagram of a compound forming alloy (Ni/Mg system) containing both eutectic and peritectic points. Melting point of Ni = 2648°F (1453°C) and melting point of Mg = 1202°F (650°C).

11.4.7 Other eutectic systems

11.4.7.1 Aluminium-Silica system

The binary phase diagram is shown in the Figure 11.20. α-phase appears as long crystals of dentritic structure, whereas β-phase appears as short crystals of a definite geometrical shape. This difference of appearance is mainly due to mixing of pronounced metallic characteristics of Al with Si which is predominantly non-metallic. The crystals present in the α-phase and β-phase are known as primary α- and primary β-crystals respectively. The α-crystal, present in the eutectic mixture is known as *secondary* α-crystal and similarly, β-crystal, present in the eutectic mixture, is known as *secondary* β-crystal.

Eutectic point is at *E*: temperature is 1000°F, composition is 12% Si and 88% α-phase.

Let us consider an Al/Si alloy-melt at the point *G* in the Figure 11.20. The melt is allowed to cool and Goo' represents the cooling line. Draw two tie lines, aoa' and $bo'b'$ through the points

o and o' respectively. Applying lever rule along aoa' we have:

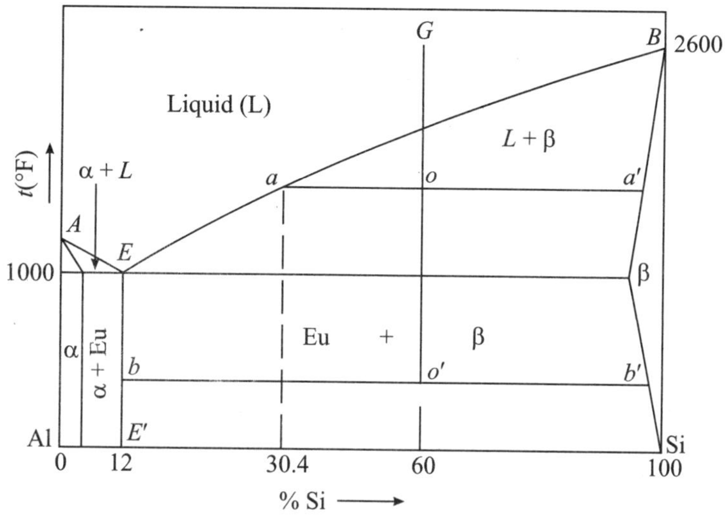

Fig 11.20 Binary phase diagram (temperature vs. composition) of Al/Si system.

Primary β-phase, containing 98% Si and liquid phase, containing 30.4% Si.

$$\% \text{ primary } \beta\text{-phase} = \frac{100(ao)}{(aa')} = \frac{100(60-30.4)}{(98-30.4)} = 43.8\% \text{ and}$$

$$\% \text{ liquid phase (or melt)} = \frac{100(oa')}{(aa')} = \frac{100(98-60)}{(98-30.4)} = 56.2\%$$

Along bob', we have: primary β-phase, containing 99% Si and eutectic phase, containing 12% Si.

$$\% \text{ primary } \beta\text{-phase} = \frac{100(bo')}{(bb')} = \frac{100(60-12)}{(99-12)} = 55.2\% \text{ and}$$

$$\% \text{ eutectic mixture} = \frac{100(o'b')}{(bb')} = \frac{100(99-60)}{(99-12)} = 44.8\%$$

11.4.7.2 Lead-Magnesium system

The binary phase diagram is shown in the Figure 11.21. Pb/Mg system forms an intermetallic compound Mg_2Pb, having a congruent melting point of around 1000°F. Intermetallic compound, having a congruent melting point, melts in the same manner as a pure metal, whereas intermetallic compound, having incongruent melting point, instead of melting decomposes into two phases, usually one solid and one liquid phase.

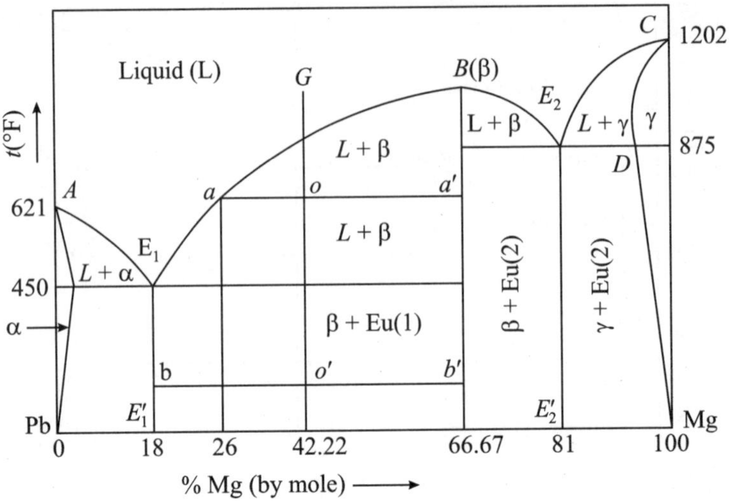

Fig. 11.21 Binary phase diagram (temperature vs. composition) of Pb/Mg system.

Eutectic point at E_1 temperature is 450°F, composition is 26.67% (by mole) $Mg_2Pb(s)$ and 73.33% (by mole) α-phase. At the point *A*, Pb-content is 100% and at the point *B*, Mg_2Pb-content is 100%. Reaction is given by:

$$\text{Melt} \underset{\text{Heating}}{\overset{\text{Cooling}}{\rightleftharpoons}} \text{Pb(s) (}\alpha\text{-phase)} + \text{Mg}_2\text{Pb(s) (}\beta\text{-phase)}$$

Eutectic point at E_2 Temperature is 875°F, composition: 52.78% γ-phase and 47.22% $Mg_2Pb(s)$-phase. At the point *B* $Mg_2Pb(s)$-content is 100%, at the point *C* Mg(s)-content is 100% and at the point *D* γ-phase is 100%. Reaction is given by:

$$\text{Melt} \underset{\text{Heating}}{\overset{\text{Cooling}}{\rightleftharpoons}} \text{Mg(s) (}\gamma\text{-phase)} + \text{Mg}_2\text{Pb(s) (}\beta\text{-phase)}$$

γ-phase at the point *D* is a homogeneous solid solution, containing 93% Mg with 7% Pb by mole basis. Consider the Pb/Mg alloy melt at the point *G* and it is allowed to cool along *Goo′*. Draw two tie lines *aoa′* and *bo′b′* through the points *o* and *o′* respectively.

At *a′* and *b′*: $Mg_2Pb(s)$ (β-phase) is 100 % (by mole basis).

At *o* and *o′*: $Mg_2Pb(s)$ (β-phase) is 57.78 % (by mole basis).

At *a*: $Mg_2Pb(s)$ (β-phase) is 39 % (by mole basis).

At *b*: $Mg_2Pb(s)$ (β-phase) is 26.67 % (by mole basis).

Consider the lever *aoa′*:

$$\% \ \beta\text{-phase} = \frac{100(ao)}{(aa')} = \frac{100(57.78 - 39)}{(100 - 39)} = 30.79\% \text{ and}$$

$$\% \text{ melt} = \frac{100(oa')}{(aa')} = \frac{100(100 - 57.78)}{(100 - 39)} = 69.21\%$$

Consider the lever $bo'b'$:

$$\% \ \beta\text{-phase} = \frac{100(bo')}{(bb')} = \frac{100(57.78 - 26.67)}{(100 - 26.67)} = 42.42\% \text{ and}$$

$$\% \text{ eutectic} = \frac{100(o'b')}{(bb')} = \frac{100(100 - 57.78)}{(100 - 26.67)} = 57.78\%$$

11.5 Solid state reactions

As we have seen earlier at eutectic point a liquid phase gives rise two solid phases. When all the three phases at the eutectic point are solid phases, the invariant point is called *eutectoid*. Thus the eutectoid reaction is given by:

$$\gamma(s) \underset{\text{Heating}}{\overset{\text{Cooling}}{\rightleftharpoons}} \alpha(s) + \beta(s)$$

Similarly, when all the three phases at the peritectic point are solid phases the invariant point is called *peritectoid*. Thus peritectoid reaction is given by:

$$\alpha(s) + \beta(s) \underset{\text{Heating}}{\overset{\text{Cooling}}{\rightleftharpoons}} \gamma(s)$$

Thus solid state reactions differ from liquid state reactions in two different ways: (a) solid state reactions occur very slowly with greater amount of super-cooling with respect to liquid state reactions and hence hardly reach to the equilibrium conditions, (b) during solid state reactions a definite orientation of crystal lattices occur to form a new crystal lattice of different structure and phase. Thus the crystal structure of new phase has a definite orientation relationship to the crystal structure of the parent phase.

11.5.1 Precipitation reaction

In most cases it has been observed that one component of binary system is partially soluble in another component to form a homogeneous solid solution. However, solubility decreases with the decrease in temperature and hence soluble component is slowly precipitated out from the homogeneous solid solution. One such example is Al/Ag system, phase diagram of which is shown in the Figure 11.22.

Consider the case of Al/Ag alloy, having composition of 15% Ag (the point G in the Figure 11.22). The alloy is slowly cooled to the point o where only homogeneous α-phase, having 48% Ag exists. So, there is no change in solubility of Ag in α-phase during cooling along

Go. If the mixture is further cooled to o', % Ag in α-phase is reduced to only 1% and the remaining Ag is slowly precipitated out to form β-phase. Draw a tie line bb' through o':

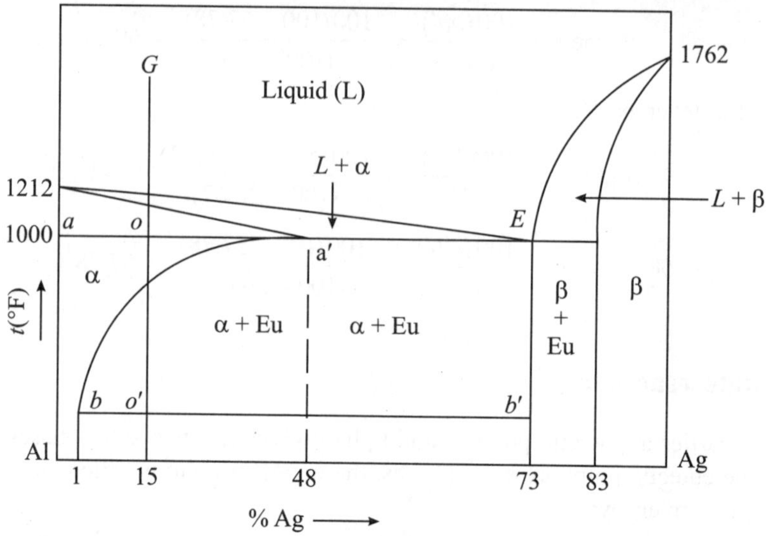

Fig. 11.22 Binary phase diagram (temperature vs. composition) of Al/Ag system. Eutectic point: 1000°F and 73% Ag.

$$\% \text{ α-phase} = \frac{100(o'b')}{(bb')} = \frac{100(73-15)}{(73-1)} = 80.56\% \text{ and}$$

$$\% \text{ eutectic} = \frac{100(bo')}{(bb')} = \frac{100(15-1)}{(73-1)} = 19.44\%$$

Thus the solubility of Ag decreases from 48% to only 1% on decreasing the temperature from 1000°F to 180°F. As the temperature goes down β-grains are separated out from α-grains and get precipitated as plates on α-grains.

11.6 Ternary systems

Usually ternary systems are investigated at constant temperature and pressure. Phase diagram is represented by an equilateral triangle with the components located at the vertices as shown in the Figure 11.20. Vertical distance from the point A to its base BC is taken as 100% of A. similarly the vertical distance from the point B to its base AC and that the point C to its base AB are taken as 100% of B and 100% of C respectively. All the three vertical distances are same and is taken as h. Consider an alloy at the point o.

$$\% A = \frac{100(oa)}{h} = 68.57\%, \quad \% B = \frac{100(ob)}{h} = 21.43\%, \quad \% C = \frac{100(oc)}{h} = 10\%$$

Consider the line *DE*, parallel to the base line *BC*. Along *DE* concentration of *A* remains unchanged and hence *DE* represents the change in composition of *B* and *C* in a given alloy at a fixed concentration of *A*. Similarly the line HI represents the change in composition of *A* and *C* at a fixed concentration of *B* and the line GF represents the change in composition of *A* and *B* at a fixed concentration of *C*.

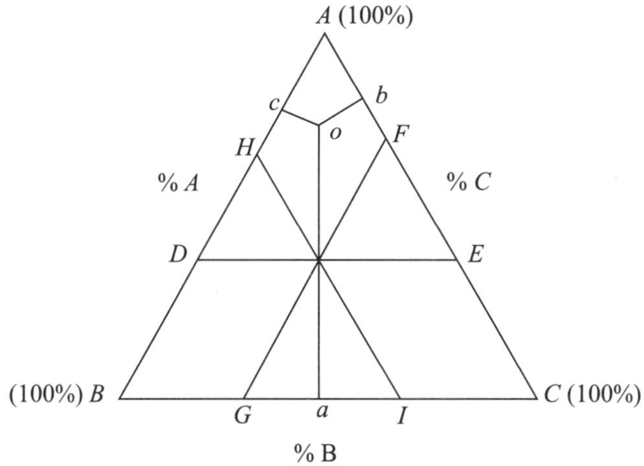

Fig. 11.23 Ternary phase diagram of a 3-component alloy at constant temperature and pressure.

11.7 Iron-Carbon diagram

Study of iron-carbon system is valuable since, (i) it helps explain the properties of steel, (ii) several solid state reactions can be studied by means of Fe/C phase diagram, which is shown in the Figure 11.24.

The iron-carbon alloy has three invariant points, *S*, *B* and *R*.

(i) At *S*, the following reaction occurs:

$$\gamma(s) \underset{\text{Heating}}{\overset{\text{Cooling}}{\rightleftharpoons}} \alpha(s) + \text{Ce}(s)$$

All the three phases are in solid state and hence the mixture is called eutectoid mixture. Ce represents cementite, having formula Fe_3C, an intermetallic compound. Eutectic temperature is 700°C or 1292°F. The carbon content at the eutectoid point is 0.85%.

(ii) At *B*, the following reaction occurs:

$$\text{Liquid} \underset{\text{Heating}}{\overset{\text{Cooling}}{\rightleftharpoons}} \gamma(s) + \text{Ce}(s)$$

Here liquid phase breaks into two solid phases and the resultant mixture is called eutectic mixture. Eutectic temperature is 1130°C or 2066°F. The carbon content at the eutectic point *B* is 4.3%.

(iii) At *R*, the following reaction occurs:

$$\text{Liquid} + \delta(s) \underset{\text{Heating}}{\overset{\text{Cooling}}{\rightleftharpoons}} \gamma(s)$$

Here two phases disappear to form a new phase and hence the mixture is called peritectic mixture. Peritectic temperature is 1493°C or 2719°F. The carbon content at the peritectic point *R* is 0.16%.

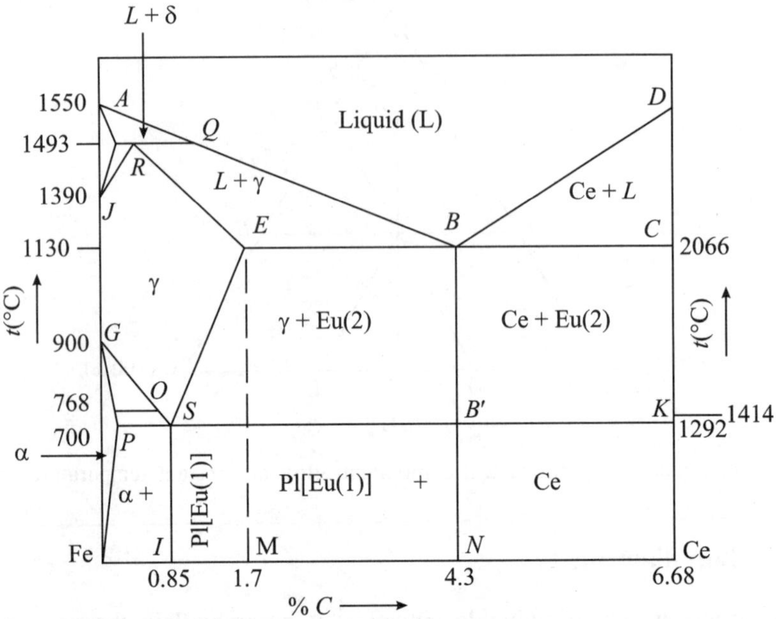

Fig. 11.24 Partial binary phase diagram of iron-carbon alloys. Ce = cementite, represented by Fe_3C, Pl = Pearlite or Eutectic(1) or Eu(1)

11.7.1 Allotropy

Iron exists in different forms at different temperatures. All theses forms are called allotropes of iron.

(i) γ-iron Characteristic features: (a) Its crystal structure is FCC, (b) it is non-magnetic, (c) it exists above 700°C. The area under the curve JRESG in the Figure 11.24 constitutes γ-iron, which is often called austenite, (d) the line SE in the Figure 11.24 represents solubility of carbon in austenite. Maximum carbon content in austenite is 1.7%, (e) γ-iron is denser than α-iron, (f) its grain size depends on temperature, time and working.

(ii) α-iron Characteristic features: (a) Its crystal structure is BCC, (b) it is lighter than γ-iron, (c) it is popularly known as ferrite and exists from room temperature to 900°C, (d) its maximum carbon content is 0.015%, (e) it is magnetic, (f) forging of α-iron (or ferrite) causes distortion of grain size, (g) its grain size does not change with time and temperature.

(iii) δ-iron Characteristic features: (a) It appears above 1390°C, (b) its crystal structure is BCC, (c) its maximum carbon content is 0.10%, (d) it is non-magnetic.

(iv) β-iron At temperature above 768°C (or 1414°F) crystal atoms of α-iron undergoes some rearrangements, as a result of which, properties like volume, internal energy and electrical conductivity change in small scale but a remarkable change in magnetic properties is observed. However, crystal structure remains unaltered during this rearrangement. This form of iron, which exists above 768°C, possesses same BCC structure of α-iron but unlike α-iron does not exhibit any magnetic characteristics, is known as β-iron. Interestingly it has been observed that a temperature little below 768°C, iron becomes strongly magnetic and is known as α-iron.

11.7.2 Composition of Fe/C alloy

11.7.2.1 Ferrite solubility curve

The line *JG* in the Figure 11.24 indicates that pure γ-iron exists at the point *J* and slowly transforms into α-iron at the point *G* on cooling. The α-iron, known as *ferrite*, contains a very little amount of carbon (0.015%) and hence is capable of holding considerable amount of various elements, such as, nickel, silicon, phosphorus etc. This α-iron can be considered as a solvent in a solid solution. The line GOS in the Figure 11.24 is the demarcation line above which all α-iron gets converted to γ-iron. The temperature of conversion from α-iron to γ-iron is maximum at *G* (900°C) and decreases with increase in carbon content, ultimately reaches a minimum at *S* (700°C). This conversion of α-iron to γ-iron is known as solubility of α-iron to γ-iron and the line GOS is often called *ferrite solubility curve*.

11.7.2.2 Cementite solubility curve

The eutectic temperature 1130°C (the point *E* in the Figure 11.24) is the temperature where γ-iron holds 1.7% carbon. As the temperature decreases below 1130°C (or 2066°F), γ-iron is unable to retain its carbon content and hence excess carbon is converted to cementite. As cooling proceeds along *ES*, cementite phase slowly gets precipitated out from γ-phase. Thus the line *SE* represents solubility of cementite in austenite at various temperatures and is often called cementite solubility curve.

11.7.2.3 Eutectoid mixture and pearlite

The point *S* in the Figure 11.24 is the eutectoid point, where carbon content is only 0.85%. The temperature at the point *S* is 700°C (or 1292°F). This is the minimum temperature at which 0.85% carbon remains in soluble form in the γ-iron matrix. If the temperature is further lowered below 700°C, two changes take place simultaneously. First γ-phase of FCC structure is converted slowly to α-phase of BCC structure. Secondly, the excess carbon slowly diffuses out of the γ-phase to form an intermetallic compound, cementite, having 6.67% carbon. Thus the resultant eutectoid mixture contains both ferrite and cementite at a definite proportion (88.6% ferrite and 11.4% cementite). The eutectoid mixture is often called pearlite because its appearance is similar to that of mother-of-pearl.

11.7.2.4 Ledeburite

Consider an alloy melt, containing carbon content 4.3%. It is cooled continuously until the eutectic point *B* (see the Figure 11.24) is reached. At this stage both γ-iron and cementite start separating out from the liquid melt and the separation continues till the temperature reaches to 700°C. This mixture, thus separated at the eutectic temperature (1130°C) along *BB′* (see the Figure 11.24), is known as *ledeburite*.

11.7.2.5 Hypo-eutectoid steel

Consider an alloy, containing 0.25% carbon at the point *G* in the Figure 11.25. The alloy is allowed to cool along *GHIo*. At the point *H*, α-phase starts separating out from γ-phase and reaches the point *I* on the eutectic line. Further cooling leads to a separation of ferrite and pearlite mixture. Draw a tie line *aoa′* through the point *o*. Thus,

$$\%\text{ ferrite} = \frac{100(oa')}{(aa')} = \frac{100(0.85-0.25)}{(0.85-0.010)} = 71.43\% \text{ and}$$

$$\%\text{ pearlite} = \frac{100(ao)}{(aa')} = \frac{100(0.25-0.010)}{(0.85-0.010)} = 28.57\%$$

When carbon content lies within 0.85%, the steel contains a mixture of ferrite and pearlite and it is known as hypo-eutectoid steel. The microstructure of hypo-eutectoid steel (see the Figure 11.26) shows that free ferrite forms network of crystal grains around the pearlite grains.

11.7.2.6 Hyper-eutectoid steel

Steel, having composition between 0.85% and 1.7% *C*, is known as hyper-eutectoid steel, also known as *tool steel*. Consider an alloy containing 1.2% carbon at the point G′ in the Figure 11.25. The phase is γ-phase and all the carbon remains dissolved in the iron matrix. As the alloy is cooled slowly the point *H′* is reached where the temperature is around 844°C (or 1550°F). γ-iron, at this stage, is unable to hold that amount of carbon and the cementite phase starts separating out. The separation of cementite phase continues till the point *I′* (700°C) is reached. If the alloy is cooled further austenite (γ-iron) splits into pearlite and cementite. A tie line *a′o′b* is drawn through the point *o′* (Figure 11.25).

$$\text{Thus }\%\text{ pearlite} = \frac{100(o'b)}{(a'b)} = \frac{100(6.67-1.2)}{(6.67-0.85)} = 93.99\% \text{ and}$$

$$\%\text{ cementite} = \frac{100(a'o')}{(a'b)} = \frac{100(1.2-0.85)}{(6.67-0.85)} = 6.01\%$$

Steel, containing carbon content within 0.85% to 1.7%, is a mixture of pearlite and cementite. It is known as hyper-eutectoid steel. The hyper-eutectoid steel consists of pearlite crystals surrounded by a network of free cementite crystals, shown in the Figure 11.27.

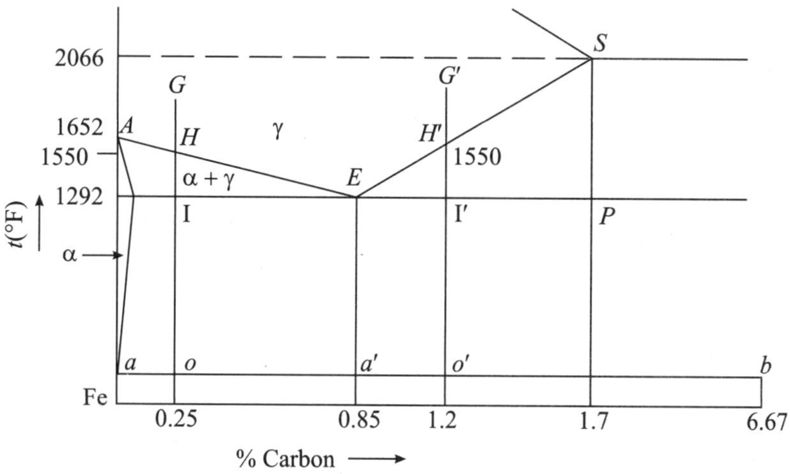

Fig. 11.25 Partial phase diagram of Fe/C alloy.

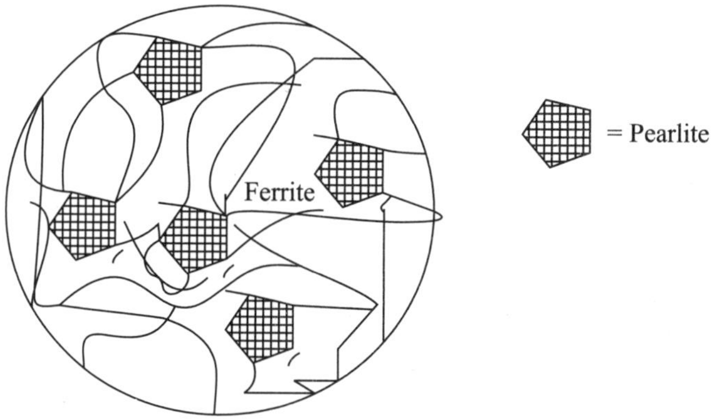

Fig. 11.26 Microstructure of hypo-eutectoid steel, containing 0.25% carbon.

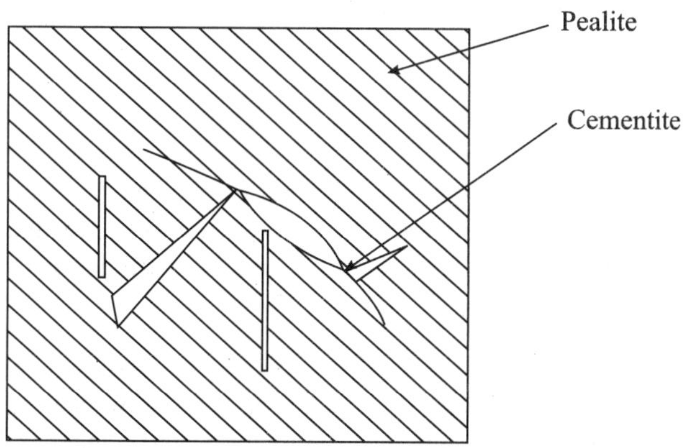

Fig. 11.27 Microstructure of hyper-eutectoid steel, containing 1.2% carbon.

11.7.3 Heating and cooling of steel

During heating and cooling of an iron-carbon alloy several changes in microstructure take place in different phases. Sufficient time must be provided to execute each structural change, which usually requires several minutes. Thus cooling or heating rate should be slow enough to execute the structural changes or in other words normal change cannot be obtained unless the proper rate of cooling/heating is employed. During continuous heating/cooling grain size of the crystal changes continuously. During solidification of alloy melt the following processes take place: (a) nucleation/crystallization, (b) crystal growth. During phase change recrystallization occurs.

Nucleation/Crystallization In the nucleation/crystallization process the liquid molecules cluster together to form numerous nuclei, which are converted into large dendritic grains. These large dendritic grains are broken down by hot working process.

Crystal growth It usually occurs at high temperature but below the solidus curve. Particles, surrounding a nucleus, join to the nucleus to increase the grain size. The phenomenon is called grain growth or crystal growth. Crystal growth is a very fast process and accelerated as the temperature is increased. Due to crystal growth number of grains decreases but the grain size increases and ultimately a *single crystal* is formed. In the case of a single crystal, cross-section of a metal piece contains only one grain. In the case of α-iron, grain growth is hardly observed below 1200°F but above 1292°F and above *GOSK* line (see the Figure 11.24) grains of γ-iron grow rapidly.

Recrystallization In iron-carbon diagram different forms of iron exist at different temperatures with different crystal structures, e.g., α-iron, β-iron, γ-iron, δ-iron etc. During allotropic transformation or phase change at eutectic/eutectoid points, crystals are totally destroyed to form new crystals of different structure. The phenomenon is known as *recrystallization*. Below the line PSK (see the Figure 11.24), grain size is big and coarse but continuous heating to above GOSE line (see the Figure 11.24), recrystallization occurs, giving rise a very fine austenite grains.

11.7.3.1 Change in grain size during cooling/heating

Cooling Consider hypo-eutectoid steel, containing 0.25% carbon at around 100°C (the point *G* in the Figure 11.25). The alloy is slowly cooled to room temperature (25°C) along *GHIo* (see the Figure 11.25). On reaching the point *H* (around 800°C) α-phase (ferrite iron) starts separating out as small crystal of ferrite (see the Figure 11.28). As the temperature decreases volume and size of ferrite crystals increase with simultaneous decrease in austenite grain size. The process continues till the point *I* (eutectic temperature 700°C) is reached (see the Figure 11.25). Slow cooling below 700°C results in recrystallization of austenite grains to pearlite grains (Figure 11.28). If the cooling rate is very slow, balls or spheroids of cementite particles are formed and distributed throughout the ferrite matrix.

Heating The same steel, if heated from 25°C to 1000°C, the resultant micro-structural change is shown in the Figure 11.28. It has been observed that upon slow heating of the steel, pearlite phase exists up to 732°C, at which point recrystallization starts and conversion of pearlite to austenite begins (the point *I′* in the Figure 11.28), which means that in heating process, eutectic temperature appears at 732°C instead of 700°C. This difference in eutectoid temperature by 32°C or 50°F is observed if slow cooling and heating rates are employed. Above 732°C α-phase and γ-phase exist but as the temperature increases volume and grain size of ferrite phase decrease. At around 845°C, recrystallization is complete, i.e., all ferrite grains recrystallize to austenite grains and the grain size of austenite phase increases as the temperature increases (see the Figure 11.28).

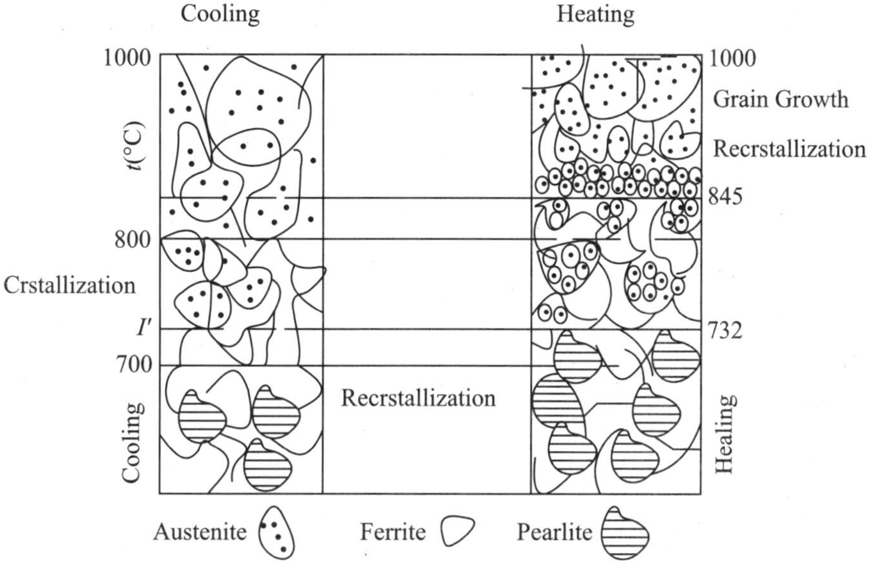

Fig. 11.28 Changes in micro-structure during heating and cooling of hypo-eutectoid steel, containing 0.25% carbon.

11.7.3.2 Critical temperature diagram

The Figure 11.25 shows the micro-structural change during heating and cooling of a hypo-eutectoid Fe/C-alloy. The corresponding phase diagram is shown in the Figure 11.26. The point *O* represents eutectoid temperature (700°C), achieved during cooling and the point *O′* represents the eutectoid temperature (732°C), achieved during heating the alloy. The line *OPQ* represents ferrite solubility curve (*Ar*-γ) during cooling operation. The same solubility curve is shifted towards the right during heating operation (curve *Ac*-γ, *O′P′Q′*). The point *P* or *P′* represents the allotropic β-temperature, above which α-iron is converted to β-iron. Interestingly it has been observed that allotropic temperature of α-form to β-form remains unchanged during cooling or heating process. The line OR represents cementite solubility curve during cooling operation (Ar-Cm) and the line O′R′ represents cementite solubility curve during heating operation (Ac-Cm).

Lag All the above temperatures and curves, shown in the Figure 11.29 are applicable only at equilibrium conditions, i.e., at a very slow heating/cooling rate. In actual practice the heating/cooling rate is fast and hence eutectic points appear at different positions from those shown in the Figure 11.29 (points *O* and *O′*). All the points on the heating *Ac-curve* (*Q′P′O′R′*) and on the cooling *Ar-curve* (*QPOR*) are in equilibrium. The equilibrium Ac-curve is shifted upwards if the heating rate is fast and the equilibrium Ar-curve is shifted downwards if the cooling rate is fast. The Ac-temperatures can be raised by about 300°F from their normal occurrence by fast heating and the Ar-temperatures can be lowered by about 1000°F or can be brought to below room temperature by rapid quenching of thin sections.

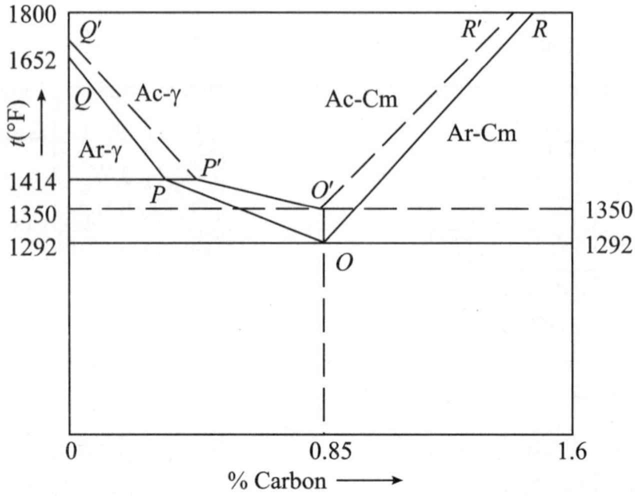

Fig. 11.29 Effect of cooling and heating on critical temperature of a eutectoid steel specimen.

11.7.3.3 Effect of impurities on Fe/C-alloy

Presence of foreign elements other than carbon has significant influence on critical temperatures or transformation temperatures from α-form to β-form to γ-form. For example for each addition of 1% Ni, Ac-critical temperatures are lowered by about 10°C.

Alloying elements are of different types

(i) Some alloying elements are totally dissolved in iron matrix to form solid solution.
(ii) Some alloying elements react with carbon to form different carbides.
(iii) Alloying element reacts with iron itself to form inter-metallic compounds like, FeCr, also known as *sigma phase*, Fe_3W_2.
(iv) Some alloying elements alter the critical temperature of the iron-carbon diagram. For example, All the Ac points are lowered by 10°C by addition of 1% nickel, while addition of 12% chromium raises the Ac_1 temperature (eutectoid temperature) to about 800°C.
(v) Carbon content of the eutectoid mixture changes in presence of alloying element, e.g., in absence of any alloying element eutectoid mixture (pearlite) contains about 0.85% C but in presence of 12% Cr, carbon content of the pearlite decreases to only 0.33% with the change in microstructure. Alloy steel containing 3% Ni and 0.3% C has the same properties and appearance with that of 0.5% carbon steel only.

(vi) Manganese and zirconium readily form sulphides and hence prevent forming dangerous ferrous sulphide during hot working process.

(vii) Presence of vanadium in the steel keeps the steel free from slag inclusions.

(viii) Cr, Al, Si and Cu readily form adherent protective film on the steel surface, imparting corrosion resistance and oxidation resistance at elevated temperatures.

(ix) Presence of Mo enhances the creep strength of the steel by forming fine molybdenum carbide.

Classification of alloying elements

Alloying elements in steel may be classified on the basis of the stability of carbides and allotropes.

(a) There are some alloying elements which form strong carbides, e.g., Cr, Ti, V, Nb, W, Mo etc.

(b) There are some elements which break down iron carbides, e.g., Al, Ni, Co, Si, etc.

(c) ***Austenite stabiliser*** There are some elements which stabilise austenite zone by raising the Ae_4 temperature and lowering the Ae_3 and Ae_1 temperatures. Examples are: Mn, Ni, Co, Cu etc. These elements have *FCC* structure like austenite and hence more soluble in austenite rather than in ferrite.

(d) ***Ferrite stabiliser*** There are some elements which stabilise ferrite zone by lowering the Ae_4 temperature and raising the Ae_3 and Ae_1 temperatures. Examples are: Cr, W, V, Mo, Si etc. These elements are more soluble in ferrite rather than in austenite. γ-loop is shortened significantly in presence of these elements.

11.7.4 Iron-Carbon alloys

Iron-Carbon alloys can be broadly classified into four main categories: (a) Ingot iron, (b) Wrought iron, (c) Steel and (d) Cast iron.

11.7.4.1 Ingot iron

Iron can be made in very pure form (99.99% pure). However, commercial pure iron contains about 0.1% impurities, including 0.01% carbon. Commercial iron of this purity is known as ingot iron. The superior ductility and electrical conductivity as well as very high corrosion resistance and magnetic permeability are some important properties of ingot iron. This iron is practically useful in various research problems.

11.7.4.2 Wrought iron

During manufacturing of iron, 3% slag particles are uniformly distributed in the iron matrix. This form of iron is known as wrought iron. When wrought iron is subjected to hot working operation, slag particles become elongated along machining direction, giving rise slag fibres. These slag fibres, due to having pronounced directional properties, impart excellent fatigue resistance and corrosion resistance properties to the iron matrix.

11.7.4.3 Steels

Steels are broadly classified into two main categories:

(i) Hypo-eutectoid steel, containing carbon content less than 0.85%.
(ii) Hyper-eutectoid steel, containing carbon content more than 0.85% and less than 1.7%.

According to end-uses of steel products steels can be classified into three main categories:

(iii) Low carbon steel, containing carbon content up to 0.2%.
(iv) Medium carbon steel, containing carbon content range from 0.2% to 0.5%.
(v) High carbon steel, containing carbon content range 0.5% to 1.7%.

Low carbon steels possess poor mechanical properties, high ductility, low cost and usually are not hardened by heat treatment. It is mainly used to produce press-moulded articles.

Medium carbon steels are characterized by good mechanical properties, sufficient ductility and can be hardened by heat treatment operation. It is mainly used as structural materials.

High carbon steels possess no ductility with high hardness and superior mechanical properties. High carbon steels are always subjected to heat treatment operation. It is mainly used in the manufacture of cutting tools.

There are several processes in manufacturing steel products. One such process is casting. In most steel castings, medium carbon steel is used. In this method the alloy is melted and the molten alloy is allowed to flow in a prescribed mould, until the mould cavity is filled up. Molten alloy takes the shape of desired article on subsequent cooling. The microstructure of the article consists of pearlite particles surrounded by large ferrite plates. Cast articles of this type of structure possess poor mechanical properties and hence need heat treatment operation, in which process, cast article is heated to about 900°C to change the entire micro-structure to fine austenite grains, followed by slow cooling operation in air medium. The ultimate structure consists of fine mixture of ferrite and pearlite particles with better mechanical properties. This particular heat treatment operation is known as *normalizing operation.*

In order to have better mechanical properties cast ingot is subjected to hot-working process. One of the cheapest and common hot-working processes is hot-rolling operation. During hot-rolling operation the temperature of the alloy reaches to austenite region. After hot-rolling the product is allowed to cool in air medium. The product alloy is known as *wrought alloy*, which is characterized by superior mechanical properties. In wrought alloy, low, medium and high carbon content are frequently used. Wrought steel is often subjected to cold-worked condition, i.e., the wrought steel is allowed to deform plastically, as a consequence of which, (i) machinability of steel is sufficiently improved with marked decrease in ductility, (ii) steel is produced with a good surface finish and close dimensional tolerances, (iii) strength of steel is sufficiently increased as in the case of cold-drawn wires.

Alloy steels

According to composition, steel can be divided into three broad categories: (a) Plain carbon steel, where only two compositions, namely iron and carbon, are present in the matrix, (b) Carbon steel, where only alloying element silicon is present in the matrix in a very minute quantity (less than 0.5%), (c) Alloy steel, where several alloying elements, like copper (Cu), nickel (Ni), cobalt

(Co) etc., are present in varying compositions. Alloying element usually shows any one of the following characteristics:

(i) It (alloying element) forms solid solution with iron. Phosphorus (P), silicon (Si), aluminium (Al), nickel (Ni) and to some extent cobalt (Co) fall in this category. Formation of solid solution with Fe/C matrix results in appreciably increase in hardness and mechanical strength of the alloy steel. However, corrosion resistance properties, electrical properties and hardenability are adversely affected in presence of solid solution.

(ii) It (alloying element) reacts with free carbon to form carbides within Fe/C matrix. Among the alloying elements, titanium (Ti) is the strongest carbide forming element followed by vanadium (V), tungsten (W), molybdenum (Mo), chromium (Cr) and manganese (Mn). Cobalt is the weakest carbide-forming element and so does manganese. Only one-fourth of total manganese is available for carbide formation and the rest of manganese remains as solid solution in Fe/C matrix. Due to high hardness of carbide particles, carbide-forming elements usually impart high wear resistance and hardness, increase hardenability as well as prevent grain coarsening during transformation to austenite steel. However, machinability of steel decreases and mechanical strength is only slightly improved in presence of carbide particles.

11.7.4.4 Cast iron

The iron-carbon alloy, containing carbon content more than 2% but less than 4.5%, is known as cast iron. The main constituent of cast iron is cementite (Fe_3C), which dissociates according to the following reaction, known as *graphitization*:

$$Fe_3C \rightarrow 3Fe + C \text{ (graphite)}.$$

However, cementite is stable at low temperature and hence graphitization occurs at a slow rate. On the basis of extent of graphitization cast iron can be divided into two categories: (a) white cast iron and (b) gray cast iron.

(a) White cast iron During forming process temperature is so high that cast iron is melted and the molten alloy is rapidly cooled to eutectic temperature 2066°F, where austenite phase appears as dentritic forms. Further rapid cooling to 1292°F, eutectic mixture austenite and cementite, called *ledeburite*, is separated out along with the austenite. Further rapid cooling below 1292°F results in decomposition of austenite into pearlite and cementite. These extra cementite particles do not have time to take part in graphitization reaction due to rapid cooling and prefer building up on cementite particles already present. So if cast alloy is moulded in metal mould, heat transfer occurs at fast rate, rendering fast cooling rate, which prevents graphitization of cementite. The ultimate cast alloy product appears white and hence the name white cast iron. The cast iron product, thus produced, is often reheated to about 1600°F to allow the graphitization reaction to occur, producing a graphite layer on the surface. The graphite layer is known as *temper carbon.* This temper carbon imparts wear resistance to the final product and the resulting alloy, thus obtained after tempering, is known as *malleable cast iron.* The malleable iron castings are widely used in industrial and farm machinery. Graphitization of cementite is favoured at the following conditions: (1) slow cooling rate, (2) high temperature and (3) presence of silicon and nickel as an alloying element, *accelerates* the graphitization rate, whereas chromium and molybdenum *retard* the rate of graphitization. Thus in the manufacture of white cast iron silicon content is restricted to a maximum of 1%.

(b) Gray cast iron During forming process if slow cooling rate is maintained graphitization starts when the temperature drops below the eutectic temperature, i.e., 1130°C. Formation of graphite flakes continues as the temperature of the mould goes down till the eutectoid temperature (700°C) is reached. When temperature drops below 700°C austenite splits into pearlite and cementite. The latter undergoes graphitization to produce more graphite flakes. If the cooling rate is very slow pearlite further dissociates into ferrite and cementite. Thus gray cast iron can be obtained in two forms depending on the cooling rate: (1) at slow cooling rate, grey cast iron, thus produced, consists of pearlite matrix into which graphite flakes are dispersed. It is known as *pearlitic gray cast iron*, (2) at very slow cooling rate, ferrite matrix, dispersed with graphite flakes, is obtained. It is known as *ferritic gray cast iron.*

Graphite flake size Graphite flake size is an important parameter since coarse flakes have low strength. Various measures have been considered to have finer flakes in the matrix. (a) Presence of alloying elements reduces the flake size. (b) Antioxidants or inoculants, e.g., ferrosilicon, calcium-silicon are frequently added to the molten alloy to protect the latter against oxidation. These inoculants increase the nucleation rate, giving rise finer flake size. (c) Sometimes alloy is superheated just before casting in order to form finer flakes of graphite.

Nodular cast iron During manufacturing of cast iron, molten alloy is sometimes treated with magnesium or cerium. Presence of magnesium or cerium tends to form nodules rather than flakes, as shown in the Figure 11.30. This is known as nodular cast iron, possessing superior strength and ductility. Mechanical strengths of different iron-carbon alloys are given in the Table 11.1.

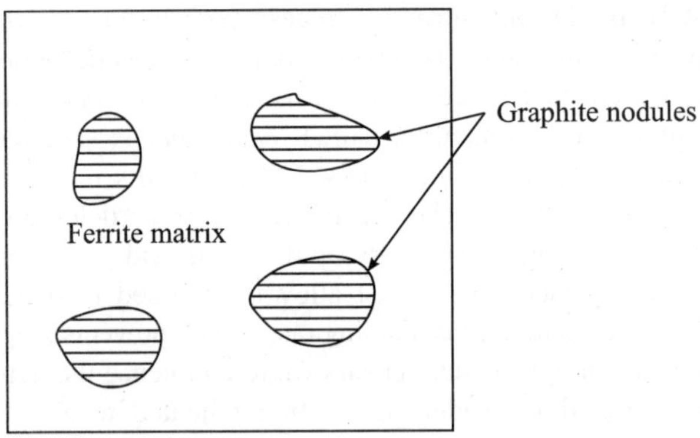

Fig. 11.30 The microstructure of nodular cast iron.

Table 11.1 Yield strength and tensile strength of different iron-carbon alloys.

Alloy	*Yield strength* (MPa)	*Tensile strength* (MPa)
Ingot iron	207	310
Wrought iron	207	345
Malleable cast iron	241	379
Ferrite cast iron	—	172
Pearlitic cast iron	—	310
Nodular cast iron	345	552

12

CORROSION

12.1 Definition

It is a process by which a metal surface undergoes deterioration through chemical or electrochemical reaction with environment.

$$\text{Metal } (G_1) \xrightarrow{\text{Reaction}} \text{Compound } (G_2)$$

If $\Delta G = (G_2 - G_1) < 0$, metals have a tendency to form compounds and hence are susceptible to corrosion.

Effects of corrosion (a) loss of usual properties, (b) decrease in efficiency and hence production rate, (c) increase in maintenance and production cost, (d) contamination of product.

12.2 Classification

Corrosion may occur in different forms. All these forms can be classified into the following groups: (a) chemical/dry corrosion, (b) electrochemical/wet corrosion, (c) other forms of corrosion.

12.2.1 Chemical or dry corrosion

Three different types of dry corrosion are usually observed: (i) oxidation corrosion, (ii) corrosion by other gases, (iii) liquid metal corrosion.

(i) **Oxidation corrosion** In this type of corrosion metals are oxidized to their oxides by natural process but oxides, thus formed, are found four different types: (1) stable oxide, (2) unstable oxide, (3) volatile oxide and (4) porous oxide.

(1) **Stable oxide** In the cases of aluminium (Al), copper (Cu) etc, oxides, developed on the surface, are stable and impervious. The oxide appears as tightly adhering film to protect the underneath metal. Oxides are formed according to the following reactions:

$$Al + O_2 \rightarrow Al_2O_3 \quad \text{and} \quad Cu + O_2 \rightarrow CuO$$

Both the alumina (Al_2O_3) and copper oxide (CuO) act as protective coating.

(2) **Unstable oxide** In some cases, metal oxides, formed on the metal surface, decompose back to the metals and oxygen. Silver (Ag), gold (Au) and platinum (Pt) oxides are highly unstable and hence they do not undergo oxidation corrosion. Thus silver, gold and platinum are highly stable and not susceptible to oxidation corrosion.

(3) **Volatile oxide** Some metal oxides are volatile and hence oxidation continues to take place until total metal is converted to the corresponding metal oxide. One such example is oxidation of molybdenum (Mo).

$$2Mo + 3O_2 \rightarrow 2MoO_3 \text{ (volatile).}$$

(4) **Porous oxide** If specific volume of oxide, formed on the surface is less than that of metal, the oxide film, thus produced on the metal surface becomes porous, allowing continuous aeration through pores and thereby, helps continue the corrosion till all the metal is exhausted. Specific volume of the pore is defined as the ratio of the volumes of metal oxide to the volume of the metal as shown below:

$$\text{Specific volume of pore} = \frac{\text{Volume of metal oxide}}{\text{Volume of metal}}$$

Pilling-Bedworth Rule This rule states that smaller the specific volume greater is the rate of oxidation corrosion. Specific volume of tungsten (W), chromium (Cr) and nickel (Ni) are 3.6, 2.0 and 1.6 respectively. The rate of corrosion is least in the case of tungsten (W). Actually tungsten is stable even at high temperature.

Significances of Pilling-Bedworth Rule

(i) If volume of metal oxide is far greater than that of metal, the oxide film is non-porous. Examples are Al/Al_2O_3, Cu/CuO etc. So these metals are the least susceptible to oxidation corrosion.

(ii) If the volume of oxide is far less than that of metal, oxide layers are sufficiently stressed or strained, leading to the formation of cracks and pores. Examples are Li, Na, K etc. $M + O_2 \rightarrow M_2O$ or MO or M_2O_y. These metals are highly susceptible to corrosion. M^{y+} is small in size and has a tendency to diffuse towards the surface at fast rate. O^{2-} is large in size and hence inward diffusion of O^{2-} through oxide layer is slow. There are several possibilities of diffusion of ions, e.g., (i) metal ions may migrate outwards, (ii) oxide ions may migrate inwards, (iii) molecular oxygen may penetrate to the metal/oxide interface, (iv) both the processes (i) and (ii) occur simultaneously.

(ii) Corrosion by other gases, e.g., chlorine (Cl_2), hydrogen sulphide (H_2S) etc.:

(a) Silver (Ag) undergoes corrosion in presence of Cl_2 according to the following reaction:

$$2Ag + Cl_2 = 2AgCl$$

However AgCl film is protective and prevents further attack of chlorine on silver.

(b) Tin (Sn) reacts with chlorine to form $SnCl_4$, which is volatile and hence accelerates the corrosion of tin metal.

$$Sn + 2Cl_2 \rightarrow SnCl_4$$

(iii) Liquid metal corrosion When a liquid metal is flowing over a solid metal surface through different temperature zones there is a possibility that solid metal gets dissolved in the liquid metal at high temperature zone but the same gets deposited again at low temperature zone. For example, sodium metal leads to corrosion of cadmium in nuclear reactor.

Wagner's theory of oxidation of metals According to this theory corrosion may occur by any one of the following processes: (i) metal ions may migrate outwards, (ii) oxide ions may migrate inwards, (iii) molecular oxygen may penetrate to the metal/oxide interface, (iv) both the processes (i) and (ii) occur simultaneously.

Mechanism of growth of oxide films (a) There are some oxides, where deficient in metals is observed. These oxides are called *p*-type oxides, e.g., NiO, Cu_2O etc., (b) there are some metal oxides where excess of metal ions is observed. These metal oxides are called *n*-type oxides, e.g., ZnO, CdO etc. Both these type of oxides are shown in the Figure 12.1. In *n*-type ZnO crystals, Zn^{2+} ions, generated at the interface, move towards the bulk and occupy interstitial voids. To neutralize these excess cations electrons will flow inwards. Thus the chemical formula of zinc oxide should be Zn_xO, where, $x > 1$. Metal oxides, formed on the metal surface, are usually non-stoichiometric and hence possess either deficient or excess of cations. These are called semi-conducting oxides.

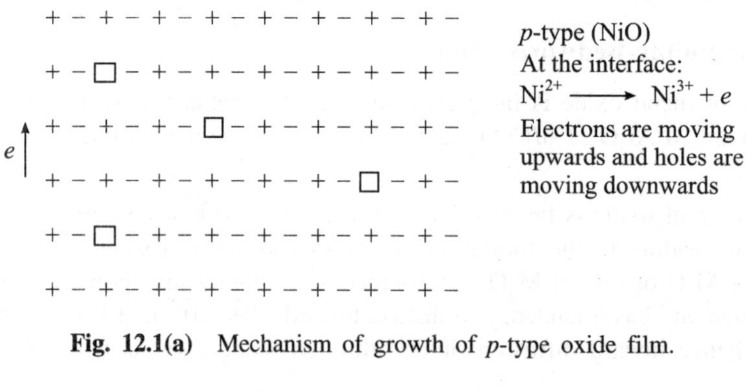

Fig. 12.1(a) Mechanism of growth of *p*-type oxide film.

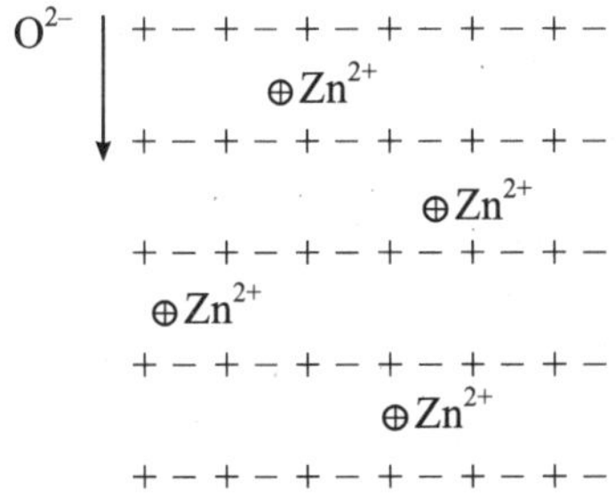

n-type (ZnO)

At the interface: $\frac{1}{2}O_2 + 2e \longrightarrow O^{2-}$

$Zn \longrightarrow Zn^{2+} + 2e$

Electrons are moving from the surface to the bulk of the crystal to neutralize excess Zn^{2+}

Fig. 12.1(b) Mechanism of growth of *n*-type oxide film.

Growth of oxide film usually follows any one of the following laws:

(i) Parabolic law

$\left(\frac{dx}{dt}\right) \propto \left(\frac{1}{x}\right)$, or, $xdx = kdt$. On integration, we get, $x^2 = 2kt + C$,
where, x is the thickness of the oxide layer and C is a constant. Co, Fe, Cr, Ni, Cu and alkaline earth metals usually follow this law.

(ii) Linear law

$\left(\frac{dx}{dt}\right) = k$ (a constant). On integration we get, $x = kt + C$.
At first a protective layer of definite thickness is formed. As the thickness increases oxide film breaks down. Usually alkali metals show this type of behaviour.

(iii) Logarithmic law

$\left(\frac{dx}{dt}\right) \propto \left(\frac{1}{t}\right)$. On integration we get, $x = k\ln(at + 1)$, since, at $t = 0$, $x = 0$.
a is a constant. Usually Al, Cu, Zn and Be show this type of behaviour.
Let us consider a general form of oxidation reaction:

$$xM + \frac{1}{2}(yO_2) \rightarrow M_xO_y$$

Or,
$$\left(\frac{2x}{y}\right)M + O_2 \rightarrow \left(\frac{2}{y}\right)M_xO_y.$$

The free energy change is given by: $\Delta G = \Delta G^\circ + RT\ln\left[\frac{a_{M_xO_y}^{2/y}}{a_M^{2x/y} \cdot a_{O_2}}\right]$, where ΔG° is the standard free energy change. When ΔG is negative, oxidation reaction occurs. Magnesium (Mg) has more negative free energy change compared to aluminium (Al) and hence MgO formation occurs more readily compared to Al_2O_3. It has been observed that oxide films that are formed on surfaces of Na, P, Zn and Mg metals are porous while oxides of Al, Cu and Ni are dense and adherent to the metal surfaces.

12.2.2 Electrochemical corrosion

When corrosion occurs through the formation of galvanic cells, the phenomenon is called electrochemical corrosion. To complete the cell circuit flow of ions is essential and hence a medium, called electrolyte, is needed. Moisture or water often acts as an electrolyte. So electrochemical corrosion always occurs in wet condition. The galvanic cell has two electrodes, cathode and anode. The anode is short of electrons and hence oxidation occurs at anode as follows: $M \rightarrow M^{n+} + ne$. Thus corrosion occurs at anode. At cathode electrons are in excess and reduction occurs. Reduction may occur either in presence of oxygen or in absence of oxygen. The mechanism of reduction process is given below:

(i) $4H^+ + O_2 + 4e \rightarrow 2H_2O$

(ii) $2H_2O + O_2 + 4e \rightarrow 4OH^-$

(iii) $O_2 + 4e \rightarrow 2O^{2-}$.

All the above reactions occur in presence of oxygen.

However, in absence of oxygen reduction process may proceed as follows:

(i) $2H^+ + 2e \rightarrow H_2$

(ii) $2H_2O + 2e \rightarrow 2OH^- + H_2$

All the reactions occur in absence of oxygen.

The cell e.m.f. is given by: $E = E_c - E_a$, where, E_c and E_a are the oxidation potentials of cathode and anode respectively. The free energy change is given by:

$$\Delta G = -nFE \tag{12.1}$$

where, n is number of electrons, F = Faraday's constant = 96,500 C/mole, E is the cell e.m.f. and ΔG is the free energy change.

Again, we know that,

$$\Delta G = \Delta G^\circ + RT \ln K_a \tag{12.2}$$

where K_a is the equilibrium constant for the following reaction:

$aA + bB \rightarrow lC + mD$, where, A & B are the reactants and C & D are the products.

$$K_a = \frac{a_C^l \cdot a_D^m}{a_A^a \cdot a_B^b}$$

$$\Delta G^\circ = (lG^\circ_C + mG^\circ_D) - (aG^\circ_A + bG^\circ_B) = -nFE^\circ \tag{12.3}$$

Combining Equations 12.1, 12.2 and 12.3, we get, $E = E^\circ - \left(\frac{RT}{nF}\right) \ln K_a$

Characteristics features

(a) Numerous galvanic cells are set up in presence of a conducting medium like water.
(b) Oxidation of metals occurs at the anode and the metal ions flow towards the cathode.
(c) OH^- and O^{2-} ions are formed at the cathode and these ions move towards the anode.
(d) The product, i.e., metal oxide is formed somewhere between cathode and anode.

Polarization of electrodes The potential difference between cathode and anode is the driving force of any electrochemical process. In some cases concentration difference mainly occurs at anode and the corrosion current, thus developed, is mainly controlled by anodic reaction rate. The phenomenon is called anodic control polarization. In some cases the corrosion current is governed by cathodic reaction rate. The phenomenon is called cathodic control polarization. When both cathodic and anodic reaction rates have significant contributions in developing corrosion current the phenomenon is called mixed control polarization.

Initially the current density at cathode is different from that of anode and hence the potential difference between cathode and anode changes. Finally a steady-state potential, where anodic current density equals to cathodic current density, is reached. That steady-state potential is called corrosion potential and the corresponding current is called corrosion current. Corrosion current mainly depends on (i) corrosion potential, (ii) diffusion of corrosive electrolytic species, (iii) temperature, (iv) ratio of anodic area to cathodic area. The diffusion of corrosive electrolytic species is the primary requirement to continue corrosion. The diffusion process is often slowed down due to deposition of products on anode and cathode plates and hence corrosion rate decreases.

With the increase in temperature corrosion rate increases due to increase in diffusion rate. The corrosion current (I_o) is given by:

$$I_o = I_a - I_c \tag{12.4}$$

where, I_a and I_c are the anodic and cathodic current densities respectively. Thus if cathodic area is larger, anodic current density is significantly high and hence corrosion rate increases as I_a is much higher than I_c. The potential versus current diagram is commonly known as Evans polarization diagram which is shown in the following figure.

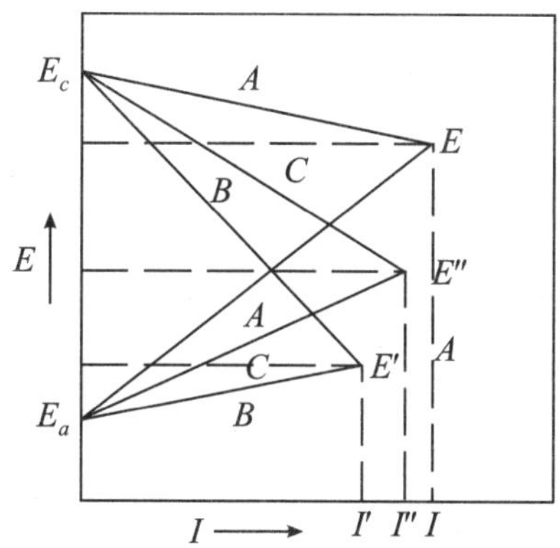

Fig. 12.2 Evans polarization diagram.

The curve *A* shows that anodic reaction rate is the controlling factor in corrosion and hence known as anodic control potential diagram. Similarly the curve *B* represents cathodic potential diagram. In the curve *C* both cathodic and anodic reaction rates control the corrosion and hence it is called *mixed control potential diagram.*

Differences between chemical and electrochemical corrosion

Chemical corrosion	*Electrochemical corrosion*
1. It occurs in dry condition	1. It occurs in wet condition
2. Corrosion occurs through chemical reaction with environment	2. Corrosion occurs through the formation of numerous galvanic cells
3. It is slow and uniform process	3. It is a fast and non-uniform process
4. It occurs both on homogeneous and heterogeneous surfaces	4. It occurs only on heterogeneous surfaces

Rusting of iron The chemical reactions responsible for rusting of iron are given below:

Anodic reaction: $2Fe \rightarrow 2Fe^{2+} + 4e$

Cathodic reaction: $O_2 + 2H_2O + 4e \rightarrow 4OH^-$

$$2Fe + 2H_2O \rightarrow 2Fe^{2+} + 4OH^-$$

$$2Fe^{2+} + 2H_2O + O_2 \rightarrow 2Fe(OH)_2$$

$Fe(OH)_2$ gets oxidized to $Fe(OH)_3$, which exists as $Fe_2O_3.xH_2O$. Usually, the value of x is 3. Thus the chemical formula of rust is $Fe_2O_3.3H_2O$.

$$2Fe(OH)_2 + H_2O + \frac{1}{2}O_2 \rightarrow 2Fe(OH)_3 \rightarrow Fe_2O_3.3H_2O.$$

The Figure 12.3 illustrates the formation of rust.

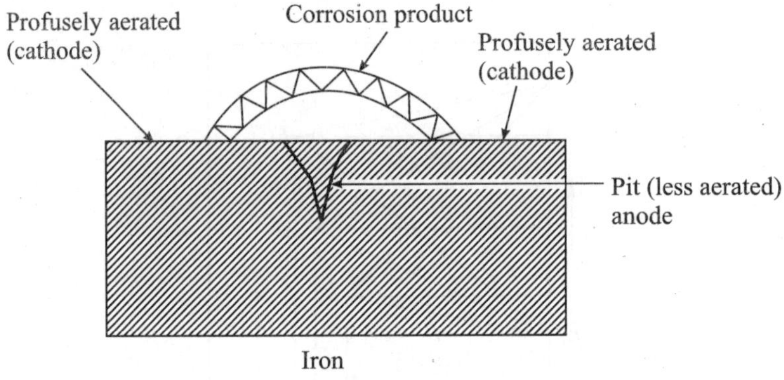

Fig. 12.3 Schematic representation of formation of rust.

12.2.3 Other forms of corrosion

Corrosion may be classified in the following types according to the process of corrosion: (i) uniform corrosion, (ii) galvanic corrosion, (iii) crevice corrosion, (iv) pitting corrosion, (v) intergranular corrosion, (vi) selective leaching, (vii) erosion corrosion, (viii) stress corrosion, (ix) biocorrosion, (x) soil corrosion and (xi) stray current corrosion.

(i) Uniform corrosion When the electrochemical corrosion occurs continuously and uniformly the corrosion is called uniform corrosion. Examples are, corrosion of iron sheets, tarnishing of silver plates etc.

(ii) Galvanic corrosion When two dissimilar metals are coupled and jointly exposed to the atmosphere, one metal acts as cathode which is protected while other metal acts as anode which undergoes corrosion. One example is Zn/Cu system.

At anode:

$$Zn \rightarrow Zn^{2+} + 2e$$

$$2Zn^{2+} + O_2 \rightarrow 2ZnO - 4e$$

$$2Zn + O_2 \rightarrow 2ZnO \text{ [oxidation \& corrosion]}$$

At cathode:

$$CuO \rightarrow Cu^{2+} + O^{2-}$$

$$Cu^{2+} + 2e \rightarrow Cu$$

$$CuO + 2e \rightarrow Cu + O^{2-} \text{ [protected]}$$

$$Zn^{2+} + O^{2-} \rightarrow ZnO$$

Or,

$$Zn + O^{2-} \rightarrow ZnO + 2e \text{ [corrosion]}$$

Passivity or passivation The phenomenon, in which a metal or an alloy exhibits much higher corrosion resistance than expected from its position in the electrochemical series, is known as passivity or passivation. Formation of a very thin protective and invisible film on the surface of the metal or an alloy makes the latter noble (about 0.0004 mm thick). One example is steel containing Ni and Cr. Chromium (Cr) forms a protective layer of Cr_2O_3 on steel making it passive in oxidizing environment.

(iii) Crevice corrosion It is usually observed in rivet joints, weld joints etc. It is a localized corrosion attack occurred due to concentration difference of oxygen. One of such crevice corrosion is shown in the Figure 12.4.

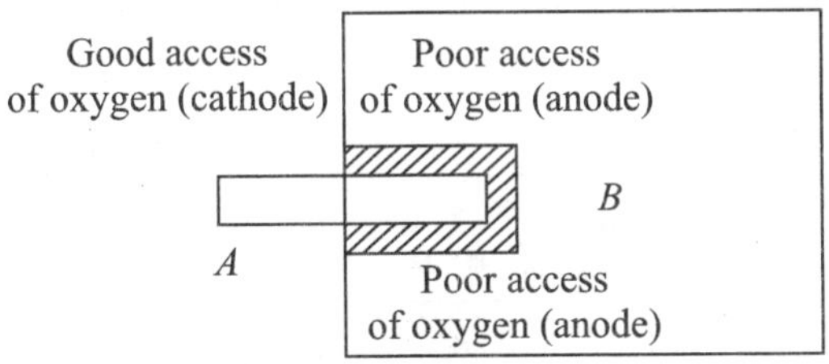

Fig. 12.4 Crevice corrosion at riveted joint.

The hatched area is scarcely aerated whereas free part of *A* is adequately aerated and hence concentration cell sets up. Scarcely aerated portion becomes anode and is susceptible to corrosion whereas the free part becomes cathode and is protected.

Anodic reaction: $M \rightarrow M^{n+} + ne$

Cathodic reaction: $O_2 + 2H_2O + 4e \rightarrow 4OH^-$

Crevice reaction can be minimized by (a) using welded joints instead of riveted or bolted joints, (b) closing crevices by welding, (c) using non-absorbent gaskets.

(iv) Pitting corrosion It is a localized accelerated attack resulting in the formation of cavities. Thus pitting corrosion occurs surrounding pits and cavities in the metal. Pitting is usually the result of the breakdown or cracking of the protective film on a metal at a specific joints, giving rise a small anodic area with a large cathodic area as shown in the Figure 12.3. Presence of impurities on metal surface also causes oxygen concentration difference, giving rise intense corrosion. Once a small pit is formed the rate of corrosion increases. Sometimes acid concentration difference in a localized region causes the pitting. Stainless steels are susceptible to be injured by pitting corrosion. Pitting can be suppressed by adding a small amount of (<2 weight %) Mo to the alloy. The Figure 12.5(a) and (b) illustrate the formation of pits in presence of liquid and solid impurities respectively.

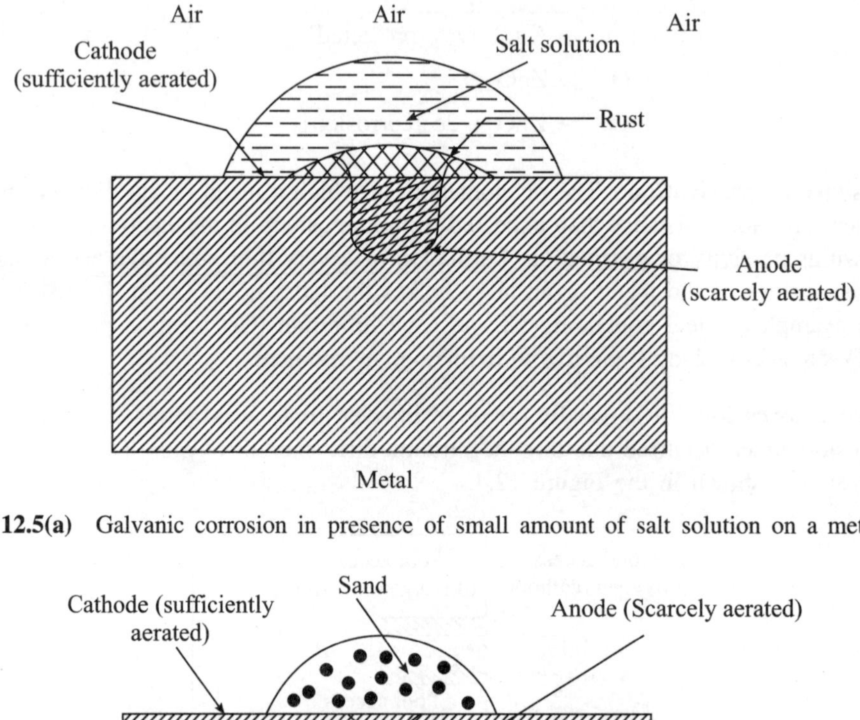

Fig. 12.5(a) Galvanic corrosion in presence of small amount of salt solution on a metal surface.

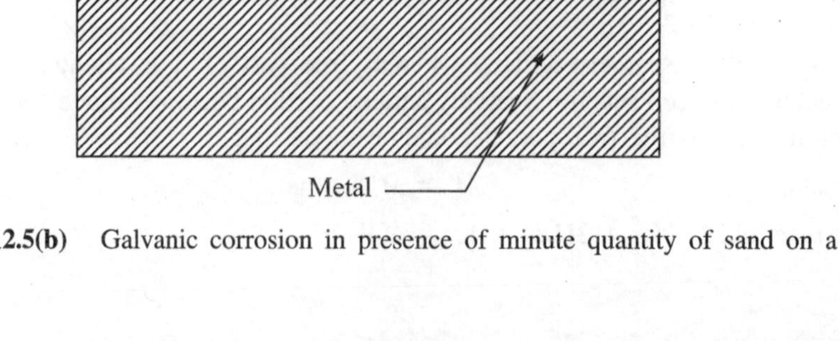

Fig. 12.5(b) Galvanic corrosion in presence of minute quantity of sand on a metal surface.

(v) Intergranular corrosion Intergranular corrosion usually occurs along the grain boundaries. For example, in steel containing Cr and Ni, Cr forms carbides ($Cr_{23}C_6$) along the grain boundaries as the temperature reaches between 500°C to 800°C, as a result of which, chromium concentration is less at grain boundaries than at the bulk of the grain. This is well illustrated in the Figure 12.6.

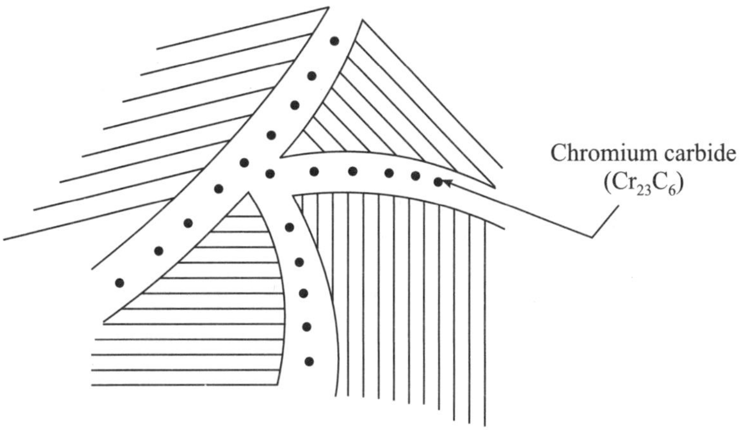

Fig. 12.6 Inter-granular corrosion in steel, containing 18% Cr and 8% Ni.

Thus precipitated region becomes anodic and is vulnerable to corrosion. This process in stainless steel is known as sensitization and the phenomenon is known as intergranular corrosion. This type of inter-granular corrosion is frequently observed during welding process, where chromium (Cr) readily forms the chromium carbide at this high temperature, rendering failure of the material during working. This type of failure is commonly known as *weld-decay*. Cr-activated inter-granular corrosion can be avoided either (a) by the addition of strong carbide forming elements like, Ti, Nb etc or (b) by lowering carbon content. Inter-granular corrosion is also observed in many non-ferrous metals, e.g., in precipitation hardened duralumin [Al/Cu (4 weight %)].

(vi) Selective leaching It is the removal of one element from a solid alloy by the process of corrosion, resulting in the loss of luster and surface texture. One example is removal of zinc from brass. The process of corrosion is known as dezincification. However, the zinc-loss can be prevented by adding a small amount of tin. In graphitization process ferrite is formed continuously from cast iron, giving rise gray cast iron. However, cementite formation prevents the graphitization process.

(vii) Erosion corrosion When material loss occurs by the combined action of corrosion and wear the phenomenon is called erosion corrosion. At first a corrosive product is formed and then removed by turbulent fluid velocity, particle impingement etc. It mainly occurs in the pipelines, turbine blades, pumps, propellers, valves, centrifugals, ducts etc. This form of corrosion can be minimized by (a) changing the design of the component as well as environment, (b) providing barrier wear resistant coatings and (c) usage of more wear/corrosion resistant materials.

(viii) Stress corrosion Intensive corrosion occurs under the joint action of stress and corrosion, leading to a cracking of a component. The phenomenon is called Stress Corrosion Cracking (SCC). SCC may also occur through hydrogen permeation in different metals (Al, Ti, Ni etc.) and alloys. Under certain conditions hydrogen atoms permeate into the metal lattices to form hydrogen gas, which builds up local pressure inside the lattice. In some cases metal hydrides are formed and entrapped within the lattice, inducing embrittlement in the metal/alloy. The phenomenon is called hydrogen embrittlement. Some engineering components are subjected to corrosion under the application of cyclic stress. This type of corrosion is often called fatigue corrosion. In all cases failure of the component occurs. For a given material the favourable condition for occurring stress corrosion is the development of tensile stress in presence of corrosive environment. For example, both copper (Cu) and zinc (Zn) are susceptible to corrosion in ammonia solution since ammonia dissolves out Cu and Zn as $[Cu(NH_3)_4]^{2+}$ and $[Zn(NH_3)_4]^{2+}$ respectively, creating tensile stress on the metal surface. In another case steels are sufficiently corroded in presence of caustic alkalis, which dissolves out iron as Na-ferrate. Na-ferrate decomposes to Fe_3O_4, giving rise strong tensile stress on the steel surface. The following reactions may occur:

$$Fe + NaOH \rightarrow Na_2FeO_2$$

$$3Na_2FeO_2 + 4H_2O \rightarrow 6NaOH + Fe_3O_4 + H_2$$

$$6Na_2FeO_2 + 6H_2O + O_2 \rightarrow 12NaOH + 2Fe_3O_4.$$

The regenerated caustic alkali helps further dissolution of iron. Stress corrosion can be prevented by employing inhibitors or by applying protective coatings on the surface.

(ix) Biocorrosion The deterioration of materials through the involvement of microorganisms is known as biocorrosion. Different micro- and macro-organisms adhere to a material surface in presence of an aqueous medium to form undesirable organic and inorganic deposites on the surface. Biocorrosion can be prevented by the addition of inhibitors. Microorganisms play an important role in extracting minerals through bioleaching. Thiobacillus ferrooxidant is widely used organism for leaching sulphide minerals.

(x) Soil corrosion It may happen in several ways, e.g., underground pipes and cables may be sources of corrosion. As the underground pipe or cable is very long, one part of it is sufficiently aerated whereas the access of oxygen is restricted to another part, as a result of which concentration cells are developed. These concentration cells not only corrode pipes/cables but also deteriorate the quality of the soil. This type of corrosion is frequently observed in soil, rich in gravel and sand. In many cases soil contains numerous air pockets, bringing about severe corrosion due to differential aeration. In water-logged areas anaerobic bacteria becomes active, causing severe deterioration of soil in that area. In presence of excess pesticides, germicides or a large amount of organic matter, metals form soluble complexes, resulting in sufficient concentration difference in metal ions. Thus several concentration cells are developed, bringing about severe soil corrosion. Presence of sufficient quantity of acidic matter also helps setting up numerous concentration cells, which cause soil corrosion.

(xi) Stray-current corrosion In electric traction systems, such as in electric trains, electric current can leak into adjacent conducting structures. The phenomenon is known as stray-current corrosion. This type of corrosion can lead to pitting and serious damages to underground structures.

12.3 Factors influencing corrosion

Corrosion of a metal surface mainly depends on two factors: (a) nature of metal and (b) nature of the corroding environment.

12.3.1 Nature of metal

(i) Position on galvanic series When two metals are in electrical contact the more active metal immediately forms anode in presence of an electrolyte and suffers corrosion. The rate of severity of corrosion depends upon the difference in their positions. Greater is the difference, faster is the corrosion of the anodic metal/alloy.

(ii) Areas of cathode and anode The corrosion of the anode is directly proportional to the ratio of areas of the cathode to anode. Rapid and severe corrosion is observed if the anodic area is small due to heavy current density on the anodic area.

(iii) Purity of metal Presence of any impurity only encourages the formation of local electrochemical cells. The rate and extent of corrosion increases with the increasing exposure and extent of impurities or in other words corrosion resistance of a metal may be improved by increasing its purity.

(iv) Nature of the corrosion product If the corrosion product is oxide, the rate of corrosion mostly depends on the specific volume ratio. Greater the specific volume ratio, lesser the oxide corrosion rate. If the corrosion product is soluble in the corroding medium, corrosion proceeds at a faster rate, or in other words, if the corrosion product is insoluble in the corroding medium, the corrosion product forms a film, which is a barrier to further corrosion. If the corrosion product is volatile underlying metal surface will be exposed readily and corrosion occurs at a faster rate.

(v) Passivity or passivation It is a phenomenon by which a metal or an alloy shows high corrosion resistance due to the formation of a very thin protective film [around 0.4 mm thick] on the surface. Passivity or noble behaviour is exhibited when metal or alloy acts as cathode. The elements or alloys can be formatted in a series with decreasing tendency of anode formation as shown below:

Na > Mg & Mg-alloys > Zn > Al > Cd > Fe > Steel and cast iron > Pb > Sn > Cu > Ni > Cr > Stainless steel > Ag > Ti > Au > Pt.

Gold (Au) and platinum (Pt) are chemically very inert and hence show superior corrosion resistance properties.

12.3.2 Nature of the corroding environment

(i) **Effect of temperature** The rate of corrosion increases with rise in temperature since rate of chemical and electrochemical reactions increases with temperature. Stress corrosion and inter-granular corrosion are usually observed at high temperatures.

(ii) **Dissolved oxygen and turbulence in liquid** As the dissolved oxygen increases corrosion occurs at faster rate. The corrosivity of water decreases with temperature since dissolved oxygen content decreases with temperature. Turbulence in liquid increases the corrosion since the corrosive product film is removed by turbulence and free metal surface appears.

(iii) **Corrosive ion concentration** Corrosion rate can be minimized through removing of corrosive ion concentration from the system, e.g., removing of chloride ions reduces the corrosion.

(iv) **Humidity** Rate and extent of corrosion increases with the increase in humidity.

(v) **pH** Corrosion, particularly electrochemical corrosion, largely depends on pH of the medium. In general acidic medium is more corrosive than alkaline or neutral medium.

(vi) **Presence of impurities in the atmosphere** Different types of suspended particles are present in the atmosphere. Some of them absorb moisture and help formation of galvanic cell. Obviously these particles increase the corrosion rate rapidly. These are called active particles. Examples are: NaCl, $(NH_4)_2SO_4$, sulphates, nitrates etc. Presence of ammonia (NH_3) increases the corrosion rate of copper (Cu) as follows:

$$Cu \rightarrow Cu^{2+} + 2e \quad \text{and} \quad Cu^{2+} + 4NH_3 \rightarrow [Cu(NH_3)_4]^{2+}.$$

As the concentration of Cu^{2+} ions decreases due to complexation more copper metal dissolves to form Cu^{2+} ions. In the similar way dissolution of zinc (Zn) metal dissolves faster in presence of ammonia (NH_3) as follows:

$$Zn \rightarrow Zn^{2+} + 2e \quad \text{and} \quad Zn^{2+} + 4NH_3 \rightarrow [Zn(NH_3)_4]^{2+}$$

Charcoals, aerosols etc. do not increase the corrosion rate directly. However, they act as catalyst and this catalytic activity increases the corrosion rate. These are called inactive particles.

In the boiler soft water is used and sodium carbonate is used in the softening process (lime-soda process). Failure of boiler may occur due to the following reason: Na_2CO_3 dissociates into caustic soda and carbon-di-oxide at high pressure according to the following reaction:

$$Na_2CO_3 \rightarrow 2NaOH + CO_2$$

So concentration of caustic soda increases and ultimately attacks the boiler to dissolve out iron as sodium ferrate. This causes embrittlement of boiler walls, causing failure of the boiler.

Thus in brief the following are the key-factors in accelerating corrosion processes:

(i) Air and moisture, (ii) electrolytes in water, (iii) presence of impurities in metal, (iv) entrapped gases like CO_2, SO_2 and NO_2 in the surrounding atmosphere, (v) internal strain in a metal, (vi) difference in oxygen concentration due to partial exposure of metal surface to the atmosphere.

12.4 Protective measure against corrosion

Protective measure is specific to a particular type, depending on the nature of the material, its environment, service conditions etc. Thus selection of material is one of the most important criteria to prevent corrosion. Furthermore, preparation of metal surface is the first step in the protective coating process. In this step the metal surface should be properly cleaned by degreasing and desealing. Degreasing is carried out by exposure to trichloroethylene or other volatile organic solvents. Desealing is carried out by sand blasting technique or by acid leaching technique or by other chemical methods. Following are some practical methods to prevent corrosion:

(i) modification of the environment
(ii) modification of the composition of the alloy
(iii) cathodic protection
(iv) use of inhibitor
(v) use of protective coatings.

12.4.1 Modification of the environment

Sometimes environment appears corrosive and in that case modification of the environment is needed to protect the material.

(a) If the humidity of the environment is high, the material should be kept over silica gel to minimize the moisture contamination.
(b) If the environment is sufficiently acidic, corrosion is prevented by treating the metal surface with lime.
(c) Salts of surrounding medium are removed by using ion-exchange resins.
(d) Reducing agents are frequently used to remove oxygen from the surrounding medium, e.g., sodium sulphite (Na_2SO_3), hydrazine etc.

$$2Na_2SO_3 + O_2 \rightarrow 2Na_2SO_4; \; N_2H_4 + O_2 \rightarrow N_2 + 2H_2O$$

12.4.2 Modification of the composition of alloy

Iron in the steel is very much susceptible to corrosion and to avoid it other alloying elements are often used. Steel, containing 12% – 18% Cr, is highly corrosion-resistant and can be used in manufacturing several instruments and equipment, including turbine brackets, heat-resisting parts etc. Other alloying elements, e.g., Al, Ni, Ti, Mo etc. are also used for this purpose. These alloying elements usually form a protective oxide layer film on the metal surface. Tungsten (W), tantalum (Ta), niobium (Nb), titanium (Ti) etc. form protective carbides on the metal surface, making the stainless steel corrosion-resistant.

Effects of alloying elements on steel

(i) Chromium (Cr) forms a protective film on iron surface but unable to protect others, e.g., copper.

(ii) 0.1% cerium (Ce) in Ni-Cr resistance wire increases life ten times through the formation of interlocking between oxide and scale. Traces of beryllium and calcium in Magnox alloy prevent burning.

(iii) An alloy element may dissolve in oxide phase of parent metal with little or no improvement.

(iv) Some alloying elements form stable oxides, which protect the underneath metal and minimize scaling, e.g., Al, Be, Si etc. form stable oxides on copper surface. Some alloying elements form low melting point oxides.

(v) Sodium and vanadium often attack the protective oxide film, giving rise low melting compounds like V_2O_5 (600°C – 900°C) and sodium sulphate etc. Formation of these low melting compounds often leads to intercrystalline failure.

(vi) Sometimes formation of carbides leads to change in alloy composition, rendering the crack formation within the matrix. One example is precipitation of chromium carbides in Ni-Cr alloys in a reducing-oxidizing atmosphere followed by oxidation through the chromium-doped matrix. This type of corrosion is called "Green decay".

(vii) When two closely fitting metal surfaces are subjected to vibration, severe damage may be resulted. This type of corrosion is known as *fretting corrosion*, which mainly arises through the formation of oxide debris like red rust. Presence of oxygen encourages the fretting corrosion. However fretting corrosion can be avoided by increasing humidity, reducing vibration and load or using a bonded coating of MoS_2 etc. The fretting corrosion is well illustrated in the Figure 12.7.

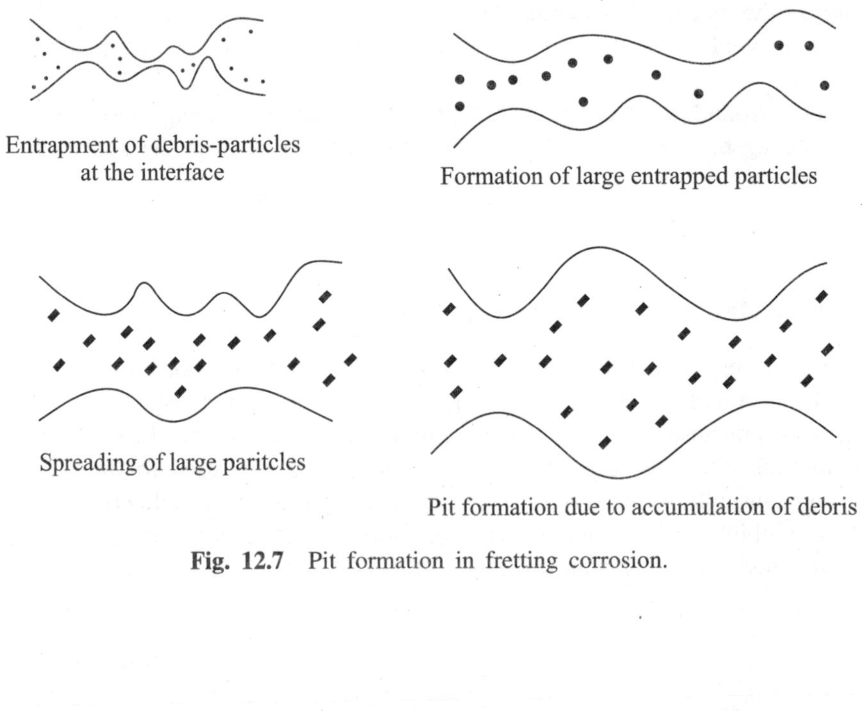

Fig. 12.7 Pit formation in fretting corrosion.

12.4.3 Metal coatings

Aluminium (Al), zinc (Zn), tin (Sn) and lead (Pb) are widely used as metal coatings on different metal surfaces since these metals usually act as anode and hence protect the underneath metal surface. Al-coating is mainly used for protecting steel and duralumin in presence of sulphurous atmospheres. In general Al protects several metals, like Cu, Ni, Pb, Cr etc. but accelerates the corrosion of magnesium (Mg). Al has a good resistance to carbonates, chromates, acetates, nitrates and sulphates in the pH-range 6.4 – 7.2. However, Al is readily attacked by dilute nitric acid (HNO_3), sulphuric acid (H_2SO_4), phosphoric acid (H_3PO_4) and hydrochloric acid (HCl). Al-alloys are attacked by alkalis and the inhibitors such as Na_2SiO_3, $Na_2Cr_2O_7$ are used to protect machines and containers, often used in soap industries. In industrial atmosphere the rate of attack of Al in the region 0 – 0.5 mg/dm^2/day and for food products the rate of attack is 0 – 5 mg/dm^2/day.

Zinc (Zn) Zinc is widely used in protection of steels against atmosphere, water, acids etc. The passivation of zinc usually depends on the formation of adherent film on its surface. Mainly oxide film, chromate film, carbonate film, hydroxide film appear as protective film on the zinc surface. However, longevity of the protective coatings mainly depends on the thickness of the film, which decreases at a very slow rate (0.0002 inch per year).

Tin (Sn) Tin has some distinct advantages over the other metals in the following respects, (a) it is physiologically inactive, (b) it is not corroded by foodstuffs in absence of oxidizers, (c) iron is cathodic to tin under certain conditions. Tin is moderately resistant to acid solutions in presence of air. In absence of air, hydrogen, which develops on the surface, increases the resistance to the flow of current and stops corrosion. Tin coating is frequently used on containers, carrying milk, since the latter has negligible effect on tin even when strongly aerated. However, fruit juices have more corrosive action on tin. Tin protects steel significantly from the attack of distilled water and atmosphere.

Lead (Pb) The corrosion resistance property of lead mainly depends on the formation of a protective layer, which is effective within the pH range 3 – 11. Thus lead is resistant to sulphurous, chromic, phosphoric acids and to the atmosphere. However, it is attacked by hydrochloric, hydrofluoric, nitric, acetic and formic acids as well as nitrate solutions.

Effect of oxide films The corrosion resistance properties of different oxide films depend upon the properties of the film, such as: (a) thickness and impermeability to media, (b) adherence to the base metal, (c) resistance to chemical attack, (d) mechanical strength, (e) the ability to repair defects, developed in the film. Many of the oxide films do not have all the above properties and hence corrosion occurs. Chloride and sulphate ions accelerate the film breakdown process whereas chromates and phosphates promote the repairing action.

12.4.4 Cathodic protection

Oxidation potential of ferrous ion to ferric ion is –0.77 V. Zinc and aluminium are placed at higher positions in the electrochemical series and hence more susceptible to be attacked by oxygen. Thus underneath iron remains bright at the cost of the cathodic metal film on its surface. This is an example of sacrificial protection. Zinc is frequently used in marine boilers, propellers and rudders. Both nickel and copper have lower oxidation potential with respect to iron. These metals protect the steel surface through the formation of protective film, which resists corrosion mechanically. However, if the coating is porous, attack occurs at accelerated rate.

There are several processes of application of metal coating on steel surface: (i) Hot dipping, (ii) Electroplating, (iii) Metal spraying, (iv) Metal cladding and (v) Cementation.

(i) Hot dipping In this process the base metal is dipped into the pool of molten metal to be deposited and kept the base metal for sufficient time. After that the base metal is withdrawn from the molten metal bath and allowed to cool it to room temperature. A thin film of protective metal is tightly adhered to the surface of the base metal. Zn-, Sn- and Pb-coatings are usually carried out by this method. The process of Zn-coating is called "Galvanizing" and the process of Sn-coating is called "Tinning".

(ii) Electroplating In this technique, the metal, to be coated, is dipped into the salt solution, containing the metal ions to be deposited. The metal plate, to be coated, acts as cathode, which attracts the metal ions of salt solution. Usually graphite acts as anode. When current is passed through the solution, metal ions are deposited as metals on the cathode plate, forming a protective film on the cathode plate. Several metals like, Ni, Cd, Sn, Au, Cu, Cr etc. may be deposited on the base metal by this technique.

(iii) Metal spraying In this technique molten metal, to be deposited, is sprayed over the surface of the base metal followed by drying. Thus a thin film of coated metal is formed on the surface of the base metal. One of the important advantages of this technique is that film of protective coating can be formed on any finished article or, any finished article can be coated with a protective film by this technique. Mostly Al- and Zn-coatings are carried out by this technique.

(iv) Metal cladding Sometimes the expensive metal or alloy is protected by covering it with a thick film of protective metal either by pressing technique or by hot rolling method. The thick cover is often called clad and the phenomenon is called cladding. The cladded material possesses both the strength of the alloy as well as corrosion resistance property of the clad. One important example is duralumin, containing 4% Cu, 0.5% Mg, 0.5% Mn with a small amount of Si and Fe. Cladding material is aluminium.

(v) Cementation In this method base metal and powdered metal, to be coated to the base metal, are heated to high temperature and kept that temperature for long time in order to allow the powdered metal to diffuse into the base metal. As a result of this, an alloy layer is formed on the base metal surface. Zn, Al, Si, Cr etc. are frequently used to form protective alloy layer on the base metal.

12.4.5 Inorganic coatings

Non-metallic coatings on metal surfaces are called inorganic coatings. Non-metallic coatings can be applied on a metal surface by different techniques: (a) through chemical reactions, (b) through anodizing, (c) through vitreous enamel coating.

(a) **Surface reactions** In this method, metal surface, to be protected, is dipped into a solution of suitable chemical, which reacts with the metal surface to form a protective coating film. This protective film not only protects the metal surface but also has some decorative value. Phosphate coatings, chromate coatings and oxide coatings can be carried out by this technique. In the phosphate coating process the base metal is treated with a solution of phosphoric acid and Zn-/Fe-/Mn-phosphate in presence of a catalyst, like, copper salt. In the chromate coating process, the base metal is immersed in a bath of sodium-di-chromate and dilute sulphuric acid. Then the base metal is further treated with a neutral chromate solution. The base metal surface is covered with a thin chromate film, such as, $Cr_2O_3 . CrO_3 . xH_2O$. Chromate coatings are frequently applied on Al, Zn, Mg-surfaces etc.

(b) **Anodized coating** In anodized oxide coating process the base metal acts as anode and it is immersed in a bath, containing sulphuric, phosphoric, chromic, oxalic or boric acid. A current is passed through the bath. The oxide coating, thus formed on the metal surface, is hardened by immersing it in boiling water. Anodizing of stainless steel is carried out by using a bath containing concentrated sodium hydroxide solution along with the oxidizing agents, like sodium nitrite or sodium chlorate at 80°C – 90°C. Oxide film, produced by anodized coatings, is sufficiently thick and hence possesses improved resistance to corrosion.

(c) **Vitreous enamel coating** In this process vitreous material is applied to the metal surface, preferably on steel and cast iron equipment. The enamel, usually called 'frit', is nothing but a fused mixture of acidic refractory substance (like, quartz, Cr_2O_3 etc.) and basic fluxes (like borax, fluorspar, soda ash, litharge etc.). High proportion of acidic refractory substance (like silica etc.) makes the metal surface acid-resistant, whereas high proportion of refractory oxides (like Cr_2O_3) protects the metal surface even at very high temperature and imparts high chemical resistance. Enamelled iron and steel have widespread usage in food and beverage industries, pharmaceuticals, refrigerators, kitchen utensils etc. Enamelled cast iron is used in the manufacture of sanitary-wares, bathroom fittings etc. Application of frit or enamel on metal surfaces may be performed in two techniques, (i) dry process and (ii) wet process. At first the metal surface is cleaned and then it is treated with an adhesive to make an undercoat. The frit material is ground and mixed with other gradients, such as floating agent, opacifier, colouring agents etc. In dry process, the metal surface is heated red hot and the mixture is sprayed over the red hot object. In wet process, the blended material, called 'slip', is suspended in water containing electrolyte, such as , soda ash or borax. The water suspended slip is then sprayed over the metal surface, which after drying, is heated to about 500°C – 650°C to produce a glossy and continuous coat on the metal surface. The slip material contains the following: (a) frit material (mixture of acidic refractory substance and basic fluxes), (b) opacifiers (TiO_2, SnO_2, Al_2O_3 etc), (c) floating agents and colourants, (d) electrolytes (Na_2CO_3, $MgSO_4$ etc.).

12.4.6 Inhibitors

In the corrosion process the function of inhibitor is to form a protective film on a metal surface under corrosive environment. When a concentration cell sets up, both cathode and anode need protection. There are some inhibitors which are readily adsorbed on the anode plate and hence the latter is automatically protected. These are known as anodic inhibitors. Examples are: phosphates, chromates, molybdates of alkali metals. These anodic inhibitors are efficient since they diminish the rate of attack by covering the anode plate. If insufficient quantity is used part of the anode plate may not be covered and pitting corrosion sets in. The latter may also occur if part of the protective film is desorbed from the surface. Thus use of anodic inhibitors is dangerous. Cathodes in the concentration cell are susceptible to corrosion in both acidic and alkaline environments. In acidic environment the corrosion proceeds through the diffusion of hydrogen ions. The diffusion process is slowed down through the formation of a protective film by different inhibitors, like lime, heavy metal soaps, amines, mercaptans, substituted ureas, thioureas etc. In neutral solution or in alkaline environment diffusion of hydroxyl ions brings about corrosion. In presence of Zn-, Ni-, Mg-salts, hydroxyl ions form insoluble hydroxide film on anodes and hence diffusion of oxygen to cathode is sufficiently retarded. Cathodic inhibitors are safe in use but at the same time it is not much efficient since corrosion rate is checked only when a protective film is formed on the cathode plate. Furthermore, as the concentration of inhibitor decreases, the protective film eroded.

12.4.7 Organic coatings

In most cases organic coating is used to protect the material. The organic coating is commonly known as paint. A good paint should have the following characteristics: (i) good adhesion to metal surfaces, (ii) sufficient fluid characteristics so that the paint can be easily spread over the metal surface, (iii) good impermeability to water, salts and gases, (iv) chemically inert to the corrosive environment.

The main constituents of a paint are: (i) pigment, (ii) vehicle, (iii) thinner, (iv) drier, (v) extender, (vi) plasticizer, (vii) anti-gelling agent.

(i) Pigment It improves opacity, mechanical strength, impermeability and colour. They are usually chemically inert and used to form durable protective film. Examples are:
white colour: zinc oxide (ZnO), titanium oxide (TiO_2), barium sulphate ($BaSO_4$), white lead etc.; **blue colour:** ultramarine blue, Prussian blue etc.; **black colour:** carbon black, graphite, Fe_2O_3 etc.; **green colour:** chromium oxide (Cr_2O_3) etc.; **red colour:** red lead, chrome red etc.; **yellow colour:** litharge etc.

(ii) Vehicle It is the film forming substance, into which pigment particles are suspended. It is mainly alkyd, i.e., glycerol ester of high molecular weight fatty acid, having the general formula: ROOC—CH_2—CH(COOR)—CH_2—COOR. The *R* represents the fatty acid part. If it is completely saturated fatty acid the corresponding alkyd is called non-drying oil. If *R* has low percentage of unsaturation, the corresponding alkyd is called semi-drying oil. If *R* has high percentage of unsaturation, the corresponding alkyd is called drying oil.

(iii) Thinner It is one of the important constituent of a paint. Its main function is to reduce the viscosity so that paint can be handled and applied comfortably. It helps pigment particles remain suspended in the vehicle. It helps drying of the film. It increases penetration power of the vehicle. Some examples of thinner: turpentine oil, kerosene, xylol, benzene, naphtha etc.

(iv) Driers It helps drying of the paint film. It reacts with vehicle to form a hard and dry film on the metal surface through oxidation, polymerization etc. Common driers are: Co-, Pb- and Mn-salts of resonates, lenoleates, naphthenates etc. Always a small quantity of drier is used in order to avoid formation of hard and brittle film.

(v) Extenders/Fillers They are mainly used to reduce the cost, to increase the durability of film as well as to increase the hardness. It also helps pigment particles distribute themselves uniformly throughout the vehicle. Examples are: china clay, Ca- and Mg-silicates, dolomites etc.

(vi) Plasticizers Its main function is to provide elasticity to the film and thus reducing the danger of cracking. Examples are: di-octyl phthalates (DOP), di-butyl phthalates (DBP), tri-crecyl phosphates (TCP) etc. Phosphate plasticizers also have fire-extinguishing property.

(vii) Anti-gelling agent This reagent is frequently used to prevent gelling and sagging of the paint film. Polyhydroxy phenols are widely used for this purpose.
Besides these above mentioned ingredients other ingredients are frequently added to the paint in order to have desired properties, e.g., (i) metal or metal oxide powders like Al, Sn, Zn, Cr_2O_3, TiO_2 etc. are used to impart high heat resistant property in a paint. Heat resistant paint can withstand temperature up to 1000°C, (ii) very often chlorinated and brominated substances like polyvinyl chloride, chlorinated rubber, substituted halogeneted organic compounds etc. as well as phosphates like Mg-, Ca- or Zn-ammonium phosphates are used to have fire-extinguishing characteristics in the final paint.

Table 12.1 Melting Point and Pilling-Bedworth Ratio (P.B.R.) of different metal oxides.

Oxide	*Melting Point* (°C)	*(P.B.R.)*
BaO	1923	1.31
CaO	2580	0.64
CoO	1935	1.74
FeO	1420	1.78
γ-Fe_2O_3	1457	2.22
NiO	1990	1.70
SiO_2	1713	2.15
SnO_2	1127	1.31
TiO_2	1830	1.76
V_2O_5	690	3.25
ZnO	1975	1.58
ZrO_2	2715	1.57
W_2O_5	850	3.12

12.5 Explain the following

(a) Corrosion occurs in a steel pipe at the steel-copper joint.

Ans. When a steel pipe, containing Fe/Cu joint is exposed in an environment, containing electrolyte, iron, placed at higher position in electrochemical series, readily forms anode and suffers from corrosion whereas the copper forms the cathode and is protected, this type of corrosion is called galvanic corrosion.

(b) Protection of iron plate by nickel coating is not an example of cathodic protection.

Ans. The oxidation potential of Fe/Fe^{2+} is 0.44 V, whereas the oxidation potential of Ni/N^{2+} is 0.242 V. In the case of cathodic protection coating material must have higher oxidation potential than the metal to be coated but here coating material, i.e., Ni has lower oxidation potential than the metal to be coated, i.e., iron. Thus protection of iron plate by Ni-coating is not an example of cathodic protection.

(c) Deposition of sand or other material on any metal surface for long period is undesirable.

Ans. Deposition of sand or any other scale on a metal surface restricts the free access of oxygen to the underneath metal surface, whereas the free metal surface enjoys the full access of oxygen. Due to this differential aeration free metal surface forms the cathode and the underneath surface forms the anode, bringing about corrosion. That is why deposition of sand or any other material on any metal surface for long period is undesirable.

(d) In a water-filled steel tank corrosion occurs at the bottom. Explain.

Ans. In a water-filled storage tank the metal surface, above the water line, is adequately aerated and acts as cathode but the bottom of the tank is suffering from insufficient aeration, forming anode. Thus a galvanic cell sets up, as a result of which, the bottom line of the water tank gets corroded.

(e) Presence of acidic oxides, like, CO_2, SO_2 and NO_2 accelerates the rusting of iron.

Ans. Iron in presence of acidic oxides readily forms the corresponding salts. The portion of iron surface, not in touch with the acidic oxides, forms the anode. The iron surface, eroded by the acidic oxides, forms the cathode, accelerating the rusting of iron.

Problems

(i) State any two causes where wet corrosion takes place.

(ii) Write down the formula of rust. If 100 kg iron is completely rusted how much rust ($Fe_2O_3.3H_2O$) is formed?

(iii) An electrochemical cell contains $Cu|CuSO_4$ solution and $Zn|ZnSO_4$ solution. Write down the cell reactions. Calculate E^0_{cell} and cell e.m.f. at 35°C. Given: $E^0_{Zn/Zn^{2+}} = +0.763$ V, $[Zn^{2+}] = 0.25$ M and $E^0_{Cu/Cu^{2+}} = -0.337$ V, $[Cu^{2+}] = 0.50$ M.

(iv) Calculate the PBR of zircona to zirconium. Given: density of Zr = 7.19 g/c.c. and density of ZrO_2 = 5.56 g/c.c.

13

ENVIRONMENT AND POLLUTION

13.1 Environmental science

Environmental chemistry is a multi-disciplinary science involving chemistry, physics, life science etc. In simpler terms it is the science of chemical phenomena in the environment. In broader sense, it is the study of sources, reactions, transport, effect and fate of chemical species in the air, water and soil. It also considers the effect of human activities on the environment.

Goals of environmental education The objective of environmental education is to throw light on the necessity and usefulness of environment in one hand and protection and conservation of environment in the other hand. It also highlights the need to restrain human activities, which lead to indiscriminate release of pollutants in the environment.

Ecology It is the science, which concerns with relationship between living organism and their physical environment.

Ecosystem It is the system of interaction through which matters and energy are exchanged between living organisms and their environment.

Characteristics of ecosystem:

(a) It contains both biotic and abiotic components.
(b) Nutrient cycles and energy cycles are completed through the interaction between biotic and abiotic components.
(c) An ecosystem has no definite boundary. It is flexible.
(d) Adaptation to local environment is a must for each biotic component, failing which, it might be destroyed.

(e) An ecosystem always passes from less complex state to more complex state. The process is known as *succession*.

(f) **Natural ecosystem** The ecosystem, which is maintained through natural cycles, is known as natural ecosystem. Examples are: desert, forest, ocean etc.

(g) **Artificial ecosystem** The ecosystem which is created and maintained by human being, is known as artificial ecosystem. Examples are: cities, industrial areas, perks, zoological gardens etc.

(h) **Incomplete ecosystem** The ecosystem, which does not possess all the components of environment, is known as incomplete ecosystem. Examples are: bottom of ocean, caves etc.

(i) **Biome** The ecosystem, encompassing a very large area, is known as biome. Examples are: desert, rainforest etc.

13.2 Pollution

(a) Sufficient alteration of natural environment, which has deleterious effect on human beings, animals, crops etc., is termed pollution.

(b) Multifarious activities of human being, like industrialization, construction, transportation, using manures, pesticides in agricultural land etc. often generate and release objectionable matters to the environment, as a result of which, environment becomes hostile to the living organisms. It is called environmental pollution.

Threshold Limit Value (TLV) This is a unique measure of toxicity, which indicates the conditions under which a worker is continuously exposed day after day without any adverse effect. It is the maximum permissible average concentration limit of a toxic pollutant under which a worker can work safely for 40 hours in a week. TLV values of some metals and gases are given below:

Element	*TLV* (μg/m³)	*Gas*	*TLV* (ppm)
As	500	CO	50
Sb	500	NH_3	50
Sn	2000	$COCl_2$	0.1
Cu	1000	O_3	0.1
Zn	500	F_2	0.1
Cd	200	Cl_2	1
Pb	200	Br_2	0.1
Hg	10	SO_2	5
		NO_2	5

Common toxic effects on human health The vital organs in a human body are: brain, heart, lungs, kidneys and liver. Exposure to toxic environment causes mal-functioning of vital organs. The effects can be categorized as follows:

(a) Asphyxiation The absence of oxygen causes cessation of breathing and may lead to death. It is called asphyxiation. Several gases are responsible to asphyxiation, e.g., CO, HCN, $COCl_2$ etc.

(b) Irritation Corrosive acids and caustic alkalis cause irritation to eyes, skin and respiratory passage. Mostly gases and organics produce trouble in respiratory passage. Examples are: NO_2, CdO, BeO, PH_3, aldehydes, alcohols and ketones.

(c) Pneumoconiosis This is a problem related to lung caused by dust and other related substances like quartz, asbestos etc.

(d) Narcosis Many organic solvents are narcotic in nature, e.g., halogeneted hydrocarbons, esters, ketones, ethers, benzene, carbon tetrachloride. Benzene has also slow poisoning effect and is a prime enemy of blood.

(e) Haematoxic effect Toxic effect to blood is called haematoxic effect. Benzene attacks the blood constituents, leading to leukemia. Lead poisoning causes anemia. Carbon monoxide (CO) breaks the haemoglobin and makes the oxygen set free which in turn leads to instanteneous death due to asphyxiation.

(f) Hepatic and renal disorder Primary function of liver is to produce blood and that of kidney is to filter toxins from the body. Compounds of Hg, Cd, Pb, Se, As and U as well as TNT, halogeneted hydrocarbons cause damage to liver and kidney. Persons are also suffering from tremor, paralysis and mental fatigue due to neurotoxic effect. Neurotoxic compounds are CH_3Br, CS_2, Mn or Hg compounds.

Some toxic metal pollutants

(a) Copper Copper is extensively used in household purposes. It is a toxic element and more toxic in ionic form. Around 1.7 ppm of copper in drinking water has deleterious effect on health. It causes fatal hemolytic anemia, fatty metamorphosis of liver, narcosis of kidney. Permissible level in potable water is 1 ppm and in soil is 0.2 ppm. The prescribed TLV is 1 mg/m^3 but 0.2 mg/m^3 is the best for all purposes. ESP and bag house filter are usually used for collection of copper dust. Wheat flour, animal feed contain considerable amount of copper. Jewellery effluents, copper mines, copper industries are the main sources of copper in the environment.

(b) Cadmium Source: Cladding industry, nuclear reactor, alkaline cells and alloy industries. Substantial amount of cadmium is released in air when rubber tyres and plastic containers are burnt.

Cadmium mainly affects kidneys, lungs, and pancreas and causes respiratory disorder. It is a cumulative poison and can reside in human body for more than 10 years.

(c) Mercury It is the king of toxic elements. Liquid mercury has no adverse effect on health but its vapour is deadly toxic. So workers should not be exposed to mercury vapours for more than 30 hours per week. Organic mercury is more dangerous than inorganic mercury. In Minamata (Japan) a paper industry released mercury wastes in sea and these mercury wastes form $(CH_3)_2Hg$ and $(C_2H_5)_2Hg$. They are consumed by fishes and finally it moved to human body through food chain. Several persons died due to mercury poisoning. It causes severe health problems like tremor of extremities, renal disorder and impairs central nervous system. Hg^{2+} readily attacks hemoglobin and finally blood cells are collapsed. Hg^{2+} is deadly dangerous to human body.

(d) Lead Source: gasoline, lead pigment industries, ceramic utensils, glass components. Auto-exhaust emissions contain considerable amount of lead.
To reduce the lead content in auto-exhaust emission catalytic converter is used. Organic lead causes intoxication, disrupts central nervous system, weakness in wrists and legs, anemia, gastrointestinal disturbance, cardiac disorders. In women it causes fertility while in men it weakens reproductive system. Inorganic lead causes abdominal pain, asthema, constipation, diarrhea, vomiting etc.

(e) Arsenic It is one of the important components in the "Big Four Elements" in the environment. Main sources are: cotton, electronic, photocopying, computer industries, pharmaceutical, paper-pulp and glass industries.
It can cause dermatitis, irritation of upper respiratory passage, ulceration and perforation of nasal septum, cutaneous cancer, jaundice and anemia, hepatic disorders, mutagenecity and carcinogenecity. Inorganic arsenic is more toxic. It causes skin abnormalities by interfering DNA synthesis process. It is a carcinogenic element. It can produce cancer in skin, lung, kidney and other parts of human body.

(f) Antimony The main source is safety match industries, drug and alloy industries. It causes violent and continuous vomiting, profuse diarrhea, muscle weakness and damage to kidneys and liver. TLV limit is 0.005 ppm in air and 0.05 ppm in water.

Carcinogenic compounds Compounds, producing cancers in human body, are known as carcinogens. Examples are: (1) Nitrosoamines. They usually contain —N=O group, causing cancer, liver damage and kidney trouble. (2) Carbocyclic aromatic amines: aniline, α-methyl amine, benzidine etc. They convert Fe(II) to Fe(III) in blood, leading to cyanosis. Benzidine causes haemolysis, bone marrow depression, kidney and liver damage. (3) MIC (Methyl Iso Cyanate) causes breathing trouble, lung damage, nausea and body pain. (4) Polycyclic aromatic hydrocarbons (PAH): naphthalene, acenaphthalene, anthracene, phenonthrine, pyrene, ϕ-pyrene, ϕ-anthracene etc. Main sources are thermal power plants, town incinarators, open air combustion, auto-exhaust emission, tobacco combustion, cigarette smoking etc. Solar radiation increases the toxicity of PAH compounds. 1-Nitropyrine and 1, 8-dinitropyrene are highly toxic, mutagenic and carcinogenic. These compounds are formed in car engine by action with NO_x. (5) Polychlorinated biphenols (PCB), dimethyl sulphate, vinyl chloride and cadmium compounds. (6) Carbon tetrachloride.

Effects of pollutants on health (1) respiratory disorder, (2) chest disease, (3) eye irritation, (4) nervous system breakdown and brain disorder, (5) chronic problem, (6) bronchitis and lung damage, (7) damage of circulatory and olfactory systems.

Effects of pollutants on vegetation and animals:

(a) It reduces crop yield and fruit yield.
(b) It affects plant growth.
(c) Acid rain depositions (HNO_3, HNO_2, H_2SO_4, HCl etc.) adversely affect vegetation, fishes and animals.
(d) Pesticides, germicides, used in agricultural land are ultimately mixed with the lake and pond water and affect badly the fishes and other aquatic animals.
(e) Heavy metals, particularly, As, Cd, Pb and Hg are deadly poisonous to all living beings.

13.3 Pollutant

Any species, having concentration more than its TLV in the environment, has deleterious effect on environment and is called pollutant.

Classification of pollution Pollution can be classified into seven broad categories: (a) air pollution, (b) water pollution, (c) soil pollution, (d) noise pollution, (e) metal pollution, (f) thermal pollution, (g) radiation pollution.

13.3.1 Acid rain

In industrial area harmful gases are often released to the environment. These gases react with atmospheric water to form corresponding acids, which return to the earth surface in the form of rain, which is known as acid rain. Some causes of acid rain are listed below:

(1) Natural processes like volcano eruptions, combustion of fossil fuel, forest fires etc. produce huge amount of sulphur and nitrogen compounds, which are released to the atmosphere, producing acid rain.
(2) Electric generation plants, smelting plants and industrial boilers release nitrogen oxides and sulphur dioxide into the atmosphere, which form acids in air.
(3) Petroleum refineries and coal combustion often release huge amount of sulphur and nitrogen oxides.
(4) Many chemical industries release huge amount of hydrochloric acid gas and chlorine gas in the atmosphere. This chlorine gas forms hydrochloric acid in the atmosphere.
(5) Automobile exhaust contains profuse amount of nitrogen oxides which form nitric acid in the atmosphere.

Mechanism of acid rain formation

(1) Industry $\rightarrow SO_2 + NO_x + HCl$
(2) $O_2 \rightarrow O_3$ (by photochemical reaction)
(3) $SO_2 + NO_x + O_3 \rightarrow H_2SO_4 + HNO_3$
(4) $SO_2 + O_3 \rightarrow SO_3 + O_2$. $SO_3 + H_2O \rightarrow H_2SO_4 \rightarrow (H_2SO_4)_n$ (aerosol droplet)
(5) $NO + O_3 \rightarrow NO_2 + O_2$
$NO_2 + O_3 \rightarrow NO_3 + O_2$

$NO_2 + NO_3 \rightarrow N_2O_5$

$N_2O_5 + H_2O \rightarrow 2HNO_3$

(6) H_2SO_4, HNO_3, HCl are entrapped within clouds and eventually return to the earth surface in the form of acid rain.

Adverse effects of acid rain

(1) Acid rain makes the leaves of plants yellow and brown. It also affects productivity of agricultural crops.
(2) It affects the properties of soil and damages the soil fertility.
(3) It also affects aquatic ecosystems.
(4) It may cause respiratory and skin diseases.
(5) It damages limestone, marble monuments and buildings.

13.3.2 Hydrocarbons

Constituents: Methane, ethane, propane, ethylene, acetylene, *n*-butane, iso-butane, toluene, xylene etc.

Sources

(1) Natural source from trees.
(2) Anaerobic decomposition of organic matter in water, sediments and soil.

$$2(CH_2O) + \text{Bacteria} \rightarrow CO_2 + H_2O$$

(3) Human activities produce large quantities of hydrocarbons in the atmosphere.
(4) Automobile exhaust is the main source of aromatic hydrocarbons.
(5) Others: Petroleum, wood, coal, fossil, solvent evaporation etc.
(6) Topography and climate also contribute certain percentage of hydrocarbon in the atmosphere.

Environmental impacts

(1) Hydrocarbons undergo photochemical reactions in presence of sunlight to form noxious pollutants.

$$CH_4 + O_2 \rightarrow CH_3^* + OH^*$$

$$CH_4 + OH^* \rightarrow CH_3^* + H_2O$$

$$CH_3\text{—}CH{=}CH_2 + OH^* \rightarrow CH_3CH^*CH_2OH \rightarrow CH_3CHO \text{ (Acetaldehyde)} + CH_3^*$$

Acetaldehyde is one of constituent of photochemical smog. All these oxidized products slowly settle on the Earth and hence polluting soil and aquatic reservoir of the Earth.

(2) Methane (CH_4) reacts with mercury (Hg) to form CH_3Hg^+, which causes damage to the central nervous system, segregation and breakage of chromosomes in cells, inhibition of cell division. Only 0.5 ppm of CH_3Hg^+ is enough to create such dreadful poisoning effect. Organomercurials, particularly, CH_3Hg^+ are soluble in fat, the lipid fraction of membranes

and brain tissue. Attachment of Hg to cell membranes is likely to inhibit active transport of sugars across the membranes and allow the passage of potassium (K) to the membrane.

(3) Hydrocarbons catalyze the oxidation of SO_2 to H_2SO_4, which is the main constituent of acid rain.

$$SO_2 + \frac{1}{2}O_2 + H_2O + C_nH_m \rightarrow H_2SO_4$$

13.3.3 Aerosols

Colloidal size solid particles with liquid dispersed in atmosphere is known as aerosols. It can be divided into four categories:

(a) **Mists** Heavy concentration of fine particles is called mists.

(b) **Dust** Heavy concentration of coarse particles is called dust.

(c) **Fog** Heavy concentration of water droplets is called fog.

(d) **Smog** Smoke associated with fog is known as smog.

Soot particle It consists of several thousands of crystallites, forming graphite like layer structure. Each layer consists of at least 100 condensed aromatic rings. Surface of soot particle adsorbs PAH compounds and other harmful elements like: Cr, Mn, Cd, Be, V etc.

Sources of particulates

(a) Natural sources, e.g., volcano eruption, mines, sea etc.
(b) Human activities, e.g., developmental work, automobile exhaust, civilization and industrial growth etc.
(c) Photochemical reactions in the atmosphere.

13.3.4 Photochemical smog

Smog, enriched with high concentration of sulphur di-oxide, is called reducing smog, also known as London smog.

Photochemical smog is oxidizing smog, enriched with high level of oxidants and organics. It appears as brown hazy fumes.

Constituents Nitrogen-di-oxide, aldehydes, free radicals, peroxybenzoyl nitrate, peroxyacyl nitrate (PAN) etc.

Source Large quantities of automobile exhausts are trapped by an inversion layer (stagnant air mass) and get oxidized while exposed to sunlight to form photochemical smog.

Effects It causes irritation to eyes and lungs and cracking of rubber, reduces visibility and damages the plant.

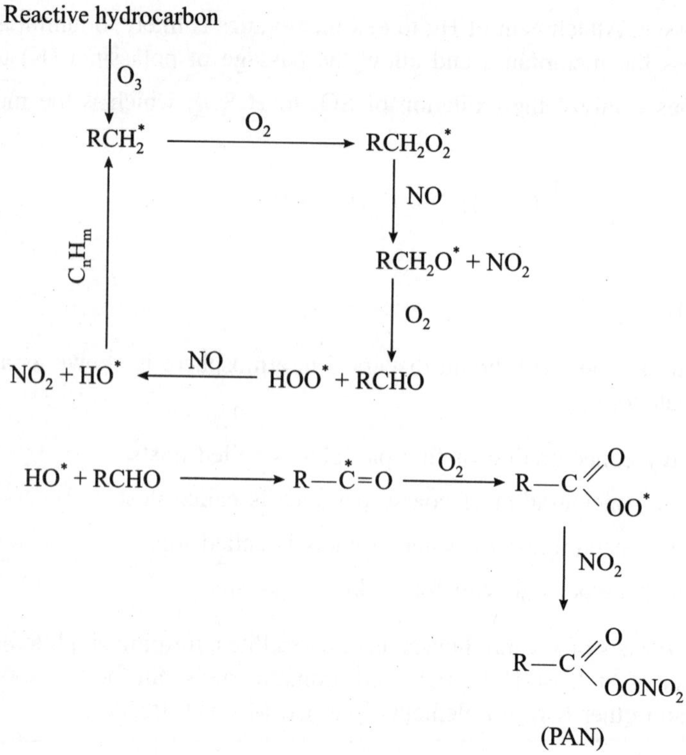

13.4 Noise pollution

Any sound source, not comfortable to human body, is called noise. The noise pollution can be classified into three categories:

(a) Transport noise It can be further divided into three categories: (1) Road traffic noise, (2) Air craft noise and (3) Rail traffic noise.

(1) Road traffic noise This noise arises from various vehicles, producing irritating sounds. The noise level increases with the increase in traffic density. Permissible sound level in different areas are given below:

Area	Day	Night
Industrial	75 dB	65 dB
Commercial	65 dB	55 dB
Residential	50 dB	45 dB
Hospitals, educational institutions, courts etc.	50 dB	40 dB

(2) **Air craft noise** Unlike road traffic noise, air craft noise is not continuous and intermittant. Permissible noise level range: 100 PNdB to 120 PNdB in day time and 90 PNdB to 100 PNdB at night, where, 1 PNdB = 1 dBA + 13.

(3) **Rail traffic noise** The frequency of rail traffic noise is lower than that of road traffic noise. So it does not appear harmful to us. However, buildings, located besides railway tracks are affected by that noise level. The introduction of diesel rail engine and electric locomotives has greatly reduced rail traffic noise.

(b) **Occupational noise** Origin of occupational noise are: (i) industrial machines and (ii) domestic machines.

Industrial machines Milling machine, lathe machine, grinding machine, news paper press, textile loom etc. produce sufficient noise, creating health hazards.

Domestic machines Vacuum cleaners, washing machines, mixers & grinders etc. can produce significant noise.

(c) **Neighbourhood noise** There are variety of noise sources which disturb and annoy the general public, e.g., TV sets, radio sets, music sets, loudspeakers etc.

Unit of noise pollution:

Sound Intensity (I): $$I = \frac{W}{A}$$

$$I = \frac{p_{\text{rms}}^2}{\rho c}$$

W = sound power in Watt, A = area normal to the direction of propagation of the sound wave.
p_{rms} = Root mean square sound pressure (Pa).
c = Velocity of sound in a specific medium (m/s). The value of c changes with temperature according to the following equation: $c = 20.05\sqrt{T}$, where, T = temperature in °K.
ρ = Density of the medium (kg/m^3), I = sound intensity in Watt/m^2.

$$L_w = 10 \log \frac{W}{10^{-12}}$$

$$L_I = 10 \log \frac{I}{10^{-12}}$$ If the intensity is 10^{-11}, sound intensity is 10 deci-Bel (dB)

$$L_P = 10 \log \frac{p_{\text{rms}}^2}{p_{\text{rms}}^{0\,2}} = 20 \log \frac{p_{\text{rms}}}{p_{\text{rms}}^0}$$ where, $p_{\text{rms}}^0 = 20\ \mu\text{Pa} \equiv 60$ dB

Noise pollution hazards For conversation within a metre 65 dBA is enough, 125 dBA gives the sensation of pain and 150 dBA might kill a human being.

Control of noise pollution

(1) Setting industries away from the residential areas.
(2) Using proper and well designed engines which produce less noise
(3) Proper lubrication and maintenance of machines.
(4) Restricting the use of loudspeakers and amplifiers.
(5) Enforcing laws to get rid of noise producing vehicles.
(6) Planting trees on both sides of roads.
(7) Using acoustic materials and mufflers which absorb the sound.
(8) Using cotton plugs and earmuffs.

13.5 Water Pollution

Hydrologic cycle: It consists of the following.

(1) Evaporation During day time huge amount of water from natural water resources like ponds, rivers, oceans etc. get evaporated in natural process.

(2) Transpiration Plants extract water from soil through their roots and this water is then transported to the leaves from which it is finally evaporated. The phenomenon is called transpiration.

(3) Condensation In winter season water vapour in atmosphere is condensed into droplets of liquid water in the form of fog and dew. The phenomenon is called condensation.

(4) Precipitation Heavy water fall occurs on the surface of the Earth in the form of rain, hail, snow, sleet etc. The phenomenon is called precipitation.

(5) Infiltration Movement of water from one aquifer to other confined water source is called infiltration.

(6) Percolation In many cases water from surface water sources moves to the surface through pores of soil. The phenomenon is called percolation.

(7) Run-off Due to melting of ice caps, down flow of water is observed from hill area to the sea level. The phenomenon is called run-off.

Causes of water pollution

(a) Dumping of wastes and chemicals in the urban and rural watersheds.
(b) Excessive use of fertilizers and pesticides in the agricultural land.
(c) Continuous discharge of effluents, containing hazardous chemicals from industries, to the nearby water reservoir.
(d) Mixing solid particles and insects with water.
(e) Leakage in pipelines and surface tanks.
(f) Sudden spill over of oils and chemicals into the water reservoir.
(g) Lack of social responsibilities.

Environmental damages due to water pollution

(a) Contaminated water threatens the health of millions of people through spreading of water-born diseases.
(b) Erosion of soil and degradation of soil quality, deforestation, resulting in huge loss in national productivity.
(c) Pollution in coastal water, rivers, lakes, ponds etc. immensely damage the aquatic life leading to a remarkable reduction in fish population.

Aerobic decomposition Microorganisms have a natural tendency to degrade or decompose organic wastes into simpler organic and inorganic substances. Some microorganisms decompose the organic wastes in presence of oxygen atmosphere. The process is known as aerobic decomposition and the corresponding bacteria is called aerobic bacteria. The products of decomposition are usually CO_2, H_2O, NO_3^-, PO_4^{3-}, SO_4^{2-} etc., which are non-objectionable.

Anaerobic decomposition Some microorganisms decompose the organic wastes in absence of oxygen atmosphere. The process is known as anaerobic decomposition. The products of decomposition are usually PH_3, H_2S, CH_4 etc., which are foul smelling and objectionable.

Biochemical Oxygen Demand (BOD) The total amount of dissolved oxygen required by aerobic bacteria to decompose all the organic wastes aerobically in a given waste water sample is called biochemical oxygen demand or BOD of that waste sample.

Significance For a given waste water sample higher BOD value indicates larger amount of decomposable organic matter present in the waste and hence the waste is said to be more polluted. So, BOD value indicates the degree of pollution in a given waste sample.

Disadvantages in BOD determination

(1) BOD results sometimes appear erroneous due to the presence of some undesirable substances.
(2) Some organic materials such as cellulose, phenols, benzene and tannic acid do not respond bio-degradation.
(3) Some organic materials, such as pesticides, industrial chemicals are non-biodegradable because they are toxic to microorganisms.
(4) Determination of BOD requires long time, at least five days.

Chemical Oxygen Demand (COD) The total amount of oxygen required to decompose all the organic wastes chemically is called chemical oxygen demand or COD. The following reagents are used for this purpose:

(1) Potassium-di-chromate ($K_2Cr_2O_7$) with 50% sulphuric acid (H_2SO_4) is used to oxidize the wastes.
(2) Trace amount of silver sulphate (Ag_2SO_4) is required to catalyze the oxidation of straight chain aliphatic compounds, aromatic hydrocarbons and pyridine.
(3) Trace amount of mercuric sulphate ($HgSO_4$) is used. Hg^{2+} ion forms complex ($HgCl_4^{2-}$) with Cl^- and hence prevents the interference of the latter during COD determination.

(4) Unreacted potassium-di-chromate is estimated with the help of standard Mohr's salt solution $[FeSO_4.(NH_4)_2SO_4.6H_2O]$. 1 ml 1 N $K_2Cr_2O_7$ solution ≡ 0.008 gm of oxygen.

Advantages of COD determination over BOD determination

(1) In BOD determination many oxidizable organic matter do not respond but in COD test both biologically oxidizable and biologically inert matters are oxidized.
(2) Determination of BOD requires at least 3 days but COD determination requires only 3 hours.
(3) In BOD test oxidation process is not at all complete but in COD test oxidation process may be assumed to be complete. As COD value for a given sample is always higher than BOD value of the same sample, COD value can be used to estimate approximate BOD value.

Dissolved oxygen (DO) Amount of oxygen remain dissolved in a given water reservoir is called dissolved oxygen or DO. Saturated value of DO is 12 to 15 mg/l and the critical value of DO is 4 to 5 mg/l. Critical DO value is the minimum DO required for survival of aquatic life.

Effects of low DO value

(1) Whenever DO level drops near or below its critical value, aquatic life becomes endangered.
(2) As DO level drops, anaerobic condition prevails, giving rise floating of sludges, bubbling of odorous gases, abnormal fungal growth etc.

Factors influencing DO level of a water reservoir:

(1) Plants add DO in daytime through photosynthesis but removes DO at night through respiration.
(2) Microorganisms remove DO due to their respiration and biodegradation of organics.
(3) Water, flowing from other sources are directly flowing into river or pond, supplies oxygen to the latter.
(4) Increase in temperature reduces the solubility of oxygen etc.

Water Pollutants Following are the major water pollutants: (a) oxygen demanding wastes, (b) excess nutrients, (c) total dissolved solids, (d) heavy metals, (e) pesticides, (f) volatile organic compounds, (g) pathogens, (h) thermal pollution.

(a) Oxygen Demanding Wastes Organics, present in the water, are oxidized by dissolved oxygen, thereby reducing the dissolved oxygen (DO) content. As DO level drops, aquatic life is in danger and undesirable colours, odours and tastes are developed in water, reducing its acceptability as a domestic supply. The saturated value of DO is DO_s, which is around 12 – 15 mg/l, whereas the critical DO level is only 4 mg/l.

(b) Excess Nutrients Major nutrients for survival of life are nitrogen, phosphorus, carbon, sulphur, Ca, K, Fe, Mn, B, Co etc. However, excess nutrients stimulate the growth of algae, the phenomenon is called eutrophication. Whenever the eutrophication starts, reservoir becomes shallower, warmer and is ultimately converted to a marsh. Thus excess nutrients must be avoided as far as practicable.

(c) Total Dissolved Solids (TDS) Concentration of total dissolved solids in a given sample of water is known as salinity of the sample. It mainly contains some acid radicals and basic radicals.

Acid radicals: Cl^-, SO_4^{2-}, HCO_3^-, NO_3^- etc.

Basic radicals: Na^+, K^+, Mg^{2+}, Ca^{2+} etc.

In fresh water TDS value usually lies within 1500 mg/l, in brackish water TDS value usually lies within 5000 mg/l, in saline water TDS value is more than 5000 mg/l, in sea water TDS value usually lies within 34000 mg/l and in drinking water TDS value must lie within 500 mg/l.

(d) Heavy Metals In chemistry metals, having specific gravity greater than 4 or 5, are known as heavy metals. However, in environmental science, toxic metals are usually called heavy metals. Examples are: As, Cd, Pb, Hg (deadly poisonous), Cu, Mn, Ni, Cr and Fe. Chromium and iron are the essential nutrients in human being but they are extremly toxic if used in higher doses.

(e) Pesticides These are often used in agriculture to kill insects. The scientist Muller first invented dichloro-diphenyl-trichloro-ethane (DDT). He was awarded Noble prize for that in 1946. Other pesticides are benzene hexa chloride (BHC), dichloro-diphenyl-dichloro-ethene, heptachlor, aldrine, dieldrin, endrin etc. DDT is soluble in lipids and easily get accumulated in fatty tissue. Pesticides are stable for a long period of time in the environment and have bad impact on food chains.

(f) Volatile Organic Compounds (VOC) There are several volatile organic compounds, which are partly soluble in water. Out of them five are extremly toxic: (i) vinyl chloride or chloroethene. It is the raw material to manufacture Polyvinyl Chloride or PVC. It is a proven carcinogenic compound to human being. (ii) Tetrachloroethylene. It is mainly used in heat transfer medium and also in the manufacture of CFCs. (iii) Trichloroethylene (TCE). It is commonly used to clean everything from electronic parts to jet engines as well as septic tanks. (iv) 1, 2-Dichloroethane. It is commonly used in the manufacture of vinyl chloride, tetraethyl lead etc. It causes injury to the central nervous system, liver and kidneys. (v) Carbon tetrachloride. It is a common house-hold cleaning agent, fire extinguishers and solvents. It is very toxic and also a carcinogenic compound.

(g) Pathogens Water-born disease producing bacteria are called pathogens or coliform bacteria. They are responsible for cholera, dysentery, typhoid etc. Some viruses are responsible for heaptitis. Sewage water contains approximately 500 viruses in 100 ml of sample and 46×10^6 coliform bactreria in 100 ml of sample. Polluted surface water contains 1 virus in 100 ml of sample and 5×10^4 coliform bactreria in 100 ml of sample.

(h) Thermal Pollution Several thermal power plants often release huge amount of heat in the nearby river or lake, resulting in sharp rise in temperature. So DO level decreases with this rise in temperature. The following chart illustrates how DO level drops with rise in temperature.

At 0°C —— DO = 14.6 mg/l
At 10°C —— DO = 11.6 mg/l
At 20°C —— DO = 9.1 mg/l
At 30°C —— DO = 7.6 mg/l

The following water quality standard is prescribed by *World Health Organization* (WHO):

Parameter	Standard
Zn	5000 ppb
Cu	1000 ppb
Fe	200 ppb
Mn	50 ppb
As	50 ppb
Pb	50 ppb
Cd	10 ppb
Hg	1 ppb
Chlorides	< 250 ppm
Sulphates	< 250 ppm
TDS	< 500 ppm
PH	7 to 8.5
Turbidity	5 to 10 NTU
Odour	≤ 3 TON
Colour	< 10 cobalt scale
Coliform bacteria	zero per 100 ml of drinking water sample.

ppb = Parts per billion

13.6 Municipal waste water treatment

Municipal waste water is treated before discharging to the natural water sources to avoid water pollution. The treatment is performed in three different stages. The waste water plant or effluent treatment plant may be divided into three major sections as given below:

(i) Primary treatment plant.
(ii) Secondary treatment plant.
(iii) Tertiary treatment plant.

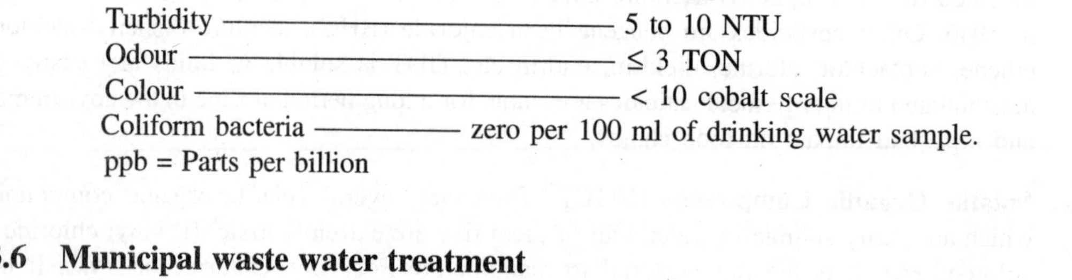

Fig. 13.1 Schemetic diagram of Municipal waste water treatment method. (a) P = Primary treatment plant, (b) S1 = Trickling filter method, (c) S2 = Activated sludge method, (d) T = Tertiary treatment palnt.

13.6.1 Primary treatment plant

This section consists of two chambers: (a) screening chamber and (b) sedimentation chamber. Raw waste water is first sent to the screening chamber, consisting of several screens of different mesh sizes. While passing through these screens large particles are separated.

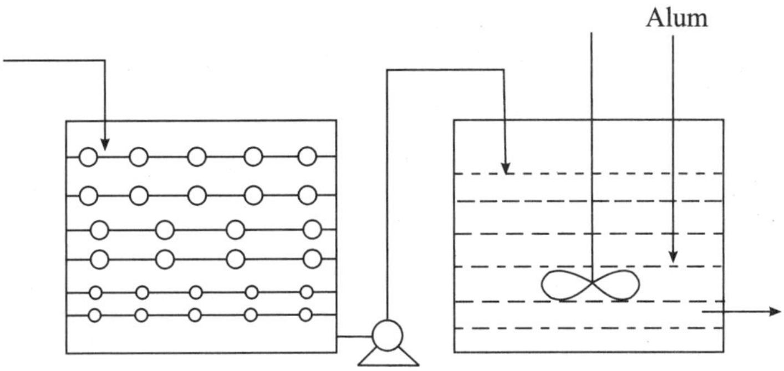

Fig. 13.2 Primary treatment plant, consisting of screening chamber and sedimentation chamber.

The effluent of screening chamber is then sent to the sedimentation chamber where it is treated with alum, namely, $Al_2(SO_4)_3$ or $FeCl_3$ or $FeSO_4$ under anaerobic condition. Alum forms bulky hydroxide precipitates or flocs, floating in the solution. These flocs absorb small colloidal size particles from the effluent and slowly get precipitated at the bottom of the chamber. In presence of air oxidative decomposition sets in, thereby preventing the formation of flocs.

The effluent of sedimentation chamber is then sent to the secondary treatment plant.

13.6.2 Secondary treatment plant

The secondary treatment of waste water may be carried out by following any one of the methods, described below.

13.6.2.1 Trickling filter method

The effluent of the primary treatment plant is first sent to a rotary drum with the help of a pump. The effluent is then spread over a filter bed through multiple outlets of the rotary drum. The filter bed is sufficiently porous and aerated so as to allow the effluent flowing through the pores under full aerobic condition. While passing through the pores microbes, distributed in the filter bed, slowly eat up the organics of the effluent. Finally the effluent is taken out from the bottom of the filter bed and sent to a sedimentation chamber where it is allowed to settle for about 12 hours. The heavy sludge is disposed off and the clear effluent is then sent to the tertiary treatment plant. However, the method has some disadvantages.

(i) The method is costly.

(ii) The biodegradation process is temperature dependent.

(iii) Short filter bed life. When the pores are filled up, the filter bed should be replaced by a new one.

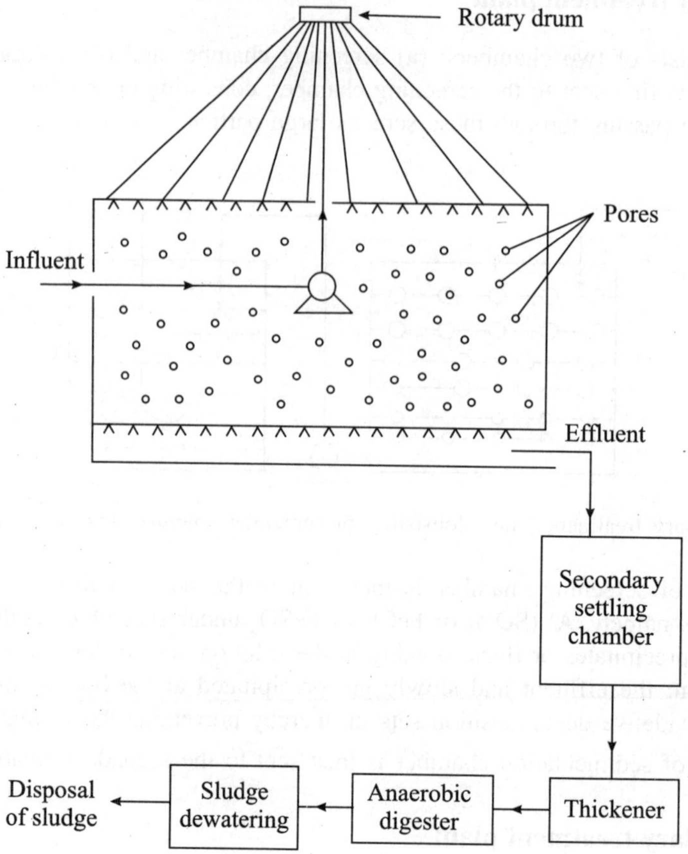

Fig. 13.3 Trickling filter method of secondary treatment plant.

13.6.2.2 Rotating Biological Contactor (RBC) method

The effluent of the primary treatment plant is sent to an open chamber, where a rotating drum is kept half-immersed in the waste. The drum, having diameter 3.6 m, is rotating at constant speed. While rotating a biomass film is developed surrounding the drum surface. Air is pumped into the waste time to time to maintain full aerobic condition during biodegradation process. Aerobic bacteria in biomass film, while immersed in the waste, absorb organics from the waste and start decomposing the organics when the biomass film remains in open atmosphere during rotation. The rotating drum is called *rotating biological contactor* or *RBC*. An agitator is often attached with the chamber in order to expedite the decomposition process. The process is continued for six to eight hours. After that waste sample is allowed to flow in a settling tank where it is allowed to settle for about 12 hours. Most of the solids are separated as sludge during this period. Clear effluent is taken out from the top and sent to the tertiary treatment plant. The heavy sludge is taken out from the bottom.

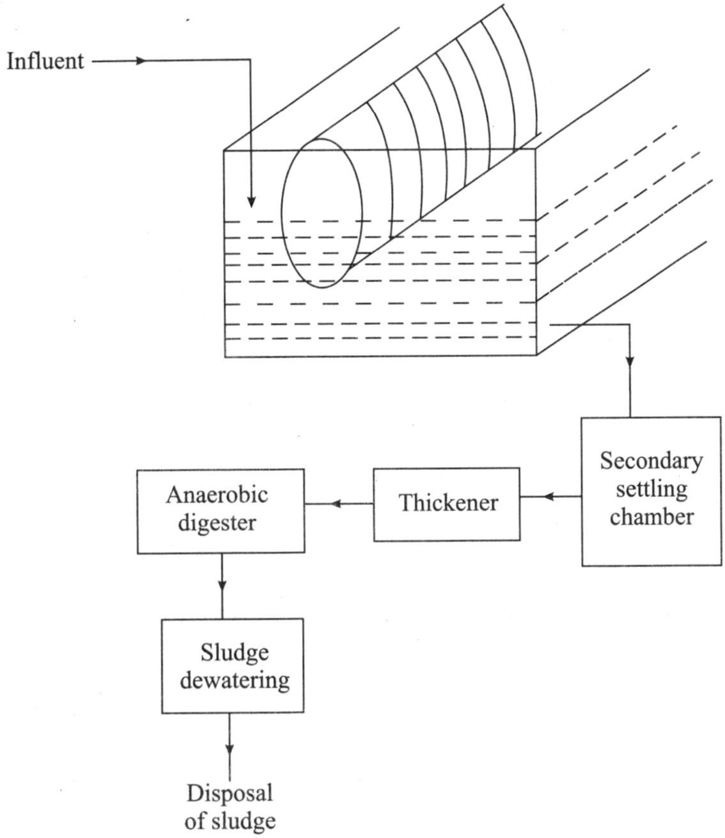

Fig. 13.4 Rotating Biological Contactor method of secondary treatment plant.

Low cost is the main advantage of this process. Furthermore the process is almost temperature independent. However, this process also has some disadvantages.

(i) The thickness of the biomass film increases with time and hence the *RBC* needs cleaning regularly.

(ii) Air is pumped into the biodegradation chamber regularly to maintain full aerobic condition.

13.6.2.3 Activated sludge method

In this method, effluent of primary treatment plant is sent to an aeration tank, where air is pumped to the tank to maintain full aerobic condition. The waste water is continuously agitated for 6 to 8 hours. Then it is sent to a secondary settling tank, where it is allowed to settle for 12 hours. Heavy sludge is taken out from the bottom and sent to a thickener. Part of this heavy sludge is mixed with the fresh batch to expedite the biodegradation process since the sludge contains aerobic bacteria. The recycled sludge is called activated sludge and hence the name of the process. The clear water is taken out from the top of the settling chamber.

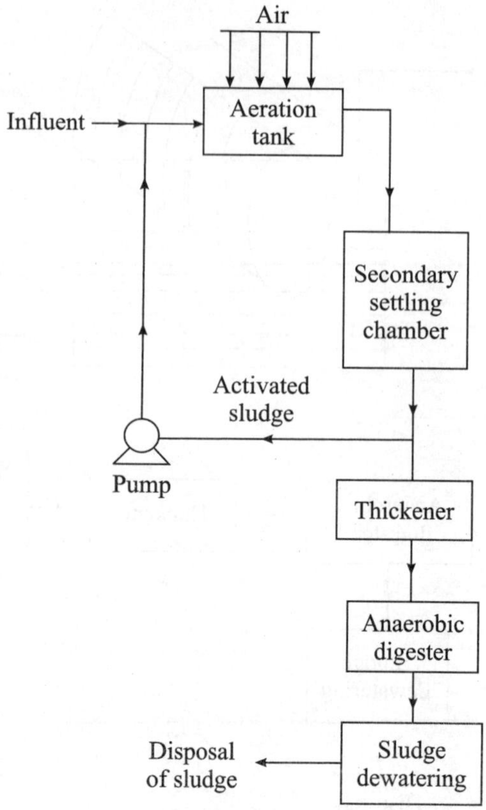

Fig. 13.5 Activated sludge method of secondary treatment plant.

13.6.3 Tertiary treatment plant

The effluent of the secondary plant is first sent to a lime chamber, where the effluent is treated with dry lime (CaO), which reacts with phosphates to form insoluble calcium phosphate [$Ca_3(PO_4)_2$], appearing as fine precipitate at the bottom. Thus phosphates are removed from the effluent.

The clear effluent is then sent to an absorption column from the top. The air is passed from the bottom. While passing through the column the air strips off ammonia from the effluent. The absorption plates are arranged in the column in such a way to have maximum contact between air and effluent. Thus ammonia is removed from the effluent.

The clear effluent is then sent to a disinfectant chamber, where the effluent is treated with bleaching powder or chlorine gas to kill all the bacteria.

Finally the effluent is sent to a pH-controller to maintain pH of the effluent near about 8. The acidic waste is treated with slaked lime [$Ca(OH)_2$] and the alkaline waste is treated with carbon-di-oxide gas to maintain pH near about 8. An agitator is attached to the chamber to expedite the process. The clear effluent is then discharged to the nearby reservoir.

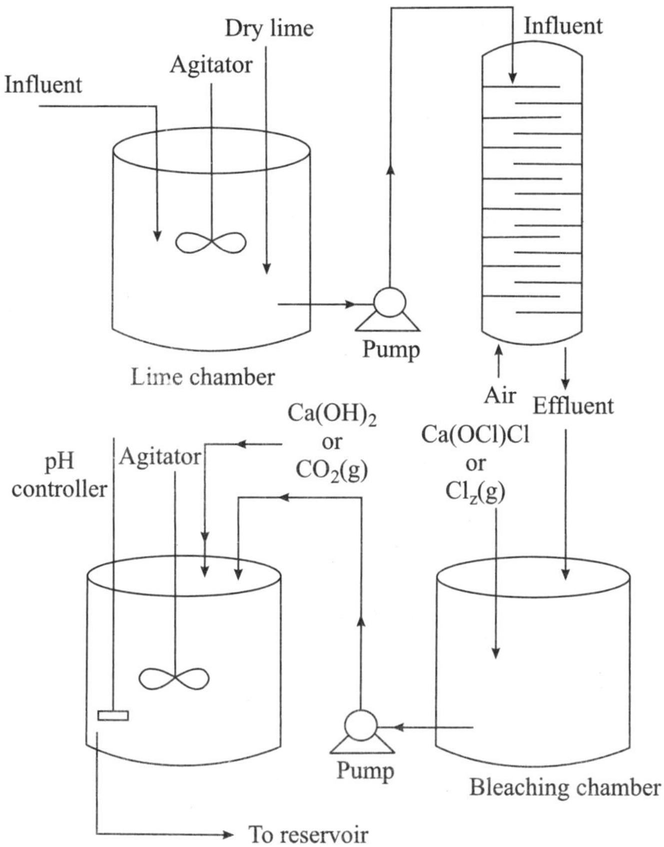

Fig. 13.6 Tertiary treatment plant.

13.7 Lithosphere

The earth crust can be divided into four regions: (a) Top-soil, (b) Sub-soil, (c) Weathered rock, (d) Bedrock.

Top-soil It mainly contains bioorganisms and organic matters. So this part of soil is very important for agriculture.

Sub-soil It receives organics, salts, nutrients from top-soil as well as inorganic minerals from weathered rock.

Weathered rock It mainly contains oxides of iron, magnesium, nickel, silicon etc.

Bedrock Inorganic minerals remain in liquid state in this region. Iron and nickel are the majority components.

Hard core This part of the earth crust mainly contains pure iron, remaining in solid state.

Productive soil contains about 5% organic matter and 95% inorganic matter. Soil has numerous capillaries for the passage of water and air. About 35% volume of soil is filled up by air. Oxygen content in soil is 15% instead of 21% since biodegradation of organics consumes some oxygen.

Solid wastes

Definition The wastes which are not liquid or gaseous such as food scraps, electrical and electronic component scraps, container and packaging wastes etc. are called solid wastes.

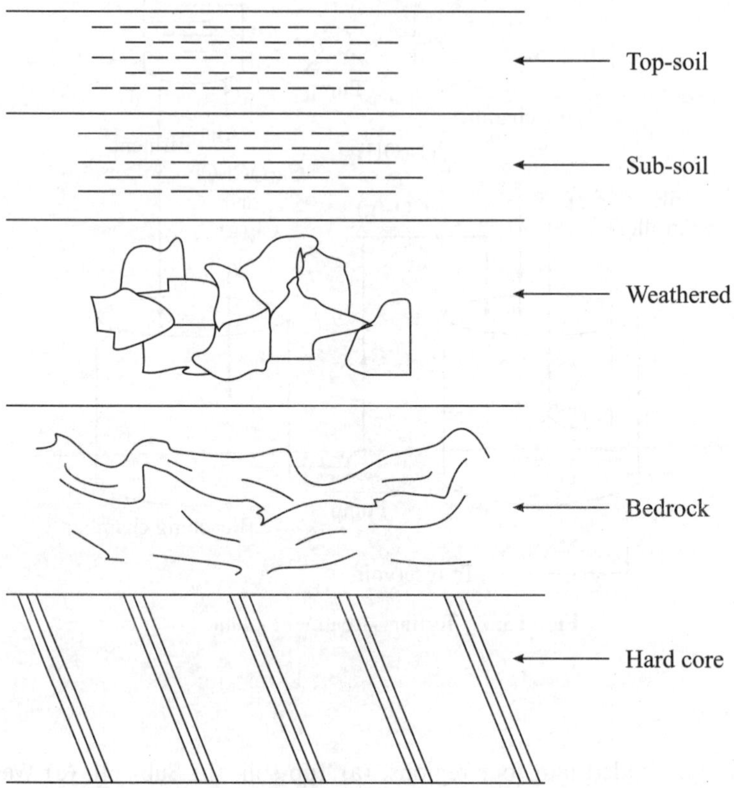

Types of solid wastes (a) Municipal solid wastes, (b) Industrial solid wastes, (c) Hazardous wastes.

Municipal solid wastes Solid wastes, obtained from residential, commercial, institutional and industrial sources, are called municipal solid wastes.

Hazardous wastes The wastes, which cause danger to the living community, are known as hazardous wastes. It may have any of the following characteristics: (a) ignitable, (b) corrosive, (c) reactive and (d) toxic.

Classification Hazardous wastes may be classified into following categories: (i) radioactive wastes, (ii) chemical wastes, (iii) biological wastes, (iv) flammable wastes, (v) explosives.

Sources Following are the major sources of hazardous wastes: (i) automobile industries, (ii) chemical industries, (iii) coal industries, (iv) metallurgical industries, (v) leather industries, (vi) textile industries.

Factors affecting the solid wastes generation The following factors are important in generation of solid wastes:

(i) Geographic location of the area.
(ii) Frequency of waste collection.
(iii) Population.
(iv) Use of home appliances.
(v) Extent of recycling.

To minimize the solid waste generation the following steps are to be taken: (i) waste reduction at source, (ii) recycling, (iii) product modification.

Some causes of soil degradation

(a) Production of excess food grains to meet the fast growth of population.
(b) Excessive cultivation of lands in low rainfall areas.
(c) Excessive use of fertilizers and pesticides.
(d) Unscientific expansion of irrigation projects, waterlogging and salinization reduce the fertility of soil to a great extent.

Environmental damages due to soil degradation

(a) Desertification, (b) water logging, (c) salinization, (d) erosion of soil and degradation of soil quality, (e) accumulation of silt in rivers, canals, dams etc.

Salinization leads to infertility in agricultural lands. Fertility in agricultural lands can be improved by several methods like agro forestry, integrated crops and life stock management systems.

13.8 Green house effect

Fraction of solar radiation, entering into the Earth's atmosphere, is reflected back to the space by the atmosphere and the Earth's surface, is known as *albedo*, represented by the symbol α. The value of α is 0.30 for the Earth. Remaining fraction of solar radiation is absorbed by the Earth's surface followed by reemission of the same. The emitted radiation has wide range of spectrum. Part of the spectrum, having wavelength range 4 – 30 μm is absorbed by some gases like CO_2, H_2O-vapour, CH_4, CFC, N_2O etc. These absorbed radiation eventually returns to the Earth's surface, resulting in the increase in Earth's surface temperature. This phenomenon is known as Green House Effect.

Green House Gases (GHG) The gases, which absorb emitted radiation from the Earth's surface, are called Green House Gases or GHG in short. Some GHG are CO_2, CH_4, N_2O, CFC etc. Besides that SO_2, H_2O-vapour, O_3 also absorb some emitted radiation.

Beneficial effect Green House gases acts as a thermal blanket surrounding the Earth to maintain the average temperature of the Earth's surface at around 15°C. So this thermal blanket is very much essential for survival of life on the Earth.

Harmful effects

(1) With the increase in population GHG concentration increases, as a result of which Earth's surface temperature increases continuously. It has been observed that atmospheric temperaturre rises by 3°C with doubling the CO_2-concentration.
(2) With the increase in GHG concentration global climate changes continuously, which brings about fierce storms, blizzards, cyclones, drought, incessant rainfall etc. in different part of the Earth's surface.
(3) The agricultural productivity is adversely affected with the increase in temperature even by 1°C. Thus increase in GHG concentration in the atmosphere hamperes the agricultural productivity. Furthermore, dissolved oxygen in aquatic reservoir decreases with the increase in temperature. Thus aquatic life is endangered with the increase in GHG concentration.
(4) As we know temperature of the Earth increases with the increase in GHG concentration, ice caps start melting, as a result of which, water level rises continuously. It is apprehended that water level will be as high as 6 metres above the present water level by 2100. If that be the case many coastal areas including many island nations will be submerged.
(5) The calculated surface temperature of the planet Venus is –41°C but the actual surface temperature is 427°C due to heavy concentration of GHG in the atmosphere of Venus. If GHG concentration in the Earth's atmosphere continues to increase, the Earth will face the fate of Venus.

13.9 Ozone layer depletion

Ozone in *tropospheric* region has deleterious effect on environment. Hence it is often called *bad ozone*. Some harmful effects of bad ozone are:

(i) It damages trees and reduces growth rates.
(ii) It reduces yields of major agricultural crops such as corn, wheat etc.
(iii) It causes irritation to the eyes and respiratory tracts of human beings. Exposure to 50 ppm of ozone for several hours will lead to death due to accumulation of fluid in the lungs, leading to severe damage to lung capillaries. Younger ones are more susceptible to toxic effects than older ones.

Ozone in *stratospheric* region has remarkable importance due to following reasons:

(i) Ozone acts as protective shield for living organisms on the Earth from various injurious effects of UV rays (220 – 230) nm, which is proven to be responsible for human sunburn and cancer.
(ii) Through absorption of UV-rays ozone layer supplies necessary heat to develop positive lapse rate in the stratosphere due to which stratospheric region is quite stable. Pollutant, if enters into stratospheric region, is entrapped locally instead of diffusion.
Prime enemies of stratospheric ozone are: CFC, NO. OH^-, Pb etc.

$$CFCl_3 \rightarrow CFCl_2 + Cl^{\bullet}$$
$$Cl^{\bullet} + O_3 \rightarrow ClO^{\bullet} + O_2$$
$$ClO^{\bullet} + O \rightarrow Cl^{\bullet} + O_2$$
$$\ldots\ldots\ldots\ldots\ldots\ldots \text{ Chain reaction continues}$$
$$NO + O_3 \rightarrow NO_2 + O_2$$
$$NO_2 + O_3 \rightarrow NO_3 + O_2$$
$$NO_2 + NO_3 \rightarrow N_2O_5$$
$$N_2O_5 + H_2O \rightarrow HNO_3 \text{ (Returned as acid rain)}$$

Ozone Depleting Potential (ODP) It is the amount of ozone layer destroyed by an ozone depleting substance with respect to that of CFC-11. It is usually expressed in *Dobson unit.*

Montreal Protocol An international agreement was made in 1987 at a conference in Montreal and signed by 34 countries. According to this agreement all countries agreed to ban CFC so that % release of CFC in atmosphere would be reduced to 50% with respect to 1987 level by the end of 20th century. Even after Montreal Protocol ozone hole continues to widen at a faster rate. Another conference was convened at London in June 1990 where the developed countries agreed to ban CFC 100% by 2000 A.D. The London amendment to the Montreal Protocol was ratified in May 1992 and came into effect on August 10, 1992.

14

HARDNESS OF WATER

14.1 Water hardness

Detergent forms lather or foam in presence of water. The main constituent of any detergent is the Na – or K –salt of fatty acids. The salt exists as colloids. These colloidal particles form micelles, each of which has two parts. One part remains dissolved on water, known as hydrophilic part and the other part repels water, known as hydrophobic part. This hydrophobic part forms lather and becomes negatively charged. These negatively charged lather attracts the dust and dirt from clothes and hence clothes are cleaned. Thus formation of lather is an important factor in washing clothes. Primary constituent of a micelle is $C_{17}H_{35}COONa$ or $C_{17}H_{35}COOK$. However, in presence of Ca^{2+}/Mg^{2+} ions the fatty acid salts get precipitated according to the following reactions

$$2C_{17}H_{35}COONa + CaCl_2 \longrightarrow (C_{17}H_{35}COO)_2Ca \downarrow + 2NaCl$$

$$2C_{17}H_{35}COONa + MgCl_2 \longrightarrow (C_{17}H_{35}COO)_2Mg \downarrow + 2NaCl$$

As a result of this cleaning efficiency of the detergent decreases. The phenomenon is known as hardening and this property of water is called hardness of water.

Definition of hardness of water It is the property of water to prevent forming lather or foam in presence of detergent. Besides Ca^{2+} and Mg^{2+} ions, other ions like Fe^{3+} etc. also prevent forming lather but only Ca^{2+} and Mg^{2+} ions are present in significant amount in surface water and hence hardness of water mainly arises due to the presence of Ca^{2+} and Mg^{2+} ions.

14.2 Types of hardness

Hardness of water may be broadly classified into following categories:

(i) Temporary hardness.

(ii) Permanent hardness.

As water inherently contains both type of hardness, total hardness of water is given by:

Total hardness = Temporary hardness + Permanent hardness

14.2.1 Temporary hardness

14.2.1.1 Boiling method

When water sample is boiled, calcium and magnesium bi-carbonates are precipitated as carbonates and hence hardness due to Ca – and Mg – bi-carbonates is removed. This is called temporary hardness. So this temporary hardness mainly arises due to the presence of $Ca(HCO_3)_2$ and $Mg(HCO_3)_2$. These bi-carbonates remain soluble in cold water but are removed as carbonates when boiled according to the following equation.

$$Ca(HCO_3)_2 \xrightarrow{Heat} CaCO_3 \downarrow + CO_2 + 2H_2O$$

$$Mg(HCO_3)_2 \xrightarrow{Heat} MgCO_3 \downarrow + CO_2 + H_2O$$

14.2.1.2 Clark's method

In this method temporary hardness can be removed through treatment of sample water by slaked lime, i.e., $Ca(OH)_2$. The following reactions occur during removal of temporary hardness.

$$Ca(HCO_3)_2 + Ca(OH)_2 \xrightarrow{Heat} 2CaCO_3 \downarrow + 2H_2O$$

$$Mg(HCO_3)_2 + Ca(OH)_2 \xrightarrow{Heat} CaCO_3 \downarrow + MgCO_3 \downarrow + 2H_2O$$

14.2.2 Permanent hardness

That fraction of hardness, which remains unaltered even after boiling, is known as permanent hardness. Presence of soluble salt of Ca^{2+} and Mg^{2+} ions other than bicarbonates, is the main source of permanent hardness.

14.2.3 Unit of hardness

The following units are used to express hardness of water

(i) *ppm* (ii) *mg/l*

(iii) $\left(\frac{M}{100}\right)$ (iv) °Clarks = °Cl

(v) °French = °Fr

Out of these 5 different units, *ppm* is usually preferred in many cases. Other units can be expressed in terms of *ppm*. Hardness of water is generally expressed as parts per million or simply *ppm* of $CaCO_3$ equivalent. Molecular weight of $CaCO_3$ is 100. Taking this advantage hardness of water is usually expressed as $CaCO_3$ equivalent. So, water hardness 1 *ppm* means 10^6 ml sample water contains 1 gm $CaCO_3$ equivalent.

$$1\ ppm = \frac{1\text{gm } CaCO_3 \text{ equivalent}}{10^6 \text{ml water}}$$

$$= \frac{10^3 \text{mg } CaCO_3 \text{ equivalent}}{10^3 \text{lit. of water}} = 1 \text{ mg/lit} = 10^{-3} \text{ gm/lit}$$

Thus, $$1000\ ppm = \frac{1\text{gm } CaCO_3 \text{ equivalent}}{1 \text{ lit. of water}}$$

$$= \frac{\left(\frac{1}{100}\right)\text{moles of } CaCO_3 \text{ equivalent}}{1 \text{ lit. of water}}$$

[Since, Molecular Weight of $CaCO_3$ is 100]

1M = Molarity of $CaCO_3$ equivalent

$$= \frac{1\text{mole of } CaCO_3 \text{ equivalent}}{1 \text{ lit. of water}}$$

Thus, $$1000\ ppm = 1 \times \left(\frac{M}{100}\right) CaCO_3 \text{ equivalent}$$

and $$1\ ppm = 10^{-3}\left(\frac{M}{100}\right) CaCO_3 \text{ equivalent}$$

So, water hardness can also be expressed as $\left(\frac{M}{100}\right) CaCO_3$ equivalent.

The different units of hardness can be expressed in terms of *ppm* as follows:

$$1\ ppm = \frac{1\text{gm } CaCO_3 \text{ equivalent}}{10^6 \text{ml of water}}$$

$$1\ mg/l = 1\ ppm = \frac{1\text{gm } CaCO_3 \text{ equivalent}}{10^6 \text{ml of water}}$$

$$1 \times \left(\frac{M}{100}\right) = 10^3\ ppm = \frac{1\text{gm } CaCO_3 \text{ equivalent}}{1 \text{ litre of water}}$$

$$1^\circ\ Cl = 0.07\ ppm = \frac{1\text{gm CaCO}_3\text{ equivalent}}{70\text{ litres of water}}$$

$$1^\circ\ Fr = 10\ ppm = \frac{1\text{gm CaCO}_3\text{ equivalent}}{10^5\text{ ml of water}}$$

14.2.4 Estimation of hardness of water using complexometric titration

14.2.4.1 Theory

Hardness of water arises mainly due to Ca^{2+} and Mg^{2+} salts. Estimation of hardness of a given sample water can be determined by using disodium salt of Ethylene Diamine Tetraacetic Acid, represented by Na_2EDTA. The latter is a strong complexing agent and easily forms complexes with Ca^{2+} and Mg^{2+}. The indicator, used for this purpose, is Eriochrome Black T (EBT), which itself is a complexing agent. When EBT is added to a given water sample, metal cations M^{2+} (Ca^{2+} and Mg^{2+}) form complexes with EBT according to the following equation

$$M^{2+} + \text{EBT} \longrightarrow [\text{M} - \text{EBT}]\text{, which appears as wine red complex.}$$

As this complex is stable at $pH \geq 10$, ammonium chloride-ammonium hydroxide buffer solution is used to maintain the pH at around 10.

During titration with EDTA, EBT of the wine red complex ([M – EBT]) is replaced by EDTA to form ([M – EDTA]) since [M – EDTA] is more stable than [M – EBT]. The following reaction takes place

$$[\text{M} - \text{EBT}] + \text{EDTA} \longrightarrow [\text{M} - \text{EDTA}] + \text{EBT (sky blue).}$$

Thus the change of wine red colour to distinct sky blue colour marks the end point of titration. Calculation of hardness from titration results gives the total hardness.

To determine Ca –hardness separately, titration is to be carried out in presence of NaOH so that Mg^{2+} is bound as $Mg(OH)_2$, leaving behind free Ca^{2+} ions. Titration is to be carried out with Na_2EDTA as before by using murexide indicator instead of EBT indicator.

The formula of Ethylene Diamine Tetraacetic Acid as below

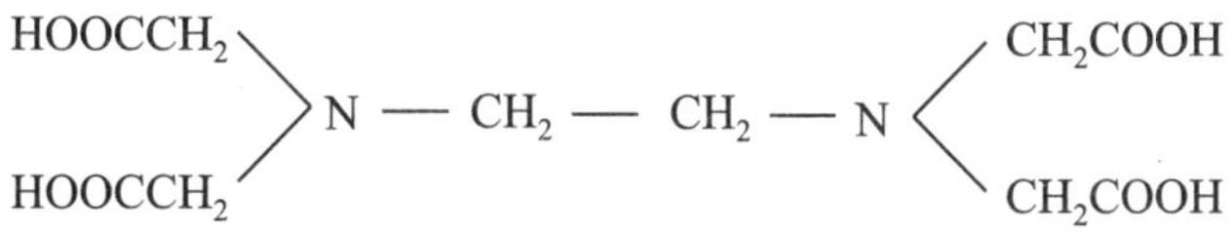

14.2.4.2 Procedure

(a) $Zn(CH_3COO)_2.2H_2O$ (MW = 219) solution of strength $0.5\left(\frac{M}{100}\right)$ is supplied.

(b) To prepare $0.5\left(\frac{M}{100}\right)$ Na_2EDTA solution around 200 mg Na_2EDTA crystals is dissolved in 100 ml distilled water in a 100 ml volumetric flask.

(c) Na_2EDTA solution is taken in a burette. 10 ml standard zinc acetate solution is taken in a 250 ml conical flask. 2 ml buffer solution is added to it followed by a pinch of indicator. The solution appears wine red, which is then titrated against Na_2EDTA solution until the colour changes to sky blue. Initial and final readings are noted and three such readings are recorded.

(d) In the next step 10 ml sample water is taken in a 250 ml conical flask. 2 ml buffer solution is added to it followed by a pinch of indicator. The solution appears wine red, which is then titrated against Na_2EDTA solution until the colour changes to sky blue. Initial and final readings are noted and three such readings are recorded.

(e) In the next step 10 ml sample water is taken in a 250 ml conical flask. 1 ml 2(N) NaOH solution is added to it to raise the *pH* to around 12. A pinch of murexide indicator is added and the solution appears pink. The solution is then titrated against Na_2EDTA solution until the colour changes to purple. Initial and final readings are noted and three such readings are recorded.

14.2.4.3 Experimental results

Table 1 Standardization of Na_2EDTA solution

No. of observations	*Volume of standard zinc acetate solution taken (ml)*	*Volume of Na_2EDTA consumed (ml)*			
		Initial	*Final*	*Difference*	*Mean volume*
1	10				
2	10				V_2
3	10				

Calculation

$V_1S_1 = V_2S_2$, where,

V_1 = Volume of standard hard water taken = 10 ml.

S_1 = Strength of zine acetate solution = $0.5\left(\frac{M}{100}\right)$

V_2 = Mean burette reading of Table 1, in ml.

S_2 = Strength of Na_2EDTA solution.

Table 2 Estimation of total hardness of supplied hard water

No. of observations	*Volume of supplied hard water taken (ml)*	*Volume of Na_2EDTA consumed (ml)*			
		Initial	*Final*	*Difference*	*Mean volume*
1	10				
2	10				V_2
3	10				

Calculation

$V_1S_1 = V_2S_2$, where,

V_1 = Volume of sample hard water taken = 10 ml.

S_1 = Strength of supplied hard water = $x\left(\dfrac{M}{100}\right)$

V_2 = Mean burette reading of Table 2, in ml.

S_2 = Strength of Na_2EDTA solution, already obtained from Table 1.

Water sample of strength $1 \times \left(\dfrac{M}{100}\right)$ is equivalent to water hardness of 1000 *ppm*.

So, water sample of strength $x\left(\dfrac{M}{100}\right)$ is equivalent to water hardness of 1000*x ppm*.

So, total hardness = 1000*x ppm*.

Table 3 Estimation of Ca –hardness of the supplied hard water

No. of observations	*Volume of supplied hard water taken (ml)*	*Volume of Na_2EDTA consumed (ml)*			
		Initial	*Final*	*Difference*	*Mean volume*
1	10				
2	10				
3	10				

Calculation

$V_1S_1 = V_2S_2$

V_1 = Volume of sample hard water taken = 10 ml.

S_1 = Strength of supplied hard water = $y\left(\dfrac{M}{100}\right)$

V_2 = Mean burette reading = …… ml

S_2 = Strength of Na_2EDTA solution, already obtained from Table 1.

Water sample of strength $1 \times \left(\dfrac{M}{100}\right)$ is equivalent to water hardness of 1000 *ppm*.

So, water sample of strength $y \times \left(\dfrac{M}{100}\right)$ is equivalent to water hardness 1000*y ppm*.

So, total Ca –hardness = 1000*y ppm*.

So, total Mg –hardness = 1000 $(x - y)$ *ppm*.

14.2.4.4 Estimation of permanent hardness

200 ml of hard water sample is taken in a 500 ml beaker and heated to boiling till the volume is reduced to one-half, i.e., 100 ml. The water is then cooled to room temperature. 50 ml of that

boiled water is taken in a 250 ml volumetric flask and fill it up to the mark by distilled water. Then perform the procedure (d) above by taking 'boiled sample water' instead of 'sample water', calculate as before and find out the permanent hardness. Subtraction of permanent hardness from total hardness gives temporary hardness.

14.3 Treatment of hard water

Hard water should be treated in two steps:

14.3.1 External treatment

14.3.1.1 Lime-Soda process

The chemicals, used in this process, is a mixture of soda, i.e., Na_2CO_3 and slaked lime, i.e., $Ca(OH)_2$. The hard water and chemical mixture are heated to around 100°C with continuous stirring in a reaction chamber, where the following reactions are believed to take place.

$$CaCl_2 + Na_2CO_3 \longrightarrow CaCO_3 \downarrow + 2NaCl$$

$$CaSO_4 + Na_2CO_3 \longrightarrow CaCO_3 \downarrow + Na_2SO_4$$

$$MgSO_4 + Ca(OH)_2 \longrightarrow Mg(OH)_2 \downarrow + CaCl_2$$

$$MgSO_4 + Na_2CO_3 + Ca(OH)_2 \longrightarrow Mg(OH)_2 \downarrow + CaCO_3 \downarrow + Na_2SO_4$$

As the process is carried out at high temperature, it is often called hot lime-soda process. Hardness of soft water usually lies within 30 *ppm*.

14.3.1.2 Zeolite process or Permutit process

In this process zeolite is used to remove the hardness of water. Its prime constituent is Na^+ ions and hence it is best represented as Na_2Ze. Its chemical formula is given below:

$$\text{Zeolite} \equiv Na_2Ze = Na_2O \cdot Al_2O_3 \cdot 2SiO_2 \cdot 6H_2O$$

When hard water is passed over this Na_2Ze bed, Ca^{2+} and Mg^{2+} ions react with Na_2Ze, precipitating CaZe and MgZe. The following reactions are believed to take place.

$$Na_2Ze + Ca(HCO_3)_2 \longrightarrow CaZe + 2NaHCO_3$$

$$Na_2Ze + Mg(HCO_3)_2 \longrightarrow MgZe + 2NaHCO_3$$

$$Na_2Ze + CaCl_2 \longrightarrow CaZe + 2NaCl$$

$$Na_2Ze + CaSO_4 \longrightarrow CaZe + Na_2SO_4$$

$$Na_2Ze + MgCl_2 \longrightarrow MgZe + 2NaCl$$

$$Na_2Ze + MgSO_4 \longrightarrow MgZe + Na_2SO_4$$

$$Na_2Ze + Fe(HCO_3)_2 \longrightarrow FeZe + 2NaHCO_3$$

$$Na_2Ze + Mn(HCO_3)_2 \longrightarrow MnZe + 2NaHCO_3$$

The process is illustrated in the Fig.14.1.

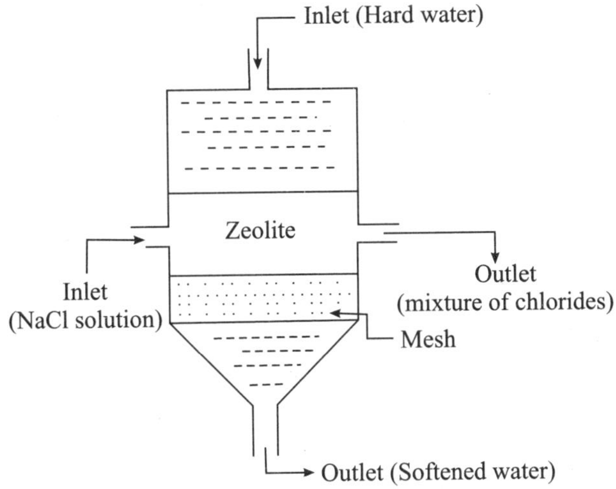

Fig. 14.1 Zeolote treatment of hard water.

The zeolite bed is regenerated by treating it with concentrated NaCl or brine solution. CaZe and MgZe react with NaCl to regenerate Na_2Ze.

Hardness of softened water lies < 10 *ppm*.

14.3.1.3 Demineralization or deionization process

In this process the raw hard water is passed through two successive filter beds. One is made of the resin, represented as $RSO_3^-H^+/RCOO^-H^+/RO^-H^+$ etc., which, when comes in contact with raw water, absorbs Ca^{2+}, Mg^{2+} or other cations with the release of H^+ ions.

$$nRSO_3^-H^+ + M^{n+} \longrightarrow (RSO_3)_nM + nH^+$$

This $RSO_3^-H^+$ is known as cation exchange resin and has the following chemical structure:

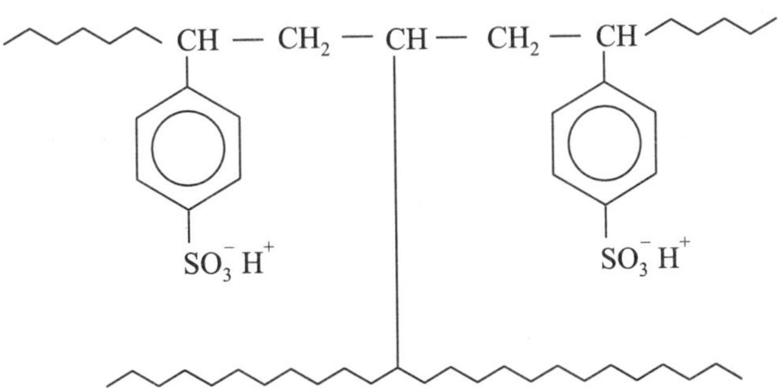

The effluent of cation exchange bed is sent to an anion exchange bed chamber, where all the anions are absorbed within the bed with the release of OH^- ions. The anion exchange resin is represented as $[RN(CH_3)_3]^+OH^-$. The chemical structure of the anion exchange resin is

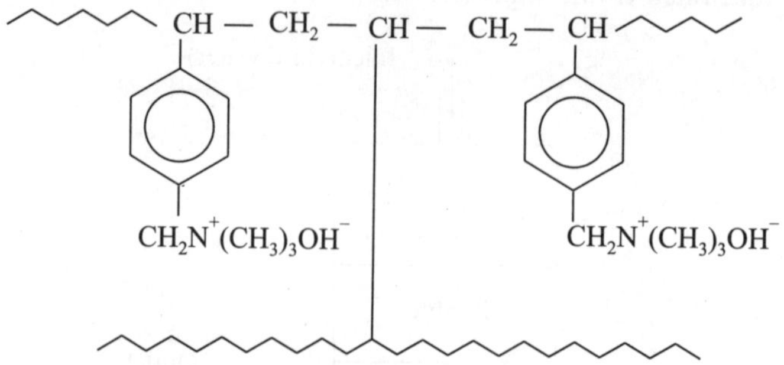

The following reaction is believed to take place during exchange of ions

$$n[RN(CH_3)_3]^+ OH^- + A^{n-} \longrightarrow [RN(CH_3)_3]_n A + nOH^-$$

Finally H^+ and OH^- ions recombine to form pure water, which is known as demineralised water or deionised water.

$$H^+ + OH^- \longrightarrow H_2O$$

The hardness of deionised water is very low, as low as 1 *ppm*. The ion exchange beds are exhausted after certain time and then the beds can be regenerated by treating acid with cation exchange bed and alkali with anion exchange bed.

$$(RSO_3)_n M + nH^+ \longrightarrow nRSO_3^- H^+ + M^{n+}$$

$$[RN(CH_3)_3]_n A + nOH^- \longrightarrow n[RN(CH_3)_3]^+ OH^- + A^{n-}$$

The process is widely used to soften any type of hard water. However, installation cost for setting up ion exchange plant is very high.

14.3.2 Internal treatment

Externally treated water is fed to the boiler but this water still possesses many undesired cations, particularly Ca^{2+} and Mg^{2+} ions, which form undesired scale and sludge. To eliminate these undesired cations the boiler feed water is treated again within the boiler. It is called internal treatment. It may be carried out in different ways as discussed below.

14.3.2.1 Phosphate conditioning

Sodium phosphate (Na_3PO_4) is used to remove undesired cations, i.e., Ca^{2+} and Mg^{2+}, from the boiler feed water. Both Ca^{2+} and Mg^{2+} ions form insoluble precipitates in presence of PO_4^{3-} ions. The following reactions are believed to take place:

$$3CaCl_2 + 2Na_3PO_4 \longrightarrow Ca_3(PO_4)_2 \downarrow + 6NaCl$$

$$3MgCl_2 + 2Na_3PO_4 \longrightarrow Mg_3(PO_4)_2 \downarrow + 6NaCl$$

$$3CaSO_4 + 2Na_3PO_4 \longrightarrow Ca_3(PO_4)_2 \downarrow + 3Na_2SO_4$$

$$3MgSO_4 + 2Na_3PO_4 \longrightarrow Mg_3(PO_4)_2 \downarrow + 3Na_2SO_4$$

However, the above precipitation reactions occur at high alkaline medium (at $pH \geq 9.5$). As Na_3PO_4 is highly alkaline, it is the most suitable reagent for treatment of water, having low alkalinity. Sometimes buffer solution (NH_4OH/NH_4Cl) is used to maintain pH at around 10. In that case Na_2HPO_4 and NaH_2PO_4 can be used instead of Na_3PO_4.

14.3.2.2 Calgon conditioning

Sodium hexametaphosphate is known as calgon and its chemical formula is $Na_2[Na_4(PO_3)_6]$. Calgon dissociates in the solution, generating the complex ion $[Na_4(PO_3)_6]^{2-}$. This complex absorbs all types of cations, particularly Ca^{2+} and Mg^{2+} to form soluble complex $Ca[Na_4(PO_3)_6]$ or $Mg[Na_4(PO_3)_6]$ respectively, thereby preventing scale and sludge formation.

14.4 Alkalinity of water

Concentration of acid required to neutralize a given sample water is known as alkalinity of water. It is expressed in terms of *ppm* of $CaCO_3$. It mainly arises due to the presence of HCO_3^-, CO_3^{2-} and OH^-. These radicals are neutralized in presence of acid according to the following reactions:

$$OH^- + H^+ \longrightarrow H_2O \quad (1)$$

$$CO_3^{2-} + H^+ \longrightarrow HCO_3^- \quad (2)$$

$$HCO_3^- + H^+ \longrightarrow H_2O + CO_2 \quad (3)$$

As HCO_3^- is alkaline in nature reactions (1) and (2) can be estimated by using phenolphthalein indicator. However, the reaction (3) can be estimated by using methyl orange only. Actually methyl orange indicator can be used to estimate all the above three reactions. Methyl orange is written as MO in short and phenolphthalein is written as P. CO_3^{2-} is neutralized in two steps [reactions (2) and (3)]. So reaction (2) indicates half neutralization of CO_3^{2-}. Thus we can write:

$$P = [OH^-] + \frac{1}{2}[CO_3^{2-}]$$

and

$$MO = [OH^-] + [CO_3^{2-}] + [HCO_3^-] \quad (4)$$

Where, P = Concentration of acid required in *ml* to neutralize a given sample water by using phenolphthalein as indicator.

MO = Concentration of acid required in *ml* to neutralize a given sample water by using methyl orange as indicator.

Several cases may arise depending on relations between P and MO.

Case – I: P = MO. It indicates, using equation (4), that $[CO_3^{2-}] = [HCO_3^-] = 0$. Sample water contains only OH^- ions and $[OH^-]$ = P = MO.

Case – II: P = 0. It indicates, using equation (4), that $[CO_3^{2-}] = [OH^-] = 0$. Sample water contains only HCO_3^- ions and $[HCO_3^-]$ = MO.

Case – III: P = $\frac{1}{2}$ MO. It indicates, using equation (4), that $[OH^-] = [HCO_3^-] = 0$. Sample water contains only CO_3^{2-} ions and $[CO_3^{2-}]$ = 2P = MO.

Case – IV: $P > \frac{1}{2}MO$. It indicates, using equation (4), that $[HCO_3^-] = 0$. Sample water contains OH^- and CO_3^{2-} ions. Thus we can write that:

$$P = [OH^-] + \frac{1}{2}[CO_3^{2-}]$$

$$\frac{1}{2}MO = \frac{1}{2}OH^- + \frac{1}{2}[CO_3^{2-}]$$

$$P - \frac{1}{2}MO = \frac{1}{2}[OH^-]$$

Or, $[OH^-] = 2P - MO$

Again, $P = [OH^-] + \frac{1}{2}[CO_3^{2-}]$

$$MO = [OH^-] + [CO_3^{2-}]$$

$$MO - P = \frac{1}{2}[CO_3^{2-}]$$

$$[CO_3^{2-}] = 2(MO - P)$$

Case – V: $P < \frac{1}{2}MO$. It indicates, using equation (4), that $[OH^-] = 0$. Sample water contains both HCO_3^- and CO_3^{2-} ions.

$$MO = [CO_3^{2-}] + [HCO_3^-]$$

$$P = \frac{1}{2}[CO_3^{2-}]$$

$$\frac{1}{2}MO - P = [HCO_3^-]$$

$$[CO_3^{2-}] = 2P$$

Alkaline water attacks the boiler plate. Thus boiler feed water should be free from alkalinity. Alkalinity is usually expressed in terms of $CaCO_3$ equivalent.

Boiler feed water requirement

(a) Hardness should be less than 0.20 *ppm*.
(b) Alkalinity due to OH^- ion should be less than 0.40 *ppm*.
(c) Alkalinity due to CO_3^{2-} ion should be less than 1 *ppm*.

14.5 Problems on boiler

The following problems are frequently encountered with boiler

(a) Corrosion
(b) Caustic embrittlement
(c) Scale and sludge
(d) Carry over

14.5.1 Boiler corrosion

During steam formation at high pressure and temperature, boiler plate gets corroded or decayed due to several reasons. The following are believed to be the most responsible for boiler corrosion.

(i) Dissolved oxygen
(ii) Carbon-di-oxide
(iii) Mineral acids, specially, HCl

(i) Dissolved oxygen Boiler feed water always contains some dissolved oxygen. Its concentration ranges from 7 *ppm* to 9 *ppm* or more. As solubility of oxygen decreases at high temperature, free oxygen is available near the boiler plate at high temperature. This free oxygen attacks the boiler plate to form rust on it.

$$\text{Boiler plate} + [O_2] \longrightarrow Fe(OH)_2 \longrightarrow Fe_2O_3.xH_2O \text{ (Rust)}$$

Thus free oxygen poses a threat to the boiler. To avoid this rust formation a reducing agent, preferably sodium sulphite (Na_2SO_3) is added. Whenever, dissolved oxygen sets free, Na_2SO_3 reacts with O_2 to form Na_2SO_4, thereby protecting the boiler plate. The following reaction is believed to occur

$$2Na_2SO_3 + O_2 \longrightarrow 2Na_2SO_4$$

However, the excess Na_2SO_4 is also dangerous since it gives off corrosive sulphur-di-oxide (SO_2) gas at high pressure. Thus very low concentration (as low as 1 *mg/l*) of Na_2SO_3 is used for this purpose.

(ii) Carbon-di-oxide CO_2 is often evolved during steam formation in boiler. At high pressure at boiler, bi-carbonates are dissociated into carbonates with the release of CO_2. This CO_2 reacts with water to form corrosive H_2CO_3, which accelerates scale formation on the boiler plates. To avoid the formation of corrosive H_2CO_3, small amount of NH_4OH is added. This NH_4OH reacts with CO_2 to form ammonium carbonate $[(NH_4)_2CO_3]$, which is soft and non-corrosive.

(iii) Mineral acids If boiler feed water contains significant amount of $MgCl_2$, the latter reacts with water to release corrosive HCl along with $Mg(OH)_2$, which forms scales on the boiler plates. HCl accelerates rust formation on the boiler plates. To avoid rust formation small amount of NH_4OH is added. This NH_4OH absorbs HCl before rust formation.

14.5.2 Caustic embrittlement

Boiler plates, after several hours of working, get corroded by caustic soda (NaOH) and become brittle. The phenomenon is called caustic embrittlement. This phenomenon is usually observed when water, softened by lime-soda process, is used at the boiler. This soft water often contains sufficient amount of sodium carbonate (Na_2CO_3), which at high pressure hydrolyses to give sodium hydroxide (NaOH) and carbon-di-oxide (CO_2). If the boiler plate contains multiple hair-cracks, NaOH flows through the hair-cracks and dissolves out iron from the boiler plate in the form of ferrates. Iron, wetted with NaOH solution, acts as anode and gets corroded continuously. CO_2 forms corrosive acid H_2CO_3 which is also dangerous to boiler plate.

$$Na_2CO_3 + H_2O \longrightarrow 2NaOH + CO_2$$

As a result of this, stress corrosion develops on the boiler plate, leading to boiler failure.

Use of zeolite-softened soft water also leads to caustic embrittlement since $NaHCO_3$ is produced in huge amount during zeolite softening process. This $NaHCO_3$ at high temperature decomposes to sodium hydroxide (NaOH) and carbon-di-oxide (CO_2).

$$NaHCO_3 \longrightarrow NaOH + CO_2$$

Prevention It is evident that caustic embrittlement occurs due to carbonates and bicarbonates. Thus sample water should be boiled for long time to remove temporary hardness. After that zeolite is used to remove permanent hardness. Finally small amount of Na_2SO_4 is added. This Na_2SO_4 fills up pores and cracks of boiler plates so that NaOH does not get in contact with boiler plate.

If Na_3PO_4 is used to remove permanent hardness, the following reactions may occur

$$2CaSO_4 + 2Na_3PO_4 \longrightarrow 2Na_2SO_4 + Ca_3(PO_4)_2 \downarrow$$

$$3MgSO_4 + 2Na_3PO_4 \longrightarrow 2Na_3SO_4 + Mg_3(PO_4)_2 \downarrow$$

In this case there is no need to add Na_2SO_4 in boiler feed water, since Na_2SO_4 is already produced in the reactions.

14.5.3 Scale and sludge

The function of a boiler is to supply steam continuously. Water is continuously supplied to the boiler, at which water is converted to steam at high temperature and pressure. During boiling, impurities, present in water, gets deposited on the boiler plate continuously. Some precipitates are very fine and remain within boiler in the form of colloids. This is called *sludge*.

The main constituents of scale are $CaCO_3$, $CaSO_4$, $Mg(OH)_2$, $CaSiO_3$, and $MgSiO_3$.

Some precipitates get precipitated in the form of coarse particles, which remain adhered to the inner wall of the boiler plate as hard deposit. This is called *scale*.

The main constituents of scale are $CaCl_2$, $MgCl_2$, $MgCO_3$, $MgSO_4$ etc.

Disadvantages of scale and sludges formation:

(a) Usually scales and sludges are poor conductor of heat and hence excess fuel is often required to achieve the desired temperature. Thus fuel wastage is the immediate effect of scale and sludge formation.

(b) Boiler efficiency decreases.

(c) Corrosion of boiler plate.

(d) Overheating of boiler plate. This is a serious problem since due to overheating high heat is developed on the boiler plate, as a result of which, metal becomes weaker, leading to the formation of cracks on the boiler plate.

(e) As thermal conductivity of boiler is adversely affected due to scale and sludge formation, some portion of boiler remains over-heated whereas other portion remains under-heated. As a result of this, boiler tube gets distorted, thereby decreasing the boiler efficiency and threatening the danger of explosion.

Prevention To avoid the scale and sludge formation, soft water, having following specifications, is recommended for the use in the boiler.

(a) Hardness should be less than 0.20 *ppm*.
(b) Alkalinity due to OH^- ion should be less than 0.40 *ppm*.
(c) Alkalinity due to CO_3^{2-} ion should be less than 1 *ppm*.
(d) Very pure form of soft water.

14.5.4 Carry over

During steam generation in a boiler, two phenomena, called priming and foaming, happen to take place very often.

Water droplets are often associated with the steam, generated in the boiler. This steam is called wet steam and the phenomenon of forming wet steam is called priming. The water droplets in the wet steam may possess impurities, which decrease the efficiency of the boiler and also lead to corrosion. The impurities are carried by the steam to the turbine and get deposited there. This undesired deposits damage the turbine.

The boiler feed water often contains oil and alkali, which react with each other during steam generation at high temperature and pressure, generating foam. The phenomenon is called foaming. Foaming is highly undesirable since it generates numerous air pockets and changes the boiler pressure. Addition of small amount of castor oil and antifoaming agent, e.g., $NaAlO_2$, $Al_2(SO_4)_3$ etc. prevents the foaming.

14.6 Numerical problems on water softening by lime-soda process

$Ca(HCO_3)_2 + L \longrightarrow 2CaCO_3 \downarrow + H_2O$ [Only Lime is required] L

$Mg(HCO_3)_2 + 2L \longrightarrow 2CaCO_3 \downarrow + Mg(OH)_2 \downarrow + 2H_2O$ [Only Lime is required] 2L

Hardness of water

$MgCl_2/MgSO_4 + L \longrightarrow Mg(OH)_2 \downarrow + Ca^{2+}$

$Ca^{2+} + S \longrightarrow CaCO_3 \downarrow + 2Na^+$ [Both Lime and Soda required] (L + S)

$CaCl_2/CaSO_4 + S \longrightarrow CaCO_3 \downarrow + 2Na^+$ [Only Soda is required] S

$CO_2 + L \longrightarrow CaCO_3 \downarrow + H_2O$ [Only Lime is required] L

$2H^+ + L \longrightarrow Ca^{2+} \downarrow + 2H_2O$

$Ca^{2+} + S \longrightarrow CaCO_3 \downarrow + 2Na^+$ [Both Lime and Soda are required] (L + S)

$2HCO_3^- + L \longrightarrow CaCO_3 \downarrow + S + H_2O$ [Soda is produced] (L – S)

$FeSO_4 + L \longrightarrow Fe(OH)_2 + CaSO_4 \longrightarrow Fe(OH)_3 \downarrow$

$CaSO_4 + S \longrightarrow CaCO_3 \downarrow + 2Na^+$ [Both Lime and Soda are required] (L + S)

$Al_2(SO_4)_3 + 3L \longrightarrow 2Al(OH)_3 \downarrow + 3CaSO_4$

$3CaSO_4 + 3S \longrightarrow 3CaCO_3 \downarrow + 3Na_2SO_4$ [Both Lime and Soda are required] (3L + 3S)

Chemical formula	*Molecular weight*
Lime [$Ca(OH)_2$]	74
Soda [Na_2CO_3]	106
$MgSO_4$	120
$MgCO_3$	84
$Mg(HCO_3)_2$	146
$MgCl_2$	95
$Ca(HCO_3)_2$	162
$CaCO_3$	100
$CaSO_4$	136
$CaCl_2$	111

P1. Analysis of a sample water gave the following result

H_2SO_4 = 200 *ppm*, $MgSO_4$ = 30 *ppm*, $CaSO_4$ = 300 *ppm*, NaCl = 35 *ppm*

Calculate the amount of lime and soda required to soften 150 × 10^6 *l* of water.

Solution

$$\text{Lime requirement} = \frac{74}{100}\{[H_2SO_4] + [MgSO_4]\}CaCO_3 \text{ eq. } (mg/l) \times 150 \times 16^6\ l$$

$$= \frac{74}{100}\left\{200\times\frac{100}{98}+30\times\frac{100}{120}\right\}\times 150 \text{ kg}$$

= 25428 kg (approx.).

$$\text{Soda requirement} = \frac{106}{100}\{[H_2SO_4] + [MgSO_4] + [CaSO_4]\}CaCO_3 \text{ eq. } (mg/l) \times 150 \times 16^6\ l$$

$$= \frac{106}{100}\left\{200\times\frac{100}{98}+30\times\frac{100}{120}+300\times\frac{100}{136}\right\}\times 150 \text{ kg}$$

= 71497.5 kg

P2. Calculate the amount of lime and soda required for softening 2 × 10^6 *l* of water containing the following impurities:

[HCl] = 2 *pm*, [$Ca(HCO_3)_2$] = 400 *ppm*, [$MgCl_2$] = 10 *ppm*, [$CaSO_4$] = 25 *ppm*, [$FeSO_4.7H_2O$] = 5 *ppm*

P3. Calculate the amount of lime and soda required for softening 3 × 10^6 *l* of water containing the following impurities:

[HCl] = 4 *ppm*, [$FeCl_3$] = 10 *ppm*, [$MgSO_4$] = 6 *ppm*, [$Mg(HCO_3)_2$] = 3 *ppm*, [$Ca(HCO_3)_2$] = 3 *ppm*, [$CaSO_4$] = 18 *ppm*

P4. Calculate the amount of lime and soda required for softening 4×10^6 *l* of water containing the following impurities:

$[Ca^{2+}] = 180$ *ppm*, $[HCO_3^-] = 395$ *ppm*, $[Mg^{2+}] = 100$ *ppm*, $[CO_3^{2-}] = 200$ *ppm*, $[CO_2] = 30$ *ppm*

P5. Calculate the quantity of lime and soda required for treatment of 2×10^6 *l* of water, having the following impurities:

$[Ca(HCO_3)_2] = 40$ *ppm*, $[MgSO_4] = 28$ *ppm*, $[Mg(HCO_3)_2] = 42$ *ppm*, $[CaSO_4] = 32$ *ppm*, $[CaCl_2] = 24$ *ppm*, $[NaCl] = 12$ *ppm*

P6. Calculate the amount of lime and soda required for softening 2×10^6 *l* of hard water, which was analyzed as follows:

Substance	*Concentration (mg/l)*
$Ca(HCO_3)_2$	243
$Mg(HCO_3)_2$	73
$CaSO_4$	102
$MgCl_2$	95
NaCl	500
$FeSO_4.7H_2O$	139

Purity of lime is 94% and that of soda is 86%. 10% chemicals are used in excess in order to complete the reaction quickly.

14.7 Numerical problems on alkalinity

P1. 100 ml of a water sample consumes 7 ml $\left(\frac{N}{50}\right) H_2SO_4$ up to the end point, using phenolphthalein as indicator but 100 ml of same water sample consumes 14 ml $\left(\frac{N}{50}\right) H_2SO_4$ up to the end point, using methyl orange as indicator. Determine the type of alkalinity and calculate the alkalinity of the water sample.

Solution

Case – I Titration with phenolphthalein indicator

We know:

$V_1S_1 = V_2S_2$, where,

V_1 = Volume of water sample = 100 ml.

S_1 = Strength of water sample, which is to be determined.

V_2 = Volume of H_2SO_4 consumed = 7 ml

S_2 = Strength of $H_2SO_4 = \left(\frac{N}{50}\right)$

Hence, $S_1 = 0.07\left(\frac{N}{50}\right) = 0.07\left(\frac{M}{100}\right)$ equivalent $CaCO_3$.

since Mol. wt. of $CaCO_3$ = 2 × Eq. wt. of $CaCO_3$.

Or, 1N $CaCO_3 = \left(\frac{M}{2}\right) CaCO_3$.

Again, $1 \times \left(\frac{M}{100}\right) = 1000\ ppm$

So, $P = 0.07\left(\frac{M}{100}\right) = 70\ ppm$

Case – II Titration with methyl orange indicator

We know:

$V_1S_1 = V_2S_2$, where,

V_1 = Volume of water sample = 100 ml.

S_1 = Strength of water sample, which is to be determined.

V_2 = Volume of H_2SO_4 consumed = 14 ml

S_2 = Strength of $H_2SO_4 = \left(\frac{N}{50}\right)$

Hence, $S_1 = 0.14\left(\frac{N}{50}\right) = 0.14\left(\frac{M}{100}\right)$ equivalent $CaCO_3$.

since Mol. wt. of $CaCO_3$ = 2 × Eq. wt. of $CaCO_3$.

Or, 1N $CaCO_3 = \left(\frac{M}{2}\right) CaCO_3$.

Again, $1 \times \left(\frac{M}{100}\right) = 1000\ ppm$

So, $MO = 0.14\left(\frac{M}{100}\right) = 140\ ppm$

Thus, $P = \frac{1}{2} MO$

So, only CO_3^{2-} ions are present in the given water sample.

P2. 100 ml of a water sample consumes 6 ml $\left(\frac{N}{50}\right)$ HCl up to the end point, using phenolphthalein as indicator but 100 ml of same water sample consumes 16 ml $\left(\frac{N}{50}\right)$ HCl up to the end point, using methyl orange as indicator. Determine the type of alkalinity and calculate the alkalinity of the water sample.

P3. A water sample is alkaline and responds both to phenolphthalein and methyl orange indicators. 50 ml of this water sample requires 15 ml $\left(\frac{N}{50}\right)H_2SO_4$ up to the end point, using phenolphthalein as indicator and another 10 ml for complete neutralization. Calculate the type of alkalinity.

P4. A water sample is alkaline but does not respond to phenolphthalein. 100 ml of this sample requires 18 ml $\left(\frac{N}{50}\right)HCl$ up to the end point, using methyl orange as indicator. Determine the type of alkalinity and amount of alkalinity present in the sample.

P5. 100 ml of a water sample consumes 6 ml $\left(\frac{N}{50}\right)H_2SO_4$ up to the end point, using phenolphthalein as indicator. 100 ml of same water sample consumes 6 ml $\left(\frac{N}{50}\right)H_2SO_4$ up to the end point, using methyl orange as indicator. Determine the type of alkalinity and calculate the alkalinity of the water sample.

P6. 200 ml of a water sample consumes 20 ml $\left(\frac{N}{50}\right)HCl$ using methyl orange as indicator. Another 200 ml of same sample requires 8 ml $\left(\frac{N}{50}\right)HCl$ using phenolphthalein as indicator. Calculate the alkalinity of the water sample in terms of $CaCO_3$ equivalent in *mg/l*.

APPENDIX – I

Table 1 Physical Constants

Avogadro's Number	N	6.023×10^{23}/g-mole or 6.023×10^{26}/kg-mole
Molar Gas Constant	R	8.314 J/(mole-°K) = 1.98 cal/(mole-°K) = 0.082 lit-atm/(mole-°K)
Boltzmenn's Constant	$k_B = \frac{R}{N}$	1.38×10^{-23} J/°K
Faraday's Constant	F	96,500 C/gm-equivalent of element
Electronic Charge	$e = \frac{F}{N}$	1.602×10^{-19} C = 4.806×10^{-10} e.s.u of charge.
Atomic Mass Unit	amu = $\frac{1}{N}$	1.66×10^{-24} gm = 1.66×10^{-27} kg
Velocity of Light	c	3.0×10^{8} m/s.
Electron Rest Mass	m_0	9.11×10^{-31} kg.
Bohr Magneton (B.M.)	μ_B	9.273×10^{-24} A-m^2
Permittivity of Free Space	ε_0	8.854×10^{-12} F/m
Acceleration due to Gravity	g	9.81 m/s^2
Dipole Moment (1 Debye)	D	10^{-18} e.s.u.-cm
Gravitational Constant	G	6.6732×10^{-11} N-m^2/kg^2
Stefan–Boltzmann Constant	σ	5.67×10^{-8} W/(m^2-°K^4)
Rydberg Constant	R_B	109737 cm^{-1}

SI unit prefixes: 1 kilo = 10^3 unit, 1 mega = 10^6 unit, 1 giga = 10^9 unit, 1 tera = 10^{12} unit, 1 milli = 10^{-3} unit, 1 micro = 10^{-6} unit, 1 nano = 10^{-9} unit, 1 pico = 10^{-12} unit, 1 femto = 10^{-15} unit, 1 atto = 10^{-18} unit, 1 Å = 10^{-10} m.

Table 2 Conversion Factors

Length:	1 m = 3.28 ft = 1.094 yd. 1 in = 2.54 cm, 1 ft = 30.48 cm. 1 mile = 1.6 km.
Area:	1 m^2 = 10.764 ft^2 1 in^2 = 645.16 mm^2 1 hectare = 100 ares = 10,000 m^2 = 2.471 acres.
Force:	1 N = 0.2248 lb_f, 1 N = 10^5 dyne 1 kg_f = 9.81 N
Energy:	1 J = 0.738 ft-lb_f = 10^7 ergs 1 eV = 1.602×10^{-19} J 1 cal = 4.19 J 1 J = 0.2392 Cal = 6.242×10^{18} eV 1 B. T. U. = 252 Cal.
Volume:	1 ft^3 = 28316.85 cm^3 or c.c. = 7.48 gallon (U.S.) 1 gallon = 3786 c.c. 1 m^3 = 35.288 ft^3 1 barrel = 42 US-gallon
Pressure:	1 torr = 1 mm Hg at 0°C 1 atm = 101.325 kPa = 14.7 psi = 1.01325 bar = 760 mm Hg at 0°C 1 bar = 100 kPa = 10^5 Pa 1 kg_f/cm^2 = 9.810×10^4 Pa = 0.981 bar = 0.968 atm 1 MPa = 145.1 psi = 1 N/mm^2 = 9.87 atm = 10.06 kg_f/cm^2 1 tsi = 15.2 MPa
Temperature:	°F = 1.8 × t°C + 32 $T(R)$ = t°F + 459.67 $T(K)$ = t°C + 273 $T(R) = 1.8\, T(K)$
Power:	1 kW = 239.2 cal/s = 1.341 hp. 1 hp = 745 W = 0.745 kW 1 lumen (lm) = 1.496×10^{-10} W 1 lm/ft^2 = 1 foot-candle (fc) = 1.609×10^{-12} W/m^2 1 lm/ft^2 = Luminous flux = 1 foot–candle (fc)
Mass and Density:	1 kg = 1000 gm = 2.2046 lb. 1 g/cm^2 = 1000 kg/m^3 = 62.43 lb/ft^3
Coefficient of Viscosity:	1 poise (p) = 100 centi-poise (cp) = 0.1 Pa-sec. 1 centi-poise(cp) = 10^{-3} Pa-sec = 6.7197×10^{-4} lb/ft-s = 2.4191 lb/ft-h.

Table 3(a) Standard Oxidation Electrode Potentials (E^0_{298}) of Selected Redox Systems in Aqueous Solution

Reduced Form	*Oxidized Form*	*No. of Electron (s) Released*	E^0_{298} (V)
Li	Li^+	1*e*	+ 3.05
K	K^+	1*e*	+ 2.936
Ca	Ca^{2+}	2*e*	+ 2.87
Na	Na^+	1*e*	+ 2.71
Mg	Mg^{2+}	2*e*	+ 2.36
Al	Al^{3+}	3*e*	+ 1.67
Mn	Mn^{2+}	2*e*	+ 1.18
$\frac{1}{2}H_2(g) + OH^-$	H_2O	1*e*	+ 0.828
Zn	Zn^{2+}	2*e*	+ 0.76
Fe	Fe^{2+}	2*e*	+ 0.44
Cd	Cd^{2+}	2e	+ 0.40
$Pb + 2I^-$	PbI_2	2*e*	+ 0.365
$Pb + SO_4^{2-}$	$PbSO_4$	2*e*	+ 0.356
Sn	Sn^{2+}	2*e*	+ 0.14
Pb	Pb^{2+}	2*e*	+0.126
Fe	Fe^{3+}	3*e*	+0.04
$\frac{1}{2}D_2$	D^+	1*e*	+ 0.01
$\frac{1}{2}H_2$	H^+	1*e*	0.00
Cu	Cu^{2+}	2*e*	– 0.34
Cu	Cu^+	1*e*	–0.518

Table 3(b) Standard Oxidation Electrode Potentials (E^0_{298}) of Selected Redox Systems in Aqueous Solution

Reduced Form	*Oxidized Form*	*No. of Electron (s) Change*	E^0_{298} (V)
I^-	$\frac{1}{2} I_2$	1*e*	–0.54
$2\,Hg(l) + SO_4^{2-}$	Hg_2SO_4	2*e*	–0.615
Fe^{2+}	Fe^{3+}	1*e*	–0.77
Ag	Ag^+	1*e*	–0.80
Hg	Hg^{2+}	2*e*	–0.85
$NO + 2H_2O$	$NO_3^- + 4H^+$	3*e*	–0.96
Br^-	$\frac{1}{2} Br_2$	1*e*	–1.07
$Mn^{2+} + 2H_2O$	$MnO_2 + 4H^+$	2*e*	–1.23
H_2O	$\frac{1}{2} O_2(g) + 2H^+$	2*e*	–1.23
$2Cr^{3+} + 7H_2O$	$Cr_2O_7^{2-} + 14H^+$	6*e*	–1.33
Cl^-	$\frac{1}{2} Cl_2$	1*e*	–1.36
$Cl^- + 3H_2O$	$ClO_3^- + 6H^+$	6*e*	–1.45
$Mn^{2+} + 4H_2O$	$MnO_4^- + 8H^+$	5*e*	–1.51
Ce^{3+}	Ce^{4+}	1*e*	–1.61
F^-	$\frac{1}{2} F_2$	1*e*	–2.87
$Ag + Br^-$	AgBr	1*e*	–0.073
$SO_3^{2-} + H_2O$	$SO_4^{2-} + 2H^+$	2*e*	–0.22
$Ag + Cl^-$	AgCl	1*e*	–0.2222
$2Hg(l) + 2Cl^-$	Hg_2Cl_2	2*e*	–0.2680

Table 4 Typical Oxyacids with their Acid Strength

Acid	*Ka*	*pKa*	*Acid Strength*
HOBr	2.06×10^{-9}	8.7	Very Weak
HOCl	3.2×10^{-8}	7.5	Very Weak
HOI	2.3×10^{-11}	10.64	Very Weak
H_3AsO_3	6.0×10^{-10}	9.2	Very Weak
H_3BO_3	5.8×10^{-10}	9.24	Very Weak
H_4SiO_4	2.0×10^{-10}	9.7	Very Weak
H_6TeO_6	2.0×10^{-8}	7.7	Very Weak
$HClO_2$	1.1×10^{-2}	1.97	Weak
HNO_2	4.0×10^{-4}	3.40	Weak
H_2CO_3	1.32×10^{-4}	3.88	Weak
H_2SO_3	1.3×10^{-2}	1.87	Weak
H_3AsO_4	6.46×10^{-3}	2.19	Weak
H_5IO_6	3.1×10^{-2}	1.57	Weak
H_3PO_4	7.25×10^{-3}	2.12	Weak
$HBrO_3$	2.0×10^{-1}	0.7	Strong
$HClO_3$	10	–1	Strong
HIO_3	1.7×10^{-1}	0.77	Strong
HNO_3	43.6	–1.64	Strong
H_2MnO_4	10^{-1}	1.0	Strong
H_2SO_4	1.0×10^{3}	–3.0	Very Strong
$HClO_4$	10^{10}	–10.0	Very Strong
$HMnO_4$	2.0×10^{2}	–2.3	Very Strong

Table 5(a) Coefficient of Viscosities (η) of Liquids in Centi-poise (*cp*) at 20°C

Liquid	η	*Liquid*	η
Water	1.005	Ethyl ether	0.245
Ethanol	1.194	Chloroform	0.563
Methanol	0.593	Chlorobenzene	0.637
Acetone	0.331	Benzene	0.647
Carbon tetrachloride	0.958	Acetic acid	1.222
Nitrobenzene	1.98	Glycerin	10.69

Table 5 (b) Coefficient of Viscosities (η) of Several Liquid Metals and Alloys

Material	*Temperature* (°C)	*Viscosity* (η) (*cp*)
Aluminium	700	3.0
Copper	1200	3.2
Iron	1600	6.2
White cast iron	1300	2.4
Steel	> M.P.	~ 3.0
Magnesium	680	1.2
Zinc	500	3.7

Table 6 Solubility Products of Some Barely Soluble Compounds in Water at 25°C

Compound	*Solubility Product* (K_{sp})	*Compound*	*Solubility Product* (K_{sp})
AgBr	5.3×10^{-13}	AgCN	1.4×10^{-16}
AgCl	1.78×10^{-10}	Ag_2CrO_4	2.5×10^{-12}
AgI	8.30×10^{-17}	PbS	2.5×10^{-27}
$CaSO_4$	9.1×10^{-6}	CdS	4.0×10^{-29}
$SrSO_4$	2.8×10^{-7}	CuS	3.0×10^{-42}
$BaSO_4$	1.1×10^{-10}	HgS	3.0×10^{-54}
$PbCl_2$	2.4×10^{-4}	CoS	2.0×10^{-27}
$PbBr_2$	7.9×10^{-5}	ZnS	1.0×10^{-23}
PbI_2	8.7×10^{-9}	NiS	1.4×10^{-24}
$CaCO_3$	4.8×10^{-9}	MnS	2.0×10^{-15}
$BaCO_3$	5.1×10^{-9}	$Mg(OH)_2$	6.0×10^{-13}
$Ca_3(PO_4)_2$	2.0×10^{-29}	CaF_2	4.0×10^{-11}
$AlPO_4$	5.75×10^{-19}	CuI	5.0×10^{-12}
CuBr	1.60×10^{-11}	CuCl	1.0×10^{-6}

Table 7 Surface Tension Values of Liquid Metals and Other Liquids

Material	*Temperature* (°C)	*Surface Tension* (g) (Dynes/cm)
Aluminium	750	520
Copper	1200	1160
Iron	1600	1360
Lead	350	453
Magnesium	681	563
Mercury	20	465
Zinc	600	770
Fused salts	400 to 930	35 to 180
Water	20	74

Table 8 Values of C_p, C_v and γ for Ideal Gases at 25°C

Gas	*Chemical Formula*	C_p (kJ/kg–°K)	C_v (kJ/kg–°K)	γ
Air	—	1.0035	0.7165	1.4
Oxygen	O_2	0.922	0.622	1.393
Nitrogen	N_2	1.042	0.745	1.4
Argon	Ar	0.520	0.312	1.667
Hydrogen	H_2	14.209	10.085	1.409
Helium	He	5.193	3.116	1.667
Carbon monoxide	CO	1.041	0.744	1.4
Carbon di-oxide	CO_2	0.842	0.653	1.289
Methane	CH_4	2.254	1.736	1.299
Ethane	C_2H_6	1.7662	1.4897	1.86
Propane	C_3H_8	1.6794	1.4909	1.126
Isobutene	C_4H_{10}	1.716	1.573	1.091
Acetylene	C_2H_2	1.712	1.394	1.23
Water vapour (Steam)	H_2O	1.8723	1.4108	1.327

Table 9 Some Reference Electrodes

Electrode	*Reaction*	E°_{ox} at 25°C
K_2SO_4 (sat) | Hg_2SO_4 | Hg	$Hg_2SO_4 + 2e \rightarrow 2Hg + SO_4^{2-}$	– 0.64 V
KCl (sat) | AgCl, Ag	$AgCl + e \rightarrow Ag + Cl^-$	– 0.1989 V
KCl (1.0 M) | AgCl, Ag	$AgCl + e \rightarrow Ag + Cl^-$	– 0.237 V
KCl (0.1 M) | AgCl, Ag	$AgCl + e \rightarrow Ag + Cl^-$	– 0.290 V

Table 10 '*K*' and '*a*' Values for Some Polymer-solvent Systems at Different Temperatures

Polymer	*Solvent*	θ(°C)	$K \times 10^5$	*a*
Polyvinyl acetate	Acetone	25	18.8	0.69
Polyvinyl acetate	Acetone	30	10.2	0.72
Polyvinyl acetate	Chloroform	25	20.3	0.72
Polyvinyl alcohol	Water	25	20	0.76
Polymethylmethacrylate	Acetone	25	7.5	0.70
Polymethylmethacrylate	Acetone	30	7.0	0.70
Polymethylmethacrylate	Toluene	25	7.5	0.71
Polymethylmethacrylate	Toluene	30	7.0	0.72
Polystyrene	Benzene	25	10.2	0.74
Polystyrene	Benzene	30	11.0	0.73
Polyisobutylene	Benzene	30	61.0	0.56
Natural Rubber	Benzene	30	18.5	—

Table 11 Average Bond Energy Values in kJ/mole

Bond	*Bond Energy*	*Bond*	*Bond Energy*
C—H	415	C—Cl	328
C—C	344	C—Br	276
C—O	350	C—I	240
C—N	292	F—F	158
C—S	259	N—H	391
C—F	441	O—H	463
S—H	368	S—S	266
N—O	175	O—O	143
N—N	159	N—Cl	200
H—H	436	C=C	615
Cl—Cl	243	C=O	725
C=N	615	N=N	418
O≡O	498	C≡C	812
C≡N	890	N≡N	946

Table 12 Dipole Moments (μ) of Different Bonds

Bond	*m*	*Bond*	*m*
H—O	1.5 D	C—O	0.8 D
H—N	1.3 D	C=O	2.5 D
H—C	0.4 D	C—N	0.5 D
C—Cl	1.5 D	C≡N	3.5 D
C—Br	1.4 D		

Note: 1 D = 10^{-18} e.s.u.-cm. Dipole moment of F—C bond is assumed to be 1.0 D.

APPENDIX – II

NUMERICAL PROBLEMS

I. THERMODYNAMICS

1. 0.1 mole of a perfect gas having $C_{V,\, m} = 1.50$ R, which is independent of temperature, undergoes the reversible cyclic process shown in the figure. Calculate q, W and ΔU for each step and for the complete cycle.
Given $R = 2$ cal/mole °K.

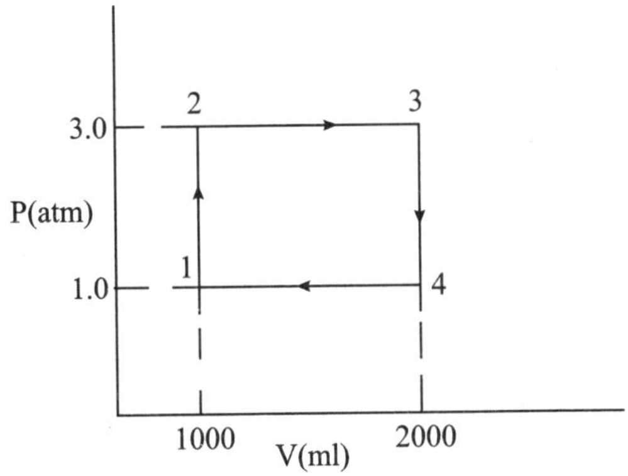

Ans. $W_{1\to2} = 0$, $W_{2\to3} = 304$ J, $W_{3\to4} = 0$, $W_{4\to1} = -101$ J
$q_{1\to2} = 304$ J, $q_{2\to3} = 761$ J, $q_{3\to4} = -608.5$ J, $q_{4\to1} = -253.5$ J
$\Delta U_{1\to2} = 304$ J, $\Delta U_{2\to3} = 457$ J, $\Delta U_{3\to4} = -608.5$ J, $\Delta U_{4\to1} = -152.5$ J

2. A cylinder fitted with a frictionless piston contains 3 moles of He gas at 1 atm. The pressure is reversibly and isothermally increased to 5 atm. Find W, q and ΔU for this process. Temperature is 127°C.

Ans. $W = -16.1$ kJ, $q = -16.1$ kJ, $\Delta U = 0$.

3. Classify each of the following properties as intensive or extensive and give the SI units of each: (a) density, (b) U, (c) H, (d) C_V, (e) C_P, (f) P, (g) molar mass, (h) T.

4. The $\bar{C}_P$ of oxygen is related to temperature as follows:

$$\bar{C}_P = a + bT, \quad a = 6.15 \text{ cal/mol-°K and } b = 0.00310 \text{ cal/mol-°K}^2$$

(a) Calculate q, W, ΔU and ΔH when 2 moles of oxygen is reversibly heated from 27 to 127°C at constant 1 atm. pressure. Assume perfect gas behaviour.

(b) Calculate the same at constant V.

Initial pressure = 1 atm., Initial temperature = 27°C, Final temperature = 127°C.

Ans. (a) $q = 1447$ cal, $W = +397$ cal, $\Delta U = 1050$ cal, $\Delta H = 1447$ cal

5. (a) Calculate q, W, ΔU and ΔH for the reversible isothermal expansion at 300K of 5 moles of a perfect gas from 500 c.c. to 1500 c.c., (b) What would ΔU and W be if the expansion occurs between the same initial and final states as in (a) but is done by having the perfect gas expanded into vacuum?

Ans. (a) 13.7 kJ, –13.7 kJ, 0, 0

6. One mole of He gas with $C_{V,m} = (3R/2)$ expands reversibly from 24.6 L and 300 K to 49.2 L. Calculate the final pressure and temperature if the expansion is (a) isothermal, (b) adiabataic. (c) Sketch these two processes on a P-V diagram.

Ans. (a) 300 K, 0.5 atm, (b) 189 K, 0.315 atm.

7. For N_2(g), $C_{P,m} = 3.5R = 29.1$ J/mole-°K. (a) Calculate q, W, ΔU and ΔH for the reversible adiabatic compression of 1.12 g of N_2(g) from 400 torr and 1 litre to 250 ml. Assume perfect gas behaviour, (b) Cool N_2(g) from 25°C and 1 atm. to 100 K using a reversible adiabatic expansion. What should the final pressure be?

Ans. (a) 0, 98.9 J, 98.9 J, 138.5 J.

8. Find q, W, ΔU and ΔH if 2.00 g of He (g) with $C_{V,m} = \left(\dfrac{3R}{2}\right)$ undergoes (a) a reversible expansion from 20 dm³ to 40 dm³ at constant pressure of 0.8 bar, (b) a reversible heating at constant V of 15 dm³ from 0.6 bar to 0.9 bar.

1 atm = 101325 Pa, 1 bar = 10^5 Pa = 750.062 torr.

1 torr = (1/760) atm = 133.322 Pa.

Ans. (a) q = 975.6 cal, W = 387.2 cal, ΔU = 585.36 cal, ΔH = 975.60 cal.
(b) q = 164.61 cal, W = 0 cal, ΔU = 164.61 cal, ΔH = 274.35 cal.

9. C_P of hydrogen, oxygen and water vapour in the temperature range from 25°C to 100°C are as follows:

$$H_2(g) = 6.92,\ O_2(g) = 7.04,\ H_2O(g) = 8.03 \text{ cal/°C-mole.}$$

The heat of reaction at 25° is – 57.8 kcal/mole. Calculate the standard heat of formation of water vapour at 100°C.

Ans. – 57,981 cal/mole.

10. Gibb's free energy of a reaction at 27°C and 37°C are –29.0 kcal and –30.0 kcal respectively. Determine ΔS and ΔH in this temperature range.

Ans. (i) 100 cal/°C, (ii) 1.0 kcal.

11. From the following bond enthalpies calculate the ΔH for the reaction:

$$H_2 + Br_2 \rightarrow 2HBr$$

Bond energy: H—H = 435.24 kJ/mole, Br—Br = 192.50 kJ/mole,
H—Br = 368.30 kJ/mole.

Ans. – 108.86 kJ.

12. When 1 g of liquid naphthalene ($C_{10}H_8$) solidifies, 149 J of heat is evolved. Calculate the enthalpy of fusion of naphthalene.

Ans. 19.072 kJ/mole.

13. Calculate the equilibrium constant for the following reaction at 298 K and 1 atm. pressure.

$$CO(g) + 3H_2(g) \rightleftharpoons CH_4(g) + H_2O(l)$$

Given: ΔS^0 for the reaction at 298 K = –333.3 J/°K-mole, R = 8.314 J/°K-mole.
ΔH_f^0 at 298 K for CO(g) = – 110.50 kJ/mole, CH_4(g) = – 74.80 kJ/mole, H_2O(l) = –286 kJ/mole.

Ans. $K_{eq.} = 2.970 \times 10^{26}$

14. Given the heats of formation of CO, CO_2 and H_2O(g) are: – 110.50, –393.80 and – 241.80 kJ/mole respectively, calculate heat of the reaction:

$$CO_2 + H_2 \rightarrow CO + H_2O$$

Ans. 41.50 kJ.

15. Calculate the heat of formation of ethane if the heats of combustion of ethane, carbon and hydrogen are –372.8, –94.5 and –68.4 kcal respectively.

Ans. –21.4 kcal/mole.

16. At what temperature will water boil when the atmospheric pressure is 528 mm of Hg? Latent heat of vaporisation of water is 545.5 cal/g.

Ans. 363 K.

17. Calculate ΔH, ΔS, ΔG and ΔU when 1 mole of water is vaporized at 100°C and 1 atm. pressure. The latent heat of vaporization of water is 540 cal/g, density of water is 1 g/c.c. Assume perfect gas behaviour of water vapour.

Ans. $\Delta H = 9720$ cal, $\Delta S = 26.06$ cal/°K or e.u., $\Delta G = 0$, $\Delta U = 8979.5$ cal.

18. 200 g of gold [$C_p = 0.0313$ cal/g-°C] at 120°C is dropped into 25 g of water at 10°C. The system is allowed to reach in equilibrium in an adiabatic container. Find (a) the final temperature, (b) ΔS_{Au}, (c) ΔS_{H_2O}, (d) $\Delta S_{Au} + \Delta S_{H_2O}$.

Ans. (a) 32°C, (b) –1.59 cal/°K, (c) 1.87 cal/°K, (d) 0.28 cal/°K.

19. Calculate ΔS for the mixing of 10 g of He at 120°C and 1.5 bar with 10 g O_2 at 120°C and 1.5 bar.

Ans. 8.14 J/°K.

20. A certain perfect gas has $C_{V,m} = a + bT$, where, $a = 25$ J/mole-°K and $b = 0.0300$ J/mole-°K^2. Let 4 moles of this gas move from 300 K and 2 atm to 500 K and 3 atm. Calculate each of the following quantities for this change of state. If it is impossible to calculate a quantity from the given information – state this: (a) q, (b) W, (c) ΔU, (d) ΔH, (e) ΔS.

21. Classify each of these processes as reversible or irreversible:

(a) freezing of water at 0°C and 1 atm.
(b) freezing of suppercooled water at –10°C and 1 atm.
(c) burning of carbon in O_2 to give CO_2 at 800 K and 1 atm.
(d) rolling a ball on a floor with friction
(e) the Joule-Thomson experiment.
(f) adiabatiac expansion of a gas into vacuum (the Joule-experiment).
(g) use of a frictionless piston to infinitely slowly increase the pressure on an equilibrium mixture of N_2, H_2 and NH_3, thereby, shifting the equilibrium.

Ans. (a) Reversible, (b) irreversible, (c) irreversible etc.

22. For each of the following pairs of systems state which system (if either) has the greater U and which has the greater S.

(a) 5 g of Fe at 20°C and 1 atm. or 10 g of Fe at 20°C and 1 atm.
(b) 2 g of water at 20°C and 1 atm. or 2 g of water vapour at 20°C and 20 torr.
(c) 2 g of benzene at 20°C and 1 bar or 2 g of benzene at 40°C and 1 bar.
(d) a system consisting of 2 g of metal M at 300 K and 1 bar and 2 g of metal M at 310 K and 1 bar or a system consisting of 4 g of metal M at 305 K and 1 bar. Assume the

specific heat of M is constant over the 300 to 310 K range and the volume change of M is negligible over this range.

(e) 1 mole of a perfect gas at 0°C and 1 atm or 1 mole of the same perfect gas at 0°C and 5 atm.

Ans. (a) 10 g has greater U and greater S.

23. Which of these cyclic integrals must, vanish for a closed system with P-V work only?

(a) $\oint PdV$, (b) $\oint (PdV + VdP)$, (c) $\oint VdP$, (d) $\oint (dq_{rev}/T)$, (e) $\oint HdT$, (f) $\oint dU$, (g) $\oint dq_{rev}$, (h) $\oint dq_p$, (i) $\oint dW_{rev}$, (j) $\oint (dW_{rev}/P)$.

24. A real gas obeys the relation $PV = RT + \left(\frac{a}{V}\right)$, where $a = 3.6$ atm-lit^2/mole^2. Calculate the work done when one mole of the gas expands from 0.224 litres to 22.4 litres at 127°C.

Ans. 33.56 kJ.

25. The heat of solution of anhydrous $CuSO_4$ and of $CuSO_4.5H_2O$ are –15.8 kcal and 2.750 kcal/mole respectively. Find the heat of hydration in the reaction:

$$CuSO_4(s) + 5H_2O = CuSO_4.5H_2O(s)$$

26. A mole of an ideal gas (C_V = 8.0 cal) was initially at 0°C and 10 atm. The external pressure is suddenly reduced to 1 atm. and the gas expands adiabatically against pressure. What is the final temperature? Final volume?

Ans. 223.8°K, 18.357 lit.

27. The heat of combustion of liquid ethanol into CO_2 and water is –327 kcal at constant pressure. Calculate q_P. The temperature is maintained at 27°C.

Ans. –326.4 kcal.

28. The specific heat of benzene is 0.39 cal/g, °K. Find the entropy change when 1 kg of benzene at 10°C is mixed with 4 kg of benzene at 30°C.

Ans. 0.7175 e.u.

29. For ammonia, $C_P = 6.2 + 7.9\ T$ cal/mole. Assuming ideal behaviour estimate the entropy change in heating 34 gm of the gas from a volume of 100 litres at 300 K to a volume of 60 litres at 800 K.

Ans. 7906 e.u.

30. A Carnot engine using 1 mole of a perfect gas (C_P = 7 cal/mole) works between 900 K and 300 K. The work done in the isothermal expansion is 900 cal at 900 K. What will be the work available per cycle of operation? Estimate also the work done on the gas and the increment in enthalpy during adiabatic compression.

Ans. $W = 600$ cal, $W_3 = -300$ cal, $\Delta H_4 = 4200$ cal.

31. 3 moles of an ideal monatomic gas initially at 27°C and 20 atm. pressure expands to a final pressure of 2 atm. Calculate ΔS, ΔA and ΔG if the process is performed (a) reversibly and isothermally, (b) by expansion into a vacuum.

Ans. (a) $\Delta S = 13.8$ e.u., $\Delta A = -4145$ cal, $\Delta G = -4145$ cal, (b) Same values.

32. Evaluate: $\left(\frac{\partial T}{\partial S}\right)_P$ and $\left(\frac{\partial T}{\partial S}\right)_V$

Ans. $\frac{T}{C_P}$ and $\frac{T}{C_V}$

33. The molar heat capacity at constant pressure of ammonia gas is expressed by:

$$C_P = (6.189 + 7.887 \times 10^{-3}\,T - 7.28 \times 10^{-7}\,T^2) \text{ cal/mole-°K}.$$

Calculate the increase in entropy when 1 mole of ammonia is heated from 25°C to 125°C at constant pressure.

Ans. $\Delta S_P = 2.52$ cal/mole-°K.

34. Calculate the change in entropy when 100 g of water at 100°C are mixed under adiabatic conditions and at constant atmospheric pressure with

(a) 100 g of water at 0°C.

(b) 100 g of ice at 0°C.

Given: C_P of H_2O (l) = 1.00 cal/g-°C, Heat of fusion of ice = 1436 cal/mole.

Ans. (a) 2.42 cal/mole-°C, (b) 5.20 cal/mole-°C.

35. Evaluate: $\left(\frac{\partial T}{\partial P}\right)_S$ for an ideal gas.

Ans. $\frac{V}{C_P}$

36. Determine ΔS when 2 moles of SO_2(g) undergoes a change of state from 25°C and 1 atm. to 325°C and 20 atm. pressure.

Given: $C_P = (6.077 + 23.54 \times 10^{-3}\,T - 9.69 \times 10^{-6}\,T^2)$ cal/mole-°K.

Ans. $\Delta S = 8.06$ e.u.

37. A steel cylinder of 10 litre capacity contains 20 moles of oxygen at 27°C. An imperfection in the valve closing the cylinder allows the gas to leak out slowly until the cylinder is empty. Assuming the atmospheric pressure to have remained constant at 760 mm of Hg during this time, calculate the work done by the gas.

Ans. 11,680 cal.

38. A current of 5 amp at 115 volts is passed through a heater immersed in water at 100°C. If the heater is operated for an hour,

(a) how much water will be evaporated

(b) how much work will be done and

(c) what will be the change in internal energy?

Given: The heat of vaporization of water at 100°C is 539.6 cal/g.
The density of water vapour is 0.5977 g/l and that of water is 0.9584 g/ml.

Ans. (a) 917 g, (b) 37 kcal, (c) 458 kcal.

39. 200 g of carbon-di-oxide is heated from 27°C to 527°C at a constant pressure of 1 atm. It is then compressed isothermally to 100 atm. Calculate the work done assuming that carbon-di-oxide behaves as an ideal gas.

Ans. –29 kcal.

40. A steel cylinder of 4-litre capacity contains 20 moles of helium at 25°C. A leak in the valve allows the gas to escape but so slowly that the temperature remains constant. Calculate how much heat must have been absorbed from the surroundings when the cylinder gauge reads empty.

Ans. 57 kcal.

41. Two bodies A and B of equal mass of the same material are at temperatures T_1 and T_2 ($T_1 > T_2$). When these two are brought in contact, what would be the entropy change at equilibrium?

Ans. $\Delta S = 2n\, C_P \ln \dfrac{T_1 + T_2}{2\sqrt{T_1 T_2}}$

42. The entropy of a liquid at 0.05 atm. and 32°C is 0.52 cal/deg-mole. Assuming that the vapour of the liquid behaves ideally, calculate the entropy of vapour per mole at 32°C and 0.52 atm. Given: L_V = 2422 cal/mole.

Ans. 3.78 e.u.

43. Calculate ΔS when 1 kg water at 25°C is converted into superheated steam at 200°C under constant atmospheric pressure.

Specific heat of water = 4180 J/kg.

Specific heat of steam = (1670 + 0.49 T) J/kg at T°K, $L_V = 23 \times 10^5$ J/kg.

Ans. 7550 J/°K.

44. One mole of an ideal gas occupying a volume of 2.24 litres at 0°C expands isothermally into a connecting evacuated vessel until the final volume is 22.4 litres. Calculate ΔG and ΔS.

Ans. $\Delta G = -1257$ cal, $\Delta S = 4.6$ e.u.

II. ENGINEERING THERMODYNAMICS

1. The compression ratio in an air-standard Otto cycle is 10. At the beginning of the compression stroke the pressure is 1 atm. and the temperature is 20°C. The heat transfer to the air per cycle is 1900 kJ/kg. Determine:
   (a) Pressure and temperature at the point of each process of the cycle.
   (b) The thermal efficiency.
   (c) The mean effective pressure.

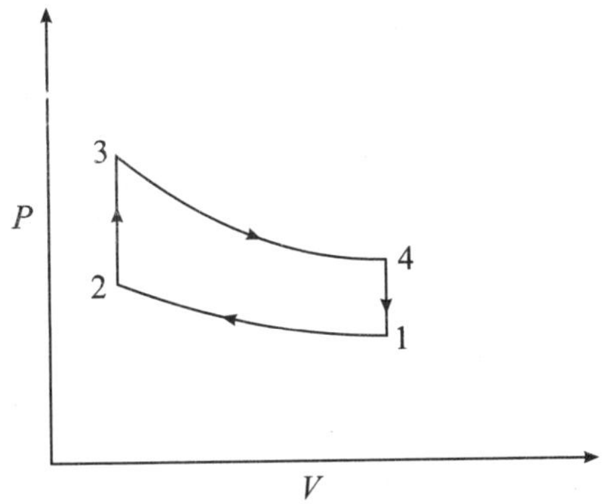

P_1 = 1 atm., t_1 = 20°C, r_c = 10, q_H = 1900 kJ/kg

Assume ideal gas behaviour. C_V = 0.72 kJ/kg-°K, R = 0.286 kJ/kg-°K (for air).

[Hint: $q_H = C_V(T_3 - T_2)$ and $\eta = 1 - \frac{1}{r_c^{\gamma-1}}$, where, $\gamma = \frac{C_P}{C_V}$]

Mean effective pressure = mep = $\frac{W_{net}}{(V_1 - V_2)}$

$TV^{\gamma-1}$ = Constant for the process 1 → 2 and 3 → 4]

2. For an ideal refrigeration cycle the temperature of the refrigerant (R – 12) in the evaporator and in the condenser are –20°C and 40°C respectively. The refrigerant is circulated at the rate of 100 kg/hr. Determine the coefficient of performance and the capacity of the plant.
   Given: At 40°C, P_2 = 0.96 MPa, h_4 = 74.53 kJ/kg, $S_2 = S_1$ = 0.71 kJ/kg-°K,
   h_1 = 178.61 kJ/kg and h_2 = 211.38 kJ/kg.

[Hint: $W_{comp} = (h_2 - h_1)$

$h_3 = h_4$ (Isenthalpic expansion process)

$$\text{COP} = \frac{Q_{condenser}}{W_{comp}}$$]

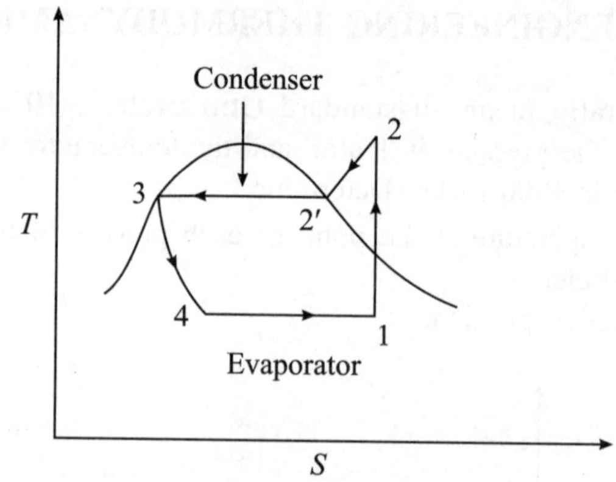

3. Consider the air standard refrigeration cycle as shown in the following figure. Air enters the compressor at 0.1 MPa, −20°C and leaves at 0.5 MPa. Air enters the expander at 15°C. Determine: (a) COP of the cycle, (b) The rate at which air must enter the compressor to provide 1.5 kW of refrigeration.
Given: C_p = 1.006 kJ/kg-°K.

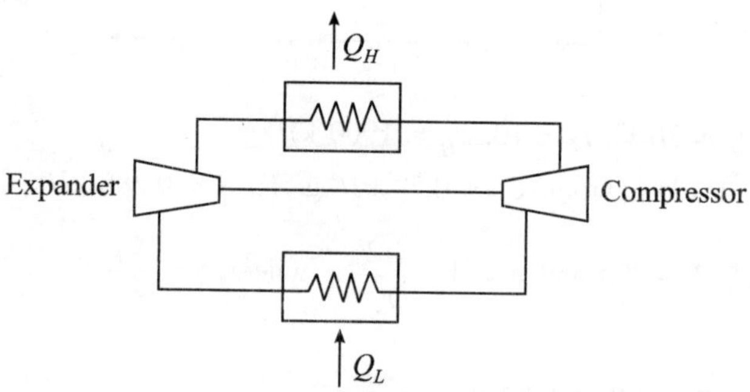

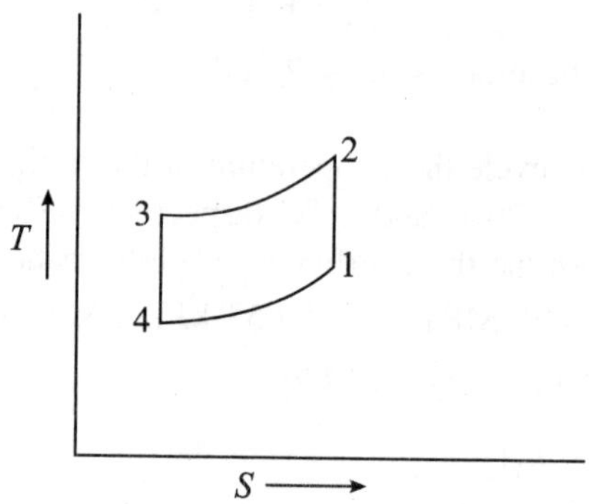

[*Hint:* $T.P^{\frac{1-\gamma}{\gamma}}$ = Constant along 1 → 2 and 3 → 4 processes]

$|W_1| = |W_{comp}| = C_P(T_2 - T_1)$

$$|W_2| = |W_{\text{turbine}}| = C_P(T_3 - T_4)$$
$$W_{\text{net}} = |W_1| - |W_2|$$
$$\text{COP} = \frac{q_L}{W_{\text{net}}}$$
$$q_L \times \dot{m} = Q_L$$
$$Q_L = 1.5 \text{ kW}]$$

4. Determine the efficiency of a Rankine cycle using steam as the working fluid in which the condenser pressure is 15 kPa. The boiler pressure is 30 bar. The steam leaves the boiler as saturated vapour. Using steam table find work done at pump, work output at turbine, heat absorbed at boiler, heat rejected at condenser and efficiency.

Hint:

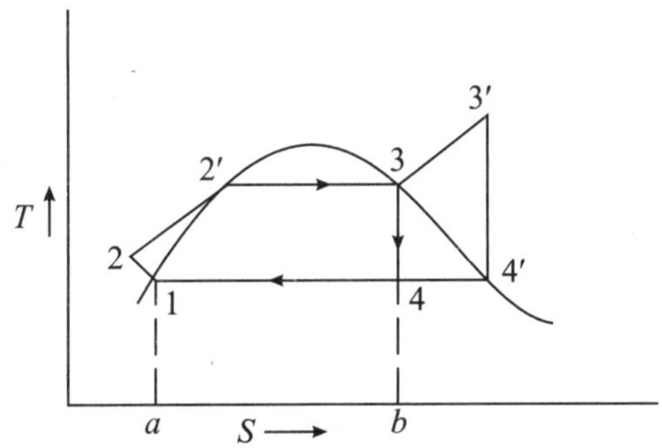

[1-2-2′-3-3′-4′-1] → Rankine cycle

Heat transferred to the working fluid = Area under the curve *a*-1-2-2′-3-4-*b*-*a*.
Heat rejected by the working fluid = Area under the curve *a*-1-4-b-*a*.
Work output = Difference of two areas = 1-2-2′-3-4-1.

At pump: Inlet: Saturated liquid. P_1 is known. Exit: P_2 is known.

Work input: $W_i = H_2 - H_1$

$$dQ = 0 \Rightarrow S_2 = S_1.$$
$$dH = dQ + VdP$$

Or, $dH = VdP$

Or, $$\int_{H_1}^{H_2} dH = \int_{P_1}^{P_2} VdP$$

As the fluid is liquid, V is constant.

At boiler: $dP = 0$

Or, $$\int_{H_2}^{H_3} dH = \int dQ$$

Or, $$Q_1 = H_3 - H_2$$

Inlet: P_2, H_2. Outlet: P_3 = Vapour pressure of saturated vapour.

At turbine: Inlet: state 3. Outlet: P_4. Work output = $W_0 = H_3 - H_4$.

$$dQ = 0 \Rightarrow S_3 = S_4 = 6.3409.$$

At state 4: $S_f = 0.6492$ kJ/kg-°K and $S_{fg} = 7.5010$ kJ/kg-°K

$h_f = 191.81$ kJ/kg and $h_{fg} = 2392.82$ kJ/kg.

So, $6.3409 = 0.6492 + x \times 7.5010$, where, x = weight fraction of vapour in the mixture.

Heat rejected = $|Q_2| = (h_4 - h_1)$

$$\eta = \frac{|Q_1| - |Q_2|}{|Q_1|}$$

5. A steam power cycle has a high pressure of 3.5 MPa and a condenser exit temperature of 45°C. The turbine efficiency is 85% and other cycle components are ideal. If the boiler superheats to 800°C, find the cycle thermal efficiency.

6. A disel engine has a compression ratio of 20:1 with an inlet of 95 kPa, 290 K with volume 5 litres (state 1). The maximum cycle temperature is 1800 K. Find the maximum pressure, the net specific work and the thermal efficiency.

Ans. 11.8 MPa, 420 kJ/kg, 70%.

7. 1 kg of air in a cylinder fitted with a piston at a pressure of 400 kPa and a temperature of 600 K. The air is expanded to 150 kPa in a reversible adiabatic process. Calculate the work done by the air. $\gamma = 1.40$.

[*Hint:* $$W = \frac{P_2V_2 - P_1V_1}{1-\gamma} = \frac{nR(T_2 - T_1)}{1-\gamma}$$

n = Number of moles]

8. In a reversible process nitrogen is compressed in a cylinder from 100 kPa, 20°C to 500 kPa. During this compression process the following relation holds good. $PV^{1.3}$ = Constant. Calculate the work and heat transfer per kg and show this process on $P - V$ and $T - S$ diagrams.

Hint:

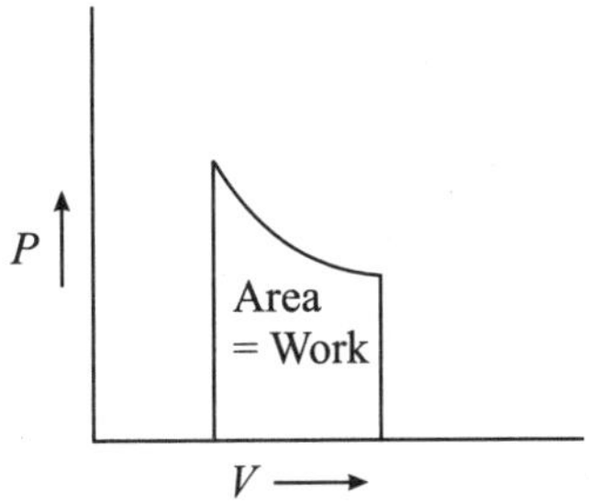

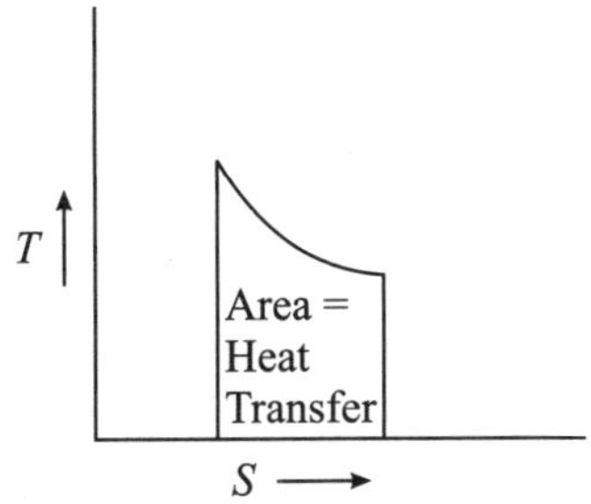

9. The centrifugal air compressor of a gas turbine receives air from the ambient atmosphere where the pressure is 1 bar and the temperature is 300 K. At the discharge of the compressor the pressure is 4 bar, the temperature is 480 K and the velocity is 100 m/s. The mass flow rate into the compressor is 15 kg/s. Determine the power required to drive the compressor.

Hint:

$$h_i = h_e + \frac{V_e^2}{2} + W$$

$$dQ = 0$$

$$h_i = \text{Inlet enthalpy}$$

$$h_e = \text{Outlet enthalpy}$$

Neglect inlet kinetic energy and change in potential energy.

C_P = 1.0035 kJ/kg-°K. $(h_e - h_i) = C_P(T_e - T_i)$. V_e = 100 m/s.

10. Steam at 0.6 MPa, 200°C enters an insulated nozzle with a velocity of 50 m/s. It leaves at a pressure of 0.15 MPa and a velocity of 600 m/s. Determine the final temperature if the steam is superheated in the final state.

$$C_P = 1.872 \text{ kJ/kg-°K.}$$

[*Hint:* $h_e = C_P(T_e - T_i)$ and $h_i + \frac{v_i^2}{2} = h_e + \frac{v_e^2}{2}$]

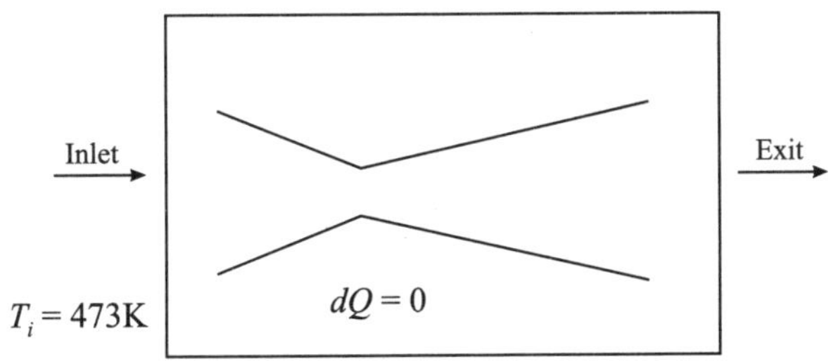

11. In an aqua-ammonia absorption refrigerator system heat is supplied to the generator by condensing steam at 0.2 MPa, 90% quality. The temperature to be maintained in the

refrigerator is –10°C and the ambient temperature is 30°C. Estimate the maximum COP of the refrigerator.
If the actual COP is 40% of the maximum COP and the refrigeration load is 20 tonnes, what will the required steam flow rate be?

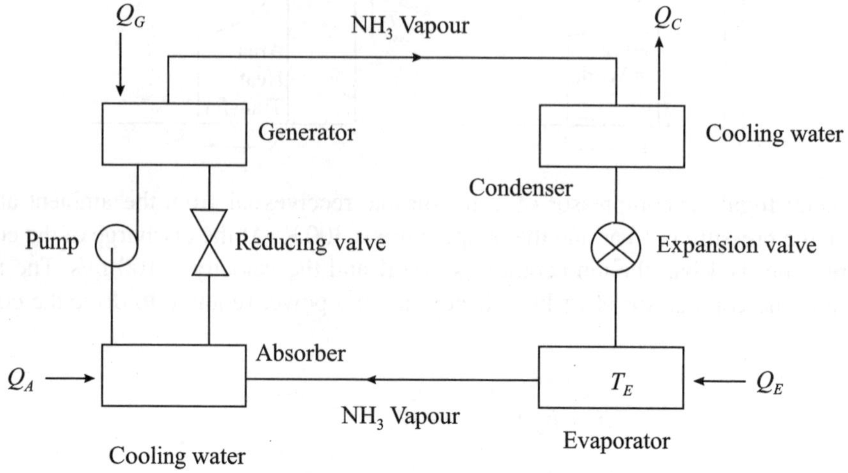

T = Condenser and absorber temperature = Ambient temperature

= (30 + 273)K = 303 K

T_E = (–10 + 273) K = 263 K.

T_G = Standard temperature of steam at 0.2 MPa = (120 + 273)K = 393 K.

$$(\text{COP})_{\max} = \left(\frac{T_E}{T - T_E}\right) \times \left(\frac{T_G - T}{T_G}\right)$$

Again,
$$\text{COP} = \frac{Q_E}{Q_G}$$

$$\text{COP} = 0.4 \times (\text{COP})_{\max}$$

$$Q_E = \frac{20 \times 14{,}000}{3600} \text{ kJ/s}$$

1 ton of refrigeration = 14,000 kJ/h

Or,
$$Q_G = \frac{20 \times 14{,}000}{3600 \times \text{COP}} \text{ kJ/s}$$

At 120°C and 0.2 MPa latent heat of evaporation = h_{fg} = 2202 kJ/kg.
Quality = 0.9, which means that 90% of the steam is considered.

12. A Diesel engine has a compression ratio of 14 and cut-off takes place at 6% of the stroke. Find the air standard efficiency.

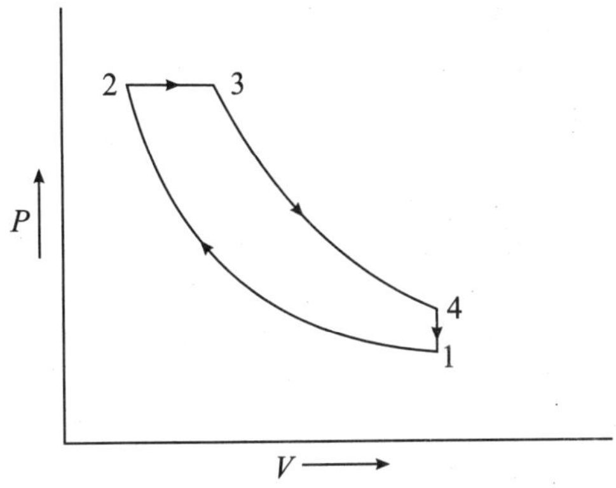

$$r_c = \frac{V_1}{V_2} = 14 \quad \text{and} \quad (V_3 - V_2) = 0.06\,(V_1 - V_2)$$

$$r_e = \frac{V_4}{V_3} = \frac{V_1}{V_3} \ \text{Cut-off-ratio} = \frac{V_3}{V_2} = k$$

$$r_c = r_e \times k \text{ and } r_e = \frac{r_c}{k}$$

$$\text{Efficiency} = \eta_{\text{diesel}} = 1 - \frac{1}{\gamma} \times \frac{1}{r_c^{(\gamma-1)}} \left(\frac{K^{\gamma} - 1}{K - 1} \right) \text{ and } \gamma = 1.4$$

13. (a) Freon-12 refrigerator, working with the evaporator and condensor temperatures of –15°C and 40°C respectively, operates on vapour compression cycle. Determine: (i) condenser and evaporator pressures, (ii) compressor discharge temperature and enthalpy, (iii) ratio of COP of the cycle to that of Carnot cycle.

(b) If the capacity of the refrigerator is $\left(\frac{1}{8}\right)$th ton of refrigeration, what is the mass flow rate of the refrigerator? Also determine the power consumption and heat rejected in the compressor.

Hint:

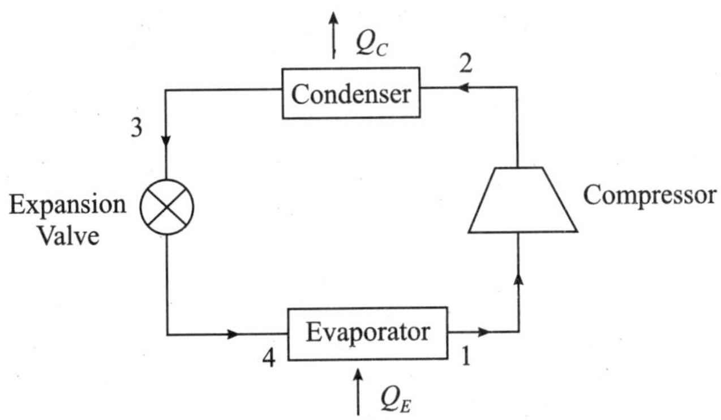

$1 \rightarrow 2$: Compressor: $(h_2 - h_1) = |W|$, $dS = 0$.

$2 \rightarrow 3$: Condenser: $dP = 0$, $dT = 0$, $|Q_C| = (h_2 - h_3)$

$3 \rightarrow 4$: Expansion valve: $dH = 0$, $h_4 = h_3$, x = quality of vapour. $h_3 = h_f + x\, h_{fg}$

$4 \rightarrow 1$: Evaporator: $dP = 0$, $dT = 0$, $|Q_E| = (h_1 - h_4)$

$$\text{COP} = \frac{Q_E}{W}$$

In the case of Carnot refrigerator: $dS = 0$ at the expansion process. So, $S_4 = S_3$ and hence $S_4 = S_{f4} + x_4\, S_{fg4}$.

At the compression process also: $S_2 = S_1$ and $S_1 = S_{f1} + x_1\, S_{fg1}$.

$$h_1 = h_f + x_1\, h_{fg},\ |W_{\text{Comp}}| = (h_2 - h_1).$$

$$h_4 = h_f + x_4\, h_{fg},\ |W_{\text{Exp}}| = (h_3 - h_4).$$

$$|W_{\text{net}}| = |W_{\text{Comp}}| - |W_{\text{Exp}}|$$

$$|Q_C| = (h_2 - h_3) \text{ and } |Q_E| = (h_1 - h_4)$$

$$\text{COP}|_{\text{Carnot}} = \frac{|Q_E|}{|W_{\text{net}}|}$$

$$\text{Capacity of the refrigerator} = \frac{1}{8} \text{ ton refrigeration}$$

$$\dot{Q}_E = \frac{1}{8} \times 3.5167 \text{ kW}$$

$$\dot{m} = \frac{\dot{Q}_E}{Q_E}\left(\frac{\text{kW}}{\text{kJ/kg}}\right) = \frac{\dot{Q}_E}{Q_E} \text{ kg/s}$$

Power consumption of the compressor $= \dot{m} \times W_{\text{Comp}}$

14. A diesel engine has a compression ratio of 20. The pressure and temperature of air at the beginning of compression are 100 kPa and 300 K. The maximum temperature attained in the engine is 2800 K. Determine the thermal efficiency and mean effective pressure of the air standard cycle.

Hint:

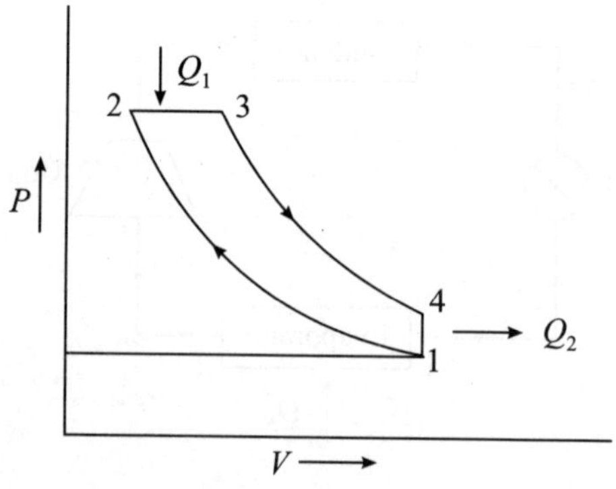

$1 \rightarrow 2$ $$dQ = 0,\ T.V^{\gamma-1} = \text{Constant}$$

$$P.V^{\gamma} = \text{Constant and } \gamma = 1.4$$

$2 \rightarrow 3$ $$Q_1 = (h_3 - h_2) = C_P(T_3 - T_2)$$

$$C_P = 1.0035 \text{ kJ/kg.°K}$$

$$T_3 = T_{max} = 2800 \text{ K}$$

$$P_2 = P_3 = P_{max}$$

$3 \rightarrow 4$ $$dQ = 0$$

$$T.V^{\gamma-1} = \text{Constant}$$

$$P.V^{\gamma} = \text{Constant}$$

$$\gamma = 1.4$$

$$V_3 = \frac{RT_3}{P_3} \text{ m}^3\text{/kg}$$

$$V_4 = \frac{RT_4}{P_4} \text{ m}^3\text{/kg}$$

$$Q_2 = (u_1 - u_4) = C_V(T_1 - T_4)$$

since, $$dP = 0$$

$$C_V = 0.7165 \text{ kJ/kg.°K}$$

$$W = (Q_1 + Q_2) \text{ kJ/kg}$$

$$\eta = \frac{W}{Q_1}$$

$$\text{mep} = P_m = \frac{W}{(V_1 - V_2)}$$

15. A petrol engine operates on the air-standard Otto cycle. The compression ratio is 8. The pressure and temperature at the beginning of compression are 100 kPa and 300 K. The heat added to air is 1800 kJ/kg of air. Determine: (i) pressure and temperature at each point in the cycle, (ii) thermal efficiency, (iii) mean effective pressure.

16. A fluid of 0.5 kg at 1.0021 MPa, 180°C expands in a reversible constant pressure process until its volume is doubled. Find the temperature after expansion, change in entropy, work done and heat transferred if the fluid is: (a) wet steam with initial quality of 0.8, (b) air.

[*Hint:* (a) Steam: Initial state:

$v_1 = v_{f1} + x_1 v_{fg1},\ h_1 = h_{f1} + x_1 h_{fg1},\ s_1 = s_{f1} + x_1 s_{fg1}.$

Final state:

$P_2 = P_1 = 1.0021$ MPa, $v_2 = 2v_1$].

17. 10 kg/s chilled water for air conditioning enters a tall building with a velocity of 50 m/s at an elevation of 30 m from the ground. The water leaves the system with a velocity of 10 m/s at an elevation of 60 m. The enthalpies of water entering in and leaving out are 21 kJ/kg and 43 kJ/kg respectively. The rate of work done by a pump in the line is 35 kW. Calculate the rate at which heat is removed from the building.

Hint:

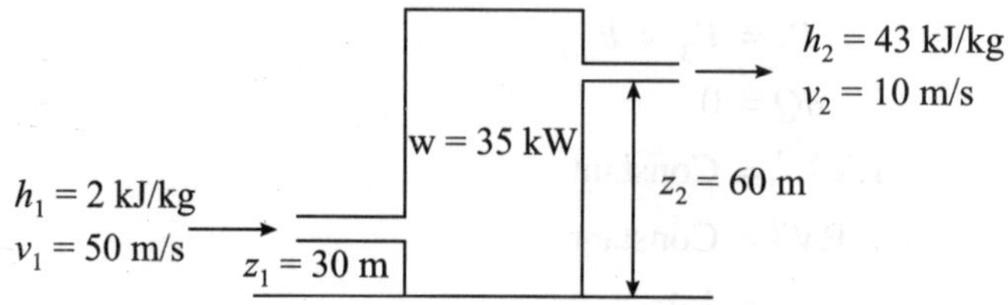

$$\dot{Q} = \dot{W} + \dot{m}(h_2 - h_1) + \frac{1}{2}\dot{m}(v_2^2 - v_1^2) + \dot{m}g(z_2 - z_1)$$

18. In an aircraft engine, compressed air at 3 bar and 450 K enters a combustion chamber in which heat is added at constant pressure to the air by the combustion of a fuel. Air is thus heated to a temperature of 1250 K. It then enters a turbine with a negligible velocity. It expands in the turbine until its temperature falls to 1000 K. The velocity of air leaving the turbine is 50 m/s. Air then enters a convergent-divergent nozzle wherein it expands until its temperature drops to 800 K. Flow through both turbine and nozzle may be considered as reversible adiabatic. Determine:

(i) heat added in the combustion chamber per kg of air.
(ii) Work done in the turbine per kg of air.
(iii) Velocity of air leaving the nozzle.
(iv) Pressure of air leaving the turbine and at exit from the nozzle.

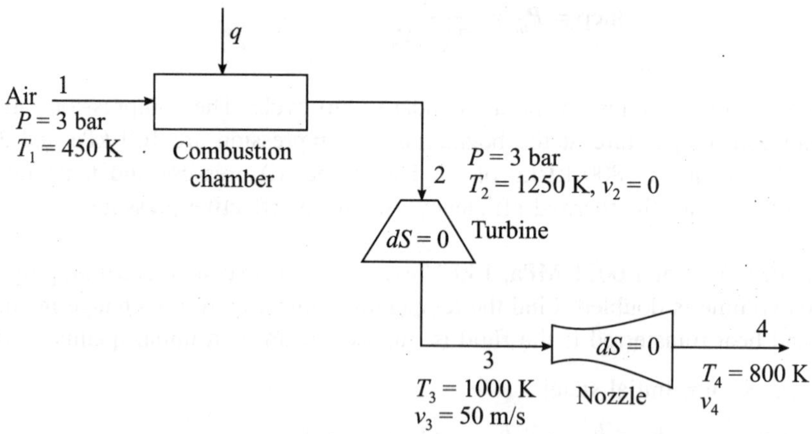

Combustion chamber: $dP = 0$

$$q = (h_2 - h_1) = C_P(T_2 - T_1)$$

At turbine: $$h_2 = h_3 + \frac{v_3^2}{2} + w$$

$$T_2 P_2^{\frac{1-\gamma}{\gamma}} = T_3 P_3^{\frac{1-\gamma}{\gamma}}$$

At nozzle: $$h_3 + \frac{v_3^2}{2} = h_4 + \frac{v_4^2}{2}$$

$$T_3 P_3^{\frac{1-\gamma}{\gamma}} = T_4 P_4^{\frac{1-\gamma}{\gamma}}$$

19. Atmospheric air is sucked in an internal combustion engine cylinder. After suction $p = 1$ bar, $t = 75°C$, $V = 800$ c.c. The air is then reversibly compressed to 15 bar and $\frac{1}{8}$th of its volume. Heat of combustion of fuel is then added to air at constant volume until its pressure reaches to 50 bar.

(a) Determine the index of compression process.
(b) Calculate the change in entropy of air during each process.
(c) Determine if there is any heat exchange between air and cylinder walls during compression. State the direction of heat flow.

Hint:

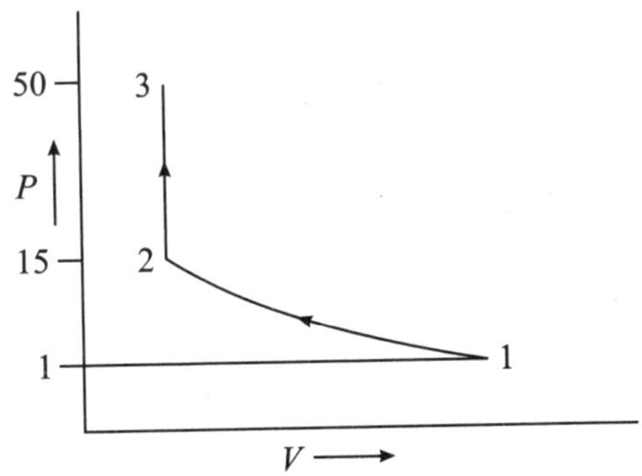

$$P_1 V_1^n = P_2 V_2^n$$

n = Compression index

$$PV = \frac{m}{M} RT$$

$$(S_2 - S_1) = m\left[C_V \ln\frac{T_2}{T_1} + R\ln\frac{V_2}{V_1}\right]$$

$$T_1 P_1^{\frac{1-n}{n}} = T_2 P_2^{\frac{1-n}{n}}$$

$$(S_3 - S_2) = m\,C_V \ln \frac{T_3}{T_2}$$

$$\frac{P_2V_2}{T_2} = \frac{P_3V_3}{T_3} \quad \text{and} \quad V_3 = V_2$$

Heat transfer: $$Q = m\,C_V\left(\frac{\gamma - n}{1 - n}\right)(T_2 - T_1) \quad \text{and} \quad \gamma = 1.4$$

20. A Carnot refrigerator is working between reservoirs of – 35 and 45°C. What is the coefficient of performance? If the actual refrigerator has a coefficient of performance 60% of the Carnot value estimate the refrigerating effect per kW of power consumption.

21. 0.1 kg of water at 300 kPa and 200°C is contained within a cylinder and piston. During a process 7.7 kJ of heat is removed and 17.5 kJ of work is done on water. Determine the entropy change of water if the pressure at the end of the process is 1.5 MPa.

III. CHEMICAL KINETICS

1. A second order reaction, where $a = b$, is 20% complete in 500s. How long will it take the reaction to go to 60% completion?

Ans. 3000s.

2. The gas phase decomposition of dimethyl ether follows first order kinetics.

$$CH_3—O—CH_3(g) \rightarrow CH_4(g) + H_2(g) + CO(g)$$

The reaction is carried out in a constant volume container at 500°C and has a half life of 14.5 min. Initially only dimethyl ether is present at a pressure of 0.40 atm. What is the total pressure of the system after 12 min? Assume ideal gas behaviour.

Ans. 0.749 atm.

3. In a second order reaction the initial concentration of reactants is 0.1 mole/l. The reaction is found to be 20% complete in 40 min. Calculate (i) the rate constant, (ii) half-life period, (iii) time required to complete 75% of the reaction.

Ans. (i) 0.0625 l/(mole-min), (ii) 160 min, (iii) 480 min.

4. The optical rotation of sucrose in 0.5 N hydrochloric acid at 35°C and at various time intervals are given below. Find out the order of the reaction.

Time (min)	0	10	20	30	40	∞
Rotation (Degree)	+32.4	+28.8	+25.5	+22.4	+19.6	–11.1

Ans. First order.

5. The 1st order decomposition of N_2O_4 into NO_2 has a k-value of 4.5×10^3 s^{-1} at 1°C and an activation energy of 58 kJ/mole. At what temperature its half-life would be 6.93×10^{-5} s [R = 2 cal/(mole-°K)].

Ans. 287.9 K.

6. Rate = $k[A]^2[B]$ for a particular reaction. What is the effect of (i) doubling the concentration of A, (ii) doubling the concentration of A and decreasing the concentration of B to one-fourth?

Ans. (i) Rate becomes four times, (ii) rate remains unaltered.

7. At 100°C the gaseous reaction $A \rightarrow 2B + C$ is observed to be 1st order. On starting with pure A it is found that at the end of 10 min the total pressure of the system is 176 mm of Hg and after a long time it is 270 mm of Hg. Calculate the half-life period of the reaction.

Ans. 10.7 min.

8. (a) Integrate the rate equation:

$$-\frac{dC}{dt} = k.C^{1/2}$$

(b) Determine the expression for $t_{1/2}$ in terms of k and C_0.

(c) What is the expression of k?

9. The 1st order decomposition of N_2O_5 to give NO_2 and O_2 has a specific rate constant of $4.30 \times 10^{-5}\ s^{-1}$ at 300 K and $6.97 \times 10^{-2}\ s^{-1}$ at 500 K. Calculate the activation energy for this reaction. Given: $R = 2$ cal/(mole-°K).

Ans. 11.1 kcal.

10. Two substances A and B undergo a bimolecular reaction step. The following table gives the concentration of A at various times for an experiment carried out at constant temperature of 17°C.

Conc. of $A \times 10^4$ moles/dm^3	10.00	7.94	6.31	5.01	3.98
Time (min)	0	10	20	30	40

The initial concentration of B is 2.5 moles/dm^3. Calculate the value of 2nd order rate constant.

Ans. 1.54×10^{-4} dm^3/(mole-s).

11. Two 2nd order reactions A and B have identical frequency factors. The activation energy of A exceeds that of B by 10.46 kJ/mole. At 100°C the reaction A is 30% complete after 1 hour when the initial concentration of the reactant is 0.1 moles/dm^3. How long will it take for the reaction B to reach 70% completion at the same temperature for an initial concentration of 0.05 moles/dm^3?

Ans. 1340 s.

12. In gas phase kinetics pressures instead of concentrations are used in rate laws. Suppose that for the reaction: $aA \rightarrow$ Products, one finds that:

$$-\frac{dP_A}{dt} = \frac{1}{a} k_P P_A^n$$

where, k_P is a constant and P_A is the partial pressure of A.

(a) Show that: $k_P = k.(RT)^{1-n}$

(b) Is that valid for any nth order reaction?

13. The 1st order reaction $2A \rightarrow 2B + C$ is 35% complete after 325 s.

(a) find k and k_A, where, $k_A = 2k$, (b) how long will it take for the reaction to be 70% complete? and 90% complete?

Ans. (a) 0.00066 s^{-1}, 0.00133 s^{-1}, (b) 905 s, 1731 s.

14. If the reaction $A \rightarrow$ products is zero order. Sketch $[A]$ versus t.

15. For the reaction $A \rightarrow$ products, data for a run with $[A]_0$ = 0.6 moles/dm^3 are:

$t(s)$	$[A]/[A]_0$	$t(s)$	$[A]/[A]_0$
0	1	400	0.511
100	0.829	600	0.385
200	0.688	1000	0.248
300	0.597		

(a) Find the order of the reaction, (b) Find the rate constant.

16. The gas-phase reaction: $2NO_2Cl \rightarrow 2NO_2 + Cl_2$ has $r = k[NO_2Cl]$. Devise two mechanisms consistent with this rate law.

17. The reaction $2Cr^{2+} + Tl^{3+} \rightarrow 2Cr^{3+} + Tl^{+}$ in aqueous solution has $r = k[Cr^{2+}]\,[Tl^{3+}]$. Devise two mechanisms consistent with this rate law.

18. The gas-phase reaction $2NO_2 + F_2 \rightarrow 2NO_2F$ has $r = k[NO_2]\,[F_2]$. Devise a mechanism consistent with this rate law.

19. For the reaction $2HI \rightarrow H_2 + I_2$ values of k are 1.2×10^{-3} and 3.0×10^{-5} dm^3/(mole-s) at 700 K and 629 K respectively. Estimate E_a and A.

Ans. 45.5 kcal/mole, 1.9×10^{11} lit/(mole-s).

20. What value of k is predicted by the Arrhenius equation for $T \rightarrow \infty$? Is this result physically reasonable?

21. The gas-phase reaction $2N_2O_5 \rightarrow 4NO_2 + O_2$ has $k = 2.05 \times 10^{13}.\ e^{\left(\frac{-E}{RT}\right)}$ s^{-1} and E = 24.65 kcal/mole. (a) Find k at 0°C, (b) Find $t_{1/2}$ at −50°C, 0°C and +50°C.

Ans. (a) 5.07×10^{-7} s^{-1}, (b) 3.4×10^{10} s, 1.37×10^{6} s, 1260 s.

22. A sample of 0.420 mg of $^{233}UF_6$ shows an activity of 9.88×10^{4} counts/s. Find $t_{1/2}$ of ^{233}U, assuming the radioactive reaction follows 1st order kinetics.

Ans. 1.62×10^{5} years.

23. (a) The half-life of ^{3}H is 12.4 years. Calculate the activity of 20.0 g of HNO_3 containing 0.200 mole percent $^{3}HNO_3$. (b) Find the activity after 6.20 years.

Ans. (a) 6.77×10^{11} s^{-1}, (b) 4.79×10^{11} s^{-1}.

24. For a system with the two competing mechanisms in:
Show that the observed activation energy is:

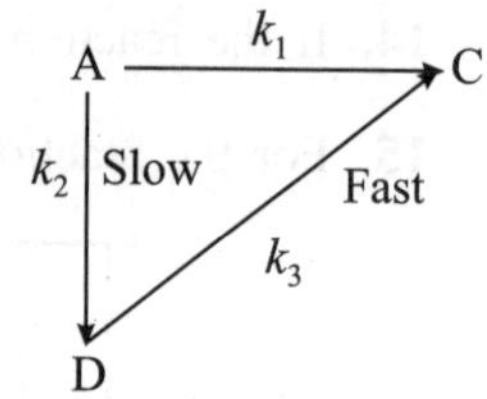

$$E_a = \frac{k_1(T)E_{a,1} + k_2(T)E_{a,2}}{k_1(T) + k_2(T)}$$

25. Half-life period of decomposition of a compound was found to be 50 min. When the concentration to start with is halved the half-life period becomes 100 min. What is the order of the reaction?

Ans. Second order.

26. The rate constant for $H^+ + OH^- = H_2O$ has been found be 1.3×10^{11} litre/(mole-s). Calculate the half-value period for the neutralization of a strong acid by a strong base, both at a concentration of 10^{-4} N.

Ans. 7.7×10^{-8} sec.

27. The decomposition of ammonia on tungsten at 1200 K is a zero order reaction. Starting with ammonia in a rigid vessel at 200 mm pressure, the total pressure was 300 mm for all gases at 160 min. What will be the partial pressure of ammonia after 1 hour if we start with ammonia at an initial pressure of 150 mm?

Ans. 112.5 mm.

28. The decomposition of a gas at an initial pressure of 600 mm was studied in a closed vessel at T°K. The gas was found to be 50% decomposed in 30 min, and 75% in 90 min. Show that the reaction is of the second order and its rate constant is 5.55×10^{-5} mm^{-1}/min.

29. For the reaction: $2HgCl_2 + K_2C_2O_4 \rightarrow 2KCl + 2CO_2 + Hg_2Cl_2$ the following results were obtained:

$HgCl_2$ (mole/litre)	$K_2C_2O_4$ (mole/litre)	Time (t) in sec.	Hg_2Cl_2 (ppt.) in gm.
0.0836	0.404	65	0.0068
0.0836	0.202	120	0.0031
0.0418	0.404	60	0.0032

Find the order of the overall reaction.

Ans. Third order.

30. The temperature coefficient of a reaction rate of a reaction is 2.8. How many times will the reaction rate increase when temperature is raised from 20°C to 75°C?

Ans. $k_{75°C}/k_{20°C} = 287$.

IV. SOLID STATE CHEMISTRY

1. Compute the limiting radius ratio $\left(\frac{r_+}{r_-}\right)$ for (a) a square planar crystal structure in which B^- ions are at the corners of the square and A^+ ions are at the centre, (b) an equilateral triangular crystal structure with B^- ions are at the apices and A^+ ions at the centre.

Ans. (a) $\left(\frac{r_+}{r_-}\right) = 0.414$, (b) $\left(\frac{r_+}{r_-}\right) = 0.155$.

2. Potassium crystallizes with a BCC lattice and has a density of 0.856 g/c.c. Calculate the length of the side of the unit cell, i.e. a and the distance between (200), (110) and (222) planes.

Ans. $a = 5.33$ Å, $d_{200} = 2.665$ Å, $d_{110} = 3.77$ Å, $d_{222} = 1.54$ Å.

3. The element chromium exists as a BCC lattice whose unit cell edge is 2.88 Å. The density of chromium is 7.20 g/c.c. How many atoms does 52.0 g of chromium contain?

Ans. 6.04×10^{23} atoms.

4. Calculate the electron affinity of iodine, from the following data:
$U_{\text{NaI}} = -165.4$ kcal/mole, $\Delta H_{f,\,\text{NaI}} = -64.8$ kcal/mole, $\Delta H_{\text{Sub, NaI}} = 25.9$ kcal/mole, $(\text{I.P.})_{\text{Na}} = 118.4$ kcal/mole, $\Delta H_{\text{S, Na}} = 26$ kcal/mole, $(\frac{1}{2}\Delta H_{\text{Sub}, I_2} + \frac{1}{2}\Delta H_{\text{Diss}, I_2}) = 25.5$ kcal/mole.

Ans. 73.22 kcal/mole.

5. Prove that $\left(\frac{c}{a}\right) = \left(\frac{2\sqrt{2}}{\sqrt{3}}\right)$ for the hexagonal closed packed structure.

6. The metal nickel crystallizes in a FCC structure. Its density is 8.90 g/c.c. Calculate: (i) the radius of a nickel atom, (ii) edge length of the unit cell and (iii) What volume do they occupy?

Ans. (i) 1.24 Å, (ii) 3.52 Å, (iii) 4.38×10^{-23} c.c.

7. Metallic gold crystallizes in a FCC structure which is 4.078 Å on one edge. The density of the gold is 19.30 g/c.c. at 25°C. Calculate: (i) the atomic radius of gold and (ii) Avogadro Number.

Ans. (i) 1.442 Å, (ii) 6.020×10^{23}/g.mole.

8. Rubidium iodide (RbI) crystallizes in a BCC structure with each Rb^+ ion surrounded by eight I^- ions and vice versa. The ionic radius of Rb^+ is 1.49 Å and the ionic radius of I^- is 2.17 Å. Find the length of one side of RbI unit cell.

Ans. 4.23 Å.

9. The solid KF has FCC structure. The density of KF at 20°C is 2.48 g/c.c. Calculate the unit cell edge length and the minimum distance between two neighbours in KF at 20°C.

Ans. (a) 5.38 Å, (b) 2.69 Å.

10. Determine the resistivity of an intrinsic Si specimen at 300 K if electron mobility μ_e = 1300 $cm^2/(V\text{-}s)$ and hole mobility μ_h = 500 $cm^2/(V\text{-}s)$. Intrinsic Si exhibits a carrier electron (or hole) density of 1.48×10^{10} electrons (or holes)/cm^3.

Ans. 2.35×10^5 Ω-cm.

11. If a Si-specimen is doped with *n*-type material to the concentration of 1 ppm atoms, determine the resulting conductivity. Given lattice parameter of SC crystal of Si = 5.431×10^{-8} cm, electron mobility of Si at 300 K = 1300 $cm^2/(V\text{-}s)$.

Ans. 10.4 Ω^{-1} cm^{-1}.

12. The resistivity of intrinsic germanium at 27°C is 0.43 Ω–m. Calculate the intrinsic carrier density at 27°C. Assuming the number of electrons near the top of the valence band available for thermal excitation is $5 \times 10^{25}/m^3$, calculate the energy gap for germanium.

Ans. $2.5 \times 10^{19}/m^3$, 0.75 eV.

13. Indium phosphide (InP) has an energy gap of 1.29 eV. The electron and hole mobilities for this semiconducting compound are 0.46 and 0.015 $m^2/(V\text{-}s)$ respectively at 300 K. Calculate its conductivity.

Ans. 5×10^{-5} Ω^{-1} m^{-1}.

14. 5 g of *n*-type silicon doped with arsenic of conductivity 10 Ω^{-1} m^{-1} is melted with 5 g of *p*-type silicon doped with aluminium having the same conductivity. What is the conductivity of the resulting material?

Ans. 3.2 Ω^{-1} m^{-1}.

V. CHEMICAL BONDING

1. The electron dipole moment of the NH_3 molecule is 1.47D and the dipole moment of the NF_3 molecule is 0.21D. Explain why the dipole moment of NF_3 is so small in spite of the large difference in electronegativity between *N* and *F*.

2. Calculate the percent ionic character of HCl and HF molecules using the following data: $\mu_{HCl} = 1.03D$, $r_0 = 1.27$ Å, $\mu_{HF} = 2.487D$, $r_0 = 1.15$ Å, $e = 1.602 \times 10^{-19}$ C.

Ans. 17% for HCl and 45% for HF.

3. Which one of the following combination should favour highest lattice energy in ionic compounds? (i) small ion and high charge, (ii) small ion and low charge, (iii) large ion and high charge, (iv) large ion and low charge.

Ans. (i).

4. Which one of the following is not a condition for a molecule to be a dipole? (i) there must be a bond with ionic character, (ii) there must be at least one double bond in the molecule, (iii) the molecule must possess a favourable shape, (iv) there must be at least two different elements in the molecule.

Ans. (ii).

5. Predict the shape of each of the following: (a) NH_4^+, (b) $BeCl_2$, (c) H_2Se, (d) PCl_3, (e) NH_2^-, (f) AsF_5, (g) IF_2^-, (h) SF_5^-, (i) XeF_3^-, (j) ClF_4^-, (k) XeF_4, (l) IF_3, (m) ICl_2^+.

Ans. (a) tetrahedral, (b) linear, (c) angular, (d) pyramidal, (e) angular, (f) trigonal bipyramid, (g) linear, (h) square pyramidal, (i) *T*-shaped, (j) square planar, (k) square planar, (l) *T*-shaped, (m) angular.

6. Calculate the dipole moment of the following compounds:

(a) 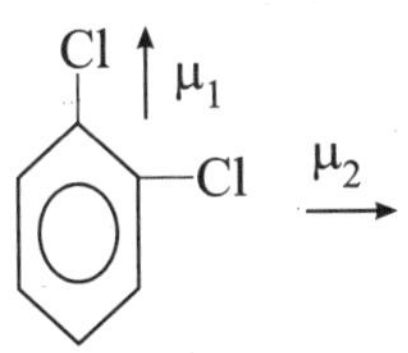

$\mu_1 = \mu_2 = 0.7D$

(b) 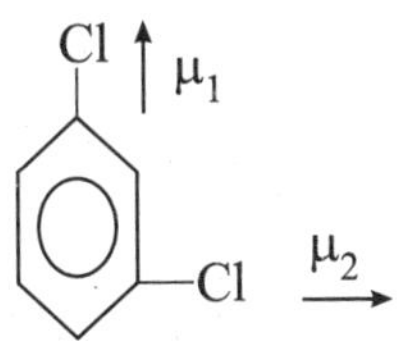

$\mu_1 = \mu_2 = 0.5D$

7. What is metallic bonding? Explain the following metallic properties with the help of metallic bonding: (a) Metallic lustre, (b) High electrical and thermal conductivity, (c) Ductility and malleability, (d) Photoelectric effect and thermionic emission of electrons.

VI. NUMERICAL ON CHEMICAL BONDING

1. The experimental lattice energy of SnO_2 is –11596 kJ/mole Calculate ΔH_f of SnO_2.
Given: ΔH_s = +292 kJ/mole, ΔH_d = +454 kJ/mole,
I = +8990 kJ/mole, E = +636 kJ/mole

2. Calculate the lattice energy of NaCl crystal from the following data:
ΔH_s = +108.7 kJ/mole, ΔH_d = +225.9 kJ/mole
I = +489.5 kJ/mole, E = –351.4 kJ/mole, ΔH_f = –414.2 kJ/mole

3. Calculate the lattice energy of *KF* crystal from the following data:
ΔH_s = +87.8 kJ/mole, ΔH_d = +158.9 kJ/mole,
I = +414.2 kJ/mole, E = –334.7 kJ/mole, ΔH_f = –560.8 kJ/mole

4. Calculate the lattice energy of MgF_2 crystal from the following data:
ΔH_s = +146.4 kJ/mole, ΔH_d = +158.9 kJ/mole,
I = +2184 kJ/mole, E = –334.7 kJ/mole, ΔH_f = –1102.6 kJ/mole

5. The inter-nuclear distance in *HI* is 1.62 Å (assuming complete transfer of electron). Calculate % ionic character and feractional charge on each ion. The measured dipole moment is 0.38*D*.

6. Calculate the lattice energy of AgCl using Born-Meyer equation. Given:
r = 0.322 nm, A = 1.748, r = 0.0345 nm, $\varepsilon_0 = 8.854 \times 10^{-12}$ F/m

7. The dipole moment of water is 6.17×10^{-30} C – m. The H – O – H angle is 104° and the O – H distance is 96 pm. Calcualte the %ionic character of the O – H bond.

VII. ELECTROCHEMICAL CELLS

1. Calculate E^0 at 25°C for the cell:

$$Cd|Cd^{2+}||Cu^{2+}|Cu$$

and determine the cell reaction and its equilibrium constant.
Given: $E^0_{Cu^{2+}}|Cu = 0.337$ V, $E^0_{Cd^{2+}}|Cd = -0.403$ V

Ans. $E^0 = 0.740$ V, $K = 1.04 \times 10^{25}$.

2. The voltage of the cell

$$Pb|PbSO_4|Na_2SO_4.10H_2O(s)|Hg_2SO_4(s)|Hg$$

is 0.9647 V at 25°C. The temperature coefficient of the cell is 1.74×10^{-4} V/°K. (a) What is the cell reaction ? (b) Calculate the values of ΔG, ΔS and ΔH.

Ans. (b) $\Delta G = -44.492$ kcal/mole, $\Delta S = 8.02$ e.u., $\Delta H = -44.21$ kcal/mole.

3. Calculate the electromotive force of

$Li(s)|LiCl(l)|Cl_2(g)$ at 900 K for $P_{Cl_2} = 1$ atm. $\Delta G^0_f = -80.10$ kcal/mole.

Ans. $E^0 = 3.474$ V.

4. Calculate the potential for each half-cell and the total emf at 25°C for the cell represented schematically as:

$$Pb|Pb^{2+}(0.0010\text{ M})|Pt,\ Cl_2(1\text{ atm})|Cl^-\ (0.10\text{ M})$$

$$E^0_{Pb|Pb^{2+}} = 0.1126\text{ V}$$

$$E^0_{Cl_2|2Cl^-} = 1.358\text{ V}$$

Ans. $E = 1.64$ V.

5. The emf of the cell

$$Zn(s)|ZnCl_2\ (0.05\text{ M})|AgCl(s),\ Ag(s)$$

is 1.015 V at 298 K, the silver electrode being positive, whilst the temperature coefficient of the cell is –0.000492 V/°K. Write down the equation for the reaction occurring when the cell is allowed to discharge and calculate the changes in (a) ΔG, (b) ΔH and (c) ΔS accompanying this reaction at 298 K.

Ans. (a) $\Delta G = -196$ kJ/mole, (b) $\Delta H = -224$ kJ/mole,
(c) $\Delta G = -95$ J/°K-mole.

6. The equivalence point in the titration of 40 ml solution of a weak monoprotic acid occurs when 35 ml of a 0.1 N NaOH solution has been added. The pH of the solution is 5.75 after 20 ml of the NaOH solution has been added. What is the dissociation constant of the acid?

Ans. $K = 2.4 \times 10^{-6}$.

7. A 50 ml sample of 0.1 M $FeSO_4$ solution is titrated with 0.1 M ceric sulphate. (a) What is the equilibrium constant for the reaction, (b) What are the concentrations of ions in equilibrium at the equivalence point?

Given: $E^0_{Ce^{3+}|Ce^{4+}} = -1.61$ V and $E^0_{Fe^{2+}|Fe^{3+}} = -0.77$ V

Ans. (a) $K = 1.6 \times 10^4$, (b) $[Fe^{3+}] = [Ce^{3+}] = 5.0 \times 10^{-2}$ M and $[Fe^{2+}] = [Ce^{4+}] = 4.0 \times 10^{-9}$ M

8. (a) Calculate the potential for the cell

$$Ni(s)|Ni^{2+}(0.01\ M)||Cl^-(2\ M)|Cl_2(1\ atm),\ Pt$$

(b) What is the value of ΔG for the cell reaction?

Given: $E^0_{Ni|Ni^{2+}} = 0.250$ V and $E^0_{Cl^-|Cl_2} = -1.36$ V

Ans. (a) 1.71 V, (b) –79 kcal.

9. What is the concentration of Cd^{2+} in the cell?

$$Cd(s)|Cd^{2+}(c)||Ni^{2+}(2\ M)|Ni(s)$$

The emf of this cell is 0.205 V at 25°C.

Given: $E^0_{Ni|Ni^{2+}} = 0.250$ V and $E^0_{Cd|Cd^{2+}} = 0.408$ V

Ans. 0.050 M.

10. (a) Calculate the e.m.f. of the following cell at 25°C.

$$Zn(s)|Zn^{2+}\ (a = 0.004)||Cd^{2+}\ (a = 0.2)|Cd(s)$$

Given: $E^0_{Zn|Zn^{2+}} = 0.763$ V and $E^0_{Cd|Cd^{2+}} = 0.408$ V

(b) Write the cell reaction,

(c) Calculate ΔG for the cell reaction.

Ans. (a) 0.4398 V, (c) $\Delta G = -20.287$ cal.

11. For the following reaction

$$Zn(s) + 2AgCl(s) = ZnCl_2(s)\ (0.555\ m) + 2Ag(s)$$

Calculate ΔG, ΔH and ΔS of the reaction.

Given: $E^0_{Zn|Zn^{2+}} = 0.763$ V and $E^0_{Ag(s)|AgCl(s),\ Cl^-} = -0.2223$ V

$$E_{0°C} = 1.015 \text{ V and } \left(\frac{\partial E}{\partial T}\right)_P = -4.02 \times 10^{-4} \text{ V/°K}$$

Ans. $\Delta S = -18.85$ e.u.

12. The emf of the cell

$$Hg(l)|Hg_2Cl_2(s),\ KCl(aq.sat)||KCl(aq.\ 0.1\ mole/kg),\ AgCl(s)|Ag(s)$$

is 0.047 V at 298 K. The reduction potential of the saturated calomel electrode is 0.2415 V and the standard reduction potential of the Ag(s)|AgCl(s), Cl^- is 0.223 V. Estimate the activity coefficient for the Cl^- ions in 0.1 mole/kg aq. KCl.

Ans. $\gamma_{Cl^-} = 0.77$.

13. Calculate the transport number of the copper ions in dilute copper sulphate solution, given that the e.m.f. (at 298 K) of the cell with transference is 0.0268 V.

$$Cu(s)|CuSO_4(aq.\ 0.001\ mole/kg),\ CuSO_4\ (aq.\ 0.01\ mole/kg)|Cu(s)$$

The mean activity coefficients of 0.001 mole/kg and 0.01 mole/kg $CuSO_4$ solutions are 0.74 and 0.44 respectively.

Ans. $t_{Cu^{2+}} = 0.415$.

14. The specific conductance at 25°C of a saturated aqueous solution of $SrSO_4$ is $1.482 \times 10^{-2}\ \Omega^{-1}\ cm^{-1}$, while that of water used is $1.5 \times 10^{-3}\ \Omega^{-1}\ cm^{-1}$. Determine at 25°C the solubility in gms/l of $SrSO_4$ in water. Given that equivalent conductivities of Sr^{2+} and SO_4^{2-} at infinite dilution are 58 and 80 $\Omega^{-1}\ cm^2$ respectively.

Ans. 0.0967 g/l.

15. A conductivity cell, having a resistance of 9950 Ω with $\left(\frac{N}{10}\right)$ acetic acid at 25°C, had a resistance of 3735 Ω with $\left(\frac{N}{100}\right)$ KCl at 25°C. The specific resistance of $\left(\frac{N}{100}\right)$ KCl at 25°C is 721 Ω-cm. Calculate the specific conductance and equivalent conductance of acetic acid.

Ans. $\kappa = 5.21 \times 10^{-4}\ \Omega^{-1}\ cm^{-1}$, $\Lambda_N = 5.21\ \Omega^{-1}\ cm^2$.

16. At 18°C the equivalent conductivities at infinite dilution of NH_4Cl, NaOH and NaCl are 129.8, 217.4 and 108.9 $\Omega^{-1}\ cm^2$ respectively. If the equivalent conductivity of $\left(\frac{N}{100}\right)$ solution of NH_4OH is 9.33 $\Omega^{-1}\ cm^2$, calculate the degree of dissociation of NH_4OH at this dilution.

Ans. 3.92%.

17. The specific conductance of a solution of a weak acid is 2.1×10^{-4} mho/cm. At the same temperature the ionic conductivities of the two ions H^+ and A^- are 349 and 76 $\Omega^{-1}\ cm^2$ respectively.

$$\lambda^0_{H^+} = 349\ \Omega^{-1}\ cm^2 \text{ and } \lambda^0_{A^-} = 76\ \Omega^{-1}\ cm^2$$

where *HA* is the weak acid. Calculate the degree of dissociation of the acid in the above solution of $\left(\frac{N}{10}\right)$ concentration.

Ans. 4.97×10^{-3}.

18. Solution *A* consists of 50 ml of 0.02 *M* NaOH and solution *B* consists of 50 ml of 0.0446 *M* HCl. The λ_0 values for H^+, Na^+, OH^-, and Cl^- are 350, 50, 200 and 75 Ω^{-1} cm^2 respectively. The two solutions are now mixed. Calculate the transference number of Na^+ in the mixture and also the specific conductance of the mixture.

Ans. $t_{Na^+} = 0.097$ and $\kappa = 6.9 \times 10^{-3}\ \Omega^{-1}\ cm^{-1}$.

19. (a) Formulate a cell in which the following reaction could occur:

$$Fe^{2+} + Br_2 = Fe^{3+} + Br^-$$

(b) Determine E^0, ΔG^0 and K for the reaction at 25°C.
(c) What would be the corresponding values for the following reaction?

$$2Fe^{2+} + \frac{1}{2}Br_2 = 2Fe^{3+} + 2Br^-$$

Ans. (a) $Pt|Fe^{2+}, Fe^{3+}|Br_2(l)|Br^-, Pt$
(b) 0.294 V, –6.8 kcal, 9.25×10^4
(c) 0.294 V, –13.6 kcal, 8.55×10^9.

20. The equilibrium constant for the reaction: $2Cu^+ = Cu^{2+} + Cu(s)$ at 25°C is 1.646×10^6. (a) Construct a cell in which the reaction could occur. (b) Calculate the standard E.M.F. for the cell. (c) Evaluate the standard potential of the $Cu|Cu^+$ electrode. Given: E^0 of $Cu|Cu^{2+}$ is –0.153 V.

Ans. (a) $Pt|Cu^+, Cu^{2+}|Cu$, (b) 0.368, (c) –0.521 V.

21. A cell consists of a Pt-electrode immersed in the redox system: $A^{x+} = A^{(x+n)+} + ne$, combined with a satured calomel electrode. From the following data at 30°C, determine the value of *n*.

% Reduced form	24.4	48.8	73.2	85.4
E.M.F. (Volts)	0.101	0.116	0.129	0.139

The oxidation potential of the calomel electrode is at 30°C is –0.238 V.

Ans. $n = 2$.

VIII. IONIC EQUILIBRIUM

1. To 35 c.c. of 0.05 mole/litre acetic acid, 10 c.c. of 0.05 mole/litre sodium hydroxide was added. What is the pH of the solution? pK_a of acetic acid = 4.74.

Ans. pH = 5.14.

2. $AgNO_3$ is slowly added to a 0.020 M solution of NaOAC. (a) Will a precipitate of AgOAC form when $[Ag^+] = 1.0 \times 10^{-2}$ M? (b) At what concentration of Ag^+ will a precipitate start forming? The solubility product of AgOAC is 2.0×10^{-3}.

Ans. (a) No, (b) $[Ag^+] = 0.10$ M.

3. The solubility product of AgCl is 2.8×10^{-10} at 25°C. Calculate the solubility of the salts in (i) pure water and (ii) 0.1 M $AgNO_3$ solution.

Ans. (i) 1.673×10^{-5} M (ii) 2.8×10^{-9} M.

4. Equal volumes of 1.0×10^{-2} M $CaCl_2$ and 4.0×10^{-4} M Na_2SO_4 are mixed. Show with steps, whether a precipitate will be formed. Solubility product of $CaSO_4$ is 2.4×10^{-5}. Calculate the solubility of $CaSO_4$.

Ans. 0.30 M.

5. A saturated solution of Ag_3AsO_4 was found to contain 1.0×10^{-6} mole of AsO_4^{3-} and 3.0×10^{-6} mole of Ag^+ ions per litre of solution. What is the value of solubility product of Ag_3AsO_4?

Ans. 2.7×10^{-23}.

6. A solution is saturated with respect to $BaSO_4$ ($K_{SP} = 1.1 \times 10^{-10}$) and BaF_2 ($K_{SP} = 1.7 \times 10^{-6}$). Find $[SO_4^{2-}]$ in the solution if $[F^-] = 4.10 \times 10^{-2}$ M.

Ans. 1.09×10^{-7} M.

7. The hydrolysis of adinosine triphosphate (ATP) to adinosine diphosphate (ADP) and inorganic phosphate at pH 8 and 25°C, is given by:

$$ATP^{4-} + H_2O = ADP^{3-} + HPO_4^{2-} + H^+$$

The standard enthalpy change for the above reaction is (–3 kcal). The standard enthalpy changes of acid dissociation of $HATP^{3-}$, $HADP^{2-}$, $H_2PO_4^-$ are (–2, 0 and +2 kcal/mole) respectively. Calculate the standard enthalpy change for the reaction:

$$HATP^{3-} + H_2O = HADP^{2-} + H_2PO_4^-$$

Ans. –7 kcal.

8. 1 litre of a buffer (pH = 5.20) is to be prepared from NH_4Cl and NH_4OH. If 0.10 mole of NH_4Cl is used, how many moles of NH_3 should be added to this solution? Assume that activity coefficients of components are all unity. pK_b values of NH_4OH = 4.749.

Ans. 8901.6×10^{-9} moles of NH_3.

9. What is the pH of a 0.05 M solution of sodium borate at 25°C? What would be the pH of the same solution at 5°C? The pK_a of boric acid is 9.44 at 5°C and pK_a =14.73 at 5°C. Given: ΔH = 13.56 kcal/mole and 2.303 R = 4.576 cal/mole/°K

Ans. pH = 10.76 at 25°C and 11.44 at 5°C.

10. To 100 c.c. of 0.1 mole/litre phosphate buffer of pH 7.1 the following were added : (a) 1 c.c. 1 M NaOH, (b) 5 c.c. 5 M NaOH, (c) 5 c.c. 1 M HCl. Calculate of pH of the resulting solution in each case. The apparent pK_a values of phosphoric acid are 1.96, 6.8 and 12.

Ans. (a) 7.24, (b) 11.3, (c) 6.5.

IX. PROBLEMS OF ORGANIC CHEMISTRY

1. Complete the following reaction with mechanism:

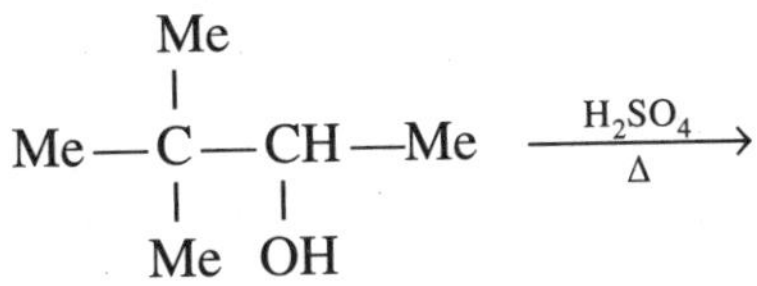

2. Explain with mechanism which one of the following is the major product:

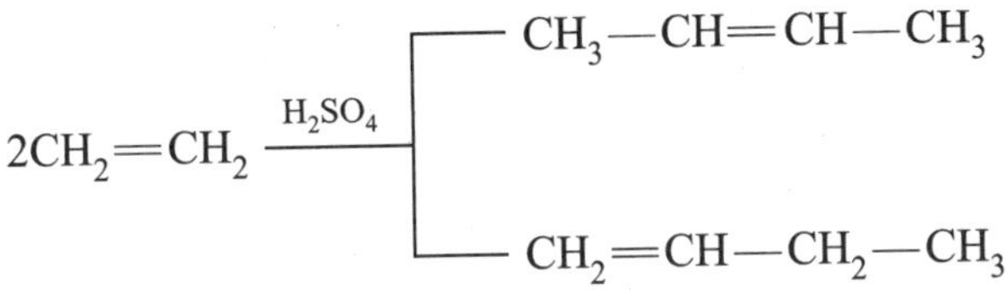

3. Complete the following reaction with mechanism:

4. Between acetophenone ($PhCOCH_3$) and acetone (CH_3COCH_3) which one is more reactive and why?

5. Explain the following order of basic strength of aromatic amines:

$$PhNH_2 > Ph_2NH > Ph_3N$$

6. Explain the following reaction with mechanism:

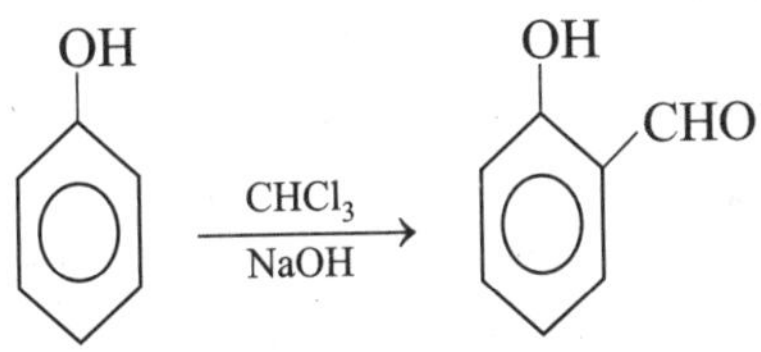

7. (a) Explain $E1_{CB}$ mechanism with an example:

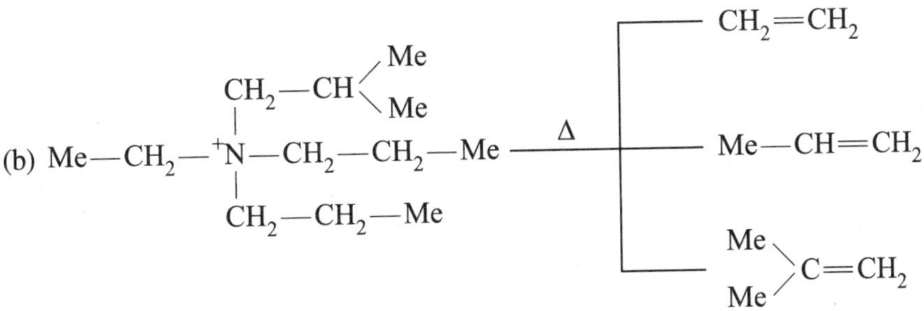

Which one is the major product and why?

8. Explain the following reactions with mechanism:

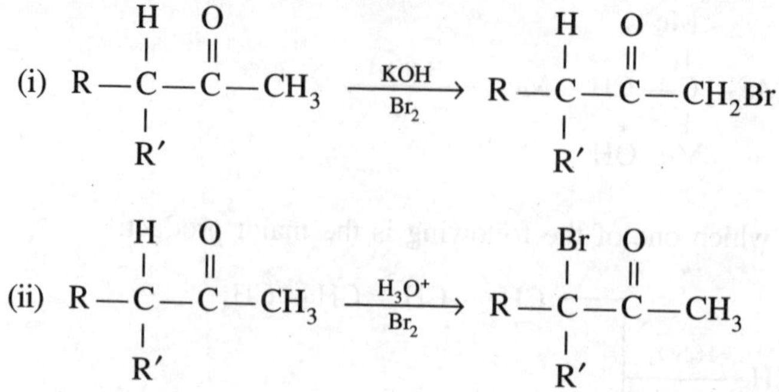

9. Explain the following:

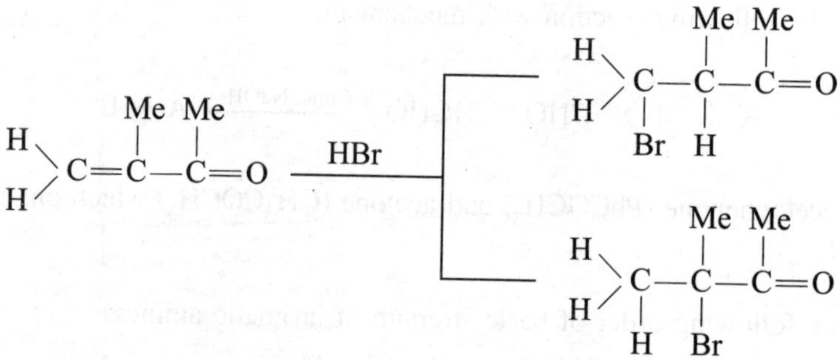

Which one is the major product and why?

10. Complete the following reaction with mechanism:

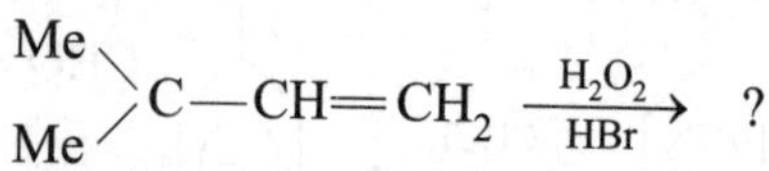

11. Explain the following reaction with mechanism:

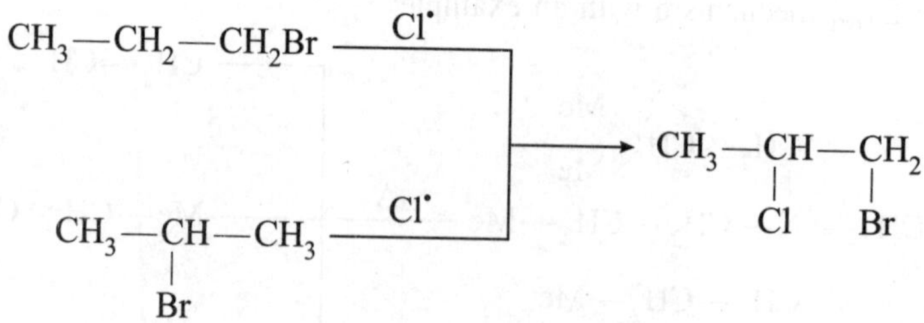

12. Explain the following reaction with mechanism:

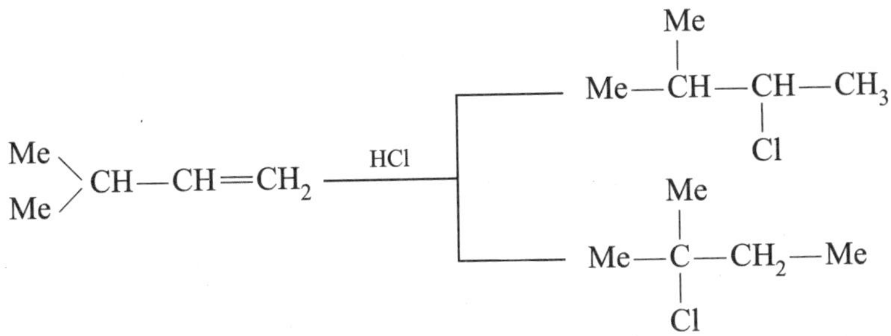

13. Which one is true and why? Increasing rate of reactivity towards HBr:

(i) $CH_2{=}CH_2 > CH_2{=}CH{-}CH{=}CH_2 > Me{-}CH{=}CH{-}Me$

(ii) $Me{-}CH{=}CH{-}Me > CH_2{=}CH_2 > CH_2{=}CH{-}CH{=}CH_2$

(iii) $CH_2{=}CH{-}CH{=}CH_2 > Me{-}CH{=}CH{-}Me > CH_2{=}CH_2$

14. Arrange the following with increasing order of acidity with explanation:

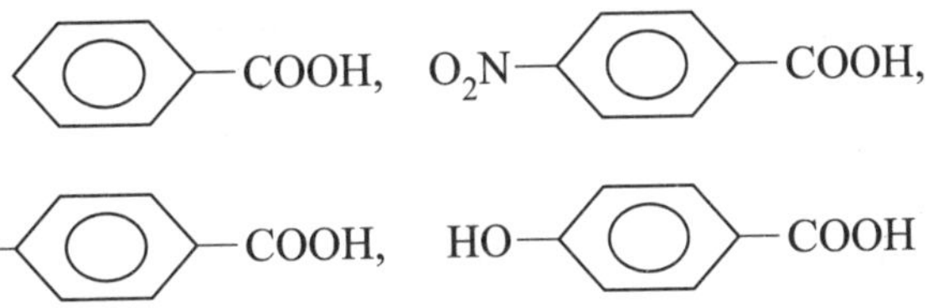

15. Arrange the following with decreasing order of reactivity towards *E*2 dehydrohalogenetion:

$CH_3CH_2CH_2CH_2CH_2Br$, $CH_3CH_2CH(Br)CH_2CH_3$, $CH_3CH_2{-}C(Br)(CH_3)CH_3$

16. Explain the following order of reactivity towards S_N1 substitution:

$$ClCH_2CH{=}CH_2 > CH_3CH_2CH_2Cl > CH_3CH{=}CHCl$$

17. Arrange the following with increasing order of basicity and explain:

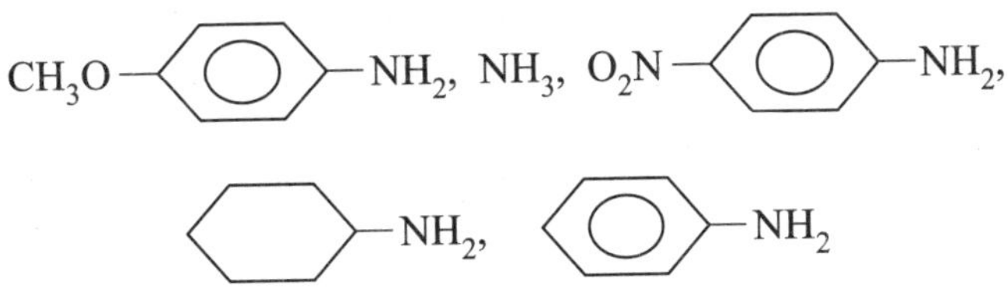

18. Offer an explanation for the fact that *p*-nitrophenol is much stronger acid than *m*-nitrophenol, whereas the acidity of *p*-nitrobenzoic acid is only slightly more than that of *m*-nitrobenzoic acid.

19. Classify the following reactions as nucleophilic or electrophilic. In each case identify the reagent as electrophile or nucleophile.

(i) $C_6H_6 + HNO_3 \rightarrow C_6H_5NO_2 + H_2O$

(ii) $(CH_3)_2C{=}CH_2 + HCl \rightarrow (CH_3)_3CCl$

(iii) $(CH_3)_3CI + H_2O \rightarrow (CH_3)_3COH$

(iv) $CH_3CHO + HCN \rightarrow CH_3CH(OH)CN$

20. Arrange the carbonium ions in order of increasing stability:

$$C_6H_5CH_2^+,\ (C_6H_5)_2CH^+,\ (CH_3)_2CH^+,\ (CH_3)_3C^+,\ CH_3CH_2^+,\ CH_3^+$$

21. Predict the order of reactivity of the following halides with (a) sodium iodide in acetone and (b) aqueous alcoholic silver nitrate.

CH_3Br, CH_3CH_2Br, 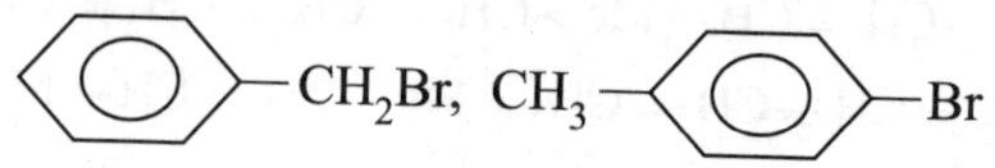

(A) (B) (C) (D)

22. Write the products of the following reactions with mechanism:

(i) 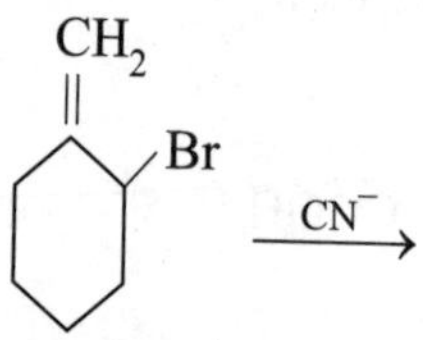

(ii) $CH_3CH_2CH{=}CH{-}CH{=}CH{-}CH_2Br \xrightarrow{SH^-}$

(iii) $CH_2{=}CH{-}C(Me)_2{-}Cl \xrightarrow{PhS^-}$

(iv) $CH_2{=}CH{-}CH{=}CH_2 \xrightarrow{Br_2}$

23. Which of the two alkyl chlorides will undergo faster hydrolysis in an S_N1 process? Suggest reasons.

$$(Me)_3C{-}C(C(Me)_3)_2{-}Cl \quad \text{and} \quad Me{-}C(Me)_2{-}Cl$$

24. (i) Arrange the following compounds in order of reactivity towards sodium iodide in acetone:

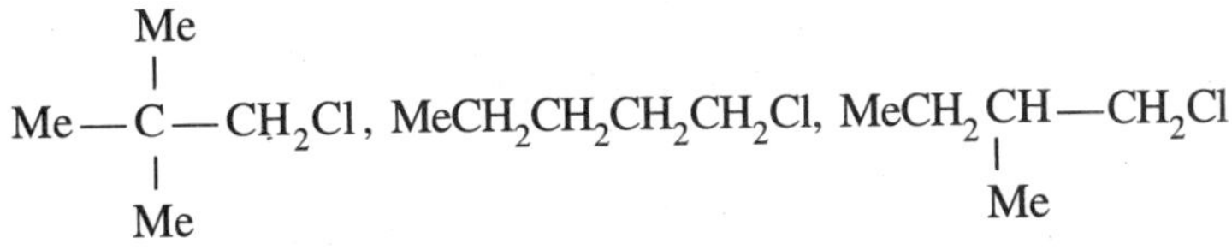

(A) (B) (C)

(ii) Arrange the following compounds in order of reactivity towards aqueous formic acid.

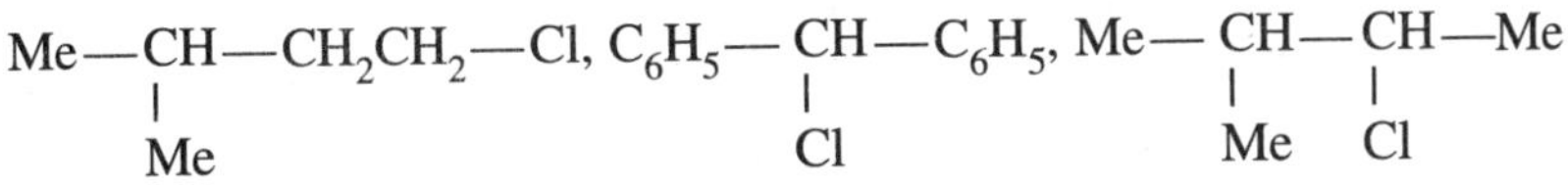

(A) (B) (C)

25. Consider radicals *A* and *B*:

$$CH_3CH_2CH_2\dot{C}H_2 \quad \text{and} \quad CH_3-\dot{C}H-CH_3$$

(A) (B)

Predict the coupling and disproportionate products of *A* and *B*.

MODEL QUESTIONS

Thermodynamics

Short Answer Type Questions (5 Marks)

1. Explain the terms: (i) diathermal wall and (ii) adiabatic wall.
2. State and explain 1st law of thermodynamics. Write the mathematical form of the 1st law of thermodynamics. How can you prove 1st law of thermodynamics from the law of conservation of energy.
3. (a) Discuss the physical significance of entropy.

   (b) Prove that: $(dS_r) > \left(\frac{dQ}{T}\right)_{ir}$
4. Show that work done in a reversible process is greater than that in an irreversible process.
5. $C_p - C_v = T\left(\frac{\partial P}{\partial T}\right)_V \left(\frac{\partial V}{\partial T}\right)_P$ (Symbols have usual significances)
6. Prove that for 1 mole of an ideal gas PV^γ = constant for adiabatic reversible process.
7. Prove that: Joule-Thomson process is an isenthalpic process.
8. State the thermodynamic condition necessary to occur a chemical reaction. Hence discuss the conditions of spontaneity of the following:

   (i) A chemical reaction for which ΔH and ΔS are positive. (ii) A chemical reaction for which ΔH and ΔS are negative. (iii) A chemical reaction for which ΔH is positive and ΔS is negative. (iv) A chemical reaction for which ΔH is negative and ΔS is positive.
9. An ideal gas is compressed adiabatically from pressure (P_1), volume (V_1) to pressure (P_2), volume (V_2). Show that:

   Work done $= W = \frac{P_2V_2 - P_1V_1}{\gamma - 1}$, where, $\gamma = \frac{C_P}{C_V}$
10. Prove that:

    $$dH = C_P dT - T\left(\frac{\partial V}{\partial T}\right)_P dP + VdP$$
11. Determine the value of ΔU and ΔH for the reversible isothermal evaporation of 90 g of water at 100°C. Assume that water vapour behaves as an ideal gas the latent heat of evaporation of water is 540 cal/g.
12. (a) What do you mean by internal energy?

    (b) Define enthalpy. How is it related with internal energy?
13. (a) State the relation between C_p and C_V for n moles of an ideal gas.

    (b) Define state function with an example. Is entropy is a state function?
14. Derive the expression of efficiency of Carnot cycle.

Long Answer Type Questions (15 Marks)

1. (a) Define internal energy. What is called state function? "Enthalpy is a state function but work is not a state function"—explain.

(b) Discuss the entropy change for the following processes:
(i) Melting of solid ice, (ii) water vapour is converted to liquid water, (iii) water is cooled from 30°C to 5°C.

2. (a) State 2nd law of thermodynamics. Deduce the expression for the efficiency of Carnot engine. State the characteristics of Carnot Heat Engine. Explain how efficiency of a Carnot engine can be improved.

(b) What is absolute temperature?

3. Show that:

(a) $C_p - C_v = nR$ for n moles of an ideal gas.

(b) $\left[\frac{\partial C_P}{\partial P}\right]_T = 0$ for an ideal gas.

4. (a) Define Joule-Thomson coefficient. What is its unit? State its significances.

(b) Show that: $\mu_J = -\frac{1}{C_P}\left[T\left(\frac{\partial V}{\partial T}\right)_P - V\right]$, where, symbols have usual significances.

(c) Show that temperature of an ideal gas remains unchanged after execution of Joule-Thomson expansion. What is called inversion temperature?

5. (a) Calculate the change in entropy when 100 gm of water is reversibly heated from 25°C to 50°C at 1 atm. pressure. Given: $C_P = 1$ cal/g – °K.

[**Ans.** 8.06 cal/°K]

(b) Calculate the work done during the synthesis of ammonia in which the volume contracts from 8.6 l to 4.3 l at a constant external pressure of 44 atm.

[**Ans.** 1.9×10^4 J]

(c) 1 l of an ideal gas A and 4 l of another gas B are mixed in a 4 l vessel at constant temperature (T) and pressure. Calculate the entropy change.

[**Ans.** 0.99 cal/mole]

6. (a) Deduce Kirchoff's equation. State its significance.

(b) State law of constant heat summation (Hess's law). Discuss one application of it.

7. Write short notes on the following:
(i) Clausius – Clapeyron equation
(ii) Gibb's-Duhem relations
(iii) Gibb's-Helmholtz equation (any of its standard form)
(iv) Gibb's free energy

8. (a) Prove that:

(i) $\left(\frac{\partial T}{\partial V}\right)_S = -\left(\frac{\partial P}{\partial S}\right)_V$, (ii) $\left(\frac{\partial P}{\partial T}\right)_V = -\left(\frac{\partial S}{\partial V}\right)_T$,

(iii) $\left(\frac{\partial T}{\partial P}\right)_S = -\left(\frac{\partial V}{\partial S}\right)_P$, (iv) $\left(\frac{\partial V}{\partial T}\right)_P = -\left(\frac{\partial S}{\partial P}\right)_T$

(b) Distinguish between intensive and extensive properties with examples.

9. (a) Prove that for 1 mole of an ideal gas, the adiabatic $P - V$ curve would be steeper than the corresponding isothermal $P - V$ curve.

(b) Show that an isothermal mixing of two ideal gases at constant pressure is a spontaneous process.

(c) Calculate the work done when one mole of an ideal gas at 0°C expands adiabatically and reversibly from 10 atm. to 1 atm. Given: $\gamma = 1.667$.

10. (a) Find the expression of $\left(\frac{\partial P}{\partial T}\right)_V$ for 1 mole of a Van-der-Waals gas.

(b) Water boils at 100°C at 1 atm. pressure but the boiling point drops to 90°C at 528 mm of Hg. Calculate latent heat of water. Given: $R = 8.314\ JK^{-1}$ mole^{-1}.

Hint: $\ln\frac{P_2}{P_1} = -\frac{L}{R}\left[\frac{1}{T_2} - \frac{1}{T_1}\right]$

(c) At N.T.P. 2.8 l of oxygen is mixed with 19.6 l of hydrogen. Calculate the increase in entropy.

Hint: $\Delta S^M = -nR(x_1 + x_2 \ln x_2)$

11. Prove that:

(i) $\left(\frac{\partial A}{\partial T}\right)_V = -S$ (ii) $\left(\frac{\partial A}{\partial V}\right)_T = -P$ (iii) $\left(\frac{\partial G}{\partial T}\right)_P = -S$ (iv) $\left(\frac{\partial G}{\partial P}\right)_T = V$

Chemical Kinetics

Short Answer Type Questions (5 Marks)

1. (a) Define activation energy. State its significance.

(b) Define specific reaction rate. State its unit for 2nd order reaction.

2. Explain how specific reaction rate changes with temperature.

3. The 1st order decomposition of N_2O_5 to give NO_2 and O_2 has a specific rate constant of $4.30 \times 10^{-5}\ s^{-1}$ at 300 K and $6.97 \times 10^{-2}\ s^{-1}$ at 500 K. Calculate the activation energy for this reaction. Given: R = 2 cal/(mole-°K).

4. The 1st order reaction: $2A \rightarrow 2B + C$ is 35% complete after 325 s. (a) Find k and k_A, where, $k_A = 2k$. (b) How long will it take for the reaction to be 70% complete and 90% complete?

Long Answer Type Questions (15 Marks)

1. (a) State the differences between Molecularity and Order.
   (b) What is called pseudounimolecular reaction? Give two examples.
   (c) Define half-life period. State its significances.

2. (a) Explain how reaction rate is influenced in presence of a catalyst.
   (b) Derive the rate expression of an irreversible 2nd order reaction where number of moles of both the reactants are same.
   (c) In a 1st order reaction, the initial concentration of the reactant is 0.1 mole/l. The reaction is found to be 20% complete in 40 min. Calculate: (i) the rate constant, (ii) the half-life period, (iii) the time required to complete 75% of the reaction.

3. (a) Show that in case of a 1st order reaction, the time required to complete 99.9% reaction is 10 times the time required to complete 50% reaction.
   (b) The half-life period of decomposition of a compound is 50 min. If the initial concentration is halved, the half-life period is reduced to 25 min. Find the order of the reaction.
   (c) For an irreversible reaction, the rate (r) is given by: $r = k[A]^2[B]$.
      (i) What will be the value of r when concentration of A is doubled?
      (ii) What will be the value of r when concentration of A is doubled and that of B is reduced to one-fourth?

4. (a) What is meant by activation energy of a chemical reaction? State the significance of it. How it varies with temperature?
   (b) The rate constant for the first order reaction is described by the equation:

   $$\ln k = 14.34 - \frac{1.25\times10^4}{T}$$

   What is the energy of activation for this reaction?
   What will be the rate constant at 500 K temperature?
   (c) The rate constant of a reaction is $1.5 \times 10^7\ s^{-1}$ at 500°C and $4.5 \times 10^7\ s^{-1}$ at 1000°C. Evaluate the Arrhenious parameters A and E_a.

5. (a) Define catalyst of a chemical reaction. State the characteristics of it.
   (b) Define homogeneous and heterogeneous catalysts with examples.
   (c) Show that rate of a homogeneous catalytic reaction depends on catalyst concentration.

Electrochemistry

Short Answer Type Questions (5 Marks)

1. State Hittorf's rule and hence explain transport number of an ion.
2. Establish the relationship: $\lambda_+ = t_+ . \Lambda_\infty$, where symbols have usual significances.
3. The standard potentials of Cu/Cu^{2+} and Cu/Cu^+ systems are 0.337V and 0.530V respectively. Find out the equilibrium constant for the reaction: $2Cu \rightleftharpoons Cu^{2+} + Cu$.

4. State Ostwald's dilution law. Discuss the factors influencing degree of dissociation of weak electrolytes.

5. Derive the expression of Nernst's equation for single electrode potential for M/M^{n+} electrode.

6. Explain standard cell with an example. State three applications of single electrode potential.

7. (a) What do you mean by 'Transport number'?
   (b) State and explain the "Kohlrausch's law of independent migration of ions".
   (c) What is the S.I. unit of specific conductance?

8. (a) The equivalent conductance of 0.05(N) KCl, NaCl and K_2SO_4 are 149.9, 126.5 and 153.3 $ohm^{-1}cm^2$ respectively. Calculate an approximate value for the equivalent conductance of Na_2SO_4 of the same concentration.

   [Ans. 106.5 $ohm^{-1}cm^2$]

   (b) State whether the statement is true or false:
   Transport number of ions are independent of ion conductance.

9. (a) Explain why H+ and OH^- ions have very large ion conductance in water.
   (b) Arrange the following according to increasing order of ionic conductance:
   K^+, Na^+, Cs^+, Li^+

10. (a) Define specific conductance. What do you mean by ionic mobility?
    (b) Discuss the advantages of the conductometric titration method over volumetric method.

11. (a) Give an example of a fuel cell. Write down the cell reaction of the fuel cell.
    (b) What are the advantages of fuel cells over ordinary batteries.

Long Answer Type Questions (15 Marks)

1. (a) Explain the following titration curves:
   (i) Titration of a weak acid (say acetic acid) by a strong base (say NaOH).
   (ii) Titration of NaCl by $AgNO_3$.
   (b) The value of equilibrium constant (K) of the reaction:

   $$H_2 + Cl_2\ 2H_2O \rightleftharpoons 2H_3O^+ + 2Cl^-$$

   is 10^{46} at 25°C. Calculate E^0 at 25°C for the cell: (Pt)H_2/HCl(aq.)/Cl_2(Pt)

2. (a) Explain how ionic mobility varies with (i) size, (ii) concentration and (iii) temperature.
   (b) State the advantages of conductometric titration over volumetric titration.
   (c) What is called molar conductance? How is it related with equivalent conductance?

3. (a) Distinguish between electrolytic and electrochemical cells.
   (b) Establish the relationship between entropy and e.m.f. of an electrolytic cell. What is called temperature coefficient of a cell?
   (c) What is meant by single electrode potential? Explain reference electrode with three examples.

4. (a) State Kohlrausch's law. Discuss three applications of it.

(b) What is glass electrode? Write down the expression of glass electrode potential and hence explain the principle of glass electrode to measure *pH* of a sample solution. State the advantages of using glass electrode.

(c) What is quinhydrone electrode? Derive the expression of quinhydrone electrode potential and hence explain the principle of quinhydrone electrode to measure *pH* of a sample solution. State the advantages of using quinhydrone electrode.

Solid State Chemistry

Short Answer Type Questions (5 Marks)

1. Explain point defects. State the differences between Schottky defect and Frenkel defect.
2. Discuss Schottky defect with a neat sketch.
3. Distinguish between intrinsic and extrinsic semiconductors.

Long Answer Type Questions (15 Marks)

1. (a) Explain the following with examples:
(i) Stoichiometric defect, (ii) Non-stoichiometric defect, (iii) Doping in semiconductor.
(b) Establish the law of mass action.
(c) What is called *band gap*? State its significance.

2. (a) Explain in detail how resistance of an intrinsic semiconductor changes with temperature.
(b) Discuss the effect of temperature on conductivity of an extrinsic semiconductor.
(c) What are the differences between *p*- type and *n*- type semiconductors?

3. (a) Explain why ZnO appears white in cold but yellow when hot.
(b) Determine the resistivity of an intrinsic Si specimen at 300 K. Intrinsic Si exhibits a carrier concentration of 1.48×10^{10} electrons/cm^3.
Given: $\mu_e = 1300$ cm^2/V – s and $\mu_h = 500$ cm^2/V – s
[**Ans.** $2.35 \times 10^5 \Omega$ – cm]
(c) *InP* has an energy gap of 1.29 eV. The electron and hole mobilities for the semi-conducting compound are 0.46 and 0.015 m^2/v – S respectively at 300 K. Calculate its conductivity.

Chemical Bonding

Short Answer Type Questions (5 Marks)

1. Explain why $AlCl_3$ is covalent in anhydrous form but ionic in solution.
2. Explain why p –nitrophenol has much higher boiling point than o –nitrophenol although both have same molecular weight.

3. Explain why Salicylic acid is 17 times stronger acid than benzoic acid.
4. Bond angle of NH_3 is 107° but the same for other hydrides, PH_3, AsH_3, SbH_3 and BiH_3 is 90°, although hybridization in all the cases is sp^3............explain.
5. State the postulates of VSEPR theory and hence explain the geometry of ClF_3 molecule.
6. Calculate the % ionic character of HCl and HF molecules using the following data:
μ_{HCl} = 1.03D, μ_{HF} = 2.487D, R_{HCl} = 1.27Å, R_{HF} = 1.15Å, e = 1.602×10^{-19}C.

Long Answer Type Questions (15 Marks)

1. Explain the following:
   (a) Electronegativity of halogens decreases in the following order: F > Cl > Br > l, but acid strength of halogen acids follows the reverse order: HF < HCl < HBr < Hl.
   (b) Ammonia (NH_3) is far stronger Lewis base than phosphine (PH_3) although nitrogen is more electronegative than phosphorus.
   (c) Hydrolysis of NCl_3 gives ammonia but hydrolysis of PCl_3 gives phosphoric acid.
2. (a) Define hybridisation. State some characteristics of it.
   (b) Predict the shape of the following compounds:
   BF_3, BrF_5, PCl_5, SF_6, XeF_4.
3. (a) State the postulates of VSEPR theory and hence explain the geometry of ClF_3 molecule.
   (b) Explain inter-molecular and intra-molecular coordinate bonds with examples.
4. (a) What is metallic bonding? Explain the following properties with the help of metallic bonding: (i) Metallic lustre, (ii) Ductility and malleability, (iii) Photoelectric effect.
   (b) Give reasons for the following:
   (i) Ionic solids are hard, brittle and high melting point.
   (ii) Ionic compounds are non-conductors in solid state but good conductors in molten state or in solution state.

Coordination Chemistry

Short Answer Type Questions (5 Marks)

1. Which of the following should possess higher conductivity in 1M solution and why?
   (i) $[Co(NH_3)_5SO_4]Cl$ and $[Co(NH_3)_4(SO_4)(Cl)]$.
   (ii) $[Cr(NH_3)_5NO_2]Br_2$ and $[Cr(NH_3)_3(NO_2)_3]$.
2. Explain the geometry and magnetic behaviour of $[FeF_6]^{3-}$ and $[Co(CN)_6]^{3-}$.
3. Explain chelates with two examples.
4. Distinguish between inner orbital complex and outer orbital complex.
5. Explain why $[NiCl_4]^{2-}$ is strongly paramagnetic but $[Ni(CN)_4]^{2-}$ is diamagnetic.

Long Answer Type Questions (15 Marks)

1. (a) State the postulates of Werner's theory. (b) State the limitations of Werner's theory.
   (c) Define ligands. Classify ligands with examples.
2. (a) Explain the following with examples:
   (i) Coordination number, (ii) Primary and secondary valencies.
   (b) Explain the following with examples:
   (i) Linkage isomerism, (ii) Coordination isomerism, (iii) Stereoisomerism,
   (iv) Ionisation isomerism.
3. (a) Write the chemical formulae of the following compounds:
   (i) Pent-aqua-fluoro-chromium (II) chloride, (ii) Potassium hexafluoro cobaltate (III),
   (iii) Tetraamino-chloro-cyano-sulphato cobalt (III).
   (b) Write the IUPAC name of the following:
   (i) $[CoCl_2(en)_2]^2$ (ii) $[Cr(NH_3)(F)(CN)_2(H_2O)_2]$, (iii) $K_2[Ni(CN)_4]$,
   (iv) $K[Fe(Cl)(en)(NO_2)_2(N_2O)]$, (v) $Na_2[Fe(CN)_5(NO)]$.
4. (a) Discuss the factors influencing the complex formation.
   (b) Both $[Ni(CO)_4]$ and $[Ni(CN)_4]^{2-}$ are diamagnetic but the former is tetrahedral and the latter is square planar—explain.

Fuel and Combustion

Short Answer Type Questions (5 Marks)

1. (a) Define high speed, medium speed and low speed diesel.
   (b) Why is ethylene-di-bromide used along with TEL in petrol engine?
2. (a) What is unleaded petrol?
   (b) Distinguish between water gas and carbureted water gas.
3. State some uses of gaseous fuel.
4. What are the information obtained by performing proximate analysis and ultimate analysis of a coal sample?
5. What is called cracking of a petrol? State the catalyst used and necessary conditions favourable for cracking process.
6. (a) What is called reforming of gasoline?
   (b) What is called synthetic gasoline?
7. (a) What are the constituents of producer gas? Which is 'blue gas'?
   (b) What are the uses of L.P.G.?

Long Answer Type Questions (15 Marks)

1. (a) Define calorific value of a fuel. Distinguish between GCV and NCV.

(b) What is called knocking of a petrol sample?
(c) Why is delay in combustion observed in diesel engine?
(d) Explain the following: (i) Octane No., (ii) Cetane No. State their significances.

2. (a) State the differences between coal and coke. Distinguish between HTC and LTC.
(b) What is called straight-run gasoline? State the demerits of straight-run gasoline.

Organic Chemistry

Short Answer Type Questions (5 Marks)

1. Explain the following order of reactivity towards S_N1 substitution:
$CH_2 = CH - CH_2Cl > CH_3CH_2CH_2Cl > CH_3CH = CHCl$
2. Discuss the factors influencing S_N1 and S_N2 reactions.
3. Discuss the conditions favourable for elimination reaction.
4. Explain $E1_{CB}$ reaction mechanism with an example.
5. Explain anti-Markownikoff's rule with an example.
6. (a) Distinguish between electrophiles and nuclophiles.
(b) Explain why only HBr but not HCl or HI responds addition reactions in presence of peroxide.

Long Answer Type Questions (15 Marks)

1. (a) Explain Inductive Effect and Electromeric Effect with examples.
(b) State some characteristics features of (i) Resonance and (ii) Hyperconjugative effect.

Polymer Chemistry

Short Answer Type Questions (5 Marks)

1. Define a polymer. What is polydispersity index? State its significance.
2. Differentiate the following:
(i) Thermoplasts and Thermosets.
(ii) Addition polymerisation and condenzation polymerization.
3. What is called copolymers? Hence define block copolymers with two examples.
4. (a) What is called fibre?
(b) Explain conducting polymers with three examples.
5. Define tacticity of polymers. Classify polymers with examples on the basis of tacticity.

6. Write the chemical structures of the repeat units of Nylon 6,6 and Nylon 6. Why are they so named?

7. Write down the chemical structure of NR. In what form is it available in nature? How does it differ from Gutta-Percha?

8. (a) What are the monomer units of NR and PVC? (1 + 1)
   (b) What is copolymer? Give two examples. (1 + 2)

Long Answer Type Questions (15 Marks)

1. (a) What is called free volume? State its significance.
   (b) Explain the term "glass transition temperature, T_g" of a polymer. State its significance.
   (c) Write down the chemical structures of repeat unit of the following rubbers:
   (i) NR (ii) SBR, (iii) Neoprene, (iv) NBR.

2. (a) What is called vulcanisation of rubber? What are the reagents used for this purpose? State why vulcanisation is essential to rubbers.
   (b) Write down the repeating units and uses of the following polymers: PE and PP.

3. (a) What is called Bakelite? State the chemical reactions, involved during manufacture of Bakelite.
   (b) Write down the name of the lightest polymer. State the differences between HDPE and LDPE.

4. (a) Write down the name of an inorganic polymer. Write down the chemical structure of the repeat unit.
   (b) Write down the chemical structure, properties and uses of silicone rubber.
   (c) What are called high temperature polymers? Give one example.

Spectroscopy

Short Answer Type Questions (5 Marks)

1. What do you mean by molecular spectroscopy? Write the names of different methods of spectroscopic studies of organic compounds.

2. Write down the mathematical form of Lambert-Beers law. State its significance.

Long Answer Type Questions (15 Marks)

1. (a) Discuss different types of molecular vibrations related to IR spectra.
   (b) State different factors having strong influence in shifting of IR spectra of different functional groups of a molecule.

2. Discuss the applications of the following
   (i) Ultraviolet Spectroscopy, (ii) Infrared Spectroscopy, (iii) Nuclear Magnetic Spectroscopy, (iv) Mass Spectroscopy.

MULTIPLE CHOICE QUESTIONS

Choose the correct answer/alternative from the following

Thermodynamics

1. Which one of the following is true for melting of ice at 0ºC
(a) $\Delta G < 0$, (b) $\Delta G > 0$, (c) $\Delta G = 0$, (d) $\Delta G \geq 0$.

2. In a reaction both ΔH and ΔS are positive. The reaction is spontaneous if
(a) $|\Delta H| < |T\Delta S|$, (b) $|\Delta H| > |T\Delta S|$,
(c) $|\Delta H| = |T\Delta S|$, (d) $|\Delta H| = |\Delta S$.

3. Change in internal energy is equal to heat change for
(a) Isochoric process, (b) Isothermal process,
(c) Isobaric process, (d) Isenthalpic process.

4. If enthalpy of a reactant is less than that of the product, the following is true
(a) The reaction is exothermic (b) The reaction is endothermic
(c) The reaction does not occur at all (d) Net free energy change is zero

5. Joule-Thomson expansion of a gas is an
(a) Isobaric process, (b) Isothermal process,
(c) Isentropic process, (d) Isenthalpic process.

6. In which of the following case does entropy decrease?
(a) Melting of solid ice, (b) Expansion of a gas,
(c) Dissolution of sugar in water, (d) Polymerization process.

7. Which one of the following is true?
(a) $\Delta H = \Delta U + \Delta nRT$, (b) $\Delta H = \Delta U + \frac{\Delta n}{RT}$,
(c) $\Delta H = \Delta U + (RT)^{\Delta n}$, (d) $\Delta H = \Delta U - (RT)^{\Delta n}$.

8. Work function is defined as
(a) $A = U + PV$, (b) $A = U + TS$,
(c) $A = U + PT$, (d) $A = U - TS$.

9. For an irreversible process entropy of the system
(a) decreases, (b) increases, (c) remains constant, (d) zero.

10. Which one of the following is an intensive property?
(a) ΔU, (b) ΔQ, (c) ΔV, (d) ΔP.

11. For an ideal gas undergoing free expansion
(a) $\Delta H = 0$ and $\Delta S > 0$, (b) $\Delta T = 0$ and $\Delta S = 0$,
(c) $\Delta T < 0$ and $\Delta S > 0$, (d) $\Delta T < 0$ and $\Delta S = 0$.

12. 2 moles of an ideal gas expands spontaneously into a vacuum. The work done is
(a) 2J, (b) 4J, (c) 0, (d) ∞.

13. According to Maxwell's relation

(a) $\left(\frac{\partial S}{\partial P}\right)_T = -\left(\frac{\partial V}{\partial T}\right)_P$, (b) $\left(\frac{\partial S}{\partial P}\right)_T = -\left(\frac{\partial S}{\partial T}\right)_V$,

(c) $\left(\frac{\partial S}{\partial P}\right)_T = +\left(\frac{\partial V}{\partial T}\right)_P$, (d) $\left(\frac{\partial S}{\partial P}\right)_T = +\left(\frac{\partial S}{\partial T}\right)_V$

14. For a real gas, subjected to $J - T$ expansion, inversion temperature is that temperature

(a) At which cooling effect is just observed
(b) At which heating effect is just observed
(c) Above which cooling effect is observed
(d) Above which heating effect is observed

15. We know: $dH = TdS + VdP$. Then which one of the following is true?

a) $\left(\frac{\partial S}{\partial V}\right)_P = \left(\frac{\partial P}{\partial T}\right)_S$, (b) $\left(\frac{\partial V}{\partial S}\right)_P = \left(\frac{\partial T}{\partial P}\right)_S$,

(c) $\left(\frac{\partial V}{\partial P}\right)_S = \left(\frac{\partial T}{\partial S}\right)_P$, (d) $\left(\frac{\partial S}{\partial P}\right)_V = \left(\frac{\partial T}{\partial V}\right)_S$.

16. For an ideal gas system if decrease in internal energy is same as the work done by the system, the process is said to be

(a) Isobaric, (b) Adiabatic, (c) Isothermal, (d) Isochoric.

Electrochemistry

1. An electrolyte is one

(a) which ionizes only when electric current is passed through it.
(b) which does not ionize at all.
(c) which dissociates into ions by dissolving it in a suitable solvent.
(d) which carries current through free electrons.

2. Strong electrolytes are those, which

(a) dissolves only at high dilution.
(b) completely dissociate into ions at all dilutions.
(c) dissolve slightly in water.
(d) does not conduct electricity in solution.

3. An electrochemical cell is used to convert

(a) electrical energy to chemical energy.
(b) chemical energy to electrical energy.
(c) chemical energy to mechanical energy.
(d) chemical energy to heat energy.

4. The unit of specific conductance is

(a) *Ohm – cm*, (b) *Ohm/cm*, (c) *mho*, (d) $mho - cm^{-1}$.

5. The conductance of weak electrolyte
   (a) decreases with increase in concentration.
   (b) increases with increase in concentration.
   (c) remains unchanged with increase in concentration.
   (d) initially decreases and then increases with increase in concentration.

6. The quantity of electricity needed to deposit 1 gm-eq. of an element is
   (a) 96,500 e.m.u. (b) 96,500 Amp,
   (c) 96,500 Coulombs, (d) 96,500 Faradays.

7. Standard hydrogen electrode has been assigned a potential of
   (a) 100 V, (b) 0 V, (c) 1 V, (d) 10 V.

8. Stronger the oxidizing agent
   (a) greater the oxidation potential, (b) faster the reaction rate,
   (c) greater the reduction potential, (d) slower the reaction rate.

9. Conductivity of an electrolyte is directly proportional to
   (a) current density, (b) volume of solution,
   (c) dilution, (d) number of ions.

10. The cell reaction is spontaneous if the cell potential is
    (a) zero, (b) positive, (c) negative, (d) infinity.

11. Equivalent conductance of a solution of weak electrolyte increases
    (a) with increase in concentration, (b) with increase in dilution,
    (c) with decrease in dilution, (d) with decrease in temperature.

12. According to Kohlrausch's law
    (a) Equivalent conductance of all strong electrolytes are same.
    (b) Equivalent conductance of an electrolytic solution is independent of concentration.
    (c) Specific conductance of an electrolytic solution is the sum of specific conductance of both positive and negative ions at infinite dilution.
    (d) Equivalent conductance of an electrolytic solution is the sum of equivalent conductance of both positive and negative ions at infinite dilution.

13. Transport number of an ion
    (a) increases with increase in valency of ion.
    (b) changes with change in type of salt.
    (c) increases with increase in temperature.
    (d) increases with increase in salt concentration.

14. Which one of the following is not true for conductometric titration?
    (a) Indicator is not required for this type of titration.
    (b) Conductometric titration is far more accurate than volumetric titration.
    (c) Precipitation type of titration can be performed by this technique.
    (d) Weak acid-weak base type of titration cannot be performed by this technique.

15. In highly concentrated NaCl solution conductivity decreases because
(a) mobility of ions decreases
(b) viscosity decreases
(c) number of ions decreases
(d) transport number of ions decreases

Chemical Kinetics and Catalysis

1. The unit of specific reaction rate of a zero order reaction is
(a) *mole/lit – s*, (b) *lit/mole – s*, (c) s^{-1}, (d) *mole/s*.

2. For a chemical reaction A → B, it is found that the reaction rate is doubled up when concentration of *A* is increased four times. So, order of the reaction is
(a) 2, (b) 1, (c) 1/2, (d) 0.

3. According to collision theory of reaction rate
(a) Each collision is a successful collision, leading to chemical reaction,
(b) Rate of reaction is proportional to velocity of molecules,
(c) All reactions, occurring in gaseous phase, are zero order reactions,
(d) Rate of reaction is proportional to number of collisions per second.

4. The minimum energy, required to produce successful collisions, is known as
(a) free energy,
(b) kinetic energy,
(c) activation energy,
(d) internal energy.

5. For a chemical reaction A → B, it is found that the reaction rate increases by a factor 100 when the concentration of *A* is increased by 10 fold. The order of the reaction is
(a) 10, (b) 1, (c) 4, (d) 2.

6. The temperature coefficient of most of the reactions lies between
(a) 1 and 3, (b) 2 and 3, (c) 1 and 4, (d) 2 and 4.

7. In any equilibrium chemical reaction which one of the following will happen in presence of a positive catalyst?
(a) The value of equilibrium constant increases,
(b) Forward reaction rate increases,
(c) Enthalpy of the products decreases,
(d) Activation energy decreases.

8. Reaction rate constant of any chemical reaction
(a) increases with increase in temperature,
(b) decreases with increase in temperature,
(c) is independent of catalyst concentration,
(d) is independent of temperature.

9. Decomposition of N_2O_2 is retarded in presence of
(a) NaOH, (b) N_3PO_4, (c) HNO_3, (d) Na_2SO_4.

10. Reaction rate constant of acid catalysed hydrolysis of ester
(a) increases with increase in water content,
(b) increases with increases in acid concentration,
(c) decreases with increase in acid concentration,
(d) is independent of acid concentration.

Solid State Chemistry

1. Which of the following defects arises due to misplace of ions in a crystal lattice?
(a) Schottky defect,
(b) Frenkel defect,
(c) Metal excess defect,
(d) Non-stoichiometric defect.

2. Which one of the following is true for Schottky defect?
(a) Density increases,
(b) Density decreases,
(c) Density remain unchanged,
(d) Density exponentially increases.

3. Which one of the following is not correct?
(a) Band gap of a conductor is less than that of a semiconductor.
(b) Conductivity of a semiconductor increases linearly with increase in temperature.
(c) Conductivity of a conductor decreases with increase in temperature.
(d) At 0ºC an intrinsic semiconductor behaves as an insulator.

4. Colour centres are found in some crystals due to
(a) Electron-trapping in cation vacancies.
(b) Electron-trapping in anion vacancies.
(c) Presence of interstitial cation.
(d) Missing of pair of ions.

5. *In*-doped Si is a
(a) *p*-type semiconductor,
(b) *n*-type semiconductor,
(c) *pn*-type semiconductor,
(d) Intrinsic semiconductor.

6. Semiconductor diode with a *pn*-junction generally acts as
(a) a conductor, (b) a rectifier, (c) an amplifier, (d) a insulator.

7. Frenkel defect is found in
(a) AgBr, (b) NaBr, (c) $CuBr_2$, (d) TlBr.

8. Which one of the following is an example of insulator?
(a) Diamond, (b) Graphite, (c) Silicon, (d) Lead.

9. The unit of mobility is
(a) m^2/V, (b) $m/(V - s)$, (c) $m/(V - s^2)$, (d) $m^2/(V - s)$.

10. PRV rating of a *Si*-diode is
(a) higher than that of *Ge*-diode,
(b) lower than that of *Ge*-diode,
(c) same as that of *Ge*-diode,
(d) slightly lower than that of *Ge*-diode.

Chemical Bonding

1. Van-der-Waals type of bond is formed
   (a) by sharing of electrons,
   (b) by transferring of electrons from one atom to the other,
   (c) by sharing of electrons by one atom only,
   (d) weak electrostatic force of attraction among fluctuating dipoles.

2. Which of the following does not follow octet rule?
   (a) N_2O, (b) PH_3, (c) PCl_5, (d) CH_4.

3. Diffusion of electron cloud is expected to be maximum for
   (a) Small cation and large anion, (b) small cation and small anion,
   (c) large cation and large anion, (d) large cation and small anion.

4. Which one of the following is an ionic compound?
   (a) $AlCl_3$, (b) AlF_3, (c) $AlBr_3$, (d) All of these.

5. Lattice energy of an ionic crystal
   (a) decreases with decrease in inter-ionic distance,
   (b) decreases with increase in ionic charge,
   (c) increases with decrease in ionic charge,
   (d) increases with decrease in inter-ionic distance.

6. % *s*-character in *sp* hybrid orbital is
   (a) 50%, (b) 25%, (c) 75%, (d) 100%.

7. Shape of water molecule is
   (a) trigonal bi-pyramid, (b) tetrahedral,
   (c) square planar, (d) triangular.

8. The dipole moment of NH_3
   (a) is less than NCl_3, (b) is greater than NCl_3,
   (c) is same as that of NCl_3, (d) cannot be compared with that of NCl_3.

9. Hydrogen bond is expected in which of the following?
   (a) Diethyl ether (b) Ethyl chloride (c) Ethanol (d) Triethyl amine

Coordination Chemistry

1. The shape of $[Cu(NH_3)_4]^{2+}$ is square planar, then the hybridization of Cu^{2+} is
   (a) sp^3. (b) sp^3d. (c) sp^3d^2. (d) dsp^2.

2. Which of the following shows ionization isomerism?
   (a) $[Pt(NH_3)_2Cl_2]$ and $[Pt(NH_3)_4][PtCl_4]$
   (b) $[Co(NH_3)_5Br]SO_4$ and $[Co(NH_3)_5SO_4]Br$
   (c) $[Co(NH_3)_6][Cr(CN)_6]$ and $[Cr(NH_3)_6][Co(CN)_6]$
   (d) $[Co(NH_3)_5(ONO)]^{2+}$ and $[Co(NH_3)_2]^{2+}$

3. Which one of the following should have maximum magnetic moment?
(a) $[Fe(CN)_6]^{4-}$, (b) $[Fe(CN)_6]^{3-}$, (c) $[Co(CN)_6]^{3-}$, (d) $[FeF_6]^{3-}$

4. IUPAC name of $K_4Fe(CN)_6$ is
(a) Prussian blue (b) Potassium ferrocyanide
(c) Potassium hexacyano ferrate (II) (d) Potassium hexacyano iron (II)

5. Valency of Cr in $[Cr(H_2O)_4Cl_2]^+$ is
(a) 5 (b) 7 (c) 3 (d) 2

6. Which one of the following has square planar structure?
(a) $[Ni(CN)_4]^{2-}$, (b) $[NiCl_4]^{2+}$, (c) $Ni(CO)_4$, (d) $[Ni(H_2O)_6]^{2+}$

7. An example of an hexadendate ligand is
(a) ethylene diamine (b) oxalato
(c) ethylene diamine tetraacetic acid (d) ammonia

8. Which of the following is an example of chelating agent?
(a) Ethylene diamine, (b) Oxalato,
(c) Acetyl acetate, (d) All of these.

9. EAN value of Fe in $[Fe(CN)_6]^{3-}$ is
(a) 26 (b) 23, (c) 32, (d) 35.

10. Which one of the following shows linkage isomerism?
(a) Cyano, (b) Sulphato, (c) Oxalato, (d) Thiocyanato.

11. Which one of the following complexes shows geometrical isomerism?
(a) $[Ma_6]$, (b) $[Ma_5b]$, (c) $[Ma_4b_2]$, (d) $[Ma_3a_3]$.

Fuel and Combustion

1. In petrol TEL is used to
(a) ignite the petrol, (b) improve knocking characteristics
(c) improve anti-knocking characteristics, (d) catalyse the oxidation process.

2. Which of the following fuels has the best ignition quality
(a) Alkenes, (b) Aromatics, (c) Naphthalenes, (d) *n*-Alkanes.

3. Which one of the following has the highest calorific value?
(a) Producer gas, (b) LPG, (c) Blue gas, (d) Natural gas.

4. Gasoline is that fraction of petroleum, having boiling point range
(a) 40ºC – 120ºC, (b) 10ºC – 30ºC,
(c) 120ºC – 180ºC, (d) 180ºC – 250ºC.

5. The calorific value of natural gas is about
(a) 4,000 kcal/mole, (b) 14,000 kcal/mole,
(c) 3,000 kcal/mole, (d) 27,000 kcal/mole.

6. The carbon content in anthracite coal is
(a) 60%, (b) 78%, (c) 83%, (d) 93%.

Organic Chemistry

1. According to Saytzeff's rule the major product of an elimination reaction is
   (a) the least substituted alkene, (b) the most substituted alkane,
   (c) the least substituted alkane, (d) the most substituted alkene.

2. Which one of the following is true for S_N1 reaction?
   (a) Two species are involved in the rate determining step.
   (b) Carbonium ion is formed in the transition state.
   (c) Steric hindrance occurs in the rate determining step.
   (d) Carbanion is formed in the transition state.

3. Which one of the following is not true for S_N2 reaction?
   (a) Two species are involved in the rate determining step.
   (b) Primary alkyl halides usually undergo S_N2 reaction.
   (c) Racimised product is obtained when a dextro-rotatory compound undergoes S_N2 reaction.
   (d) Steric hindrance plays an important role in S_N2 reaction.

4. Which one of the following is not true for resonance?
   (a) Energy content of all resonating structures is widely different from one another.
   (b) Resonating structure, containing maximum number of covalent bonds, is the most stable structure.
   (c) Identical resonating structures have equal contributions in the actual molecule.
   (d) Resonating structures of a molecule cannot be separated.

Polymer Chemistry

1. Nylon is a
   (a) metal, (b) rubber,
   (c) semi-synthetic polymer, (d) fibre.

2. Natural rubber is polymer of
   (a) ethylene, (b) isobutylene, (c) isoprene, (d) propylene.

3. Bakelite is an example of
   (a) metal, (b) thermoplastic, (c) rubber, (d) thermoset.

4. Which one of the following is an example of addition polymer?
   (a) Nylon 6,6, (b) Bakelite,
   (c) Polyvinyl chloride, (d) Polyethylene terephthalate.

5. For a given polymer sample higher the polydispersity index,
   (a) lower the molecular weight,
   (b) narrower the molecular weight distribution,
   (c) higher the molecular weight,
   (d) broader the molecular weight distribution.

6. A polymer is said to be plastic when its T_g is
   (a) much higher than room temperature,
   (b) same as room temperature,
   (c) 0ºC,
   (d) much lower than room temperature.

7. Which one of the following is not a plastic?
   (a) Polyethylene, (b) Polypropylene,
   (c) Polyvinyl chloride, (d) Polyisoprene.

8. Which one of he following is a conducting polymer?
   (a) Polystyrene, (b) Polyethylene,
   (c) Poly(*p*-phenylene), (d) PVC.

9. Which one of the following undergoes vulcanization process?
   (a) HDPE, (b) PVC, (c) Nylon 6,6, (d) NR.

10. $\bar{M}_n$ and $\bar{M}_w$ are the number average and weight average molecular weights of a polymer. Then which one of the following is true?
    (a) $\bar{M}_w < \bar{M}_n$, (b) $\bar{M}_w = \bar{M}_n$,
    (c) $\bar{M}_w > \bar{M}_n$, (d) cannot be correlated.

Spectroscopy

1. Bathochromic shift is related with
   (a) Low intensity → High intensity, (b) High intensity → Low intensity,
   (c) Low λ_{max} → High λ_{max}, (d) High λ_{max} → Low λ_{max}.

2. Which one of the following is the highest energy transition in UV-spectrum:
   (a) $\pi \rightarrow \pi^*$, (b) $n \rightarrow \pi^*$, (c) $n \rightarrow \sigma^*$, (d) $\sigma \rightarrow \sigma^*$.

3. Fingerprint region is:
   (a) $1400 cm^{-1} - 1700 cm^{-1}$, (b) $1400 cm^{-1} - 650 cm^{-1}$,
   (c) $3600 cm^{-1} - 2300 cm^{-1}$, (d) $1600 cm^{-1} - 2200 cm^{-1}$.

4. The intra- and inter-molecular hydrogen bonding can be distinguished by:
   (a) UV spectroscopy, (b) IR spectroscopy,
   (c) ^{1}H-NMR spectroscopy, (d) Both IR and ^{1}H-NMR spectroscopy.

5. Essential condition for a molecule to be IR active is that the
   (a) molecule should be polar,
   (b) molecule should have permanent dipole moment,
   (c) molecule should have oscillating dipole moment,
   (d) molecule should be non-polar.

6. In NMR spectrum of toluene, signal for methyl group appears as
   (a) Singlet, (b) Doublet, (c) Triplet, (d) Quartet.

Miscellaneous

1. Example of an electrophile is
(a) $AlCl_3$, (b) NH_3, (c) CH_3OH, (d) CN^-

2. Example of nucleophile is
(a) NH_3, (b) H_2O, (c) C_2H_5OH, (d) All of these.

3. Electromeric effect occurs due to
(a) permanent partial charge separation, (b) delocalization of σ-electrons,
(c) distortion of electron cloud, (d) shifting of π-electrons.

4. Which one of the following responds anti-Markownikoff's rule?
(a) HBr, (b) HI, (c) HCl, (d) H_2O

5. α-elimination is that elimination reaction where, two atoms are
(a) eliminated simultaneously,
(b) eliminated from two consecutive carbon atoms,
(c) eliminated from the same carbon atom,
(d) eliminated to form a bridged cation.

6. Which one of the following favours the S_N2 reaction mechanism?
(a) Combination of weak nucleophile and weak leaving group.
(b) Presence of protic solvent.
(c) Combination of weak nucleophile and strong leaving group.
(d) Combination of strong nucleophile and weak leaving group.

7. Octane number of aviation gasoline is
(a) 40, (b) 100, (c) 80, (d) 75.

8. The compound with cetane number of zero is
(a) Benzene, (b) Naphthalene,
(c) 2-Methyl naphthalene, (d) Anthracene.

9. The compound with octane number zero is
(a) *n*-pentane, (b) *n*-hexane, (c) *n*-heptane, (d) *n*-octane.

10. The range of hydrocarbon fraction in kerosene oil is
(a) $C_{20} - C_{28}$, (b) $C_{10} - C_{16}$, (c) $C_5 - C_9$, (d) $C_{17} - C_{30}$.

11. One British Thermal Unit is equal to
(a) 252 J, (b) 252 kcal, (c) 252 kJ, (d) 252 cal.

12. Volatile matter content of a coal sample is determined at a temperature of
(a) 650°C, (b) 750°C, (c) 850°C, (d) 950°C.

13. The temperature of the system decreases in an
(a) adiabatic expansion, (b) isothermal expansion,
(c) isothermal compression, (d) adiabatic compression.

14. Variation of heat of reaction with temperature is known as
(a) Vant Hoff isotherm,
(b) Vant Hoff isochore,
(c) Kirchoffs equation,
(d) Clasius-Clapeyron equation.

15. $H_2 + Cl_2 \rightarrow 2HCl$. $\Delta H = 44$ *kcals* So heat of formation of 1 mole of HCl is
(a) –44 *kcal*, (b) 44 *kcal*, (c) – 22 *kcal*, (d) 22 *kcal*.

16. Which one of the following is true for endothermic reaction?
(a) $\Delta G > 0$,
(b) $\Delta H < 0$,
(c) $\Delta G < 0$,
(d) $\Delta H > 0$ and $\Delta S > 0$.

NUMERICAL PROBLEMS

1. Calculate activation energy of a reaction whose rate constant is doubled when T is increased from 300 K to 310 K.

[**Ans:** 12.9 *Kcal*]

[*Hint:* $\log \frac{K_1}{K_2} = -\frac{E}{2.303R}\left[\frac{1}{T_1} - \frac{1}{T_2}\right]$]

2. At NTP 3 dm^3 of oxygen was mixed with 2 dm^2 of hydrogen. Calculate the increase in entropy.

[**Ans:** 0.81 *cal/K*]

[*Hint:* $\Delta S^M = -nR(x_1 \ln x_1 + x_2 \ln x_2)$]

3. The half-life period of a 1st order reaction is 15 minutes. Calculate the rate constant and the time taken to complete 80% of the reaction.

[**Ans:** 34.84 min.]

[*Hint:* $k = \frac{1}{t}\ln\frac{a}{(a-x)}$ and $t_{\frac{1}{2}} = \frac{0.693}{k}$]

4. A gas expands against a variable external pressure given by: $P = 10/V$ atm., where V is the volume at each stage of expansion. In expanding from 10 to 100 litres, the gas undergoes a change in internal energy ΔU = 100 *cal*. How much heat has been absorbed?

[*Hint:* $\Delta W = \int_{10}^{100} PdV$, $\Delta Q = \Delta W + \Delta U$]

5. Calculate the mass of air needed for complete combustion of 5 kg of coal containing 80% Carbon, 15% Hydrogen and the rest oxygen.

6. Calculate the amount of work done when 5 moles of an ideal gas expands isothermally and reversibly at 27°C from 5 litres to 50 litres.

[**Ans:** 28.7 *KJ*]

[*Hint:* Work done = nRT $\ln\frac{V_2}{V_1}$]

7. A 1st order reaction is 20% complete in 10 minutes. Calculate the time taken by the reaction to go to 80% completion.

[**Ans:** 72.2 min]

[*Hint:* $k = \frac{1}{t}\ln\frac{a}{(a-x)}$]

8. Calculate the e.m.f. of the following cell:

Ni|Ni^{2+} (1M)|Pb^{2+} (1M)|Pb at 25°C
Given: Standard potentials of Ni and Pb are –0.24*V* and –0.13*V* respectively at 25°C.

[Ans: 0.114 *V*]

[*Hint:* $E^0_{ox} = E^0_L - E^0_R$]

9. The standard reduction potential for the electrode $Sn^{4+}|Sn^{2+}$ is 0.16*V* and that of $Fe^{3+}|Fe^{2+}$ electrode is 0.77*V*. Write the cell reaction and calculate the value of the equilibrium constant of the reaction at 25°C.

[Ans: $K_a = 2.51 \times 10^{20}$]

[*Hint:* $\Delta G^o = - -nE^0_{ox} = -RT \ln K_a$]

10. Show that for an ideal gas undergoing reversible adiabatic expansion or compression, PV^γ = constant. A diatomic ideal gas (γ = 1.4), initially at 600 K and 10 atm undergoes reversible adiabatic expansion till the final pressure becomes 2 atm. Find out its final volume.

[Ans: 15.5 litres]

[*Hint:* $P_1V_1^\gamma = P_2V_2^\gamma$]

11. Rate constants of a reaction at 300 K and 310 K are $4.5 \times 10^{-5}\ s^{-1}$ and $9 \times 10^{-5}\ s^{-1}$ respectively. Evaluate the activation energy and the pre-exponential factor (frequency factor) of the reaction. What is the order of the reaction?

[Ans: $A = 90 \times 10^4\ s^{-1}$, Order = 1]

[*Hint:* $\log\frac{K_1}{K_2} = -\frac{E}{2.303R}\left[\frac{1}{T_1} - \frac{1}{T_2}\right]$]

BIBLIOGRAPHY

1. Y. A. Gerasimov (Ed.), *Physical Chemistry*, English translation by Mir Publishers (1974).
2. S. M. Walas, *Reaction Kinetics for Chemical Engineers*, New York (1959).
3. S. Glasstone, K. J. Laidler and H. Eyring, *The Theory of Rate Processes*, New York (1941).
4. S. W. Benson, *The Formulation of Chemical Kinetics*, New York (1960).
5. G. W. Castellan, *Physical Chemistry*, Reading Mass. (1964).
6. J. O'M Bockris (Ed.), *Modern Aspects of Electrochemistry*, New York (1966).
7. V. A. Kireev, *Physical Chemistry*, Moscow (1972).
8. S. Glasstone and D. Lewis, *Elements of Physical Chemistry*, The Macmillan Press Limited (1960).
9. Darrow (Concept of entropy), Amer. J. Phys., 12, 183 (1944).
10. A. K. Galwey, *Chemistry of Solids*, Science Paperbacks and Chapman and Hall Ltd., London (1967).
11. N. N. Greenwood, *Ionic Crystals*, Lattice Defects and Non-Stoichiometry, Butterworths (1968).
12. Ira N. Levine, *Physical Chemistry*, Tata McGraw-Hill Publishing Company Limited, New Delhi (1995).
13. J. H. Espenson, *Chemical Kinetics and Reaction Mechanisms*, McGraw-Hill (1981).
14. G. S. Van Wylen and R. E. Sonntag, *Fundamentals of Classical Thermodynamics* (3rd edition), Wiley (1985).
15. P. K. Nag, *Engineering Thermodynamics* (2nd edition), Tata McGraw-Hill Publishing Company Limited (2004).
16. R. E. Sonntag, C. Borgnakke and G. J. Van Wylen, *Fundamentals of Thermodynamics* (5th edition), John Wiley & Sons Inc. (2000).
17. P. C. Rakshit, *Physical Chemistry*, Sarat Book Distributors (2001).

18. W. C. Reynolds and H. C. Perkins, *Engineering Thermodynamics*, McGraw-Hill (1977).
19. I. L. Finar, Chemistry (Volume 2), *Stereochemistry and the Chemistry of Natural Products* (5th Edition), The English Language Book Society and Longman Group Limited (1975).
20. E. L. Eliel, *Stereochemistry of Carbon Compounds*, McGraw-Hill Book Co. (1962).
21. Newman (Ed.), *Steric Effects in Organic Chemistry*, Wiley (1956).
22. De La Mare and Ridd, *Aromatic Substitution: Nitration and Halogenetion*, Butterworths (1959).
23. I. L. Finar, Organic Chemistry Volume 1: *The Fundamental Principles* (6th edition), The English Language Book Society and Longman Group Limited (1980).
24. Wynberg, (The Reimer-Tiemann Reaction), Chem. Rev., 60, 169 (1960).
25. Shine, Aromatic Rearrangements, Elsevier (1967).
26. Wassermann, Diels-Alder Reactions, Elsevier (1965).
27. S. M. Mukherjee and S. P. Singh, *Reaction Mechanism in Organic Chemistry*, Macmillan India Limited (1986).
28. J. March, *Advanced Organic Chemistry* (2nd edition), McGraw-Hill Book Co., New York (1977).
29. O.T. Benfey, *Introduction to Organic Reaction Mechanism*, Mcgraw-Hill Book Co., New York (1970).
30. R. A. Jackson, *Mechanism: An Introduction to the Study of Organic Reactions*, Clarendon Press, Oxford (1972).
31. R. Stewart, *The Investigation of Organic Reactions*, Prentice Hall Inc., Englewood Cliffs, N. J. (1966).
32. C. K. Ingold, *Structure and Mechanism in Organic Chemistry* (2nd edition), Bell, London (1969).
33. F. A. Carey and R. J. Sundberg, *Advanced Organic Chemistry*, Plenum Press, New York (1977).
34. D. Benthorpy, *Elimination Reactions*, Elsevier Publishing Co., New York (1963).
35. P. de Mayo (Ed.), *Molecular Rearrangements* (Volumes 1 and 2), Interscience Publishers Inc., New York (1963).
36. R. T. Morrison and R. N. Boyd, *Organic Chemistry* (6th edition), Prentice Hall of India Private Limited, New Delhi (2001).
37. Peter Sykes, *A Guidebook to Mechanism in Organic Chemistry* (6th edition), Pearson Education Limited (1986).
38. R. W. Alder, R. Baker and J. M. Brown, *Mechanism in Organic Chemistry*, Wiley-Interscience (1976).

39. C. K. Ingold, *Structure and Mechanism in Organic Chemistry* (2nd edition), Cornell University, Ithaca (1969).

40. Fred W. Billmeyer, Jr., *Textbook of Polymer Science* (3rd edition), John Wiley & Sons., New York (1984).

41. P. J. Flory, *Principles of Polymer Chemistry*, Cornell University Press, Ithaca, New York (1953).

42. Bibla, *Interpretation of NMR Spectra*, Plenum Press (1965).

43. Jackman and Sternhell, *Applications of Nuclear Magnetic Spectroscopy in Organic Chemistry* (2nd edition), Pergamon Press (1969).

44. Cross and Jones, *Introduction to Practical Infrared Spectroscopy* (3rd edition), Butterworths, (1969).

45. G. Socrates, *Infrared Characteristics Group Frequencies*, Wiley Interscience Publication, John Wiley & Sons (1980).

46. John R. Dyer, *Applications of Absorption Spectroscopy of Organic Compounds*, Prentice Hall of India Private Limited (1987).

47. H. Gunther, NMR Spectroscopy, Wiley (1980).

48. B. P. Straughan and S. Walker (eds.), *Spectroscopy*, Volumes 1–3 (2nd edition), Chapman and Hall (1976).

49. F. A. Cotton and G. Wilkinson, *Advanced Inorganic Chemistry* (3rd edition), Wiley (1972).

50. J. D. Lee, *Concise Inorganic Chemistry* (3rd edition), English Language Book Society and Nostrand Reinhold Company Limited, London (1979).

51. M. Karplus and R. N. Porter, *Atoms and Molecules*, Benjamin (1964).

52. J. R. Partington, *General and Inorganic Chemistry* (4th edition), Macmillan (1966).

53. H. Krebs, *Fundamentals of Inorganic Crystal Chemistry*, Mcgraw-Hill (1968).

54. H. B. Gray, *Electrons and Chemical Bonding*, Benjamin (1964).

55. J. E. Ferguson, *Stereochemistry and Bonding in Inorganic Chemistry*, Prentice Hall (1974).

56. L. Pauling, *The Nature of the Chemical Bond* (3rd edition), Oxford University Press (1961).

57. T. Moeller, *Inorganic Chemistry*, Wiley (1952).

58. R. E. Reed-Hill and R. Abbaschian, *Physical Metallurgy Principles* (3rd edition), Thomson Asia Pvt. Ltd., Singapore (2003).

59. Albert G. Guy, *Elements of Physical Metallurgy* (2nd edition), Addison-Wesley Publishing Company Inc. (USA) (1962).

60. M. Hansen and K. Anderko, *Constitution of Binary Alloys* (2nd edition), New York (1958).

61. J. E. Ricci, The Phase Rule and Heterogeneous Equilibrium, Dover (1966).

The Periodic Table

	Gr-2	Gr-3	Gr-4	Gr-5	Gr-6	Gr-7	Gr-8	Gr-9
1 H 1.008 $1s^1$								
3 Li 6.941 $2s^1$	4 Be 9.012 $2s^2$							
11 Na 22.99 $3s^1$	12 Mg 24.31 $3s^2$							
19 K 39.10 $4s^1$	20 Ca 40.08 $4s^2$	21 Sc 44.96 $3d^1 4s^2$	22 Ti 47.90 $3d^2 4s^2$	23 V 50.94 $3d^3 4s^2$	24 Cr. 52 $3d^5 4s^1$	25 Mn 54.94 $3d^5 4s^2$	26 Fe 55.85 $3d^6 4s^2$	27 Co 58.93 $3d^7 4s^2$
37 Rb 85.47 $5s^1$	38 Sr 87.62 $5s^2$	39 Y 88.91 $4d^1 5s^2$	40 Zr 91.22 $4d^2 5s^2$	41 Nb 92.91 $4d^4 5s^1$	42 Mo 95.94 $4d^5 5s^1$	43 Tc 98 $4d^6 5s^1$	44 Ru 101.1 $4d^7 5s^1$	45 Rh 102.9 $4d^8 5s^1$
55 Cs 132.9 $6s^1$	56 Ba 137.3 $6s^2$	57 La 138.9 $4f^0 5d^1 6s^2$	72 Hf 178.5 $4f^{14} 5d^2 6s^2$	73 Ta 180.95 $4f^{14} 5d^3 6s^2$	74 W 183.9 $4f^{14} 5d^4 6s^2$	75 Re 186.2 $4f^{14} 5d^5 6s^2$	76 Os 190.2 $4f^{14} 5d^6 6s^2$	77 Ir 192.2 $4f^{14} 5d^9 6s^0$
87 Fr 223 $7s^1$	88 Ra 226 $7s^2$	89 Ac 227 $6d^1 7s^2$	104 Rf 261 $5f^{14} 6d^2 7s^2$	105 Db 262 $5f^{14} 6d^3 7s^2$	106 Sg 266 $5f^{14} 6d^5 7s^1$	107 Bh 264 $5f^{14} 6d^5 7s^2$	108 Hs 277 $5f^{14} 6d^6 7s^2$	109 Mt 268 $5f^{14} 6d^7 7s^2$

1st transition series, known as Lanthanides ($4f^1$ to $4f^{14}$)

58	59	60	61	62	63	64
Ce 140.1 $4f^2 6s^2$	Pr 140.9 $4f^3 6s^2$	Nd 144.3 $4f^4 6s^2$	Pm 147 $4f^5 6s^2$	Sm 150.3 $4f^6 6s^2$	Eu 152.0 $4f^7 6s^2$	Gd 157.3 $4f^7 5d^1 6s^2$

2nd transition series, known as Actinides ($5f^1$ to $5f^{14}$)

90	91	92	93	94	95	96
Th 232 $6d^2 7s^2$	Pa 231 $5f^2 6d^1 7s^2$	U 238 $5f^3 6d^1 7s^2$	Np 237 $5f^5 7s^2$	Pu 244 $5f^6 7s^2$	Am 243 $5f^7 7s^2$	Cm 247 $5f^7 6d^1 7s^2$

Gr-10	Gr-11	Gr-12	Gr-13	Gr-14	Gr-15	Gr-16	Gr-17	
								2 He 4 $1s^2$
			5 B 10.81 $2s^2 2p^1$	6 C 12.01 $2s^2 2p^2$	7 N 14.01 $2s^2 2p^3$	8 O 16 $2s^2 2p^4$	9 F 19 $2s^2 2p^5$	10 Ne 20.18 $2s^2 2p^6$
			13 Al 26.98 $3s^2 3p^1$	14 Si 28.09 $3s^2 3p^2$	15 P 30.97 $3s^2 3p^3$	16 S 32.06 $3s^2 3p^4$	17 Cl 35.45 $3s^2 3p^5$	18 Ar 39.95 $3s^2 3p^6$
28 Ni 58.71 $3d^8 4s^2$	29 Cu 63.54 $3d^{10} 4s^1$	30 Zn 65.37 $3d^{10} 4s^2$	31 Ga 69.72 $4s^2 4p^1$	32 Ge 72.59 $4s^2 4p^2$	33 As 74.92 $4s^2 4p^3$	34 Se 78.96 $4s^2 4p^4$	35 Br 79.91 $4s^2 4p^5$	36 Kr 83.80 $4s^2 4p^6$
46 Pd 106.4 $4d^{10} 5s^0$	47 Ag 107.9 $4d^{10} 5s^1$	48 Cd 112.4 $4d^{10} 5s^2$	49 In 114.8 $5s^2 5p^1$	50 Sn 118.7 $5s^2 5p^2$	51 Sb 121.7 $5s^2 5p^3$	52 Te 127.6 $5s^2 5p^4$	53 I 126.9 $5s^2 5p^5$	54 Xe 131.3 $5s^2 5p^6$
78 Pt 195.1 $4f^{14} 5d^9 6s^1$	79 Au 197 $4f^{14} 5d^{10} 6s^1$	80 Hg 200.6 $4f^{14} 5d^{10} 6s^2$	81 T1 204.4 $6s^2 6p^1$	82 Pb 207.2 $6s^2 6p^2$	83 Bi 209 $6s^2 6p^3$	84 Po 210 $6s^2 6p^4$	85 At 211 $6s^2 6p^5$	86 Rn 222 $6s^2 6p^6$
110 Ds 269 $5f^{14} 6d^8 7s^2$	111 Rg 272 $5f^{14} 6d^9 7s^2$	112 Uub 277 $5f^{14} 6d^{10} 7s^2$						

65 Tb 158.9 $4f^8 5d^1 6s^2$	66 Dy 162.5 $4f^{10} 6s^2$	67 Ho 164.9 $4f^{11} 6s^2$	68 Er 167.3 $4f^{12} 6s^2$	69 Tm 168.9 $4f^{13} 6s^2$	70 Yb 173.0 $4f^{14} 6s^2$	71 Lu 175.0 $4f^{14} 5d^1 6s^2$

97 Bk 247 $5f^8 6d^1 7s^2$	98 Cf 251 $5f^{10} 7s^2$	99 Es 254 $5f^{11} 7s^2$	100 Fm 257 $5f^{12} 7s^2$	101 Md 256 $5f^{13} 7s^2$	102 No 254 $5f^{14} 7s^2$	103 Lr 257 $5f^{14} 6d^1 7s^2$

The Chemical Elements

Symbol	Name	Symbol	Name	Symbol	Name
Ac	Actinium	Gd	Gadolinium	Po	Polonium
Ag	Silver (Argentum)	Ge	Germanium	Pr	Praseodymium
Al	Aluminium, Aluminum	H	Hydrogen	Pt	Platinum
Am	Americium	He	Helium	Pu	Plutonium
Ar	Argon	Hf	Hafnium	Ra	Radium
As	Arsenic	Hg	Mercury (Hydrargyrum)	Rb	Rubidium
At	Astatine	Ho	Holmium	Re	Rhenium
Au	Gold (Aurum)	Hs	Hassium	Rf	Rutherfordium
B	Boron	I	Iodine	Rg	Roentgenium
Ba	Barium	In	Indium	Rh	Rhodium
Be	Beryllium	Ir	Iridium	Rn	Radon
Bh	Bohrium	K	Potassium (Kalium)	Ru	Ruthenium
Bi	Bismuth	Kr	Krypton	S	Sulfur
Bk	Berkelium	La	Lanthanum	Sb	Antimony (Stibium)
Br	Bromine	Li	Lithium	Sc	Scandium
C	Carbon	Lr	Lawrencium	Se	Selenium
Ca	Calcium	Lu	Lutetium	Sg	Seaborgium
Cd	Cadmium	Md	Mendelevium	Si	Silicon
Ce	Cerium	Mg	Magnesium	Sm	Samarium
Cf	Californium	Mn	Manganese	Sn	Tin (Stannum)
Cl	Chlorine	Mo	Molybdenum	Sr	Strontium
Cm	Curium	Mt	Meitnerium	Ta	Tantalum
Co	Cobalt	N	Nitrogen	Tb	Terbium
Cr	Chromium	Na	Sodium (Natrium)	Tc	Technetium
Cs	Caesium, Cesium	Nb	Niobium	Te	Tellurium
Cu	Copper (Cuprum)	Nd	Neodymium	Th	Thorium
Db	Dubnium	Ne	Neon	Ti	Titanium
Ds	Darmstadtium	Ni	Nickel	Tl	Thallium
Dy	Dysprosium	No	Nobelium	Tm	Thulium
Er	Erbium	Np	Neptunium	U	Uranium
Es	Einsteinium	O	Oxygen	V	Vanadium
Eu	Europium	Os	Osmium	W	Tungsten (Wolfram)
F	Fluorine	P	Phosphorus	Xe	Xenon
Fe	Iron (Ferrum)	Pa	Protactinium	Y	Yttrium
Fm	Fermium	Pb	Lead (Plumbum)	Yb	Ytterbium
Fr	Francium	Pd	Palladium	Zn	Zinc
Ga	Gallium	Pm	Promethium	Zr	Zirconium

INDEX

C

D

E

F

U

V

W

Z